A **Singular** Introduction to Commutative Algebra

T0237364

Gert-Martin Greuel · Gerhard Pfister

A **Singular** Introduction to Commutative Algebra

With contributions by
Olaf Bachmann, Christoph Lossen and Hans Schönemann

Second, Extended Edition
With 49 Figures

 Springer

Gert-Martin Greuel
Gerhard Pfister

University of Kaiserslautern
Department of Mathematics
67663 Kaiserslautern
Germany
greuel@mathematik.uni-kl.de
pfister@mathematik.uni-kl.de

Mathematics Subject Classi cation (2000): 13-XX, 13-01, 13-04, 13P10, 14-XX, 14-01, 14-04, 14QXX

Additional material to this book can be downloaded from http:// extras .springer .com .

ISBN 978-3-642-44254-4   Springer Berlin Heidelberg New York

Springer is a part of Springer Science+Business Media
springer.com
© Springer-Verlag Berlin Heidelberg 2002, 2008
Softcover re-print of the Hardcover 2nd edition 2008

Typesetting: by the authors and Integra, India using a Springer LATEX macro package
Cover Design: KünkelLopka, Heidelberg

Printed on acid-free paper   SPIN: 12073818   5 4 3 2 1 0

To Ursula, Ursina, Joscha, Bastian, Wanja, Grischa
G.–M. G.

To Marlis, Alexander, Jeannette
G. P.

Preface to the Second Edition

The first edition of this book was published 5 years ago. When we have been asked to prepare another edition, we decided not only to correct typographical errors, update the references, and improve some of the proofs but also to add new material, some appearing in printed form for the first time.

The major changes in this edition are the following:

(1) A new section about non–commutative Gröbner basis is added to chapter one, written mainly by Viktor Levandovskyy.
(2) Two new sections about characteristic sets and triangular sets together with the corresponding decomposition–algorithm are added to chapter four.
(3) There is a new appendix about polynomial factorization containing univariate factorization over $\mathbb{F}_p$ and $\mathbb{Q}$ and algebraic extensions, as well as multivariate factorization over these fields and over the algebraic closure of $\mathbb{Q}$.
(4) The system SINGULAR has improved quite a lot. A new CD is included, containing the version 3-0-3 with all examples of the book and several new SINGULAR–libraries.
(5) The appendix concerning SINGULAR is rewritten corresponding to the version 3-0-3. In particular, more examples on how to write libraries and about the communication with other systems are given.

We should like to thank many readers for helpful comments and finding typographical errors in the first edition. We thank the Singular Team for the support in producing the new CD. Special thanks to Anne Frühbis–Krüger, Santiago Laplagne, Thomas Markwig, Hans Schönemann, Oliver Wienand, for proof–reading, Viktor Levandovskyy for providing the chapter on non–commutative Gröbner bases and Petra Bäsell for typing the manuscript.

Kaiserslautern, July, 2007

Gert–Martin Greuel
Gerhard Pfister

Preface to the First Edition

> In theory there is no difference
> between theory and practice.
> In practice there is.
>
> Yogi Berra

A SINGULAR Introduction to Commutative Algebra offers a rigorous introduction to commutative algebra and, at the same time, provides algorithms and computational practice. In this book, we do not separate the theoretical and the computational part. Coincidentally, as new concepts are introduced, it is consequently shown, by means of concrete examples and general procedures, how these concepts are handled by a computer. We believe that this combination of theory and practice will provide not only a fast way to enter a rather abstract field but also a better understanding of the theory, showing concurrently how the theory can be applied.

We exemplify the computational part by using the computer algebra system SINGULAR, a system for polynomial computations, which was developed in order to support mathematical research in commutative algebra, algebraic geometry and singularity theory. As the restriction to a specific system is necessary for such an exposition, the book should be useful also for users of other systems (such as *Macaulay2* and *CoCoA*) with similar goals. Indeed, once the algorithms and the method of their application in one system is known, it is usually not difficult to transfer them to another system.

The choice of the topics in this book is largely motivated by what we believe is most useful for studying commutative algebra with a view toward algebraic geometry and singularity theory. The development of commutative algebra, although a mathematical discipline in its own right, has been greatly influenced by problems in algebraic geometry and, conversely, contributed significantly to the solution of geometric problems. The relationship between both disciplines can be characterized by saying that algebra provides rigour while geometry provides intuition.

In this connection, we place computer algebra on top of rigour, but we should like to stress its limited value if it is used without intuition.

During the past thirty years, in commutative algebra, as in many parts of mathematics, there has been a change of interest from a most general theoretical setting towards a more concrete and algorithmic understanding. One of the reasons for this was that new algorithms, together with the development of fast computers, allowed non–trivial computations, which had been intractable before. Another reason is the growing belief that algorithms can contribute to a better understanding of a problem. The human idea of "understanding", obviously, depends on the historical, cultural and technical status of the society and, nowadays, understanding in mathematics requires more and more algorithmic treatment and computational mastering. We hope that this book will contribute to a better understanding of commutative algebra and its applications in this sense.

The algorithms in this book are almost all based on Gröbner bases or standard bases. The theory of Gröbner bases is by far the most important tool for computations in commutative algebra and algebraic geometry. Gröbner bases were introduced originally by Buchberger as a basis for algorithms to test the solvability of a system of polynomial computations, to count the number of solutions (with multiplicities) if this number is finite and, more algebraically, to compute in the quotient ring modulo the given polynomials. Since then, Gröbner bases have played an important role for any symbolic computations involving polynomial data, not only in mathematics. We present, right at the beginning, the theory of Gröbner bases and, more generally, standard bases, in a somewhat new flavour.

Synopsis of the Contents of this Book

From the beginning, our aim is to be able to compute effectively in a polynomial ring as well as in the localization of a polynomial ring at a maximal ideal. Geometrically, this means that we want to compute globally with (affine or projective) algebraic varieties and locally with its singularities. In other words, we develop the theory and tools to study the solutions of a system of polynomial equations, either globally or in a neighbourhood of a given point.

The first two chapters introduce the basic theories of rings, ideals, modules and standard bases. They do not require more than a course in linear algebra, together with some training, to follow and do rigorous proofs. The main emphasis is on ideals and modules over polynomial rings. In the examples, we use a few facts from algebra, mainly from field theory, and mainly to illustrate how to use SINGULAR to compute over these fields.

In order to treat Gröbner bases, we need, in addition to the ring structure, a total ordering on the set of monomials. We do not require, as is the case in usual treatments of Gröbner bases, that this ordering be a well–ordering. Indeed, non–well–orderings give rise to local rings, and are necessary for a computational treatment of local commutative algebra. Therefore, we introduce, at an early stage, the general notion of localization. Having this, we introduce the notion of a (weak) normal form in an axiomatic way. The standard basis algorithm, as we present it, is the same for any monomial ordering,

only the normal form algorithm differs for well–orderings, called global orderings in this book, and for non–global orderings, called local, respectively
mixed, orderings.

A standard basis of an ideal or a module is nothing but a special set of
generators (the leading monomials generate the leading ideal), which allows
the computation of many invariants of the ideal or module just from its
leading monomials. We follow the tradition and call a standard basis for a
global ordering a Gröbner basis. The algorithm for computing Gröbner bases
is Buchberger's celebrated algorithm. It was modified by Mora to compute
standard bases for local orderings, and generalized by the authors to arbitrary
(mixed) orderings. Mixed orderings are necessary to generalize algorithms
(which use an extra variable to be eliminated later) from polynomial rings to
local rings. As the general standard basis algorithm already requires slightly
more abstraction than Buchberger's original algorithm, we present it first in
the framework of ideals. The generalization to modules is then a matter of
translation after the reader has become familiar with modules. Chapter 2 also
contains some less elementary concepts such as tensor products, syzygies and
resolutions. We use syzygies to give a proof of Buchberger's criterion and,
at the same time, the main step for a constructive proof of Hilbert's syzygy
theorem for the (localization of the) polynomial ring. These first two chapters
finish with a collection of methods on how to use standard bases for various
computations with ideals and modules, so–called "Gröbner basics".

The next four chapters treat some more involved but central concepts of
commutative algebra. We follow the same method as in the first two chapters,
by consequently showing how to use computers to compute more complicated
algebraic structures as well. Naturally, the presentation is a little more condensed, and the verification of several facts of a rather elementary nature are
left to the reader as an exercise.

Chapter 3 treats integral closure, dimension theory and Noether normalization. Noether normalization is a cornerstone in the theory of affine algebras, theoretically as well as computationally. It relates affine algebras, in a
controlled manner, to polynomial algebras. We apply the Noether normalization to develop the dimension theory for affine algebras, to prove the Hilbert
Nullstellensatz and E. Noether's theorem that the normalization of an affine
ring (that is, the integral closure in its total ring of fractions) is a finite extension. For all this, we provide algorithms and concrete examples on how to
compute them. A highlight of this chapter is the algorithm to compute the
non–normal locus and the normalization of an affine ring. This algorithm is
based on a criterion due to Grauert and Remmert, which had escaped the
computer algebra community for many years, and was rediscovered by T. de
Jong. The chapter ends with an extra section containing some of the larger
procedures, written in the SINGULAR programming language.

Chapter 4 is devoted to primary decomposition and related topics such
as the equidimensional part and the radical of an ideal. We start with the

usual, short and elegant but not constructive proof, of primary decomposition of an ideal. Then we present the constructive approach due to Gianni, Trager and Zacharias. This algorithm returns the primary ideals and the associated primes of an ideal in the polynomial ring over a field of characteristic 0, but also works well if the characteristic is sufficiently large, depending on the given ideal. The algorithm, as implemented in SINGULAR is often surprisingly fast. As in Chapter 3, we present the main procedures in an extra section.

In contrast to the relatively simple existence proof for primary decomposition, it is extremely difficult to actually decompose even quite simple ideals, by hand. The reason becomes clear when we consider the constructive proofs which are all quite involved, and which use many non–obvious results from commutative algebra, field theory and Gröbner bases. Indeed, primary decomposition is an important example, where we learn much more from the constructive proof than from the abstract one.

In Chapter 5 we introduce the Hilbert function and the Hilbert polynomial of graded modules together with its application to dimension theory. The Hilbert polynomial, respectively its local counterpart, the Hilbert–Samuel polynomial, contains important information about a homogeneous ideal in a polynomial ring, respectively an arbitrary ideal, in a local ring. The most important one, besides the dimension, is the degree in the homogeneous case, respectively the multiplicity in the local case. We prove that the Hilbert (–Samuel) polynomial of an ideal and of its leading ideal coincide, with respect to a degree ordering, which is the basis for the computation of these functions. The chapter finishes with a proof of the Jacobian criterion for affine K–algebras and its application to the computation of the singular locus, which uses the equidimensional decomposition of the previous chapter; other algorithms, not using any decomposition, are given in the exercises to Chapter 7.

Standard bases were, independent of Buchberger, introduced by Hironaka in connection with resolution of singularities and by Grauert in connection with deformation of singularities, both for ideals in power series rings. We introduce completions and formal power series in Chapter 6. We prove the classical Weierstraß preparation and division theorems and Grauert's generalization of the division theorem to ideals, in formal power series rings. Besides this, the main result here is that standard bases of ideals in power series rings can be computed if the ideal is generated by polynomials. This is the basis for computations in local analytic geometry and singularity theory.

The last chapter, Chapter 7, gives a short introduction to homological algebra. The main purpose is to study various aspects of depth and flatness. Both notions play an important role in modern commutative algebra and algebraic geometry. Indeed, flatness is the algebraic reason for what the ancient geometers called "principle of conservation of numbers", as it guarantees that certain invariants behave continuously in families of modules, respectively varieties. After studying and showing how to compute Tor–modules, we use Fitting ideals to show that the flat locus of a finitely presented module is open. Moreover, we present an algorithm to compute the non–flat locus and, even further, a flattening stratification of a finitely presented module. We study, in some detail, the relation between flatness and standard bases, which is somewhat subtle for mixed monomial orderings. In particular, we use flatness to show that, for any monomial ordering, the ideal and the leading ideal have the same dimension.

In the final sections of this chapter we use the Koszul complex to study the relation between the depth and the projective dimension of a module. In particular, we prove the Auslander–Buchsbaum formula and Serre's characterization of regular local rings. These can be used to effectively test the Cohen–Macaulay property and the regularity of a local K–algebra.

The book ends with two appendices, one on the geometric background and the second one on an overview on the main functionality of the system SINGULAR.

The geometric background introduces the geometric language, to illustrate some of the algebraic constructions introduced in the previous chapters. One of the objects is to explain, in the affine as well as in the projective setting, the geometric meaning of elimination as a method to compute the (closure of the) image of a morphism. Moreover, we explain the geometric meaning of the degree and the multiplicity defined in the chapter on the Hilbert Polynomial (Chapter 5), and prove some of its geometric properties. This appendix ends with a view towards singularity theory, just touching on Milnor and Tjurina numbers, Arnold's classification of singularities, and deformation theory. All this, together with other concepts of singularity theory, such as Puiseux series of plane curve singularities and monodromy of isolated hypersurface singularities, and many more, which are not treated in this book, can be found in the accompanying libraries of SINGULAR.

The second appendix gives a condensed overview of the programming language of SINGULAR, data types, functions and control structure of the system, as well as of the procedures appearing in the libraries distributed with the system. Moreover, we show by three examples (Maple, Mathematica, MuPAD), how SINGULAR can communicate with other systems.

How to Use the Text

The present book is based on a series of lectures held by the authors over the past ten years. We tried several combinations in courses of two, respectively four, hours per week in a semester ($12 - 14$ weeks). There are at least four aspects on how to use the text for a lecture:

(A) Focus on computational aspects of standard bases, and syzygies.

A possible selection for a two–hour lecture is to treat Chapters 1 and 2 completely (possibly omitting 2.6, 2.7). In a four–hour course one can treat, additionally, $3.1 - 3.5$ together with either $4.1 - 4.3$ or 4.1 and $5.1 - 5.3$.

(B) Focus on applications of methods based on standard basis, respectively syzygies, for treating more advanced problems such as primary decomposition, Hilbert functions, or flatness (regarding the standard basis, respectively syzygy, computations as "black boxes").

In this context a two–hour lecture could cover Sections 1.1 – 1.4 (only treating global orderings), 1.6 (omitting the algorithms), 1.8, 2.1, Chapter 3 and Section 4.1. A four–hour lecture could treat, in addition, the case of local orderings, Section 1.5, and selected parts of Chapters 5 and 7.

(C) Focus on the theory of commutative algebra, using SINGULAR as a tool for examples and experiments.

Here a two–hour course could be based on Sections 1.1, 1.3, 1.4, 2.1, 2.2, 2.4, 2.7, 3.1 – 3.5 and 4.1. For a four–hour lecture one could choose, additionally, Chapter 5 and Sections 7.1 – 7.4.

(D) Focus on geometric aspects, using SINGULAR as a tool for examples.

In this context a two–hour lecture could be based on Appendix A.1, A.2 and A.4, together with the needed concepts and statements of Chapters 1 and 3. For a four–hour lecture one is free to choose additional parts of the appendix (again together with the necessary background from Chapters 1 – 7).

Of course, the book may also serve as a basis for seminars and, last but not least, as a reference book for computational commutative algebra and algebraic geometry.

Working with SINGULAR

The original motivation for the authors to develop a computer algebra system in the mid eighties, was the need to compute invariants of ideals and modules in local rings, such as Milnor numbers, Tjurina numbers, and dimensions of modules of differentials. The question was whether the exactness of the Poincaré complex of a complete intersection curve singularity is equivalent to the curve being quasihomogeneous. This question was answered by an early version of SINGULAR: it is not [190]. In the sequel, the development of SINGULAR was always influenced by mathematical problems, for instance, the famous Zariski conjecture, saying that the constancy of the Milnor number in a family implies constant multiplicity [111]. This conjecture is still unsolved.

Enclosed in the book one finds a CD with folders EXAMPLES, LIBRARIES, MAC, MANUAL, UNIX and WINDOWS. The folder EXAMPLES contains all SINGULAR Examples of the book, the procedures and the links to Mathematica, Maple and MuPAD. The other folders contain the SINGULAR binaries for the respective platforms, the manual, a tutorial and the SINGULAR libraries. SINGULAR can be installed following the instructions in the INSTALL_<platform>.html (or INSTALL_<platform>.txt) file of the respective folder. We also should like to refer to the SINGULAR homepage `http://www.singular.uni-kl.de`

which always offers the possibility to download the newest version of SINGU-LAR, provides support for SINGULAR users and a discussion forum. Moreover, one finds there a lot of useful information around SINGULAR, for instance, more advanced examples and applications than provided in this book.

Comments and Corrections

We should like to encourage comments, suggestions and corrections to the book. Please send them to either of us:

Gert–Martin Greuel   greuel@mathematik.uni-kl.de
Gerhard Pfister   pfister@mathematik.uni-kl.de

We also encourage the readers to check the web site for *A SINGULAR Introduction to Commutative Algebra*,

`http://www.singular.uni-kl.de/Singular-book.html`

This site will contain lists of corrections, respectively of solutions for selected exercises.

Acknowledgements

As is customary for textbooks, we use and reproduce results from commutative algebra, usually without any specific attribution and reference. However, we should like to mention that we have learned commutative algebra mainly from the books of Zariski–Samuel [238], Nagata [183], Atiyah–Macdonald [6], Matsumura [159] and from Eisenbud's recent book [66]. The geometric background and motivation, present at all times while writing this book, were laid by our teachers Egbert Brieskorn and Herbert Kurke. The reader will easily recognize that our book owes a lot to the admirable work of the above–mentioned mathematicians, which we gratefully acknowledge.

There remains only the pleasant duty of thanking the many people who have contributed in one way or another to the preparation of this work. First of all, we should like to mention Christoph Lossen, who not only substantially improved the presentation but also contributed to the theory as well as to proofs, examples and exercises.

The book could not have been written without the system SINGULAR, which has been developed over a period of about fifteen years by Hans Schönemann and the authors, with considerable contributions by Olaf Bachmann. We feel that it is just fair to mention these two as co–authors of the book, acknowledging, in this way, their contribution as the principal creators of the SINGULAR system.[1]

---

[1] *"Software is hard. It's harder than anything else I've ever had to do."* (Donald E. Knuth)

Further main contributors to SINGULAR include: W. Decker, A. Frühbis-Krüger, H. Grassmann, T. Keilen, K. Krüger, V. Levandovskyy, C. Lossen, M. Messollen, W. Neumann, W. Pohl, J. Schmidt, M. Schulze, T. Siebert, R. Stobbe, M. Wenk, E. Westenberger and T. Wichmann, together with many authors of SINGULAR libraries mentioned in the headers of the corresponding library.

Proofreading was done by many of the above contributors and, moreover, by Y. Drozd, T. de Jong, D. Popescu, and our students M. Brickenstein, K. Dehmann, M. Kunte, H. Markwig and M. Olbermann. Last but not least, Pauline Bitsch did the LaTeX–typesetting of many versions of our manuscript and most of the pictures were prepared by Thomas Keilen.

We wish to express our heartfelt[2] thanks to all these contributors.

The book is dedicated to our families, especially to our wives Ursula and Marlis, whose encouragement and constant support have been invaluable.

Kaiserslautern, March, 2002   Gert–Martin Greuel
                                                                Gerhard Pfister

---

[2] The heart is displayed by using the programme surf, see SINGULAR Example A.1.1.

Contents

1. Rings, Ideals and Standard Bases

1.1 Rings, Polynomials and Ring Maps

The concept of a ring is probably the most basic one in commutative and non–commutative algebra. Best known are the ring of integers $\mathbb{Z}$ and the polynomial ring $K[x]$ in one variable x over a field K.

We shall now introduce the general concept of a ring with special emphasis on polynomial rings.

Definition 1.1.1.

(1) A *ring* is a set A together with an addition $+ : A \times A \to A$, $(a, b) \mapsto a + b$, and a multiplication $\cdot : A \times A \to A$, $(a, b) \mapsto a \cdot b = ab$, satisfying
   a) A, together with the addition, is an abelian group; the neutral element being denoted by 0 and the inverse of $a \in A$ by $-a$;
   b) the multiplication on A is associative, that is, $(ab)c = a(bc)$ and the distributive law holds, that is, $a(b+c) = ab + ac$ and $(b+c)a = ba + ca$, for all $a, b, c \in A$.
(2) A is called *commutative* if $ab = ba$ for $a, b \in A$ and has an *identity* if there exists an element in A, denoted by 1, such that $1 \cdot a = a \cdot 1$ for all $a \in A$.

In this book, except for chapter 1.9, a *ring always means a commutative ring with identity*. Because of (1) a ring cannot be empty but it may consist only of one element 0, this being the case if and only if $1 = 0$.

Definition 1.1.2.

(1) A subset of a ring A is called a *subring* if it contains 1 and is closed under the ring operations induced from A.
(2) $u \in A$ is called a *unit* if there exists a $u' \in A$ such that $uu' = 1$. The set of units is denoted by A^*; it is a group under multiplication.
(3) A ring is a *field* if $1 \neq 0$ and any non–zero element is a unit, that is $A^* = A - \{0\}$.
(4) Let A be a ring, $a \in A$, then $\langle a \rangle := \{af \mid f \in A\}$.

Any field is a ring, such as $\mathbb{Q}$ (the rational numbers), or $\mathbb{R}$ (the real numbers), or $\mathbb{C}$ (the complex numbers), or $\mathbb{F}_p = \mathbb{Z}/p\mathbb{Z}$ (the finite field with p elements where p is a prime number, cf. Exercise 1.1.3) but $\mathbb{Z}$ (the integers) is a ring which is not a field.

$\mathbb{Z}$ is a subring of $\mathbb{Q}$, we have $\mathbb{Z}^* = \{\pm 1\}$, $\mathbb{Q}^* = \mathbb{Q} \smallsetminus \{0\}$. $\mathbb{N} \subset \mathbb{Z}$ denotes the set of nonnegative integers.

Definition 1.1.3. Let A be a ring.

(1) A *monomial* in n variables (or indeterminates) $x_1, \ldots, x_n$ is a power product

$$x^\alpha = x_1^{\alpha_1} \cdot \ldots \cdot x_n^{\alpha_n}, \qquad \alpha = (\alpha_1, \ldots, \alpha_n) \in \mathbb{N}^n.$$

The set of monomials in n variables is denoted by

$$\mathrm{Mon}(x_1, \ldots, x_n) = \mathrm{Mon}_n := \{x^\alpha \mid \alpha \in \mathbb{N}^n\}.$$

Note that $\mathrm{Mon}(x_1, \ldots, x_n)$ is a semigroup under multiplication, with neutral element $1 = x_1^0 \cdot \ldots \cdot x_n^0$.

We write $x^\alpha \mid x^\beta$ if x^α *divides* x^β, which means that $\alpha_i \leq \beta_i$ for all i and, hence, $x^\beta = x^\gamma x^\alpha$ for $\gamma = \beta - \alpha \in \mathbb{N}^n$.

(2) A *term* is a monomial times a coefficient (an element of A),

$$ax^\alpha = ax_1^{\alpha_1} \cdot \ldots \cdot x_n^{\alpha_n}, \qquad a \in A.$$

(3) A *polynomial over A* is a finite A–linear combination of monomials, that is, a finite sum of terms,

$$f = \sum_\alpha a_\alpha x^\alpha = \sum_{\alpha \in \mathbb{N}^n}^{\text{finite}} a_{\alpha_1 \ldots \alpha_n} x_1^{\alpha_1} \cdot \ldots \cdot x_n^{\alpha_n},$$

with $a_\alpha \in A$. For $\alpha \in \mathbb{N}^n$, let $|\alpha| := \alpha_1 + \cdots + \alpha_n$.

The integer $\deg(f) := \max\{|\alpha| \mid a_\alpha \neq 0\}$ is called the *degree* of f if $f \neq 0$; we set $\deg(f) = -1$ for f the zero polynomial.

(4) The *polynomial ring* $A[x] = A[x_1, \ldots, x_n]$ in n variables over A is the set of all polynomials together with the usual addition and multiplication:

$$\sum_\alpha a_\alpha x^\alpha + \sum_\alpha b_\alpha x^\alpha := \sum_\alpha (a_\alpha + b_\alpha) x^\alpha,$$

$$\left(\sum_\alpha a_\alpha x^\alpha \right) \cdot \left(\sum_\beta b_\beta x^\beta \right) := \sum_\gamma \left(\sum_{\alpha + \beta = \gamma} a_\alpha b_\beta \right) x^\gamma.$$

$A[x_1, \ldots, x_n]$ is a commutative ring with identity $1 = x_1^0 \cdot \ldots \cdot x_n^0$ which we identify with the identity element $1 \in A$. Elements of $A \subset A[x]$ are called *constant polynomials*, they are characterized by having degree ≤ 0. A is called the *ground ring* of $A[x]$, respectively the *ground field*, if A is a field.

Note that any monomial is a term (with coefficient 1) but, for example, 0 is a term but not a monomial. For us the most important case is the polynomial ring $K[x] = K[x_1, \ldots, x_n]$ over a field K. By Exercise 1.3.1 only the non–zero constants are units of $K[x]$, that is, $K[x]^* = K^* = K \smallsetminus \{0\}$.

If K is an infinite field, we can identify polynomials $f \in K[x_1, \ldots, x_n]$ with their associated *polynomial function*

$$\tilde{f} : K^n \longrightarrow K, \quad (p_1, \ldots, p_n) \longmapsto f(p_1, \ldots, p_n),$$

but for finite fields $\tilde{f}$ may be zero for a non–zero f (cf. Exercise 1.1.4).

Any polynomial in $n-1$ variables can be considered as a polynomial in n variables (where the n–th variable does not appear) with the usual ring operations on polynomials in n variables. Hence, $A[x_1, \ldots, x_{n-1}] \subset A[x_1, \ldots, x_n]$ is a subring and it follows directly from the definition of polynomials that

$$A[x_1, \ldots, x_n] = (A[x_1, \ldots, x_{n-1}])[x_n].$$

Hence, we can write $f \in A[x_1, \ldots, x_n]$ in a unique way, either as

$$f = \sum_{\alpha \in \mathbb{N}^n}^{\text{finite}} a_\alpha x^\alpha, \ a_\alpha \in A$$

or as

$$f = \sum_{\nu \in \mathbb{N}}^{\text{finite}} f_\nu x_n^\nu, \quad f_\nu \in A[x_1, \ldots, x_{n-1}].$$

The first representation of f is called *distributive* while the second is called *recursive*.

Remark 1.1.4. Both representations play an important role in computer algebra. The practical performance of an implemented algorithm may depend drastically on the internal representation of polynomials (in the computer). Usually the distributive representation is chosen for algorithms related to Gröbner basis computations while the recursive representation is preferred for algorithms related to factorization of polynomials.

Definition 1.1.5. A *morphism* or *homomorphism* of rings is a map $\varphi : A \to B$ satisfying $\varphi(a + a') = \varphi(a) + \varphi(a')$, $\varphi(aa') = \varphi(a)\varphi(a')$, for all $a, a' \in A$, and $\varphi(1) = 1$. We call a morphism of rings also a *ring map*, and B is called an *A–algebra*.[1]

We have $\varphi(a) = \varphi(a \cdot 1) = \varphi(a) \cdot 1$. If φ is fixed, we also write $a \cdot b$ instead of $\varphi(a) \cdot b$ for $a \in A$ and $b \in B$.

---

[1] See also Example 2.1.2 and Definition 2.1.3.

Lemma 1.1.6. *Let $A[x_1, \ldots x_n]$ be a polynomial ring, $\psi : A \to B$ a ring map, C a B–algebra, and $f_1, \ldots, f_n \in C$. Then there exists a unique ring map*

$$\varphi : A[x_1, \ldots, x_n] \longrightarrow C$$

satisfying $\varphi(x_i) = f_i$ for $i = 1, \ldots, n$ and $\varphi(a) = \psi(a) \cdot 1 \in C$ for $a \in A$.

Proof. Given any $f = \sum_\alpha a_\alpha x^\alpha \in A[x]$, then a ring map φ with $\varphi(x_i) = f_i$, and $\varphi(a) = \psi(a)$ for $a \in A$ must satisfy (by Definition 1.1.5)

$$\varphi(f) = \sum_\alpha \psi(a_\alpha) \varphi(x_1)^{\alpha_1} \cdot \ldots \cdot \varphi(x_n)^{\alpha_n} .$$

Hence, φ is uniquely determined. Moreover, defining $\varphi(f)$ for $f \in A[x]$ by the above formula, it is easy to see that φ becomes a homomorphism, which proves existence. $\square$

We shall apply this lemma mainly to the case where C is the polynomial ring $B[y_1, \ldots, y_m]$.

In SINGULAR one can define polynomial rings over the following fields:

(1) the field of rational numbers $\mathbb{Q}$,
(2) finite fields $\mathbb{F}_p$, p a prime number ≤ 32003,
(3) finite fields $\mathrm{GF}(p^n)$ with p^n elements, p a prime, $p^n \leq 2^{15}$,
(4) transcendental extensions of $\mathbb{Q}$ or $\mathbb{F}_p$,
(5) simple algebraic extensions of $\mathbb{Q}$ or $\mathbb{F}_p$,
(6) simple precision real floating point numbers,
(7) arbitrary prescribed real floating point numbers,
(8) arbitrary prescribed complex floating point numbers.

For the definitions of rings over fields of type (3) and (5) we use the fact that for a polynomial ring $K[x]$ in one variable x over a field and $f \in K[x] \smallsetminus \{0\}$ the quotient ring $K[x]/\langle f \rangle$ is a field if and only if f is *irreducible*, that is, f cannot be written as a product of two polynomials of lower degree (cf. Exercise 1.1.5). If f is irreducible and monic, then it is called the *minimal polynomial* of the field extension $K \subset K[x]/\langle f \rangle$ (cf. Example 1.1.8).

Remark 1.1.7. Indeed, the computation over the above fields (1) – (5) is exact, only limited by the internal memory of the computer. Strictly speaking, floating point numbers, as in (6) – (8), do not represent the field of real (or complex) numbers. Because of rounding errors, the product of two non–zero elements or the difference between two unequal elements may be zero (the latter case is the more serious one since the individual elements may be very big). Of course, in many cases one can trust the result, but we should like to emphasize that this remains the responsibility of the user, even if one computes with very high precision.

In SINGULAR, field elements have the type *number* but notice that one *can define and use numbers only in a polynomial ring with at least one variable* and a specified monomial ordering. For example, if one wishes to compute with arbitrarily big integers or with exact arithmetic in $\mathbb{Q}$, this can be done as follows:

SINGULAR Example 1.1.8 (computation in fields).
In the examples below we have used the degree reverse lexicographical ordering dp but we could have used any other monomial ordering (cf. Section 1.2). Actually, this makes no difference as long as we do simple manipulations with polynomials. However, more complicated operations on ideals such as the std or groebner command return results which depend very much on the chosen ordering.

(1) Computation in the field of *rational numbers*:

```
ring A = 0,x,dp;
number n = 12345/6789;
n^5;                    //common divisors are cancelled
//-> 11799100858126071875/59350279669807543
```

Note: Typing just 123456789^5; will result in integer overflow since 123456789 is considered as an integer (machine integer of limited size) and not as an element in the field of rational numbers; however, also correct would be number(123456789)^5;.

(2) Computation in *finite fields*:

```
ring A1 = 32003,x,dp;   //finite field Z/32003
number(123456789)^5;
//-> 8705

ring A2 = (2^3,a),x,dp; //finite (Galois) field GF(8)
                        //with 8 elements
number n = a+a2;        //a is a generator of the group
                        //GF(8)-{0}
n^5;
//-> a6
minpoly;                //minimal polynomial of GF(8)
//-> 1*a^3+1*a^1+1*a^0

ring A3 = (2,a),x,dp;   //infinite field Z/2(a) of
                        //characteristic 2
minpoly = a20+a3+1;     //define a minimal polynomial
                        //a^20+a^3+1
                        //now the ground field is
                        //GF(2^20)=Z/2[a]/<a^20+a^3+1>,
number n = a+a2;        //a finite field
```

```
            //with 2^20 elements
   n^5;     //a is a generator of the group
            //GF(2^20)-{0}
//-> (a10+a9+a6+a5)
```

Note: For computation in finite fields $\mathbb{Z}/p\mathbb{Z}$, $p \leq 32003$, respectively $GF(p^n)$, $p^n \leq 2^{15}$, one should use rings as **A1** respectively **A2** since for these fields SINGULAR uses look–up tables, which is quite fast. For other finite fields a *minimal polynomial* as in **A3** must be specified. A good choice are the *Conway polynomials* (cf. [126]). SINGULAR does not, however, check the irreducibility of the chosen minimal polynomial. This can be done as in the following example.

```
ring tst = 2,a,dp;
factorize(a20+a2+1,1);
//-> _[1]=a3+a+1    //not irreducible! We have two factors
//-> _[2]=a7+a5+a4+a3+1
factorize(a20+a3+1,1);   //irreducible
//-> _[1]=a20+a3+1
```

To obtain the multiplicities of the factors, use `factorize(a20+a2+1);`.

(3) Computation with *real* and *complex floating point numbers*, 30 digits precision:

```
ring R1 = (real,30),x,dp;
number n = 123456789.0;
n^5;                //compute with a precision of 30 digits
//-> 0.286797186029971810723376143809e+41
```

Note: n^5 is a number whose integral part has 41 digits (indicated by e+41). However, only 30 digits are computed.

```
ring R2 = (complex,30,I),x,dp;//I denotes imaginary unit
number n = 123456789.0+0.0001*I;
n^5;                //complex number with 30 digits precision
//-> (0.286797186029971810723374262133e+41
        +I*1161528613991129622075046746710)
```

(4) Computation with rational numbers and *parameters*, that is, in $\mathbb{Q}(a, b, c)$, the quotient field of $\mathbb{Q}[a, b, c]$:

```
ring R3 = (0,a,b,c),x,dp;
number n = 12345a+12345/(78bc);
n^2;
//->(103021740900a2b2c2+2641583100abc+16933225)/(676b2c2)
n/9c;
//-> (320970abc+4115)/(234bc2)
```

We shall now show how to define the polynomial ring in n variables $x_1, \ldots, x_n$ over the above mentioned fields K. We can do this for any n, but we have to specify an integer n first. The same remark applies if we work with transcendental extensions of degree m; we usually call the elements $t_1, \ldots, t_m$ of a transcendental basis (free) *parameters*. If g is any non–zero polynomial in the parameters $t_1, \ldots, t_m$, then g and $1/g$ are numbers in the corresponding ring.

For further examples see the SINGULAR Manual [116].

SINGULAR Example 1.1.9 (computation in polynomial rings).

Let us create *polynomial rings* over different fields. By typing the name of the *ring* we obtain all relevant information about the ring.

```
ring A = 0,(x,y,z),dp;
poly f = x3+y2+z2;           //same as x^3+y^2+z^2
f*f-f;
//-> x6+2x3y2+2x3z2+y4+2y2z2+z4-x3-y2-z2
```

SINGULAR understands short (e.g., 2x2+y3) and long (e.g., 2*x^2+y^3) input. By default the short output is displayed in rings without parameters and with one–letter variables, whilst the long output is used, for example, for indexed variables. The command short=0; forces all output to be displayed in the long format.

Computations in polynomial rings over other fields follow the same pattern. Try ring R=32003,x(1..3),dp; (finite ground field), respectively ring R=(0,a,b,c),(x,y,z,w),dp; (ground field with parameters), and type R; to obtain information about the ring. The command setring allows switching from one ring to another, for example, setring A4; makes A4 the basering.

We use Lemma 1.1.6 to define ring maps in SINGULAR. Indeed, one has three possibilities, fetch, imap and map, to define ring maps by giving the name of the preimage ring and a list of polynomials $f_1, \ldots, f_n$ (as many as there are variables in the preimage ring) in the current basering. The commands fetch, respectively imap, map an object directly from the preimage ring to the basering whereas fetch maps the first variable to the first, the second to the second and so on (hence, is convenient for renaming the variables), while imap maps a variable to the variable with the same name (or to 0 if it does not exist), hence is convenient for inclusion of sub–rings or for changing the monomial ordering.

Note: All maps go from a predefined ring to the basering.

SINGULAR Example 1.1.10 (methods for creating ring maps).

$$\text{map: preimage ring} \longrightarrow \text{basering}$$

(1) General definition of a *map*:

```
ring A = 0,(a,b,c),dp;
poly f = a+b+ab+c3;

ring B = 0,(x,y,z),dp;
map F  = A, x+y,x-y,z;//map F from ring A (to basering B)
                      //sending a -> x+y, b -> x-y, c -> z
poly g = F(f);        //apply F
g;
//-> z3+x2-y2+2x
```

(2) Special maps (imap, fetch):

```
ring A1 = 0,(x,y,c,b,a,z),dp;
imap(A,f);             //imap preserves names of variables
//-> c3+ba+b+a
fetch(A,f);            //fetch preserves order of variables
//-> c3+xy+x+y
```

Exercises

1.1.1. The set of units A^* of a ring A is a group under multiplication.

1.1.2. The *direct sum of rings* $A \oplus B$, together with component–wise addition and multiplication is again a ring.

1.1.3. Prove that, for $n \in \mathbb{Z}$, the following are equivalent:

(1) $\mathbb{Z}/\langle n \rangle$ is a field.
(2) $\mathbb{Z}/\langle n \rangle$ is an integral domain.
(3) n is a prime number.

1.1.4. Let K be a field and $f \in K[x_1, \ldots, x_n]$. Then f determines a *polynomial function* $\tilde{f} : K^n \to K$, $(p_1, \ldots, p_n) \mapsto f(p_1, \ldots, p_n)$.

(1) If K is infinite then f is uniquely determined by $\tilde{f}$.
(2) Show by an example that this is not necessarily true for K finite.
(3) Let K be a finite field with q elements. Show that each polynomial $f \in K[x_1, \ldots, x_n]$ of degree at most $q - 1$ in each variable is already determined by the polynomial function $\tilde{f} : K^n \to K$.

1.1.5. Let $f \in K[x]$ be a non–constant polynomial in one variable over the field K. f is called *irreducible* if $f \notin K$ and if it is not the product of two polynomials of strictly smaller degree. Prove that the following are equivalent:

(1) $K[x]/\langle f \rangle$ is a field.
(2) $K[x]/\langle f \rangle$ is an integral domain.
(3) f is irreducible.

1.1.6. An irreducible polynomial $f = a_n x^n + \cdots + a_1 x + a_0 \in K[x]$, K a field, is called *separable*, if f has only simple roots in $\overline{K}$, the algebraic closure of K.

An algebraic field extension $K \subset L$ is called *separable* if any element $a \in L$ is separable over K, that is, the minimal polynomial of a over K is separable.

(1) Show that $f \neq 0$ is separable if and only if f and its formal derivative $Df := n a_n x^{n-1} + \cdots + a_1$ have no common factor of degree ≥ 1.
(2) A finite separable field extension $K \subset L$ is generated by a *primitive element*, that is, there exists an irreducible $f \in K[x]$ such that $L \cong K[x]/\langle f \rangle$.
(3) K is called a *perfect field* if every irreducible polynomial $f \in K[x]$ is separable. Show that finite fields, algebraically closed fields and fields of characteristic 0 are perfect.

1.1.7. Which of the fields in SINGULAR, (1) – (5), are perfect, which not?

1.1.8. Compute (10!)^5 with the help of SINGULAR.

1.1.9. Declare in SINGULAR a polynomial ring in the variables x(1), x(2), x(3), x(4) over the finite field with eight elements

1.1.10. Declare in SINGULAR the ring $A = \mathbb{Q}(a, b, c)[x, y, z, w]$ and compute f^2/c^2 for $f = (ax^3 + by^2 + cz^2)(ac - bc)$.

1.1.11. Declare in SINGULAR the rings $A = \mathbb{Q}[a, b, c]$ and $B = \mathbb{Q}[a]$. In A define the polynomial $f = a + b + ab + c^3$. Try in B the commands imap(A,f) and fetch(A,f).

1.1.12. Declare in SINGULAR the ring $\mathbb{Q}(i)[u, v, w]$, $i^2 = -1$, and compute $\left((i + i^2 + 1)(uvw)\right)^3$.

1.1.13. Write a SINGULAR procedure, depending on two integers p, d, with p a prime, which returns all polynomials in $\mathbb{F}_p[x]$ of degree d such that the corresponding polynomial function vanishes. Use the procedure to display all $f \in (\mathbb{Z}/5\mathbb{Z})[x]$ of degree ≤ 6 such that $\tilde{f} = 0$.

1.2 Monomial Orderings

The presentation of a polynomial as a linear combination of monomials is unique only up to an order of the summands, due to the commutativity of the addition. We can make this order unique by choosing a total ordering on the set of monomials. For further applications it is necessary, however, that the ordering is compatible with the semigroup structure on Mon_n.

Definition 1.2.1. A *monomial ordering* or *semigroup ordering* is a total (or linear) ordering $>$ on the set of monomials $\mathrm{Mon}_n = \{x^\alpha \mid \alpha \in \mathbb{N}^n\}$ in n variables satisfying

$$x^\alpha > x^\beta \implies x^\gamma x^\alpha > x^\gamma x^\beta$$

for all $\alpha, \beta, \gamma \in \mathbb{N}^n$. We say also $>$ is a *monomial ordering on* $A[x_1, \ldots, x_n]$, A any ring, meaning that $>$ is a monomial ordering on Mon_n.

We identify Mon_n with $\mathbb{N}^n$, and then a monomial ordering is a total ordering on $\mathbb{N}^n$, which is compatible with the semigroup structure on $\mathbb{N}^n$ given by addition. A typical, and important, example is provided by the lexicographical ordering on $\mathbb{N}^n$: $x^\alpha > x^\beta$ if and only if the first non–zero entry of $\alpha - \beta$ is positive. We shall see different monomial orderings later.

Monomial orderings provide an extra structure on the set of monomials and, hence, also on the polynomial ring. Although they have been used in several places to prove difficult mathematical theorems they are hardly part of classical commutative algebra. Monomial orderings, however, can be quite powerful tools in theoretical investigations (cf. [98]) but, in addition, they are indispensable in many serious and deeper polynomial computations.

From a practical point of view, a monomial ordering $>$ allows us to write a polynomial $f \in K[x]$ in a unique ordered way as

$$f = a_\alpha x^\alpha + a_\beta x^\beta + \cdots + a_\gamma x^\gamma,$$

with $x^\alpha > x^\beta > \cdots > x^\gamma$, where no coefficient is zero (a sparse representation of f). Moreover, this allows the representation of a polynomial in a computer as an ordered list of coefficients, making equality tests very simple and fast (assuming this is the case for the ground field). Additionally, this order does not change if we multiply f with a monomial. For highly sophisticated presentations of monomials and polynomials in a computer see [10]. There are many more and deeper properties of monomial orderings and, moreover, different orderings have different further properties.

Definition 1.2.2. Let $>$ be a fixed monomial ordering. Write $f \in K[x]$, $f \neq 0$, in a unique way as a sum of non–zero terms

$$f = a_\alpha x^\alpha + a_\beta x^\beta + \cdots + a_\gamma x^\gamma, \quad x^\alpha > x^\beta > \cdots > x^\gamma,$$

and $a_\alpha, a_\beta, \ldots, a_\gamma \in K$. We define:

(1) $\mathrm{LM}(f) := \mathtt{leadmonom(f)} := x^\alpha$, the *leading monomial* of f,
(2) $\mathrm{LE}(f) := \mathtt{leadexp(f)} := \alpha$, the *leading exponent* of f,
(3) $\mathrm{LT}(f) := \mathtt{lead(f)} := a_\alpha x^\alpha$, the *leading term* or *head* of f,
(4) $\mathrm{LC}(f) := \mathtt{leadcoef(f)} := a_\alpha$, the *leading coefficient* of f
(5) $\mathrm{tail}(f) := f - \mathtt{lead(f)} = a_\beta x^\beta + \cdots + a_\gamma x^\gamma$, the *tail* of f.

Let us consider an example with the lexicographical ordering. In SINGULAR every polynomial belongs to a ring which has to be defined first. We define the ring $A = \mathbb{Q}[x, y, z]$ together with the lexicographical ordering.

A is the name of the ring, 0 the characteristic of the ground field $\mathbb{Q}$, x, y, z are the names of the variables and 1p defines the lexicographical ordering with $x > y > z$, see Example 1.2.8.

SINGULAR Example 1.2.3 (leading data).

```
ring A = 0,(x,y,z),lp;
poly f = y4z3+2x2y2z2+3x5+4z4+5y2;
f;                          //display f in a lex-ordered way
//-> 3x5+2x2y2z2+y4z3+5y2+4z4
leadmonom(f);               //leading monomial
//-> x5
leadexp(f);                 //leading exponent
//-> 5,0,0
lead(f);                    //leading term
//-> 3x5
leadcoef(f);                //leading coefficient
//-> 3
f - lead(f);                //tail
//-> 2x2y2z2+y4z3+5y2+4z4
```

The most important distinction is between global and local orderings.

Definition 1.2.4. Let $>$ be a monomial ordering on $\{x^\alpha \mid \alpha \in \mathbb{N}^n\}$.

(1) $>$ is called a *global ordering* if $x^\alpha > 1$ for all $\alpha \neq (0, \ldots, 0)$,
(2) $>$ is called a *local ordering* if $x^\alpha < 1$ for all $\alpha \neq (0, \ldots, 0)$,
(3) $>$ is called a *mixed ordering* if it is neither global nor local.

Of course, if we turn the ordering around by setting $x^\alpha >' x^\beta$ if $x^\beta > x^\alpha$, then $>'$ is global if and only if $>$ is local. However, local and global (and mixed) orderings have quite different properties. Here are the most important characterizations of a global ordering.

Lemma 1.2.5. *Let $>$ be a monomial ordering, then the following conditions are equivalent:*

(1) $>$ is a well–ordering.
(2) $x_i > 1$ for $i = 1, \ldots, n$.
(3) $x^\alpha > 1$ for all $\alpha \neq (0, \ldots, 0)$, that is, $>$ is global.
(4) $\alpha \geq_{nat} \beta$ and $\alpha \neq \beta$ implies $x^\alpha > x^\beta$.

The last condition means that $>$ is a refinement of the natural partial ordering on $\mathbb{N}^n$ defined by

$$(\alpha_1, \ldots, \alpha_n) \geq_{\text{nat}} (\beta_1, \ldots, \beta_n) :\Longleftrightarrow \alpha_i \geq \beta_i \text{ for all } i.$$

Proof. $(1) \Rightarrow (2)$: if $x_i < 1$ for some i, then $x_i^p < x_i^{p-1} < 1$, yielding a set of monomials without smallest element (recall that a well–ordering is a total ordering on a set such that each non–empty subset has a smallest element).

$(2) \Rightarrow (3)$: write $x^\alpha = x^{\alpha'} x_j$ for some j and use induction. For $(3) \Rightarrow (4)$ let $(\alpha_1, \ldots, \alpha_n) \geq_{\text{nat}} (\beta_1, \ldots, \beta_n)$ and $\alpha \neq \beta$. Then $\gamma := \alpha - \beta \in \mathbb{N}^n \setminus \{0\}$, hence $x^\gamma > 1$ and, therefore, $x^\alpha = x^\beta x^\gamma > x^\beta$.

$(4) \Rightarrow (1)$: Let M be a non–empty set of monomials. By Dickson's Lemma (Lemma 1.2.6) there is a finite subset $B \subset M$ such that for each $x^\alpha \in M$ there is an $x^\beta \in B$ with $\beta \leq_{\text{nat}} \alpha$. By assumption, $x^\beta < x^\alpha$ or $x^\beta = x^\alpha$, that is, B contains a smallest element of M with respect to $>$.   □

Lemma 1.2.6 (Dickson, 1913). *Let $M \subset \mathbb{N}^n$ be any subset. Then there is a finite set $B \subset M$ satisfying*

$$\forall\, \alpha \in M \;\exists\, \beta \in B \text{ such that } \beta \leq_{nat} \alpha.$$

B is sometimes called a Dickson basis *of M.*

Proof. We write $\geq$ instead of $\geq_{\text{nat}}$ and use induction on n. For $n = 1$ we can take the minimum of M as the only element of B.

For $n > 1$ and $i \in \mathbb{N}$ define

$$M_i = \{\alpha' = (\alpha_1, \ldots, \alpha_{n-1}) \in \mathbb{N}^{n-1} \mid (\alpha', i) \in M\}$$

and, by induction, M_i has a Dickson basis B_i.

Again, by induction hypothesis, $\bigcup_{i \in \mathbb{N}} B_i$ has a Dickson basis B'. B' is finite, hence $B' \subset B_1 \cup \cdots \cup B_s$ for some s.

We claim that

$$B := \{(\beta', i) \in \mathbb{N}^n \mid 0 \leq i \leq s,\ \beta' \in B_i\}$$

is a Dickson basis of M.

To see this, let $(\alpha', \alpha_n) \in M$. Then $\alpha' \in M_{\alpha_n}$ and, since B_{α_n} is a Dickson basis of M_{α_n}, there is a $\beta' \in B_{\alpha_n}$ with $\beta' \leq \alpha'$. If $\alpha_n \leq s$, then $(\beta', \alpha_n) \in B$ and $(\beta', \alpha_n) \leq (\alpha', \alpha_n)$. If $\alpha_n > s$, we can find a $\gamma' \in B'$ and an $i \leq s$ such that $\gamma' \leq \beta'$ and $(\gamma', i) \in B_i$. Then $(\gamma', i) \in B$ and $(\gamma', i) \leq (\alpha', \alpha_n)$.   □

Remark 1.2.7. If A is an $n \times n$ integer matrix with only non–negative entries and determinant $\neq 0$, and if $>$ is a monomial ordering, we can define a *matrix ordering* $>_{(A,>)}$ by setting

$$x^\alpha >_{(A,>)} x^\beta :\Longleftrightarrow x^{A\alpha} > x^{A\beta}$$

where α and β are considered as column vectors. By Exercise 1.2.6 (2), $>_{(A,>)}$ is again a monomial ordering. We can even use matrices $A \in \mathrm{GL}(n, \mathbb{R})$ with real entries to obtain a monomial ordering by setting

$$x^\alpha >_A x^\beta :\Longleftrightarrow A\alpha > A\beta,$$

where $>$ on the right–hand side is the lexicographical ordering on $\mathbb{R}^n$.

Robbiano proved in [196], that every monomial ordering arises in this way from the lexicographical ordering on $\mathbb{R}^n$. However, we do not need this fact (cf. Exercise 1.2.9).

Important examples of monomial orderings are:

Example 1.2.8 (monomial orderings).
In the following examples we fix an enumeration $x_1, \ldots, x_n$ of the variables, any other enumeration leads to a different ordering.

(1) GLOBAL ORDERINGS

    (i) *Lexicographical ordering* $>_{lp}$ (also denoted by lex):

$$x^\alpha >_{lp} x^\beta \;:\Longleftrightarrow\; \exists\, 1 \le i \le n : \alpha_1 = \beta_1, \ldots, \alpha_{i-1} = \beta_{i-1}, \alpha_i > \beta_i\,.$$

    (ii) *Degree reverse lexicographical ordering* $>_{dp}$ (denoted by degrevlex):

$$x^\alpha >_{dp} x^\beta \;:\Longleftrightarrow\; \deg x^\alpha > \deg x^\beta$$
$$\text{or } \big(\deg x^\alpha = \deg x^\beta \text{ and } \exists\, 1 \le i \le n :$$
$$\alpha_n = \beta_n, \ldots, \alpha_{i+1} = \beta_{i+1}, \alpha_i < \beta_i\big)\,,$$

    where $\deg x^\alpha = \alpha_1 + \cdots + \alpha_n$.

  (iii) *Degree lexicographical ordering* $>_{Dp}$ (also denoted by deglex):

$$x^\alpha >_{Dp} x^\beta \;:\Longleftrightarrow\; \deg x^\alpha > \deg x^\beta$$
$$\text{or } \big(\deg x^\alpha = \deg x^\beta \text{ and } \exists\, 1 \le i \le n :$$
$$\alpha_1 = \beta_1, \ldots, \alpha_{i-1} = \beta_{i-1}, \alpha_i > \beta_i\big).$$

In all three cases $x_1, \ldots, x_n > 1$. For example, we have $x_1^3 >_{lp} x_1^2 x_2^2$ but $x_1^2 x_2^2 >_{dp,Dp} x_1^3$. An example where dp and Dp differ: $x_1^2 x_2 x_3^2 >_{Dp} x_1 x_2^3 x_3$ but $x_1 x_2^3 x_3 >_{dp} x_1^2 x_2 x_3^2$.

Given a vector $w = (w_1, \ldots, w_n)$ of integers, we define the *weighted degree* of x^α by

$$\text{w–deg}(x^\alpha) := \langle w, \alpha \rangle := w_1 \alpha_1 + \cdots + w_n \alpha_n\,,$$

that is, the variable x_i has degree w_i. For a polynomial $f = \sum_\alpha a_\alpha x^\alpha$, we define the weighted degree,

$$\text{w–deg}(f) := \max\big\{\text{w–deg}(x^\alpha) \,\big|\, a_\alpha \ne 0\big\}\,.$$

Using the weighted degree in (ii), respectively (iii), with all $w_i > 0$, instead of the usual degree, we obtain the *weighted reverse lexicographical ordering*, wp$(w_1, \ldots, w_n)$, respectively the *weighted lexicographical ordering*, Wp$(w_1, \ldots, w_n)$.

(2) LOCAL ORDERINGS

    (i) *Negative lexicographical ordering* $>_{ls}$:

$$x^\alpha >_{ls} x^\beta \; :\Longleftrightarrow \; \exists \, 1 \leq i \leq n, \alpha_1 = \beta_1, \ldots, \alpha_{i-1} = \beta_{i-1}, \alpha_i < \beta_i \,.$$

    (ii) *Negative degree reverse lexicographical ordering* $>_{ds}$:

$$x^\alpha >_{ds} x^\beta \; :\Longleftrightarrow \; \deg x^\alpha < \deg x^\beta, \text{ where } \deg x^\alpha = \alpha_1 + \cdots + \alpha_n,$$
$$\text{or } \big(\deg x^\alpha = \deg x^\beta \text{ and } \exists \, 1 \leq i \leq n:$$
$$\alpha_n = \beta_n, \ldots, \alpha_{i+1} = \beta_{i+1}, \, \alpha_i < \beta_i \big).$$

    (iii) *Negative degree lexicographical ordering* $>_{Ds}$:

$$x^\alpha >_{Ds} x^\beta \; :\Longleftrightarrow \; \deg x^\alpha < \deg x^\beta,$$
$$\text{or } \big(\deg x^\alpha = \deg x^\beta \text{ and } \exists \, 1 \leq i \leq n:$$
$$\alpha_1 = \beta_1, \ldots, \alpha_{i-1} = \beta_{i-1}, \, \alpha_i > \beta_i \big).$$

Similarly, as above, we can define weighted versions $\mathtt{ws}(w_1, \ldots, w_n)$ and $\mathtt{Ws}(w_1, \ldots, w_n)$ of the two last local orderings.

(3) PRODUCT OR BLOCK ORDERINGS

Now consider $>_1$, a monomial ordering on $\mathrm{Mon}(x_1, \ldots, x_n)$, and $>_2$, a monomial ordering on $\mathrm{Mon}(y_1, \ldots, y_m)$. Then the *product ordering* or *block ordering* $>$, also denoted by $(>_1, >_2)$ on $\mathrm{Mon}(x_1, \ldots, x_n, y_1, \ldots, y_m)$, is defined as

$$x^\alpha y^\beta > x^{\alpha'} y^{\beta'} \; :\Longleftrightarrow \; x^\alpha >_1 x^{\alpha'}$$
$$\text{or } \big(x^\alpha = x^{\alpha'} \text{ and } y^\beta >_2 y^{\beta'} \big).$$

If $>_1$ is a global ordering then the product ordering has the property that monomials which contain an x_i are always larger than monomials containing no x_i. If the special orderings $>_1$ on $\mathrm{Mon}(x_1, \ldots, x_n)$ and $>_2$ on $\mathrm{Mon}(y_1, \ldots, y_m)$ are irrelevant, for a product ordering on $\mathrm{Mon}(x_1, \ldots, x_n, y_1, \ldots, y_m)$ we write just $x \gg y$.

If $>_1$ and $>_2$ are global (respectively local), then the product ordering is global (respectively local) but the product ordering is mixed if one of the orderings $>_1$ and $>_2$ is global and the other local. This is how mixed orderings arise in a natural way.

Definition 1.2.9. A monomial ordering $>$ on $\{x^\alpha \mid \alpha \in \mathbb{N}^n\}$ is called a *weighted degree ordering* if there exists a vector $w = (w_1, \ldots, w_n)$ of non–zero integers such that

$$\mathrm{w\text{–}deg}(x^\alpha) > \mathrm{w\text{–}deg}(x^\beta) \implies x^\alpha > x^\beta \,.$$

It is called a *global* (respectively *local*) *degree ordering* if the above holds for $w_i = 1$ (respectively $w_i = -1$) for all i.

Remark 1.2.10. Consider a matrix ordering defined by $A \in \mathrm{GL}(n, \mathbb{R})$. Since the columns of A are lexicographically greater than the 0–vector if and only if the variables are greater than 1, it follows that a matrix ordering $>_A$ is a well–ordering if and only if the first non–zero entry in each column of A is positive. It is a (weighted) degree ordering if and only if all entries in the first row of A are non–zero.

Of course, different matrices can define the same ordering. For examples of matrices defining the above orderings see the SINGULAR Manual.

Although we can represent any monomial ordering $>$ as a matrix ordering $>_A$ for some $A \in \mathrm{GL}(n, \mathbb{R})$, it turns out to be useful to represent $>$ just by one weight vector. This is, in general, not possible on the set of all monomials (cf. Exercise 1.2.10) but it is possible, as we shall see, for finite subsets.

For this purpose, we introduce the set of differences

$$D := \{\alpha - \beta \mid x^\alpha > x^\beta\} \subset \mathbb{Z}^n$$

associated to a monomial ordering on $\mathrm{Mon}(x_1, \ldots, x_n)$. D has the following properties,

- $0 \notin D$,
- $\gamma_1, \gamma_2 \in D \Longrightarrow \gamma_1 + \gamma_2 \in D$.

The last property follows from the fact that $>$ is a semigroup ordering. Namely, if $\gamma_1 = \alpha_1 - \beta_1$, $\gamma_2 = \alpha_2 - \beta_2 \in D$, then $x^{\alpha_1} > x^{\beta_1}$ implies that $x^{\alpha_1 + \alpha_2} > x^{\beta_1 + \alpha_2}$, and $x^{\alpha_2} > x^{\beta_2}$ implies that $x^{\beta_1 + \alpha_2} > x^{\beta_1 + \beta_2}$, therefore $x^{\alpha_1 + \alpha_2} > x^{\beta_1 + \beta_2}$ and $\gamma_1 + \gamma_2 = (\alpha_1 + \alpha_2) - (\beta_1 + \beta_2) \in D$.

It follows that $\sum_{i=1}^{k} n_i \gamma_i \in D$ for $n_i \in \mathbb{N} \setminus \{0\}$ and $\gamma_i \in D$, and, hence, $\sum_{i=1}^{k} r_i \gamma_i \neq 0$ for any finite linear combination of elements of D with $r_i \in \mathbb{Q}_{>0}$. In particular, no convex combination $\sum_{i=1}^{k} r_i \gamma_i$, $r_i \in \mathbb{Q}_{\geq 0}$, $\sum_{i=1}^{k} r_i = 1$, yields 0, that is, 0 is not contained in the convex hull of D. This fact will be used in the following lemma.

Lemma 1.2.11. *Let $>$ be a monomial ordering and $M \subset \mathrm{Mon}(x_1, \ldots, x_n)$ a finite set. Then there exists some $w = (w_1, \ldots, w_n) \in \mathbb{Z}^n$ such that $x^\alpha > x^\beta$ if and only if $\langle w, \alpha \rangle > \langle w, \beta \rangle$ for all $x^\alpha, x^\beta \in M$. Moreover, w can be chosen such that $w_i > 0$ for $x_i > 1$ and $w_i < 0$ if $x_i < 1$.*

The integer vector w is called a *weight–vector* and we say that w induces $>$ on M.

Proof. Since $\langle w, \alpha \rangle > \langle w, \beta \rangle$ if and only if $\langle w, \alpha - \beta \rangle > 0$, we have to find $w \in \mathbb{Z}^n$ such that $\langle w, \gamma \rangle > 0$ for all

$$\gamma \in D_M := \{\alpha - \beta \in D \mid x^\alpha, x^\beta \in M, \ x^\alpha > x^\beta\}.$$

This means that D_M should be in the positive half–space defined by the linear form $\langle w, - \rangle$ on $\mathbb{Q}^n$. Since 0 is not contained in the convex hull of D_M and since D_M is finite, we can, indeed, find such a linear form (see, for example, [221], Theorem 2.10).

To see the last statement, include 1 and x_i, $i = 1, \ldots, n$, into M. Then $w_i > 0$ if $x_i > 1$ and $w_i < 0$ if $x_i < 1$. $\square$

Example 1.2.12. A weight vector for the lexicographical ordering 1p can be determined as follows. For $M \subset \mathrm{Mon}_n$ finite, consider an n–dimensional cube spanned by the coordinate axes containing M. Choose an integer v larger than the side length of this cube. Then $w = (v^{n-1}, v^{n-2}, \ldots, v, 1)$ induces 1p on M.

We shall now define in SINGULAR the same ring $\mathbb{Q}[x, y, z]$ with different orderings, which are considered as different rings in SINGULAR. Then we map a given polynomial f to the different rings using imap and display f as a sum of terms in decreasing order, the method by which f is represented in the given ring.

SINGULAR Example 1.2.13 (monomial orderings).

Global orderings are denoted with a p at the end, referring to "polynomial ring" while local orderings end with an s, referring to "series ring". Note that SINGULAR stores and outputs a polynomial in an ordered way, in decreasing order.

(1) Global orderings:

```
ring A1 = 0,(x,y,z),1p;       //lexicographical
poly f = x3yz + y5 + z4 + x3 + xy2; f;
//-> x3yz+x3+xy2+y5+z4

ring A2 = 0,(x,y,z),dp;       //degree reverse lexicographical
poly f = imap(A1,f); f;
//-> y5+x3yz+z4+x3+xy2

ring A3 = 0,(x,y,z),Dp;       //degree lexicographical
poly f = imap(A1,f); f;
//-> x3yz+y5+z4+x3+xy2

ring A4 = 0,(x,y,z),Wp(5,3,2);//weighted degree
                              //lexicographical
poly f = imap(A1,f); f;
//-> x3yz+x3+y5+xy2+z4
```

(2) Local orderings:

```
ring A5 = 0,(x,y,z),1s;       //negative lexicographical
poly f = imap(A1,f); f;
//-> z4+y5+xy2+x3+x3yz
```

```
ring A6 = 0,(x,y,z),ds;        //negative degree reverse
                               //lexicographical
poly f = imap(A1,f); f;
//-> x3+xy2+z4+y5+x3yz

ring A7 = 0,(x,y,z),Ws(5,3,2);//negative weighted degree
                               //lexicographical
poly f = imap(A1,f); f;
//-> z4+xy2+x3+y5+x3yz
```

(3) Product and matrix orderings:

```
ring A8 = 0,(x,y,z),(dp(1),ds(2)); //mixed product ordering
poly f = imap(A1,f); f;
//-> x3+x3yz+xy2+z4+y5

intmat A[3][3] = -1, -1, -1, 0, 0, 1, 0, 1, 0;
print(A);
//->      -1    -1    -1
//->       0     0     1
//->       0     1     0
```

Now define your own matrix ordering using A:

```
ring A9 = 0,(x,y,z),M(A); //a local ordering
poly f = imap(A1,f); f;
//-> xy2+x3+z4+x3yz+y5
```

Exercises

1.2.1. Show that lp, dp, Dp, wp(w(1..m)), Wp(w(1..n)), respectively ls, ds, Ds, ws(w(1..m)), Ws(w(1..n)), as defined in Example 1.2.8 are indeed global, respectively local, monomial orderings.

1.2.2. Determine the names of the orderings given by the following matrices:

$$\begin{pmatrix} 1 & 1 \\ 0 & -1 \end{pmatrix}, \quad \begin{pmatrix} 1 & 0 \\ 0 & 1 \end{pmatrix}, \quad \begin{pmatrix} -1 & -1 \\ 0 & -1 \end{pmatrix}, \quad \begin{pmatrix} -1 & 0 \\ 0 & -1 \end{pmatrix}, \quad \begin{pmatrix} 1 & 2 \\ 0 & -1 \end{pmatrix}, \quad \begin{pmatrix} 1 & 1 & 0 & 0 \\ 0 & -1 & 0 & 0 \\ 0 & 0 & -1 & -1 \\ 0 & 0 & 0 & -1 \end{pmatrix}.$$

1.2.3. Order the polynomial $x^4 + z^5 + x^3z + yz^4 + x^2y^2$ with respect to the orderings dp,Dp,lp,ds,Ds,ls,wp(5,3,4),ws(5,5,4).

1.2.4. Compute the leading term and the leading coefficient

$$f = 4xy^2z + 4z^2 - 5x^3 + 7xy^2 - 7y^4$$

with respect to the orderings lp on $\mathbb{Q}[x,y,z]$, lp on $\mathbb{Q}(x)[z,y]$, lp on $\mathbb{Q}[z,y,x]$, Dp on $(\mathbb{Z}/2\mathbb{Z})[z,y,x]$, ls on $\mathbb{Q}[x,y,z]$, wp(w(1..3)) on $(\mathbb{Z}/2\mathbb{Z})[x,y,z]$, where wp(w(1..3)) is given by w–deg$(x^\alpha y^\beta z^\gamma) := 3\alpha + 2\beta + \gamma$.

1.2.5. Determine matrices defining the orderings dp, Dp, lp, ds, Ds, ls, wp(5,3,4), ws(5,5,4).

1.2.6. Let $>$ be any monomial ordering on $\mathrm{Mon}(x_1, \ldots, x_n)$.

(1) Let $w = (w_1, \ldots, w_n) \in \mathbb{R}^n$ be arbitrary. Show that

$$x^\alpha >_w x^\beta \ :\Longleftrightarrow \ \langle w, \alpha \rangle > \langle w, \beta \rangle \text{ or } \langle w, \alpha \rangle = \langle w, \beta \rangle \text{ and } x^\alpha > x^\beta$$

defines a monomial ordering on $\mathrm{Mon}(x_1, \ldots, x_n)$.

Note that the ordering $>_w$ is a (weighted) degree ordering. It is a global ordering if $w_i > 0$ for all i and a local ordering if $w_i < 0$ for all i.

(2) Let A be an $n \times n$ integer matrix with non–negative entries, which is invertible over $\mathbb{Q}$. Show that

$$x^\alpha >_{(A,>)} x^\beta \Leftrightarrow x^{A\alpha} > x^{A\beta}$$

defines a monomial ordering on $\mathrm{Mon}(x_1, \ldots, x_n)$.

1.2.7. (1) Prove the claim made in Example 1.2.12.

(2) Consider a matrix ordering $>_A$ for some matrix $A \in \mathrm{GL}(n, \mathbb{Q})$ and $M \subset \mathrm{Mon}_n$ a finite set. Use (1) and the fact that $x^\alpha >_A x^\beta$ if and only if $A\alpha >_{\mathrm{lex}} A\beta$ to determine a weight vector which induces $>_A$ on M.

1.2.8. (1) Determine weight vectors w which induce dp, respectively ds, on $M = \{x^i y^j z^k \mid 1 \leq i, j, k \leq 5\}$.

(2) Check your result, using SINGULAR, in the following way: create a polynomial f, being the sum of all monomials of degree ≤ 5 in the rings with ordering dp, respectively ds, and convert f to a string. Then do the same in the rings with ordering wp(w), respectively ws(-w), ((a(w),lp), respectively (a(-w),lp)), and compare the respective strings.

1.2.9. Show that any monomial ordering $>$ can be defined as $>_A$ by a matrix $A \in \mathrm{GL}(n, \mathbb{R})$.

(Hint: You may proceed as follows: first show that a semigroup ordering on $(\mathbb{Z}^n_{\geq 0}, +)$ extends in a unique way to a group ordering on $(\mathbb{Q}^n, +)$. Then show that, for any $\mathbb{Q}$–subvector space $V \subset \mathbb{Q}^n$ of dimension r, the set

$$V_0 := \left\{ z \in \mathbb{R}^n \ \middle| \ \begin{array}{c} \forall \varepsilon > 0 \ \exists \ z_+(\varepsilon), z_-(\varepsilon) \in U_\varepsilon(z) \cap V \\ \text{such that } z_+(\varepsilon) > 0, z_-(\varepsilon) < 0 \end{array} \right\}$$

is an $\mathbb{R}$–subvector space in $\mathbb{R}^n$ of dimension $r - 1$. Use this to construct, successively, the rows of A.)

1.2.10. Let $w_1, \ldots, w_n \in \mathbb{R}$ be linearly independent over $\mathbb{Q}$ and define $>$ by setting $x^\alpha < x^\beta$ if $\sum_{i=1}^n w_i \alpha_i < \sum_{i=1}^n w_i \beta_i$. Prove that $>$ is a monomial ordering. Show that there is no matrix $A \in \mathrm{GL}(n, \mathbb{Q})$ defining this ordering.

1.3 Ideals and Quotient Rings

Ideals are in the centre of commutative algebra and algebraic geometry. Here we introduce only the basic notions related to them.

Let A be a ring, as always, commutative and with 1.

Definition 1.3.1. A subset $I \subset A$ is called an *ideal* if it is an additive subgroup which is closed under scalar multiplication, that is,

$$f, g \in I \implies f + g \in I$$
$$f \in I, a \in A \implies af \in I.$$

Definition 1.3.2.

(1) Let $I \subset A$ be an ideal. A family $(f_\lambda)_{\lambda \in \Lambda}$, Λ any index set, and $f_\lambda \in I$, is called a *system of generators* of I if every element $f \in I$ can be expressed as a finite linear combination $f = \sum_\lambda a_\lambda f_\lambda$ for suitable $a_\lambda \in A$. We then write

$$I = \langle f_\lambda \mid \lambda \in \Lambda \rangle_A = \langle f_\lambda \mid \lambda \in \Lambda \rangle = \sum_{\lambda \in \Lambda} f_\lambda A$$

or, if $\Lambda = \{1, \ldots, k\}$,

$$I = \langle f_1, \ldots, f_k \rangle_A = \langle f_1, \ldots, f_k \rangle.$$

(2) I is called *finitely generated* if it has a finite system of generators; it is called *principal* if it can be generated by one element.

(3) If $(I_\lambda)_{\lambda \in \Lambda}$ is a family of ideals, then $\sum_{\lambda \in \Lambda} I_\lambda$ denotes the ideal generated by $\bigcup_{\lambda \in \Lambda} I_\lambda$.

(4) If I_1, I_2 are ideals, then $I_1 I_2$ (or $I_1 \cdot I_2$) denotes the ideal generated by the set $\{ab \mid a \in I_1, b \in I_2\}$.

Note that the union of ideals is, in general, not an ideal (but the intersection is). We have

$$\sum_{\lambda \in \Lambda} I_\lambda = \left\{ \sum_{\lambda \in \Lambda} a_\lambda \;\middle|\; a_\lambda \in I_\lambda, \; a_\lambda = 0 \text{ for almost all } \lambda \right\}.$$

Because the empty sum is defined to be 0, the 0–ideal is generated by the empty set (but also by 0). The expression $f = \sum_\lambda a_\lambda f_\lambda$ as a linear combination of the generators is, in general, by no means unique. For example, if $I = \langle f_1, f_2 \rangle$ then we have the *trivial relation* $f_1 f_2 - f_2 f_1 = 0$, hence $a_1 f_1 = a_2 f_2$ with $a_1 = f_2$, $a_2 = f_1$. Usually there are also further relations, which lead to the notion of the module of syzygies (cf. Chapter 2).

Ideals occur in connection with ring maps. If $\varphi : A \to B$ is a ring homomorphism and $J \subset B$ an ideal, then the *preimage*

$$\varphi^{-1}(J) = \{a \in A \mid \varphi(a) \in J\}$$

is an ideal. In particular,

$$\mathrm{Ker}\,\varphi = \{a \in A \mid \varphi(a) = 0\}$$

is an ideal in A. On the other hand, the *image*

$$\varphi(I) = \{\varphi(a) \mid a \in I\}$$

of an ideal $I \subset A$ is, in general, not an ideal. In particular, $\mathrm{Im}\,\varphi = \varphi(A) \subset B$ is not, generally, an ideal (for example, consider $\mathbb{Z} \subset \mathbb{Q}$, then no non–zero ideal in $\mathbb{Z}$ is an ideal in $\mathbb{Q}$). All these statements are very easy to check.

φ is called *injective* if $\mathrm{Ker}\,\varphi = 0$, and *surjective* if $\mathrm{Im}\,\varphi = B$. A *bijective*, that is injective and surjective, morphism is called an *isomorphism*, an isomorphism from A to A an *automorphism*.

SINGULAR contains the built–in command **preimage** which can be used to compute the kernel of a ring map.

If a ring map $\varphi : K[x_1, \ldots, x_k] \to K[y_1, \ldots, y_m]$ is given by $f_1, \ldots, f_k$, that is, $\varphi(x_i) = f_i$, then φ is surjective if and only if $y_1, \ldots, y_m$ are contained in the subring $\mathrm{Im}\,\varphi = K[f_1, \ldots, f_m]$ of $K[y_1, \ldots, y_m]$. This fact is used in SINGULAR to check surjectivity.

We shall explain the algorithms for checking injectivity, surjectivity, bijectivity of a ring map in Chapter 2. Here we just apply the corresponding procedures from `algebra.lib`.

SINGULAR Example 1.3.3 (properties of ring maps).

(1) Checking injectivity:

```
ring S = 0,(a,b,c),lp;
ring R = 0,(x,y,z),dp;
ideal i = x, y, x2-y3;
map phi = S,i;       //a map from S to R, a->x, b->y, c->x2-y3
LIB "algebra.lib"; //load algebra.lib
```

By default, SINGULAR displays the names and paths of those libraries which are used by `algebra.lib` and which are also loaded. We suppress this message.

We test injectivity using the procedure `is_injective`, then we compute the kernel by using the procedure `alg_kernel` (which displays the kernel, an object of the preimage ring, as a string).

```
is_injective(phi,S);
//-> 0                         // phi is not injective
```

```
ideal j = x, x+y, z-x2+y3;
map psi = S,j;                   // another map from S to R
is_injective(psi,S);
//-> 1                          // psi is injective

alg_kernel(phi,S);
//-> b^3-a^2+c                  // <b^3-a^2+c> = Ker(phi)
alg_kernel(psi,S);
//-> 0
```

(2) Computing the *preimage*:

Using the **preimage** command, we must first go back to S, since the preimage is an ideal in the preimage ring.

```
ideal Z;                        //the zero ideal in R
setring S;
preimage(R,phi,Z);              //computes kernel of phi in S
//-> _[1]=a2-b3-c                //kernel of phi = preimage of Z
```

(3) Checking *surjectivity* and *bijectivity*.

```
setring R;
is_surjective(psi,S);
//-> 1
is_bijective(psi,S);            //faster than is_injective,
                                //is_surjective
//-> 1
```

Definition 1.3.4. A ring A is called *Noetherian* if every ideal in A is finitely generated.

It is a fundamental fact that the polynomial ring $A[x_1, \ldots, x_n]$ over a Noetherian ring A is again Noetherian; this is the content of the Hilbert basis theorem. Since a field is obviously a Noetherian ring, the polynomial ring over a field is Noetherian. It follows that the kernel of a ring map between Noetherian rings is finitely generated. An important point of the SINGULAR Example 1.3.3 is that we can explicitly compute a finite set of generators for the kernel of a map between polynomial rings.

Theorem 1.3.5 (Hilbert basis theorem). *If A is a Noetherian ring then the polynomial ring $A[x_1, \ldots, x_n]$ is Noetherian.*

For the proof of the Hilbert basis theorem we use

Proposition 1.3.6. *The following properties of a ring A are equivalent:*

(1) A is Noetherian.
(2) Every ascending chain of ideals

$$I_1 \subset I_2 \subset I_3 \subset \ldots \subset I_k \subset \ldots$$

becomes stationary (that is, there exists some j_0 such that $I_j = I_{j_0}$ for all $j \geq j_0$).
(3) Every non–empty set of ideals in A has a maximal element (with regard to inclusion).

Condition (2) is called the *ascending chain condition* and (3) the *maximality condition*. We leave the proof of this proposition as Exercise 1.3.9.

Proof of Theorem 1.3.5. We need to show the theorem only for $n = 1$, the general case follows by induction.

We argue by contradiction. Let us assume that there exists an ideal $I \subset A[x]$ which is not finitely generated. Choose polynomials

$$f_1 \in I, \quad f_2 \in I \setminus \langle f_1 \rangle, \quad \ldots, \quad f_{k+1} \in I \setminus \langle f_1, \ldots, f_k \rangle, \quad \ldots$$

of minimal possible degree. If $d_i = \deg(f_i)$,

$$f_i = a_i x^{d_i} + \text{lower terms in } x,$$

then $d_1 \leq d_2 \leq \ldots$ and $\langle a_1 \rangle \subset \langle a_1, a_2 \rangle \subset \ldots$ is an ascending chain of ideals in A. By assumption it is stationary, that is, $\langle a_1, \ldots, a_k \rangle = \langle a_1, \ldots, a_{k+1} \rangle$ for some k, hence, $a_{k+1} = \sum_{i=1}^{k} b_i a_i$ for suitable $b_i \in A$. Consider the polynomial

$$g = f_{k+1} - \sum_{i=1}^{k} b_i x^{d_{k+1}-d_i} f_i = a_{k+1} x^{d_{k+1}} - \sum_{i=1}^{k} b_i a_i x^{d_{k+1}} + \text{lower terms}.$$

Since $f_{k+1} \in I \setminus \langle f_1, \ldots, f_k \rangle$, it follows that $g \in I \setminus \langle f_1, \ldots, f_k \rangle$ is a polynomial of degree smaller than d_{k+1}, a contradiction to the choice of f_{k+1}. $\qquad\square$

Definition 1.3.7. Let I be any ideal in the ring A. We define the *quotient ring* or *factor ring* A/I as follows.

(1) A/I is the set of co–sets $\{[a] := a + I \mid a \in A\}^2$ with addition and multiplication defined via representatives:

$$[a] + [b] := [a + b],$$
$$[a] \cdot [b] := [a \cdot b].$$

---

[2] $a + I := \{a + f \mid f \in I\}$.

It is easy to see that the definitions are independent of the chosen representatives and that $(A/I, +, \cdot)$ is, indeed, a ring. Moreover, A/I is not the zero ring if and only if $1 \notin I$.

(2) The *residue map* or *quotient map* is defined by

$$\pi : A \longrightarrow A/I, \quad a \longmapsto [a].$$

π is a surjective ring homomorphism with kernel I.

The following lemma is left as an easy exercise.

Lemma 1.3.8. *The map $J \mapsto \pi(J)$ induces a bijection*

$$\{ideals\ in\ A\ containing\ I\} \longrightarrow \{ideals\ in\ A/I\}$$

with $J' \mapsto \pi^{-1}(J')$ being the inverse map.

Definition 1.3.9.

(1) An element $a \in A$ is called a *zerodivisor* if there exists an element $b \in A \smallsetminus \{0\}$ satisfying $ab = 0$; otherwise a is a *non–zerodivisor*.

(2) A is called an *integral domain* if $A \neq 0$ and if A has no zerodivisors except 0.

(3) A is a *principal ideal ring* if every ideal in A is principal; if A is, moreover, an integral domain it is called a *principal ideal domain*.

Polynomial rings over a field are integral domains (Exercise 1.3.1 (4)). This is, however, not generally true for quotient rings $K[x_1, \ldots, x_n]/I$. For example, if $I = \langle f \cdot g \rangle$ with $f, g \in K[x_1, \ldots, x_n]$ polynomials of positive degree, then $[f]$ and $[g]$ are zerodivisors in $K[x_1, \ldots, x_n]/I$ and not zero.

A ring A, which is isomorphic to a factor ring $K[x_1, \ldots, x_n]/I$, is called an *affine ring* over K.

Definition 1.3.10. Let $I \subset A$ be an ideal.

(1) I is a *prime ideal* if $I \neq A$ and if for each $a, b \in A : ab \in I \Rightarrow a \in I$ or $b \in I$.

(2) I is a *maximal ideal* if $I \neq A$ and if it is maximal with respect to inclusion (that is, for any ideal $I' \subsetneqq A$ and $I \subset I'$ implies $I = I'$).

(3) The set of prime ideals is denoted by $\mathrm{Spec}(A)$ and the set of maximal ideals by $\mathrm{Max}(A)$.

The set of prime ideals $\mathrm{Spec}(A)$ of a ring A is made a topological space by endowing it with the so–called Zariski topology, creating, thus, a bridge between algebra and topology. We refer to the Appendix, in particular A.3, for a short introduction. In many cases in the text we use $\mathrm{Spec}(A)$ just as a set. But, from time to time, when we think we should relax and enjoy geometry, then we consider the affine space $\mathrm{Spec}(A)$ instead of the ring A and the variety $V(I) \subset \mathrm{Spec}(A)$ instead of the ideal I. Most of the examples deal with affine rings over a field K.

Lemma 1.3.11.

(1) $I \subset A$ is a prime ideal if and only if A/I is an integral domain.
(2) $I \subset A$ is a maximal ideal if and only if A/I is a field.
(3) Every maximal ideal is prime.

Proof. Let $I \subsetneq A$. For $a, b \in A$ we have $ab \in I \iff [ab] = [a] \cdot [b] = 0$ in A/I, which implies (1). By Lemma 1.3.8, A/I has only the trivial ideals 0 and A/I, if and only if I and A are the only ideals of A which contain I, which implies (2). Finally, (3) follows from (2) and (1), since a field is an integral domain. $\qquad\square$

If $\varphi : A \to B$ is a ring map and $I \subset B$ is a prime ideal, then $\varphi^{-1}(I)$ is a prime ideal (an easy check). However, the preimage of a maximal ideal need not be maximal. (Consider $\mathbb{Z} \subset \mathbb{Q}$, then 0 is a maximal ideal in $\mathbb{Q}$ but not in $\mathbb{Z}$.)

Lemma 1.3.12. *Let A be a ring.*

(1) Let $P, I, J \subset A$ be ideals with P prime. Then $I \not\subset P$, $IJ \subset P$ implies $J \subset P$.
(2) Let $I_1, \ldots, I_n$, $P \subset A$ be ideals with P prime and $\bigcap_{i=1}^{n} I_i \subset P$ (respectively $\bigcap_i I_i = P$), then $P \supset I_i$ (respectively $P = I_i$) for some i.
(3) (Prime avoidance) Let $P_1, \ldots, P_n$, $I \subset A$ be ideals with P_i prime and $I \subset \bigcup_{i=1}^{n} P_i$, then $I \subset P_i$ for some i.

Proof. For (1) let $J = \langle f_1, \ldots, f_n \rangle$ and $x \in I$ such that $x \notin P$. By assumption, we have $xf_i \in P$ for all i. Now P is prime and, therefore, $f_i \in P$ for all i. This implies that $J \subset P$.

To prove (2) assume that $\bigcap I_i \subset P$. Then $\prod I_i \subset P$ and, therefore, using (1), $I_k \subset P$ for some k. If, additionally, $\bigcap I_i = P$, then $P = I_k$.

To prove (3) we use induction on n. The case $n = 1$ is trivial. Assume (3) is true for $n - 1$ prime ideals. If $I \subset \bigcup_{j \neq i} P_j$ for some i, then $I \subset P_k$ for some k.

We may assume now that $I \not\subset \bigcup_{j \neq i} P_j$ for all $i = 1, \ldots, n$ and choose $x_1, \ldots, x_n \in I$ such that $x_i \notin \bigcup_{j \neq i} P_j$. This implies especially that $x_i \in P_i$ because $x_i \in I \subset \bigcup P_j$.

Now consider the element $x_1 + x_2 \cdot \ldots \cdot x_n \in I$. Since $I \subset \bigcup P_j$, there exists a k such that $x_1 + x_2 \cdot \ldots \cdot x_n \in P_k$. If $k = 1$ then, since $x_1 \in P_1$, we obtain $x_2 \cdot \ldots \cdot x_n \in P_1$. This implies that $x_\ell \in P_1$ for some $\ell > 1$ which is a contradiction to the choice of $x_\ell \notin \bigcup_{j \neq \ell} P_j$. If $k > 1$ then, since $x_2 \cdot \ldots \cdot x_n \in P_k$, we obtain $x_1 \in P_k$ which is again a contradiction to the choice of $x_1 \notin \bigcup_{j \neq 1} P_j$. $\qquad\square$

Many of the concepts introduced so far in this section can be treated effectively using SINGULAR. We define a quotient ring and test equality and the zerodivisor property in the quotient ring.

SINGULAR Example 1.3.13 (computation in quotient rings).

(1) Define a *quotient ring*:

```
ring R = 32003,(x,y,z),dp;
ideal I = x2+y2-z5, z-x-y2;
qring Q = groebner(I);    //defines the quotient ring Q = R/I
Q;
//-> //   characteristic : 32003
//-> //   number of vars : 3
//-> //          block   1 : ordering dp
//-> //                    : names     x y z
//-> //          block   2 : ordering C
//-> // quotient ring from ideal
//-> _[1]=y2+x-z
//-> _[2]=z5-x2+x-z
```

(2) *Equality test* in quotient rings:

Equality test in quotient rings is difficult. The test `f==g` checks only formal equality of polynomials, it does not work correctly in quotient rings. Instead, we have to compute a normal form of the difference $f - g$. Why and how this works, will be explained in Section 1.6 on standard bases.

```
poly f = z2+y2;
poly g = z2+2x-2z-3z5+3x2+6y2;
reduce(f-g,std(0));    //normal form, result is 0 iff f=g in Q
//-> 0
```

The same can be tested without going to the quotient ring.

```
setring R;
poly f = z2+y2;    poly g = z2+2x-2z-3z5+3x2+6y2;
reduce(f-g,groebner(I)); //result is 0 iff f-g is in I
//-> 0
```

(3) *Zerodivisor test* in quotient rings:

```
setring Q;
ideal q = quotient(0,f);//this defines q = <0>:<f>
q = reduce(q,std(0));    //normal form of ideal q in Q
size(q);                 //the number of non-zero generators
//-> 0                   //hence, f is a non-zerodivisor in Q
```

Testing primality of a principal ideal $\langle f \rangle$ in a polynomial ring is easily achieved by using `factorize(f);`. For an arbitrary ideal this is much more involved. One can use `primdecGTZ` or `primdecSY` from `primdec.lib`, as will be explained in Chapter 4.

(4) Computing the *inverse* in quotient rings:

If $I \subset K[x] = K[x_1, \ldots, x_n]$ is a maximal ideal, then the quotient ring $K[x]/I$ is a field. To be able to compute effectively in the field $K[x]/I$ we need, in addition to the ring operations, the inverse of a non–zero element. The following example shows that we can effectively compute in all fields of finite type over a prime field.

If the polynomial f is invertible, then the command `lift(f,1)[1,1]` gives the inverse (`lift` checks whether $1 \in \langle f \rangle$ and then expresses 1 as a multiple of f):

```
ring R=(0,x),(y,z),dp;
ideal I=-z5+y2+(x2),-y2+z+(-x);
I=std(I);
qring Q=I;
```

We shall now compute the inverse of z in $Q = R/I$.

```
poly p=lift(z,1)[1,1];
p;
//->1/(x2-x)*z4-1/(x2-x)
```

We make a test for p being the inverse of z.

```
reduce(p*z,std(0));
//->1
```

The ideal I is a maximal ideal if and only if R/I is a field. We shall now prove that, in our example, I is a maximal ideal.

```
ring R1=(0,x),(z,y),lp;
ideal I=imap(R,I);
I=std(I);
I;
//-> I[1]=y10+(5x)*y8+(10x2)*y6+(10x3)*y4+(5x4-1)*y2+(x5-x2)
//-> I[2]=z-y2+(-x)
```

Since $\mathbb{Q}(x)[z, y]/\langle z - y^2 - x \rangle \cong \mathbb{Q}(x)[y]$, we see that

$$R/I \cong \mathbb{Q}(x)[y]/\langle y^{10} + 5xy^8 + 10x^2y^6 + 10x^3y^4 + (5x^4 - 1)y^2 + x^5 - x^2 \rangle.$$

```
factorize(I[1]);
//-> [1]:
//->     _[1]=1
//->     _[2]=y10+(5x)*y8+(10x2)*y6+(10x3)*y4+(5x4-1)*y2
//->           +(x5-x2)
//-> [2]:
//->     1,1
```

The polynomial is irreducible and, therefore, R/I is a field and I a maximal ideal.

Definition 1.3.14. Let A be a ring and $I, J \subset A$ ideals.

(1) The *ideal quotient* of I by J is defined as

$$I : J := \{a \in A \mid aJ \subset I\}.$$

The *saturation of I with respect to J* is

$$I : J^{\infty} = \{a \in A \mid \exists\, n \text{ such that } aJ^n \subset I\}.$$

(2) The *radical* of I, denoted by $\sqrt{I}$ or $\mathrm{rad}(I)$ is the ideal

$$\sqrt{I} = \{a \in A \mid \exists\, d \in \mathbb{N} \text{ such that } a^d \in I\},$$

which is an ideal containing I. I is called *reduced* or a *radical ideal* if $I = \sqrt{I}$.

(3) $a \in A$ is called *nilpotent* if $a^n = 0$ for some $n \in \mathbb{N}$; the minimal n is called *index of nilpotency*. The set of nilpotent elements of A is equal to $\sqrt{\langle 0 \rangle}$ and called the *nilradical* of A.

(4) The ring A itself is called *reduced* if it has no nilpotent elements except 0, that is, if $\sqrt{\langle 0 \rangle} = \langle 0 \rangle$. For any ring, the quotient ring

$$A_{\mathrm{red}} = A/\sqrt{\langle 0 \rangle}$$

is called the *reduction* of A or the *reduced ring associated to A*.

The ideal quotient $I : J$ is an ideal in A which is very useful. In SINGULAR the command `quotient(I,J);` computes generators of this ideal. In particular,

$$\langle 0 \rangle : J = \mathrm{Ann}_A(J)$$

is the *annihilator* of J and, hence, $\langle 0 \rangle : \langle f \rangle = \langle 0 \rangle$ if and only if f is a non–zerodivisor of A.

It is clear that A_{red} is reduced and that $A = A_{\mathrm{red}}$ if and only if A is reduced. Any integral domain is reduced.

Computing the radical is already quite involved (cf. Chapter 4). The radical membership problem is, however, much easier (cf. Section 1.8.6).

SINGULAR Example 1.3.15 (computing with radicals).

(1) Compute the radical of an ideal:

```
ring R = 0,(x,y,z),dp;
poly p = z4+2z2+1;
LIB "primdec.lib";          //loads library for radical
```

```
radical(p);                    //squarefree part of p
//-> _[1]=z2+1


ideal I = xyz, x2, y4+y5;  //a more complicated ideal
radical(I);
//-> _[1]=x
//-> _[2]=y2+y                 //we see that I is not reduced
```

(2) Compute the *index of nilpotency* in a quotient ring:

Since $y^2 + y$ is contained in the radical of I, some power of $y^2 + y$ must be contained in I. We compute the minimal power k so that $(y^2 + y)^k$ is contained in I by using the normal form as in Example 1.3.13. This is the same as saying that $y^2 + y$ is nilpotent in the quotient ring R/I and then k is the index of nilpotency of $y^2 + y$ in R/I.

```
ideal Is = groebner(I);
int k;
while (reduce((y2+y)^k,Is) != 0 ) {k++;}
k;
//-> 4             //minimal power (index of nilpotency) is 4
```

Exercises

1.3.1. Let A be a ring and $f = \sum_{|\alpha| \geq 0} a_\alpha x^\alpha \in A[x_1, \ldots, x_n]$. Prove the following statements:

(1) f is nilpotent if and only if a_α is nilpotent for all α.

   (Hint: choose a monomial ordering and argue by induction on the number of summands.)

   In particular: $A[x_1, \ldots, x_n]$ is reduced if and only if A is reduced.

(2) f is a unit in $A[x_1, \ldots, x_n]$ if and only if $a_{0,\ldots,0}$ is a unit in A and a_α are nilpotent for $\alpha \neq 0$.

   (Hint: Remember the geometric series for $1/(1 - g)$ and use (1).)

   In particular: $(A[x_1, \ldots, x_n])^* = A^*$ if and only if A is reduced.

(3) f is a zerodivisor in $A[x_1, \ldots, x_n]$ if and only if there exists some $a \neq 0$ in A such that $af = 0$. Give two proofs: one by induction on n, the other by using a monomial ordering.

   (Hint: choose a monomial ordering and $g \in A[x_1, \ldots, x_n]$ with minimal number of terms so that $f \cdot g = 0$, consider the biggest term and conclude that g must be a monomial.)

(4) $A[x_1, \ldots, x_n]$ is an integral domain if and only if $\deg(fg) = \deg(f) + \deg(g)$ for all $f, g \in A[x_1, \ldots, x_n]$.

   In particular: $A[x_1, \ldots, x_n]$ is an integral domain if and only if A is an integral domain.

1.3.2. Let $\varphi : A \to B$ be a ring homomorphism, I an ideal in A and J an ideal in B. Show that:

(1) $\varphi^{-1}(J) \subset I$ is an ideal.
(2) $\varphi(I)$ is a subring of B, not necessarily with 1, but, in general, not an ideal.
(3) If φ is surjective then $\varphi(I)$ is an ideal in B.

1.3.3. Prove the following statements:

(1) $\mathbb{Z}$ and the polynomial ring $K[x]$ in one variable over a field are principal ideal domains [use division with remainder].
(2) Let A be any ring, then $A[x_1, \ldots, x_n]$, $n > 1$, is not a principal ideal domain.

1.3.4. Let A be a ring. A non–unit $f \in A$ is called *irreducible* if $f = f_1 f_2$, $f_1, f_2 \in A$, implies that f_1 or f_2 is a unit. f is called a *prime element* if $\langle f \rangle$ is a prime ideal. Prove Exercise 1.1.5 with $K[x]$ replaced by any principal ideal domain A. Moreover, prove that the conditions $(1)-(3)$ of Exercise 1.1.5 are equivalent to

(4) The ideal $\langle f \rangle$ is a prime ideal.
(5) The ideal $\langle f \rangle$ is a maximal ideal.

1.3.5. Let R be a principal ideal domain. Use Exercise 1.3.4 to prove that every non–unit $f \in R$ can be written in a unique way as a product of finitely many prime elements. Unique means here modulo permutation and multiplication with a unit.

1.3.6. The quotient ring of a principal ideal ring is a principal ideal ring. Show, by an example, that the quotient ring of an integral domain (respectively a reduced ring) need not be an integral domain.

1.3.7. (1) If A, B are principal ideal rings, then, also $A \oplus B$.
(2) $A \oplus B$ is never an integral domain, unless A or B are trivial.
(3) How many ideals has $K \oplus F$ if K and F are fields?

1.3.8. Prove the following statements:

(1) Let $n > 1$, then $\mathbb{Z}/n\mathbb{Z}$ is reduced if and only if n is a product of pairwise different primes.
(2) Let K be a field, and let $f \in K[x_1, \ldots, x_n]$ be a polynomial of degree ≥ 1. Then $K[x_1, \ldots, x_n]/\langle f \rangle$ is reduced (respectively an integral domain) if and only if f is a product of pairwise different irreducible polynomials (respectively irreducible).

1.3.9. Prove Proposition 1.3.6.

1.3.10. Prove Lemma 1.3.8.

1.3.11. Let A be a Noetherian ring, and let $I \subset A$ be an ideal. Prove that A/I is Noetherian.

1.3.12. Let A be a Noetherian ring, and let $\varphi : A \to A$ be a surjective ring homomorphism. Prove that φ is injective.

1.3.13. (*Chinese remainder theorem*) Let A be a ring, and let $I_1, \ldots, I_s$ be ideals in A. Assume that $\bigcap_{j=1}^s I_j = \langle 0 \rangle$ and $I_j + I_k = A$ for $j \neq k$. Prove that the canonical map

$$A \longrightarrow \bigoplus_{j=1}^s A/I_j, \quad a \longmapsto (a + I_1, \ldots, a + I_s),$$

is an isomorphism of rings[3]

1.3.14. Let K be a field and A a K–algebra. Then A is called an *Artinian K–algebra* if $\dim_K(A) < \infty$. Prove the following statements:

(1) An Artinian K–algebra is Noetherian.
(2) A is an Artinian K–algebra if and only if each descending chain of ideals

$$I_1 \supset I_2 \supset I_3 \supset \ldots \supset I_k \supset \ldots$$

becomes stationary (that is, there exists some j_0 such that $I_j = I_{j_0}$ for all $j \geq j_0$). [4]

1.3.15. Show that $\mathbb{Q}[x]/\langle x^2 + 1 \rangle$ is a field and compute in this field the quotient $(x^3 + x^2 + x)/(x^3 + x^2 + 1)$, first by hand and then by using SINGULAR as in Example 1.3.13. Alternatively use the method of Example 1.1.8 (in characteristic 0), defining a `minpoly`.

1.3.16. Let $f = x^3 + y^3 + z^3 + 3xyz$, and let I be the ideal in $\mathbb{Q}[x, y, z]$, respectively $\mathbb{F}_3[x, y, z]$, generated by f and its partial derivatives. Moreover, let $R := \mathbb{Q}[x, y, z]/I$ and $S := \mathbb{F}_3[x, y, z]/I$.

(1) Is xyz a zerodivisor in R, respectively in S?
(2) Compute the index of nilpotency of $x + y + z$ in R, respectively S.

(Hint: type `?diff;` or `?jacob;` to see how to create the ideal I.)

1.4 Local Rings and Localization

Localization of a ring means enlarging the ring by allowing denominators, similar to the passage from $\mathbb{Z}$ to $\mathbb{Q}$. The name, however, comes from the geometric interpretation. For example, localizing $K[x_1, \ldots, x_n]$ at $\langle x_1, \ldots, x_n \rangle$

---

[3] If $A = \mathbb{Z}$ the theorem can be reformulated as follows: Let $a_1, \ldots, a_s \in \mathbb{Z}$ such that $\gcd(a_i, a_j) = 1$ for $i \neq j$ and $a = \prod_{i=1}^s a_i$. Then for given $x_1, \ldots, x_s \in \mathbb{Z}$ the congruences $x \equiv x_i \mod a_i, 1 \leq i \leq s$ have a solution which is uniquely determined modulo a. The procedure `chineseRem` of the library `crypto.lib` computes this solution.

[4] This is the usual way to define an *Artinian ring*.

means considering rational functions f/g where f and g are polynomials with $g(0) \neq 0$. Of course, any polynomial $f = f/1$ is of this form but, as g may have zeros arbitrary close to 0, f/g is defined only locally, in an arbitrary small neighbourhood of 0 (cf. Appendix A.8).

Definition 1.4.1. A ring A is called *local* if it has exactly one maximal ideal $\mathfrak{m}$. $A/\mathfrak{m}$ is called the *residue field* of A. Rings with finitely many maximal ideals are called *semi–local*. We denote local rings also by $(A, \mathfrak{m})$ or $(A, \mathfrak{m}, K)$ where $K = A/\mathfrak{m}$.

Fields are local rings. A polynomial ring $K[x_1, \ldots, x_n]$ with $n \geq 1$ over a field K is, however, never local. To see this, consider for any $(a_1, \ldots, a_n) \in K^n$ the ideal $\mathfrak{m}_a := \langle x_1 - a_1, \ldots, x_n - a_n \rangle$. Since $\varphi : K[x_1, \ldots, x_n] \to K[x_1, \ldots, x_n]$, $\varphi(x_i) := x_i - a_i$, is an isomorphism sending $\mathfrak{m}_0 = \langle x_1, \ldots, x_n \rangle$ to $\mathfrak{m}_a$, it follows that $K[x_1, \ldots, x_n]/\mathfrak{m}_a \cong K$ is a field, hence $\mathfrak{m}_a$ is a maximal ideal. Since K has at least two elements, K^n has at least two different points and, hence, $K[x_1, \ldots, x_n]$ has at least as many maximal ideals as K^n points (those of type $\mathfrak{m}_a$). If K is algebraically closed, then the ideals $\mathfrak{m}_a$, $a \in K^n$ are all maximal ideals of $K[x_1, \ldots, x_n]$ (this is one form of Hilbert's Nullstellensatz).

A typical local ring is the formal power series ring $K[[x_1, \ldots, x_n]]$ with maximal ideal $\mathfrak{m} = \langle x_1, \ldots, x_n \rangle$, that is, all power series without constant term. That this ring is local follows easily from Lemma 1.4.3. We shall treat power series rings in Chapter 6. Other examples are localizations of polynomial rings at prime ideals, cf. Example 1.4.6.

Theorem 1.4.2. *Every ring $A \neq 0$ contains at least one maximal ideal. If $I \subsetneq A$ is an ideal, then there exists a maximal ideal $\mathfrak{m} \subset A$ such that $I \subset \mathfrak{m}$.*

Proof. The first statement follows from the second with $I = 0$. If I is not maximal there exists an $f_1 \in A$ such that $I \subsetneq I_1 := \langle I, f_1 \rangle \subsetneq A$. If I_1 is not maximal there is an f_2 such that $I_1 \subsetneq I_2 = \langle I_1, f_2 \rangle \subsetneq A$. Continuing in this manner, we obtain a sequence of strictly increasing ideals $I \subsetneq I_1 \subsetneq I_2 \subsetneq \ldots$ which must become stationary, say $I_m = I_n$ for $m \geq n$ if A is Noetherian by Proposition 1.3.6. Thus, I_n is maximal and contains I. In general, if A is not Noetherian, $\bigcup_{n \geq 1} I_n$ is an ideal containing I, and the result follows from Zorn's lemma.[5] $\qquad \square$

Lemma 1.4.3. *Let A be a ring.*

(1) A is a local ring if and only if the set of non–units is an ideal (which is then the maximal ideal).

(2) Let $\mathfrak{m} \subset A$ be a maximal ideal such that every element of the form $1 + a$, $a \in \mathfrak{m}$ is a unit. Then A is local.

---

[5] Zorn's Lemma says: let S be a non–empty system of sets such that for each chain $I_1 \subset I_2 \subset \ldots \subset I_n \subset \ldots$ in S, the union of the chain elements belong to S. Then any element of S is contained in a maximal element (w.r.t. inclusion) of S. This "lemma" is actually an axiom, equivalent to the axiom of choice.

Proof. (1) is obvious. To see (2) let $u \in A \setminus \mathfrak{m}$. Since $\mathfrak{m}$ is maximal $\langle \mathfrak{m}, u \rangle = A$ and, hence, $1 = uv + a$ for some $v \in A$, $a \in \mathfrak{m}$. By assumption $uv = 1 - a$ is a unit. Hence, u is a unit and $\mathfrak{m}$ is the set of non–units. The claim follows from (1). □

Localization generalizes the construction of the quotient field: if A is an integral domain, then the set

$$\mathrm{Quot}(A) := Q(A) := \left\{ \frac{a}{b} \;\middle|\; a, b \in A, \; b \neq 0 \right\},$$

together with the operations

$$\frac{a}{b} + \frac{a'}{b'} = \frac{ab' + a'b}{bb'}, \qquad \frac{a}{b} \cdot \frac{a'}{b'} = \frac{aa'}{bb'}$$

is a field, the *quotient field* or *field of fractions* of A. Here a/b denotes the class of (a, b) under the equivalence relation

$$(a, b) \sim (a', b') :\Longleftrightarrow ab' = a'b.$$

The map $A \to Q(A)$, $a \mapsto a/1$ is an injective ring homomorphism and we identify A with its image. Since $a/b = 0$ if and only if $a = 0$, every element $a/b \neq 0$ has an inverse b/a and, therefore, $Q(A)$ is a field.

The denominators in $Q(A)$ are the elements of the set $S = A \setminus \{0\}$ and S satisfies
(1) $1 \in S$,
(2) $a \in S$, $b \in S \implies ab \in S$.

This notion can be generalized as follows.

Definition 1.4.4. Let A be a ring.

(1) A subset $S \subset A$ is called *multiplicative* or *multiplicatively closed* if conditions (1) and (2) above hold.
(2) Let $S \subset A$ be multiplicatively closed. We define the *localization* or the *ring of fractions* $S^{-1}A$ of A with respect to S as follows:

$$S^{-1}A := \left\{ \frac{a}{b} \;\middle|\; a \in A, \; b \in S \right\}$$

where a/b denotes the equivalence class of $(a, b) \in A \times S$ with respect to the following equivalence relation:

$$(a, b) \sim (a', b') :\Longleftrightarrow \exists \, s \in S \text{ such that } s(ab' - a'b) = 0.$$

Moreover, on $S^{-1}A$ we define an addition and multiplication by the same formulas as for the quotient field above.

The following proposition is left as an exercise.

Proposition 1.4.5.

(1) The operations $+$ and $\cdot$ on $S^{-1}A$ are well–defined (independent of the chosen representatives) and make $S^{-1}A$ a ring (commutative and with $1 = 1/1$).

(2) The map $j : A \to S^{-1}A$, $a \mapsto a/1$ is a ring homomorphism satisfying

   a) $j(s)$ is a unit in $S^{-1}A$ if $s \in S$,

   b) $j(a) = 0$ if and only if $as = 0$ for some $s \in S$,

   c) j is injective if and only if S consists of non–zerodivisors,

   d) j is bijective if and only if S consists of units.

(3) $S^{-1}A = 0$ if and only if $0 \in S$.

(4) If $S_1 \subset S_2$ are multiplicatively closed in A and consist of non–zerodivisors, then $S_1^{-1}A \subset S_2^{-1}A$.

(5) Every ideal in $S^{-1}A$ is generated by the image of an ideal in A under the map j. Moreover, the prime ideals in $S^{-1}A$ are in one–to–one correspondence with the prime ideals in A which do not meet S.

Examples 1.4.6.

(1) $A \smallsetminus P$ is multiplicatively closed for any prime ideal $P \subset A$. The localization of A with respect to $A \smallsetminus P$ is denoted by A_P and

$$A_P = \left\{ \frac{a}{b} \;\middle|\; a, b \in A,\ b \notin P \right\}$$

is called the *localization* of A *at the prime ideal* P.
The set

$$PA_P = \left\{ \frac{a}{b} \;\middle|\; a \in P,\ b \notin P \right\}$$

is clearly an ideal in A_P. Any element $a/b \in A_P \smallsetminus PA_P$ satisfies $a \notin P$, hence, $b/a \in A_P$ and, therefore, a/b is a unit.
This shows that A_P is a local ring with maximal ideal PA_P by Lemma 1.4.3. In particular, if $\mathfrak{m} \subset A$ is a maximal ideal then $A_\mathfrak{m}$ is local with maximal ideal $\mathfrak{m}A_\mathfrak{m}$.

(2) For any $f \in A$, the set $S := \{ f^n \mid n \geq 0 \}$ is multiplicatively closed (with $f^0 = 1$). We use the special notation

$$A_f := S^{-1}A = \left\{ \frac{a}{f^n} \;\middle|\; a \in A, n \geq 0 \right\},$$

not to be confused with $A_{\langle f \rangle}$, if $\langle f \rangle \subset A$ is a prime ideal.

(3) The set S of all non–zerodivisors of A is multiplicatively closed. For this S, $S^{-1}A =: Q(A) =: \mathrm{Quot}(A)$ is called the *total ring of fractions* or the *total quotient ring* of A. If A is an integral domain, this is just the quotient field of A.

Two special but important cases are the following: if $K[x_1, \ldots, x_n]$ is the polynomial ring over a field, then the quotient field is denoted by $K(x_1, \ldots, x_n)$,

$$K(x_1, \ldots, x_n) := Q(K[x_1, \ldots, x_n]),$$

which is also called the *function field* in n variables; the x_i are then also called *parameters*. For computing with parameters cf. SINGULAR-Example 1.1.8.

The localization of $K[x] = K[x_1, \ldots, x_n]$ with respect to the maximal ideal $\langle x \rangle = \langle x_1, \ldots, x_n \rangle$ is

$$K[x]_{\langle x \rangle} = \left\{ \frac{f}{g} \;\middle|\; f, g \in K[x],\ g(0) \neq 0 \right\}.$$

It is an important fact that we can compute in this ring without explicit denominators, just by defining a suitable monomial ordering on $K[x]$ (cf. Section 1.5). More generally, we can compute in $K[x]_{\mathfrak{m}_a}$, $\mathfrak{m}_a = \langle x_1 - a_1, \ldots x_n - a_n \rangle$, for any $a = (a_1, \ldots, a_n) \in K^n$, by translating our polynomial data to $K[x]_{\langle x \rangle}$ via the ring map $x_i \mapsto x_i + a_i$.

Proposition 1.4.7. *Let $\varphi : A \to B$ be a ring homomorphism, $S \subset A$ multiplicatively closed, and $j : A \to S^{-1}A$ the canonical ring homomorphism $a \mapsto a/1$.*

(1) Assume
  (i) $\varphi(s)$ is a unit in B for all $s \in S$.
  Then there exists a unique ring homomorphism $\psi : S^{-1}A \to B$ such that the following diagram commutes:

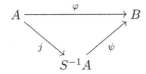

(2) Assume moreover
  (ii) $\varphi(a) = 0$ implies $sa = 0$ for some $s \in S$,
  (iii) every element of B is of the form $\varphi(a)\varphi(s)^{-1}$.
  Then ψ is an isomorphism.

Property (1) is called the *universal property of localization*.

Proof. (1) Since $\varphi(a) = \psi(a/1)$ for $a \in A$, we obtain, for any $a/s \in S^{-1}A$, that $\psi(a/s) = \psi(a/1) \cdot \psi(1/s) = \psi(a/1)\psi(s/1)^{-1} = \varphi(a)\varphi(s)^{-1}$. In particular, ψ is unique if it exists. Now define $\psi(a/s) := \varphi(a)\varphi(s)^{-1}$ and check that ψ is well–defined and a ring homomorphism.
(2) (ii) implies that ψ is injective and (iii) that ψ is surjective.   □

Lemma 1.4.8. *Let $S \subset A$ be multiplicatively closed and $j : A \to S^{-1}A$ the canonical ring homomorphism $a \mapsto a/1$.*

(1) If $J \subset S^{-1}A$ is an ideal and $I = j^{-1}(J)$ then $IS^{-1}A = J$. In particular, if $f_1, \ldots, f_k$ generate I over A then $f_1, \ldots, f_k$ generate J over $S^{-1}A$.
(2) If A is Noetherian, then $S^{-1}A$ is Noetherian.

Proof. (1) If $f/s \in J$ then $f/1 = s \cdot f/s \in J$, hence $f \in I = j^{-1}(J)$ and, therefore, $f/s = f \cdot 1/s \in IS^{-1}A$. The other inclusion is clear. Statement (2) follows directly from (1). $\qquad\square$

To define the local ring $K[x]_{\langle x \rangle} = K[x_1, \ldots, x_n]_{\langle x_1, \ldots, x_n \rangle}$ in SINGULAR, we have to choose a local ordering such as `ds`, `Ds`, `ls` or a weighted local ordering. This is explained in detail in the next section. We shall now show the difference between local and global rings by some examples. Note that objects defined in the local ring $K[x]_{\langle x \rangle}$ contain geometric information (usually only) about a Zariski neighbourhood of $0 \in K^n$ (cf. A.2, page 454), while objects in $K[x]$ contain geometric information which is valid in the whole affine space K^n.

Consider the ideal $I = \langle y(x-1), z(x-1) \rangle \subset \mathbb{Q}[x, y, z]$ and consider the common zero–set of all elements of I,

$$V(I) = \{(x, y, z) \in \mathbb{C}^3 \mid f(x, y, z) = 0 \;\forall\, f \in I\}$$
$$= \{(x, y, z) \in \mathbb{C}^3 \mid y(x-1) = z(x-1) = 0\}.$$

The real picture of $V(I)$ is displayed in Figure 1.1.

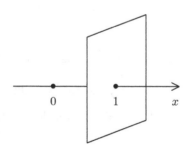

Fig. 1.1. The real zero–set of $\langle y(x-1), z(x-1) \rangle$.

Although we treat dimension theory later, it should be intuitively clear from the picture that the (local) dimension of $V(I)$ is 1 at the point $(0, 0, 0)$ and 2 at the point $(1, 0, 0)$.

We compute the global dimension of $V(I)$ (which is the maximum of the dimensions at each point) and then the dimension of $V(I)$ in the points $(0, 0, 0)$ and $(1, 0, 0)$. As we shall see in Section 3.3, we always have to compute a standard basis of the ideal with respect to the given ordering first and then apply the command `dim`.

SINGULAR Example 1.4.9 (global versus local rings).

(1) Compute the dimension of $V(I)$, that is, compute $\dim(I)$, the Krull dimension of S/I, $S = \mathbb{Q}[x, y, z]$ (cf. Chapter 3, Section 3.3).

```
ring S = 0,(x,y,z),dp;
ideal I = y*(x-1), z*(x-1);
ideal J = std(I);      //compute a standard basis J of I in S
J;                     //J = <z(x-1),y(x-1)>
//-> J[1]=xz-z
//-> J[2]=xy-y
dim(J);                //the (global) dimension of V(I) is 2
//-> 2

reduce(y,J);           //y is not in I
                       //(result is 0 iff y is in I)
//-> y
```

(2) Compute the dimension of $V(I)$ at $0 = (0,0,0)$, that is, compute $\dim(I)$, the Krull dimension of R/IR, $R = \mathbb{Q}[x, y, z]_{\langle x,y,z \rangle}$.

```
ring R = 0,(x,y,z),ds;
ideal I = fetch(S,I);//fetch I from S to basering
ideal J = std(I);      //compute a standard basis J of I in R
J;
//-> J[1]=y                //J = <y,z> since x-1 is a unit in R
//-> J[2]=z
dim(J);
//-> 1                     //(local) dimension of V(I) at 0 is 1
reduce(y,J);
//-> 0                     //now y is in IR = JR
```

(3) Compute the dimension of $V(I)$ at $(1,0,0)$, that is, compute $\dim(I_1)$, the Krull dimension of $R/I_1 R$ in $R = \mathbb{Q}[x, y, z]_{\langle x,y,z \rangle}$ where I_1 is the translation of I to $(1,0,0)$.

```
map trans = S, x+1,y,z;  //replace x by x+1 and leave
                         //y,z fixed, i.e., translate
                         //(0,0,0) to (1,0,0)
ideal I1 = trans(I);
I1;
//-> I1[1]=xy
//-> I1[2]=xz
dim(std(I1));            //dimension of V(I) at (1,0,0) is 2
//-> 2
```

(4) Compute the (global) dimension of $V(I)$ after translation.

```
setring S;                //go back to global ring S
map trans = S, x+1,y,z;
ideal I1 = trans(I);      //translate I, as in (3)
I1;
//-> I1[1]=xy
//-> I1[2]=xz
dim(std(I1));             //(global) dimension of translated
//-> 2                    //variety has not changed

kill S,R;
```

The above computation illustrates what is intuitively clear from the picture in Figure 1.1: the dimension of the local rings varies. Dimension theory is treated in detail in Chapter 3, Section 3.3. For this example, it is enough to have an intuitive feeling for the dimension as it is visualized in the real picture of $V(I)$.

Exercises

1.4.1. Prove Proposition 1.4.5.

1.4.2. Let A be a ring, $I \subset A$ an ideal and $f \in A$.
Prove that $IA_f \cap A$ is the saturation of I with respect to f, that is, equal to $I : \langle f \rangle^\infty = \{g \in A \mid \exists n \text{ such that } gf^n \in I\}$.

1.4.3. Let $(A, \mathfrak{m})$ be a local K–algebra, K a field, and $I \subset A$ an ideal such that $\dim_K(A/I) < \infty$. Show that $\mathfrak{m}^n \subset I$ for some n.

1.4.4. Let A be a ring and $J(A)$ the intersection of all maximal ideals of A, which is called the *Jacobson radical* of A. Prove that for all $x \in J(A)$, $1 + x$ is a unit in A.

1.4.5. Let S and T be two multiplicatively closed sets in the ring A. Show that ST is multiplicatively closed and that $(ST)^{-1}(A)$ and $T^{-1}(S^{-1}A)$ are isomorphic, if T denotes also the image of T in $S^{-1}A$.

In particular, if $S \subset T$ then $T^{-1}A \cong T^{-1}(S^{-1}A)$. Hence, for $Q \subset P$ two prime ideals we obtain $A_Q \cong (A_P)_{QA_P}$.

1.4.6. Let $S \subset A$ be the set of non–zerodivisors. Show the following statements about the total ring of fractions $\text{Quot}(A) = S^{-1}A$:

(1) S is the biggest multiplicatively closed subset of A such that $A \to S^{-1}A$ is injective.
(2) Each element of $\text{Quot}(A)$ is either a unit or a zerodivisor.
(3) A ring A, such that each non–unit is a zerodivisor, is equal to its total ring of fractions, that is, $A \to \text{Quot}(A)$ is bijective.

1.4.7. (1) Consider the two rings

$$A = \mathbb{C}[x,y]/\langle x^2 - y^3 \rangle \text{ and } B = \mathbb{C}[x,y]/\langle xy \rangle$$

and the multiplicative sets:
- S the set of non–zerodivisors of A, respectively of B, and
- $T := A \smallsetminus \langle x,y \rangle_A$, respectively $T := B \smallsetminus \langle x,y \rangle_B$.

Determine the localizations of A and B with respect to T and S.
(2) Are any two of the six rings $A, B, S^{-1}A, S^{-1}B, T^{-1}A, T^{-1}B$ isomorphic?

1.4.8. Let A be a ring and $B = A/(P_1 \cap \cdots \cap P_r)$ with $P_i \subset A$ prime ideals. Show that the rings $\mathrm{Quot}(B)$ and $\bigoplus_{i=1}^{r} \mathrm{Quot}(A/P_i)$ are isomorphic. In particular, $\mathrm{Quot}(B)$ is a direct sum of fields.
(Hint: use Exercise 1.3.13.)

1.4.9. Let A be a *unique factorization domain* (that is, A is a domain and every non–unit of A can be written as a product of irreducible elements such that the factors are uniquely determined up to multiplication with units) and $S \subset A$ multiplicatively closed. Show that $S^{-1}A$ is a unique factorization domain.

(Hint: enlarge, if necessary, S to a multiplicative system $\widetilde{S}$ such that
(1) $\widetilde{S}^{-1}A = S^{-1}A$ and
(2) if $s \in \widetilde{S}$ and $s = s_1 s_2$ then $s_1, s_2 \in \widetilde{S}$.)

1.4.10. Let A be an integral domain. Then, for any prime ideal $P \subset A$, we consider the localization A_P as a subring of the quotient field $\mathrm{Quot}(A)$ and, hence, we can consider their intersection. Show that

$$A = \bigcap_{P \in \mathrm{Spec}\,A} A_P = \bigcap_{\mathfrak{m} \in \mathrm{Max}\,A} A_{\mathfrak{m}}.$$

1.4.11. Let $I = I_1 I_2 I_3 \subset Q[x,y,z]$ be the product of the ideals $I_1 = \langle z - x^2 \rangle$, $I_2 = \langle y, z \rangle$ and $I_3 = \langle x \rangle$. Compute, as in SINGULAR Example 1.4.9, the dimension of $V(I)$ at the points $(0,0,0)$, $(0,0,1)$, $(1,0,0)$ and $(1,1,1)$. Draw a real picture of $V(I)$ and interpret your results geometrically.

1.5 Rings Associated to Monomial Orderings

In this section we show that non–global monomial orderings lead to new rings which are localizations of the polynomial ring. This fact has far–reaching computational consequences. For example, choosing a local ordering, we can, basically, do the same calculations in the localization of a polynomial ring as with a global ordering in the polynomial ring itself. In particular, we can effectively compute in $K[x_1, \ldots, x_n]_{\langle x_1, \ldots, x_k \rangle}$ for $k \leq n$ (by Lemma 1.5.2 (3) and Example 1.5.3).

Let $>$ be a monomial ordering on the set of monomials $\mathrm{Mon}(x_1, \ldots, x_n) = \{x^\alpha \mid \alpha \in \mathbb{N}^n\}$, and $K[x] = K[x_1, \ldots, x_n]$ the polynomial ring in n variables over a field K. Then the leading monomial function LM has the following properties for polynomials $f, g \in K[x] \setminus \{0\}$:

(1) $\mathrm{LM}(gf) = \mathrm{LM}(g)\,\mathrm{LM}(f)$.
(2) $\mathrm{LM}(g + f) \leq \max\{\mathrm{LM}(g), \mathrm{LM}(f)\}$ with equality if and only if the leading terms of f and g do not cancel.

In particular, it follows that

$$S_> := \{u \in K[x] \setminus \{0\} \mid \mathrm{LM}(u) = 1\}$$

is a multiplicatively closed set.

Definition 1.5.1. For any monomial ordering $>$ on $\mathrm{Mon}(x_1, \ldots, x_n)$, we define

$$K[x]_> := S_>^{-1} K[x] = \left\{ \frac{f}{u} \;\middle|\; f, u \in K[x],\, \mathrm{LM}(u) = 1 \right\},$$

the localization of $K[x]$ with respect to $S_>$ and call $K[x]_>$ the *ring associated to $K[x]$ and $>$*.

Note that $S_> = K^*$ if and only if $>$ is global and $S_> = K[x] \setminus \langle x_1, \ldots, x_n \rangle$ if and only if $>$ is local.

Lemma 1.5.2. *Let K be a field, $K[x] = K[x_1, \ldots, x_n]$, and let $>$ be a monomial ordering on $\mathrm{Mon}(x_1, \ldots, x_n)$. Then*

(1) $K[x] \subset K[x]_> \subset K[x]_{\langle x \rangle}$.
(2) The set of units in $K[x]_>$ is given by

$$(K[x]_>)^* = \left\{ \frac{v}{u} \;\middle|\; u, v \in K[x],\, \mathrm{LM}(v) = \mathrm{LM}(u) = 1 \right\},$$

and satisfies $(K[x]_>)^* \cap K[x] = S_>$.
(3) $K[x] = K[x]_>$ if and only if $>$ is a global ordering and $K[x]_> = K[x]_{\langle x \rangle}$ if and only if $>$ is a local ordering.
(4) $K[x]_>$ is a Noetherian ring.
(5) $K[x]_>$ is factorial.

We shall see later (Corollary 7.4.6) that the inclusions of Lemma 1.5.2 (1) are flat ring morphisms.

Proof. (1) The first inclusion is clear by Proposition 1.4.5 (2), the second follows from Proposition 1.4.5 (4) since $\mathrm{LM}(u) = 1$ implies $u \notin \langle x \rangle$.
(2) If f/u is a unit in $K[x]_>$, there is a h/v such that $(f/u) \cdot (h/v) = 1$. Hence, $fh = uv$ and $\mathrm{LM}(f)\,\mathrm{LM}(h) = 1$, which implies $\mathrm{LM}(f) = 1$.

(3) By Proposition 1.4.5 (2), $K[x] = K[x]_>$ if and only if $S_>$ consists of units of $K[x]$, that is if and only if $S_> \subset K^*$, which is equivalent to $>$ being global. The second equality follows since $K[x] \setminus \langle x \rangle$ consists of units in $K[x]_>$ if and only if every polynomial with non–zero constant term belongs to $S_>$ which is equivalent to $>$ being local.

(4) Follows from Lemma 1.4.8.

(5) Since $K[x]$ is factorial, this follows from Exercise 1.4.9.   □

Examples 1.5.3. We describe some familiar and some less familiar rings, associated to a polynomial ring and a monomial ordering.

(1) Let $K[x, y] = K[x_1, \ldots, x_n, y_1, \ldots, y_m]$ and consider the product ordering $> = (>_1, >_2)$ on $\mathrm{Mon}(x_1, \ldots, x_n, y_1, \ldots, y_m)$, where $>_1$ is global on $\mathrm{Mon}(x_1, \ldots, x_n)$ and $>_2$ is local on $\mathrm{Mon}(y_1, \ldots, y_m)$. Then

$$x^\alpha y^\gamma > 1 > y^\beta \text{ for all } \alpha, \beta \neq 0, \text{ all } \gamma$$

and hence $S_> = K^* + \langle y \rangle \cdot K[y]$. It follows that

$$K[x, y]_> = (K[y]_{\langle y \rangle})[x],$$

which equals $K[y]_{\langle y \rangle} \otimes_K K[x]$ (cf. Section 2.7 for the tensor product).

(2) Now let $>_1$ be local and $>_2$ global, $> = (>_1, >_2)$, then

$$x^\alpha y^\gamma < 1 < y^\beta \text{ for all } \alpha, \beta \neq 0, \text{ all } \gamma$$

and hence $S_> = K^* + \langle x \rangle K[x, y]$. We obtain strict inclusions

$$(K[x]_{\langle x \rangle})[y] \subsetneq K[x, y]_> \subsetneq K[x, y]_{\langle x \rangle},$$

since $1/(1 + xy)$ is in the second but not in the first and $1/y$ is in the third but not in the second ring.

(3) If $>_1$ is global, $>_2$ arbitrary and $> = (>_1, >_2)$ then $S_>$ consists of elements $u \in K[y]$ satisfying $\mathrm{LM}_{>_2}(u) = 1$. Hence,

$$K[x, y]_> = (K[y]_{>_2})[x]$$

(cf. Exercise 2.7.12). This ordering has the following *elimination property* for $x_1, \ldots, x_n$:

$$f \in K[x, y], \ \mathrm{LM}(f) \in K[y] \Rightarrow f \in K[y].$$

(4) Let $>$ be a local ordering on $\mathrm{Mon}(x_1, \ldots, x_n)$ and $K(y)$ the quotient field of $K[y] = K[y_1, \ldots, y_m]$. It is not difficult to see that

$$K(y)[x]_> = K[x, y]_{\langle x \rangle}$$

(Exercise 1.5.6). Hence, we can effectively compute in the localization $K[x_1, \ldots, x_n]_P$, where P is a prime ideal generated by a subset of the variables.

Definition 1.5.4. A monomial ordering $>$ on $K[x_1, \ldots, x_n]$ having the elimination property for $x_1, \ldots, x_s$ (cf. Example 1.5.3 (3)) is called an *elimination ordering* for $x_1, \ldots, x_s$.

An elimination ordering need not be a product ordering but must satisfy $x_i > 1$ for $i = 1, \ldots, s$ (since, if $x_i < 1$ then $\mathrm{LM}(1 + x_i) = 1$ but $1 + x_i \notin K[x_{s+1}, \ldots, x_n]$), that is, an elimination ordering for $x_1, \ldots, x_s$ must be global on $\mathrm{Mon}(x_1, \ldots, x_s)$. Since the lexicographical ordering is the product of the degree orderings on $\mathrm{Mon}(x_i)$ for $i = 1, \ldots, n$, it is an elimination ordering for $x_1, \ldots, x_j$, for $j = 1, \ldots, n$. (Cf. Section 1.8.2 for applications of elimination orderings.)

We now extend the leading data to $K[x]_>$.

Definition 1.5.5. Let $>$ be any monomial ordering:

(1) For $f \in K[x]_>$ choose $u \in K[x]$ such that $\mathrm{LT}(u) = 1$ and $uf \in K[x]$. We define

$$\mathrm{LM}(f) := \mathrm{LM}(uf),$$
$$\mathrm{LC}(f) := \mathrm{LC}(uf),$$
$$\mathrm{LT}(f) := \mathrm{LT}(uf),$$
$$\mathrm{LE}(f) := \mathrm{LE}(uf),$$

and $\mathrm{tail}(f) = f - \mathrm{LT}(f)$.

(2) For any subset $G \subset K[x]_>$ define the ideal

$$L_>(G) := L(G) := \langle \mathrm{LM}(g) \mid g \in G \setminus \{0\} \rangle_{K[x]}.$$

$L(G) \subset K[x]$ is called the *leading ideal* of G.

Remark 1.5.6.

(1) The definitions in 1.5.5 (1) are independent of the choice of u.

(2) Since $K[x]_> \subset K[x]_{\langle x \rangle} \subset K[[x]]$, where $K[[x]]$ denotes the formal power series ring (cf. Section 6.1), we may consider $f \in K[x]_>$ as a formal power series. It follows easily that $\mathrm{LM}(f)$, respectively $\mathrm{LT}(f)$, corresponds to a unique monomial, respectively term, in the power series expansion of f. Hence $\mathrm{tail}(f)$ is the power series of f with the leading term deleted.

(3) Note that if I is an ideal, then $L(I)$ is the ideal generated by all leading monomials of all elements of I and not only by the leading monomials of a given set of generators of I.

Example 1.5.7.

(1) Consider $\mathbb{Q}[x]$ with a local ordering (in one variable all local, respectively global, orderings coincide). For $f = 3x/(1 + x) + x$ we have $\mathrm{LM}(f) = x$, $\mathrm{LC}(f) = 4$, $\mathrm{LT}(f) = 4x$, $\mathrm{LE}(f) = 1$ and $\mathrm{tail}(f) = -3x^2/(1 + x)$.

(2) Let $G = \{f, g\}$ with $f = xy^2 + xy$, $g = x^2y + x^2 - y \in \mathbb{Q}[x, y]$ and monomial ordering dp. If $I = \langle f, g \rangle$ then $L(G) \subsetneq L(I)$, since $L(G) = \langle xy^2, x^2y \rangle$,
but $xf - yg = y^2$. Thus, $y^2 \in L(I)$, but $y^2 \notin L(G)$.

Ring maps between rings associated to a monomial ordering are almost as
easy as ring maps between polynomial rings.

Lemma 1.5.8. *Let $\psi : K \to L$ be a morphism of fields and $>_1$, $>_2$, mono-
mial orderings on $\mathrm{Mon}(x_1, \ldots, x_n)$ and on $\mathrm{Mon}(y_1, \ldots, y_m)$. Let $f_1, \ldots, f_n \in$
$L[y_1, \ldots, y_m]_{>_2}$ and assume that, for all $h \in S_{>_1}$, $h(f_1, \ldots, f_n) \in S_{>_2}$. Then
there exists a unique ring map*

$$\varphi : K[x_1, \ldots, x_n]_{>_1} \to L[y_1, \ldots, y_m]_{>_2}$$

satisfying $\varphi(x_i) = f_i$ for $i = 1, \ldots, n$, and $\varphi(a) = \psi(a)$ for $a \in K$.

Proof. By Lemma 1.1.6, there is a unique ring map $\widetilde{\varphi} : K[x] \to L[y]_{>_2}$ with
$\widetilde{\varphi}(x_i) = f_i$ and $\widetilde{\varphi}(a) = \psi(a)$, $a \in K$. The assumption says that $\widetilde{\varphi}(u)$ is a unit
in $L[y]_{>_2}$ for each $u \in S_{>_1}$. Hence, the result follows from the universal prop-
erty of localization (Proposition 1.4.7). $\square$

In particular, if $>_1$ is global, there is no condition on the f_i and any elements
$f_1, \ldots, f_n \in K[y_1, \ldots, y_m]_>$ define a unique map

$$K[x_1, \ldots, x_n] \longrightarrow K[y_1, \ldots, y_m]_>, \quad x_i \longmapsto f_i,$$

for any monomial ordering $>$ on $\mathrm{Mon}(y_1, \ldots, y_m)$.

Remark 1.5.9. With the notations of Lemma 1.5.8, the condition "$h \in S_{>_1}$
implies $h(f_1, \ldots, f_n) \in S_{>_2}$" cannot be replaced by "$1 >_2 \mathrm{LM}(f_i)$ for those i
where $1 >_1 x_i$". Consider the following example: let $>_1$, respectively $>_2$, be
defined on $K[x, y]$ by the matrix $\left(\begin{smallmatrix} -2 & 1 \\ 1 & 0 \end{smallmatrix}\right)$, respectively $\left(\begin{smallmatrix} -1 & 2 \\ 1 & 0 \end{smallmatrix}\right)$. Then $x <_1 y$
and $x <_2 y$ but $xy + 1 \in S_{<_1}$ and $xy + 1 \notin S_{>_2}$.

SINGULAR Example 1.5.10 (realization of rings).
We show how to create the rings of Examples 1.5.3. Note that SINGULAR
sorts the monomials with respect to the monomial ordering, the greatest
being first. Hence, the position of 1 in the output shows which monomials
are greater, respectively smaller, than 1.

```
int n,m=2,3;
ring A1 = 0,(x(1..n),y(1..m)),(dp(n),ds(m));
poly f   = x(1)*x(2)^2+1+y(1)^10+x(1)*y(2)^5+y(3);
f;
//-> x(1)*x(2)^2+x(1)*y(2)^5+1+y(3)+y(1)^10

1>y(1)^10;   //the monomial 1 is greater than y(1)^10
```

```
//-> 1

ring A2 = 0,(x(1..n),y(1..m)),(ds(n),dp(m));
fetch(A1,f);
//-> y(1)^10+y(3)+1+x(1)*y(2)^5+x(1)*x(2)^2

x(1)*y(2)^5<1;
//-> 1

ring A3 = 0,(x(1..n),y(1..m)),(dp(n),ds(2),dp(m-2));
fetch(A1,f);
//-> x(1)*x(2)^2+x(1)*y(2)^5+y(3)+1+y(1)^10
```

Exercises

1.5.1. Prove Remark 1.5.6.

1.5.2. Give one possible realization of the following rings within SINGULAR:

(1) $\mathbb{Q}[x, y, z]$,
(2) $\mathbb{F}_5[x, y, z]$,
(3) $\mathbb{Q}[x, y, z]/\langle x^5 + y^3 + z^2 \rangle$,
(4) $\mathbb{Q}(i)[x, y]$, $i^2 = -1$,
(5) $\mathbb{F}_{27}[x_1, \ldots, x_{10}]_{\langle x_1, \ldots, x_{10} \rangle}$,
(6) $\mathbb{F}_{32003}[x, y, z]_{\langle x,y,z \rangle}/\langle x^5 + y^3 + z^2, xy \rangle$,
(7) $\mathbb{Q}(t)[x, y, z]$,
(8) $(\mathbb{Q}[t]/(t^3 + t^2 + 1))[x, y, z]_{\langle x,y,z \rangle}$,
(9) $(\mathbb{Q}[t]_{\langle t \rangle})[x, y, z]$,
(10) $\mathbb{F}_2(a, b, c)[x, y, z]_{\langle x,y,z \rangle}$.

(Hint: see the SINGULAR Manual for how to define a quotient ring modulo some ideal.)

1.5.3. What are the units in the rings of Exercise 1.5.2?

1.5.4. Write a SINGULAR procedure, having as input a polynomial f and returning 1 if f is a unit in the basering and 0 otherwise.

(Hint: type `?procedures;`.)

Test the procedure by creating, for each ring of Exercise 1.5.2, two polynomials, one a unit, the other not.

1.5.5. Write a SINGULAR procedure, having as input a polynomial f and an integer n, which returns the power series expansion of the inverse of f up to terms of degree n if f is a unit in the basering and 0 if f is not a unit.

(Hint: remember the geometric series.)

1.5.6. (1) Let $>$ be a local ordering on $\mathrm{Mon}(x_1, \ldots, x_n)$. Show that
$$K[x_1, \ldots, x_n, y_1, \ldots, y_m]_{\langle x_1, \ldots, x_n \rangle} = K(y_1, \ldots, y_m)[x_1, \ldots, x_n]_>.$$
(2) Implement the ring $\mathbb{Q}[x, y, z]_{\langle x, y \rangle}$ inside SINGULAR.

1.6 Normal Forms and Standard Bases

In this section we define standard bases, respectively Gröbner bases, of an ideal $I \subset K[x]_>$ as a set of polynomials of I such that their leading monomials generate the leading ideal $L(I)$. The next section gives an algorithm to compute standard bases. For global orderings this is Buchberger's algorithm, which is a generalization of the Gaussian elimination algorithm and the Euclidean algorithm. For local orderings it is Mora's tangent cone algorithm, which itself is a variant of Buchberger's algorithm. The general case is a variation of Mora's algorithm, which is due to the authors and implemented in SINGULAR since 1990.

The leading ideal $L(I)$ contains a lot of information about the ideal I, which often can be computed purely combinatorially from $L(I)$, because the leading ideal is *generated by monomials*. Standard bases have turned out to be the fundamental tool for computations with ideals and modules. The idea of standard bases is already contained in the work of Gordan [93]. Later, monomial orderings were used by Macaulay [157] and Gröbner [117] to study Hilbert functions of graded ideals, and, more generally, to find bases of zero–dimensional factor rings. The notion of a standard basis was introduced later, independently, by Hironaka [123], Grauert [98] (for special local orderings) and Buchberger [32] (for global orderings).

In the following, special emphasis is made to axiomatically characterize normal forms, respectively weak normal forms, which play an important role in the standard basis algorithm. They generalize division with remainder to the case of ideals, respectively finite sets of polynomials.

In the case of a global ordering, for any polynomial f and any ideal I, there is a unique normal form $\mathrm{NF}(f \mid I)$ of f with respect to I, such that no monomial of $\mathrm{NF}(f \mid I)$ is in the leading ideal $L(I)$. This can be used to decide, for instance, whether f is in the ideal I (if the normal form is 0).

In the general case, the above property turns out to be too strong. Hence, the requirements for a normal form have to be weakened. For instance, for the decision whether a polynomial is in an ideal or not, only the leading term of a normal form $\mathrm{NF}(f \mid I)$ is important. Thus, for this purpose, it is enough to require that $\mathrm{NF}(f \mid I)$ is either 0 or has a leading term, which is not in $L(I)$. After weakening the requirements, there is no more uniqueness statement for the normal form.

Our intention is to keep the definition of a normal form as general as possible. Moreover, our presentation separates the normal form algorithm from a general standard basis algorithm and shows that different versions of standard basis algorithms are due to different normal forms.

The axiomatic definition of a normal form as presented in this section has been introduced in [111, 112], although its properties have already commonly been used before. It seems to be the minimal requirement in order to carry through standard basis theory in the present context.

Let $>$ be a fixed monomial ordering and let, in this section,

$$R = K[x_1, \ldots, x_n]_>$$

be the localization of $K[x] = K[x_1, \ldots, x_n]$ with respect to $>$. Recall that $R = S_>^{-1} K[x]$ with $S_> = \{u \in K[x] \smallsetminus \{0\} \mid \mathrm{LM}(u) = 1\}$, and that $R = K[x]$ if $>$ is global and $R = K[x]_{\langle x \rangle}$ if $>$ is local. In any case, R may be considered as a subring of the ring $K[[x]]$ of formal power series (cf. Section 6.1).

Definition 1.6.1. Let $I \subset R$ be an ideal.

(1) A finite set $G \subset R$ is called a *standard basis* of I if

$$G \subset I, \text{ and } L(I) = L(G).$$

That is, G is a standard basis, if the leading monomials of the elements of G generate the leading ideal of I, or, in other words, if for any $f \in I \smallsetminus \{0\}$ there exists a $g \in G$ satisfying $\mathrm{LM}(g) \mid \mathrm{LM}(f)$.

(2) If $>$ is global, a standard basis is also called a *Gröbner basis*.

(3) If we just say that G is a standard basis, we mean that G is a standard basis of the ideal $\langle G \rangle_R$ generated by G.

Let $>$ be any monomial ordering on $\mathrm{Mon}(x_1, \ldots, x_n)$. Then each non–zero ideal $I \subset K[x]_>$ has a standard basis. To see this, choose a finite set of generators $m_1, \ldots, m_s$ of $L(I) \subset K[x]$, which exists, since $K[x]$ is Noetherian (Theorem 1.3.5). By definition of the leading ideal, these generators are leading monomials of suitable elements $g_1, \ldots, g_s \in I$. By construction, the set $\{g_1, \ldots, g_s\}$ is a standard basis for I.

Definition 1.6.2. Let $G \subset R$ be any subset.

(1) G is called *interreduced* if $0 \notin G$ and if $\mathrm{LM}(g) \nmid \mathrm{LM}(f)$ for any two elements $f \neq g$ in G. An interreduced standard basis is also called *minimal*.

(2) $f \in R$ is called *(completely) reduced with respect to G* if no monomial of the power series expansion of f is contained in $L(G)$.

(3) G is called *(completely) reduced* if G is interreduced and if, for any $g \in G$, $\mathrm{LC}(g) = 1$ and $\mathrm{tail}(g)$ is completely reduced with respect to G.

Remark 1.6.3.

(1) If $>$ is a global ordering, then any finite set G can be transformed into an interreduced set: for any $g \in G$ such that there exists an $f \in G \smallsetminus \{g\}$ with $\mathrm{LM}(f) \mid \mathrm{LM}(g)$ replace g by $g - mf$, where m is a term with $\mathrm{LT}(g) = m \mathrm{LT}(f)$. The result is called the *(inter)reduction* of G; it generates the same ideal as G.

(2) Every standard basis G can be transformed into an interreduced one by just deleting elements of G: delete zeros and then, successively, any g such that $\mathrm{LM}(g)$ is divisible by $\mathrm{LM}(f)$ for some $f \in G \smallsetminus \{g\}$. The result is again a standard basis. Thus, G is interreduced if and only if G is minimal (that is, we cannot delete any element of G without violating the property of being a standard basis).

(3) Let $G \subset R$ be an interreduced set and $g \in G$. If $\mathrm{tail}(g)$ is not reduced with respect to G, the power series expansion of $\mathrm{tail}(g)$ has a monomial which is either divisible by $L(g)$ or by $L(f)$ for some $f \in G \smallsetminus \{g\}$. If $>$ is global, then no monomial of $\mathrm{tail}(g)$ is divisible by $L(g)$ since $>$ refines the natural partial ordering on $\mathbb{N}^n$, that is, $\mathrm{tail}(g)$ is reduced with respect to $\{g\}$. For local or mixed orderings, however, it is possible to reduce $\mathrm{tail}(g)$ with g and we actually have to do this.

(4) It follows that a Gröbner basis $G \subset K[x]$, which consists of monic polynomials, is (completely) reduced if for any $f \neq g \in G$, $\mathrm{LM}(g)$ does not divide any monomial of f.

We shall see later that reduced Gröbner bases can always be computed[6] (cf. the remark after Algorithm 1.6.10), and are unique (Exercise 1.6.1), but reduced standard bases are, in general, not computable (in a finite number of steps).

The following two definitions are crucial for our treatment of standard bases.

Definition 1.6.4. Let $\mathcal{G}$ denote the set of all finite lists $G \subset R$.

$$\mathrm{NF} : R \times \mathcal{G} \to R, \ (f, G) \mapsto \mathrm{NF}(f \mid G),$$

is called a *normal form* on R if, for all $G \in \mathcal{G}$,

(0) $\mathrm{NF}(0 \mid G) = 0$,

and, for all $f \in R$ and $G \in \mathcal{G}$,

(1) $\mathrm{NF}(f \mid G) \neq 0 \implies \mathrm{LM}\big(\mathrm{NF}(f \mid G)\big) \notin L(G)$.

(2) If $G = \{g_1, \ldots, g_s\}$, then $f - \mathrm{NF}(f|G)$ (or, by abuse of notation, f) has a *standard representation* with respect to $\mathrm{NF}(- \mid G)$, that is,

$$f - \mathrm{NF}(f \mid G) = \sum_{i=1}^{s} a_i g_i, \ a_i \in R, \ s \geq 0,$$

satisfying $\mathrm{LM}(\Sigma_{i=1}^{s} a_i g_i) \geq \mathrm{LM}(a_i g_i)$ for all i such that $a_i g_i \neq 0$.
NF is called a *reduced normal form*, if, moreover, $\mathrm{NF}(f \mid G)$ is reduced with respect to G.

---

[6] The SINGULAR command `std` can be forced to compute reduced Gröbner bases G (up to normalization) using `option(redSB)`. To normalize G, one may use the SINGULAR command `simplify(G,1)`.

Definition 1.6.5.

(1) A map NF $: R \times \mathcal{G} \to R$, as in Definition 1.6.4, is called a *weak normal form* on R if it satisfies (0),(1) of 1.6.4 and, instead of (2),

   (2') for all $f \in R$ and $G \in \mathcal{G}$ there exists a unit $u \in R^*$ such that uf has a standard representation with respect to $\mathrm{NF}(- \mid G)$.

(2) A weak normal form NF is called *polynomial* if, whenever $f \in K[x]$ and $G \subset K[x]$, there exists a unit $u \in R^* \cap K[x]$ such that uf has a standard representation with $a_i \in K[x]$.

Remark 1.6.6.

(1) The notion of weak normal forms is only interesting for non–global orderings since for global orderings we have $R = K[x]$ and, hence, $R^* = K^*$. Even in general, if a weak normal form NF exists, then, theoretically, there exists also a normal form $\widetilde{\mathrm{NF}}$

$$(f, G) \longmapsto \frac{1}{u} \mathrm{NF}(f \mid G) =: \widetilde{\mathrm{NF}}(f \mid G)$$

for an appropriate choice of $u \in R^*$ (depending on f and G). However, we are really interested in polynomial normal forms, and for non–global orderings $1/u$ is, in general, not a polynomial but a power series. Note that $R^* \cap K[x] = S_>$.

(2) Consider $f = y$, $g = (y - x)(1 - y)$, and $G = \{g\}$ in $R = K[x, y]_{\langle x, y \rangle}$ with local ordering $\mathtt{ls}$. Assume $h := \mathrm{NF}(f \mid G) \in K[x, y]$ is a polynomial normal form of f with respect to G. Since $f \notin \langle G \rangle_R = \langle y - x \rangle_R$, we have $h \neq 0$, hence, $\mathrm{LM}(h) \notin L(G) = \langle y \rangle$. Moreover $h - y = h - f \in \langle y - x \rangle_R$, which implies $\mathrm{LM}(h) < 1$. Therefore, we obtain $h = xh'$ for some h' (because of the chosen ordering $\mathtt{ls}$). However, $y - xh' \notin \langle (y - x)(1 - y) \rangle_{K[x,y]}$ (substitute $(0, 1)$ for (x, y)) and, therefore no polynomial normal form of (f, G) exists. On the other hand, setting $u = (1 - y)$ and $h = x(1 - y)$ then $uy - h = (y - x)(1 - y)$ and, hence, h is a polynomial weak normal form.

(3) For applications (weak) normal forms are most useful if G is a standard basis of $\langle G \rangle_R$. We shall demonstrate this with a first application in Lemma 1.6.7.

(4) $f = \sum_i a_i g_i$ being a standard representation means that no cancellation of leading terms $> \mathrm{LM}(f)$ between the $a_i g_i$ can occur and that $\mathrm{LM}(f) = \mathrm{LM}(a_i g_i)$ for at least one i.

(5) We do not distinguish strictly between lists and (ordered) sets. Since, in the definition of normal form, we allow repetitions of elements in G we need lists, that is, sequences of elements, instead of sets. We assume a given set G to be ordered (somehow) when we apply $\mathrm{NF}(- \mid G)$.

(6) The existence of a normal form resp. a polynomial weak normal form with respect to $G \subset K[x]$ which we prove in Algorithm 1.6.10 resp. Algorithm 1.7.6 says:

For any $f \in R$ there exist polynomials $u, a_1, \ldots, a_s \in K[x]$ such that

$$uf = \sum_{i=1}^{s} a_i g_i + h, \ \mathrm{LM}(u) = 1,$$

satisfying:

(1) If $h \neq 0$ then $\mathrm{LM}(h)$ is not divisible by $\mathrm{LM}(g_i)$, $i = 1, \ldots, s$,
(2) $\mathrm{LM}(\sum_{i=1}^{s} a_i g_i) \geq \mathrm{LM}(a_i g_i)$ for all i with $a_i g_i \neq 0$ (and hence equality holds for at least one i).

Moreover, if $>$ is global then the unit u can be chosen as 1.

Thus the existence of a (weak) normal form is a *division theorem* where f (resp. uf) is divided by $G = \{g_1, \ldots, g_s\}$ with main part $\sum_{i=1}^{s} a_i g_i$ and *remainder* $h = \mathrm{NF}(f|G)$.

The SINGULAR command *reduce* or *NF* resp. *division* returns the remainder h resp. h together with the a_i and the unit u.

Lemma 1.6.7. *Let $I \subset R$ be an ideal, $G \subset I$ a standard basis of I and $\mathrm{NF}(- \mid G)$ a weak normal form on R with respect to G.*

(1) For any $f \in R$ we have $f \in I$ if and only if $\mathrm{NF}(f \mid G) = 0$.
(2) If $J \subset R$ is an ideal with $I \subset J$, then $L(I) = L(J)$ implies $I = J$.
(3) $I = \langle G \rangle_R$, that is, the standard basis G generates I as R–ideal.
(4) If $\mathrm{NF}(- \mid G)$ is a reduced normal form, then it is unique.[7]

Proof. (1) If $\mathrm{NF}(f \mid G) = 0$ then $uf \in I$ and, hence, $f \in I$. If $\mathrm{NF}(f \mid G) \neq 0$, then $\mathrm{LM}(\mathrm{NF}(f \mid G)) \notin L(G) = L(I)$, hence $\mathrm{NF}(f \mid G) \notin I$, which implies $f \notin I$, since $\langle G \rangle_R \subset I$. To prove (2), let $f \in J$ and assume that $\mathrm{NF}(f \mid G) \neq 0$. Then $\mathrm{LM}(\mathrm{NF}(f \mid G)) \notin L(G) = L(I) = L(J)$, contradicting $\mathrm{NF}(f \mid G) \in J$. Hence, $f \in I$ by (1).

(3) follows from (2), since $L(I) = L(G) \subset L(\langle G \rangle_R) \subset L(I)$, in particular, G is also a standard basis of $\langle G \rangle_R$. Finally, to prove (4), let $f \in R$ and assume that h, h' are two reduced normal forms of f with respect to G. Then no monomial of the power series expansion of h or h' is divisible by any monomial of $L(G)$ and, moreover, $h - h' = (f - h') - (f - h) \in \langle G \rangle_R = I$. If $h - h' \neq 0$, then $\mathrm{LM}(h - h') \in L(I) = L(G)$, a contradiction, since $\mathrm{LM}(h - h')$ is a monomial of either h or h'. $\square$

Remark 1.6.8. The above properties are well–known for Gröbner bases with $R = K[x]$. For local or mixed orderings it is quite important to work rigorously with R instead of $K[x]$. We give an example showing that none of the

---

[7] In the case of a global ordering, we shall see below that such a reduced normal form exists. Then we also write $\mathrm{NF}(- \mid I)$ for $\mathrm{NF}(- \mid G)$, G any standard basis of I, and call it the *normal form with respect to I*.

above properties (1)–(3) holds for $K[x]$, if they make sense, that is, if the input data are polynomial.

Let $f_1 := x^{10} - x^9 y^2$, $f_2 := y^8 - x^2 y^7$, $f_3 := x^{10} y^7$, and consider the (local) ordering $\mathtt{ds}$ on $K[x,y]$. Then $R = K[x,y]_{\langle x,y \rangle}$, $(1 - xy)f_3 = y^7 f_1 + x^9 y f_2$, and we set

$$I := \langle f_1, f_2 \rangle_R = \langle f_1, f_2, f_3 \rangle_R\,, \quad I' := \langle f_1, f_2 \rangle_{K[x,y]}\,, \quad J' := \langle f_1, f_2, f_3 \rangle_{K[x,y]},$$

and $G := \{f_1, f_2\}$. Then G is a reduced standard basis of I (since we must multiply f_1 at least with y^8 and f_2 with x^{10} to produce new monomials, but $L(G) \supset \langle x, y \rangle^{17}$). If $\mathrm{NF}(-\mid G)$ is any weak normal form on R, then $\mathrm{NF}(f_3 \mid G) = 0$, since $f_3 \in I$. Hence, we have in this case

(1) $\mathrm{NF}(f_3 \mid G) = 0$, but $f_3 \notin I'$,
(2) $I' \subset J'$, $L(I') = L(J')$, but $I' \neq J'$,
(3) $G \subset J'$, but $\langle G \rangle_{K[x]} \neq J'$.

Note that J' is even $\langle x, y \rangle$–primary (for a definition cf. Chapter 4).

We concentrate first on well–orderings, Gröbner bases and *Buchberger's algorithm*. To describe Buchberger's normal form algorithm, we need the notion of an s–polynomial, due to Buchberger.

Definition 1.6.9. Let $f, g \in R \setminus \{0\}$ with $\mathrm{LM}(f) = x^\alpha$ and $\mathrm{LM}(g) = x^\beta$, respectively. Set

$$\gamma := \mathrm{lcm}(\alpha, \beta) := \big(\max(\alpha_1, \beta_1), \ldots, \max(\alpha_n, \beta_n)\big)$$

and let $\mathrm{lcm}(x^\alpha, x^\beta) := x^\gamma$ be the *least common multiple* of x^α and x^β. We define the *s–polynomial* (*spoly*, for short) of f and g to be

$$\mathrm{spoly}(f,g) := x^{\gamma - \alpha} f - \frac{\mathrm{LC}(f)}{\mathrm{LC}(g)} \cdot x^{\gamma - \beta} g\,.$$

If $\mathrm{LM}(g)$ divides $\mathrm{LM}(f)$, say $\mathrm{LM}(g) = x^\beta$, $\mathrm{LM}(f) = x^\alpha$, then the s–polynomial is particularly simple,

$$\mathrm{spoly}(f,g) = f - \frac{\mathrm{LC}(f)}{\mathrm{LC}(g)} \cdot x^{\alpha - \beta} g\,,$$

and $\mathrm{LM}\big(\mathrm{spoly}(f,g)\big) < \mathrm{LM}(f)$.

For the normal form algorithm, the s–polynomial will only be used in the second form, while for the standard basis algorithm we need it in the general form above. In order to be able to use the same expression in both algorithms, we prefer the above definition of the s–polynomial and not the symmetric form $\mathrm{LC}(g)x^{\gamma - \alpha} f - \mathrm{LC}(f)x^{\gamma - \beta} g$. Both are, of course, equivalent, since we work over a field K. However, in connection with pseudo standard bases (Exercise 2.3.6) we have to use the symmetric form.

Algorithm 1.6.10 (NFBUCHBERGER($f \mid G$)).
Assume that $>$ is a global monomial ordering.

Input: $f \in K[x]$, $G \in \mathcal{G}$

Output: $h \in K[x]$, a normal form of f with respect to G.

- $h := f$;
- while ($h \neq 0$ and $G_h := \{g \in G \mid \mathrm{LM}(g) \text{ divides } \mathrm{LM}(h)\} \neq \emptyset$)
    choose any $g \in G_h$;
    $h := \mathrm{spoly}(h, g)$;
- return h;

Note that each specific choice of "any" can give a different normal form function.

Proof. The algorithm terminates, since in the i–th step of the while loop we create (setting $h_0 := f$) an s–polynomial

$$h_i = h_{i-1} - m_i g_i, \quad \mathrm{LM}(h_{i-1}) > \mathrm{LM}(h_i),$$

where m_i is a term such that $\mathrm{LT}(m_i g_i) = \mathrm{LT}(h_{i-1})$, and $g_i \in G$ (allowing repetitions).

Since $>$ is a well–ordering, $\{\mathrm{LM}(h_i)\}$ has a minimum, which is reached at some step m. We obtain

$$h_1 = f - m_1 g_1$$
$$h_2 = h_1 - m_2 g_2 = f - m_1 g_1 - m_2 g_2$$
$$\vdots$$
$$h_m = f - \sum_{i=1}^{m} m_i g_i,$$

satisfying $\mathrm{LM}(f) = \mathrm{LM}(m_1 g_1) > \mathrm{LM}(m_i g_i) > \mathrm{LM}(h_m)$. This shows that $h := h_m$ is a normal form with respect to G.

Moreover, if $h \neq 0$, then $G_h = \emptyset$ and, hence, $\mathrm{LM}(h) \notin L(G)$ if $h \neq 0$. This proves correctness, independent of the specific choice of "any" in the while loop. $\square$

It is easy to extend NFBUCHBERGER to a reduced normal form. Either we do tail–reduction during NFBUCHBERGER, that is, we set

$$h := \mathrm{spoly}(h, g);$$
$$h := \mathrm{LT}(h) + \mathrm{NFBUCHBERGER}(\mathrm{tail}(h) \mid G);$$

in the while loop, or do tail–reduction after applying NFBUCHBERGER, as in Algorithm 1.6.11. Indeed, the argument holds for any normal form with respect to a global ordering.

Algorithm 1.6.11 (REDNFBUCHBERGER($f \mid G$))**.**
Assume that $>$ is a global monomial ordering.

Input: $f \in K[x]$, $G \in \mathcal{G}$
Output: $h \in K[x]$, a reduced normal form of f with respect to G

- $h := 0$, $g := f$;
- while ($g \neq 0$)
    $g := $ NFBUCHBERGER $(g \mid G)$;
    if ($g \neq 0$)
       $h := h + \mathrm{LT}(g)$;
       $g := \mathrm{tail}(g)$;
- return $h/\mathrm{LC}(h)$;

Since $\mathrm{tail}(g)$ has strictly smaller leading term than g, the algorithm terminates, since $>$ is a well–ordering. Correctness follows from the correctness of NFBUCHBERGER.

Example 1.6.12. Let $>$ be the ordering dp on $\mathrm{Mon}(x, y, z)$,

$$f = x^3 + y^2 + 2z^2 + x + y + 1\,, \quad G = \{x, y\}\,.$$

NFBUCHBERGER proceeds as follows:

$\mathrm{LM}(f) = x^3$, $G_f = \{x\}$,
$h_1 = \mathrm{spoly}(f, x) = y^2 + 2z^2 + x + y + 1$,
$\mathrm{LM}(h_1) = y^2$, $G_{h_1} = \{y\}$,
$h_2 = \mathrm{spoly}(h_1, y) = 2z^2 + x + y + 1$, $G_{h_2} = \emptyset$.

Hence, NFBUCHBERGER$(f \mid G) = 2z^2 + x + y + 1$. For the reduced normal form in Algorithm 1.6.11 we obtain:

$g_0 = $ NFBUCHBERGER$(f \mid G) = 2z^2 + x + y + 1$, $\mathrm{LT}(g_0) = 2z^2$,
$h_1 = 2z^2$, $g_1 = \mathrm{tail}(g_0) = x + y + 1$,
$g_2 = $ NFBUCHBERGER$(g_1 \mid G) = 1$, $\mathrm{LT}(g_2) = 1$,
$h_2 = 2z^2 + 1$, $g_3 = \mathrm{tail}(g_2) = 0$.

Hence, REDNFBUCHBERGER$(f \mid G) = z^2 + 1/2$.

SINGULAR Example 1.6.13 (normal form).
Note that $\mathrm{NF}(f \mid G)$ may depend on the sorting of the elements of G. The function `reduce` computes a normal form.

```
ring A   = 0,(x,y,z),dp;  //a global ordering
poly f   = x2yz+xy2z+y2z+z3+xy;
poly f1  = xy+y2-1;
poly f2  = xy;
ideal G  = f1,f2;
```

```
ideal S = std(G);          //a standard basis of <G>
S;
//-> S[1]=xy
//-> S[2]=y2-1

reduce(f,G);
//** G is no standardbasis
//-> y2z+z3                //NF w.r.t. a non-standard basis

G=f2,f1;
reduce(f,G);
//** G is no standardbasis
//-> y2z+z3-y2+xz+1        //NF for a different numbering in G

reduce(f,S,1);             //NFBuchberger
//-> z3+xy+z

reduce(f,S);               //redNFBuchberger
//-> z3+z
```

Remark 1.6.14. There exists also the notion of a *standard basis over a ring*. Namely, let R be Noetherian and $R[x] = R[x_1, \ldots, x_n]$. The leading data of $f \in R[x_1, \ldots, x_n] \setminus \{0\}$ with respect to a monomial ordering $>$ on $\mathrm{Mon}(x_1, \ldots, x_n)$ are defined as in Definition 1.2.2. If $I \subset R[x]$ is an ideal and $G \subset I$ a finite set, then G is a *standard basis* of I if

$$\langle \mathrm{LT}(f) \mid f \in I \rangle = \langle \mathrm{LT}(g) \mid g \in G \rangle.$$

Note that we used leading terms and not leading monomials (which is, of course, equivalent if R is a field). The normal form algorithm over rings is more complicated than over fields. For example, if $>$ is a global ordering, the algorithm NFBUCHBERGER has to be modified to

$h := f$;
while ($h \neq 0$ and $G_h = \{g_1, \ldots, g_s\} \neq \emptyset$ and $\mathrm{LT}(h) \in \langle \mathrm{LT}(g) \mid g \in G_h \rangle$)
    choose $c_i \in R \setminus \{0\}$ and monomials m_i with $m_i \mathrm{LM}(g_i) = \mathrm{LM}(h)$ such that
        $\mathrm{LT}(h) = c_1 m_1 \mathrm{LT}(g_1) + \cdots + c_s m_s \mathrm{LT}(g_s)$;
    $h := h - \sum_{i=1}^{s} c_i m_i g_i$;
return h;

The determination of the c_i requires the solving of linear equations over R and not just a divisibility test for monomials as for s–polynomials. With this normal form, standard bases can be computed as in the next section. For details see [1], [90], [129].

In practice, however, this notion is not frequently used so far and there seems to be no publicly available system having this implemented. A weaker concept are the *comprehensive Gröbner bases* of Weispfenning [232], which are Gröbner bases depending on parameters and which specialize to a Gröbner basis for all possible fixed values of the parameters.

For a simple criterion, when the *specialization of a standard basis* is again a standard basis, see Exercises 2.3.7, 2.3.8, where we introduce *pseudo standard bases*.

Exercises

Let $>$ be any monomial ordering and $R = K[x_1, \ldots, x_n]_>$.

1.6.1. Let $I \subset R$ be an ideal. Show that if I has a reduced standard basis, then it is unique.

1.6.2. Let $>$ be a local or mixed ordering. Prove that Algorithm 1.6.11 computes, theoretically, (possibly in infinitely many steps) for $f \in R$ and $G \subset R$ a reduced normal form. Hence, it can be used to compute, for local degree orderings, a normal form which is completely reduced up to a finite, but arbitrarily high order.

1.6.3. Show by an example, with f and G consisting of polynomials and $>$ not global, that a completely reduced normal form of f with respect to G does not exist in R. (Note that Exercise 1.6.2 only says that it exists as formal power series.)

1.6.4. Apply NFBUCHBERGER to $(f, G, >)$ without using SINGULAR:

(1) $f = 1$, $G = \{x - 1\}$ and ordering lp, respectively ls.
(2) $f = x^4 + y^4 + z^4 + xyz$, $G = \{\partial f/\partial x, \partial f/\partial y, \partial f/\partial z\}$ and ordering dp.

1.6.5. Give a direct argument that the set G in Exercise 1.6.4 (2) is a standard basis with respect to dp.

1.6.6. Write a SINGULAR procedure, having two polynomials f, g as input and returning spoly(f, g) as output.

1.6.7. Write your own SINGULAR procedure, having a polynomial f and an ideal I as input and NFBUCHBERGER $(f \mid I)$ as output by always choosing the first element from G_h. (Note that an ideal is given by a list of polynomials.)

1.6.8. Implement, as SINGULAR procedures, the two ways described in the text to compute a reduced normal form. (The first method is a good exercise in recursive programming.)

Check your procedures with the SINGULAR Example 1.6.13.

1.6.9. Let $R = K[t_1, \ldots, t_n]$, K a field. Write a SINGULAR procedure which computes the normal form NFBUCHBERGER over the ring R, as explained in Remark 1.6.14.

(Hint: use the SINGULAR command lift.)

1.7 The Standard Basis Algorithm

Let $>$ be a fixed monomial ordering and let, in this section,

$$R = K[x_1, \ldots, x_n]_>$$

be the localization of $K[x]$, $x = (x_1, \ldots, x_n)$, with respect to $>$. Recall that $R = S_>^{-1} K[x]$ with $S_> = \{u \in K[x] \setminus \{0\} \mid \mathrm{LM}(u) = 1\}$, and that $R = K[x]$ if $>$ is global and $R = K[x]_{\langle x \rangle}$ if $>$ is local. In any case, R may be considered as a subring of the ring $K[[x]]$ of formal power series.

The idea of many standard basis algorithms may be formalized as follows:

Algorithm 1.7.1 (STANDARD(G,NF)).
Let $>$ be any monomial ordering, and $R := K[x_1, \ldots, x_n]_>$.

Input: $G \in \mathcal{G}$, NF an algorithm returning a weak normal form.
Output: $S \in \mathcal{G}$ such that S is a standard basis of $I = \langle G \rangle_R \subset R$

- $S := G$;
- $P := \{(f, g) \mid f, g \in S, f \neq g\}$, the pair–set;
- while $(P \neq \emptyset)$
    choose $(f, g) \in P$;
    $P := P \setminus \{(f, g)\}$;
    $h := \mathrm{NF}\big(\mathrm{spoly}(f, g) \mid S\big)$;
    if $(h \neq 0)$
      $P := P \cup \{(h, f) \mid f \in S\}$;
      $S := S \cup \{h\}$;
- return S;

To see termination of STANDARD, note that if $h \neq 0$ then $\mathrm{LM}(h) \notin L(S)$ by property (i) of NF. Hence, we obtain a strictly increasing sequence of monomial ideals $L(S)$ of $K[x]$, which becomes stationary as $K[x]$ is Noetherian. That is, after finitely many steps, we always have $\mathrm{NF}\big(\mathrm{spoly}(f, g) \mid S\big) = 0$ for $(f, g) \in P$, and, again after finitely many steps, the pair-set P will become empty. Correctness follows from applying Buchberger's fundamental standard basis criterion below.

Remark 1.7.2. If NF is a reduced normal form and if G is reduced, then S, as returned by STANDARD(G,NF), is a reduced standard basis if we delete elements whose leading monomials are divisible by a leading monomial of another element in S. If G is not reduced, we may apply a reduced normal form afterwards to $(f, S \setminus \{f\})$ for all $f \in S$ in order to obtain a reduced standard basis.

Theorem 1.7.3 (Buchberger's criterion). *Let $I \subset R$ be an ideal and $G = \{g_1, \ldots, g_s\} \subset I$. Let $\mathrm{NF}(- \mid G)$ be a weak normal form on R with respect to G. Then the following are equivalent:*[8]

---

[8] Usually, the implication (4) $\Rightarrow$ (1) is called Buchberger's criterion. But with our concept of (weak) normal forms, we need, indeed, the implication (5) $\Rightarrow$ (1) to prove the correctness of the standard basis algorithm.

(1) G is a standard basis of I.
(2) $\mathrm{NF}(f \mid G) = 0$ *for all* $f \in I$.
(3) Each $f \in I$ *has a standard representation with respect to* $\mathrm{NF}(- \mid G)$.
(4) G generates I and $\mathrm{NF}(\mathrm{spoly}(g_i, g_j) \mid G) = 0$ *for* $i, j = 1, \ldots, s$.
(5) G generates I and $\mathrm{NF}(\mathrm{spoly}(g_i, g_j) \mid G_{ij}) = 0$ *for a suitable subset* $G_{ij} \subset$
   G *and* $i, j = 1, \ldots, s$.

Proof. The implication (1) $\Rightarrow$ (2) follows from Lemma 1.6.7, (2) $\Rightarrow$ (3) is trivial. To see (3) $\Rightarrow$ (4), note that $h := \mathrm{NF}(\mathrm{spoly}(g_i, g_j) \mid G) \in I$ and, hence, either $h = 0$ or $\mathrm{LM}(h) \in L(G)$ by (3), a contradiction to property (i) of NF. The fact that G generates I follows immediately from (3). (4) $\Rightarrow$ (5) is trivial.

Finally, the implication (5) $\Rightarrow$ (1) is the important Buchberger criterion which allows the checking and construction of standard bases in finitely many steps. Our proof uses syzygies and is, therefore, postponed to the next chapter.   □

Example 1.7.4. Let $>$ be the ordering **dp** on $\mathrm{Mon}(x, y)$, NF=NFBUCHBER-GER and $G = \{x^2 + y, xy + x\}$. Then we obtain as initialization
   $S = \{x^2 + y, xy + x\}$
   $P = \{(x^2 + y, xy + x)\}$.
The while–loop gives, in the first turn,
   $P = \emptyset$
   $h = \mathrm{NF}(-x^2 + y^2 \mid S) = y^2 + y$
   $P = \{(y^2 + y, x^2 + y), (y^2 + y, xy + x)\}$
   $S = \{x^2 + y, xy + x, y^2 + y\}$.
In the second turn
   $P = \{(y^2 + y, xy + x)\}$
   $h = \mathrm{NF}(-x^2 y + y^3 \mid S) = 0$.
In the third turn
   $P = \emptyset$
   $h = \mathrm{NF}(0 \mid S) = 0$.
The algorithm terminates and $S = \{x^2 + y, xy + x, y^2 + y\}$ is a standard basis.

We present now a general normal form algorithm, which works for any monomial ordering. The basic idea is due to Mora [176], but our algorithm is more general, with a different notion of ecart. It has been implemented in SINGU-LAR since 1990, the first publication appeared in [97], [111].

Before turning to the details, let us first analyze Buchberger's algorithm in the case of a non–global ordering. We may assume that in $K[x, y]$ we have $x_1, \ldots, x_n < 1, y_1, \ldots, y_m > 1$ $(m \geq 0)$.

Look at the sequence $m_i = c_i x^{\alpha_i} y^{\beta_i}$, $i \geq 1$, of terms constructed in the algorithm NFBUCHBERGER. If $\deg_x(m_i)$ is bounded, then, since $>$ induces a well–ordering on $K[y]$, the algorithm stops after finitely many steps.

On the other hand, if the degree of m_i in x is unbounded, then, for each fixed factor x^{α_i}, there can only be finitely many cofactors y^{β_j} and, hence, $\sum_{i \geq 1} m_i$ converges in the $\langle x \rangle$–adic topology (cf. Definition 6.1.6), that is,

$\sum_{i\geq1} m_i \in K[y][[x]]$. If $G = \{g_1, \ldots, g_s\}$ we may gather the factors m_j of any g_i, obtaining thus in NFBUCHBERGER an expression

$$h = f - \sum_{i=1}^{s} a_i g_i, \quad h, a_i \in K[y][[x]],$$

which holds in $K[y][[x]]$. However, this process does not stop.

The standard example is in one variable x, with $x < 1$, $f := x$ and $G := \{g = x - x^2\}$. Using NFBUCHBERGER we obtain

$$x - \left(\sum_{i=0}^{\infty} x^i\right)(x - x^2) = 0$$

in $K[[x]]$, which is true, since $\sum_{i=0}^{\infty} x^i = 1/(1-x)$ in $K[[x]]$. However, the algorithm constructs a power series $\sum_{i=0}^{\infty} x^i$ having infinitely many terms and not the finite expression $1/(1-x)$.

In order to avoid infinite power series, we have to allow a wider class of elements for the reduction in order to create a standard expression of the form

$$uf = \sum_{i=1}^{s} a_i g_i + \text{NF}(f \mid G),$$

where u is a unit in R, and u, a_i and $\text{NF}(f \mid G)$ are polynomials in the case when the input data f and $G = \{g_1, \ldots, g_s\}$ are polynomials. In the previous example we arrive at an expression

$$(1 - x)x = x - x^2$$

instead of $x = (\sum_{i=0}^{\infty} x^i)(x - x^2)$.

Definition 1.7.5. For $f \in K[x] \smallsetminus \{0\}$ we define the *ecart* of f as

$$\text{ecart}(f) := \deg f - \deg \text{LM}(f).$$

Note that, for a homogeneous polynomial f, we have $\text{ecart}(f) = 0$.

If $w = (w_1, \ldots, w_n)$ is any tuple of positive real numbers, we can define the *weighted ecart* by $\text{ecart}_w(f) := \text{w–deg}(f) - \text{w–deg}(\text{LM}(f))$. In the following normal form algorithm NFMORA, we may always take ecart_w instead of ecart, the algorithm works as well. It was noted in [94] that, for certain examples, the algorithm can become much faster for a good choice of w.

Another description of $\text{ecart}(f)$ turns out to be quite useful. Let f^h denote the homogenization of f with respect to a new variable t (such that all monomials of f are of the same degree, cf. Exercise 1.7.4). Define on $\text{Mon}(t, x_1, \ldots, x_n)$ an ordering $>_h$ by $t^p x^\alpha >_h t^q x^\beta$ if $p + |\alpha| > q + |\beta|$ or if $p + |\alpha| = q + |\beta|$ and $x^\alpha > x^\beta$. Equivalently, $>_h$ is given by the matrix

$$\begin{pmatrix} 1\ 1\ \dots\ 1 \\ 0 \\ \vdots \quad\quad A \\ 0 \end{pmatrix}$$

where A is a matrix defining the ordering on $K[x]$. This defines a well-ordering on $\mathrm{Mon}(t, x)$.

For $f \in K[x]$ we have

$$\mathrm{LM}_{>_h}(f^h) = t^{\mathrm{ecart}(f)}\, \mathrm{LM}_>(f)\,,$$

in particular, $\mathrm{ecart}(f) = \deg_t \mathrm{LM}_{>_h}(f^h)$.

Algorithm 1.7.6 ($\mathrm{NFMORA}(f \mid G)$).

Let $>$ be any monomial ordering.

Input: $f \in K[x]$, G a finite list in $K[x]$
Output: $h \in K[x]$ a polynomial weak normal form of f with respect to G.

- $h := f$;
- $T := G$;
- while($h \neq 0$ and $T_h := \{g \in T \mid \mathrm{LM}(g) \mid \mathrm{LM}(h)\} \neq \emptyset$)
   choose $g \in T_h$ with $\mathrm{ecart}(g)$ minimal;
   if ($\mathrm{ecart}(g) > \mathrm{ecart}(h)$)
    $T := T \cup \{h\}$;
   $h := \mathrm{spoly}(h, g)$;
- return h;

Example 1.7.7. Let $>$ be the ordering ds on $\mathrm{Mon}(x, y, z)$, $f = x^2 + y^2 + z^3 + x^4 + y^5$, $G = \{x, y\}$. Then $\mathrm{NFMORA}\,(f \mid G) = z^3 + x^4 + y^5$.

If the input is homogeneous, then the ecart is always 0 and NFMORA is equal to $\mathrm{NFBUCHBERGER}$. If $>$ is a well–ordering, then $\mathrm{LM}(g) \mid \mathrm{LM}(h)$ implies that $\mathrm{LM}(g) \leq \mathrm{LM}(h)$, hence, even if h is added to T during the algorithm, it cannot be used in further reductions. Thus, NFMORA is the same as $\mathrm{NFBUCHBERGER}$, but with a special selection strategy for the elements from G.

Proof of Algorithm 1.7.6. Termination is most easily seen by using homogenization: start with $h := f^h$ and $T := G^h = \{g^h \mid g \in G\}$. The while loop looks as follows (see Exercise 1.7.9):

- while ($h \neq 0$ and $T_h := \{g \in T \mid \mathrm{LM}(g)$ divides $t^\alpha \mathrm{LM}(h)$ for some $\alpha\} \neq \emptyset$)
   choose $g \in T_h$ with $\alpha \geq 0$ minimal;
   if ($\alpha > 0$)
    $T := T \cup \{h\}$;
   $h := \mathrm{spoly}(t^\alpha h, g)$;
   $h := (h|_{t=1})^h$;
- return $h|_{t=1}$;

Since R is Noetherian, there exists some positive integer N such that $L(T_\nu)$ becomes stable for $\nu \geq N$, where T_ν denotes the set T after the ν–th turn of the while loop. The next h, therefore, satisfies $\mathrm{LM}(h) \in L(T_N) = L(T)$, whence, $\mathrm{LM}(g)$ divides $\mathrm{LM}(h)$ for some $g \in T$ and $\alpha = 0$. That is, T_ν itself becomes stable for $\nu \geq N$ and the algorithm continues with fixed T. Then it terminates, since $>$ is a well–ordering on $K[t,x]$.

To see correctness, consider the ν–th turn in the while loop of Algorithm 1.7.6. There we create (with $h_0 := f$) $h_\nu := \mathrm{spoly}(h_{\nu-1}, g'_\nu)$ for some $g'_\nu \in T_{\nu-1}$ such that $\mathrm{LM}(g'_\nu) \mid \mathrm{LM}(h_{\nu-1})$. Hence, there exists some term $m_\nu \in K[x]$, $\mathrm{LT}(m_\nu g'_\nu) = \mathrm{LT}(h_{\nu-1})$, such that

$$h_\nu = h_{\nu-1} - m_\nu g'_\nu, \quad \mathrm{LM}(h_{\nu-1}) = \mathrm{LM}(m_\nu g'_\nu) > \mathrm{LM}(h_\nu),$$

Now for g'_ν we have two possibilities:

(1) $g'_\nu = g_i \in G = \{g_1, \ldots, g_s\}$ for some i, or
(2) $g'_\nu \in T \smallsetminus G \subset \{h_0, h_1, \ldots, h_{\nu-2}\}$.

Suppose, by induction, that in the first $\nu - 1$ steps ($\nu \geq 1$) we have constructed standard representations

$$u_j f = \sum_{i=1}^{s} a_i^{(j)} g_i + h_j, \quad u_j \in S_>, \quad a_i^{(j)} \in K[x],$$

$0 \leq j \leq \nu - 1$, starting with $u_0 := 1$, $a_i^{(0)} := 0$.

Consider this standard representation for $j = \nu - 1$. In case (1), we replace $h_{\nu-1}$ on the right–hand side by $h_\nu + m_\nu g_i$, hence, obtaining

$$u_\nu f = \sum_{i=1}^{s} a_i^{(\nu)} g_i + h_\nu$$

with $u_\nu := u_{\nu-1}$ and some $a_i^{(\nu)} \in K[x]$.

In case (2), we have to substitute $h_{\nu-1}$ by

$$h_\nu + m_\nu h_j = h_\nu - m_\nu \left(\sum_{i=1}^{s} a_i^{(j)} g_i - u_j f \right)$$

with $j < \nu - 1$. Hence, we obtain an expression

$$(u_{\nu-1} - m_\nu u_j) f = \sum_{i=1}^{s} a_i^{(\nu)} g_i + h_\nu, \quad a_i^{(\nu)} \in K[x].$$

Since $\mathrm{LM}(m_\nu) \cdot \mathrm{LM}(h_j) = \mathrm{LM}(m_\nu h_j) = \mathrm{LM}(h_{\nu-1}) < \mathrm{LM}(h_j)$, we obtain that $\mathrm{LM}(m_\nu) < 1$ and, hence, $u_\nu = u_{\nu-1} - m_\nu u_j \in S_>$. $\qquad \square$

It is clear that, with a little extra storage, the algorithm does also return $u \in S_>$. Moreover, with quite a bit of bookkeeping one obtains the a_i.

Now, the standard basis algorithm for arbitrary monomial orderings formally looks as follows:

Algorithm 1.7.8 (STANDARDBASIS(G)).
Let $>$ be any monomial ordering, $R = K[x]_>$.

Input: $G = \{g_1, \ldots, g_s\} \subset K[x]$
Output: $S = \{h_1, \ldots, h_t\} \subset K[x]$ such that S is a standard basis of the ideal $\langle G \rangle_R \subset R$.

- $S := $ STANDARD(G,NFMORA);
- return S;

The following corollary shows that the property of being a standard basis depends only on the ordering of finitely many monomials. This property is used in our study of flatness and standard bases (Section 7.5).

Corollary 1.7.9 (finite determinacy of standard bases). *Let $I \subset K[x]$ be an ideal and $G \subset K[x]$ be a standard basis of I with respect to an arbitrary monomial ordering $>$. Then there exists a finite set $F \subset \mathrm{Mon}(x)$ with the following properties:*
Let $>_1$ be any monomial ordering on $\mathrm{Mon}(x)$ coinciding with $>$ on F, then

(1) $\mathrm{LM}_>(g) = \mathrm{LM}_{>_1}(g)$ for all $g \in G$,
(2) G is a standard basis of I with respect to $>_1$.

Proof. We apply Theorem 1.7.3 with NF = NFMORA.
    Let $G = \{g_1, \ldots, g_s\}$, and let F be the set of all monomials occurring in all polynomials during the reduction process of $\mathrm{spoly}(g_i, g_j)$ to 0 in NFMORA. Then $\mathrm{NF}\big(\mathrm{spoly}(g_i, g_j) \mid G\big) = 0$ also with respect to $>_1$, and the result follows, using Theorem 1.7.3 (4).   □

SINGULAR Example 1.7.10 (standard bases).
The same generators for an ideal give different standard bases with respect to different orderings:

```
ring A  = 0,(x,y),dp;   //global ordering: degrevlex
ideal I = x10+x9y2,y8-x2y7;
ideal J = std(I);
J;
//-> J[1]=x2y7-y8    J[2]=x9y2+x10    J[3]=x12y+xy11
//-> J[4]=x13-xy12   J[5]=y14+xy12    J[6]=xy13+y12

ring A1 = 0,(x,y),lp;   //global ordering: lex
ideal I = fetch(A,I);
```

```
ideal J = std(I);
J;
//-> J[1]=y15-y12 J[2]=xy12+y14 J[3]=x2y7-y8 J[4]=x10+x9y2

ring B  = 0,(x,y),ds;    //local ordering: local degrevlex
ideal I = fetch(A,I);
ideal J = std(I);
J;
//-> J[1]=y8-x2y7    J[2]=x10+x9y2

ring B1 = 0,(x,y),ls;    //local ordering: negative lex
ideal I = fetch(A,I);
ideal J = std(I);
J;
//-> J[1]=y8-x2y7    J[2]=x9y2+x10    J[3]=x13

intmat O[3][3]=1,1,1,0,-1,-1,0,0,-1;
ring  C = 0,(t,x,y),M(O);    //global ordering: matrix O
ideal I = homog(imap(A,I),t); //gives a standard basis for
                              //local degrevlex
ideal J = std(I);            //cf. Exercise 1.7.5
      J = subst(J,t,1);
J;
//-> J[1]=-x2y7+y8    J[2]=x9y2+x10    J[3]=x12y7+x9y10
                                       //already J[1],J[2] is a
                                       //standard basis
```

We finish this section with the so–called highest corner, a notion which is computationally extremely useful for 0–dimensional ideals in local rings. Moreover, the highest corner is tightly connected with the determinacy of an isolated hypersurface singularity (cf. A.9).

Definition 1.7.11. Let $>$ be a monomial ordering on $\mathrm{Mon}(x_1,\ldots,x_n)$ and let $I \subset K[x_1,\ldots,x_n]_>$ be an ideal. A monomial $m \in \mathrm{Mon}(x_1,\ldots,x_n)$ is called the *highest corner* of I (with respect to $>$), denoted by $\mathrm{HC}(I)$, if

(1) $m \notin L(I)$;
(2) $m' \in \mathrm{Mon}(x_1,\ldots,x_n)$, $m' < m \implies m' \in L(I)$.

Note that for a global ordering the highest corner is 1 if I is a proper ideal (and does not exist if $1 \in I$). Since, by definition $\mathrm{HC}(I) = \mathrm{HC}\big(L(I)\big)$, it can be computed combinatorially from a standard basis of I.

SINGULAR has a built–in function `highcorner` which returns, for a given set of generators $f_1,\ldots,f_k$ of I, the highest corner of the ideal $\langle \mathrm{LM}(f_1),\ldots,\mathrm{LM}(f_k)\rangle$, respectively 0, if the highest corner does not exist.

SINGULAR Example 1.7.12 (highest corner).

```
ring A = 0,(x,y),ds;
ideal I = y4+x5,x3y3;
highcorner(I);
//-> // ** I is not a standard basis
//-> 0                    //no highest corner for <y4,x3y3>

std(I);
//-> _[1]=y4+x5   _[2]=x3y3   _[3]=x8
highcorner(std(I));
//-> x7y2
```

The highest corner of I is x^7y^2, as can be seen from Figure 1.2.

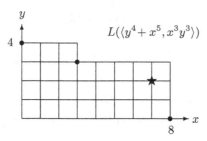

Fig. 1.2. $L(\langle y^4 + x^5, x^3y^3 \rangle)$ is generated by the monomials y^4, x^3y^3, x^8 (marked by a •). The highest corner is x^7y^2 (marked by a ★).

Lemma 1.7.13. *Let $>$ be a monomial ordering on $\mathrm{Mon}(x_1, \dots, x_n)$ and $I \subset K[x_1, \dots, x_n]_>$ be an ideal. Let m be a monomial such that $m' < m$ implies $m' \in L(I)$. Let $f \in K[x_1, \dots, x_n]$ such that $\mathrm{LM}(f) < m$. Then $f \in I$.*

Proof. Let $r = \mathrm{NFMora}\,(f \mid G)$, G a standard basis for I. If $r \neq 0$, then $\mathrm{LM}(r) < \mathrm{LM}(f) < m$ and, therefore, $\mathrm{LM}(r) \in I$ which is a contradiction to the properties of the normal form. □

Lemma 1.7.14. *Let $>$ be a monomial ordering on $\mathrm{Mon}(x_1, \dots, x_n)$ and denote by $z_1, \dots, z_r$ the variables < 1 from $\{x_1, \dots, x_n\}$ and by $y_1, \dots, y_s$ the variables > 1 ($0 \leq r, s, r + s = n$). Assume that the restriction of $>$ to $\mathrm{Mon}(z_1, \dots, z_r)$ is a weighted degree ordering. The following are equivalent for an ideal $I \subset K[x_1, \dots, x_n]_>$:*

(1) $\mathrm{HC}(I)$ exists,
(2) $\langle z_1, \dots, z_r \rangle^N \subset L(I)$ for some $N \geq 0$,
(3) $\langle z_1, \dots, z_r \rangle^M \subset I$ for some $M \geq 0$.

Moreover, $\mathrm{HC}(I) \in \mathrm{Mon}(z_1, \dots, z_r)$ if it exists.

Proof. To see that (1) implies (2), let $m := \mathrm{HC}(I)$. If $m = 1$, then $1 \notin I$ and $z_1, \ldots, z_r \in L(I)$ by definition of the highest corner. If $m \neq 1$ and if we write $m = x_i m'$ for some monomial m' then $x_i < 1$ (otherwise, $m' < m$, which would imply $m' \in L(I)$, hence, $m \in L(I)$, a contradiction), and it follows that $m \in \mathrm{Mon}(z_1, \ldots, z_r)$. Since $>$ is a weighted degree ordering on $\mathrm{Mon}(z_1, \ldots, z_r)$, the definition of the highest corner implies (2).

Conversely, if $\langle z_1, \ldots, z_r \rangle^N \subset L(I)$ then there are only finitely many monomials in $\mathrm{Mon}(z_1, \ldots, z_r)$ which are not in $L(I)$. This finite set has a minimum m. If $m' = z^\alpha y^\beta < m$ then $z^\alpha < m$ which implies $z^\alpha \in L(I)$ and, hence, $m' \in L(I)$.

The implication $(3) \Rightarrow (2)$ being trivial, it remains only to show that (2) implies (3). Let $M \geq N$. Since $z_i^M \in L(I)$, we have $z_i^M + h_i \in I$ for some h_i with $\mathrm{LM}(h_i) = z^\alpha y^\beta < z_i^M$, in particular, $z^\alpha < z_i^M$. Let $m := \mathrm{HC}(I)$, which exists by the equivalence of (2) and (1) proven before, and enlarge M, if necessary, such that $z_i^M \leq m$. Then $z^\alpha < m$ implies $z^\alpha \in L(I)$ and, hence, $\mathrm{LM}(h_i) \in L(I)$. Now we apply Lemma 1.7.13 and obtain $h_i \in I$. Therefore, $z_i^M \in I$ for $i = 1, \ldots, r$, and (3) follows.   □

Remark 1.7.15. As a direct consequence, for a local weighted degree ordering, we have

$$\mathrm{HC}(I) \text{ exists} \iff \dim_K(K[x_1, \ldots, x_n]_>/I) < \infty$$
$$\iff \dim_K\big(K[x_1, \ldots, x_n]/L(I)\big) < \infty.$$

Indeed, we show in Section 7.5 that, for any monomial ordering,

$$\dim_K(K[x_1, \ldots, x_n]_>/I) = \dim_K\big(K[x_1, \ldots, x_n]/L(I)\big).$$

(see also Corollary 5.3.14).

Remark 1.7.16. The implications $(2) \Leftrightarrow (3) \Rightarrow (1)$ in Lemma 1.7.14 hold without any assumption on the ordering $>$. This is a consequence of Lemma 1.2.11.

The implication $(1) \Rightarrow (2)$ is wrong in general: let $>$ be the negative lexicographical ordering $\mathtt{ls}$ and $I = \langle xy, x^2 \rangle$, then $\mathrm{HC}(I) = x$.

Lemma 1.7.17. *Let $>$ be a weighted degree ordering on $\mathrm{Mon}(x_1, \ldots, x_n)$. Moreover, let $f_1, \ldots, f_k$ be a set of generators of the ideal $I \subset K[x_1, \ldots, x_n]_>$ such that $J := \langle \mathrm{LM}(f_1), \ldots, \mathrm{LM}(f_k) \rangle$ has a highest corner $m := \mathrm{HC}(J)$, and let $f \in K[x_1, \ldots, x_n]_>$. Then the following holds:*

(1) $\mathrm{HC}(I)$ exists, and, moreover, $\mathrm{HC}(I) \geq \mathrm{HC}(J)$ and $\mathrm{HC}(I) = \mathrm{HC}(J)$ if $f_1, \ldots, f_k$ is a standard basis of I.

(2) If $\mathrm{LM}(f) < \mathrm{HC}(J)$ then $f \in I$.

(3) For a fixed monomial $m' < \mathrm{HC}(J)$ set $M = \{i \mid \mathrm{LM}(f_i) \leq m'\}$ and define

$$\hat{f}_i := \begin{cases} f_i, & \text{if } i \in M \\ f_i + a_i \cdot m', & \text{if } i \notin M \end{cases}$$

where $a_i \in K$ is arbitrary. Then $I = \langle \hat{f}_1, \ldots, \hat{f}_k \rangle$.

Proof. (1) Since $J \subset L(I)$ and $J = L(I)$ if $f_1, \ldots, f_k$ is a standard basis, the claim follows from Lemma 1.7.14.

(2) $\mathrm{LM}(f) < m$ implies $\mathrm{LM}(f) \in L(J) \subset L(I)$. The assertion is a consequence of Lemma 1.7.13.

(3) Since $m' < m$, $m' \in I$ by (2) and, therefore, $I = \langle \hat{f}_1, \ldots, \hat{f}_k, m' \rangle$. We have to show $m' \in \hat{I} = \langle \hat{f}_1, \ldots, \hat{f}_k \rangle$. Since $\mathrm{LM}(f_i) = \mathrm{LM}(\hat{f}_i)$ for all i, we can apply (2) to $\hat{I}$ instead of I with the same J and m and, therefore, $m' \in \hat{I}$. $\square$

The lemma shows that we can delete from f_i all terms $a \cdot m'$, $a \in K$, with $m' < \min\{m, \mathrm{LM}(f_i)\}$, still keeping a set of generators of I. This is used in SINGULAR during standard basis computations in local orderings to keep the polynomials sparse and to have *early termination* if, in the reduction process, the leading monomial becomes smaller than the highest corner.

Exercises

1.7.1. Prove the *Product Criterion*: let $f, g \in K[x_1, \ldots, x_n]$ be polynomials such that $\mathrm{lcm}(\mathrm{LM}(f), \mathrm{LM}(g)) = \mathrm{LM}(f) \cdot \mathrm{LM}(g)$, then

$$\mathrm{NF}\big(\mathrm{spoly}(f, g) \mid \{f, g\}\big) = 0.$$

(Hint: It is sufficient to prove the statement for $\mathrm{NF} = \mathrm{NFMora}$. Assume that $\mathrm{LC}(f) = \mathrm{LC}(g) = 1$ and claim that $\mathrm{spoly}(f, g) = -\mathrm{tail}(g)f + \mathrm{tail}(f)g$. Moreover, assume that, after some steps in NFMora, $u \cdot \mathrm{spoly}(f, g)$ (u a unit) is reduced to $hf + kg$. If $\mathrm{LT}(hf) + \mathrm{LT}(kg) = 0$ then $\mathrm{LT}(h) = m \cdot \mathrm{LM}(g)$ and $\mathrm{LT}(k) = -m \, \mathrm{LM}(f)$ for a suitable term m, and $(u - m) \, \mathrm{spoly}(f, g)$ is reduced to $\mathrm{tail}(h)f + \mathrm{tail}(k)g$. If $\mathrm{LT}(hf) + \mathrm{LT}(kg) \neq 0$ then assume $\mathrm{LM}(hf + kg) = \mathrm{LM}(hf)$, and $hf + kg$ reduces to $\mathrm{tail}(h)f + kg$.)

1.7.2. Let $I := \langle x^3 y^2 + x^4, x^2 y^3 + y^4 \rangle \subset K[x, y]$ (resp. $I := \langle x^3 + y^2, y^4 + x \rangle$). Compute (without using SINGULAR) a standard basis of I with respect to the degree lexicographical ordering (respectively lexicographical ordering).

1.7.3. Which of the following orderings are elimination orderings: `lp`, `ls`, `(lp(n),ls(m))`, `(ls(n),lp(m))`, `(a(1,...,1,0,...0),dp)`?

Compute a standard basis of the ideal $\langle x - t^2, y - t^3, z - t^4 \rangle$ for all those orderings.

1.7.4. For an arbitrary polynomial $g \in K[x_1, \ldots, x_n]$ of degree d, let

$$g^h(x_0, x_1, \ldots, x_n) := x_0^d g\left(\frac{x_1}{x_0}, \ldots, \frac{x_n}{x_0}\right) \in K[x_0, \ldots, x_n]$$

be the *homogenization* of g (with respect to x_0). For an ideal $I \subset K[x_1, \ldots, x_n]$ let $I^h := \langle f^h \mid f \in I \rangle \subset K[x_0, \ldots, x_n]$.

Let $>$ be a global degree ordering, and let $\{f_1, \ldots, f_m\}$ be a Gröbner basis of I. Prove that

$$I^h = \langle f_1^h, \ldots, f_m^h \rangle .$$

1.7.5. For $w = (w_1, \ldots, w_n) \in \mathbb{Z}^n$, $w_i \neq 0$ for $i = 1, \ldots, n$, and a polynomial $g \in K[x_1, \ldots, x_n]$ with w–deg$(g) = d$, let

$$g^h(x_0, x_1, \ldots, x_n) := x_0^d g\left(\frac{x_1}{x_0^{w_1}}, \ldots, \frac{x_n}{x_0^{w_n}}\right) \in K[x_0, \ldots, x_n]$$

be the *(weighted) homogenization* of g (with respect to x_0). For an ideal $I \subset K[x_1, \ldots, x_n]$ let $I^h := \langle f^h \mid f \in I \rangle \subset K[x_0, \ldots, x_n]$.

Let $>$ be a weighted degree ordering with weight vector w, and let $\{f_1, \ldots, f_m\}$ be a Gröbner basis of I. Prove that

$$I^h K[x, t]_{>_h} = \langle f_1^h, \ldots, f_m^h \rangle K[x, t]_{>_h} ,$$

where $>_h$ denotes the monomial ordering on $\mathrm{Mon}(x_0, \ldots, x_n)$ defined by the matrix

$$\begin{pmatrix} 1 & w_1 & \cdots & w_m \\ 0 & & & \\ \vdots & & A & \\ 0 & & & \end{pmatrix}$$

with $A \in \mathrm{GL}(n, \mathbb{R})$ a matrix defining $>$ on $\mathrm{Mon}(x_1, \ldots, x_n)$.

1.7.6. Let $A \in \mathrm{GL}(n, \mathbb{Q})$ be a matrix defining, on $\mathrm{Mon}(x_1, \ldots, x_n)$, the ordering $>$ and let $I = \langle f_1, \ldots, f_m \rangle \subset K[x_1, \ldots, x_n]$ be an ideal. Consider the ordering $>_h$ on $\mathrm{Mon}(t, x_1, \ldots, x_n)$ defined by the matrix

$$\begin{pmatrix} 1 & 1 & \cdots & 1 \\ 0 & & & \\ \vdots & & A & \\ 0 & & & \end{pmatrix}$$

(cf. the remark after Definition 1.7.5) and let $\{G_1, \ldots, G_s\}$ be a homogeneous standard basis of $\langle f_1^h, \ldots, f_m^h \rangle$, f_i^h the homogenization of f_i with respect to t. Prove that $\{G_1|_{t=1}, \ldots, G_s|_{t=1}\}$ is a standard basis for I.

1.7.7. Let $I = \langle f_1, \ldots, f_m \rangle \subset K[x_1, \ldots, x_n]$ be an ideal, and consider on $\mathrm{Mon}(t, x_1, \ldots, x_n)$ the ordering dp (respectively Dp). Let $\{G_1, \ldots, G_s\}$ be a standard basis of $\langle f_1^h, \ldots, f_m^h \rangle$, f_i^h the homogenization of f_i with respect to t. Prove that $\{G_1|_{t=1}, \ldots, G_s|_{t=1}\}$ is a standard basis for I with respect to the ordering ls on $\mathrm{Mon}(x_n, x_{n-1}, \ldots, x_1)$ (respectively Ds on $\mathrm{Mon}(x_1, \ldots, x_n)$).

1.7.8. Let $I = \langle f_1, \ldots, f_m \rangle \subset K[x_1, \ldots, x_n]$ be an ideal, and consider on $\mathrm{Mon}(x_1, \ldots, x_n, t)$ the ordering dp. Let $\{G_1, \ldots, G_s\}$ be a standard basis of $\langle f_1^h, \ldots, f_m^h \rangle$, f_i^h the homogenization of f_i with respect to t. Prove that $\{G_1|_{t=1}, \ldots, G_s|_{t=1}\}$ is a standard basis for I with respect to the ordering dp on $\mathrm{Mon}(x_1, \ldots, x_n)$.

1.7.9. Prove that the while loops in Algorithm 1.7.6 and at the beginning of its proof give the same result.

1.7.10. Check (by hand) whether the following polynomials f are contained in the respective ideals I:

(1) $f = xy^3 - z^2 + y^5 - z^3$, $I = \langle -x^3 + y, x^2y - z \rangle$ in $\mathbb{Q}[x, y, z]$,
(2) $f = x^3z - 2y^2$, $I = \langle yz - y, xy + 2z^2, y - z \rangle$ in $\mathbb{Q}[x, y, z]$,
(3) f and I as in (2) but in $\mathbb{Q}[x, y, z]_{\langle x,y,z \rangle}$.

1.7.11. Verify your computation in 1.7.10 by using SINGULAR.

1.7.12. Compute a standard basis of

(1) $\langle x^3, x^2y - y^3 \rangle$ with respect to ls and lp.
(2) $\langle x^3 + xy, x^2y - y^3 \rangle$ with respect to ds and dp.

1.7.13. Determine all solutions in $\mathbb{C}^2$ of the system of polynomial equations

$$xy - x - 2y + 2 = 0, \quad x^2 + xy - 2x = 0.$$

(Hint: compute first a lexicographical Gröbner basis of the two polynomials.)

1.7.14. Use SINGULAR to determine all points in $\mathbb{C}^3$ lying on the variety V given by:

(1) $V = V(xz - y, xy + 2z^2, y - z)$,
(2) $V = V(x^2 + y^2 + z^2 - 1, y^2 - z, x^2 + y^2)$.

1.7.15. Consider $f(x, y) := x^2 - y^3 - \frac{3}{2}y^2$.

(1) Compute all critical points of f (that is, points where $\partial f/\partial x$ and $\partial f/\partial y$ vanish).
(2) Which of the critical points are local minima, maxima, saddle points?
(3) Do the same for $g(x, y) = f(x, y) \cdot (y - 1)$.

1.7.16. Let K be a field, let $\mathfrak{m} \subset K[x_1, \ldots, x_n]$ be a maximal ideal, and let $L := K[x_1, \ldots, x_n]/\mathfrak{m}$. Moreover, let $I = \langle f_1, \ldots, f_m \rangle \subset L[y_1, \ldots, y_s]$ be an ideal, and let $J := \langle F_1, \ldots, F_m \rangle \subset K[x_1, \ldots, x_n, y_1, \ldots, y_s]$ be generated by representatives F_i of the f_i, $i = 1, \ldots, m$.

Finally, let $\{H_1, \ldots, H_t\}$ be a standard basis of J with respect to a block ordering $>= (>_1, >_2)$ on $\mathrm{Mon}(y_1, \ldots, y_s, x_1, \ldots, x_n)$ with $>_1, >_2$ global.

(1) Prove that $\{H_1 \bmod \mathfrak{m}, \ldots, H_t \bmod \mathfrak{m}\}$ is a standard basis of I with respect to $>_1$.
(2) Write a SINGULAR procedure to compute a minimal standard basis in $L[y_1, \ldots, y_s]$ (where K is one of the base fields of SINGULAR) such that the leading coefficients are 1.

1.7.17. Let $I \subset K[x_1, \ldots, x_n]$ be an ideal and $>$ a monomial ordering. Then there exists a weight vector $w = (w_1, \ldots, w_n) \in \mathbb{Z}^n$, with $w_i > 0$ if $x_i > 1$ and $w_i < 0$ if $x_i < 1$, such that the weighted degree lexicographical ordering defined by w and the given ordering $>$ yield the same leading ideal $L(I)$.

(Hint: use Lemma 1.2.11.)

1.7.18. (1) Let $I \subset K[x_1, \ldots, x_n]_>$ be an ideal, and let $>$ denote the negative lexicographical ordering $\mathtt{ls}$. Moreover, let x^α, $\alpha = (\alpha_1, \ldots, \alpha_n)$ denote the highest corner of I. Show that, for $i = 1, \ldots, n$,

$$\alpha_i = \max\{p \mid x_1^{\alpha_1} \cdot \ldots \cdot x_{i-1}^{\alpha_{i-1}} x_i^p \notin L(I)\}.$$

(2) Compute the highest corner of $I = \langle x^2 + x^2 y, y^3 + xy^3, z^3 - xz^2 \rangle$ with respect to the orderings $\mathtt{ls}$ and $\mathtt{ds}$.

(This can be done by hand; you may check your results by using the SINGULAR function $\mathtt{highcorner}$.)

1.7.19. Let K be a field, x one variable and $>$ the well–ordering on $K[x]$.

(1) Prove that the standard basis algorithm is the Euclidean algorithm.
(2) Use SINGULAR to compute for $f = (x^3 + 5)^2 (x - 2)(x^2 + x + 2)^4$ and $g = (x^3 + 5)(x^2 - 3)(x^2 + x + 2)$ the $\gcd(f, g)$. Try $\mathtt{std(ideal(f,g))}$ and $\mathtt{gcd(f,g)}$.

1.7.20. Let K be a field, $x = (x_1, \ldots, x_n)$ and $>$ the lexicographical ordering on $K[x]$.

(1) Prove that the standard basis algorithm is the Gaussian elimination algorithm if it is applied to linear polynomials.
(2) Use SINGULAR to solve the following linear system of equations:

$$\begin{aligned} 22x + 77y + z &= 3 \\ x + y + z &= 77 \\ x - y - z &= -11. \end{aligned}$$

With $\mathtt{option(redSB)}$ the complete reduction of the standard basis can be forced. Try both possibilities.

1.7.21. Prove that the equivalence of (2) and (3) in Lemma 1.7.14 holds for any monomial ordering.

1.7.22. Let $>$ be an arbitrary monomial ordering on $\mathrm{Mon}(x_1, \ldots, x_n)$, and let $I \subset K[x]$ be an ideal. Let $G \subset K[x]$ be a standard basis of I with respect to $>$. Assume, moreover that $\dim_K(K[x]/L(I)) < \infty$. Prove that there exists a standard basis $G' \supset G$ such that REDNFBUCHBERGER$(- \mid G')$ terminates. (Hint: Denote by $z = (z_1, \ldots, z_r)$ the variables < 1. Use Exercise 1.7.21 to choose M such that $\langle z \rangle^M \subset I$. Enlarge G by adding all monomials in z of degree M.)

1.7.23. Let $>$ be an arbitrary monomial ordering on $\mathrm{Mon}(x_1, \ldots, x_n)$, and let $I \subset K[x]_>$ be an ideal. Denote by $z = (z_1, \ldots, z_r)$ the variables < 1, and assume that $\langle z \rangle^m \subset I$ for some positive integer m. Prove that the canonical injection $K[x]/(I \cap K[x]) \hookrightarrow K[x]_>/I$ is an isomorphism.

1.7.24. Use Remark 1.6.14 and Exercise 1.6.9 for writing a SINGULAR procedure which computes standard bases over a polynomial ring $K[t_1, \ldots, t_s]$, K a field.

1.8 Operations on Ideals and Their Computation

The methods developed so far already allow some interesting applications to basic ideal operations.

In general, we assume we have given a finite set of ideals, each is given by a finite set of polynomial generators. We want to either affirmatively answer a specific question about the ideals or to compute a specific operation on these ideals, that is, compute a finite set of generators for the result of the operation.

1.8.1 Ideal Membership

Let $K[x] = K[x_1, \ldots, x_n]$ be the polynomial ring over a field K, $>_0$ an arbitrary monomial ordering and $R = K[x]_{>_0}$ the ring associated to $K[x]$ and $>_0$. Recall that $K[x] \subset R \subset K[x]_{\langle x \rangle}$, and that $R = K[x]_{\langle x \rangle}$ if and only if $>_0$ is local (cf. Section 1.5).

Let NF denote a weak normal form and redNF a reduced normal form (cf. Section 1.6). We do not need any further assumptions about NF, respectively redNF, however, we may think of NFBUCHBERGER (1.6.10), respectively REDNFBUCHBERGER (1.6.11), if $>_0$ is global, and NFMORA (1.7.6) in the general case. These are also the normal forms implemented in SINGULAR.

Problem: Given $f, f_1, \ldots, f_k \in K[x]$, and let $I = \langle f_1, \ldots, f_k \rangle_R$. We wish to decide whether $f \in I$, or not.

Solution: We choose any monomial ordering $>$ such that $K[x]_> = R$ and compute a standard basis $G = \{g_1, \ldots, g_s\}$ of I with respect to $>$. If NF is any weak normal form, then $f \in I$ if and only if $\mathrm{NF}(f \mid G) = 0$. Correctness follows from Lemma 1.6.7. $\qquad\qquad\square$

Since the result is independent of the chosen NF, we should use, for reasons of efficiency, a non–reduced normal form. If $>_0$ is global, we usually choose dp and, if $>_0$ is local, then ls or ds are preferred.

SINGULAR Example 1.8.1 (ideal membership).

(1) Check inclusion of a polynomial in an ideal

```
ring A   = 0,(x,y),dp;
ideal I = x10+x9y2,y8-x2y7;
ideal J = std(I);
poly f   = x2y7+y14;
reduce(f,J,1);       //3rd parameter 1 avoids tail reduction
//-> -xy12+x2y7      //f is not in I
     f   = xy13+y12;
reduce(f,J,1);
//-> 0               //f is in I
```

(2) Check inclusion and equality of ideals.

```
ideal K = f,x2y7+y14;
reduce(K,J,1);         //normal form for each generator of K

//-> _[1]=0  _[2]=-xy12+x2y7  //K is not in I

K=f,y14+xy12;
size(reduce(K,J,1));          //result is 0 iff K is in I

//-> 0
```

Now assume that $f \in I = \langle f_1, \ldots, f_k \rangle_R$. Then there exist $u \in K[x] \cap R^*$, $a_1, \ldots, a_k \in K[x]$ such that

$$uf = a_1 f_1 + \cdots + a_k f_k. \tag{*}$$

If $\{f_1, \ldots, f_k\}$ is a standard basis of I, then, in principle, the normal form algorithm NFMORA provides u and the a_i. However, it is also possible to express f as a linear combination of arbitrary given generators $f_1, \ldots, f_k$, by using the lift or division command. How this can be done is explained in Chapter 2, Section 2.8.1.

If the ordering is global, then we can choose $u = 1$ in the above expression (*). This is illustrated in the following example.

SINGULAR Example 1.8.2 (linear combination of ideal members).
We exemplify the SINGULAR commands lift and division:

```
ring A  = 0,(x,y),dp;
ideal I = x10+x9y2,y8-x2y7;
poly f  = xy13+y12;
matrix M=lift(I,f);         //f=M[1,1]*I[1]+...+M[r,1]*I[r]
M;
//-> M[1,1]=y7
//-> M[2,1]=x7y2+x8+x5y3+x6y+x3y4+x4y2+xy5+x2y3+y4
```

Hence, f can be expressed as a linear combination of $I[1]$ and $I[2]$ using M:

```
f-M[1,1]*I[1]-M[2,1]*I[2]; //test
//-> 0
```

In a local ring we can, in general, only express uf as a polynomial linear combination of the generators of I if $f \in I$:

```
ring R = 0,(x,y,z),ds;
poly f = yx2+yx;
ideal I = x-x2,y+x;
```

```
list L = division(f,I);            //division with remainder
L;
//-> [1]:                 [2]:              [3]:
//->    _[1,1]=y-y2            _[1]=0             _[1,1]=1+y
//->    _[2,1]=2xy
```

```
matrix(f)*L[3] - matrix(I)*L[1] - matrix(L[2]);   //test
//-> _[1,1]=0
```

Hence $(1 + y)f = (x - x^2)(y - y^2) + (y + x)(2xy)$, the remainder being 0.

1.8.2 Intersection with Subrings (Elimination of variables)

This is one of the most important applications of Gröbner bases. The problem may be formulated as follows (we restrict ourselves for the moment to the case of the polynomial ring):

Problem: Given $f_1, \ldots, f_k \in K[x] = K[x_1, \ldots, x_n]$, $I = \langle f_1, \ldots, f_k \rangle_{K[x]}$, we should like to find generators of the ideal

$$I' = I \cap K[x_{s+1}, \ldots, x_n], \quad s < n.$$

Elements of the ideal I' are said to be obtained from $f_1, \ldots, f_k$ by *eliminating* $x_1, \ldots, x_s$.

In order to treat this problem, we need a global elimination ordering for $x_1, \ldots, x_s$. We can use the lexicographical ordering lp which is an elimination ordering (Definition 1.5.4) for each s, but lp is, in almost all cases, the most expensive choice. A good choice is, usually, $(\mathrm{dp(s)}, \mathrm{dp(n-s)})$, the product ordering of two degrevlex orderings. But there is another way to construct an elimination ordering which is often quite fast.

Let $>$ be an arbitrary ordering and let $a_1, \ldots, a_s$ be positive integers. Define $>_a$ by

$$x^\alpha >_a x^\beta :\Longleftrightarrow a_1\alpha_1 + \cdots + a_s\alpha_s > a_1\beta_1 + \cdots + a_s\beta_s$$
$$\text{or } a_1\alpha_1 + \cdots + a_s\alpha_s = a_1\beta_1 + \cdots + a_s\beta_s \text{ and } x^\alpha > x^\beta.$$

Then $>_a$ is an elimination ordering and $a = (a_1, \ldots, a_s)$ is called an *extra weight vector*.

If $>$ is an arbitrary elimination ordering for $x_1, \ldots, x_s$, then

$$K[x_1, \ldots, x_n]_> = (K[x_{s+1}, \ldots, x_n]_{>'})[x_1, \ldots, x_s],$$

since the units in $K[x]_>$ do not involve $x_1, \ldots, x_s$ (we denote, by $>'$, the ordering on $\mathrm{Mon}(x_{s+1}, \ldots, x_n)$ induced by $>$). Hence, $f \in K[x_{s+1}, \ldots, x_n]_{>'}$ for any $f \in K[x_1, \ldots, x_n]_>$ such that $\mathrm{LM}(f) \in K[x_{s+1}, \ldots, x_n]$.

The following lemma is the basis for solving the elimination problem.

Lemma 1.8.3. *Let $>$ be an elimination ordering for $x_1, \ldots, x_s$ on the set of monomials $\mathrm{Mon}(x_1, \ldots, x_n)$, and let $I \subset K[x_1, \ldots, x_n]_>$ be an ideal. If $S = \{g_1, \ldots, g_k\}$ is a standard basis of I, then*

$$S' := \{g \in S \mid \mathrm{LM}(g) \in K[x_{s+1}, \ldots, x_n]\}$$

is a standard basis of $I' := I \cap K[x_{s+1}, \ldots, x_n]_{>'}$. In particular, S' generates the ideal I'.

Proof. Given $f \in I' \subset I$ there exists $g_i \in S$ such that $\mathrm{LM}(g_i)$ divides $\mathrm{LM}(f)$, since S is a standard basis of I. Since $f \in K[x_{s+1}, \ldots, x_n]_>$, we have $\mathrm{LM}(f) \in K[x_{s+1}, \ldots, x_n]$ and, hence, $g_i \in S'$ by the above remark. Finally, since $S' \subset I'$, S' is a standard basis of I'. $\qquad\square$

The general elimination problem can be posed, for any ring associated to a monomial ordering, as follows. Recall that the ordering on the variable to be eliminated must be global.

Problem: Given polynomials $f_1, \ldots, f_k \in K[x_1, \ldots, x_n]$, let $I := \langle f_1, \ldots, f_k \rangle_R$ with $R := (K[x_{s+1}, \ldots, x_n]_>)[x_1, \ldots, x_s]$ for some monomial ordering $>$ on $\mathrm{Mon}(x_{s+1}, \ldots, x_n)$. Find generators for the ideal $I' := I \cap K[x_{s+1}, \ldots, x_n]_>$.

Solution: Choose an elimination ordering for $x_1, \ldots, x_s$ on $\mathrm{Mon}(x_1, \ldots, x_n)$, which induces the given ordering $>$ on $\mathrm{Mon}(x_{s+1}, \ldots, x_n)$, and compute a standard basis $S = \{g_1, \ldots, g_k\}$ of I. By Lemma 1.8.3, those g_i, for which $\mathrm{LM}(g_i)$ does not involve $x_1, \ldots, x_s$, generate I' (even more, they are a standard basis of I').

A good choice of an ordering on $\mathrm{Mon}(x_1, \ldots, x_n)$ may be $(\mathrm{dp(s)}, >)$, but instead of $>$ we may choose any ordering $>'$ on $\mathrm{Mon}(x_{s+1}, \ldots, x_n)$ such that $K[x_{s+1}, \ldots, x_n]_{>'} = K[x_{s+1}, \ldots, x_n]_>$. For any global ordering $>$ on $\mathrm{Mon}(x_{s+1}, \ldots, x_n)$, we have, thus, a solution to the elimination problem in the polynomial ring, as stated at the beginning of this section. $\qquad\square$

SINGULAR Example 1.8.4 (elimination of variables).

```
ring A =0,(t,x,y,z),dp;
ideal I=t2+x2+y2+z2,t2+2x2-xy-z2,t+y3-z3;

eliminate(I,t);
//-> _[1]=x2-xy-y2-2z2    _[2]=y6-2y3z3+z6+2x2-xy-z2
```

Alternatively choose a product ordering:

```
ring A1=0,(t,x,y,z),(dp(1),dp(3));
ideal I=imap(A,I);
ideal J=std(I);
J;
//-> J[1]=x2-xy-y2-2z2    J[2]=y6-2y3z3+z6+2x2-xy-z2
//-> J[3]=t+y3-z3
```

We can also choose the *extra weight vector* $a = (1, 0, 0, 0)$ to obtain an elimination ordering:

```
ring A2=0,(t,x,y,z),(a(1),dp);
ideal I=imap(A,I);
ideal J=std(I);
J;
//-> J[1]=x2-xy-y2-2z2     J[2]=y6-2y3z3+z6+2x2-xy-z2
//-> J[3]=t+y3-z3
```

By Lemma 1.8.3, the elements of J which do not involve t (here J[1] and J[2]), are a standard basis of $I \cap K[x, y, z]$.

1.8.3 Zariski Closure of the Image

Here we study the geometric counterpart of elimination. The reader who is not familiar with the geometrical background should read Section A.1 first. In this section we assume K to be algebraically closed.

Suppose $\varphi : K[x] = K[x_1, \ldots, x_n] \to K[t] = K[t_1, \ldots, t_m]$ is a ring map given by $f_1, \ldots, f_n \in K[t]$ such that $\varphi(x_i) = f_i$. Let $I = \langle g_1, \ldots, g_k \rangle \subset K[t]$ and $J = \langle h_1, \ldots, h_l \rangle \subset K[x]$ be ideals such that $\varphi(J) \subset I$. Then φ induces a ring map $\bar{\varphi} : K[x]/J \to K[t]/I$ and, hence, we obtain a commutative diagram of morphism of affine schemes (cf. Section A.1)

$$
\begin{array}{ccc}
X := V(I) & \xrightarrow{\ f = \bar{\varphi}^{\#}\ } & V(J) =: Y \\
\cup & & \cup \\
\mathbb{A}^m & \xrightarrow{\ \varphi^{\#}\ } & \mathbb{A}^n .
\end{array}
$$

We cannot compute the image $f(X)$, since it is, in general, not closed. However, we can compute the (Zariski) closure $\overline{f(X)}$.

Problem: The problem is to find polynomials $p_1, \ldots, p_r \in K[x]$ such that

$$\overline{f(X)} = V(p_1, \ldots, p_r) \subset \mathbb{A}^n .$$

Solution: Define the ideal

$$N = \langle g_1(t), \ldots, g_k(t), x_1 - f_1(t), \ldots, x_n - f_n(t) \rangle_{K[t,x]}$$

and eliminate $t_1, \ldots, t_m$ from N, that is, compute generators $p_1, \ldots, p_r \in K[x]$ of $N \cap K[x]$. Then $V(p_1, \ldots, p_r) = \overline{f(X)}$.

Hence, we can proceed as in Section 1.8.2. We choose a global ordering which is an elimination ordering for $t_1, \ldots, t_m$ on $\mathrm{Mon}(t_1, \ldots, t_m, x_1, \ldots, x_n)$, compute a Gröbner basis G of N and select those elements $p_1, \ldots, p_r$ from G which do not depend on t. Correctness follows from Lemma A.3.10. $\qquad\square$

Since K is algebraically closed (with $f = (f_1, \ldots, f_n) : K^m \to K^n$), Lemma A.2.18 implies

$$\overline{f(X)} = \overline{\{x \in K^n \mid \exists\, t \in K^m \text{ such that } f(t) = x\}}.$$

The following example shows that the question whether $f(X)$ is closed or not may depend on the field.

Example 1.8.5. Consider the ring map $\varphi : K[x] \to K[x, y]/\langle x^2 + y^2 - 1\rangle$ given by $\varphi(x) := x$, and the induced morphism

$$
\begin{array}{ccc}
X := V(\langle x^2 + y^2 - 1\rangle) & \xrightarrow{\;f = \bar{\varphi}^{\#}\;} & V(\langle 0\rangle) \\[4pt]
\Big\downarrow & & \Big\| \\[4pt]
\mathbb{A}^2 & \xrightarrow{\;\varphi^{\#}\;} & \mathbb{A}^1 \,.
\end{array}
$$

It is easy to see that, if K is algebraically closed, then f is surjective, and hence, $f(X)$ is closed. However, if $K = \mathbb{R}$, then $f(X)$ is a segment but the Zariski closure of the segment is the whole line.

Now we treat the problem of computing the closure of the image of a map between spectra of local rings. More generally, let $>_1$, respectively $>_2$, be monomial orderings on $\mathrm{Mon}(x_1, \ldots, x_n)$, respectively $\mathrm{Mon}(t_1, \ldots, t_m)$, and let $\varphi : K[x]_{>_1} \to K[t]_{>_2}$ be a ring map defined by $\varphi(x_i) = f_i(t) \in K[t]$ (cf. Lemma 1.5.8).

Let $I \subset K[t]$ and $J \subset K[x]$ be ideals as above, satisfying $\varphi(J) \subset I$, and

$$\bar{\varphi} : K[x]_{>_1}/J \to K[t]_{>_2}/I$$

the induced map.

Problem: We want to compute equations for $\overline{f(X)} \subset Y$ for the map

$$f = \bar{\varphi}^{\#} : X = \mathrm{Spec}(K[t]_{>_2}/I) \longrightarrow Y = \mathrm{Spec}(K[x]_{>_1}/J)\,.$$

We claim that the following algorithm solves the problem:

Solution: Choose any ordering $>$ on $\mathrm{Mon}(t_1, \ldots, t_m, x_1, \ldots, x_n)$ which is an elimination ordering for $t_1, \ldots, t_m$ and satisfies $K[x]_{>'} \subset K[x]_{>_1}$ where $>'$ is the ordering on $\mathrm{Mon}(x_1, \ldots, x_n)$ induced by $>$.[9] Compute a standard basis G of the ideal

$$N := \langle I, J, x_1 - f_1(t), \ldots, x_n - f_n(t)\rangle$$

as above with respect to this ordering. Select those elements $p_1, \ldots, p_r$ from G which do not depend on t. Then $\overline{f(X)} = V(\langle p_1, \ldots, p_r\rangle_{K[x]_{>_1}})$.   □

---

[9] We could choose, for example, $>$ to be $(\mathrm{dp}(m), \mathrm{dp}(n))$ or $>$ to be $(\mathrm{dp}(m), >_1)$.

The only problem in seeing correctness results from the fact that we only assume $K[x]_{>'} \subset K[x]_{>_1}$ but no other relation between $>$ and $>_1, >_2$.

The graph construction from Appendix A.2, applied to the localized rings, shows that $\overline{f(X)}$ is the zero–set of the ideal

$$N \cdot (K[t]_{>_2} \otimes_K K[x]_{>_1}) \cap K[x]_{>_1} .$$

Now the above algorithm computes polynomial generators of the intersection $(N \cdot K[t,x]_>) \cap K[x]_{>'}$. We have $K[t,x]_> = K[t] \otimes K[x]_{>'}$, $K[x]_{>'} \subset K[x]_{>_1}$ and an inclusion of rings

$$K[t,x] \subset R_1 := K[t,x]_> \subset R_2 := K[t]_{>_2} \otimes K[x]_{>_1} \subset R_3 := K[t,x]_{(>_2, >_1)} ,$$

where $(>_2, >_1)$ is the product ordering on $\mathrm{Mon}(t_1, \ldots, t_m, x_1, \ldots, x_n)$.

Moreover, by Lemma 1.4.8 (1), we have $(N \cdot R_3) \cap K[t,x] = N$, hence, $(N \cdot R_i) \cap K[t,x] = N$ for $i = 1, 2$ and, therefore,

$$(N \cdot R_1) \cap K[x] = N \cap K[x] = (N \cdot R_2) \cap K[x] .$$

Again, by Lemma 1.4.8 (1), $(N \cdot R_2) \cap K[x]_{>_1} = (N \cdot R_2 \cap K[x]) \cdot K[x]_{>_1}$ and $(N \cdot R_1) \cap K[x]_{>'} = (N \cdot R_1 \cap K[x]) \cdot K[x]_{>'}$. Altogether, we have

$$\begin{aligned}
(N \cdot R_2) \cap K[x]_{>_1} &= \big((N \cdot R_1) \cap K[x]\big) \cdot K[x]_{>_1} \\
&= \big((N \cdot R_1) \cap K[x]_{>'}\big) \cdot K[x]_{>_1},
\end{aligned}$$

where the left–hand side defines $\overline{f(X)}$ and generators for the right–hand side are computed. □

Thus, we have many choices for orderings on $\mathrm{Mon}(x_1, \ldots, x_n)$ for computing $\overline{f(X)} \subset Y$. In particular, we can always choose a global ordering.

SINGULAR Example 1.8.6 (Zariski closure of the image).
Compute an implicit equation for the surface defined parametrically by the map $f : \mathbb{A}^2 \to \mathbb{A}^3$, $(u,v) \mapsto (uv, uv^2, u^2)$.

```
ring A =0,(u,v,x,y,z),dp;
ideal I=x-uv,y-uv2,z-u2;
ideal J=eliminate(I,uv);
J;
//-> J[1]=x4-y2z              //defines the closure of f(X)
```

Note that the image does not contain the y–axis, however, the closure of the image contains the y–axis. This surface is called the *Whitney umbrella*.

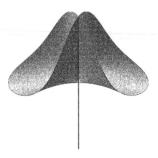

Fig. 1.3. Whitney Umbrella.

1.8.4 Solvability of Polynomial Equations

Problem: Given $f_1, \ldots, f_k \in K[x_1, \ldots, x_n]$, we want to assure whether the system of polynomial equations

$$f_1(x) = \cdots = f_k(x) = 0$$

has a solution in $\overline{K}^n$, where $\overline{K}$ is the algebraic closure of K.

Let $I = \langle f_1, \ldots, f_k \rangle_{K[x]}$, then the question is whether the algebraic set $V(I) \subset \overline{K}^n$ is empty or not.

Solution: By Hilbert's Nullstellensatz, $V(I) = \emptyset$ if and only if $1 \in I$. We compute a Gröbner basis G of I with respect to any global ordering on $\mathrm{Mon}(x_1, \ldots, x_n)$ and normalize it (that is, divide every $g \in G$ by $\mathrm{LC}(g)$). Since $1 \in I$ if and only if $1 \in L(I)$, we have $V(I) = \emptyset$ if and only if 1 is an element of a normalized Gröbner basis of I. Of course, we can avoid normalizing, which is expensive in rings with parameters. Since $1 \in I$ if and only if G contains a non–zero constant polynomial, we have only to look for an element of degree 0 in G. □

1.8.5 Solving Polynomial Equations

A fundamental task with countless applications is to solve a system of polynomial equations, $f_1(x) = 0, \ldots, f_k(x) = 0$, $f_i \in K[x] = K[x_1, \ldots, x_n]$. However, what is actually meant by "solving" very much depends on the context. For instance, it could mean to determine one (respectively some, respectively all) points of the solution set $V(f_1, \ldots, f_k)$, either considered as a subset of K^n or of $\overline{K}^n$, where $\overline{K}$ is the algebraic closure of K (for notations cf. Section A.1).

Here, we consider only the case where the ideal $I = \langle f_1, \ldots, f_k \rangle_{K[x]}$ is 0–dimensional, that is, where $f_1 = \cdots = f_k = 0$ has only finitely many solutions in $\overline{K}^n$.

From an algebraic point of view, a primary decomposition $I = \bigcap_{i=1}^{r} Q_i$ of I with $P_i = \sqrt{Q_i} \subset K[x_1, \ldots, x_n]$ a maximal ideal, could be considered as a solution (cf. Chapter 4). At least, it provides a decomposition $V(I) = V(P_1) \cup \cdots \cup V(P_r)$ and if $p = (p_1, \ldots, p_n) \in K^n$ is a solution, then $P_j = \langle x_1 - p_1, \ldots, x_n - p_n \rangle$ for some j and we can, indeed, recover the coordinates of p from a primary decomposition of I. Moreover, for solutions $p \in \overline{K}^n$ which are not in K^n the primary decomposition provides irreducible polynomials defining a field extension $\widetilde{K}$ of K such that p has coordinates in $\widetilde{K}$ (cf. Chapter 4).

Besides the fact that primary decomposition is very expensive, the answer would be unsatisfactory from a practical point of view. Indeed, if $K = \mathbb{R}$ or $\mathbb{C}$, most people would probably interpret solving as finding approximate numerical coordinates of one (respectively some, respectively all) point(s) of $V(I)$. And this means that, at some point, we need a numerical root finder.

Numerical solving of equations (even transcendental or (partial) differential equations) is a highly developed discipline in mathematics which is very successful in applications to real life problems. However, there are inherent problems which often make it difficult, or even impossible, either to find a solution or to ensure that a detected solution is (approximately) correct. Particular problems are, for example, to find *all* solutions or to guarantee stability and convergence of algorithms in the presence of singularities. In this context symbolic methods can be useful in preparing the system by finding another set of generators for I (hence, having the same solutions) which is better suited for numerical computations.

Here we describe only how lexicographical Gröbner bases can be used to reduce the problem of multivariate solving to univariate solving.

Problem: Given $f_1, \ldots, f_k \in K[x_1, \ldots, x_n]$, $K = \mathbb{R}$ or $\mathbb{C}$, which we assume to have only finitely many solutions $p_1, \ldots, p_r \in \mathbb{C}^n$. We wish to find coordinates of all p_i in decimal format up to a given number of digits. We are also interested in locating multiple solutions.

Solution: Compute a lexicographical Gröbner basis $G = \{g_1, \ldots, g_s\}$ of I for $x_1 > x_2 > \cdots > x_n$. Then we have (cf. Exercise 1.8.6) $s \geq n$ and, after renumbering G, there are elements $g_1, \ldots, g_n \in G$ such that

$$g_1 = g_1(x_n), \qquad \mathrm{LM}(g_1) = x_n^{n_n},$$
$$g_2 = g_2(x_{n-1}, x_n), \quad \mathrm{LM}(g_2) = x_{n-1}^{n_{n-1}},$$
$$\vdots$$
$$g_n = g_n(x_1, \ldots, x_n), \, \mathrm{LM}(g_n) = x_1^{n_1}.$$

Now use any numerical univariate solver (for example Laguerre's method) to find all complex solutions of $g_1(x_n) = 0$ up to the required number of digits. Substitute these in g_2 and for each substitution solve g_2 in x_{n-1}, as before. Continue in this way up to g_n. Thus, we computed all coordinates of all solutions of $g_1 = \cdots = g_n = 0$. Finally, we have to discard those solutions for which one of the remaining polynomials $g_{n+1}, \ldots, g_s$ does not vanish. $\square$

We should like to mention that this is not the best possible method. In particular, the last step, that is, discarding non–solutions, may lead to numerical problems. A better method is to use *triangular sets*, either in the spirit of Lazard [148] or Möller [170] (cf. [102] for experimental results and a comparison to resultant based methods). Triangular sets are implemented in the SINGULAR library `triang.lib`.

SINGULAR Example 1.8.7 (solving equations).

```
ring   A=0,(x,y,z),lp;
ideal I=x2+y+z-1,
        x+y2+z-1,
        x+y+z2-1;
ideal J=groebner(I);      //the lexicographical Groebner basis
J;
//-> J[1]=z6-4z4+4z3-z2      J[2]=2yz2+z4-z2
//-> J[3]=y2-y-z2+z          J[4]=x+y+z2-1
```

We use the multivariate solver based on triangular sets, due to Möller and Hillebrand [170], [123], and the univariate Laguerre–solver.

```
LIB"solve.lib";
list s1=solve(I,6);
//-> // name of new current ring: AC
s1;
//-> [1]:          [2]:      [3]:            [4]:      [5]:
//->    [1]:          [1]:      [1]:            [1]:      [1]:
//->        0.414214      0        -2.414214       1         0
//->    [2]:          [2]:      [2]:            [2]:      [2]:
//->        0.414214      0        -2.414214       0         1
//->    [3]:          [3]:      [3]:            [3]:      [3]:
//->        0.414214      1        -2.414214       0         0
```

If we want to compute the zeros with multiplicities then we use 1 as a third parameter for the `solve` command:

```
setring A;
list s2=solve(I,6,1);
s2;
//-> [1]:                       [2]:
//->    [1]:                        [1]:
//->        [1]:                        [1]:
//->            [1]:                        [1]:
//->                -2.414214                   0
//->            [2]:                        [2]:
//->                -2.414214                   1
//->        [3]:                        [3]:
```

```
//->                 -2.414214                    0
//->        [2]:                        [2]:
//->          [1]:                        [1]:
//->             0.414214                     1
//->          [2]:                        [2]:
//->             0.414214                     0
//->          [3]:                        [3]:
//->             0.414214                     0
//->        [2]:                        [3]:
//->           1                          [1]:
//->                                        0
//->                                      [2]:
//->                                        0
//->                                      [3]:
//->                                        1
//->                          [2]:
//->                            2
```

The output has to be interpreted as follows: there are two zeros of multiplicity 1 and three zeros $((0,1,0), (1,0,0), (0,0,1))$ of multiplicity 2.

Note that a possible way to check whether a system of polynomial equations $f_1 = \cdots = f_k = 0$ has finitely many solutions in $\overline{K}$, is to compute a Gröbner basis G of $I = \langle f_1, \ldots, f_k \rangle$ with respect to any ordering (usually dp is the fastest). Then $V(I)$ is finite if and only if dim(G)=0 or, equivalently, lead(G) contains $x_i^{n_i}$ for $i = 1, \ldots, n$ and some n_i (and then the number of solutions is $\leq n_1 \cdot \ldots \cdot n_n$). The number of solutions, counting multiplicities, in $\overline{K}^n$ is equal to vdim(G)$= \dim_K K[x]/I$ (cf. Exercises 1.8.6 to 1.8.8).

1.8.6 Radical Membership

Problem: Let $f_1, \ldots, f_k \in K[x]_>$, $>$ a monomial ordering on $\mathrm{Mon}(x_1, \ldots, x_n)$ and $I = \langle f_1, \ldots, f_k \rangle_{K[x]_>}$. Given some $f \in K[x]_>$ we want to decide whether $f \in \sqrt{I}$. The following lemma, which is sometimes called *Rabinowich's trick*, is the basis for solving this problem. [10]

Lemma 1.8.8. *Let A be a ring, $I \subset A$ an ideal and $f \in A$. Then*

$$f \in \sqrt{I} \iff 1 \in \tilde{I} := \langle I, 1 - tf \rangle_{A[t]}$$

where t is an additional new variable.

Proof. If $f^m \in I$ then $t^m f^m \in \tilde{I}$ and, hence,

$$1 = t^m f^m + (1 - t^m f^m) = t^m f^m + (1 - tf)(1 + tf + \cdots + t^{m-1} f^{m-1}) \in \tilde{I}.$$

---

[10] We can even compute the full radical $\sqrt{I}$ as is shown in Section 4.5, but this is a much harder computation.

Conversely, let $1 \in \tilde{I}$. Without loss of generality, we may assume that f is not nilpotent since, otherwise, f is clearly in $\sqrt{I}$.

By assumption, there are $f_1, \ldots, f_k \in I$ and $a_i(t) = \sum_{j=0}^{d_i} a_{ij} t^j \in A[t]$, $i = 0, \ldots, k$ such that

$$1 = \sum_{i=1}^{k} a_i(t) f_i + a_0(t)(1 - tf) \, .$$

Since f is not nilpotent we can replace t by $1/f$ and obtain

$$1 = \sum_i a_i \left(\frac{1}{f}\right) f_i = \sum_{i,j} a_{ij} f^{-j} f_i$$

in the localization A_f, see Section 1.4. Multiplying with f^m, for m sufficiently large, we obtain $f^m = \sum_{i,j}(a_{ij} f^{m-j}) f_i \in I$ (even in A, not only in A_f). $\square$

Solution: By Lemma 1.8.8, we have $f \in \sqrt{I}$ if and only if

$$1 \in J := \langle f_1, \ldots, f_k, 1 - tf \rangle (K[x]_>)[t] \, ,$$

where t is a new variable.

To solve the problem, we choose on $\mathrm{Mon}(t, x_1, \ldots, x_n)$ an elimination ordering for t inducing $>'$ on $\mathrm{Mon}(x_1, \ldots, x_n)$ such that $K[x]_{>'} = K[x]_>$ (for example, take (lp(1),>)) and compute a standard basis G of J. Then $f \in \sqrt{I}$ if and only if G contains an element g with $\mathrm{LM}(g) = 1$. $\square$

SINGULAR Example 1.8.9 (radical membership).

```
ring A =0,(x,y,z),dp;
ideal I=x5,xy3,y7,z3+xyz;
poly f =x+y+z;

ring B =0,(t,x,y,z),dp; //need t for radical test
ideal I=imap(A,I);
poly f =imap(A,f);
I=I,1-t*f;
std(I);
//-> _[1]=1              //f is in the radical

LIB"primdec.lib";   //just to see, we compute the radical
setring A;
radical(I);
//-> _[1]=z    _[2]=y       _[3]=x
```

1.8.7 Intersection of Ideals

Problem: Given $f_1, \ldots, f_k, h_1, \ldots, h_r \in K[x]$ and $>$ a monomial ordering. Let $I_1 = \langle f_1, \ldots, f_k \rangle K[x]_>$ and $I_2 = \langle h_1, \ldots, h_r \rangle K[x]_>$. We wish to find generators for $I_1 \cap I_2$.

Consider the ideal $J := \langle tf_1, \ldots, tf_k, (1-t)h_1, \ldots, (1-t)h_r \rangle (K[x]_>)[t]$.

Lemma 1.8.10. *With the above notations, $I_1 \cap I_2 = J \cap K[x]_>$.*

Proof. Let $f \in J \cap K[x]_>$, then

$$f(x) = t \cdot \sum_{i=1}^{k} a_i(t,x) f_i(x) + (1-t) \sum_{j=1}^{r} b_j(t,x) h_j(x).$$

Since the polynomial f is independent of t, we have $f = \sum_{i=1}^{k} a_i(1,x) f_i \in I_1$ and $f = \sum_{j=1}^{r} b_j(0,x) h_j \in I_2$, hence $f \in I_1 \cap I_2$. Conversely, if $f \in I_1 \cap I_2$, then $f = tf + (1-t)f \in J \cap K[x]_>$. □

Solution: We choose an elimination ordering for t on $\mathrm{Mon}(t, x_1, \ldots, x_n)$ inducing $>'$ on $\mathrm{Mon}(x_1, \ldots, x_n)$ such that $K[x]_{>'} = K[x]_>$ (for example, take (lp(1),>)). Then we compute a standard basis of J and get generators for $J \cap K[x]_>$ as in Section 1.8.2. □

A different solution, using syzygies, is described in Chapter 2, Section 2.8.3.

SINGULAR Example 1.8.11 (intersection of ideals).

```
ring A=0,(x,y,z),dp;
ideal I1=x,y;
ideal I2=y2,z;
intersect(I1,I2);          //the built-in SINGULAR command
//-> _[1]=y2     _[2]=yz     _[3]=xz

ring B=0,(t,x,y,z),dp;      //the way described above
ideal I1=imap(A,I1);
ideal I2=imap(A,I2);
ideal J=t*I1+(1-t)*I2;
eliminate(J,t);
//-> _[1]=yz  _[2]=xz   _[3]=y2
```

1.8.8 Quotient of Ideals

Problem: Let I_1 and $I_2 \subset K[x]_>$ be as in Section 1.8.7. We want to compute

$$I_1 : I_2 = \{ g \in K[x]_> \mid gI_2 \subset I_1 \}.$$

Since, obviously, $I_1 : \langle h_1, \ldots, h_r \rangle = \bigcap_{i=1}^{r} (I_1 : \langle h_i \rangle)$, we can compute $I_1 : \langle h_i \rangle$ for each i and then apply SINGULAR Example 1.8.11. The next lemma shows a way to compute $I_1 : \langle h_i \rangle$.

Lemma 1.8.12. *Let $I \subset K[x]_>$ be an ideal, and let $h \in K[x]_>$, $h \neq 0$. Moreover, let $I \cap \langle h \rangle = \langle g_1 \cdot h, \ldots, g_s \cdot h \rangle$. Then $I : \langle h \rangle = \langle g_1, \ldots, g_s \rangle_{K[x]_>}$.*

Proof. Any set of generators of $I \cap \langle h \rangle$ is of the form $\{g_1 h, \ldots, g_s h\}$. Therefore, $h \langle g_1, \ldots, g_s \rangle \subset I$, hence $\langle g_1, \ldots, g_s \rangle \subset I : \langle h \rangle$. Conversely, if $g \in I : \langle h \rangle$, then $hg \in I \cap \langle h \rangle$ and $hg = h \cdot \sum_i a_i g_i$ for some a_i. Since $K[x]_>$ has no zero-divisors and $h \neq 0$, we have $g = \sum_i a_i g_i$ which proves the claim. □

Solution 1: We can compute $I_1 : I_2$ by computing, for $i = 1, \ldots, r$, $I_1 \cap \langle h_i \rangle$ according to Section 1.8.7, divide the generators by h_i getting $I_1 : \langle h_i \rangle$ and compute the intersection $\bigcap_i (I_1 : \langle h_i \rangle)$, according to Section 1.8.7. □

Instead of computing $\bigcap_i (I_1 : \langle h_i \rangle)$, we can define

$$h := h_1 + t_1 h_2 + \cdots + t_{r-1} h_r \in K[t_1, \ldots, t_{r-1}, x_1, \ldots, x_n]$$

and obtain

$$I_1 : I_2 = \Big(I_1(K[x]_>)[t] : \langle h \rangle \Big) \cap K[x]_> .$$

This holds, since $g(x) \in I_1 : \langle h \rangle$ if and only if

$$g(x)\big(h_1(x) + t_1 h_2(x) + \cdots + t_{r-1} h_r(x) \big) = \sum_{i=1}^{k} a_i(x, t) f_i(x)$$

for some $a_i \in (K[x]_>)[t]$, which is equivalent to $g(x) h_j(x) \in \langle f_1, \ldots, f_k \rangle_{K[x]_>}$ for all j (set $t_i := 0$ for all i, and then $t_j := 1$ and $t_i = 0$ for $i \neq j$).

Solution 2: Define h as above. We can compute $I_1 : \langle h \rangle$ by Lemma 1.8.12 and then $I_1 : I_2$ by eliminating $t_1, \ldots, t_{r-1}$ from $I_1 : \langle h \rangle$ according to Section 1.8.2.

The same procedure works with $h := h_1 + t h_2 + t^2 h_3 + \cdots + t^{r-1} h_r \in (K[x]_>)[t]$ with just one new variable t (Exercise 1.8.2). □

SINGULAR Example 1.8.13 (quotient of ideals).

```
ring A=0,(x,y,z),dp;
ideal I1=x,y;
ideal I2=y2,z;
quotient(I1,I2);              //the built-in SINGULAR command
//-> _[1]=y    _[2]=x
```

Now let us proceed as described in Lemma 1.8.12:

```
ideal J1=intersect(I1,ideal(I2[1]));
ideal J2=intersect(I1,ideal(I2[2]));
J1;
//-> J1[1]=y2
```

$J1/I2[1]=1$ implies $I1:I2[1]=A$.

```
J2;
//-> J2[1]=yz    J2[2]=xz
```

J2/I2[2]=<x,y> implies I1:I2[2]=<x,y> and all together we obtain
I1:I2=<x,y>:

```
ideal K1=J1[1]/I2[1];
ideal K2=J2[1]/I2[2],J2[2]/I2[2];
intersect(K1,K2);
//-> _[1]=y    _[2]=x
```

1.8.9 Saturation

Let $I_1, I_2 \subset K[x]_>$ be as in Section 1.8.7. We consider the quotient of I_1 by powers of I_2

$$I_1 = I_1 : I_2^0 \subset I_1 : I_2^1 \subset I_1 : I_2^2 \subset I_1 : I_2^3 \subset \ldots \subset K[x]_> .$$

Since $K[x]_>$ is Noetherian, there exists an s such that $I_1 : I_2^s = I_1 : I_2^{s+i}$ for all $i \geq 0$. Such an s satisfies

$$I_1 : I_2^\infty := \bigcup_{i \geq 0} I_1 : I_2^i = I_1 : I_2^s ,$$

and $I_1 : I_2^s$ is called the *saturation of I_1 with respect to I_2*.

The minimal such s is called the *saturation exponent*. If I_1 is radical, then the saturation exponent is 1.

Problem: Given ideals $I_1, I_2 \subset K[x]_>$, we want to compute generators for $I_1 : I_2^\infty$ and the saturation exponent.

Solution: Set $I^{(0)} = I_1$ and compute successively $I^{(j+1)} = I^{(j)} : I_2$, $j \geq 0$, by any of the methods of Section 1.8.8. In each step check whether $I^{(j+1)} \subset I^{(j)}$, by using Section 1.8.1. If s is the first j when this happens, then $I^{(s)} = I_1 : I_2^\infty$ and s is the saturation exponent. □

Correctness follows from $I^{(j)} = I_1 : I_2^j$, which is a consequence of Lemma 1.8.14 (1). The above method is usually much faster than computing $I_1 : I_2^j$, since I_2^j can become quite large.

To provide a geometric interpretation of ideal–quotient and saturation, we state the following:

Lemma 1.8.14. *Let A be a ring and I_1, I_2, I_3 ideals in A.*

(1) a) $(I_1 \cap I_2) : I_3 = (I_1 : I_3) \cap (I_2 : I_3)$, in particular
        $$I_1 : I_3 = (I_1 \cap I_2) : I_3 \quad \text{if } I_3 \subset I_2,$$
     b) $(I_1 : I_2) : I_3 = I_1 : (I_2 \cdot I_3)$.
(2) If I_1 is prime and $I_2 \not\subset I_1$, then $I_1 : I_2^j = I_1$ for $j \geq 1$.
(3) If $I_1 = \bigcap_{i=1}^r J_i$ with J_i prime, then $I_1 : I_2^\infty = I_1 : I_2 = \bigcap_{I_2 \not\subset J_i} J_i$.

Proof. (1) is an easy exercise.

(2) $I_1 \subset I_1 : I_2^j$ is clear. Let $gI_2^j \in I_1$. Since $I_2 \not\subset I_1$ and I_1 is radical, $I_2^j \not\subset I_1$ and we can find an $h \in I_2^j$ such that $h \notin I_1$ and $gh \in I_1$. Since I_1 is prime, we have $g \in I_1$.

(3) follows from (1) and (2) since $I_2^s \not\subset J_i$ if and only if $I_2 \not\subset J_i$:

$$\left(\bigcap_{j=1}^r J_i\right) : I_2^s = \left(\bigcap_{I_2^s \not\subset J_i} (J_i : I_2^s)\right) \cap \left(\bigcap_{I_2^s \subset J_i} J_i : I_2^s\right) = \bigcap_{I_2^s \not\subset J_i} J_i. \qquad \square$$

We shall see in Chapter 3 that in a Noetherian ring each radical ideal I_1 has a prime decomposition $I_1 = \bigcap_{i=1}^r J_i$ with J_i prime. For the *geometric interpretation of* the *ideal quotient* and the *saturation*, we use the notations of Appendix A.2, respectively A.3. We have

$$V(I_1) = \bigcup_{i=1}^r V(J_i).$$

Moreover, we have $I_2 \subset J_i$ if and only if $V(J_i)$ is a closed subscheme of $V(I_2)$. Hence, the variety defined by $I_1 : I_2$ is

$$V(I_1 : I_2) = \bigcup_{V(J_i) \not\subset V(I_2)} V(J_i).$$

In other words, if I_1 is a radical ideal, then $V(I_1 : I_2)$ is the Zariski closure of $V(I_1) \smallsetminus V(I_2)$.

Note that $V(\langle 0 \rangle : I) = \mathrm{supp}(I) := \{P \in \mathrm{Spec}(A) \mid P \supset \mathrm{Ann}_A(I)\}$, due to Lemma 2.1.41 below. More generally, for finitely generated ideals I_1, I_2,

$$V(I_1 : I_2) = \mathrm{supp}((I_2 + I_1)/I_1) \subset \mathrm{Spec}(A/I_1).$$

Here is another example, where we do not know a priori whether we are dealing with a radical ideal or not: given an ideal $I \subset K[x_1, \ldots, x_n]$ and some point $a = (a_1, \ldots, a_n) \in V(I)$ such that $V' := V(I) \smallsetminus \{a\} \subset \mathbb{A}^n$ is Zariski closed. We wish to know equations for V', that is, some ideal I' such that $V' = V(I')$. At the moment, we only know that there exist such ideal I' and an ideal $J \subset K[x]$ satisfying $I = I' \cap J$ and $V(J) = \{a\}$, but we neither know I' nor J. Now, since $V(J) = \{a\}$, some power of the maximal ideal $\mathfrak{m}_a = \langle x_1 - a_1, \ldots, x_n - a_n \rangle$ is contained in J (see Lemma A.2.3). Now, since $I' \not\subset \mathfrak{m}_a$, it is not difficult to show that $I' : \mathfrak{m}_a = I'$, using the existence of a primary decomposition (Theorem 4.1.4) and Exercise 4.1.3. Hence, again using Exercise 4.1.3, we can conclude that $I : \mathfrak{m}_a^\infty = I' : \mathfrak{m}_a^\infty = I'$, that is,

$$V(I) \smallsetminus \{a\} = V(I : \mathfrak{m}_a^\infty).$$

In general, however, a geometric interpretation of $I_1 : I_2$ is more difficult and requires a careful study of the primary decompositions of I_1 and I_2. $I_1 : I_2$, or even $I_1 : I_2^\infty$, may not kill a whole component of $V(I_1)$, it may just reduce part of the structure. For example, if $I_1 = \langle xy^2, y^3 \rangle$, $I_2 = \langle x, y \rangle$, then

$$I_1 : I_2 = I_1 : I_2^\infty = \langle xy^2, y^3 \rangle : \langle x, y \rangle = \langle y^2 \rangle,$$

hence, $V(I_1 : I_2)$ is set-theoretically the same as $V(I_1)$ (namely, the x–axis), just with a slightly reduced structure (indicated in Figure 1.4 by the small arrow pointing in y–direction).

$$V(I_1) \qquad\qquad\qquad\qquad V(I_1 : I_2)$$

Fig. 1.4. Symbolic pictures of $V(\langle xy^2, y^3 \rangle)$ and $V(\langle xy^2, y^3 \rangle : \langle x, y \rangle)$.

Saturation is an important tool in computational projective geometry, cf. Appendix A.5, in particular, Lemma A.5.2 and the subsequent discussion.

SINGULAR Example 1.8.15 (saturation).

```
ring A  =0,(x,y,z),dp;
ideal I1=x5z3,xyz,yz4;
ideal I2=z;
LIB"elim.lib";
sat(I1,I2);                //the SINGULAR procedure
//-> [1]:                  //the result
//->    _[1]=y
//->    _[2]=x5
//-> [2]:
//->    4                  //the saturation exponent

ideal J=quotient(I1,I2);   //the way described above
int k;
while(size(reduce(J,std(I1)))!=0)
{
   k++;
   I1=J;
   J=quotient(I1,I2);
}
J;
//-> J[1]=y    J[2]=x5
k;
//-> 4                     //we needed to take the quotient 4 times
```

1.8.10 Kernel of a Ring Map

Let $\varphi : R_1 := (K[x]_{>_1})/I \to (K[y]_{>_2})/J =: R_2$ be a ring map defined by polynomials $\varphi(x_i) = f_i \in K[y] = K[y_1, \ldots, y_m]$ for $i = 1, \ldots, n$ (and assume that the monomial orderings satisfy $1 >_2 \mathrm{LM}(f_i)$ if $1 >_1 x_i$, cf. Lemma 1.5.8).

Define $J_0 := J \cap K[y]$, and $I_0 := I \cap K[x]$. Then φ is induced by

$$\tilde{\varphi} : K[x]/I_0 \to K[y]/J_0, \quad x_i \mapsto f_i,$$

and we have a commutative diagram

$$
\begin{array}{ccc}
K[x]/I_0 & \xrightarrow{\tilde{\varphi}} & K[y]/J_0 \\
\downarrow & & \downarrow \\
R_1 & \xrightarrow{\quad\varphi\quad} & R_2.
\end{array}
$$

Problem: Let I, J and φ be as above. Compute generators for $\mathrm{Ker}(\varphi)$.

Solution: Assume that $J_0 = \langle g_1, \ldots, g_s \rangle_{K[y]}$ and $I_0 = \langle h_1, \ldots, h_t \rangle_{K[x]}.$[11] Set $H := \langle h_1, \ldots, h_t, g_1, \ldots, g_s, x_1 - f_1, \ldots, x_n - f_n \rangle \subset K[x, y]$, and compute $H' := H \cap K[x]$ by eliminating $y_1, \ldots, y_m$ from H (cf. Section 1.8.2). Then H' generates $\mathrm{Ker}(\varphi)$ by the following lemma. $\square$

Lemma 1.8.16. *With the above notations, $\mathrm{Ker}(\varphi) = \mathrm{Ker}(\tilde{\varphi})R_1$ and*

$$\mathrm{Ker}(\tilde{\varphi}) = \big(I_0 + \langle g_1, \ldots, g_s, x_1 - f_1, \ldots, x_n - f_n \rangle_{K[x,y]} \cap K[x]\big) \ mod \ I_0.$$

In particular, if $>_1$ is global, then $\mathrm{Ker}(\varphi) = \mathrm{Ker}(\tilde{\varphi})$.

Proof. Obviously $\mathrm{Ker}(\tilde{\varphi})R_1 \subset \mathrm{Ker}(\varphi)$. On the other hand, let $h \in \mathrm{Ker}(\varphi)$, where $h = (h_1/h_2) + I$ for some $h_1 \in K[x]$, $h_2 \in S_{>_1}$, then $h_1 + I_0 \in \mathrm{Ker}(\tilde{\varphi})$. We conclude that $\mathrm{Ker}(\varphi) = \mathrm{Ker}(\tilde{\varphi})R_1$.

Now let $h \in K[x]$ satisfy $\tilde{\varphi}(h + I_0) = 0$, in other words, there exist polynomials $a_1, \ldots, a_s \in K[y]$ such that

$$h(f_1, \ldots, f_n) + \sum_{j=1}^{s} a_j g_j = 0.$$

Applying Taylor's formula to the polynomial $h(x)$, we obtain

---

[11] Let $>$ be any monomial ordering on $K[x]$, let $f_1, \ldots, f_m \in K[x]$, and let I be the ideal generated by $f_1, \ldots, f_m$ in $K[x]_>$. Then, for global orderings we have $\langle f_1, \ldots, f_m \rangle_{K[x]} = I_0$ but, if $>$ is non–global, the inclusion $\langle f_1, \ldots, f_m \rangle_{K[x]} \subset I_0 := I \cap K[x]$ may be strict. To compute I_0 in this case, one may proceed as follows: compute a primary decomposition $Q_1 \cap \ldots \cap Q_s$ of $\langle f_1, \ldots, f_m \rangle_{K[x]}$ (see Chapter 4). Assume that $Q_i K[x]_> \subsetneq K[x]_>$ iff $1 \leq i \leq t$. Then $I_0 = Q_1 \cap \ldots \cap Q_t$.

$$h(x) = h(f_1, \ldots, f_n) + \sum_{i=1}^{n} \frac{\partial h}{\partial x_i}(f_1, \ldots, f_n) \cdot (x_i - f_i) + \ldots .$$

This implies that, for suitable $b_i \in K[x, y]$,

$$h(x) + \sum_{i=1}^{n} b_i(x, y) \cdot (x_i - f_i(y)) + \sum_{j=1}^{s} a_j(y) g_j(y) = 0 .$$

This implies that $h \in \langle g_1, \ldots, g_s, x_1 - f_1(y), \ldots, x_n - f_n(y) \rangle_{K[x,y]} \cap K[x]$.
Conversely, let $h \in I_0 + \langle g_1, \ldots, g_s, x_1 - f_1(y), \ldots, x_n - f_n(y) \rangle_{K[x,y]} \cap K[x]$,

$$h = h_1 + \sum_{i=1}^{s} a_i g_i + \sum_{i=1}^{n} b_i(x_i - f_i), \quad h_1 \in I_0 .$$

Substituting x_i by f_i we obtain

$$h(f_1, \ldots, f_n) = h_1(f_1, \ldots, f_n) + \sum_{i=1}^{s} a_i(f_1, \ldots, f_n, y) g_i .$$

But $h_1(f_1, \ldots, f_n) \in J_0$ and $g_1, \ldots, g_s \in J_0$, hence, $h(f_1, \ldots, f_n) \in J_0$, which proves the claim. □

Remark 1.8.17. Given a ring map $\tilde{\varphi} : A \to B$, and $J \subset B$ an ideal, then $\tilde{\varphi}$ induces a ring map $\varphi : A \to B/J$ and $\mathrm{Ker}(\varphi) = \tilde{\varphi}^{-1}(J)$. Hence, the same method for computing the kernel can be used to compute preimages of ideals. Since $\mathrm{Ker}(\varphi) = \varphi^{-1}(0)$, to compute kernels or preimages is equivalent. SINGULAR has the built–in command `preimage`.

SINGULAR Example 1.8.18 (kernel of a ring map).

```
ring A=0,(x,y,z),dp;
ring B=0,(a,b),dp;
map phi=A,a2,ab,b2;
ideal zero;              //compute the preimage of 0
setring A;
preimage(B,phi,zero);    //the built-in SINGULAR command
//-> _[1]=y2-xz


ring C=0,(x,y,z,a,b), dp; //the method described above
ideal H=x-a2, y-ab, z-b2;
eliminate(H,ab);
//-> _[1]=y2-xz
```

1.8.11 Algebraic Dependence and Subalgebra Membership

Recall that a sequence of polynomials $f_1, \ldots, f_k \in K[x_1, \ldots, x_n]$ is called *algebraically dependent* if there exists a polynomial $g \in K[y_1, \ldots, y_k] \setminus \{0\}$ satisfying $g(f_1, \ldots, f_k) = 0$. This is equivalent to $\mathrm{Ker}(\varphi) \neq 0$, where $\varphi : K[y_1, \ldots, y_k] \to K[x_1, \ldots, x_n]$ is defined by $\varphi(y_i) = f_i$. $\mathrm{Ker}(\varphi)$ can be computed according to Section 1.8.10, and any $g \in \mathrm{Ker}(\varphi) \setminus \{0\}$ defines an algebraic relation between the $f_1, \ldots, f_k$. In particular, $f_1, \ldots, f_k$ are algebraically independent if and only if $\mathrm{Ker}(\varphi) = 0$ and this problem was solved in Section 1.8.10.

Related, but slightly different is the subalgebra–membership problem.

Problem: Given $f \in K[x_1, \ldots, x_n]$, we may ask whether f is an element of the subalgebra $K[f_1, \ldots, f_k] \subset K[x_1, \ldots, x_n] = K[x]$.

Solution 1: Define $\psi : K[y_0, \ldots, y_k] \to K[x]$, $y_0 \mapsto f$, $y_i \mapsto f_i$, compute $\mathrm{Ker}(\psi)$ according to Section 1.8.10 and check whether $\mathrm{Ker}(\psi)$ contains an element of the form $y_0 - g(y_1, \ldots, y_k)$. That is, we define an elimination ordering for $x_1, \ldots, x_n$ on $\mathrm{Mon}(x_1, \ldots, x_n, y_0, \ldots, y_k)$ with y_0 greater than $y_1, \ldots, y_k$ (for example, (dp(n), dp(1), dp(k))) and compute a standard basis G of $\langle y_0 - f, y_1 - f_1, \ldots y_k - f_k \rangle$. Then G contains an element with leading monomial y_0 if and only if $f \in K[f_1, \ldots, f_k]$.   □

Solution 2: Compute a standard basis of $\langle y_1 - f_1, \ldots, y_k - f_k \rangle$ for an elimination ordering for $x_1, \ldots, x_n$ on $\mathrm{Mon}(x_1, \ldots, x_n, y_1, \ldots, y_k)$ and check whether the normal form of f with respect to this standard basis does not involve any x_i. This is the case if and only if $f \in K[f_1, \ldots, f_k]$ and the normal form expresses f as a polynomial in $f_1, \ldots, f_k$.   □

We omit the proofs for these statements (cf. Exercise 1.8.10).

Note that $f \in K[f_1, \ldots, f_k]$ implies a relation $h(f, f_1, \ldots, f_k) = 0$ with $h(y_0, y_1, \ldots, y_k) = y_0 - g(y_1, \ldots, y_n)$, hence $f, f_1, \ldots, f_k$ are algebraically dependent (the converse does not need to be true).

Note further that the map $\varphi : K[y_1, \ldots, y_k] \to K[x_1, \ldots, x_n]$, $y_i \to f_i(x)$ is surjective if and only if $x_i \in K[f_1, \ldots, f_k]$ for all i. Hence, Solution 1 or Solution 2 can be used to check whether a given ring map is surjective.

SINGULAR Example 1.8.19 (algebraic dependence).

```
ring A=0,(x,y),dp;
poly f=x4-y4;
poly f1=x2+y2;
poly f2=x2-y2;
LIB"algebra.lib";
algDependent(ideal(f,f1,f2))[1];//a SINGULAR procedure
//-> 1

ring B=0,(u,v,w),dp;                    //the method described above
```

```
setring A;
ideal zero;
map phi=B,f,f1,f2;
setring B;
preimage(A,phi,zero);         //the kernel of phi
//-> _[1]=vw-u                //f=f1*f2 and hence f,f1,f2
                              //are algebraically dependent
```

SINGULAR Example 1.8.20 (subalgebra membership).

```
ring A=0,(x,y),dp;
poly f,f1,f2=x4-y4,x2+y2,x2-y2;
LIB"algebra.lib";
inSubring(f,ideal(f1,f2));//a SINGULAR procedure
//-> [1]:
//->    1                    //means f is contained in K[f1,f2]
//-> [2]:
//->    y(1)*y(2)-y(0)  //means f1*f2-f=0
```

Another SINGULAR procedure which also tests subalgebra membership is
algebra_containment.

Now let us proceed as explained in the text:

```
ring B = 0,(x,y,u,v,w),(dp(2),dp(1),dp(2)); //solution 1
ideal H=u-imap(A,f),v-imap(A,f1),w-imap(A,f2);
std(H);
//-> _[1]=u-vw   _[2]=2y2-v+w   _[3]=x2-y2-w
```

Since u appears as a leading monomial, $f \in K[f_1, f_2]$. Moreover, the existence
of $u - vw$ in H implies $f = f_1 f_2$.

```
ring C=0,(x,y,v,w),(dp(2),dp(2));  //solution 2
ideal H=v-imap(A,f1), w-imap(A,f2);
poly f=imap(A,f);
reduce(f,std(H));
//-> vw                       //again we find f=f1*f2
```

Exercises

1.8.1. Let I_1, I_2 be two ideals in $K[x]_>$ with $I_2 = \langle h_1, \ldots, h_r \rangle$, $h_i \in K[x]$.
Define $h := h_1 + th_2 + t^2 h_3 + \cdots + t^{r-1} h_r \in K[x, t]$. Prove that

$$I_1 : I_2 = (I_1 : h) \cap K[x]_> .$$

1.8.2. Let $I := \langle x^2 + 2y^2 - 3, x^2 + xy + y^2 - 3 \rangle \subset \mathbb{Q}[x, y]$. Compute the intersections $I \cap \mathbb{Q}[x]$ and $I \cap \mathbb{Q}[y]$.

1.8.3. Let $\varphi : \mathbb{Q}^2 \to \mathbb{Q}^4$ be the map defined by $(s,t) \mapsto (s^4, s^3t, st^3, t^4)$. Compute the Zariski closure of the image, $\overline{\varphi(\mathbb{Q}^2)}$, and decide whether $\varphi(\mathbb{Q}^2)$ coincides with its closure or not.

1.8.4. Compute all complex solutions of the system

$$x^2 + 2y^2 - 2 = 0$$
$$x^2 + xy + y^2 - 2 = 0.$$

1.8.5. Check whether the polynomial $x^2 + 5x$ is in the radical of the ideal $I = \langle x^2 + y^3, y^7 + x^3y^5 \rangle_{K[x,y]}$, respectively of the ideal $IK[x,y]_{\langle x,y \rangle}$.

1.8.6. Let $>$ be any monomial ordering on $\mathrm{Mon}(x_1, \ldots, x_n)$, let $I \subset K[x]_>$ be an ideal, and let G be a standard basis of I with respect to $>$. Show that the following are equivalent:

(1) $\dim_K(K[x]_>/I) < \infty$,
(2) for each $i = 1, \ldots, n$ there exists an $n_i \geq 0$ such that $x_i^{n_i}$ is a leading monomial of an element of G.

(Hint: Use Exercise 1.7.22)

1.8.7. Let $K[x]$ be the polynomial ring in one variable, and let $f \in K[x]$ decompose into linear factors, $f = (x - a_1)^{n_1} \cdot \ldots \cdot (x - a_r)^{n_r}$ for pairwise different $a_i \in K$. Show that $SK[x]/\langle f \rangle \cong K[x]/\langle x - a_1 \rangle^{n_1} \oplus \cdots \oplus K[x]/\langle x - a_r \rangle^{n_r}$ and conclude that $\dim_K K[x]/\langle f \rangle = n_1 + \cdots + n_r$.

1.8.8. Let $I = \langle f_1, \ldots, f_k \rangle \subset K[x_1, \ldots, x_n]$ be an ideal. Use a lexicographical Gröbner basis of I to show that $\dim_K(K[x]/I) < \infty$ if and only if the system of equations $f_1 = \ldots = f_k = 0$ has only finitely many solutions in $\overline{K}^n$, where $\overline{K}$ denotes the algebraic closure of K.

(Hint: use induction on n, the previous exercises and Appendix A.)

1.8.9. Prove statement (1) of Lemma 1.8.14.

1.8.10. Prove that Solutions 1 and 2 to the subalgebra–membership problem in Section 1.8.11 are correct.

1.8.11. Use SINGULAR to check whether the line defined by $x + y = 3$ (respectively $x + y = 500$) and the circle defined by $x^2 + y^2 = 2$ intersect.

1.8.12. Compute the kernel of the ring map $\mathbb{Q}[x,y,z] \to \mathbb{Q}[t]/\langle t^{12} \rangle$ defined by $x \mapsto t^5$, $y \mapsto t^7 + t^8$, $z \mapsto t^{11}$.

1.8.13. Show that the ring $\mathbb{Q}[s^4, s^3t, st^3, t^4]$ is isomorphic to

$$\mathbb{Q}[x_1, x_2, x_3, x_4]/I$$

with $I = \langle x_2x_3 - x_1x_4, x_3^3 - x_2x_4^2, x_2^3 - x_1^2x_3, x_1x_3^2 - x_2^2x_4 \rangle$.

1.8.14. Create a homogeneous polynomial p of degree 3 in three variables with random coefficients and use the `lift` command to express p as a linear combination of the partial derivatives of p.

1.9 Non–Commutative G–Algebras

SINGULAR contains a kernel extension (sometimes called PLURAL), providing Gröbner bases algorithms and implementations of Gröbner bases for ideals and modules in non–commutative G-algebras and, more generally, GR–algebras with respect to global monomial orderings. In this section we give a short introduction to the basic definitions and some of the non–commutative features of SINGULAR. For simplicity we restrict ourselves mainly to ideals, the case of modules being an immediate generalization.

In non–commutative algebras we have three kinds of ideals, namely left, right and two–sided ideals[12]. Given a finite set $F = \{f_1, \ldots, f_k\}$ from an algebra A, we denote by $_A\langle F \rangle = \{\sum_{i=1}^{k} a_i f_i | a_i \in A\}$ the left ideal, by $\langle F \rangle_A = \{\sum_{i=1}^{k} f_i a_i | a_i \in A\}$ the right ideal and by $_A\langle F \rangle_A = \{\sum_{i,j} a_i f_j b_i | a_i, b_i \in A\}$ a two–sided ideal[13], generated by F. The same notation will be used for monoid ideals in the commutative monoid $\mathbb{N}^n$.

Let $T_n = K\langle x_1, \ldots, x_n \rangle$ be the free associative K–algebra, generated by $\{x_1, \ldots, x_n\}$ over K. A K–basis of T consists of *words* $x_{i_1}^{\alpha_1} x_{i_2}^{\alpha_2} \ldots x_{i_m}^{\alpha_m}$, where $1 \le i_1, i_2, \ldots, i_m \le n$ with $m \ge 0$ and $\alpha_i \ge 0$. The elements of the form $x_{i_1}^{\alpha_1} x_{i_2}^{\alpha_2} \ldots x_{i_m}^{\alpha_m}$, with ordered indices $1 \le i_1 < i_2 < \ldots < i_m \le n$, are often called *standard words* and form a subset of the set of all words. The subvector space of T_n generated by the standard words is called vectorspace of *standard polynomials*.

Every finitely presented associative K–algebra A is isomorphic to T_n/I for some n and some two–sided ideal $I \subset T_n$. If I is given by a finite set of two–sided generators $I_1, \ldots, I_k$[14] we say that A is generated by $\{x_1, \ldots, x_n\}$ subject to the relations $\{I_1, \ldots, I_k\}$. We use the notation $A = K\langle x_1, \ldots, x_n \mid I_1 = 0, \ldots, I_k = 0 \rangle$.

A K–algebra A is said to have a *Poincaré–Birkhoff–Witt* (shortly, *PBW*) *basis*, if the set of standard words $\{x_1^{\alpha_1} x_2^{\alpha_2} \ldots x_n^{\alpha_n} \mid \alpha_k \ge 0\}$ is a K-basis of A. It is clear, that the commutative polynomial ring $K[x_1, \ldots, x_n]$ has a PBW basis and that the free associative K–algebra $K\langle x_1, \ldots, x_n \rangle$ does not have one. However, many important non–commutative algebras have a PBW basis.

Definition 1.9.1. Let $c_{ij} \in K \setminus \{0\}$ and $d_{ij} \in T_n$, $1 \le i < j \le n$, be standard polynomials. Consider the algebra

$$A = K\langle x_1, \ldots, x_n \mid x_j x_i = c_{ij} \cdot x_i x_j + d_{ij}, \ 1 \le i < j \le n \rangle.$$

---

[12] $I \subset A$ is called a *left–sided* (resp. *right–sided* resp. *two–sided*) *ideal* if I is a subset which is closed under addition and under multiplication by elements from A from the left (resp. from the right, resp. from both sides).

[13] Here is a difference to the commutative case. In the sum different terms $a_i, f_j b_i$ with the same f_j are necessary.

[14] This means that I is the set of linear combinations of the form $\sum_{i,j} a_i I_j b_i$ with $a_i, b_i \in T_n$.

A is called a G-*algebra*, if the following two conditions hold:

(1) there exists a monomial well–ordering $<$ on $\mathbb{N}^n$ such that[15]

$$\forall i < j \quad \mathrm{LM}(d_{ij}) < x_i x_j.$$

(2) For all $1 \le i < j < k \le n$, the polynomial

$$c_{ik}c_{jk} \cdot d_{ij}x_k - x_k d_{ij} + c_{jk} \cdot x_j d_{ik} - c_{ij} \cdot d_{ik}x_j + d_{jk}x_i - c_{ij}c_{ik} \cdot x_i d_{jk}$$

reduces to 0 with respect to the relations of A.

The matrices (c_{ij}) and (d_{ij}) are called *structural matrices*.

G-algebras, first introduced under this name by J. Apel in [3], were studied in [178]. They are also called *algebras of solvable type* [132], [135], [154] and *PBW algebras* [38].

Proposition 1.9.2. *Let A be a G-algebra. Then*

(1) A has a PBW basis $\{x_1^{\alpha_1} x_2^{\alpha_2} \ldots x_n^{\alpha_n} \mid \alpha_i \ge 0\}$,
(2) A is left and right Noetherian,
(3) A is an integral domain.

The proof of this proposition and a list of further important properties of G-algebras can be found e.g. in [152],[153].

Examples of G-algebras include:

- Weyl algebras [51] and their various generalizations [154],
- quasi–commutative polynomial rings (for example, the quantum plane $yx = q \cdot xy$ and the anti–commutative rings with relations $x_j x_i = -x_i x_j$),
- universal enveloping algebras of finite dimensional Lie algebras [60],
- some iterated Ore extensions [132],
- many quantum groups [38] and nonstandard quantum deformations,
- many important operator algebras [44], [154], et cetera,

(cf. the SINGULAR libraries `ncalg.lib`, `nctools.lib` and `qmatrix.lib`).

In the following example we set up several algebras with SINGULAR. According to Definition 1.9.1, we have to define two (strictly upper triangular) $n \times n$ matrices c and d. The initialization command `ncalgebra(C,D)`, executed in a commutative ring, turns the ring into a non–commutative G-algebra or returns an error message. An error message is returned, if the condition $\mathrm{LM}(d_{ij}) < x_i x_j$, for all $1 \le i < j \le n$, is not satisfied by the ordering in the given commutative ring.

---

[15] The notion of leading monomial, $\mathrm{LM}(f)$, for a standard polynomial f is defined as in the commutative case (cf. Definition 1.9.5).

SINGULAR Example 1.9.3 (enveloping algebra). Consider the universal enveloping algebra $U(\mathfrak{sl}_2)$ over the field K of characteristic 0. It is defined as $K\langle e, f, h \mid fe = ef - h, \; he = eh + 2e, \; hf = fh - 2f\rangle$. Thus, $c_{12} = c_{13} = c_{23} = 1$ and $d_{12} = -h$, $d_{13} = 2e$, $d_{23} = -2f$. The explicit definition of $U(\mathfrak{sl}_2)$ in SINGULAR is as follows:

```
ring r = 0,(e,f,h),dp;
matrix C[3][3];
C[1,2] = 1; C[1,3] = 1; C[2,3] = 1;
matrix D[3][3];
D[1,2] = -h;
D[1,3] = 2e;
D[2,3] = -2f;
ncalgebra(C,D);
```

Since the nonzero entries of the matrix C are all equal, we can execute `ncalgebra(1,D);` and obtain the same result.

Examining the properties of the ring r, we see that in addition to the data usually displayed for commutative polynomial rings, the non–commutative relations between variables are also displayed.

```
r;
//->    characteristic : 0
//->    number of vars : 3
//->          block   1 : ordering dp
//->                   : names     e f h
//->          block   2 : ordering C
//->    noncommutative relations:
//->     fe=ef-h
//->     he=eh+2e
//->     hf=fh-2f
```

The non–commutative multiplication between polynomials is carried out, as soon as the symbol $*$ is used.

```
fe; // "commutative" syntax, not correct
//-> ef
f*e; // correct non-commutative syntax
//-> ef-h
```

SINGULAR Example 1.9.4 (quantum deformation of $U(\mathfrak{so}_3)$). Consider the algebra $U_q'(\mathfrak{so}_3)$, which is a non–standard (quantum) deformation of the algebra $U(\mathfrak{so}_3)$. The quantum parameter q is invertible. It is considered to be either a free parameter (that is, we work over the transcendental field extension $K(q)$) or a primitive root of unity (then we work over the simple algebraic field extension $K[q]/\mu(q)$, where $\mu(q)$ is the corresponding minimal polynomial). Computation in both cases is possible in SINGULAR.

The algebra $U_q'(\mathfrak{so}_3)$ is defined as

$$K\langle x,y,z \mid yx = q \cdot xy - \sqrt{q}z, \ zx = \frac{1}{q} \cdot xz + \frac{1}{\sqrt{q}}y, \ zy = q \cdot yz - \sqrt{q}x\rangle.$$

Hence, we have $c_{12} = q, c_{13} = \frac{1}{q}, c_{23} = q$ and $d_{12} = -\sqrt{q}z, d_{13} = \frac{1}{\sqrt{q}}y,$ $d_{23} = -\sqrt{q}x.$

Let us consider q as a free parameter and, moreover, set $Q := \sqrt{q}.$

```
ring s   = (0,Q),(x,y,z),dp;
matrix C[3][3];
C[1,2] = Q2;
C[1,3] = 1/Q2;
C[2,3] = Q2;
matrix D[3][3];
D[1,2] = -Q*z;
D[1,3] = 1/Q*y;
D[2,3] = -Q*x;
ncalgebra(C,D);
```

We obtain the following relations in the non–commutative ring s:

```
//->     noncommutative relations:
//->     yx=(Q2)*xy+(-Q)*z
//->     zx=1/(Q2)*xz+1/(Q)*y
//->     zy=(Q2)*yz+(-Q)*x
```

Many important algebras are predefined in numerous procedures of the libraries `ncalg.lib`, `nctools.lib` and `qmatrix.lib`, distributed with SINGULAR.

In the library `ncalg.lib`, there are procedures defining the universal enveloping algebras $U(\mathfrak{sl}_n)$, $U(\mathfrak{gl}_n)$, $U(\mathfrak{so}_m)$, $U(\mathfrak{sp}_m)$, $U(\mathfrak{g}_2)$, $U(\mathfrak{f}_4)$, $U(\mathfrak{e}_6)$, $U(\mathfrak{e}_7)$, $U(\mathfrak{e}_8)$. Moreover, there are procedures for the quantized enveloping algebras $U_q(\mathfrak{sl}_2)$, $U_q(\mathfrak{sl}_3)$ and the non–standard quantum deformation $U_q'(\mathfrak{so}_3)$.

Weyl, Heisenberg, exterior algebras, as well as finite dimensional algebras (given via multiplication table between generators) are implemented in the library `nctools.lib`. The ring of quantum matrices $\mathcal{O}_q(M_n)$ can be set with the procedure `quantMat` from `qmatrix.lib`. This procedure can be used for defining the algebras $\mathcal{O}_q(GL_n)$ and $\mathcal{O}_q(SL_n)$ as factor algebras of $\mathcal{O}_q(M_n)$.

Having defined some interesting non–commutative algebras we start with describing Gröbner bases for left ideals.

Let $A = K\langle x_1, \ldots, x_n \mid x_j x_i = c_{ij} x_i x_j + d_{ij}, 1 \leq i < j \leq n\rangle$ be a G–algebra over a field K. Since A has a PBW basis, we call an element $x^\alpha = x_1^{\alpha_1} x_2^{\alpha_2} \ldots x_n^{\alpha_n}$ of this basis, i.e. a standard word, a *monomial* in A, and denote the set of monomials from A by $\text{Mon}(A)$. Note that, although A is a quotient algebra of T_n the set $\text{Mon}(A)$ coincides with Mon_n defined in Definition 1.1.3. A *term* in A is an element of A of the form cx^α with $c \in K$ and x^α a monomial.

Definition 1.9.5. Let A be a G–algebra in n variables.

(1) A total ordering $<$ on $\text{Mon}(A)$ is called a *a (global) monomial ordering* on A, if it is a global monomial ordering on Mon_n in the sense of Definition 1.2.2, that is, if the following conditions hold:
  - $<$ is a well–ordering,
  - $\forall \alpha, \beta, \gamma \in \mathbb{N}^n$ such that $x^\alpha < x^\beta$, we have $x^{\alpha+\gamma} < x^{\beta+\gamma}$.

(2) Since $\text{Mon}(A)$ is a K–basis of A, any $f \in A \setminus \{0\}$ can be written uniquely as $f = c_\alpha x^\alpha + g$ with $c_\alpha \in K \setminus \{0\}$, x^α a monomial, and $x^\beta < x^\alpha$ for any nonzero term $c_\beta x^\beta$ of g. We define

$$
\begin{aligned}
\text{LM}(f) &= x^\alpha &&\in \text{Mon}(A), &&\text{the } \textit{leading monomial} \text{ of } f, \\
\text{LC}(f) &= c_\alpha &&\in K \setminus \{0\}, &&\text{the } \textit{leading coefficient} \text{ of } f, \\
\text{LE}(f) &= \alpha &&\in \mathbb{N}^n, &&\text{the } \textit{leading exponent} \text{ of } f.
\end{aligned}
$$

(3) We say that x^α *divides* the monomial x^β, if $\alpha_i \leq \beta_i \ \forall i = 1 \ldots n$ and denote it by $x^\alpha | x^\beta$.

Example 1.9.6. Consider two exponent vectors $\alpha = (1,1)$ and $\beta = (1,2)$ from $\mathbb{N}^2$. Let A be a G–algebra in the variables $x = \{x_1, x_2\} = \{y, \partial\}$, and let $m_1 = x^\alpha = y\partial$ and $m_2 = x^\beta = y\partial^2$, hence $m_1 \mid m_2$. However, the left division of m_2 by m_1 gives various answers in different algebras. For example, in the commutative polynomial ring $R = K[y, \partial]$, we have $m_2 = \partial m_1$, whereas in the *first quantized Weyl algebra* $A_q = K(q)\langle y, \partial \mid \partial y = q^2 x \partial + 1\rangle$, we obtain $m_2 = q^{-2} \cdot \partial \cdot m_1 - q^{-2}\partial$.

Definition 1.9.7. Let S be any subset of a G–algebra A.

- We define $\mathcal{L}(S) \subseteq \mathbb{N}^n$ to be the monoid ideal (with respect to addition) in $\mathbb{N}^n$, generated by the leading exponents of the elements of S, that is

$$
\mathcal{L}(S) = {}_{\mathbb{N}^n}\langle \alpha \mid \exists s \in S \ \ \text{LE}(s) = \alpha \rangle.
$$

We call $\mathcal{L}(S)$ a *monoid ideal of leading exponents*. By Dixon's Lemma (Lemma 1.2.6), $\mathcal{L}(S)$ is finitely generated, i.e. there exist $\alpha_1, \ldots, \alpha_m \in \mathbb{N}^n$, such that $\mathcal{L}(S) = {}_{\mathbb{N}^n}\langle \alpha_1, \ldots, \alpha_m \rangle$.

- The *span of leading monomials of S* is defined to be the K–vector space spanned by the set $\{x^\alpha \mid \alpha \in \mathcal{L}(S)\} \subseteq \text{Mon}(A)$. We denote it by $L(S) := {}_K\langle \{x^\alpha \mid \alpha \in \mathcal{L}(S)\} \rangle \subseteq A$.

Definition 1.9.8. Let $<$ be a monomial ordering on the G–algebra A, $I \subset A$ a left ideal and $G \subset I$ a finite subset. G is called a *left Gröbner basis* of I if for any $f \in I \setminus \{0\}$ there exists $g \in G$ satisfying $\text{LM}(g) \mid \text{LM}(f)$.

Remark 1.9.9. In commutative rings, one usually defines Gröbner basis via leading ideals (cf. Definition 1.6.1). In general, it is impossible to adapt this definition in the context of G-algebras. One of the reasons is that for $S \subset A$, $L(S)$ is just a K-vector subspace of A and is not, in general, an ideal.

Let us define $L'(S) = {}_A\langle\{\mathrm{LM}(f) \mid f \in S\}\rangle$ to be the left leading ideal of a finite set S. Recall, that for a commutative algebra A, a finite set S is a Gröbner basis of an ideal I, if $S \subset I$ and $L'(S) = L'({}_A\langle S\rangle) = L'(I)$.

Consider the *Weyl algebra* $A = K\langle x, \partial | \partial x = x\partial + 1\rangle$, the set $S = \{x\partial + 1, x\}$ and $I = {}_A\langle S\rangle$. I is a proper left ideal and $\{x\}$ is a reduced left Gröbner basis of I. Hence, the K-vector spaces $L(I)$ and $L(\{x\})$ are equal, but $L'(I) = {}_A\langle\{x\partial, x\}\rangle = A \neq L'(\{x\}) = {}_A\langle x\rangle$.

Proposition 1.9.10. *Let $<$ be a monomial ordering on the G-algebra A, $I \subset A$ a left ideal and $G \subset I$ a finite subset. Then the following conditions are equivalent:*

- *G is a left Gröbner basis of I,*
- *$L(G) = L(I)$ as K-vector spaces,*
- *$\mathcal{L}(G) = \mathcal{L}(I)$ as monoid ideals in $\mathbb{N}^n$.*

Let us now introduce the notion of a normal form or *divison with remainder* in the non–commutative setting.

Definition 1.9.11. Denote by $\mathcal{G}$ the set of all finite ordered subsets of the G-algebra A. A map $\mathrm{NF} : A \times \mathcal{G} \to A$, $(f, G) \mapsto \mathrm{NF}(f|G)$, is called a *(left) normal form* on A if:

(1) For all $f \in A$, $G \in \mathcal{G}$
   (i) $\mathrm{NF}(0 \mid G) = 0$,
   (ii) $\mathrm{NF}(f \mid G) \neq 0 \Rightarrow \mathrm{LM}(\mathrm{NF}(f \mid G)) \notin L(G)$,
   (iii) $f - \mathrm{NF}(f \mid G) \in {}_A\langle G\rangle$.
(2) Let $G = \{g_1, \ldots, g_s\} \in \mathcal{G}$ and $f \in A$. Then

$$f - \mathrm{NF}(f \mid G) = \sum_{i=1}^{s} a_i g_i, \ a_i \in A, \ s \geq 0,$$

is either 0 or $\mathrm{LM}(\Sigma_{i=1}^{s} a_i g_i) \geq \mathrm{LM}(a_i g_i)$ for all i such that $a_i g_i \neq 0$. We say that $f - \mathrm{NF}(f|G)$ (or, by abuse of notation, f) has a *(left) standard representation* with respect to G.

Lemma 1.9.12. *Let $I \subset A$ be a left ideal, $G \subset I$ a left Gröbner basis of I and $\mathrm{NF}(\cdot \mid G)$ a left normal form on A with respect to G.*

(1) For any $f \in A$, we have $f \in I \Longleftrightarrow \mathrm{NF}(f \mid G) = 0$.
(2) If $J \subset A$ is a left ideal with $I \subset J$, then $L(I) = L(J)$ implies $I = J$. In particular, G generates I as a left ideal.
(3) If $\mathrm{NF}(\cdot \mid G)$ is a reduced left normal form, then it is unique.

Definition 1.9.13. Let $f, g \in A \setminus \{0\}$ with $\mathrm{LM}(f) = x^\alpha$ and $\mathrm{LM}(g) = x^\beta$. Set $\gamma := \mathrm{lcm}(\alpha, \beta)$ and define the *(left) s–polynomial* of f and g to be

$$\mathrm{LeftSpoly}(f, g) := x^{\gamma - \alpha} f - \frac{\mathrm{LC}(x^{\gamma - \alpha} f)}{\mathrm{LC}(x^{\gamma - \beta} g)} x^{\gamma - \beta} g.$$

Remark 1.9.14.

(1) It is easy to see that $\mathrm{LM}(\mathrm{LeftSpoly}(f, g)) < \mathrm{LM}(f \cdot g)$ holds. If $\mathrm{LM}(g) \mid \mathrm{LM}(f)$, say $\mathrm{LM}(g) = x^\beta$ and $\mathrm{LM}(f) = x^\alpha$, then the s–polynomial is especially simple:

$$\mathrm{LeftSpoly}(f, g) = f - \frac{\mathrm{LC}(f)}{\mathrm{LC}(x^{\alpha - \beta} g)} x^{\alpha - \beta} g,$$

and $\mathrm{LM}\big(\mathrm{LeftSpoly}(f, g)\big) < \mathrm{LM}(f)$ holds.

(2) Let A be a G–algebra, where all the relations are of the form $\{x_j x_i = x_i x_j + d_{ij}\}$. Then, there is an easier formula for the s–polynomial, namely

$$\mathrm{LeftSpoly}(f, g) := x^{\gamma - \alpha} f - \frac{\mathrm{LC}(f)}{\mathrm{LC}(g)} x^{\gamma - \beta} g,$$

which looks exactly like the formula in the Definition 1.6.9.

As before, we assume that A is a G–algebra and $<$ is a fixed global monomial ordering on A.

Algorithm 1.9.15. LeftNF($f \mid G$)

Input: $f \in A$, $G \in \mathcal{G}$;
Output: $h \in A$, a left normal form of f with respect to G.

- $h := f$;
- while $((h \neq 0)$ and $(G_h := \{g \in G : \mathrm{LM}(g) \mid \mathrm{LM}(h)\} \neq \emptyset))$
    choose any $g \in G_h$;
    $h := \mathrm{LeftSpoly}(h, g)$;
- return h;

Algorithm 1.9.16. LeftGröbnerBasis(G)

Input: $G \in \mathcal{G}$;
Output: $S \in \mathcal{G}$, a left Gröbner basis of the left submodule $I = {}_A\langle G \rangle \subset A$.

- $S := G$;
- $P := \{(f, g) \mid f, g \in S\} \subset S \times S$;
- while $(P \neq \emptyset)$
    choose $(f, g) \in P$;
    $P := P \setminus \{(f, g)\}$;
    $h := \mathrm{LeftNF}\big(\mathrm{LeftSpoly}(f, g) \mid S\big)$;

if ($h \neq 0$)
$\quad P := P \cup \{(h, f) \mid f \in S\}$;
$\quad S := S \cup \{h\}$;
return S;

As one can see, with the chosen setup, we are able to keep the form of the algorithms exactly as in the commutative case. However, since we use left s–polynomials, left normal forms, and compute left Gröbner bases respectively, the proofs are somewhat different.

Theorem 1.9.17 (Left Buchberger's Criterion). *Let $I \subset A$ be a left ideal and $G = \{g_1, \ldots, g_s\}$, $g_i \in I$. Let $\mathrm{LeftNF}(\cdot \mid G)$ be a left normal form on A with respect to G. Then the following are equivalent:*

(1) G is a left Gröbner basis of I,
(2) $\mathrm{LeftNF}(f \mid G) = 0$ for all $f \in I$,
(3) each $f \in I$ has a left standard representation with respect to G,
(4) $\mathrm{LeftNF}(\mathrm{LeftSpoly}(g_i, g_j) \mid G) = 0$ for $1 \leq i, j \leq s$.

The practical computation of Gröbner bases in non–commutative algebras can be extremely time and space consuming already for small examples, much more than in the commutative case. It is therefore important to know which criteria for discarding useless pairs continue to hold. We point out, that the generalization of the *chain criterion* (Lemma 2.5.10) holds in the LEFTGRÖBNERBASIS algorithm over G–algebras with no restrictions, whereas the *product criterion* (Exercise 1.7.1) does not hold in general. However, for some cases it is possible to use a weaker statement.

Lemma 1.9.18 (Generalized Product Criterion). *Let A be a G–algebra, such that $\forall 1 \leq i < j \leq n$, $c_{ij} = 1$. That is, the relations of A are of the form $\{x_j x_i = x_i x_j + d_{ij}\}$.*
Let $f, g \in A$. Suppose that $\mathrm{LM}(f)$ and $\mathrm{LM}(g)$ have no common factor, then $\mathrm{LeftNF}(\mathrm{LeftSpoly}(f, g) \mid \{f, g\}) = fg - gf$.

Example 1.9.19. Let $A = U(\mathfrak{sl}_2)$ over the field $\mathbb{Q}$, that is $A = \mathbb{Q}\langle e, f, h \mid fe = ef - h, \ he = eh + 2e, \ hf = fh - 2f\rangle$. Let $I \subset A$ be the left ideal generated by $\{e^2, f\}$. We compute a left Gröbner basis of I with respect to the **dp** ordering.

Let $p_1 := e^2$, $p_2 := f$, then $S = \{p_1, p_2\}$ and $P = \{(p_1, p_2)\}$. Since p_1, p_2 do not have common factors, we apply the generalized product criterion and obtain

- $\mathrm{spoly}(p_1, p_2) \to e^2 f - fe^2 = 2eh + 2e$. It is not reducible by the elements of S, so $p_3 := eh + e$, $S := \{p_1, p_2, p_3\}$ and $P := \{(p_1, p_3), (p_2, p_3)\}$.
- $\mathrm{spoly}(p_1, p_3) = hp_1 - ep_3 = 3e^2 = 3p_1$, hence $\mathrm{NF}(\mathrm{spoly}(p_1, p_3) \mid S) = 0$.
- $\mathrm{spoly}(p_2, p_3) \to p_3 p_2 - p_2 p_3 = 2ef - h^2 - h =: g$. $\mathrm{NF}(g \mid S) = g - 2ep_2 = -(h^2 + h)$, so $p_4 := h^2 + h$, $S := \{p_1, p_2, p_3, p_4\}$ and $P := \{(p_1, p_4), (p_2, p_4), (p_3, p_4)\}$.

- spoly$(p_1, p_4) \rightarrow p_1 p_4 - p_4 p_1 = -8e^2 h - 20e^2 =: g$, then NF$(g \mid S) = g + 8ep_3 + 12p_1 = 0$.
- spoly$(p_2, p_4) = p_2 p_4 - p_4 p_2 = 4fh - 2f =: g$, then NF$(g \mid S) = g - 4hp_2 - 6p_2 = 0$.
- spoly$(p_3, p_4) = hp_3 - ep_4 = 2(eh + e) = 2p_3$, so NF$($spoly$(p_3, p_4) \mid S) = 0$. $S = \{p_1, p_2, p_3, p_4\}$ and $P = \emptyset$.

Hence, after reordering the elements in an ascending way, we conclude that $S = \{f, h^2 + h, eh + h, e^2\}$ is a left Gröbner basis of I.

SINGULAR Example 1.9.20 (Left Gröbner basis). Let us compute the above example with SINGULAR. The command std computes a left Gröbner basis of its argument of type ideal or module.

```
LIB "ncalg.lib";
// load the library with the definition of U((sl_2)
def A = makeUsl(2); // set up U(sl_2)
setring A;
option(redSB);
option(redTail); // we wish to compute reduced bases
ideal I = e2,f;
ideal LI = std(I); LI;
//-> LI[1]=f
//-> LI[2]=h2+h
//-> LI[3]=eh+e
//-> LI[4]=e2
```

Above, we have sketched the left Gröbner basis theory. By replacing every left–sided with a right–sided action in the statements and proofs, one obtains a right Gröbner basis theory. However, it is not necessary to rewrite the algorithms, since we can compute with right ideals by using left Gröbner bases in opposite algebras.

Let A be an associative algebra over K. The *opposite algebra* A^{opp} is defined by taking the underlying vector-space of A and introducing a new "opposite" multiplication on it, by setting $f * g := g \cdot f$. Then, A^{opp} is an associative K–algebra, and $(A^{\mathrm{opp}})^{\mathrm{opp}} = A$.

Lemma 1.9.21. *Let A be a G–algebra, then A^{opp} is a G–algebra too.*

There is one–to–one correspondence between left (right) ideals of A and right (left) ideals of A^{opp}. Thus, in order to compute a right Gröbner basis of a left ideal I in A, we have to create the opposite algebra A^{opp} of A, compute the right ideal I^{opp} corresponding to I. Then, we compute a left Gröbner basis of I^{opp} and "oppose" the result back to A. This can be achieved by the following procedure:

```
proc rightStd(ideal I)
{
  def A = basering;
  def Aopp = opposite(A);
  setring Aopp;
  ideal Iopp = oppose(A,I);
  ideal Jopp = std(Iopp);
  setring A;
  ideal J = oppose(Aopp,Jopp);
  return(J);
}
```

The same principle applies to computation of right normal forms, right syzygies etc. The corresponding procedures rightStd, rightNF, rightSyz, rightModulo are implemented in the library nctools.lib.

SINGULAR Example 1.9.22 (Right Gröbner basis).

```
LIB "ncalg.lib";
def A = makeUsl(2);
setring A;
option(redSB);
option(redTail); // we wish to compute reduced bases
ideal I = e2,f;
ideal LI = std(I);
print(matrix(LI)); // a compact form of an ideal
//-> f,h2+h,eh+e,e2
ideal RI = rightStd(I);
print(matrix(RI));
//-> f,h2-h,eh+e,e2
```

As we can see, in this case the left and right bases differ only by one generator.

A *two–sided Gröbner basis* of an ideal I is a finite set of generators $F = \{f_1, \ldots, f_s\}$, such that F is a left and a right Gröbner basis of I. In particular, if F is a two–sided Gröbner basis of I, we have $_A\langle F \rangle = \langle F \rangle_A = _A\langle F \rangle_A = I$.

We use a special algorithm for computing two–sided Gröbner bases which is described in detail in [152] and which is behind the command twostd. Let us continue with the example before.

SINGULAR Example 1.9.23 (Two–sided Gröbner basis).

```
ideal I = e2,f;
ideal LI = std(I);
print(matrix(LI)); // a compact form of an ideal
//-> f,h2+h,eh+e,e2
```

```
ideal TI = twostd(I);
print(matrix(TI));
//-> h,f,e
```

Two–sided Gröbner bases are essential for computations ain factor–algebras. Let A be a G–algebra and $T \subset A$ be a nonzero two–sided ideal, then there is a factor–algebra A/T which we call a *GR–algebra*. The data type qring in the non–commutative case corresponds to GR–algebras. The ideal T must be given as a two–sided Gröbner basis.

SINGULAR Example 1.9.24 (Computations in factor algebras).

```
LIB "ncalg.lib";
def A = makeUsl2();
setring A;
ideal T = 4*e*f+h^2-2*h; // central element in U(sl_2)
T = twostd(T);
T;
//-> T[1]=4ef+h2-2h
qring Q = twostd(T);
ideal I = e2,f;
ideal LI = std(I);
LI;
//-> LI[1]=h
//-> LI[2]=f
//-> LI[3]=e
```

As we can see, the left Gröbner basis of $\{e^2, f\}$ is very different, if we pass from $U(\mathfrak{sl}_2)$ to the factor algebra $U(\mathfrak{sl}_2)/\langle 4ef + h^2 - 2h \rangle$.

Many of the SINGULAR functions are available both for G–algebras and for GR–algebras. Among them are the functions for computing left syzygy modules (syz), left transformation matrices between bases (lift), left free resolutions (nres, mres) and many others.

1.9.1 Centralizers and Centers

In many applications we need natural subalgebras of a non–commutative G–algebra A, like the *centralizer* of a finite set $S \subset A$, defined to be $C_A(S) := \{f \in A \mid fs = sf \ \forall s \in S\}$, and the *center* of A, $Z(A) := C_A(A) = \{f \in A \mid fa = af \ \forall a \in A\}$. As one can easily see, $K \subseteq Z(A) \subset C_A(S)$ for any finite subset S of A. The computation of centers and centralizers is implemented in the library central.lib (cf. [172]). For a general G–algebra we do not have any information on the number of generators of the center and of their degree. Hence, both procedures centralizer and center need extra arguments. Namely, one sets an upper bound for the degree and, optionally, a bound for the number of elements to be computed. Note, that although both procedures return data of the type `ideal`, the data consists of generators of a subalgebra.

SINGULAR Example 1.9.25 (Center and centralizer).

```
LIB "ncalg.lib";
LIB "central.lib";
def A0 = makeUsl2(); // U(sl_2) over the rationals
setring A0;
// compute the centralizer of f^2 up to degree 6
ideal C = centralizer(f^2,6); C;
// -> C[1]=f
// -> C[2]=4ef+h2-2h
ideal Z = center(5); Z;
// -> Z[1]=4ef+h2-2h
def A5 = makeUsl2(5); // U(sl_2) over Z/5Z
setring A5;
ideal Z = center(5); Z;
// -> Z[1]=ef-h2+2h
// -> Z[2]=h5-h
// -> Z[3]=f5
// -> Z[4]=e5
```

As we can see, the centralizers depend heavily on the ground field K of a given G–algebra. Let us demonstrate the computations for GR–algebras. We continue with the example above. Since the element $4ef + h^2 - 2h$ is central in $U(\mathfrak{sl}_2)$ for any K, it generates a principal two–sided ideal.

```
// we are in the algebra A5
ideal T = twostd(4ef+h2-2h);  T;
// -> T[1]=ef-h2+2h
qring Q = T;
// compute the centralizer of f^2 up to degree 6
ideal C = centralizer(f^2,6); C;
// -> C[1]=f
// -> C[2]=eh3-2eh2-eh+2e
// -> C[3]=h5-h
// -> C[4]=e5
// -> C[5]=e4h2-2e4h
```

1.9.2 Left Ideal Membership

In order to test whether a given polynomial lies in the given left ideal, we have to compute, according to Lemma 1.9.12, a left Gröbner basis of the ideal and then the left normal form of the polynomial with respect to the latter basis.

This method is also used for "canonizing" representatives of polynomials in factor algebras. Let us continue with Example 1.9.25.

The procedure bracket(a, b) returns $ab - ba$. Let us check, that C[2] and C[5] lie in the centralizer of f^2 in the algebra $U(\mathfrak{sl}_2)/\langle 4ef + h^2 - 2h\rangle$. For this, we use the left ideal membership approach by invoking NF(b,std(0)) or, alternatively, reduce(b,std(0)) for a polynomial b.

Recall, that, in a factor ring std(0) stands for the two–sided Gröbner basis of the ideal defining the factor ring, which has been constructed as qring Q in the Singular example 1.9.25.

```
poly b = bracket(C[2],f^2);  b;
// -> -2ef2h2+2fh4-ef2h-fh3-2ef2+2fh2+fh+f
NF(b,std(0));
// -> 0
b = bracket(C[5],f^2);  b;
// -> 2e4f2h-2e3fh3-e4f2-2e2h4-2e3fh-e2h3-2e2h2-e2h
reduce(b,std(0));
// -> 0
```

1.9.3 Intersection with Subalgebras (Elimination of Variables)

Let A be a G–algebra generated by $\{x_1,\ldots,x_n\}$ with structural matrices (c_{ij}) and (d_{ij}). For a fixed $r, 1 \leq r < n$, consider the subalgebra A_r, generated by the $\{x_{r+1},\ldots,x_n\}$. We say, that A_r is an *essential* subalgebra (or *admissible* for elimination), if $\forall\, i,j$ such that $r + 1 \leq i < j \leq n$, the polynomials d_{ij} involve only the variables $x_{r+1},\ldots,x_n$.

Example 1.9.26 (Essential and non–essential subalgebras). Consider $A = U(\mathfrak{sl}_2)$ (see SINGULAR Example 1.9.3). $\{f,h\}$ generate an essential subalgebra (recall the relations $he = eh + 2e$ and $hf = fh - 2f$). However, the subalgebra generated by $\{e,f\}$ is not essential, since $fe = ef - h$ and hence, h is the third generator of this subalgebra. That is, the set $\{e,f,h\}$ generates the same algebra over K as the set $\{e,f\}$, namely the whole A. As a consequence, we cannot "eliminate" h from any ideal of A, since this would require the intersection with the subalgebra generated by $\{e,f\}$, which is A, and hence this would not change anything.

The notion of *elimination* of variables in the context of non–commutative algebras means the *intersection of an ideal with an essential subalgebra.*

Recall, that an *ordering* $<_r$ for $x_1,\ldots,x_n$ (cf. Definition 1.5.4) is said to have the elimination property for $x_1,\ldots,x_r$, if, for any $f \in A$, $\mathrm{LM}(f) \in A_r$ implies $f \in A_r$; it is then called an *elimination ordering.*

The following lemma is the constructive generalization of Lemma 1.8.3 to the class of G–algebras.

Lemma 1.9.27. *Let A be a G–algebra, generated by $\{x_1, \ldots, x_n\}$ and $I \subset A$ an ideal. Suppose, that the following conditions are satisfied for a fixed r, $1 \leq r < n$:*

- *the set $\{x_{r+1}, \ldots, x_n\}$ generates an essential subalgebra A_r,*
- *there exists an admissible elimination ordering* [16] *$<_r$ for $x_1, \ldots, x_r$.*

Then, if S is a left (resp. right) Gröbner basis of I with respect to $<_r$, $S \cap A_r$ is a left (resp. right) Gröbner basis of $I \cap A_r$.

Note, that both conditions in Lemma 1.9.27 are automatically satisfied in a commutative polynomial ring as well as in a free associative algebra.

The SINGULAR command `eliminate` works along the lines of Lemma 1.9.27. At first it checks whether B is essential and, if it is the case, the check of the admissibility of the elimination ordering is performed. If one of these conditions is not satisfied, the corresponding error message is returned.

SINGULAR Example 1.9.28 (Intersection with essential subalgebras).

```
LIB "ncalg.lib";
def  U = makeUsl2(); // U(sl_2) over the rationals
setring U;
ring A = 0,(a),dp;
def  UA = U + A;
setring UA;
```

The algebra `UA` corresponds to $U(\mathfrak{sl}_2) \otimes_K K[a]$, in particular, a commutes with e, f and h in `UA`.

```
poly  p = 4*e*f+h^2-2*h - a;
// p is a central element of UA
ideal I = e^3, f^3, h^3-4*h, p;
// intersect I with the ring K[a]
ideal J = eliminate(I,e*f*h);
J;
//-> J[1]=a3-32a2+192a
```

Hence, $_{U(\mathfrak{sl}_2) \otimes_K K[a]}\langle 4ef + h^2 - 2h - a \rangle \cap K[a] = \langle a(a-8)(a-24) \rangle$. From Example 1.9.26 we know, that $\{e, f\}$ does not generate an essential subalgebra. Let us see what happens if we try to intersect it with this subalgebra (i.e. "eliminate" h).

```
eliminate(I,h);
//-> ? no elimination possible: subalgebra is not admissible
//-> ? error occurred in line 13: 'eliminate(I,h);'
```

---

[16] that is, satisfying the ordering condition in Definition 1.9.1 and having the elimination property for $x_1, \ldots, x_r$.

Since a commutes with e, f and h we can eliminate a, that is intersect I with the subalgebra $U(\mathfrak{sl}_2) \subset U(\mathfrak{sl}_2) \otimes_K K[a]$. Moreover, we can intersect I with the subalgebra $K[h] \subset U(\mathfrak{sl}_2) \otimes_K K[a]$, being achieved by eliminating e, f and a.

```
eliminate(I,e*f*a);
//-> _[1]=h3-4h
```

SINGULAR Example 1.9.29 (No elimination ordering exists). Let $A = K\langle p, q \mid qp = pq + q^2 \rangle$ be a G–algebra for a fixed ordering $<$. In particular $q^2 < pq$ and hence $q < p$ holds. An elimination ordering for q requires that $q > p$ holds, which is a contradiction to the ordering condition for the G–algebra A.

```
ring s = 0,(p,q),dp;
ncalgebra(1,q^2); // setting the relation qp = pq + q^2
ideal I = p+q, p2+q2;
eliminate(I,q);
//-> Bad ordering at 1,2
//-> ? no elimination possible: ordering condition violated
//-> ? error occurred in STDIN line 4: 'eliminate(I,q);'
```

The first line of the error message says, that the ordering condition is violated for the relation between the 1st and 2nd variable. However, we can intersect the ideal with the subalgebra $K[q]$.

```
eliminate(I,p);
//-> _[1]=q2
```

1.9.4 Kernel of a Left Module Homomorphism

Let A be a GR–algebra. Consider a left A–module homomorphism

$$\phi: \quad A^m/U \longrightarrow A^n/V \quad e_i \longmapsto \Phi_i, \quad \Phi \in Mat(n \times m, A),$$

where $U \subset A^m$ and $V \subset A^n$. The kernel of a homomorphism ϕ can be computed with the procedure modulo (compare SINGULAR Example 2.1.26).

SINGULAR Example 1.9.30 (Kernel of module homomorphism). Let $A = U(\mathfrak{sl}_2)/I$, where the two–sided ideal I is generated by $\{e^2, f^2, h^2 - 1\}$. Let us study the endomorphisms $\tau : A \to A$.

```
LIB "ncalg.lib";
def A0 = makeUsl2();   setring A0;
option(redSB);   option(redTail);
ideal I = e2,f2,h2-1;
I = twostd(I);
print(matrix(I));       // ideal in a compact form
//-> h2-1,fh-f,eh+e,f2,2ef-h-1,e2
```

From the two–sided Gröbner basis of I, we can read off that A is 4–dimensional over K with basis $\{1, e, f, h\}$.

```
qring A  = I;          // we move to a GR--algebra A
ideal Ke = modulo(e,0);
Ke = std(Ke+std(0)); // normalize Ke w.r.t. the factor ideal
Ke;
//-> Ke[1]=h-1       // the kernel of e
//-> Ke[2]=e
ideal Kh = modulo(h-1,0);
Kh = std(Kh+std(0));
Kh;                    // the kernel of h-1
//-> Kh[1]=h+1
//-> Kh[2]=f
```

Computing with more endomorphisms, we get the following information on their kernels.
For non–zero $k \in \mathbb{K}$, $\ker(\tau : 1 \mapsto e + k) = \ker(\tau : 1 \mapsto f + k) = 0$.
For $k^2 \neq 1$, $\ker(\tau : 1 \mapsto h + k) = 0$.
$\ker(\tau : 1 \mapsto e) = \ker(\tau : 1 \mapsto h + 1) = {}_A\langle e, h - 1 \rangle$.
$\ker(\tau : 1 \mapsto f) = \ker(\tau : 1 \mapsto h - 1) = {}_A\langle f, h + 1 \rangle$.

1.9.5 Left Syzygy Modules

Let A be an associative algebra and A^n the canonical free module of rank n over A. A *left* (resp. *right*) *syzygy* of elements $f_1, \ldots, f_m$ from A^n is an m–tuple $(a_1, \ldots, a_m)$, $a_i \in A$ such that $\sum_{i=1}^m a_i f_i = 0$ (resp. $\sum_{i=1}^m f_i a_i = 0$). It can be shown, that the set of all left (resp. right) syzygies forms a left (resp. right) A–module.

We can view the elements $f_i \in A^n$ as columns of a matrix $F \in Mat(n \times m, A)$. It is convenient to view a single left syzygy, which is an element of A^m, as a column in a matrix. If the left syzygy module is generated by s elements, it can be represented by a matrix $S \in Mat(m \times s, A)$ and then $S^T \cdot F^T = 0$ holds where S^T and F^T denote the transposed matrices. Similar remarks apply to right syzygies.

The command $\mathtt{syz}$ computes the first (left) syzygy module of a given set of elements. The higher syzygy modules are defined as successive syzygies of syzygies etc.

SINGULAR Example 1.9.31 (Syzygies).

Consider the algebra $U_q'(\mathfrak{so}_3)$ (Example 1.9.30), specializing the quantum parameter Q at the primitive 6th root of unity. The corresponding minimal polynomial for the algebraic field extension is $Q^2 - Q + 1$.

```
LIB "ncalg.lib";
def R = makeQso3(3);
```

```
setring R;
option(redSB); option(redTail); // for reduced output
ideal K = x+y+z,y+z,z;
module S = syz(K);          // the (left) syzygy module of K
print(S);
//-> (Q-1),          (-Q+1)*z,    (Q-1)*y,
//-> (Q)*z+(-Q+1),(Q-1)*z+(Q),(Q)*x+(-Q+1)*y,
//-> y+(-Q)*z,      x+(-Q),      (-Q)*x-1
```

The columns of the above matrix generate the (left) syzygy module of K. Let us check the property ($S^T \cdot F^T = 0$) of a syzygy matrix from above.

```
ideal tst = ideal(transpose(S)*transpose(K));
print(matrix(tst));
//-> 0,0,0
```

It is easy to see, that the (left) Gröbner basis of the ideal K is $\{x, y, z\}$. Let us compute the first syzygy module of this set of generators.

```
K = x,y,z;
S = syz(K);
print(S);
//-> (Q-1),0,               0,
//-> (Q)*z,-z2-1,           (Q)*yz+(-Q)*x,
//-> y,    (-Q)*yz+(Q)*x,y2+1
```

There are quadratic terms in the generators. It is important to mention, that the command syz does not return a left Gröbner basis of the first syzygy module. However, one can force it to return a Gröbner basis by setting the option returnSB as follows

```
option(returnSB);
S = syz(K);
print(S);
//-> (Q-1),(-Q+1)*y,-z,
//-> (Q)*z,(-Q)*x,   (-Q+1),
//-> y,    1,        (-Q)*x
```

The latter generators of the syzygy module are linear.

1.9.6 Left Free Resolutions

Computing syzygy modules of a module M iteratively, we get a free left resolution (cf. Definition 2.4.10) of M. If M is a finitely presented A–module, where A is a G–algebra in n variables, we know that this process stops at most after n steps.

The commands `nres` and `mres` compute free left resolutions and the command `minres` minimizes a given resolution also in the non–commutative setting.

SINGULAR Example 1.9.32 (Resolution with `nres` and `minres`). In the algebra $A = U(\mathfrak{sl}_2)$ we consider the ideal I, generated by $\{e^2, f\}$. Its left Gröbner basis has been computed in example 1.9.19 and SINGULAR example 1.9.20. Now we want to compute a left free resolution of the latter set of generators.

```
LIB "ncalg.lib";
def A = makeUsl2();  setring A;
option(redSB);  option(redTail);
ideal I = e2,f;
ideal J = groebner(I);
resolution F = nres(J,0);
F;
//->  1      4      4      1
//-> A <-- A <-- A <-- A
//-> 0      1      2      3
//-> resolution not minimized yet
print(matrix(F[1]));          // F[1] is the left map
//-> f,h2+h,eh+e,e2
print(matrix(F[2]));          // F[2] is the middle map
//-> 0,    h2+5h+6,eh+3e,e2,
//-> 0,    -f,     -1,   0,
//-> e,    0,      -f,   -2,
//-> -h+3,0,       0,    -f
print(matrix(F[3]));          // F[3] is the right map
//-> f2,
//-> -e,
//-> ef,
//-> -fh+f
```

With the help of `minres`, we can minimize a given resolution.

```
resolution MF = minres(F);
print(matrix(MF[1]));
//-> f,e2
print(matrix(MF[2]));
//-> e3,       e2f2-6efh-6ef+6h2+18h+12,
//-> -ef-2h+6,-f3
print(matrix(MF[3]));
//-> f2,
//-> -e
```

Applying `mres` produces the same result as the two commands `nres` and `minres` together.

1.9.7 Betti Numbers in Graded GR–algebras

A graded G–algebra in n variables is characterized by the following property: $\forall\ 1 \leq i < j \leq n$ the polynomials $x_i x_j + d_{ij}$ are weighted homogeneous. A graded GR–algebra is a factor algebra of a G–algebra modulo a two–sided ideal T, whose two–sided Gröbner basis consists of weighted homogeneous polynomials.

The Betti numbers (see Definition 2.4.10) of graded objects in graded GR–algebras can be computed with the procedure betti.

SINGULAR Example 1.9.33 (Betti numbers).

```
ring r = 0,(x,d,q),dp;
matrix D[3][3];
D[1,2]=q^2;
ncalgebra(1,D);
ideal I = x,d,q;
option(redSB); option(redTail);
resolution R = mres(I,0);
R;
//-> 1      3      3      1
//-> r <-- r <-- r <-- r
//-> 0      1      2      3
print(betti(R),"betti");
//->          0      1      2      3
//-> ------------------------------
//->     0:   1      3      3      1
//-> ------------------------------
//-> total:   1      3      3      1
```

1.9.8 Gel'fand–Kirillov Dimension

The standard SINGULAR command dim computes the Krull dimension of a module or an ideal. In the non–commutative case, the Gel'fand–Kirillov dimension GKdim [180] plays a similar important role as the Krull dimension in the commutative case. Note, that for an ideal I in the polynomial ring $K[\mathbf{x}] = K[x_1, \ldots, x_n]$, the Krull dimension $\dim(I)$ and the Gel'fand–Kirillov dimension $\mathrm{GKdim}(I)$ of $K[\mathbf{x}]/I$ coincide.

The algorithm for computing the Gel'fand–Kirillov (or, shortly, GK) dimension [38, 154] uses Gröbner bases. It is implemented in the library gkdim.lib.

SINGULAR Example 1.9.34 (Gel'fand–Kirillov Dimension).

In this example we compute the Gel'fand–Kirillov dimensions of some modules which appeared in the examples 1.9.3, 1.9.22, 1.9.28, and 1.9.30 before.

```
LIB "gkdim.lib";
LIB "ncalg.lib";
def A = makeUsl(2); // set up U(sl_2)
setring A;
ideal I  = e2,f;
ideal LI = std(I);
GKdim(LI);
//-> 0
ideal TI = twostd(I);
GKdim(TI);
//-> 0
ideal Z = 4*e*f + h^2 - 4*h;
Z = std(Z);
GKdim(Z);
//-> 2
ring B = 0,(a),dp;
def  C = A + B;
setring C;
ideal I = e^3, f^3, h^3-4*h, 4*e*f+h^2-2*h - a;
I = std(I);
GKdim(I);
//-> 0
ideal J = eliminate(I,e*f*h);
GKdim(J);
//-> 3
setring A;
resolution F = nres(LI,0); // we computed it before
GKdim(F[1]); // this is LI itself
//-> 0
GKdim(F[2]);
//-> 3
GKdim(F[3]);
//-> 3
```

2. Modules

Module theory may, perhaps, best be characterized as linear algebra over a ring. While classical commutative algebra was basically ideal theory, modules are in the centre of modern commutative algebra as a unifying approach. Formally, the notion of a module over a ring is the analogue of the notion of a vector space over a field, in the sense that a module is defined by the same axioms, except that we allow ring elements as scalars and not just field elements. Just as vector spaces appear naturally as the solution sets of systems of linear equations over a field, modules appear as solution sets of such systems over a ring. However, contrary to vector spaces, not every module has a basis and this makes linear algebra over a ring much richer than linear algebra over a field.

This chapter contains the basic definitions and constructions in connection with modules with some emphasis on syzygies and free resolutions. Modules over special rings, such as graded rings and principal ideal domains, are treated in a special section.

Again, every construction is accompanied by concrete computational examples.

2.1 Modules, Submodules and Homomorphisms

This section contains the most elementary definitions and properties of modules. As far as the theory is completely analogous to that of vector spaces, we leave the verification of such results as exercises, with a few exceptions, in order to give some examples on how to proceed.

Definition 2.1.1. Let A be a ring. A set M, together with two operations $+ : M \times M \to M$ (*addition*) and $\cdot : A \times M \to M$ (*scalar multiplication*) is called A–*module* if

(1) $(M, +)$ is an abelian group.
(2) $(a + b) \cdot m = a \cdot m + b \cdot m$
    $a \cdot (m + n) = a \cdot m + a \cdot n$
    $(ab) \cdot m = a \cdot (b \cdot m)$
    $1 \cdot m = m$
    for all $a, b \in A$ and $m, n \in M$.

Example 2.1.2.

(1) Let A be a ring, then A is an A–module with the ring operation.
(2) If $A = K$ is a field, then A–modules are just K–vector spaces.
(3) Every abelian group is a $\mathbb{Z}$–module with scalar multiplication

$$n \cdot x := \underbrace{x + \cdots + x}_{n \text{ times}} .$$

(4) Let $I \subset A$ be an ideal, then I and A/I are A–modules with the obvious addition and scalar multiplication.
(5) Let A be a ring and $A^n = \{(x_1, \ldots, x_n) \mid x_i \in A\}$ the n–fold Cartesian product of A, then A^n is, in a canonical way, an A–module (with the component–wise addition and scalar multiplication).
(6) Let K be a field and $A = K[x]$ the polynomial ring in one variable x. An A–module M can be considered as a K–vector space M together with a linear map $\varphi : M \to M$ defined by $\varphi(m) := x \cdot m$ for all $m \in M$. On the other hand, given a K–vector space M and a linear map $\varphi : M \to M$, then we can give M the structure of a $K[x]$–module defining $x \cdot m := \varphi(m)$ for all $m \in M$.
(7) Let $\varphi : A \to B$ be a ring map, and set $a \cdot b := \varphi(a) \cdot b$ for $a \in A$ and $b \in B$. This defines an A–module structure on B. The ring B together with this structure is called an A–*algebra*.

Definition 2.1.3.

(1) Let M, N be A–modules. A map $\varphi : M \to N$ is called A–*module homomorphism* (or simply homomorphism) if, for all $a \in A$ and $m, n \in M$,
   a) $\varphi(am) = a\varphi(m)$,
   b) $\varphi(m + n) = \varphi(m) + \varphi(n)$.
   We also say that φ is A–*linear* or just *linear*. If $N = M$, then φ is called an *endomorphism*.
(2) The set of all A–module homomorphisms from M to N is denoted by $\mathrm{Hom}_A(M, N)$.
(3) A bijective A–module homomorphism is called *isomorphism* (the inverse is automatically a homomorphism, Exercise 2.1.12).
(4) M is called *isomorphic* to N, denoted by $M \cong N$, if there exists an isomorphism $M \to N$.
(5) If $\varphi : A \to B$ and $\psi : A \to C$ are two ring maps then a ring map $\alpha : B \to C$ is called an A–*algebra map* or a *homomorphism of A–algebras* if it is an A–module homomorphism, that is, if $\alpha \circ \varphi = \psi$.

Lemma 2.1.4. *Define two operations on* $\mathrm{Hom}_A(M, N)$ *by*

$$(\varphi + \psi)(m) := \varphi(m) + \psi(m),$$
$$(a\varphi)(m) := a \cdot \varphi(m).$$

Then $\mathrm{Hom}_A(M, N)$ *is an A–module.*

The proof is left as Exercise 2.1.1. The module

$$M^* := \operatorname{Hom}_A(M, A)$$

is called the *dual module* of M.

Lemma 2.1.5. *Let* M, N, L *be* A*–modules and* $\varphi : M \to N$ *be an* A*–module homomorphism. Define* $\phi : \operatorname{Hom}_A(N, L) \to \operatorname{Hom}_A(M, L)$ *by* $\phi(\lambda) := \lambda \circ \varphi$ *and* $\psi : \operatorname{Hom}_A(L, M) \to \operatorname{Hom}_A(L, N)$ *by* $\psi(\lambda) := \varphi \circ \lambda$. *Then* ϕ *and* ψ *are* A*–module homomorphisms.*

Proof. This is just a formal verification of the definition. To give an example, we show that ϕ is an A–module homomorphism. The proof for ψ is similar and left to the reader.

Let $\lambda, \mu \in \operatorname{Hom}_A(N, L)$, $a \in A$ and $m \in M$ arbitrary. Then

$$\big(\phi(a\lambda)\big)(m) = \big((a\lambda) \circ \varphi\big)(m) = (a\lambda)\big(\varphi(m)\big) = a \cdot \lambda\big(\varphi(m)\big)$$
$$= \big(a \cdot (\lambda \circ \varphi)\big)(m) = \big(a\phi(\lambda)\big)(m),$$
$$\big(\phi(\lambda + \mu)\big)(m) = \big((\lambda + \mu) \circ \varphi\big)(m) = (\lambda + \mu)\big(\varphi(m)\big) = \lambda\big(\varphi(m)\big) + \mu\big(\varphi(m)\big)$$
$$= (\lambda \circ \varphi)(m) + (\mu \circ \varphi)(m) = \big(\phi(\lambda) + \phi(\mu)\big)(m).$$

Since m is arbitrary, we have $\phi(a\lambda) = a \cdot \phi(\lambda)$ and $\phi(\lambda + \mu) = \phi(\lambda) + \phi(\mu)$. $\square$

Let us first consider a homomorphism $\varphi : A^n \to A^m$. If $\{e_1, \dots, e_n\}$ denotes the *canonical basis* of A^n (that is, $e_i = (0, \dots, 1, \dots, 0)$, 1 at place i), then any $x \in A^n$ is a unique linear combination $x = x_1 e_1 + \cdots + x_n e_n$, $x_i \in A$. Hence, $\varphi(e_i)$ has a unique representation as

$$\varphi(e_i) = \sum_{j=1}^{n} M_{ji} e_j, \qquad i = 1, \dots, n.$$

By linearity of φ we obtain, if we write x as a column vector,

$$\varphi(x) = \begin{pmatrix} M_{11} & \dots & M_{1n} \\ \vdots & & \vdots \\ M_{m1} & \dots & M_{mn} \end{pmatrix} \cdot \begin{pmatrix} x_1 \\ \vdots \\ x_n \end{pmatrix} = M \cdot x$$

where $M = (M_{ij})$ is an $m \times n$–matrix with entries in A. That is, φ is given by a matrix $M \in \operatorname{Mat}(m \times n, A)$ and any such M defines a homomorphism $A^n \to A^m$. We identify these homomorphisms with matrices. This is all as for vector spaces over a field. In particular, the addition and scalar multiplication of homomorphisms correspond to addition and scalar multiplication of matrices. The composition of linear maps corresponds to matrix multiplication.

SINGULAR Example 2.1.6 (matrix operations). A matrix in SIN-
GULAR is a matrix with polynomial entries, hence they can be defined only
when a basering is active. This applies also to matrices with numbers as entries (compare SINGULAR Examples 1.1.8 and 1.1.9). A matrix is filled with
entries from left to right, row by row, spaces are allowed.

```
ring A = 0,(x,y,z),dp;
matrix M[2][3] = 1, x+y, z2,          //2x3 matrix
                 x,   0, xyz;
matrix N[3][3] = 1,2,3,4,5,6,7,8,9; //3x3 matrix

M;                                    //lists all entries of M
//-> M[1,1] = 1
//-> M[1,2]=x+y
//-> M[1,3]=z2
//-> M[2,1]=x
//-> M[2,2]=0
//-> M[2,3]=xyz

print(N);                             //displays N as usual
//-> 1,2,3,                           //if the entries are small
//-> 4,5,6,
//-> 7,8,9

print(M+M);                           //addition of matrices
//-> 2,  2x+2y,2z2,
//-> 2x,0,      2xyz

print(x*N);
//-> x, 2x,3x,                        //scalar multiplication
//-> 4x,5x,6x,
//-> 7x,8x,9x

print(M*N);                           //multiplication of matrices
//-> 7z2+4x+4y+1,8z2+5x+5y+2,9z2+6x+6y+3,
//-> 7xyz+x,    8xyz+2x,    9xyz+3x

M[2,3];                               //access to single entry
//-> xyz
M[2,3]=37;                            //change single entry
print(M);
//-> 1,x+y,z2,
//-> x,0,  37
```

Further matrix operations are contained in the library matrix.lib. There is
a procedure pmat in inout.lib which formats matrices similarly to print,

but allows additional parameters, for example to show only the first terms of each entry for big matrices.

```
LIB "matrix.lib"; LIB "inout.lib";

print(power(N,3));                //exponentiation of matrices
//-> 468, 576, 684,
//-> 1062,1305,1548
//-> 1656,2034,2412


pmat(power((x+y+z)*N,3),15);//show first 15 terms of entries
//-> 468x3+1404x2y+1 576x3+1728x2y+1 684x3+2052x2y+2
//-> 1062x3+3186x2y+ 1305x3+3915x2y+ 1548x3+4644x2y+
//-> 1656x3+4968x2y+ 2034x3+6102x2y+ 2412x3+7236x2y+


matrix K = concat(M,N);   //concatenation
print(K);
//-> 1,x+y,z2,1,2,3,
//-> x,0,  37,4,5,6,
//-> 0,0,  0, 7,8,9


ideal(M);                //converts matrix to ideal
//-> _[1]=1                //same as 'flatten' from matrix.lib
//-> _[2]=x+y
//-> _[3]=z2
//-> _[4]=x
//-> _[5]=0
//-> _[6]=37


print(unitmat(5));       //5x5 unit matrix
//-> 1,0,0,0,0,
//-> 0,1,0,0,0,
//-> 0,0,1,0,0,
//-> 0,0,0,1,0,
//-> 0,0,0,0,1,
```

Besides matrices, there are integer matrices which do not need a ring. These are mainly used for bookkeeping or storing integer results. The operations are the same as for matrices.

```
intmat I[2][3]=1,2,3,4,5,6;
I;
//-> 1,2,3,
//-> 4,5,6
```

We construct now the matrices corresponding to the linear maps of Lemma 2.1.5.

SINGULAR Example 2.1.7 (maps induced by Hom).
Let $\varphi : A^n \to A^m$ be the linear map defined by the $m \times n$–matrix $M = (M_{ij})$ with entries in A, $\varphi(x) = M \cdot x$. We want to compute the induced map

$$\varphi^* : \mathrm{Hom}(A^m, A^s) \to \mathrm{Hom}(A^n, A^s) \,.$$

To do so, we identify $\mathrm{Hom}(A^n, A^s) = A^{sn}$ and $\mathrm{Hom}(A^m, A^s) = A^{ms}$, using Exercise 2.1.14.
Let $\{e_1, \dots, e_n\}$, $\{f_1, \dots, f_m\}$, $\{h_1, \dots, h_s\}$ denote the canonical bases of A^n, A^m, A^s, respectively. Then $\varphi(e_i) = \sum_{j=1}^m M_{ji} f_j$. Moreover, if $\{\sigma_{ij}\}$, $\{\kappa_{ij}\}$ are the bases of $\mathrm{Hom}(A^m, A^s)$, respectively $\mathrm{Hom}(A^n, A^s)$, defined by $\sigma_{ij}(f_\ell) = \delta_{j\ell} h_i$,[1] respectively $\kappa_{ij}(e_\ell) = \delta_{j\ell} h_i$, then

$$\varphi^*(\sigma_{ij})(e_k) = \sigma_{ij} \circ \varphi(e_k) = \sigma_{ij} \left(\sum_{\ell=1}^m M_{\ell k} f_\ell \right) = \sum_{\ell=1}^m M_{\ell k} \delta_{j\ell} h_i$$

$$= M_{jk} h_i = \sum_{\ell=1}^n M_{j\ell} \delta_{\ell k} h_i = \sum_{\ell=1}^n M_{j\ell} \kappa_{i\ell}(e_k) \,.$$

This implies $\varphi^*(\sigma_{ab}) = \sum_{c=1}^n M_{bc} \kappa_{ac}$. To obtain the $sn \times sm$–matrix R defining φ^*, we order the basis elements σ_{ij} and κ_{ij} as follows

$$\{\sigma_{11}, \sigma_{12}, \dots, \sigma_{1m}, \sigma_{21}, \sigma_{22}, \dots\dots, \sigma_{s1}, \sigma_{s2}, \dots, \sigma_{sm}\} \,,$$
$$\{\kappa_{11}, \kappa_{12}, \dots, \kappa_{1n}, \kappa_{21}, \kappa_{22}, \dots\dots, \kappa_{s1}, \kappa_{s2}, \dots, \kappa_{sn}\} \,,$$

and set, for $a, d = 1, \dots, s$, $b = 1, \dots, m$, $c = 1, \dots, n$,

$$i := (d-1)n + c, \quad j := (a-1)m + b \,.$$

Then

$$R_{ij} = \begin{cases} 0, & d \neq a, \\ M_{bc} & d = a \,. \end{cases}$$

We program this in a short procedure: given a matrix M, defining a homomorphism $A^n \to A^m$, and an integer s, the procedure kontraHom returns a matrix defining $R : \mathrm{Hom}(A^m, A^s) \to \mathrm{Hom}(A^n, A^s)$.

```
proc kontraHom(matrix M,int s)
{
    int n,m=ncols(M),nrows(M);
    int a,b,c;
    matrix R[s*n][s*m];
    for(b=1;b<=m;b++)
    {
        for(a=1;a<=s;a++)
        {
```

---

[1] Here $\delta_{j\ell}$ is the *Kronecker symbol* ($\delta_{j\ell} = 0$ if $j \neq \ell$ and $\delta_{jj} = 1$).

```
      for(c=1;c<=n;c++)
      {
         R[(a-1)*n+c,(a-1)*m+b]=M[b,c];
      }
    }
  }
  return(R);
}
```

Let us try an example.

```
ring A=0,(x,y,z),dp;
matrix M[3][3]=1,2,3,
               4,5,6,
               7,8,9;

print(kontraHom(M,2));
//-> 1,4,7,0,0,0,
//-> 2,5,8,0,0,0,
//-> 3,6,9,0,0,0,
//-> 0,0,0,1,4,7,
//-> 0,0,0,2,5,8,
//-> 0,0,0,3,6,9
```

This procedure is contained as `kontrahom` in `homolog.lib`. Note that for $s = 1$, the dual map, that is, the transposed matrix, is computed.

Similarly, we can compute the map

$$\varphi_* : \mathrm{Hom}(A^s, A^n) \to \mathrm{Hom}(A^s, A^m).$$

If $\{\sigma_{ij}\}$ and $\{\kappa_{ij}\}$ are defined as before as bases of $\mathrm{Hom}(A^s, A^n)$, respectively $\mathrm{Hom}(A^s, A^m)$, then one checks that $\varphi_*(\sigma_{ab}) = \sum_{c=1}^m M_{ca}\kappa_{cb}$.

We obtain the following procedure: given $M : A^n \to A^m$ and s, `kohom` returns $R : \mathrm{Hom}(A^s, A^n) \to \mathrm{Hom}(A^s, A^m)$.

```
proc kohom(matrix M,int s)
{
    int n,m=ncols(M),nrows(M);
    int a,b,c;
    matrix R[s*m][s*n];
    for(b=1;b<=s;b++)
    {
        for(a=1;a<=m;a++)
        {
            for(c=1;c<=n;c++)
            {
                R[(a-1)*s+b,(c-1)*s+b]=M[a,c];
```

```
            }
         }
      }
      return(R);
}
```

As an example use the matrix defined above.

```
print(kohom(M,2));
//-> 1,0,2,0,3,0,
//-> 0,1,0,2,0,3,
//-> 4,0,5,0,6,0,
//-> 0,4,0,5,0,6,
//-> 7,0,8,0,9,0,
//-> 0,7,0,8,0,9
```

This procedure is contained as `kohom` in `homolog.lib`.

Definition 2.1.8. Let M be an A–module. A non–empty subset $N \subset M$ is called a *submodule* of M if, for all $m, n \in N$ and $a \in A$,

(1) $m + n \in N$,
(2) $a \cdot m \in N$.

Note that every submodule of an A–module is itself an A–module.

Remark 2.1.9. Every element of A^n is represented as

$$x = x_1 e_1 + \cdots + x_n e_n = \begin{pmatrix} x_1 \\ \vdots \\ x_n \end{pmatrix}, \quad x_i \in A,$$

that is, as a linear combination in terms of the canonical basis or as a (column, respectively row) vector. Both representations are used in SINGULAR. An element of A^n is called a *vector*.

SINGULAR Example 2.1.10 (submodules of A^n). We shall explain how to declare submodules of A^n in SINGULAR, where A is any ring of SINGULAR. In the same way as ideals $I \subset A$ are given by elements of A as generators, submodules are given by vectors in A^n as generators. The canonical basis elements e_i of A^n are denoted by `gen(i)` in SINGULAR.

```
ring    A=0,(x,y,z),dp;
module M=[xy-1,z2+3,xyz],[y4,x3,z2];
M;
//-> M[1]=xyz*gen(3)+xy*gen(1)+z2*gen(2)+3*gen(2)-gen(1)
//-> M[2]=y4*gen(1)+x3*gen(2)+z2*gen(3)
```

M is the submodule of $\mathbb{Q}[x, y, z]^3$ generated by the two (column) vectors $(xy - 1, z^2 + 3, xyz)$ and (y^4, x^3, z^2). The output is given as linear combination of the canonical basis. SINGULAR understands both formats as input.

```
ideal  I=x2+y2+z2;
qring  Q=std(I);          //create quotient ring mod I
module M=fetch(A,M);      //map M from A to Q
```

Here we consider M as a submodule in $(\mathbb{Q}[x, y, z]/\langle x^2 + y^2 + z^2 \rangle)^3$.

Definition 2.1.11. Let $\varphi : M \to N$ be an A–module homomorphism. The *kernel* of φ, $\mathrm{Ker}(\varphi)$ is defined by $\mathrm{Ker}(\varphi) := \{m \in M \mid \varphi(m) = 0\}$. The *image* of φ, $\mathrm{Im}(\varphi)$, is defined by $\mathrm{Im}(\varphi) := \{\varphi(m) \mid m \in M\}$.

Lemma 2.1.12. $\mathrm{Ker}(\varphi)$ *and* $\mathrm{Im}(\varphi)$ *are submodules of* M, *respectively* N.

The easy proof is left as Exercise 2.1.6.

SINGULAR Example 2.1.13 (kernel and image of a module homomorphism).

```
ring A=0,(x,y,z),(c,dp);
matrix M[2][3]=x,xy,z,x2,xyz,yz;
print(M);
//->x, xy,  z,
//->x2,xyz,yz
```

To compute the kernel of a module homomorphism means to solve a system of linear equations over a ring. The `syz` command, which is based on Gröbner basis computations is, hence, a generalization of Gaussian elimination from fields to rings (see Section 2.5).

```
module Ker=syz(M);
Ker;
//-> Ker[1]=[y2z-yz2,xz-yz,-x2y+xyz]
```

For the image, there is nothing to compute. The column vectors of M generate the image.

```
module Im=M[1],M[2],M[3];
Im;
//-> Im[1]=[x,x2]
//-> Im[2]=[xy,xyz]
//-> Im[3]=[z,yz]
```

Definition 2.1.14.

(1) Let M be an A–module and $N \subset M$ be a submodule. We define the *quotient module* or *factor module* M/N by

$$M/N := \{m + N \mid m \in M\}.$$

That is, M/N is the set of equivalence classes of elements of M, where $m, n \in M$ are equivalent if $m - n \in N$. An equivalence class is denoted by $m + N$ or by $[m]$. Each element in the class $m + N$ is called a *representative* of the class.

(2) Let $\varphi : M \to N$ be an A–module homomorphism, then

$$\mathrm{Coker}(\varphi) := N/\mathrm{Im}(\varphi)$$

is called the *cokernel* of φ.

Lemma 2.1.15. *With the canonical operations, by choosing representatives,*

$$(m + N) + (n + N) := (m + n) + N, \quad a \cdot (m + N) := am + N$$

the set M/N is an A–module. N, the equivalence class of $0 \in M$ is the 0–element in M/N. The map $\pi : M \to M/N$, $\pi(m) := m + N$ is a surjective A–module homomorphism.

The proof is left as Exercise 2.1.7. We just show that the addition is well–defined (independent of the chosen representatives). If $(m' + N)$ and $(n' + N)$ are other representatives, then $m - m', n - n' \in N$.

Hence, $(m + n) - (m' + n') = (m - m') + (n - n') \in N$, which shows that $(m + n) + N = (m' + n') + N$.

Proposition 2.1.16. *Let $\varphi : M \to N$ be an A–module homomorphism, then*

$$\mathrm{Im}(\varphi) \cong M/\mathrm{Ker}(\varphi).$$

Proof. Define a map $\lambda : M/\mathrm{Ker}(\varphi) \to \mathrm{Im}(\varphi)$ by $\lambda(m + \mathrm{Ker}(\varphi)) := \varphi(m)$. It is easy to see that λ is well–defined, that is, does not depend on the choice of the representative m. λ is surjective by definition. To see that λ is injective, let $\lambda(m + \mathrm{Ker}(\varphi)) = 0$. That is, $\varphi(m) = 0$, and, hence, $m \in \mathrm{Ker}(\varphi)$. But then $m + \mathrm{Ker}(\varphi) = \mathrm{Ker}(\varphi)$ which is the 0–element in $M/\mathrm{Ker}(\varphi)$. One can also easily check that λ is an A–module homomorphism. □

Corollary 2.1.17. *Let $L \supset M \supset N$ be A–modules, then*

$$(L/N)/(M/N) \cong L/M.$$

Proof. The inclusion $N \subset M$ induces a homomorphism $\pi : L/N \to L/M$ of A–modules. Obviously, π is surjective and $\mathrm{Ker}(\pi) = M/N$. Therefore, the corollary is a consequence of Proposition 2.1.16. □

Definition 2.1.18.

(1) Let M be an A–module and $M_i \subset M$ be submodules, $i \in I$. We define the *sum* of the M_i by

$$\sum_{i \in I} M_i := \left\{ \sum_{i \in I} m_i \ \middle| \ m_i \in M_i, \ m_i \neq 0 \text{ only for finitely many } i \right\}.$$

(2) Let $J \subset A$ be an ideal and M an A–module. We define JM by

$$JM := \left\{ \sum_{i \in I} a_i m_i \ \middle| \ I \text{ finite}, \quad a_i \in J, \quad m_i \in M \right\}.$$

(3) An A–module M is called *finitely generated* if $M = \sum_{i=1}^{n} A \cdot m_i$ for suitable $m_1, \ldots, m_n \in M$. We then write $M = \langle m_1, \ldots, m_n \rangle$, and $m_1, \ldots, m_n$ are called *generators* of M. A module generated by one element is called a *cyclic module*.

(4) Let M be an A–module. The *torsion submodule* $\mathrm{Tors}(M)$ is defined by

$$\mathrm{Tors}(M) := \{ m \in M \mid \exists \text{ a non–zerodivisor } a \in A \text{ with } am = 0 \}.$$

A module M is called *torsion free* if $\mathrm{Tors}(M) = 0$. M itself is called a *torsion module* if $\mathrm{Tors}(M) = M$.

(5) Let $N, P \subset M$ be submodules, then the *quotient* $N : P$ is defined by

$$N : P := N :_A P := \{ a \in A \mid aP \subset N \}.$$

In particular, the *annihilator* of P is

$$\mathrm{Ann}(P) := \mathrm{Ann}_A(P) := \langle 0 \rangle : P = \{ a \in A \mid aP = 0 \}.$$

Note that the module quotient is a generalization of the ideal quotient.

(6) There is still another quotient. Let $I \subset A$ be an ideal, then the *quotient of P by I in M* is

$$P :_M I := \{ m \in M \mid I \cdot m \subset P \}.$$

(7) Let M_i, $i \in I$, be A–modules. The *direct sum* $\bigoplus_{i \in I} M_i$ is defined by

$$\bigoplus_{i \in I} M_i = \{ (m_i)_{i \in I} \mid m_i \in M_i, \ m_i \neq 0 \text{ for only finitely many } i \}.$$

The *direct product* $\prod_{i \in I} M_i$ is defined by

$$\prod_{i \in I} M_i = \{ (m_i)_{i \in I} \mid m_i \in M_i \}.$$

Note that for a finite index set $I = \{1, \ldots, n\}$ the direct sum and the direct product coincide and are denoted as

$$M_1 \oplus \cdots \oplus M_n.$$

(8) Let M be an A–module. M is called *free* if $M \cong \bigoplus_{i \in I} A$. The cardinality of the index set I is called the *rank* of M. A subset $S \subset M$ is called a *basis* of M if every $m \in M$ can be written in a unique way as a finite linear combination $m = a_1 m_1 + \cdots + a_n m_n$ with $m_i \in S$ and $a_i \in A$, for some n (depending on m).

If A is a field the modules are vector spaces and every module is free. In general, this is not true. The $\mathbb{Z}$–module $\mathbb{Z}/\langle 2 \rangle$ is not free. More generally, a module M with $\mathrm{Tors}(M) \neq 0$ cannot be free (Exercise 2.1.9).

Lemma 2.1.19. *The sum of submodules of an A–module, the product of an ideal with an A–module, the direct sum and the direct product of A–modules are again A–modules. The module quotient of two submodules of an A–module is an ideal in A. The quotient of a submodule by an ideal is a submodule of M. The torsion module $\mathrm{Tors}(M)$ is a submodule of M.*

The proof is left as Exercise 2.1.8.

SINGULAR Example 2.1.20 (sum, intersection, module quotient). The sum of two modules is generated by the union of the generators, the "+" lets SINGULAR simplify the union by deleting 0's and identical generators.

```
ring A=0,(x,y,z),(c,dp);//the ordering (c,..) has the effect
module M=[xy,xz],[x,x]; //that the vectors are internally
module N=[y2,z2],[x,x]; //represented component-wise.
M+N;
//-> _[1]=[xy,xz]        //the output is, as the internal
//-> _[2]=[y2,z2]        //representation, component-wise
//-> _[3]=[x,x]
```

intersect and quotient require standard basis computations.

```
intersect(M,N);          //intersection, see Section 2.8.3
//-> _[1]=[x,x]
//-> _[2]=[xy2,xz2]


quotient(M,N);           //M:N, see Section 2.8.4
//-> _[1]=x


quotient(N,M);
//-> _[1]=y+z


qring Q=std(x5);         //quotient ring Q[x,y,z]/<x5>
module M=fetch(A,M);     //map M from A to Q
module Null;             //creates zero-module
M;
//-> M[1]=[xy,xz]
//-> M[2]=[x,x]
```

```
Null;                      //the zero-module
//-> Null[1]=0


quotient(Null,M);          //the annihilator of M
//-> _[1]=x4
```

Proposition 2.1.21. *Let M be an A-module and $N_1, N_2 \subset M$ submodules, then $(N_1 + N_2)/N_1 \cong N_2/N_1 \cap N_2$.*

Proof. The inclusion $N_2 \subset N_1 + N_2$ induces an A-module homomorphism $\pi : N_2 \to (N_1 + N_2)/N_1$. Obviously π is surjective and $\mathrm{Ker}(\pi) = N_1 \cap N_2$. Now we can use Proposition 2.1.16.   □

Lemma 2.1.22. *Let M be an A-module. M is finitely generated if and only if $M \cong A^n/L$ for a suitable $n \in \mathbb{N}$ and a suitable submodule $L \subset A^n$. Equivalently, there exists a surjective homomorphism $\varphi : A^n \twoheadrightarrow M$.*

Proof. Assume that $M = \langle x_1, \ldots, x_n \rangle$ and consider the A-module homomorphism $\varphi : A^n \to M$ defined by $\varphi(a_1, \ldots, a_n) = \sum_{i=1}^n a_i x_i$. φ is surjective because $x_1, \ldots, x_n$ generate M. Let $L := \mathrm{Ker}(\varphi)$ then Proposition 2.1.16 implies that $M \cong A^n/L$.

Now assume $M \cong A^n/L$ for some submodule $L \subset A^n$. Let $\{e_1, \ldots, e_n\}$ be a basis of A^n, then the preimages of $x_1 := e_1 + L, \ldots, x_n := e_n + L$ generate M.   □

Let M be a finitely generated A-module and $M \cong A^n/L$ for some submodule $L \subset A^n$. If L is also finitely generated, then $L \cong A^m/N$ for a suitable submodule $N \subset A^m$ and we have homomorphisms $A^m \twoheadrightarrow A^m/N \cong L \subset A^n \twoheadrightarrow A^n/L \cong M$. Therefore, M is isomorphic to the cokernel of a homomorphism $\varphi : A^m \to A^n$, the composition $A^n \to A^m/N \cong L \subset A^n$. Fixing bases in A^m and A^n, φ is given by an $n \times m$–matrix, which we also denote by φ.

Definition 2.1.23. Let M be an A-module. M is called of *finite presentation* if there exists an $n \times m$–matrix φ such that M is isomorphic to the cokernel of the map $A^m \xrightarrow{\varphi} A^n$. φ is called a *presentation matrix* of M. We write $A^m \xrightarrow{\varphi} A^n \to M \to 0$ to denote a presentation of M.

Constructive module theory is concerned with modules of finite presentation, that is, with modules which can be given as the cokernel of some matrix. All operations with modules are then represented by operations with the corresponding presentation matrices. We shall see below (Proposition 2.1.29) that every finitely generated module over a Noetherian ring is finitely presented. As polynomial rings and localizations thereof are Noetherian (Lemma 1.4.8), every finitely generated module over these rings is of finite presentation.

We shall see how we can actually compute with finitely generated modules over the rings $K[x_1, \ldots, x_n]_>$ (for any monomial ordering $>$, cf. Chapter 1), or, more generally, over quotient rings A of those. To start with, we must know how to represent a module within SINGULAR. Since any finitely generated module over $K[x]_>$ has a presentation matrix with polynomial entries, and, as we know how to define polynomial matrices, we can define arbitrary finitely generated A–modules in SINGULAR by giving a polynomial presentation matrix. In fact, for arbitrary modules, there is no other way, we have to know a presentation matrix.

However, submodules of A^n (which is a special class, since they are, for example, torsion free, see Exercise 2.1.8) can be given just by a set of generators, that is, by m vectors of A^n. Given the generators, we can *compute* the presentation matrix by using the syz command, which is based on Gröbner bases (Section 2.5). Giving m vectors in A^n is, up to numbering, the same as giving an $n \times m$–matrix over A. Since we can only give ordered lists of generators, this is indeed the same.

Thus, defining a matrix or a module in SINGULAR can be interpreted in two ways: either as the presentation matrix of the factor module of A^n or as the submodule of A^n generated by the columns of the matrix.

SINGULAR Example 2.1.24 (submodules, presentation of a module).
SINGULAR distinguishes between modules and matrices. For matrices see Example 2.1.6. A module is always given by generators, either with brackets, or as a linear combination of the canonical generators gen(1),...,gen(n) of A^n, where only the non–zero coefficients have to be given. The last (sparse) representation is internally used. Matrices, however, are represented internally non–sparse, therefore, it is recommended to use modules instead of matrices for large input.

SINGULAR assumes a module to be a submodule of A^n if, for some generator, gen(n) has a non–zero coefficient, and if, for each generator, the coefficients of gen(i) are zero for $i > n$.

```
ring A = 0,(x,y,z),dp;
module N = [xy,0,yz],[0,xz,z2];  //submodule of A^3,
N;                               //2 generators
//-> N[1]=xy*gen(1)+yz*gen(3)    //output in sparse format
//-> N[2]=xz*gen(2)+z2*gen(3)

LIB "inout.lib";                 //library for formatting output
show(N);                         //shows the generators as vectors
//-> // module, 2 generator(s)
//-> [xy,0,yz]
//-> [0,xz,z2]

print(N);                        //the corresponding matrix
//-> xy,0,
```

```
//-> 0,xz,
//-> yz,z2
```

Modules may be added and multiplied with a polynomial or an ideal. Not that addition of modules means, as for ideals, the sum of modules, which is quite different from the sum of matrices.

```
show(N+x*N);
//-> [xy,0,yz]
//-> [0,xz,z2]
//-> [x2y,0,xyz]
//-> [0,x2z,xz2]
```

There are type conversions from matrix to module, and from module to matrix: `module(matrix)` creates a module with generators the columns of the matrix, `matrix(module)` creates a matrix with columns the generators of the module. `module(matrix(module))` restores the original module and `matrix(module(matrix))` restores the original matrix.

```
module M = [xy,yz],[xz,z2]; //submodule of A^2
matrix MM = M;               //automatic type conversion,
MM;                          // same as matrix MM=matrix(M);
//-> MM[1,1]=xy
//-> MM[1,2]=xz
//-> MM[2,1]=yz
//-> MM[2,2]=z2
```

The operations on modules are operations as submodules. However, as explained above, M (or, better, `matrix(M)`) can be considered as the presentation matrix

$$A^2 \xrightarrow{\left(\begin{smallmatrix} xy & xz \\ yz & z^2 \end{smallmatrix}\right)} A^2$$

of the module A^2/M.

On the other hand, if M is considered as a submodule of A^2, then we can compute a presentation as

```
module K = syz(M);          //computes the kernel of M
show(K);
//-> K[1]=[-z,y]
```

This means that $A \xrightarrow{\left(\begin{smallmatrix} -z \\ y \end{smallmatrix}\right)} A^2 \to M \to 0$ is a presentation of M.

Lemma 2.1.25. *Let M, N be two A–modules with presentations*

$$A^m \xrightarrow{\varphi} A^n \xrightarrow{\pi} M \to 0 \text{ and } A^r \xrightarrow{\psi} A^s \xrightarrow{\kappa} N \to 0.$$

(1) Let $\lambda : M \to N$ be an A–module homomorphism, then there exist A–module homomorphisms $\alpha : A^m \to A^r$ and $\beta : A^n \to A^s$ such that the following diagram commutes:

$$
\begin{array}{ccccccc}
A^m & \xrightarrow{\varphi} & A^n & \xrightarrow{\pi} & M & \longrightarrow & 0 \\
{\scriptstyle\alpha}\downarrow & & {\scriptstyle\beta}\downarrow & & {\scriptstyle\lambda}\downarrow & & \\
A^r & \xrightarrow{\psi} & A^s & \xrightarrow{\kappa} & N & \longrightarrow & 0
\end{array}
\qquad (2.1)
$$

that is, $\beta \circ \varphi = \psi \circ \alpha$ and $\lambda \circ \pi = \kappa \circ \beta$.

(2) Let $\beta : A^n \to A^s$ be an A–module homomorphism such that $\beta\big(\mathrm{Im}(\varphi)\big) \subset \mathrm{Im}(\psi)$. Then there exist A–module homomorphisms $\alpha : A^m \to A^r$ and $\lambda : M \to N$ such that the corresponding diagram (as in (2.1)) commutes.

Proof. (1) Let $\{e_1, \ldots, e_n\}$ be a basis of A^n and choose $x_i \in A^s$ such that $\kappa(x_i) = \lambda \circ \pi(e_i)$, $i = 1, \ldots, n$. We define $\beta\left(\sum_{i=1}^n a_i e_i\right) = \sum_{i=1}^n a_i x_i$. Obviously, β is an A–module homomorphism and $\lambda \circ \pi = \kappa \circ \beta$. Let $\{f_1, \ldots, f_m\}$ be a basis of A^m. Then $\kappa \circ \beta \circ \varphi(f_i) = \lambda \circ \pi \circ \varphi(f_i) = 0$. Therefore, there exist $y_i \in A^r$ such that $\psi(y_i) = \beta \circ \varphi(f_i)$. We define $\alpha\left(\sum_{i=1}^n b_i f_i\right) = \sum_{i=1}^n b_i y_i$. Again α is an A–module homomorphism and $\psi \circ \alpha = \beta \circ \varphi$.

(2) Define $\lambda(m) = \kappa \circ \beta(f)$, for some $f \in A^n$ with $\pi(f) = m$. This definition does not depend on the choice of f, because $\mathrm{Ker}(\pi) = \mathrm{Im}(\varphi)$ and $\beta\big(\mathrm{Im}(\varphi)\big) \subset \mathrm{Im}(\psi) = \mathrm{Ker}(\kappa)$. Obviously, λ is an A–module homomorphism satisfying $\lambda \circ \pi = \kappa \circ \beta$. We can define α as in (1). $\qquad\square$

SINGULAR Example 2.1.26 (computation of Hom).

With the notations of Lemma 2.1.25 we obtain the following commutative diagram:

$$
\begin{array}{ccccc}
\mathrm{Hom}(M,N) & \longrightarrow & \mathrm{Hom}(A^n,N) & \xrightarrow{\varphi_N^*} & \mathrm{Hom}(A^m,N) \\
& & \uparrow & & \uparrow \\
& & \mathrm{Hom}(A^n,A^s) & \xrightarrow{\varphi^*} & \mathrm{Hom}(A^m,A^s) \\
& & \uparrow{\scriptstyle j} & & \uparrow{\scriptstyle i} \\
& & \mathrm{Hom}(A^n,A^r) & \longrightarrow & \mathrm{Hom}(A^m,A^r),
\end{array}
$$

the maps being defined as in Lemma 2.1.5. In particular, $\varphi_N^*(\sigma) = \sigma \circ \varphi$, $\varphi^*(\sigma) = \sigma \circ \varphi$, $i(\sigma) = \psi \circ \sigma$, and $j(\sigma) = \psi \circ \sigma$. Lemma 2.1.25 and Proposition 2.4.3 below imply that

$$
\mathrm{Hom}(M,N) = \mathrm{Ker}(\varphi_N^*) \cong \varphi^{*-1}\big(\mathrm{Im}(i)\big) \big/ \mathrm{Im}(j).
$$

Using the SINGULAR built–in command `modulo`, which is explained below, we
have (identifying, as before, $\mathrm{Hom}(A^n, A^s) = A^{sn}$ and $\mathrm{Hom}(A^m, A^s) = A^{ms}$)

$$D := \varphi^{*-1}\big(\mathrm{Im}(i)\big) = \mathrm{Ker}\big(A^{ns} \xrightarrow{\overline{\varphi^*}} A^{ms}/\mathrm{Im}(i)\big) = \text{modulo}\,(\varphi^*, i)\,,$$

which is given by a $ns \times k$–matrix with entries in A, and we can compute
$\mathrm{Hom}(M, N)$ as

$$\varphi^{*-1}\big(\mathrm{Im}(i)\big)/\mathrm{Im}(j) = A^k/\mathrm{Ker}\big(A^k \xrightarrow{\overline{D}} A^{ns}/\mathrm{Im}(j)\big) = A^k/\text{modulo}\,(D, j)\,.$$

Finally, we obtain the following procedure with $F = \varphi^*$, $B = i$, $C = j$.

```
proc Hom(matrix M, matrix N)
{
  matrix F = kontraHom(M,nrows(N));
  matrix B = kohom(N,ncols(M));
  matrix C = kohom(N,nrows(M));
  matrix D = modulo(F,B);
  matrix E = modulo(D,C);
  return(E);
}
```

Here is an example.

```
ring A=0,(x,y,z),dp;
matrix M[3][3]=1,2,3,
              4,5,6,
              7,8,9;
matrix N[2][2]=x,y,
              z,0;

print(Hom(M,N));        //a 6x6 matrix
//-> 0,0,0,0,y,x,
//-> 0,0,0,0,0,z,
//-> 1,0,0,0,0,0,
//-> 0,1,0,0,0,0,
//-> 0,0,1,0,0,0,
//-> 0,0,0,1,0,0
```

We explain the `modulo` command: let the matrices $M \in \mathrm{Mat}(m \times n, A)$, respectively $N \in \mathrm{Mat}(m \times s, A)$, represent linear maps

$$A^n \xrightarrow{M} A^m$$
$$\Big\uparrow{\scriptstyle N}$$
$$A^s\,.$$

Then $\texttt{modulo}(M, N) = \operatorname{Ker}\big(A^n \xrightarrow{\overline{M}} A^m / \operatorname{Im}(N)\big)$, where $\overline{M}$ is the map induced by M; more precisely, $\texttt{modulo}(M, N)$ returns a set of vectors in A^n which generate $\operatorname{Ker}(\overline{M})$ [2]. Hence, $\texttt{matrix(modulo(M,N))}$ is a presentation matrix for the quotient $(\operatorname{Im}(M) + \operatorname{Im}(N))/\operatorname{Im}(N)$. The computation is explained in SINGULAR Example 2.8.9.

Definition 2.1.27. Let M be an A–module. Then M is called *Noetherian* if every submodule $N \subset M$ is finitely generated.

Note that, in particular, a ring A is a Noetherian A–module if and only if it is a Noetherian ring (cf. Definition 1.3.4).

Lemma 2.1.28.

(1) Submodules and quotient modules of Noetherian modules are Noetherian.
(2) Let $N \subset M$ be A–modules, then M is Noetherian if and only if N and M/N are Noetherian.
(3) Let M be an A–module, then the following properties are equivalent:
  a) M is Noetherian.
  b) Every ascending chain of submodules

$$M_1 \subset M_2 \subset \ldots \subset M_k \subset \ldots$$

  becomes stationary.
  c) Every non–empty set of submodules of M has a maximal element (with regard to inclusion).

The proof is left as Exercise 2.1.9 (compare Proposition 1.3.6).

The following proposition relates Noetherian and finitely generated modules.

Proposition 2.1.29. *Let A be a Noetherian ring and M be a finitely generated A–module, then M is a Noetherian A–module.*

Proof. Using Lemma 2.1.22 and Lemma 2.1.28 we may assume that $M = A^n$ and prove the statement using induction on n. For $n = 1$ the statement follows by assumption. Let $n \geq 2$ and consider the projection $\pi : A^n \to A^{n-1}$, $(a_1, \ldots, a_n) \mapsto (a_1, \ldots, a_{n-1})$. Clearly, $\operatorname{Ker}(\pi) = \{(0, \ldots, 0, a) \mid a \in A\} \cong A$, and $A^{n-1} = A^n / \operatorname{Ker}(\pi)$. Hence, the result follows from Lemma 2.1.28 (2) and the induction hypothesis. $\qquad\square$

Lemma 2.1.30 (Nakayama). *Let A be a ring and $I \subset A$ an ideal which is contained in the Jacobson radical of A. Let M be a finitely generated A–module and $N \subset M$ a submodule such that $M = IM + N$. Then $M = N$. In particular, if $M = IM$, then $M = 0$.*

---

[2] Using automatic type conversion, we can apply the $\texttt{modulo}$–command to modules as well as to matrices.

Proof. By passing to the quotient module it is enough to prove the lemma in the case $N = \langle 0 \rangle$. Assume $M \neq \langle 0 \rangle$ and let $m_1, \ldots, m_n$ be a minimal system of generators of M. Since $m_n \in M = IM$, we can choose $a_1, \ldots, a_n \in I$ such that $m_n = \sum_{i=1}^{n} a_i m_i$. This implies $(1 - a_n) m_n = \sum_{i=1}^{n-1} a_i m_i$.

By Exercise 1.4.4, $(1 - a_n)$ is a unit in A, and, therefore, $m_1, \ldots, m_{n-1}$ generate M, which is a contradiction to the minimality of the chosen system of generators.   □

Corollary 2.1.31. *Let $(A, \mathfrak{m})$ be a local ring and M a finitely generated A-module. Let $m_1, \ldots, m_n \in M$ such that their classes form a system of generators for the $A/\mathfrak{m}$–vector space $M/\mathfrak{m}M$. Then $m_1, \ldots, m_n$ generate M.*

Proof. Let $N := \sum_i A m_i$ and consider the canonical map $N \to M \to M/\mathfrak{m}M$. This map is surjective, which implies $N + \mathfrak{m}M = M$. Thus, the corollary is a consequence of Lemma 2.1.30.   □

Remark 2.1.32. With the assumptions of Corollary 2.1.31, $\{m_1, \ldots, m_n\}$ is a *minimal system of generators* of M if and only if their classes form a basis of $M/\mathfrak{m}M$, and then n is the dimension of the $A/\mathfrak{m}$–vector space $M/\mathfrak{m}M$.

Definition 2.1.33. Let $(A, \mathfrak{m})$ be a local ring and M an A–module. A presentation $A^m \xrightarrow{\varphi} A^n \to M \to 0$ of M is called a *minimal presentation* if $n = \dim_{A/\mathfrak{m}}(M/\mathfrak{m}M)$.

Note that $n = \dim_{A/\mathfrak{m}}(M/\mathfrak{m}M)$ if and only if $\varphi(A^m) \subset \mathfrak{m}A^n$, that is, the entries of the presentation matrix are in $\mathfrak{m}$ (Exercise 2.1.17).

How can we make a presentation φ of a module M minimal if it is not? If an entry φ_{ij} of φ is a unit, we can perform elementary row and column operations to produce a matrix $\tilde{\varphi}$ which has, except at position (i, j), only zeros in row i and column j. Elementary row, respectively column, operations mean that we multiply φ from the left, respectively right, with an invertible matrix. Hence $\mathrm{Coker}(\varphi) \cong \mathrm{Coker}(\tilde{\varphi})$, that is, $\tilde{\varphi}$ is a presentation matrix of a module isomorphic to M. But from $\tilde{\varphi}$ we can delete the ith row and the jth column without changing the cokernel.

Doing this, successively, with every entry which is a unit, we obtain a minimal presentation of (a module isomorphic to) M.

Note that this is nothing else but a Gauß reduction with a pivot element being a unit. If A is a field, then every element $\neq 0$ is a unit and we can carry out a complete Gauß reduction. The SINGULAR command prune produces a minimal presentation matrix.

SINGULAR Example 2.1.34 (minimal presentations, prune).

```
ring A=0,(x,y,z),ds;  //local ring with max. ideal <x,y,z>
module M=[0,xy-1,xy+1],[y,xz,xz];
print(M);
//-> 0,    y,
```

```
//-> -1+xy,xz,            //we have units in the first column
//-> 1+xy, xz

print(prune(M));
//-> -y+xy2,
//-> -2xz
```

Let $A = \mathbb{Q}[x, y, z]_{\langle x,y,z \rangle}$ and $N = A^3/M$, where M denotes the submodule of A^3 generated by the vectors $(0, xy - 1, xy + 1)$, (y, xz, xz). Then

$$A^2 \xrightarrow{\begin{pmatrix} 0 & y \\ xy-1 & xz \\ xy+1 & xz \end{pmatrix}} A^3 \longrightarrow N \longrightarrow 0$$

is a presentation. We computed, using the command **prune**, a minimal presentation of N:

$$A \xrightarrow{\begin{pmatrix} y-xy^2 \\ 2xz \end{pmatrix}} A^2 \longrightarrow N \longrightarrow 0.$$

```
ring B=0,(x,y,z),dp;              //non-local ring
module M=[0,xy-1,xy+1],[y,xz,xz]; //no units as entries
print(prune(M));
//-> 0,   y,
//-> xy-1,xz,
//-> xy+1,xz

M=[0,1,xy+1],[y,xz,xz];
print(M);
//-> 0,   y,
//-> 1,   xz,
//-> xy+1,xz

print(prune(M));
//-> y,
//-> -x2yz
```

Corollary 2.1.35 (Krull's intersection theorem). *Let A be a Noetherian ring, $I \subset A$ an ideal contained in the Jacobson radical and M a finitely generated A–module. Then $\bigcap_{k \in \mathbb{N}} I^k M = \langle 0 \rangle$.*

Proof. Let $N := \bigcap_k I^k M$. N is a finitely generated A-module, since it is a submodule of the finitely generated module M over the Noetherian ring A. By Nakayama's Lemma it is sufficient to show that $IN = N$. Let

$$\mathfrak{M} := \{L \subset M \text{ submodule} \mid L \cap N = IN\}.$$

Since A is Noetherian, the set $\mathfrak{M}$ has a maximal element which we call L. It remains to prove that $I^k M \subset L$ for some k, because this implies

$N = I^k M \cap N \subset L \cap N = IN$. Since I is finitely generated, it suffices to prove that for any $x \in I$ there is some positive integer a such that $x^a M \subset L$. Let $x \in I$ and consider the chain of ideals $L :_M \langle x \rangle \subset L :_M \langle x^2 \rangle \subset \cdots$. This chain stabilizes because A is Noetherian.

Choose a with $L :_M \langle x^a \rangle = L :_M \langle x^{a+1} \rangle$. We claim that $x^a M \subset L$. By the maximality of L it is enough to prove that $(L + x^a M) \cap N \subset IN$ (note that, obviously, $IN \subset (L + x^a M) \cap N$). Let $m \in (L + x^a M) \cap N$, $m = n + x^a s$, with $n \in L$, $s \in M$. Now $xm - xn = x^{a+1} s \in IN + L = L$, which implies $s \in L :_M \langle x^{a+1} \rangle = L :_M \langle x^a \rangle$. Therefore, $x^a s \in L$ and, consequently, $m \in L$. This implies $m \in L \cap N = IN$. □

Definition 2.1.36. Let A be a ring, $S \subset A$ be a multiplicatively closed subset and M be an A–module.

(1) We define the *localization of M with respect to S*, $S^{-1}M$, as follows:

$$S^{-1}M := \left\{ \frac{m}{s} \ \middle| \ m \in M, \ s \in S \right\}$$

where m/s denotes the equivalence class of $(m, s) \in M \times S$ with respect to the following equivalence relation:

$$(m, s) \sim (m', s') :\Longleftrightarrow \exists \, s'' \in S, \text{ such that } s''(s'm - sm') = 0 \,.$$

Moreover, on $S^{-1}M$ we define an addition and multiplication with ring elements by the same formulæ as for the quotient field (see before Definition 1.4.4). We shall also use the notation M_S instead of $S^{-1}M$. If $S = \{1, f, f^2, \ldots\}$ then we write M_f instead of $S^{-1}M$. If $S = A \setminus P$, P a prime ideal, we write M_P instead of $S^{-1}M$.

(2) Let $\varphi : M \to N$ be an A–module homomorphism, then we define the induced $S^{-1}A$–module homomorphism,

$$\varphi_S : M_S \longrightarrow N_S, \qquad \frac{m}{s} \longmapsto \frac{\varphi(m)}{s} \,.$$

Note that the latter is, indeed, a well–defined $S^{-1}A$–module homomorphism (Exercise 2.1.19).

Proposition 2.1.37. *Let A be a ring, $S \subset A$ be a multiplicatively closed subset, M, N be A–modules and $\varphi : M \to N$ be an A–module homomorphism. Then*

(1) $\mathrm{Ker}(\varphi_S) = \mathrm{Ker}(\varphi)_S$.
(2) $\mathrm{Im}(\varphi_S) = \mathrm{Im}(\varphi)_S$.
(3) $\mathrm{Coker}(\varphi_S) = \mathrm{Coker}(\varphi)_S$.

In particular, localization with respect to S is an *exact functor*. That is, if $0 \to M' \to M \to M'' \to 0$ is an exact sequence of A–modules, then $0 \to M'_S \to M_S \to M''_S \to 0$ is an exact sequence of A_S–modules (cf. Definition 2.4.1).

Proof. (1) follows, since $\varphi_S(m/s) = 0$ if and only if there exists some $s' \in S$ such that $s'\varphi(m) = 0$, that is, $s'm \in \mathrm{Ker}(\varphi)$. (2) is clear by definition of φ_S. Finally, using Exercise 2.1.20, we have

$$\mathrm{Coker}(\varphi_S) = N_S/\mathrm{Im}(\varphi_S) = N_S/\mathrm{Im}(\varphi)_S = \left(N/\mathrm{Im}(\varphi)\right)_S,$$

which implies (3). □

Proposition 2.1.38. *Let A be a ring, M be an A–module. The following conditions are equivalent:*

(1) $M = \langle 0 \rangle$.
(2) $M_P = \langle 0 \rangle$ for all prime ideals P.
(3) $M_{\mathfrak{m}} = \langle 0 \rangle$ for all maximal ideals in $\mathfrak{m}$.

Proof. The only non–trivial part is to prove (3) ⇒ (1). Let $m \in M$ and assume $\mathrm{Ann}(m) \subset \mathfrak{m}$ for some maximal ideal $\mathfrak{m}$. Then $m/1 \neq 0$ in $M_{\mathfrak{m}}$ contradicting the assumption $M_{\mathfrak{m}} = \langle 0 \rangle$. This implies that the annihilator of every element $m \in M$ is A, that is, $M = \langle 0 \rangle$, since $1 \in A$. □

Corollary 2.1.39. *Let A be ring, M, N A–modules, and $\varphi : M \to N$ an A–module homomorphism. Then φ is injective (respectively surjective) if and only if $\varphi_{\mathfrak{m}}$ is injective (respectively surjective) for all maximal ideals $\mathfrak{m}$.*[3]

Proof. The corollary is an immediate consequence of Proposition 2.1.37 and Proposition 2.1.38. □

Definition 2.1.40. Let A be a ring and M an A–module. The *support* of M, $\mathrm{supp}(M)$, is defined by

$$\mathrm{supp}(M) := \{P \subset A \text{ prime ideal} \mid M_P \neq \langle 0 \rangle\}.$$

Lemma 2.1.41. *Let A be a ring and M a finitely generated A–module. Then $\mathrm{supp}(M) = \{P \subset A \text{ prime ideal} \mid P \supset \mathrm{Ann}(M)\} =: V(\mathrm{Ann}(M))$.*

Proof. Assume that $\mathrm{Ann}(M) \not\subset P$, then there exists some $s \in \mathrm{Ann}(M)$ satisfying $s \notin P$. Let $m \in M$, then $sm = 0$. This implies $m/1 = sm/s = 0$ in M_P and, therefore, $M_P = \langle 0 \rangle$.

On the other hand, if $M_P = \langle 0 \rangle$, then $A_P = \mathrm{Ann}(M_P) = \left(\mathrm{Ann}(M)\right)_P$ (Exercise 2.1.24) implies that $\mathrm{Ann}(M) \not\subset P$. □

For flatness properties of $S^{-1}M$ see Exercise 7.3.1.

---

[3] This means that injectivity (respectively surjectivity) is a local property.

Exercises

2.1.1. Prove Lemma 2.1.4.

2.1.2. Let $\varphi : M \to N$ be a bijective module homomorphism. Show that φ^{-1} is a homomorphism.

2.1.3. Let A be a ring and M an A–module. Prove that $\mathrm{Hom}_A(A, M) \cong M$ and give an example which shows that, in general, $\mathrm{Hom}_A(M, A) \not\cong M$.

2.1.4. Prove that isomorphisms between A–modules define an equivalence relation on the set of all A–modules.

2.1.5. Complete the proof of Lemma 2.1.5.

2.1.6. Prove Lemma 2.1.12.

2.1.7. Prove Lemma 2.1.15.

2.1.8. Prove Lemma 2.1.19.

2.1.9. Prove Lemma 2.1.28.

2.1.10. Let A be a ring, M an A–module and $I \subset A$ an ideal. Prove that M/IM has a canonical A/I–module structure.

2.1.11. Let A be a ring and $\varphi : A^n \to A^s$ an isomorphism of free A–modules. Prove that $n = s$.

2.1.12. Let A be a ring and M an A–module. Prove that M is, in a natural way, an $A/\mathrm{Ann}(M)$–module.

2.1.13. Let A be a ring and M an A–module. Prove that $\mathrm{Tors}(M)$ is a submodule of M.

2.1.14. Let A be a ring, and let $M = \bigoplus_{i=1}^{n} M_i$, N be A–modules. Prove that $\mathrm{Hom}_A(M, N) \cong \bigoplus_{i=1}^{n} \mathrm{Hom}_A(M_i, N)$, $\mathrm{Hom}_A(N, M) \cong \bigoplus_{i=1}^{n} \mathrm{Hom}_A(N, M_i)$. In particular, $\mathrm{Hom}(A^n, A^m) \cong A^{m \cdot n}$.

2.1.15. Let K be a field and M a $K[x]$–module which is finite dimensional as K–vector space. Prove that M is a torsion module.

2.1.16. Let A be an integral domain and $I \subset A$ be an ideal. Prove that I is a free A–module if and only if I can be generated by one element.

2.1.17. Let $(A, \mathfrak{m})$ be a local ring and M an A–module with presentation $A^m \xrightarrow{\varphi} A^n \to M \to 0$. Prove that this presentation is minimal if and only if $\varphi(A^m) \subset \mathfrak{m}A^n$.

2.1.18. Let $(A, \mathfrak{m})$ be a local ring and $A^m \cong A^s \oplus N$ for a suitable A–module N. Prove that $N \cong A^{m-s}$.

2.1.19. Prove (with the notations of Definition 2.1.36) that $S^{-1}M$ is an $S^{-1}A$–module. Prove that φ_S is well–defined and an $S^{-1}A$–module homomorphism.

2.1.20. Let A be a ring, $S \subset A$ be a multiplicatively closed subset and M, N be A–modules with $N \subset M$. Prove that $(M/N)_S \cong M_S/N_S$.

2.1.21. Prove that a module homomorphism is injective if and only if its kernel is zero.

2.1.22. Let A be a ring and $P_1, \ldots, P_m$ prime ideals. Let $\langle 0 \rangle \neq M$ be a finitely generated A–module such that $M_{P_j} \neq \langle 0 \rangle$ for all j. Prove that there exists $x \in M$ such that $x \notin P_j M_{P_j}$ for all j.

2.1.23. Let A be a ring, M an A–module and N, L submodules of M. Prove that $N = L$ if and only if $N_P = L_P$ for all prime ideals P.

2.1.24. Let A be a ring, $S \subset A$ be a multiplicatively closed subset and M a finitely generated A–module. Prove that $\mathrm{Ann}(S^{-1}M) = S^{-1}\mathrm{Ann}(M)$.

2.1.25. Compute the kernel and image of the following homomorphism: $A^3 \xrightarrow{M} A^2$ with $A = \mathbb{Q}[x, y, z]$ and $M = \left(\begin{smallmatrix} xy & xz & yz \\ x-1 & y-1 & z-1 \end{smallmatrix} \right)$.

2.1.26. Compute a minimal presentation of the A–module M with $M = A^3 / \left\langle \left(\begin{smallmatrix} 1 \\ xy-1 \\ xz \end{smallmatrix} \right), \left(\begin{smallmatrix} 0 \\ yz-1 \\ xy \end{smallmatrix} \right) \right\rangle$ and $A = \mathbb{Q}[x, y, z]_{\langle x,y,z \rangle}$.

2.1.27. Compute the support of the module of Exercise 2.1.26.

2.2 Graded Rings and Modules

Definition 2.2.1. A *graded ring* A is a ring together with a direct sum decomposition $A = \bigoplus_{\nu \geq 0} A_\nu$, where the A_ν are abelian groups satisfying $A_\nu A_\mu \subset A_{\nu+\mu}$ for all $\nu, \mu \geq 0$.

A *graded K–algebra*, K a field, is a K–algebra which is a graded ring such that A_ν is a K–vector space for all $\nu \geq 0$, and $A_0 = K$.

The A_ν are called *homogeneous components* and the elements of A_ν are called *homogeneous* elements of *degree ν*.

Remark 2.2.2. Let $A = \bigoplus_{\nu \geq 0} A_\nu$ be a graded ring, then A_0 is a subring of A. This follows since $1 \cdot 1 = 1$, hence $1 \in A_0$. For a K–algebra A, this implies already $K \subset A_0$, but to be a graded K–algebra we require even $K = A_0$.

Example 2.2.3.

(1) Let K be a field and $A = K[x_1, \ldots, x_n]$. Moreover, let $w = (w_1, \ldots, w_n)$ be a vector of positive integers, and let A_d be the K–vector space generated by all monomials x^α with w–$\deg(x^\alpha) = d$. Then $A = \bigoplus_{\nu \geq 0} A_\nu$ is a graded K–algebra. Namely, $A_0 = K$, and for each i we have $x_i \in A_{w_i}$. The elements of A_d are called *quasihomogeneous* or *weighted homogeneous* polynomials of *(weighted) degree d* with respect to the weights $w_1, \ldots, w_n$. If $w_1 = \cdots = w_n = 1$ we obtain the usual notion of homogeneous polynomials.

(2) Let A be any ring, then $A_0 := A$ and $A_\nu := 0$ for $\nu > 0$ defines a (trivial) structure of a graded ring for A.

(3) Let A be a Noetherian K–algebra, $I \subset A$ be an ideal, then

$$\mathrm{Gr}_I(A) := \bigoplus_{\nu \geq 0} I^\nu / I^{\nu+1}$$

is a graded K–algebra in a natural way. If $(A, \mathfrak{m})$ is a local ring, then all homogeneous components of $\mathrm{Gr}_\mathfrak{m}(A) = \bigoplus_{\nu \geq 0} \mathfrak{m}^\nu / \mathfrak{m}^{\nu+1}$ are finite dimensional vector spaces over $A/\mathfrak{m}$.

Definition 2.2.4. Let $A = \bigoplus_{\nu \geq 0} A_\nu$ be a graded ring. An A–module M, together with a direct sum decomposition $M = \bigoplus_{\mu \in \mathbb{Z}} M_\mu$ into abelian groups is called a *graded A–module* if $A_\nu M_\mu \subset M_{\nu+\mu}$ for all $\nu \geq 0$, $\mu \in \mathbb{Z}$.

The elements from M_ν are called *homogeneous* of *degree ν*. If $m = \sum_\nu m_\nu$, $m_\nu \in M_\nu$ then m_ν is called the *homogeneous part of degree ν* of m.

Example 2.2.5. Let $A = \bigoplus_{\nu \geq 0} A_\nu$ be a graded K–algebra and consider the free module $A^m = \bigoplus_{i=1}^{m} A e_i$, $e_i = (0, \ldots, 1, \ldots, 0)$ with 1 at the i-th place. Let $\nu_1, \ldots, \nu_m \in \mathbb{Z}$, define $\deg(e_i) := \nu_i$, and let M_ν be the A_0–module generated by all $f e_i$ with $f \in A_{\nu - \nu_i}$, then $A^m = \bigoplus_{\nu \in \mathbb{Z}} M_\nu$ is a graded A–module.

Definition 2.2.6. Let $M = \bigoplus_{\nu \in \mathbb{Z}} M_\nu$ be a graded A–module and define $M(d) := \bigoplus_{\nu \in \mathbb{Z}} M(d)_\nu$ with $M(d)_\nu := M_{\nu+d}$. Then $M(d)$ is a graded A–module, especially $A(d)$ is a graded A–module. $M(d)$ is called the *d–th twist* or the *d–th shift* of M.

Lemma 2.2.7. *Let $M = \bigoplus_{\nu \in \mathbb{Z}} M_\nu$ be a graded A–module and $N \subset M$ a submodule. The following conditions are equivalent:*

(1) N is graded with the induced grading, that is, $N = \bigoplus_{\nu \in \mathbb{Z}} (M_\nu \cap N)$.
(2) N is generated by homogeneous elements.
(3) Let $m = \sum m_\nu$, $m_\nu \in M_\nu$. Then $m \in N$ if and only if $m_\nu \in N$ for all ν.

The proof is easy and left as Exercise 2.2.1.

Definition 2.2.8. A submodule $N \subset M$, satisfying the equivalent conditions of Lemma 2.2.7, is called a *graded* (or *homogeneous*) submodule. A graded submodule of a graded ring is called a *graded ideal* or *homogeneous ideal* .

Remark 2.2.9. Let $A = \bigoplus_{\nu \geq 0} A_\nu$ be a graded ring, and let $I \subset A$ be a homogeneous ideal. Then the quotient A/I has an induced structure as graded ring: $A/I = \bigoplus_{\nu \geq 0} (A_\nu + I)/I \cong \bigoplus_{\nu \geq 0} A_\nu / (I \cap A_\nu)$.

Definition 2.2.10. Let $A = \bigoplus_{\nu \geq 0} A_\nu$ be a graded ring and $M = \bigoplus_{\nu \in \mathbb{Z}} M_\nu$, $N = \bigoplus_{\nu \in \mathbb{Z}} N_\nu$ be graded A–modules. A homomorphism $\varphi : M \to N$ is called *homogeneous* (or *graded*) of degree d if $\varphi(M_\nu) \subset N_{\nu+d}$ for all ν. If φ is homogeneous of degree zero we call φ just *homogeneous*.

Example 2.2.11. Let M be a graded A–module and $f \in A_d$ then the multiplication with f defines a graded homomorphism $M \to M$ of degree d. It also defines a graded homomorphism $M \to M(d)$ of degree 0.

Lemma 2.2.12. *Let A be a graded ring and M, N be graded A–modules. Let $\varphi : M \to N$ be a homogeneous A–module homomorphism, then $\mathrm{Ker}(\varphi)$, $\mathrm{Coker}(\varphi)$ and $\mathrm{Im}(\varphi)$ are graded A–modules with the induced grading.*

Proof. Let $M = \bigoplus_{i \in \mathbb{Z}} M_i$ and $N = \bigoplus_{i \in \mathbb{Z}} N_i$, and define $K_i := \mathrm{Ker}(\varphi) \cap M_i$. Then, clearly, $\bigoplus_{i \in \mathbb{Z}} K_i \subset \mathrm{Ker}(\varphi)$. Moreover, let $m = \sum_{i=1}^{k} m_i$, $m_i \in M_i$ and assume that $\varphi(m) = 0$. Then $\varphi(m) = \sum_i \varphi(m_i) = 0$, with $\varphi(m_i) \in N_i$. This implies $\varphi(m_i) = 0$, that is, $m_i \in K_i$ and, therefore, $\mathrm{Ker}(\varphi) = \bigoplus_{i \in \mathbb{Z}} K_i$.
The other statements can be proved similarly. $\square$

Example 2.2.13. Let A be a graded ring, $M = \bigoplus_{\nu \in \mathbb{Z}} M_\nu$ a graded A–module and $N \subset M$ a homogeneous submodule. Let $N_\nu := N \cap M_\nu$, then the quotient M/N has an induced structure as graded A–module:

$$M/N = \bigoplus_{\nu \in \mathbb{Z}} (M_\nu + N)/N \cong \bigoplus_{\nu \in \mathbb{Z}} M_\nu/N_\nu \,.$$

Lemma 2.2.14. *Let $A = \bigoplus_{\nu \geq 0} A_\nu$ be a Noetherian graded K–algebra and $M = \bigoplus_{\nu \in \mathbb{Z}} M_\nu$ be a finitely generated A–module. Then*

(1) there exist $m \in \mathbb{Z}$ such that $M_\nu = \langle 0 \rangle$ for $\nu < m$;
(2) $\dim_K M_\nu < \infty$ for all ν.

Proof. (1) is obvious because M is finitely generated and a graded A–module. To prove (2) it is enough to prove that M_ν is a finitely generated A_0–module for all ν.

By assumption M is finitely generated and we may choose finitely many homogeneous elements $m_1, \ldots, m_k$ to generate M. Assume that $m_i \in M_{e_i}$ for $i = 1, \ldots, k$, then $\sum_i A_{n-e_i} \cdot m_i = M_n$ (with the convention $A_\nu = 0$ for $\nu < 0$). This implies that M_n is a finitely generated A_0–module because the A_ν are finitely generated A_0–modules. $\square$

SINGULAR Example 2.2.15 (graded rings and modules).
We give examples here on how to work with graded rings and modules.

First we consider the ideal $\langle y^3 - z^2, x^3 - z \rangle$ in $A = \mathbb{Q}[x, y, z]$. This ideal is homogeneous if we consider $\mathbb{Q}[x, y, z]$ as a graded ring with weights $w_1 = 1$, $w_2 = 2$, $w_3 = 3$. Note that it is not homogeneous in $\mathbb{Q}[x, y, z]$ with the usual graduation.

```
ring A=0,(x,y,z),dp;
ideal I=y3-z2,x3-z;
qhweight(I);
//-> 1,2,3
```

The SINGULAR command `homog(I)` checks whether `I` is homogeneous with respect to the weights given to the variables of the basering (if no weights are assigned explicitly, all weights are assumed to be 1).

```
homog(I);
//-> 0

ring B=0,(x,y,z),wp(1,2,3);
ideal I=fetch(A,I);
homog(I);
//-> 1
```

Next we consider $B = \mathbb{Q}[x, y, z]$ as a graded ring with weights $w_1 = 1$, $w_2 = 2$, $w_3 = 3$, and a B–module $M = \left\langle \left(\begin{smallmatrix} y^3 - z^2 \\ x^3 - z \end{smallmatrix} \right), \left(\begin{smallmatrix} x^3 \\ 1 \end{smallmatrix} \right) \right\rangle$. Then M is a homogeneous submodule of B^2 if we consider B^2 as graded B–module with $\deg((1, 0)) = 0$ and $\deg((0, 1)) = 3$. This can be seen as follows:

```
module M=[y3-z2,x3-z],[x3,1];
homog(M);
//-> 1
```

M is homogeneous. SINGULAR defines internally a corresponding *attribute*.

```
attrib(M,"isHomog");        //asks for attributed weights
//-> 0,3
```

The degree of $(1,0)$ is 0 and the degree of $(0,1)$ is 3.

The grading for M being homogeneous is not uniquely determined. We can also use $\deg((1, 0)) = 4$ and $\text{degree}((0, 1)) = 7$.

```
intvec v=4,7;
attrib(M,"isHomog",v);      //sets externally an attribute,
attrib(M,"isHomog");        //without changing the module
//-> 4,7
```

Exercises

2.2.1. Prove Lemma 2.2.7.

2.2.2. Prove the remaining statements of Lemma 2.2.12.

2.2.3. Let A be a graded ring and M a graded A–module. Show that the annihilator $\text{Ann}_A(M)$ is a homogeneous ideal.

2.2.4. Let I_1, I_2 be homogeneous ideals in a graded ring. Show that $I_1 + I_2$, $I_1 \cdot I_2$, $I_1 \cap I_2$, $I_1 : I_2$ and $\sqrt{I_1}$ are homogeneous.

2.2.5. Let A be a graded ring. A homogeneous ideal $I \subset A$ is prime if and only if for any two *homogeneous* elements $f, g \in A$, $f \cdot g \in I$ implies $f \in I$ or $g \in I$.

2.2.6. Prove the homogeneous version of Nakayama's Lemma: let A be a graded K–algebra and $\mathfrak{m}$ the ideal generated by the elements of positive degree. Let M be a finitely generated, graded A–module and $N \subset M$ a graded submodule. If $N + \mathfrak{m}M = M$ then $N = M$.

2.2.7. Test whether the following ideals in $K[x, y, z]$ are homogeneous w.r.t. suitable weights, $\langle y^5 - z^2, \ x^3 - z, \ x^6 - y^5 \rangle$, $\langle y^5 - z^2, \ x^3 - z, \ x^7 - y^5 \rangle$.

2.3 Standard Bases for Modules

For our intended applications of standard bases, but also for an elegant proof of Buchberger's standard basis criterion, we have to extend the notion of monomial orderings to the free module $K[x]^r = \bigoplus_{i=1}^r K[x]e_i$, where

$$e_i = (0, \dots, 1, \dots, 0) \in K[x]^r$$

denotes the i–th canonical basis vector of $K[x]^r$ with 1 at the i–th place. We call

$$x^\alpha e_i = (0, \dots, x^\alpha, \dots, 0) \in K[x]^r$$

a *monomial (involving component i)*.

Definition 2.3.1. Let $>$ be a monomial ordering on $K[x]$. A *(module) monomial ordering* or a *module ordering* on $K[x]^r$ is a total ordering $>_m$ on the set of monomials $\{x^\alpha e_i \mid \alpha \in \mathbb{N}^n, i = 1, \dots, r\}$, which is compatible with the $K[x]$–module structure including the ordering $>$, that is, satisfying

(1) $x^\alpha e_i >_m x^\beta e_j \implies x^{\alpha+\gamma} e_i >_m x^{\beta+\gamma} e_j$,
(2) $x^\alpha > x^\beta \implies x^\alpha e_i >_m x^\beta e_i$,

for all $\alpha, \beta, \gamma \in \mathbb{N}^n$, $i, j = 1, \dots, r$.

Two module orderings are of particular practical interest:

$$x^\alpha e_i > x^\beta e_j :\Longleftrightarrow i < j \text{ or } (i = j \text{ and } x^\alpha > x^\beta),$$

giving priority to the components, denoted by $(c, >)$, and

$$x^\alpha e_i > x^\beta e_j :\Longleftrightarrow x^\alpha > x^\beta \text{ or } (x^\alpha = x^\beta \text{ and } i < j),$$

which gives priority to the monomials in $K[x]$, denoted by $(>, c)$.
Note that, by the second condition of Definition 2.3.1, each component of $K[x]^r$ carries the ordering of $K[x]$. Hence, $>_m$ is a well-ordering on $K[x]^r$ if and only if $>$ is a well-ordering on $K[x]$. We call $>_m$ *global*, respectively *local*, respectively *mixed*, if this holds for $>$ respectively.

In the case of a well–ordering it makes sense to define a module ordering without fixing a ring ordering, only requiring (1) (cf. [53]). In the general case, this could lead to standard bases which do not generate the module (Exercise 2.3.5).

Now we fix a module ordering $>_m$ and denote it also with $>$. Since any vector $f \in K[x]^r \smallsetminus \{0\}$ can be written uniquely as

$$f = cx^\alpha e_i + f^*$$

with $c \in K \smallsetminus \{0\}$ and $x^\alpha e_i > x^{\alpha^*} e_j$ for any non–zero term $c^* x^{\alpha^*} e_j$ of f^* we can define as before

$$\mathrm{LM}(f) := x^\alpha e_i\,,$$
$$\mathrm{LC}(f) := c\,,$$
$$\mathrm{LT}(f) := cx^\alpha e_i$$

and call it the *leading monomial, leading coefficient* and *leading term*, respectively, of f. $\mathrm{tail}(f) := f - \mathrm{LT}(f)$ is called the *tail* of f. Moreover, for $G \subset K[x]^r$ we call

$$L_>(G) := L(G) := \langle\, \mathrm{LM}(g) \mid g \in G \smallsetminus \{0\} \,\rangle_{K[x]} \subset K[x]^r$$

the *leading submodule* of $\langle G \rangle$. In particular, if $I \subset K[x]^r$ is a submodule, then $L_>(I) = L(I)$ is called the *leading module* of I.

As from $K[x]$ to $K[x]_>$ these definitions carry over naturally from $K[x]^r$ to $K[x]_>^r$.

Note that the set of monomials of $K[x]^r$ may be identified with $\mathbb{N}^n \times E^r \subset \mathbb{N}^n \times \mathbb{N}^r = \mathbb{N}^{n+r}$, $E^r = \{e_1, \dots, e_r\}$ where e_i is considered as an element of $\mathbb{N}^r$. The natural partial order on $\mathbb{N}^{n+r}$ induces a partial order $\geq_{\mathrm{nat}}$ on the set of monomials, which is given by

$$x^\alpha e_i \leq_{\mathrm{nat}} x^\beta e_j :\Longleftrightarrow i = j \text{ and } x^\alpha \mid x^\beta \Longleftrightarrow: x^\alpha e_i \mid x^\beta e_j$$

(we say that $x^\beta e_j$ is *divisible by* $x^\alpha e_i$ if $i = j$ and $x^\alpha \mid x^\beta$). For any set of monomials $G \subset K[x]^r$ and any monomial $x^\alpha e_i$, we have

$$x^\alpha e_i \notin \langle G \rangle_{K[x]} \Longleftrightarrow x^\alpha e_i \text{ is not divisible by any element of } G.$$

Hence, Dickson's Lemma for $\mathbb{N}^m$ (m arbitrary) is equivalent to the statement that any monomial submodule of $K[x]^r$ (r arbitrary) is finitely generated.

Let $>$ be a fixed monomial ordering. Again we write

$$R := K[x]_> = S_>^{-1} K[x]$$

to denote the localization of $K[x]$ with respect to $>$. Since $R^r \subset K[[x]]^r$, we can talk about the power series expansion of elements of R^r.

The theory of standard bases for ideals carries over to modules almost without any changes. We formulate the relevant definitions and theorems but omit the proofs, since they are practically identical to the ideal case.

We fix a module ordering on R^r. As for ideals, we define:

Definition 2.3.2.

(1) Let $I \subset R^r$ be a submodule. A finite set $G \subset I$ is called a *standard basis* of I if and only if $L(G) = L(I)$, that is, for any $f \in I \setminus \{0\}$ there exists a $g \in G$ satisfying $\mathrm{LM}(g) \mid \mathrm{LM}(f)$.

(2) If the ordering is a well–ordering then a standard basis G is called a *Gröbner basis*. In this case $R = K[x]$ and, hence, $G \subset I \subset K[x]^r$.

(3) A set $G \subset R^r$ is called *interreduced* if $0 \notin G$ and if $\mathrm{LM}(g) \notin L(G \setminus \{g\})$ for each $g \in G$. An interreduced standard basis is also called *minimal*.

(4) For $f \in R^r$ and $G \subset R^r$ we say that f is *reduced with respect to G* if no monomial of the power series expansion of f is contained in $L(G)$.

(5) A set $G \subset R^r$ is called *reduced* if $0 \notin G$ and if each $g \in G$ is reduced with respect to $G \setminus \{g\}$, $\mathrm{LC}(g) = 1$, and if, moreover, $\mathrm{tail}(g)$ is reduced with respect to G. For $>$ a well–ordering, this just means that for each $g \in G \subset K[x]^r$, $\mathrm{LM}(g)$ does not divide any monomial of any element of $G \setminus \{g\}$.

Definition 2.3.3. Let $\mathcal{G}$ denote the set of all finite ordered subsets $G \subset R^r$.

(1) A map
$$\mathrm{NF} : R^r \times \mathcal{G} \to R^r, \ (f, G) \mapsto \mathrm{NF}(f \mid G),$$

is called a *normal form* on R^r if for all $f \in R^r$ and $G \in \mathcal{G}$, $\mathrm{NF}(0 \mid G) = 0$, and

a) $\mathrm{NF}(f \mid G) \neq 0 \Rightarrow \mathrm{LM}(\mathrm{NF}(f \mid G)) \notin L(G)$,

b) If $G = \{g_1, \ldots, g_s\}$ then $f - \mathrm{NF}(f|g)$ (or f) has a *standard representation* with respect to $\mathrm{NF}(- \mid G)$, that is,

$$f - \mathrm{NF}(f \mid G) = \sum_{i=1}^{s} a_i g_i, \ a_i \in R, \ s \geq 0,$$

satisfying $\mathrm{LM}(\sum_{i=1}^{s} a_i g_i) \geq \mathrm{LM}(a_i g_i)$ for all i such that $a_i g_i \neq 0$.

NF is called a *reduced normal form* if, moreover, $\mathrm{NF}(f \mid G)$ is reduced with respect to G for all $G \in \mathcal{G}$.

(2) NF is called a *weak normal form* if, instead of b), only condition b') holds:

b') for each $f \in R^r$ and each $G \in \mathcal{G}$ there exists a unit $u \in R$ such that uf has a standard representation with respect to $\mathrm{NF}(- \mid G)$.

(3) Similarly to Definition 1.6.5 (2) *polynomial weak normal forms* are defined.

Remark 2.3.4. In the same manner as for ideals, a reduced normal form exists for global orderings. We just have to apply $\mathrm{NF}(- \mid G)$ not only to f but successively to $\mathrm{tail}(f)$ until it terminates (cf. Algorithm 1.6.11). For non–global orderings, this procedure may not terminate.

However, for an arbitrary module ordering $>$, we can always find a weak normal form NF with the following property which is stronger than 1.a from Definition 2.3.3: if, for any $f \in K[x]^r$ and $G = \{g_1, \dots, g_s\} \in \mathcal{G}$,

$$\mathrm{NF}(f \mid G) = f_1 e_1 + \cdots + f_r e_r \in K[x]^r \,,$$

then $\mathrm{LM}(f_i e_i) \notin L(G)$ for all i with $f_i \neq 0$.

Proof. We may assume that $\mathrm{LM}(f_1 e_1) = \mathrm{LM}\big(\mathrm{NF}(f \mid G)\big) \notin L(G)$ and proceed by induction on r, to show that we can successively reduce $\sum_{j=1}^r f_j e_j$ with respect to G to obtain the above property. Let $f^{(2)} := f_2 e_2 + \cdots + f_r e_r$, and let $G^{(2)} := \{g \in G \mid \mathrm{LM}(g) \notin K[x] e_1\}$. We consider the images

$$\overline{f}^{(2)} := \pi(f^{(2)}), \quad \overline{G}^{(2)} := \pi(G^{(2)})$$

under the canonical projection $\pi : \bigoplus_{i=1}^r Re_i \to \bigoplus_{i=2}^r Re_i$. Then, by induction hypothesis, we can assume that there exists a weak normal form

$$\mathrm{NF}\big(\overline{f}^{(2)} \mid \overline{G}^{(2)}\big) = f_2^{(2)} e_2 + \cdots + f_r^{(2)} e_r$$

such that $\mathrm{LM}(f_j^{(2)} e_j) \notin L(\overline{G}^{(2)}) = L(G) \cap \bigoplus_{i=2}^r K[x] e_i$ for $j = 2, \dots, r$. Let

$$u^{(2)} \overline{f}^{(2)} = \sum_{j=2}^r f_j^{(2)} e_j + \sum_{g \in G^{(2)}} a_g^{(2)} \overline{g}$$

be a standard representation with respect to $\overline{G}^{(2)}$ ($u^{(2)}$ a unit in R). Then, by construction,

$$u^{(2)} f^{(2)} - \sum_{g \in G^{(2)}} a_g^{(2)} g = f_1^{(2)} e_1 + \sum_{j=2}^r f_j^{(2)} e_j$$

for some $f_1^{(2)}$ such that either $\mathrm{LM}(f_1^{(2)} e_1) \leq \mathrm{LM}(f^{(2)})$ or $f_1^{(2)} = 0$. Now, it is easy to see that

$$(u^{(2)} f_1 + f_1^{(2)}) e_1 + \sum_{j=2}^r f_j^{(2)} e_j$$

is a weak normal form for f with respect to G with the required property. □

Lemma 2.3.5. *Let $I \subset R^r$ be a submodule, $G \subset I$ a standard basis of I and $\mathrm{NF}(- \mid G)$ a weak normal form on R^r with respect to G.*

(1) For any $f \in R^r$ we have $f \in I$ if and only if $\mathrm{NF}(f \mid G) = 0$.

(2) If $J \subset R^r$ is a submodule with $I \subset J$, then $L(I) = L(J)$ implies $I = J$.
(3) $I = \langle G \rangle_R$, that is, G generates I as an R–module.
(4) If $\mathrm{NF}(\_ \mid G)$ is a reduced normal form, then it is unique.

The proof is the same as for ideals. Also the notion of s–polynomial carries over to modules.

Definition 2.3.6. Let $f, g \in R^r \smallsetminus \{0\}$ with $\mathrm{LM}(f) = x^\alpha e_i$, $\mathrm{LM}(g) = x^\beta e_j$. Let
$$\gamma := \mathrm{lcm}(\alpha, \beta) := \left(\max(\alpha_1, \beta_1), \ldots, \max(\alpha_n, \beta_n) \right)$$
be the least common multiple of α and β and define the s–polynomial of f and g to be
$$\mathrm{spoly}(f, g) := \begin{cases} x^{\gamma - \alpha} f - \frac{\mathrm{LC}(f)}{\mathrm{LC}(g)} \cdot x^{\gamma - \beta} g, & \text{if } i = j \\ 0, & \text{if } i \neq j. \end{cases}$$

Definition 2.3.7. For a monomial $x^\alpha e_i \in K[x]^r$ set
$$\deg x^\alpha e_i := \deg x^\alpha = \alpha_1 + \cdots + \alpha_n.$$

For $f \in K[x]^r \smallsetminus \{0\}$, let $\deg f$ be the maximal degree of all monomials occurring in f. We define the *ecart* of f as
$$\mathrm{ecart}\,(f) := \deg f - \deg \mathrm{LM}(f).$$

Similarly to Definition 1.7.5 one can define the weighted ecart and interpret the ecart as $\deg_t \left(\mathrm{LM}(f^h) \right)$ for the homogenization of f with respect to a new variable t.

The Algorithms 1.6.10 (NFBUCHBERGER), 1.6.11 (REDNFBUCHBERGER) carry over verbatim to the module case if we replace $K[x]$ by $K[x]^r$. Similarly for the Algorithms 1.7.1 (STANDARD), 1.7.6 (NFMORA) and 1.7.8 (STANDARDBASIS). However, for the sake of completeness, we shall formulate them for modules, omitting the proofs.

Let $>$ be any monomial ordering on R^r and assume that a weak normal form algorithm NF on R^r is given.

Algorithm 2.3.8 (STANDARD(G,NF)).

Input: $G \in \mathcal{G}$, NF a weak normal form algorithm.
Output: $S \in \mathcal{G}$ such that S is a standard basis of $I = \langle G \rangle_R \subset R^r$.

- $S = G$;
- $P = \{(f, g) \mid f, g \in S, f \neq g\}$;

- while $(P \neq \emptyset)$
    choose $(f, g) \in P$;
    $P = P \setminus \{(f, g)\}$;
    $h = \mathrm{NF}\big(\mathrm{spoly}(f, g) \mid S\big)$;
    If $(h \neq 0)$
        $P = P \cup \{(h, f) \mid f \in S\}$;
        $S = S \cup \{h\}$;
- return S;

Let $>$ be any monomial ordering on $K[x]^r$, $R = K[x]_>$.

Algorithm 2.3.9 ($\mathrm{NFMORA}(f \mid G)$).

Input: $f \in K[x]^r$, $G = \{g_1, \ldots, g_s\} \subset K[x]^r$.

Output: $h \in K[x]^r$ a weak normal form of f with respect to G, such that there
         exists a standard representation $uf = h + \sum_{i=1}^{s} a_i g_i$ with $a_i \in K[x]$,
         $u \in S_>$.

- $h = f$;
- $T = G$;
- while$\big(h \neq 0$ and $T_h = \big\{g \in T \mid \mathrm{LM}(g) \text{ divides } \mathrm{LM}(h)\big\} \neq \emptyset\big)$
    choose $g \in T_h$ with $\mathrm{ecart}(g)$ minimal;
    if $(\mathrm{ecart}(g) > \mathrm{ecart}(h))$
        $T = T \cup \{h\}$;
    $h = \mathrm{spoly}(h, g)$;
- return h;

SINGULAR Example 2.3.10 (normal form).

```
ring A=0,(x,y,z),(c,dp);
module I=[x,y,1],[xy,z,z2];
vector f=[zx,y2+yz-z,y];
reduce(f,I);
//-> // ** I is no standardbasis
//-> [0,y2-z,y-z]

reduce(f,std(I));
//-> [0,0,z2-z]
```

We have seen in SINGULAR Example 1.6.13 that the normal form may not be
unique, in particular, if we do not have a standard basis for I.

Let $>$ be any monomial ordering on $K[x]^r$, $R = K[x]_>$.

Algorithm 2.3.11 ($\mathrm{STANDARDBASIS}(G)$).

Input: $G = \{g_1, \ldots, g_s\} \subset K[x]^r$.

Output: $S = \{h_1, \ldots, h_t\} \subset K[x]^r$ such that S is a standard basis of $I = \langle G \rangle_R \subset R^r$.

- $S = \text{STANDARD}(G, \text{NFMORA})$;
- return S;

SINGULAR Example 2.3.12 (standard bases).
The example shows the influence of different orderings to standard bases.

```
ring A=0,(x,y,z),(c,dp);
module I=[x+1,y,1],[xy,z,z2];
std(I);
//-> _[1]=[0,xy2-xz-z,-xz2+xy-z2]
//-> _[2]=[y,y2-z,-z2+y]
//-> _[3]=[x+1,y,1]

ring B=0,(x,y,z),dp;
module I=fetch(A,I);
std(I);
//-> _[1]=x*gen(1)+y*gen(2)+gen(3)+gen(1)
//-> _[2]=y2*gen(2)-z2*gen(3)+y*gen(3)+y*gen(1)-z*gen(2)

ring C=0,(x,y,z),lp;
module I=fetch(A,I);
std(I);
//-> _[1]=y2*gen(2)+y*gen(3)+y*gen(1)-z2*gen(3)-z*gen(2)
//-> _[2]=x*gen(1)+y*gen(2)+gen(3)+gen(1)

ring D=0,(x,y,z),(c,ds);
module I=fetch(A,I);
std(I);
//-> _[1]=[1+x,y,1]
//-> _[2]=[0,z+xz-xy2,-xy+z2+xz2]

ring E=0,(x,y,z),ds;
module I=fetch(A,I);
std(I);
//-> _[1]=gen(3)+gen(1)+x*gen(1)+y*gen(2)
//-> _[2]=z*gen(2)+xy*gen(1)+z2*gen(3)
```

Similarly to Chapter 1 we also have *Buchberger's criterion*.

Theorem 2.3.13. *Let $I \subset R^r$ be a submodule and $G = \{g_1, \ldots, g_s\} \subset I$. Let $\mathrm{NF}(- \mid G)$ be a weak normal form on R^r with respect to G. Then the following are equivalent:*

(1) G is a standard basis of I.

(2) $\mathrm{NF}(f \mid G) = 0$ *for all* $f \in I$.

(3) Each $f \in I$ *has a standard representation with respect to* $\mathrm{NF}(- \mid G)$.

(4) G *generates* I *and* $\mathrm{NF}\big(\mathrm{spoly}(g_i, g_j) \mid G\big) = 0$ *for* $i, j = 1, \ldots, s$.

(5) G *generates* I *and* $\mathrm{NF}\big(\mathrm{spoly}(g_i, g_j) \mid G_{ij}\big) = 0$ *for some* $G_{ij} \subset G$ *and*
$i, j = 1, \ldots, s$.

Proof. The proof of $(1) \Rightarrow (2) \Rightarrow (3) \Rightarrow (4) \Rightarrow (5)$ is similar to the proof of Theorem 1.7.3.

The proof of $(5) \Rightarrow (1)$ will be given in Section 2.5. This proof needs the equivalence of (1) and (2). We prove now that $(2) \Rightarrow (1)$. Let $f \in I$ then (2) implies $\mathrm{NF}(f \mid G) = 0$, and there exists a unit $u \in R^*$ such that uf has a standard representation $uf = \sum_i a_i g_i$ with $\mathrm{LM}(f) \geq \mathrm{LM}(a_i g_i)$. Therefore, there exists some i such that $\mathrm{LM}(f) = \mathrm{LM}(a_i g_i)$. This implies that $\mathrm{LM}(g_i) \mid \mathrm{LM}(f)$, which proves that G is a standard basis of I. $\qquad\square$

Exercises

2.3.1. Verify that the proofs for Lemma 2.3.5 and that the algorithms carry over from the ideal case of Chapter 1.

2.3.2. Let $R = K[x]$, and let $M = (m_{ij})$ be an $n \times n$–matrix with entries in R. Consider the matrix (M, E) obtained by concatenating M with the $n \times n$–unit matrix E, and let $v_1, \ldots, v_n \in R^{2n}$ be the rows of (M, E). On the free R–module $R^{2n} = \bigoplus_{i=1}^{2n} Re_i$, $e_1 = (1, 0, \ldots, 0)$, $\ldots$, $e_{2n} = (0, \ldots, 0, 1)$, consider the ordering defined by $x^\alpha e_i < x^\beta e_j$ if $i > j$ or if $i = j$ and $x^\alpha < x^\beta$.

Let $\{w_1, \ldots, w_m\} \subset R^{2n}$ be the reduced standard basis of $\langle v_1, \ldots, v_n \rangle$, with $\mathrm{LM}(w_1) > \cdots > \mathrm{LM}(w_m)$. Prove that M is invertible if and only if $m = n$ and $\mathrm{LM}(w_i) = e_i$ for $i = 1, \ldots, m$, and then $w_1, \ldots, w_m$ are the rows of (E, M^{-1}).

2.3.3. Let $I \subset K[x]^r$ be a submodule, $x = (x_1, \ldots, x_n)$, and let $>$ be a global module ordering on $K[x]^r$. Prove that

$$K[x]^r \cong I \oplus \left(\bigoplus_{m \notin L[I]} K \cdot m \right).$$

2.3.4. Compute the normal form of $\binom{x+y}{y-1}$ w.r.t. the module $M \subset K[x, y]^2$ generated by the vectors $\binom{x^2}{xy}$, $\binom{x}{y^2}$, and the ordering $(\mathtt{c}, \mathtt{dp})$.

2.3.5. Let K be a field and $R = K[x]$ the polynomial ring in one variable. Consider on $K[x]^2$ the following ordering:

$$x^\alpha e_i > x^\beta e_j :\Longleftrightarrow i > j \text{ or } (i = j = 1 \text{ and } \alpha < \beta)$$
$$\text{or } (i = j = 2 \text{ and } \alpha > \beta).$$

Prove that $\{(1+x)e_1, e_2\}$ is a standard basis of $K[x]^2$ (w.r.t. $>$), but it does not generate $K[x]^2$ (as $K[x]$-module).

The following exercises $(2.3.6) - (2.3.12)$ are related to the behaviour of *standard bases under specialization*. Let, in these exercises, R be an integral domain, $R[x] = R[x_1, \ldots, x_n]$, and let $>$ be a fixed monomial ordering on $\mathrm{Mon}(x_1, \ldots, x_n)$. The set $S_> = \{f \in R[x] \mid \mathrm{LT}(F) = 1\}$ is multiplicatively closed and the localization of $R[x]$ with respect to $S_>$ is

$$R[x]_> = S_>^{-1}R[x] = \left\{\frac{f}{g} \;\middle|\; f, g \in R[x],\, \mathrm{LT}(g) = 1\right\}.$$

If $>$ is global, then $R[x]_> = R[x]$ and if $>$ is local then $R[x]_> = R[x]_{\langle x_1,\ldots,x_n \rangle}$, the localization of $R[x]$ with respect to the prime ideal $\langle x_1, \ldots, x_n \rangle$ (check this).

2.3.6. Let $\mathcal{G}$ denote the set of finite ordered sets $G \subset R[x]_>^r$, where $>$ is a fixed module ordering on the set of monomials $\{x^\alpha e_i\}$. Call a function

$$\mathrm{NF} : R[x]_>^r \times \mathcal{G} \to R[x]_>^r,\ (f, g) \mapsto \mathrm{NF}(f \mid G),$$

a *pseudo normal form* on $R[x]_>^r$, if the following holds:

(1) $\mathrm{NF}(f \mid G) \neq 0 \Rightarrow \mathrm{LM}\big(\mathrm{NF}(f \mid G)\big)$ is not divisible by $\mathrm{LM}(g)$ for all $g \in G$.
(2) For all $f \in R[x]_>^r$ and $G = \{g_1, \ldots, g_s\} \in \mathcal{G}$ there exists a $u \in R[x]_>$ with $\mathrm{LT}(u)$ a product of leading coefficients of elements of G such that uf has a standard representation with respect to $\mathrm{NF}(- \mid G)$, that is,

$$uf = \mathrm{NF}(f \mid G) + \sum_{i=1}^{s} a_i g_i,\ a_i \in R[x]_>,$$

such that $\mathrm{LM}(r) \geq \mathrm{LM}(a_i g_i)$ for all i with $a_i g_i \neq 0$.

NF is called *polynomial* if, for $f \in R[x]^r$, $G \subset R[x]^r$, then $u, a_i \in R[x]$. Define, in this situation, the s-polynomial of $f, g \in R[x]_> \setminus \{0\}$, $\mathrm{LT}(f) = ax^\alpha e_i$, $\mathrm{LT}(g) = bx^\beta e_j$, as

$$\mathrm{spoly}\,(f, g) = \begin{cases} bx^{\gamma-\alpha}f - ax^{\gamma-\beta}g & \text{if } i = j, \\ 0 & \text{if } i \neq j, \end{cases}$$

where $\gamma = \mathrm{lcm}(\alpha, \beta)$.

Show the following: if we use this definition of s-polynomial, the algorithms NFBUCHBERGER (for global orderings), respectively NFMORA (for arbitrary orderings) from Chapter 1, Section 1.6, respectively Chapter 2, Section 2.3, define a pseudo normal form on $R[x]_>^r$ (where the element u itself from (2) is a product of leading coefficients of elements of G for NFBUCHBERGER). We call NFBUCHBERGER, respectively NFMORA, with the above s-polynomial *normal forms without division* .

2.3.7. Show the following generalization of *Buchberger's criterion*.
Let $I \subset R[x]^r_>$ be a $R[x]_>$-submodule, NF a pseudo normal form on $R[x]^r_>$ and $G = \{g_1, \ldots, g_0\} \subset I$ satisfying

(1) G generates I as $R[x]_>$-module;
(2) $\mathrm{NF}\big(\mathrm{spoly}(g_i, g_j) \mid G\big) = 0$ for all $1 \le i < j \le s$.

Then, for any maximal ideal $\mathfrak{m} \subset R$ such that $\mathrm{LC}(g_i) \notin \mathfrak{m}$ for all i, the set $\{\overline{g_1}, \ldots, \overline{g_s}\}$ is a standard basis of $I \cdot (R/\mathfrak{m})[x]_>$. Here $\overline{g_i}$ denotes the residue class of g_i in $(R/\mathfrak{m})[x]$.
(Hint: compare the proof of Theorem 2.5.9.)

We call G a *pseudo standard basis* of I if it satisfies the above conditions (1) and (2). Show that Algorithm 2.3.8, STANDARD(G,NF), returns a pseudo standard basis if NF is a pseudo normal form. (Note that a pseudo standard basis is not a standard basis over a ring, as defined in Remark 1.6.14.)

2.3.8. Let K be a field, $R = K[t] = K[t_1, \ldots, t_p]$, $K[t, x] = R[x_1, \ldots, x_n]$ and $\{g_1, \ldots, g_s\} \subset K[t, x]^r$ a pseudo standard basis of the submodule $I \subset K[t, x]^r_>$. Set $h_i = \mathrm{LC}(g_i) \in K[t]$ and $h = h_1 \cdot \ldots \cdot h_s$. Then, for any t_0 with $h(t_0) \neq 0$, show that $\{g_1(t_0, x), \ldots, g_s(t_0, x)\}$ is a standard basis of $(I|_{t=t_0})K[x]_>$.

2.3.9. Let $\{g_1, \ldots, g_s\} \subset \mathbb{Z}[x]^r = \mathbb{Z}[x_1, \ldots, x_n]^r$ be a pseudo standard basis of the submodule $I \subset \mathbb{Z}[x]^r_>$. Set $m_i = \mathrm{LC}(g_i) \in \mathbb{Z}$ and $m = m_1 \cdot \ldots \cdot m_s$. Then, for any prime number p such that $p \nmid m$, $\{\overline{g_1}, \ldots, \overline{g_s}\}$ is a standard basis of $I \cdot \mathbb{Z}/p\mathbb{Z}[x]_>$.

2.3.10. Consider $f := x^3 + y^3 + z^4 + ax^2yz + bxy^2z$, where a and b are parameters. Let $d := \dim_K K[x, y, z]/\langle f_x, f_y, f_z \rangle$, where K is the field $\mathbb{Q}$ or $\mathbb{F}_p$, p a prime. Hence, $d = d(a, b, p)$ depends on $(a, b) \in K^2$ and the characteristic p ($p = 0$ if $K = \mathbb{Q}$). Show

(1) $d(a, b, p) < \infty$ if and only if $p \notin \{2, 3\}$.
(2) For $p \notin \{2, 3\}$, we have

$$\begin{aligned} d(a, b, p) &= 24 &&\text{if } a \neq 0 \text{ and } b \neq 0 \text{ and } a^3 \neq b^3 \\ d(0, b, p) &= 21 &&\text{if } b \neq 0 \\ d(a, 0, p) &= 21 &&\text{if } a \neq 0 \\ d(a, a, p) &= 15 &&\text{if } a \neq 0 \\ d(0, 0, p) &= 12\,. \end{aligned}$$

(Hint: use the previous exercises.)

Note that SINGULAR avoids divisions in standard basis computations if the options `intStrategy` and `contentSB` are set.

2.3.11. Let $M = M(a, b) \subset \mathbb{Q}^2$ be the submodule generated by the vectors $[ax^2, (a + 3b)x^3y + z^4]$, $[(a - 2b)3y^3 + xyz, by^3]$, $[5az^4, (a + b)z^2]$ with a, b parameters. Compute a *comprehensive Gröbner basis* of M, that is, a system of generators which is a Gröbner basis of $M(a, b)$ for all $a, b \in \mathbb{Q}$ for the ordering (`c,ds`).

(Hint: start with a Gröbner basis in the ring R=(0,a,b),(x,y,z),(c,ds) and then distinguish cases, as in Exercise 2.3.10, and take the union of all Gröbner bases.)

2.3.12. Show that the system of equations

$$
\begin{aligned}
x^3 + yz + y + z &= 0 & x + y &= 1 \\
y^4 + xz + x + z &= 0 & x + z &= 1 \\
z^5 + xy + x + y &= 0 & y + z &= 1
\end{aligned}
$$

has no solution in $\overline{\mathbb{F}}_p^3$ for *all* prime numbers p, where $\overline{\mathbb{F}}_p$ is the algebraic closure of $\mathbb{F}_p$.

2.4 Exact Sequences and Free Resolutions

Definition 2.4.1. A sequence of A–modules and homomorphisms

$$
\cdots \to M_{k+1} \xrightarrow{\varphi_{k+1}} M_k \xrightarrow{\varphi_k} M_{k-1} \to \cdots
$$

is called a *complex* if $\mathrm{Ker}(\varphi_k) \supset \mathrm{Im}(\varphi_{k+1})$. It is called *exact at* M_k if

$$
\mathrm{Ker}(\varphi_k) = \mathrm{Im}(\varphi_{k+1}).
$$

It is called *exact* if it is exact at all M_k. An exact sequence

$$
0 \longrightarrow M' \xrightarrow{\varphi} M \xrightarrow{\psi} M'' \longrightarrow 0
$$

is called a *short exact sequence*.

Example 2.4.2.

(1) $0 \to M \xrightarrow{\varphi} N$ is exact if and only if φ is injective.
(2) $M \xrightarrow{\varphi} N \to 0$ is exact if and only if φ is surjective.
(3) $0 \to M \xrightarrow{\varphi} N \to 0$ is exact if and only if φ is an isomorphism.
(4) $0 \to M_1 \xrightarrow{\varphi} M_2 \xrightarrow{\psi} M_3 \to 0$ is a short exact sequence if and only if φ is injective, ψ is surjective and ψ induces an isomorphism $M_2/\mathrm{Im}(\varphi) \cong M_3$.
(5) $0 \to M_1 \xrightarrow{\varphi} M_1 \oplus M_2 \xrightarrow{\psi} M_2 \to 0$ with $\varphi(x) = (x,0)$, $\psi(x,y) = y$ is exact.

Proposition 2.4.3. *Let M', M, M'' be A–modules.*

(1) Let $M' \xrightarrow{\varphi} M \xrightarrow{\psi} M'' \to 0$ be a complex. The complex is exact if and only if, for all A–modules N, the sequence

$$
0 \to \mathrm{Hom}(M'', N) \xrightarrow{\psi^*} \mathrm{Hom}(M, N) \xrightarrow{\varphi^*} \mathrm{Hom}(M', N)
$$

is exact. Here $\psi^*(\lambda) := \lambda \circ \psi$ and $\varphi^*(\sigma) := \sigma \circ \varphi$.

(2) Let $0 \to M' \xrightarrow{\varphi} M \xrightarrow{\psi} M''$ be a complex. The complex is exact if and only if, for all A–modules N,

$$0 \to \operatorname{Hom}(N, M') \xrightarrow{\varphi_*} \operatorname{Hom}(N, M) \xrightarrow{\psi_*} \operatorname{Hom}(N, M'')$$

is exact. Here $\varphi_*(\lambda) = \varphi \circ \lambda$ and $\psi_*(\sigma) = \psi \circ \sigma$.

Proof. We prove (1). The proof of (2) is similar and left as Exercise 2.4.4. Assume that $M' \xrightarrow{\varphi} M \xrightarrow{\psi} M'' \to 0$ is exact and consider the sequence

$$0 \to \operatorname{Hom}(M'', N) \xrightarrow{\psi^*} \operatorname{Hom}(M, N) \xrightarrow{\varphi^*} \operatorname{Hom}(M', N) \,,$$

which is a complex, since $\varphi^* \circ \psi^* = (\psi \circ \varphi)^*$. We have to show that it is, indeed, exact: let $\psi^*(\lambda) = \lambda \circ \psi = 0$, then, since ψ is surjective, $\lambda = 0$. Hence, ψ^* is injective. Let $\sigma \in \operatorname{Hom}(M, N)$ with $\varphi^*(\sigma) = \sigma \circ \varphi = 0$,

and define $\bar{\sigma} : M'' \to N$ by $\bar{\sigma}(m'') = \sigma(m)$ for some $m \in M$ with $\psi(m) = m''$. $\bar{\sigma}$ is well–defined because $\operatorname{Ker}(\psi) = \operatorname{Im}(\varphi)$ and $\sigma \circ \varphi = 0$. We have $\psi^*(\bar{\sigma}) = \sigma$ and, hence, $\operatorname{Im}(\psi^*) \supset \operatorname{Ker}(\varphi^*)$.

Assume now that $0 \to \operatorname{Hom}(M'', N) \xrightarrow{\psi^*} \operatorname{Hom}(M, N) \xrightarrow{\varphi^*} \operatorname{Hom}(M', N)$ is exact for all A–modules N, that is,

- ψ^* is injective;
- $\operatorname{Im}(\psi^*) = \operatorname{Ker}(\varphi^*)$.

To prove that ψ is surjective, we consider $N := M''/\operatorname{Im}(\psi)$ and the canonical map $\pi : M'' \to N$. Then $\psi^*(\pi) = \pi \circ \psi = 0$. Because ψ^* is injective we obtain $\pi = 0$ and, therefore, $N = 0$, that is, ψ is surjective.

To prove that $\operatorname{Ker}(\psi) = \operatorname{Im}(\varphi)$ we choose $N = M/\operatorname{Im}(\varphi)$ and $\pi : M \to N$ the canonical morphism. Then $\varphi^*(\pi) = \pi \circ \varphi = 0$. Hence, $\pi \in \operatorname{Im}(\psi^*)$, that is, $\pi = \psi^*(\sigma) = \sigma \circ \psi$ for a suitable $\sigma : M'' \to N$, as shown in the diagram

This implies that $\operatorname{Im}(\varphi) \supset \operatorname{Ker}(\psi)$, the inverse inclusion being satisfied by assumption. $\qquad \square$

Remark 2.4.4. Let

$$0 \to M_n \xrightarrow{\varphi_n} M_{n-1} \xrightarrow{\varphi_{n-1}} \cdots \to M_2 \xrightarrow{\varphi_2} M_1 \xrightarrow{\varphi_1} M_0 \to 0$$

be a sequence of A–modules. The sequence is exact if and only if $\mathrm{Im}(\varphi_1) = M_0$, $\mathrm{Ker}(\varphi_n) = 0$ and $\mathrm{Ker}(\varphi_i) = \mathrm{Im}(\varphi_{i+1})$ for $i = 1, \ldots, n-1$. If the sequence is exact, this defines short exact sequences

$$0 \to \mathrm{Ker}(\varphi_1) \to M_1 \xrightarrow{\varphi_1} \mathrm{Im}(\varphi_1) \to 0$$
$$0 \to \mathrm{Ker}(\varphi_2) \to M_2 \xrightarrow{\varphi_2} \mathrm{Im}(\varphi_2) \to 0$$
$$\vdots \qquad \qquad \vdots$$
$$0 \to \mathrm{Ker}(\varphi_n) \to M_n \xrightarrow{\varphi_n} \mathrm{Im}(\varphi_n) \to 0.$$

Conversely, if short exact sequences

$$0 \to K_1 \to M_1 \to M_0 \to 0$$
$$0 \to K_2 \to M_2 \to K_1 \to 0$$
$$\vdots \qquad \qquad \vdots$$
$$0 \to 0 \to M_n \to K_{n-1} \to 0$$

are given, then they obviously lead to a long exact sequence

$$0 \to M_n \to M_{n-1} \to \cdots \to M_1 \to M_0 \to 0.$$

Definition 2.4.5. Let A be a ring and $\mathcal{C}$ be a class of A–modules. A map $\lambda : \mathcal{C} \to \mathbb{Z}$ is called *additive function* if $\lambda(M) = \lambda(M') + \lambda(M'')$ for every short exact sequence $0 \to M' \to M \to M'' \to 0$ with $M', M, M'' \in \mathcal{C}$.

Example 2.4.6.

(1) Let K be a field and $\mathcal{C}$ be the class of all finite dimensional K–vector spaces. Then $\lambda : \mathcal{C} \to \mathbb{Z}$ defined by $\lambda(V) = \dim_K(V)$ is additive.
(2) Let $\mathcal{C}$ be the class of finitely generated abelian groups, that is, finitely generated $\mathbb{Z}$–modules. It will be proved in Section 2.6 that every finitely generated $\mathbb{Z}$–module M decomposes as $M = F \oplus \mathrm{Tors}(M)$, with F a free module. Defining $\lambda(M) = \mathrm{rank}(F)$ we obtain an additive map $\lambda : \mathcal{C} \to \mathbb{Z}$.

Proposition 2.4.7. *Let $\mathcal{C}$ be a class of A–modules which contains all submodules and factor modules of each of its elements and let $\lambda : \mathcal{C} \to \mathbb{Z}$ be additive. If*

$$0 \to M_n \xrightarrow{\varphi_n} M_{n-1} \xrightarrow{\varphi_{n-1}} \cdots \to M_1 \xrightarrow{\varphi_{-1}} M_0 \to 0$$

is an exact sequence with $M_0, \ldots, M_n \in \mathcal{C}$, then $\sum_{i=0}^{n}(-1)^i \lambda(M_i) = 0$.

Proof. We consider the short exact sequences

$$0 \longrightarrow \operatorname{Ker}(\varphi_i) \longrightarrow M_i \longrightarrow \operatorname{Im}(\varphi_i) \longrightarrow 0, \quad i = 1, \ldots, n.$$

The additivity of λ implies

$$\lambda(M_1) = \lambda\big(\operatorname{Ker}(\varphi_1)\big) + \lambda(M_0),$$
$$\lambda(M_2) = \lambda\big(\operatorname{Ker}(\varphi_2)\big) + \lambda\big(\operatorname{Ker}(\varphi_1)\big),$$
$$\vdots$$
$$\lambda(M_n) = \lambda\big(\operatorname{Ker}(\varphi_n)\big) + \lambda\big(\operatorname{Ker}(\varphi_{n-1})\big).$$

Taking the alternating sum, we obtain

$$\sum_{i=1}^{n} (-1)^{i-1}\lambda(M_i) = \lambda(M_0) + (-1)^{n-1}\lambda\big(\operatorname{Ker}(\varphi_n)\big).$$

But $\operatorname{Ker}(\varphi_n) = 0$ and λ being additive implies that $\lambda\big(\operatorname{Ker}(\varphi_n)\big) = 0$. □

Lemma 2.4.8 (Snake Lemma). *Let* $0 \to M_1 \xrightarrow{\varphi_1} M_2 \xrightarrow{\varphi_2} M_3 \to 0$ *and* $0 \to N_1 \xrightarrow{\psi_1} N_2 \xrightarrow{\psi_2} N_3 \to 0$ *be short exact sequences of* A–*modules. Moreover, let* $\lambda_i : M_i \to N_i$, $i = 1, 2, 3$, *be module homomorphisms such that the induced diagram commutes, that is,* $\lambda_3 \circ \varphi_2 = \psi_2 \circ \lambda_2$ *and* $\lambda_2 \circ \varphi_1 = \psi_1 \circ \lambda_1$. *Then there is an exact sequence*

$$0 \to \operatorname{Ker}(\lambda_1) \to \operatorname{Ker}(\lambda_2) \to \operatorname{Ker}(\lambda_3)$$
$$\to \operatorname{Coker}(\lambda_1) \to \operatorname{Coker}(\lambda_2) \to \operatorname{Coker}(\lambda_3) \to 0.$$

Proof. The sequences $0 \to \operatorname{Ker}(\lambda_i) \xrightarrow{\nu_i} M_i \xrightarrow{\lambda_i} N_i \xrightarrow{\pi_i} \operatorname{Coker}(\lambda_i) \to 0$ are exact and lead to the following diagram:

It is not difficult to see that the canonically defined φ_i' (restriction of the φ_i) and ψ_i' make the diagram commutative, that is,

$$\pi_3 \circ \psi_2 = \psi_2' \circ \pi_2, \quad \pi_2 \circ \psi_1 = \psi_1' \circ \pi_1, \quad \varphi_2 \circ \nu_2 = \nu_3 \circ \varphi_2',$$
$$\nu_2 \circ \varphi_1' = \varphi_1 \circ \nu_1, \quad \lambda_3 \circ \varphi_2 = \psi_2 \circ \lambda_2 \text{ and } \lambda_2 \circ \varphi_1 = \psi_1 \circ \lambda_1.$$

We have to prove that the first and the last row are exact. This is left as an exercise.

The important part of the proof is to define the connecting homomorphism $d : \mathrm{Ker}(\lambda_3) \to \mathrm{Coker}(\lambda_1)$ and prove that $\mathrm{Ker}(d) = \mathrm{Im}(\varphi_2')$ and $\mathrm{Im}(d) = \mathrm{Ker}(\psi_1')$.

Let $x \in \mathrm{Ker}(\lambda_3)$ and choose $y \in M_2$ with $\varphi_2(y) = x$.[4] Then

$$0 = \lambda_3(x) = \lambda_3 \circ \varphi_2(y) = \psi_2 \circ \lambda_2(y).$$

The exactness of the third row of the diagram implies that there exists a unique $z \in N_1$ with $\psi_1(z) = \lambda_2(y)$. We define $d(x) := \pi_1(z)$.

We have to prove that this definition is independent of the choice of y. Let $\overline{z} \in N_1$ with $\psi_1(\overline{z}) = \lambda_2(\overline{y})$ and $\varphi_2(\overline{y}) = x$. Then $\varphi_2(y - \overline{y}) = 0$ implies that $y - \overline{y} = \varphi_1(u)$ for a suitable $u \in M_1$. Therefore,

$$\psi_1(z - \overline{z}) = \lambda_2(y - \overline{y}) = \lambda_2 \circ \varphi_1(u) = \psi_1 \circ \lambda_1(u).$$

But ψ_1 is injective, hence $z = \overline{z} + \lambda_1(u)$. It follows that $\pi_1(z) = \pi_1(\overline{z})$. This proves that d is well–defined.

It is not difficult to see that d is, indeed, a module homomorphism. We shall prove now that $\mathrm{Ker}(d) = \mathrm{Im}(\varphi_2')$.

Let $x \in \mathrm{Ker}(d)$. This implies, by definition of d, that there exist $z \in N_1$ and $y \in M_2$ such that $\psi_1(z) = \lambda_2(y)$, $\varphi_2(y) = x$ and $\pi_1(z) = 0$. Let $u \in M_1$ such that $z = \lambda_1(u)$. Then $\lambda_2(y) = \psi_1(z) = \psi_1 \circ \lambda_1(u) = \lambda_2 \circ \varphi_1(u)$, whence, $y - \varphi_1(u) \in \mathrm{Ker}(\lambda_2)$. But $\varphi_2'\big(y - \varphi_1(u)\big) = \varphi_2\big(y - \varphi_1(u)\big) = \varphi_2(y) = x$. It follows that $\mathrm{Ker}(d) \subset \mathrm{Im}(\varphi_2')$.

On the other hand, let $x \in \mathrm{Im}(\varphi_2')$. Then $x = \varphi_2(y)$, $y \in \mathrm{Ker}(\lambda_2)$ implies $\psi_1(0) = 0 = \lambda_2(y)$, that is, $d(x) = 0$ by definition of d. Thus we have shown the equality $\mathrm{Ker}(d) = \mathrm{Im}(\varphi_2')$.

To prove that $\mathrm{Im}(d) = \mathrm{Ker}(\psi_1')$ let $\overline{x} \in \mathrm{Im}(d)$, that is, $\overline{x} = \pi_1(z)$ such that there exists a $y \in M_2$ with $\lambda_2(y) = \psi_1(z)$. But

$$0 = \pi_2 \circ \lambda_2(y) = \pi_2 \circ \psi_1(z) = \psi_1' \circ \pi_1(z) = \psi_1'(\overline{z})$$

and, therefore, $\mathrm{Im}(d) \subset \mathrm{Ker}(\psi_1')$. Now let $\overline{z} \in \mathrm{Ker}(\psi_1')$ and choose a preimage $z \in N_1, \pi(z) = \overline{z}$. Then $0 = \psi_1'(\overline{z}) = \psi_1' \circ \pi_1(z) = \pi_2 \circ \psi_1(z)$. Therefore, there exists some $y \in M_2$ with $\lambda_2(y) = \psi_1(z)$. Then $\overline{z} = d\big(\varphi_2(y)\big)$, which proves that $\mathrm{Ker}(\psi_1') \subset \mathrm{Im}(d)$. $\qquad\square$

---

[4] Since ν_3 is the canonical inclusion, we simplify the notations by identifying $\nu_3(x)$ and x.

Proposition 2.4.9. *Let*

$$\cdots \to M_{k+1} \xrightarrow{\varphi_{k+1}} M_k \xrightarrow{\varphi_k} M_{k-1} \to \cdots$$

be an exact sequence of A–modules and $x \in A$ a non–zerodivisor for all M_k, then the induced sequence

$$\cdots \to M_{k+1}/xM_{k+1} \to M_k/xM_k \to M_{k-1}/xM_{k-1} \to \cdots$$

is exact.

Proof. Because of Remark 2.4.4 it is enough to prove the proposition for short exact sequences. Consider an exact sequence

$$0 \to M_1 \xrightarrow{\varphi_1} M_2 \xrightarrow{\varphi_2} M_3 \to 0$$

then the multiplication by x induces injective maps $M_i \to M_i$, $i = 1, 2, 3$. Using the Snake Lemma (Lemma 2.4.8) we obtain that the induced sequence

$$0 \to M_1/xM_1 \to M_2/xM_2 \to M_3/xM_3 \to 0$$

is exact. $\square$

Definition 2.4.10. Let A be a ring and M a finitely generated A–module. A *free resolution* of M is an exact sequence[5]

$$\cdots \longrightarrow F_{k+1} \xrightarrow{\varphi_{k+1}} F_k \longrightarrow \cdots \longrightarrow F_1 \xrightarrow{\varphi_1} F_0 \xrightarrow{\varphi_0} M \to 0$$

with finitely generated free A–modules F_i for $i \geq 0$. We say that a free resolution has *(finite) length* n if $F_k = 0$ for all $k > n$ and n is minimal with this property.

If $(A, \mathfrak{m})$ is a local ring, then a free resolution as above is called *minimal* if $\varphi_k(F_k) \subset \mathfrak{m}F_{k-1}$ for $k \geq 1$, and then $b_k(M) := \mathrm{rank}(F_k)$, $k \geq 0$, is called the k–th *Betti number* of M.

The following theorem shows that the Betti numbers of M are, indeed, well–defined.

Theorem 2.4.11. *Let $(A, \mathfrak{m})$ be a local Noetherian ring and M be a finitely generated A–module, then M has a minimal free resolution. The rank of F_k in a minimal free resolution is independent of the resolution. If M has a minimal resolution of finite length n,*

---

[5] Frequently the complex of free A–modules

$$F_\bullet : \cdots \longrightarrow F_{k+1} \xrightarrow{\varphi_{k+1}} F_k \longrightarrow \cdots \longrightarrow F_1 \xrightarrow{\varphi_1} F_0 \longrightarrow 0$$

is called a free resolution of M.

$$0 \to F_n \to F_{n-1} \to \cdots \to F_0 \to M \to 0$$

and if

$$0 \to G_m \to G_{m-1} \to \cdots \to G_0 \to M \to 0$$

is any free resolution, then $m \geq n$.

Proof. Let $m_1, \ldots, m_{s_0}$ be a minimal set of generators of M and consider the surjective map $\varphi_0 : F_0 := A^{s_0} \to M$ defined by $\varphi_0(a_1, \ldots, a_{s_0}) = \sum_{i=1}^{s_0} a_i m_i$. Because of Nakayama's Lemma, $m_1, \ldots, m_{s_0}$ induces a basis of the vector space $M/\mathfrak{m}M$. Hence, φ_0 induces an isomorphism $\overline{\varphi}_0 : F_0/\mathfrak{m}F_0 \cong M/\mathfrak{m}M$. Let K_1 be the kernel of φ_0. Then $K_1 \subset \mathfrak{m}F_0$. K_1 is a submodule of a finitely generated module over a Noetherian ring, hence finitely generated. As before, we can find a surjective map $F_1 := A^{s_1} \to K_1$, where s_1 is the minimal number of generators of K_1.

Let $\varphi_1 : F_1 \to F_0$ be defined by the composition $F_1 \to K_1 \hookrightarrow F_0$. As $K_1 \subset \mathfrak{m}F_0$, it follows that $\varphi_1(F_1) \subset \mathfrak{m}F_0$. Up to now, we have constructed the exact sequence $F_1 \xrightarrow{\varphi_1} F_0 \xrightarrow{\varphi_0} M \to 0$.

Continuing like this we obtain a minimal free resolution for M. To show the invariance of the Betti numbers, we consider two minimal resolutions of M:

$$\cdots \xrightarrow{\varphi_{n+1}} F_k \to \ldots \xrightarrow{\varphi_1} F_0 \xrightarrow{\varphi_0} M \to 0$$
$$\cdots \xrightarrow{\psi_{n+1}} G_k \to \ldots \xrightarrow{\psi_1} G_0 \xrightarrow{\psi_0} M \to 0 .$$

We have $F_0/\mathfrak{m}F_0 \cong M/\mathfrak{m}M \cong G_0/\mathfrak{m}G_0$ and, therefore, $\mathrm{rank}(F_0) = \mathrm{rank}(G_0)$. Let $\{f_1, \ldots, f_{s_0}\}$, respectively $\{g_1, \ldots, g_{s_0}\}$, be bases of F_0, respectively G_0. As $\{\psi_0(g_i)\}$ generate M, we have $\varphi_0(f_i) = \sum_j h_{ij} \cdot \psi_0(g_j)$ for some $h_{ij} \in A$. The matrix (h_{ij}) defines a map $h_1 : F_0 \to G_0$. The induced map $\overline{h}_1 : F_0/\mathfrak{m}F_0 \to G_0/\mathfrak{m}G_0$ is an isomorphism. In particular, we derive that $\det(h_{ij}) \neq 0 \mod \mathfrak{m}$. This implies that $\det(h_{ij})$ is a unit in A and h_1 is an isomorphism. Especially, h_1 induces an isomorphism $\mathrm{Ker}(\varphi_0) \xrightarrow{\sim} \mathrm{Ker}(\psi_0)$. As φ_1 and ψ_1, considered as matrices, have entries in $\mathfrak{m}$, and since we have surjections $F_1 \to \mathrm{Ker}(\varphi_0)$ and $G_1 \to \mathrm{Ker}(\psi_0)$, it follows, as before, that $\mathrm{rank}(F_1) = \mathrm{rank}(G_1)$. Now we can continue like this and obtain the invariance of the Betti numbers.

To prove the last statement, let

$$0 \to F_n \to F_{n-1} \to \cdots \to F_0 \to M \to 0$$

be a minimal free resolution with $F_n \neq \langle 0 \rangle$ and

$$0 \to G_m \to G_{m-1} \to \cdots \to G_0 \to M \to 0$$

be any free resolution. We have to prove that $m \geq n$. This can be proved in a similar way to the previous step. With the same idea, one can prove that there are injections $h_i : F_i \to G_i$ for all $i \leq n$. $\qquad\square$

2.4 Exact Sequences and Free Resolutions 153

SINGULAR Example 2.4.12 (resolution and Betti numbers).
SINGULAR has several commands, based on different algorithms, to compute free resolutions, see also Section 2.5. mres computes, for modules over local rings and for homogeneous modules over graded rings (cf. Definition 2.4.13), a minimal free resolution. More precisely, let $A = \mathtt{matrix}(I)$, then $\mathtt{mres(I,k)}$ computes a free resolution of $\mathrm{Coker}(A) = F_0/I$ [6]

$$\ldots \to F_2 \xrightarrow{\varphi_2} F_1 \xrightarrow{\varphi_1} F_0 \to F_0/I \to 0,$$

where the columns of the matrix φ_1 are a minimal set of generators of I if the basering is local or if I is homogeneous. If k is not zero then the computation stops after k steps and returns a list of modules $M_i = \mathtt{module}(\varphi_i)$, $i = 1 \ldots k$. $\mathtt{mres(I,0)}$ stops computation after, at most, $n + 2$ steps, where n is the number of variables of the basering. Note that the latter suffices to compute all non-zero modules of a minimal free resolution if the basering is not a quotient ring (cf. Theorem 2.5.15).

In some cases it is faster to use the SINGULAR commands res (or sres, or lres) and then to apply minres to minimize the computed resolution.

```
ring A=0,(x,y),(c,ds);
ideal I=x,y;
resolution Re=mres(I,0);
```

Typing Re; displays a pictorial description of the computed resolution, where the exponents are the ranks of the free modules and the lower index i corresponds to the index of the respective free module F_i in the resolution of $M = A/I$ (see Definition 2.4.10).

```
Re;
//-> 1     2      1
//-> A  <-- A  <-- A
//->
//-> 0     1      2
```

The corresponding list of matrices φ_i is displayed when typing print(Re);. More precisely, Re[i] $= \mathrm{Im}(\varphi_i)$, hence the columns of φ_i are given by the generators of Re[i].

```
print(Re);
//-> [1]:
//->    _[1]=x
//->    _[2]=y
//-> [2]:
//->    _[1]=[y,-x]
```

---

[6] To obtain a minimal free resolution of F_0/I, use mres(prune(I),0).

This is an example of the resolution of $\mathbb{Q} = A/\langle x, y \rangle$, as $A = \mathbb{Q}[x, y]_{\langle x,y \rangle}-$ module:

$$0 \to \mathbb{Q}[x,y]_{\langle x,y \rangle} \xrightarrow{\binom{y}{-x}} \mathbb{Q}[x,y]^2_{\langle x,y \rangle} \xrightarrow{(x,y)} \mathbb{Q}[x,y]_{\langle x,y \rangle} \to \mathbb{Q} \to 0 \ .$$

We can see that the Betti numbers $b_k(A/\langle x, y \rangle)$ are 1, 2 and 1. Let us compute them with SINGULAR:

```
betti(Re);            //intmat of Betti numbers
//-> 1,2,1
```

Now we consider an example of a cyclic infinite minimal resolution of the module $M = R^2 / \langle \binom{x}{0}, \binom{0}{y} \rangle$ over the local ring $R = A/\langle xy \rangle$ with $A = \mathbb{Q}[x,y]_{\langle x,y \rangle}$.

```
qring R=std(xy);
module M=[x,0],[0,y];
resolution Re=mres(M,4);      //mres(M,k) stops at F_k
Re;
//-> 2      2      2      2      2
//-> R  <-- R  <-- R  <-- R  <-- R
//->
//-> 0      1      2      3      4
```

Let us have a look at the matrices in the computed resolution:

```
print(Re);
//-> [1]:
//->    _[1]=[x]
//->    _[2]=[0,y]
//-> [2]:
//->    _[1]=[y]
//->    _[2]=[0,x]
//-> [3]:
//->    _[1]=[x]
//->    _[2]=[0,y]
//-> [4]:
//->    _[1]=[y]
//->    _[2]=[0,x]
```

Definition 2.4.13. Let K be a field, A be a graded K–algebra and M a graded A–module: a *homogeneous free resolution* of M is a resolution

$$\cdots \to F_{k+1} \xrightarrow{\varphi_{k+1}} F_k \to \ldots \xrightarrow{\varphi_1} F_0 \to M \to 0$$

such that the F_k are finitely generated free A–modules,

$$F_k = \bigoplus_{j \in \mathbb{Z}} A(-j)^{b_{j-k,k}} \ ,$$

and the φ_k are homogeneous maps (of degree 0). Such a resolution is called *minimal*, if $\varphi_k(F_k) \subset \mathfrak{m}F_{k-1}$, where $\mathfrak{m}$ is the ideal generated by all elements of positive degree. Then the numbers $b_{j,k} =: b_{j,k}(M)$ are called *graded Betti numbers* of M and $b_k(M) := \sum_j b_{j,k}(M)$ is called the *k–th Betti number* of M.

The following theorem shows that the graded Betti numbers of M are well-defined.

Theorem 2.4.14. *Let K be a field, A be a graded K–algebra and M be a finitely generated graded A–module. Then M has a minimal free resolution. The numbers $b_{j,k}$ and, in particular, the rank of F_k, in a minimal free resolution are independent of the resolution.*

The proof is similar to the proof of Theorem 2.4.11 and left as Exercise 2.4.3.

SINGULAR Example 2.4.15 (homogeneous resolution and graded Betti numbers).
We compute a minimal resolution of a homogeneous module $M = A/I$, with $A = \mathbb{Q}[w, x, y, z]$ and $I = \langle xyz, wz, x + y \rangle$, and compute its graded Betti numbers.

```
ring A = 0,(w,x,y,z),(c,dp);
ideal I = xyz, wz, x+y;
resolution Re = mres(I,0);
Re;
//-> 1      3      3      1
//-> A <-- A <-- A <-- A
//->
//-> 0      1      2      3

print(Re);        //display the matrices in the resolution
//-> [1]:
//->    _[1]=x+y
//->    _[2]=wz
//->    _[3]=y2z
//-> [2]:
//->    _[1]=[0,y2,-w]
//->    _[2]=[wz,-x-y]
//->    _[3]=[y2z,0,-x-y]
//-> [3]:
//->    _[1]=[x+y,y2,-w]
//-> [4]:
//->    _[1]=0
```

To show the correct format of the matrices, we use `print(matrix(Re[i]))`:

```
print(matrix(Re[2]));
//-> 0, wz,  y2z,
//-> y2,-x-y,0,
//-> -w,0,   -x-y
```

We compute the matrix of graded Betti numbers $\big(b_{j,k}(A/I)\big)$: [7]

```
betti(Re);
//-> 1,1,0,0,
//-> 0,1,1,0,
//-> 0,1,2,1
```

The print command allows attributes, like `"betti"`, to format the output:

```
print(betti(Re),"betti");    //display graded Betti numbers
//->            0    1    2    3
//-> ------------------------------
//->    0:      1    1    -    -
//->    1:      -    1    1    -
//->    2:      -    1    2    1
//-> ------------------------------
//-> total:     1    3    3    1
```

Hence, we have computed the following (minimal) homogeneous free resolution (written as displayed by `Re;`):

$$0 \longleftarrow A/I \longleftarrow A(0) \xleftarrow{(x+y,wz,xyz)} \begin{matrix} A(-1) \\ \oplus \\ A(-2) \\ \oplus \\ A(-3) \end{matrix} \xleftarrow{\left(\begin{smallmatrix} 0 & wz & y^2z \\ y^2 & -x-y & 0 \\ -w & 0 & -x-y \end{smallmatrix}\right)} \begin{matrix} A(-4) \\ \oplus \\ A(-3) \\ \oplus \\ A(-4) \end{matrix}$$

$$\xleftarrow{\left(\begin{smallmatrix} x+y \\ yz \\ -w \end{smallmatrix}\right)} A(-5) \longleftarrow 0.$$

Exercises

2.4.1. Let M, N be A–modules and $0 \to M \xrightarrow{\varphi} N \xrightarrow{\psi} A^s \to 0$ be an exact sequence, then this sequence splits, that is, there exists an isomorphism $\lambda : M \oplus A^s \to N$ such that the diagram

$$\begin{array}{ccccccccc}
0 & \longrightarrow & M & \xrightarrow{\varphi} & N & \xrightarrow{\psi} & A^s & \longrightarrow & 0 \\
 & & \| & & \uparrow{\scriptstyle\lambda} & & \| & & \\
0 & \longrightarrow & M & \xrightarrow{i} & M \oplus A^s & \xrightarrow{\pi} & A^s & \longrightarrow & 0
\end{array}$$

is commutative ($\lambda \circ i = \varphi$, $\psi \circ \lambda = \pi$).

---

[7] Note that $b_{j,k}(I) = b_{j-1,k+1}(A/I)$ for all $k \geq 0$.

2.4.2. Let $A = \mathbb{Q}[x,y]_{\langle x,y \rangle}/\langle xy \rangle$. Compute the Betti numbers of the A–module $M = \left\langle \left(\begin{smallmatrix} x^2 \\ y \end{smallmatrix} \right), \left(\begin{smallmatrix} x \\ y \end{smallmatrix} \right) \right\rangle$.

2.4.3. Prove Theorem 2.4.14.
(Hint: Use the fact that kernel and image of a homogeneous map are graded with the induced grading and choose in every step a minimal system of generators using Nakayama's Lemma (Exercise 2.2.6)).

2.4.4. Prove (2) of Proposition 2.4.3.

2.4.5. With the notations of the Snake Lemma (Lemma 2.4.8) prove that

(1) $0 \to \mathrm{Ker}(\lambda_1) \to \mathrm{Ker}(\lambda_2) \to \mathrm{Ker}(\lambda_3)$,
(2) $\mathrm{Coker}(\lambda_1) \to \mathrm{Coker}(\lambda_2) \to \mathrm{Coker}(\lambda_3) \to 0$,

are exact sequences.

2.4.6. Let A be a ring and $S \subset A$ a multiplicatively closed subset. Let $M' \to M \to M''$ be an exact sequence of A–modules. Prove that the induced sequence $M'_S \to M_S \to M''_S$ of $S^{-1}A$–modules is exact.

2.4.7. Let A be a ring and $\cdots \to M_{i+1} \to M_i \to M_{i-1} \to \cdots$ be a complex of A–modules. Prove that the complex is exact if and only if the induced complexes $\cdots \to (M_{i+1})_P \to (M_i)_P \to (M_{i-1})_P \to \cdots$ of A_P–modules are exact for all prime ideals $P \subset A$.

2.4.8. Compute a minimal free resolution of $\mathbb{Q}$ as $\mathbb{Q}[x_1, \ldots, x_n]$–module for small n by using SINGULAR. Do you see a pattern, at least for the Betti numbers?
(Hint: if you do not succeed, you may have a look at Sections 2.5 and 7.6.)

2.4.9. Let $A = \mathbb{Q}[x,y,z]_{\langle x,y,z \rangle}/\langle x^3 + y^3 + z^3 \rangle$. Use SINGULAR to compute several steps of a minimal free resolution of $\langle x, y, z \rangle$ as A–module, until you see a periodicity. Prove that the resolution is infinite.

2.5 Computing Resolutions and the Syzygy Theorem

Let K be a field and $>$ a monomial ordering on $K[x]^r$. Again R denotes the localization of $K[x]$ with respect to $S_>$.

We shall give a method, using standard bases, to compute syzygies and, more generally, free resolutions of finitely generated R–modules. Syzygies and free resolutions are very important objects and basic ingredients for many constructions in homological algebra and algebraic geometry. On the other hand, the use of syzygies gives a very elegant way to prove Buchberger's criterion for standard bases. Moreover, a close inspection of the syzygies of the generators of an ideal allows detection of useless pairs during the computation of a standard basis.

In the following definition R can be an arbitrary ring.

Definition 2.5.1. A *syzygy* or *relation* between k elements $f_1, \ldots, f_k$ of an R–module M is a k–tuple $(g_1, \ldots, g_k) \in R^k$ satisfying

$$\sum_{i=1}^{k} g_i f_i = 0 \,.$$

The set of all syzygies between $f_1, \ldots, f_k$ is a submodule of R^k. Indeed, it is the kernel of the ring homomorphism

$$\varphi : F_1 := \bigoplus_{i=1}^{k} R\varepsilon_i \longrightarrow M \,, \quad \varepsilon_i \longmapsto f_i \,,$$

where $\{\varepsilon_1, \ldots, \varepsilon_k\}$ denotes the canonical basis of R^k. φ surjects onto the R–module $I := \langle f_1, \ldots, f_k \rangle_R$ and

$$\mathrm{syz}(I) := \mathrm{syz}(f_1, \ldots, f_k) := \mathrm{Ker}(\varphi)$$

is called the *module of syzygies* of I with respect to the generators $f_1, \ldots, f_k$. [8]

Remark 2.5.2. If R is a local (respectively graded) ring and $\{f_1, \ldots, f_k\}$, $\{g_1, \ldots, g_k\}$ are minimal sets of (homogeneous) generators of I then

$$\mathrm{syz}(f_1, \ldots, f_k) \cong \mathrm{syz}(g_1, \ldots, g_k) \,,$$

hence, $\mathrm{syz}(I)$ is well–defined up to (graded) isomorphism (cf. Exercises 2.5.7 and 2.5.8). More generally, setting $\mathrm{syz}_0(I) := I$, the modules

$$\mathrm{syz}_k(I) := \mathrm{syz}\big(\mathrm{syz}_{k-1}(I)\big) \,,$$

$k \geq 1$, are well–defined up to (graded) isomorphisms. We call $\mathrm{syz}_k(I)$ the *k–th syzygy module* of I. [9]

Note that the k–th Betti number $b_k(I)$ is the minimal number of generators for the k–th syzygy module $\mathrm{syz}_k(I)$. Moreover, in the homogeneous case, the graded Betti number $b_{j,k}(I)$ is the minimal number of generators of the k–th syzygy module $\mathrm{syz}_k(I)$ in degree $j + k$.

---

[8] In general, the notion $\mathrm{syz}(I)$ is a little misleading, because it depends on the chosen system of generators of I. But it can be proved (cf. Exercise 2.5.6) that for $I = \langle f_1, \ldots, f_k \rangle = \langle g_1, \ldots, g_s \rangle$,

$$\mathrm{syz}(f_1, \ldots, f_k) \oplus R^s \cong \mathrm{syz}(g_1, \ldots, g_s) \oplus R^k \,.$$

For this reason, we keep using the notation $\mathrm{syz}(I)$ as long as we are not interested in a special system of generators.

[9] We also write $\mathrm{syz}^R(I)$, respectively $\mathrm{syz}_k^R(I)$, if we want to emphasize the basering R.

The following lemma provides a method to compute syzygies for submodules of R^r, $R = K[x]_>$, $x = (x_1, \ldots, x_n)$.

Lemma 2.5.3. *Let* $I = \langle f_1, \ldots, f_k \rangle_R \subset R^r = \bigoplus_{i=1}^r Re_i$, *with* $e_1, \ldots, e_r$ *the canonical basis of* R^r. *Consider the canonical embedding*

$$R^r \subset R^{r+k} = \bigoplus_{i=1}^{r+k} Re_i$$

and the canonical projection $\pi : R^{r+k} \to R^k$. *Let* $G = \{g_1, \ldots, g_s\}$ *be a standard basis of* $F = \langle f_1 + e_{r+1}, \ldots, f_k + e_{r+k} \rangle_R$ *with respect to an elimination ordering for* $e_1, \ldots, e_r$ *(for example, the ordering* $(c, <)$: $x^\alpha e_i < x^\beta e_j$ *if* $j < i$ *or* $j = i$ *and* $x^\alpha < x^\beta$). *Suppose that* $\{g_1, \ldots, g_\ell\} = G \cap \bigoplus_{i=r+1}^{r+k} Re_i$, *then*

$$\mathrm{syz}(I) = \langle \pi(g_1), \ldots, \pi(g_\ell) \rangle.$$

Proof. $G \cap \bigoplus_{i=r+1}^{r+k} Re_i$ is a standard basis of $F \cap \bigoplus_{i=r+1}^{r+k} Re_i$ (see Lemma 2.8.2, below). On the other hand, $\pi\left(F \cap \bigoplus_{i=r+1}^{r+k} Re_i\right) = \mathrm{syz}(I)$. Namely, let $h \in F \cap \bigoplus_{i=r+1}^{r+k} Re_i$, that is, $h = \sum_{\nu=r+1}^{r+k} h_\nu e_\nu = \sum_{j=1}^k b_j(f_j + e_{r+j})$ for suitable $b_j \in R$. This implies that $\sum_{j=1}^k b_j f_j = 0$ and $b_j = h_{r+j}$.

Conversely, if $h = (h_1, \ldots, h_k) \in \mathrm{syz}(I)$, that is, if $\sum_{\nu=1}^k h_\nu f_\nu = 0$, then $\sum_{\nu=1}^k h_\nu(f_\nu + e_{r+\nu}) \in F \cap \bigoplus_{i=r+1}^{r+k} Re_i$. □

Algorithm 2.5.4 ($\mathrm{SYZ}(f_1, \ldots, f_k)$).

Let $>$ be any monomial ordering on $\mathrm{Mon}(x_1, \ldots, x_n)$ and $R = K[x]_>$.

Input: $f_1, \ldots, f_k \in K[x]^r$.
Output: $S = \{s_1, \ldots, s_\ell\} \subset K[x]^k$ such that $\langle S \rangle = \mathrm{syz}(f_1, \ldots, f_k) \subset R^k$.

- $F := \{f_1 + e_{r+1}, \ldots, f_k + e_{r+k}\}$, where $e_1, \ldots, e_{r+k}$ denote the canonical generators of $R^{r+k} = R^r \oplus R^k$ such that $f_1, \ldots, f_k \in R^r = \bigoplus_{i=1}^r Re_i$;
- compute a standard basis G of $\langle F \rangle \subset R^{r+k}$ with respect to $(c, >)$;
- $G_0 := G \cap \bigoplus_{i=r+1}^{r+k} Re_i = \{g_1, \ldots, g_\ell\}$, with $g_i = \sum_{j=1}^k a_{ij}e_{r+j}$, $i = 1, \ldots, \ell$;
- $s_i := (a_{i1}, \ldots a_{ik})$, $i = 1, \ldots, \ell$;
- return $S = \{s_1, \ldots, s_\ell\}$.

SINGULAR Example 2.5.5 (syzygies).

We apply first the built-in command **syz**, while, in the second example, we proceed as in Lemma 2.5.3 (resp. Algorithm 2.5.4). The latter method gives more flexibility in choosing a faster ordering for specific examples.

```
ring R=0,(x,y,z),(c,dp);
ideal I=xy,yz,xz;
module M=syz(I);   //the module of syzygies of xy,yz,xz
```

```
M;
//-> M[1]=[0,x,-y]
//-> M[2]=[z,0,-y]

module T=[xy,1,0,0],[yz,0,1,0],[xz,0,0,1];
module N=std(T);
N;                //the first two elements give the syzygies
//-> N[1]=[0,0,x,-y]
//-> N[2]=[0,z,0,-y]
//-> N[3]=[yz,0,1]
//-> N[4]=[xz,0,0,1]
//-> N[5]=[xy,1]
```

Remark 2.5.6. Let R be a ring, $I = \langle f_1, \ldots, f_s \rangle \subset R$ an ideal, and

$$\overline{M} = \langle \overline{m}_1, \ldots, \overline{m}_k \rangle \subset (R/I)^r$$

a submodule. Then $\overline{M}$ is an R– as well as an R/I–module, and we denote by $\mathrm{syz}(\overline{M}) := \mathrm{syz}^R(\overline{m}_1, \ldots, \overline{m}_k)$ and $\mathrm{syz}^{R/I}(\overline{M}) := \mathrm{syz}^{R/I}(\overline{m}_1, \ldots, \overline{m}_k)$ the respective modules of syzygies. They can be computed as follows: let $e_1, \ldots, e_r$ be the canonical basis of R^r, and let $m_1, \ldots, m_k \in R^r$ be representatives of $\overline{m}_1, \ldots, \overline{m}_k$. Moreover, let

$$M := \langle m_1, \ldots, m_k, f_1 e_1, \ldots, f_1 e_r, \ldots, f_s e_1, \ldots, f_s e_r \rangle \subset R^r$$

and $\mathrm{syz}(M) = \{s_1, \ldots, s_\ell\}$, where $s_i = (s_{i1}, \ldots, s_{iN})$, $N = k + rs$. Then

$$\mathrm{syz}(\overline{M}) = \langle \overline{s}_1, \ldots, \overline{s}_\ell \rangle \subset R^k,$$

where $\overline{s}_i = (s_{i1}, \ldots, s_{ik})$, $i = 1, \ldots, \ell$. Now $\mathrm{syz}^{R/I}(\overline{M})$ is the image of $\mathrm{syz}(\overline{M})$ when projecting modulo I.

Successively computing syzygies of syzygies, we obtain an algorithm to compute free resolutions up to any given length.

Algorithm 2.5.7 (RESOLUTION(I, m)).
Let $>$ be any monomial ordering on $\mathrm{Mon}(x_1, \ldots, x_n)$ and $R = K[x]_>$.

Input: $f_1, \ldots, f_k \in K[x]^r$, $I = \langle f_1, \ldots, f_k \rangle \subset R^r$, and m a positive integer.
Output: A list of matrices $A_1, \ldots, A_m$ with $A_i \in \mathrm{Mat}(r_{i-1} \times r_i, K[x])$, $i = 1, \ldots, m$, such that

$$\ldots \longrightarrow R^{r_m} \xrightarrow{A_m} R^{r_{m-1}} \longrightarrow \ldots \longrightarrow R^{r_1} \xrightarrow{A_1} R^r \longrightarrow R^r/I \longrightarrow 0$$

is a free resolution of R^r/I.

- $i := 1$;
- $A_1 := \mathrm{matrix}(f_1, \ldots, f_k) \in \mathrm{Mat}(r \times k, K[x])$;

- while $(i < m)$
  $i := i + 1;$
  $A_i := \text{syz}(A_{i-1});$
- return $A_1, \ldots, A_m.$

With the notation of Definition 2.5.1, we shall now define a monomial ordering on F_1, the free R–module containing $\text{syz}(I)$, $I = \langle f_1, \ldots, f_k \rangle \subset R^r =: F_0$, $f_i \neq 0$ for all i, which behaves perfectly well with respect to standard bases of syzygies. This ordering was first introduced and used by Schreyer [204].

We define the *Schreyer ordering* as follows

$$x^\alpha \varepsilon_i >_1 x^\beta \varepsilon_j :\Longleftrightarrow \text{LM}(x^\alpha f_i) > \text{LM}(x^\beta f_j) \text{ or}$$
$$\text{LM}(x^\alpha f_i) = \text{LM}(x^\beta f_j) \text{ and } i > j.$$

The left–hand side $>_1$ is the new ordering on F_1 and the right–hand side $>$ is the given ordering on F_0. The same ordering on R is induced by $>$ and $>_1$. Note that the Schreyer ordering depends on $f_1, \ldots, f_k$.

Now we are going to prove Buchberger's criterion, which states that $G = \{f_1, \ldots, f_k\}$ is a standard basis of $I = \langle f_1, \ldots, f_k \rangle$, if, for all $i < j$, $\text{NF}(\text{spoly}(f_i, f_j) \mid G_{ij}) = 0$ for suitable $G_{ij} \subset G$. We give a proof by using syzygies, which works for arbitrary monomial orderings and which is different from Schreyer's (cf. [204], [205]) original proof (cf. also [66]), although the basic ideas are due to Schreyer. Our proof gives, at the same time, a proof of Schreyer's result that the syzygies derived from a standard representation of $\text{spoly}(f_i, f_j)$ form a standard basis of $\text{syz}(I)$ for the Schreyer ordering.

We introduce some notations. For each $i \neq j$ such that f_i and f_j have their leading terms in the same component, say $\text{LM}(f_i) = x^{\alpha_i} e_\nu$, $\text{LM}(f_j) = x^{\alpha_j} e_\nu$, we define the monomial

$$m_{ji} := x^{\gamma - \alpha_i} \in K[x],$$

where $\gamma = \text{lcm}(\alpha_i, \alpha_j)$. If $c_i = \text{LC}(f_i)$ and $c_j = \text{LC}(f_j)$ then

$$m_{ji} f_i - \frac{c_i}{c_j} m_{ij} f_j = \text{spoly}(f_i, f_j).$$

Assume that we have a standard representation

$$m_{ji} f_i - \frac{c_i}{c_j} m_{ij} f_j = \sum_{\nu=1}^k a_\nu^{(ij)} f_\nu, \quad a_\nu^{(ij)} \in R.$$

For $i < j$ such that $\text{LM}(f_i)$ and $\text{LM}(f_j)$ involve the same component, define

$$s_{ij} := m_{ji} \varepsilon_i - \frac{c_i}{c_j} m_{ij} \varepsilon_j - \sum_\nu a_\nu^{(ij)} \varepsilon_\nu.$$

Then $s_{ij} \in \text{syz}(I)$ and it is easy to see that

Lemma 2.5.8. *With the notations introduced above,* $\mathrm{LM}(s_{ij}) = m_{ji}\varepsilon_i$.

Proof. Since $\mathrm{LM}(m_{ij}f_j) = \mathrm{LM}(m_{ji}f_i)$ and $i < j$, the definition of $>_1$ gives $m_{ji}\varepsilon_i > m_{ij}\varepsilon_j$. From the defining property of a standard representation we obtain

$$\mathrm{LM}(a_\nu^{(ij)}f_\nu) \leq \mathrm{LM}\left(m_{ji}f_i - \frac{c_i}{c_j}m_{ij}f_j\right) < \mathrm{LM}(m_{ji}f_i)$$

and, hence, the claim.   □

Theorem 2.5.9. *Let* $G = \{f_1, \ldots, f_k\}$ *be a set of generators of* $I \subset R^r$. *Let*

$$M := \left\{(i,j) \,\middle|\, 1 \leq i < j \leq k,\ \mathrm{LM}(f_i), \mathrm{LM}(f_j) \text{ involve the same component}\right\},$$

and let $J \subset M$. *Assume that*

- $\mathrm{NF}\big(\mathrm{spoly}(f_i, f_j) \,\big|\, G_{ij}\big) = 0$ *for some* $G_{ij} \subset G$ *and* $(i,j) \in J$.
- $\big\langle \{m_{ji}\varepsilon_i \mid (i,j) \in J\} \big\rangle = \big\langle \{m_{ji}\varepsilon_i \mid (i,j) \in M\} \big\rangle$ *for* $i = 1, \ldots, r$.

Then the following statements hold:

(1) G *is a standard basis of* I (Buchberger's criterion).
(2) $S := \{s_{ij} \mid (i,j) \in J\}$ *is a standard basis of* $\mathrm{syz}(I)$ *with respect to the Schreyer ordering. In particular,* S *generates* $\mathrm{syz}(I)$.

Proof. We give a proof of (1) and (2) at the same time (recall the notations of Definition 2.5.1).

Take any $f \in I$ and a preimage $g \in F_1$ of f,

$$g = \sum_{i=1}^{k} a_i\varepsilon_i, \quad f = \varphi(g) = \sum_{i=1}^{k} a_i f_i.$$

This is possible as G generates I. In case (1), we assume $f \neq 0$, in case (2) $f = 0$.

Consider a standard representation of ug, u a unit,

$$ug = h + \sum_{(i,j)\in J} a_{ij}s_{ij}, \quad a_{ij} \in R,$$

where $h = \sum_j h_j\varepsilon_j \in F_1$ is a normal form of g with respect to S for some weak normal form on F_1 (we need only know that it exists). We can assume, if $h \neq 0$,

$$h = h_1\varepsilon_1 + \cdots + h_k\varepsilon_k$$

and $\mathrm{LM}(h_\nu\varepsilon_\nu) \notin \langle \mathrm{LM}(s_{ij}) \,|_{(i,j)\in J}\rangle = \langle m_{ji}\varepsilon_i \,|_{(i,j)\in J}\rangle$ by Lemma 2.5.8 and Remark 2.3.4 for all ν such that $h_\nu \neq 0$. This shows

$$m_{j\nu} \nmid \mathrm{LM}(h_\nu)$$

for all ν, j such that f_j and f_ν have the leading term in the same component. Since $ug - h \in \langle S \rangle \subset \text{syz}(I)$, we obtain

$$uf = \varphi(ug) = \varphi(h) = \sum_j h_j f_j \, .$$

Assume that for some $j \neq \nu$, $\text{LM}(h_j f_j) = \text{LM}(h_\nu f_\nu)$ and let $\text{LM}(f_\nu) = x^{\alpha_\nu} e_k$, $\text{LM}(f_j) = x^{\alpha_j} e_k$. Then $\text{LM}(h_\nu f_\nu)$ is divisible by $x^{\alpha_\nu} e_k$ and by $x^{\alpha_j} e_k$, hence,

$$\text{lcm}(x^{\alpha_\nu}, x^{\alpha_j}) \mid \text{LM}(h_\nu f_\nu) = \text{LM}(h_\nu) x^{\alpha_\nu} e_k \, .$$

But $m_{j\nu} = \text{lcm}(x^{\alpha_\nu}, x^{\alpha_j})/x^{\alpha_\nu}$. This contradicts $m_{j\nu} \nmid \text{LM}(h_\nu)$.

In case (1) we obtain $\text{LM}(f) = \text{LM}(h_\nu f_\nu) \in L(G)$ for some ν and, hence, G is a standard basis by definition. In case (2) it shows that $h \neq 0$ leads to a contradiction and S is a standard basis by Theorem 2.3.13, (2) $\Rightarrow$ (1), which was already proved. $\qquad\square$

Lemma 2.5.10 (Chain Criterion). *With the notations of Theorem 2.5.9 assume that $(i, j) \in M$ and $(j, \ell) \in M$. Let $\text{LM}(f_i) = x^{\alpha_i} e_\nu$, $\text{LM}(f_j) = x^{\alpha_j} e_\nu$ and $\text{LM}(f_\ell) = x^{\alpha_\ell} e_\nu$. If x^{α_j} divides $\text{lcm}(x^{\alpha_i}, x^{\alpha_\ell})$ then $m_{\ell i} \varepsilon_i \in \langle m_{ji} \varepsilon_i \rangle$. In particular, if $s_{i\ell}, s_{ij} \in S$ then $S \setminus \{s_{i\ell}\}$ is already a standard basis of $\text{syz}(I)$.*

Proof. $x^{\alpha_j} \mid \text{lcm}(x^{\alpha_i}, x^{\alpha_\ell})$ implies that $\text{lcm}(x^{\alpha_i}, x^{\alpha_j}) \mid \text{lcm}(x^{\alpha_i}, x^{\alpha_\ell})$. Dividing by x^{α_i} we obtain that m_{ji} divides $m_{\ell i}$. $\qquad\square$

Remark 2.5.11. The chain criterion can be used to refine the Standard Basis Algorithm 2.3.8. If (f_i, f_j), (f_i, f_ℓ) and (f_j, f_ℓ) are in the pair set P and (with the notations of the lemma) $x^{\alpha_j} \mid \text{lcm}(x^{\alpha_i}, x^{\alpha_\ell})$ then we can delete (f_i, f_ℓ) from P. For a generalization of the criterion cf. [168]. Note that the *Product Criterion* (Exercise 1.7.1) is only applicable for modules with all module components 0 except one.

We want to illustrate this with the following example.

Example 2.5.12. Let $I = \langle u^5 - v^5, v^5 - x^5, x^5 - y^5, y^5 - z^5, u^4 v + v^4 x + x^4 y + y^4 z + z^4 u \rangle \subset \mathbb{Q}[u, v, x, y, z]$. The reduced standard basis of this ideal with respect to the ordering **dp** has 149 elements.

Using Buchberger's criterion (Theorem 2.3.13), we see that during the computation of the standard basis $\binom{149}{2} = 11026$ pairs have to be considered.

In the implementation of Buchberger's algorithm in SINGULAR, the chain criterion is applied 10288 times and the product criterion 166 times. Therefore, instead of reducing 11026 s-polynomials, we only need to consider 572 (about 5 %) of them. This shows that these criteria have an enormous influence on the performance of Buchberger's algorithm.

We shall now see, as an application, that Hilbert's syzygy theorem holds for the rings $R = K[x]_>$, stating that each finitely generated R-module has a free resolution of length at most n, the number of variables.

Lemma 2.5.13. *Let* $G = \{g_1, \ldots, g_s\}$ *be a minimal standard basis of the submodule* $I \subset R^r = \bigoplus_{i=1}^r Re_i$ *such that* $\mathrm{LM}(g_i) \in \{e_1, \ldots, e_r\}$ *for all* i. *Let* J *denote the set of indices* j *such that* $e_j \notin \{\mathrm{LM}(g_1), \ldots, \mathrm{LM}(g_s)\}$. *Then*

$$I = \bigoplus_{i=1}^s Rg_i, \qquad R^r/I \cong \bigoplus_{j \in J} Re_j.$$

Proof. The set $G' := G \cup \{e_j \mid j \in J\}$ is R–linearly independent, since the leading terms are. This shows that both sums above are direct.

Let $f \in R^r$ and consider a weak normal form h of f with respect to G'. Assuming $h \neq 0$, we would have $\mathrm{LM}(h) \notin \langle e_1, \ldots, e_r \rangle$, a contradiction. Hence, $h = 0$, that is, $f \in I + \langle \{e_j \mid j \in J\} \rangle$. $\qquad\square$

Lemma 2.5.14. *Let* $G = \{g_1, \ldots, g_s\}$ *be a standard basis of* $I \subset R^r$, *ordered in such a way that the following holds: if* $i < j$ *and* $\mathrm{LM}(g_i) = x^{\alpha_i} e_\nu$, $\mathrm{LM}(g_j) = x^{\alpha_j} e_\nu$ *for some* ν, *then* $\alpha_i \geq \alpha_j$ *lexicographically. Let* s_{ij} *denote the syzygies defined above. Suppose that* $\mathrm{LM}(g_1), \ldots, \mathrm{LM}(g_s)$ *do not depend on the variables* $x_1, \ldots, x_k$. *Then the* $\mathrm{LM}(s_{ij})$, *taken with respect to the Schreyer ordering, do not depend on* $x_1, \ldots, x_{k+1}$.

Proof. Given s_{ij}, then $i < j$ and $\mathrm{LM}(g_i)$ and $\mathrm{LM}(g_j)$ involve the same component, say e_ν. By assumption, $\mathrm{LM}(g_i) = x^{\alpha_i} e_\nu$, $\mathrm{LM}(g_j) = x^{\alpha_j} e_\nu$ satisfy $\alpha_i = (0, \ldots, \alpha_{i,k+1}, \ldots)$ and $\alpha_j = (0, \ldots, \alpha_{j,k+1}, \ldots)$ with $\alpha_{i,k+1} \geq \alpha_{j,k+1}$. Therefore, $\mathrm{LM}(s_{ij}) = m_{ji}\varepsilon_i$, $m_{ji} = x^{\mathrm{lcm}(\alpha_i, \alpha_j) - \alpha_i}$, does not depend on x_{k+1}. $\qquad\square$

Applying the lemma successively to the higher syzygy modules, we obtain

Theorem 2.5.15 (Hilbert's Syzygy Theorem). *Let* $>$ *be any monomial ordering on* $K[x] = K[x_1, \ldots, x_n]$, *and let* $R = K[x]_>$ *be the associated ring. Then any finitely generated* R–*module* M *has a free resolution*

$$0 \to F_m \to F_{m-1} \to \cdots \to F_0 \to M \to 0$$

of length $m \leq n$, *where the* F_i *are free* R–*modules.*

Proof. Since R is Noetherian, M has a presentation

$$0 \to I \to F_0 \to M \to 0,$$

with $F_0 = \bigoplus_{i=1}^{r_0} Re_i$, and I being finitely generated. Let $G = \{g_1, \ldots, g_s\}$ be a standard basis of I and assume that the $\mathrm{LM}(g_i)$ do not depend on the variables $x_1, \ldots, x_k$, $k \geq 0$. By Theorem 2.5.9, the syzygies $s_{ij} =: s_{ij}^{(1)}$ are a standard basis of $\mathrm{syz}(I)$ and, by Lemma 2.5.14, we may assume that the $\mathrm{LM}(s_{ij}^{(1)})$ do not depend on $x_1, \ldots, x_{k+1}$. Hence, we obtain an exact sequence

$$0 \to \mathrm{Ker}(\varphi_1) = \mathrm{syz}(I) \to F_1 \xrightarrow{\varphi_1} F_0 \to M \to 0$$

$F_1 = \bigoplus_{i=1}^{s} R\varepsilon_i$, $\varphi_1(\varepsilon_i) = g_i$, with $\mathrm{syz}(I)$ satisfying analogous properties as I. We can, therefore, construct by induction an exact sequence

$$0 \to \mathrm{Ker}(\varphi_{n-k}) \to F_{n-k} \xrightarrow{\varphi_{n-k}} F_{n-k-1} \to \ldots \xrightarrow{\varphi_2} F_1 \to F_0 \to M \to 0$$

with F_i free of rank r_i and $\mathrm{Ker}(\varphi_{n-k})$ given by a standard basis $\{s_{ij}^{(n-k)}\}$ such that none of the variables appears in $\mathrm{LM}(s_{ij}^{(n-k)})$. By Lemma 2.5.13, the quotient $F_{n-k}/\mathrm{Ker}(\varphi_{n-k})$ is a free R–module, and replacing F_{n-k} by $F_{n-k}/\mathrm{Ker}(\varphi_{n-k})$ we obtain the desired free resolution. $\qquad\square$

Algorithm 2.5.16 (SRESOLUTION).
Let $>$ be any monomial ordering on $\mathrm{Mon}(x_1, \ldots, x_r)$, $R = K[x]_>$.

Input: $>_m$ any module ordering on R^r, $\emptyset \neq G = \{g_1, \ldots, g_k\} \subset K[x]^r$ an interreduced standard basis of $I = \langle G \rangle \subset R^r$ w.r.t. $>_m$.

Output: A list of matrices $A_1, \ldots, A_m$ with $A_i \in \mathrm{Mat}(r_{i-1} \times r_i, K[x])$, $i = 1, \ldots, m$, such that $m \leq n$ and

$$0 \longrightarrow R^{r_m} \xrightarrow{A_m} R^{r_{m-1}} \longrightarrow \ldots \longrightarrow R^{r_1} \xrightarrow{A_1} R^r \longrightarrow R^r/I \longrightarrow 0$$

is a free resolution.

- Renumber the elements $g_1, \ldots, g_k$ of G such that the following holds: if $i < j$ and $\mathrm{LM}(g_i) = x^{\alpha_i} e_\nu$, $\mathrm{LM}(g_j) = x^{\alpha_j} e_\nu$ for some ν, then $\alpha_i >_{lp} \alpha_j$.
- Set $A_1 = (g_1, \ldots, g_k) \in \mathrm{Mat}(r \times k, K[x])$.
- With the notations of Theorem 2.5.9, choose a subset $J \subset M$ such that $\langle \{m_{ji}\varepsilon_i \mid (i,j) \in J\} \rangle \supset \{m_{ji}\varepsilon_i \mid (i,j) \in M\}$ for $i = 1, \ldots, r$.
- For $(i,j) \in J$ compute a standard representation

$$u_{ij} \cdot \mathrm{spoly}(g_i, g_j) = \sum_{\nu=1}^{k} a_\nu^{(ij)} g_\nu \,,$$

$u_{ij} \in K[x] \cap R^*$, $a_\nu^{(ij)} \in K[x]$, and set[10]

$$s_{ij} := u_{ij} m_{ji} \varepsilon_i - \frac{c_i}{c_j} u_{ij} m_{ij} \varepsilon_j - \sum_{\nu=1}^{k} a_\nu^{(ij)} \varepsilon_\nu \,.$$

- Set $S := \mathrm{INTERREDUCTION}(\{s_{ij} \mid (i,j) \in J\})$.
- If, for each $s \in S$, $\mathrm{LM}(s) \in \{\varepsilon_1, \ldots, \varepsilon_k\}$ (where the leading monomial is taken w.r.t. the Schreyer ordering $>_1$)
  then
     set $J' := \{1 \leq j \leq k \mid \varepsilon_j \notin \{\mathrm{LM}(s) \mid s \in S\}\}$,

---

[10] Recall $c_i = \mathrm{LC}(g_i)$ and $m_{ji} = x^{\gamma-\alpha_i}$ with $\gamma = \mathrm{lcm}(\alpha_i, \alpha_j)$ and $\mathrm{LM}(g_i) = x^{\alpha_i} e_\nu$, $\mathrm{LM}(g_j) = x_{\alpha_j} e_\nu$.

        delete in the matrix A_1 all columns with index $j \notin J'$,
        return(A_1),
    else
        list $L = A_1$, SRESOLUTION($>_1, S$),
        return(L).

Note that the free resolution computed by SRESOLUTION($>_m, G$) is, in general, not minimal (in the local, respectively homogeneous case). But, it can be *minimized* afterwards, following the procedure described on page 127 (applying Gaussian elimination and deleting rows and columns of the corresponding matrices), with the only difference that each column operation on the matrix A_i has to be succeeded by a certain row operation on A_{i+1} (and vice versa, cf. Exercise 2.5.4).

Example 2.5.17. Let $R := K[x, y, z]$ with degree reverse lexicographical ordering $>_{dp}$, let $>_m = (c, >_{dp})$, and consider

$$G := \{yz + z^2, y^2 + xz, xy + z^2, z^3, xz^2, x^2z\}$$

(which is an interreduced standard basis of $I := \langle G \rangle \subset R$). After renumbering the elements of G such that $\mathrm{LM}(g_1) >_{lp} \mathrm{LM}(g_2) >_{lp} \cdots >_{lp} \mathrm{LM}(g_6)$, we obtain

$$A_1 := (\underbrace{x^2z}_{g_1}, \underbrace{xy + z^2}_{g_2}, \underbrace{xz^2}_{g_3}, \underbrace{y^2 + xz}_{g_4}, \underbrace{yz + z^2}_{g_5}, \underbrace{z^3}_{g_6}) \in \mathrm{Mat}(1 \times 6, K[x, y, z]).$$

The respective monomials $m_{ji}\varepsilon_i$, $1 \le i < j \le 6$, are given in the following table:

$i\backslash^j$	2	3	4	5	6
1	$y\varepsilon_1$	$z\varepsilon_1$	$y^2\varepsilon_1$	$y\varepsilon_1$	$z^2\varepsilon_1$
2	$--$	$z^2\varepsilon_2$	$y\varepsilon_2$	$z\varepsilon_2$	$z^3\varepsilon_2$
3	$--$	$--$	$y^2\varepsilon_3$	$y\varepsilon_3$	$z\varepsilon_3$
4	$--$	$--$	$--$	$z\varepsilon_4$	$z^3\varepsilon_4$
5	$--$	$--$	$--$	$--$	$z^2\varepsilon_5$

Hence, we may choose

$$J := \big\{(1,2), (1,3), (2,4), (2,5), (3,5), (3,6), (4,5), (5,6)\big\}$$

and compute

$$s_{1,2} = y\varepsilon_1 - xz\varepsilon_2 + x\varepsilon_6,$$
$$s_{1,3} = z\varepsilon_1 - x\varepsilon_3,$$
$$s_{2,4} = y\varepsilon_2 - x\varepsilon_4 + \varepsilon_1 - z\varepsilon_5 + \varepsilon_6,$$
$$s_{2,5} = z\varepsilon_2 - x\varepsilon_5 + \varepsilon_3 - \varepsilon_6,$$
$$s_{3,5} = y\varepsilon_3 - xz\varepsilon_5 + x\varepsilon_6,$$
$$s_{3,6} = z\varepsilon_3 - x\varepsilon_6,$$

$$s_{4,5} = z\varepsilon_4 - y\varepsilon_5 - \varepsilon_3 + z\varepsilon_5 - \varepsilon_6,$$
$$s_{5,6} = z^2\varepsilon_5 - y\varepsilon_6 - z\varepsilon_6.$$

The set $S := \{s_{1,2}, s_{1,3}, s_{2,4}, s_{2,5}, s_{3,5}, s_{3,6}, s_{4,5}, s_{5,6}\}$ is an interreduced standard basis for $\mathrm{syz}(I)$ (w.r.t. the Schreyer ordering $>_1$), and we are left with computing SRESOLUTION($>_1, S$). By accident, the elements of S are already ordered as needed, and we set

$$A_2 := \begin{pmatrix} y & z & 1 & 0 & 0 & 0 & 0 & 0 \\ -xz & 0 & y & z & 0 & 0 & 0 & 0 \\ 0 & -x & 0 & 1 & y & z & -1 & 0 \\ 0 & 0 & -x & 0 & 0 & 0 & z & 0 \\ 0 & 0 & -z & -x & -xz & 0 & -y+z & z^2 \\ x & 0 & 1 & -1 & x & -x & -1 & -y-z \end{pmatrix}.$$

We see that the set M of pairs (i,j), $1 \le i < j \le 8$, such that the leading monomials of the i-th and j-th element of S involve the same components consists of precisely 3 elements: $M = \{(1,2), (3,4), (5,6)\}$. We compute

$$s_{1,2}^{(1)} = z\varepsilon_1 - y\varepsilon_2 + xz\varepsilon_4 - x\varepsilon_5 - x\varepsilon_6,$$
$$s_{3,4}^{(1)} = z\varepsilon_3 - y\varepsilon_4 - \varepsilon_2 + \varepsilon_5 + x\varepsilon_7 + \varepsilon_8,$$
$$s_{5,6}^{(1)} = z\varepsilon_5 - y\varepsilon_6 + x\varepsilon_8,$$

Again, $S^{(1)} := \{s_{1,2}^{(1)}, s_{3,4}^{(1)}, s_{5,6}^{(1)}\}$ is an interreduced standard basis for $\mathrm{syz}(\langle S \rangle)$. Since the leading monomials of the elements of $S^{(1)}$ (w.r.t. the Schreyer ordering $>_2$) involve different components, SRESOLUTION($>_2, S^{(1)}$) returns

$$A_3 := \begin{pmatrix} z & 0 & 0 \\ -y & -1 & 0 \\ 0 & z & 0 \\ xz & -y & 0 \\ -x & 1 & z \\ -x & 0 & -y \\ 0 & x & x \\ 0 & 1 & x \end{pmatrix}.$$

Finally, we have computed the free resolution

$$0 \longrightarrow R^3 \xrightarrow{A_3} R^8 \xrightarrow{A_2} R^6 \xrightarrow{A_1} R \longrightarrow R/I \longrightarrow 0.$$

Note that I is a *homogeneous* ideal, hence R/I has the structure of a graded K-algebra. Now we can easily derive the *Betti numbers* of R/I from the computed (non-minimal) free resolution (without computing the minimal resolution). To do so, we consider the matrices $A_1, A_2, A_3 \bmod \mathfrak{m} = \langle x, y, z \rangle$, apply Gaussian elimination in $K = R/\mathfrak{m}$ as indicated above (for minimizing the resolution), and delete the respective rows and columns (cf. also Exercise

2.5.4). We obtain that $b_3(R/I) = 2$, $b_2(R/I) = 4$, $b_1(R/I) = 3$, $b_0(R/I) = 1$, that is, the minimal resolutions of R/I look as follows

$$0 \longrightarrow R^2 \longrightarrow R^4 \longrightarrow R^3 \longrightarrow R \longrightarrow R/I \longrightarrow 0\,.$$

The implemented algorithm `sres` in SINGULAR is a modification of SRES-OLUTION. For efficiency reasons the generators (with leading monomials in the same components) are not ordered lexicographically but with respect to $>_{dp}$. Hence, `sres` does not necessarily compute a free resolution of finite length, but stops the computation after $n + 1$ steps (n being the number of variables of the basering). Anyhow, we can obtain the Betti numbers (using the built–in command `betti`), respectively minimize the obtained resolution (using `minres`).

SINGULAR Example 2.5.18 (Schreyer resolution).

```
ring R=0,(x,y,z),(c,dp);
ideal I=yz+z2,y2+xz,xy+z2,z3,xz2,x2z;
```

First compute a minimal free resolution of R/I:

```
resolution Re=mres(I,0);
Re;
//-> 1      3      4      2
//-> R  <-- R  <-- R  <-- R
//->
//-> 0      1      2      3
```

Display the matrices in the resolution:

```
print(Re);
//-> [1]:
//->    _[1]=yz+z2
//->    _[2]=y2+xz
//->    _[3]=xy+z2
//-> [2]:
//->    _[1]=[0,xy+z2,-y2-xz]
//->    _[2]=[y2+xz,-yz-z2]
//->    _[3]=[xy+z2,0,-yz-z2]
//->    _[4]=[x2-yz,-xz+z2,-xz+yz]
//-> [3]:
//->    _[1]=[0,x-z,x-y,-y-z]
//->    _[2]=[z,z,-x,y]
```

Now apply the modification of Schreyer's algorithm to compute a free resolution for R/I and display the matrices:

```
resolution Se=sres(I,0);
Se;
//-> 1     6     8     3
//-> R  <-- R  <-- R  <-- R
//->
//-> 0     1     2     3
//-> resolution not minimized yet

print(Se);
//-> [1]:
//->    _[1]=x2z
//->    _[2]=xz2
//->    _[3]=z3
//->    _[4]=xy+z2
//->    _[5]=y2+xz
//->    _[6]=yz+z2
//-> [2]:
//->    _[1]=[z,-x]
//->    _[2]=[y,0,x,-xz]
//->    _[3]=[0,z,-x]
//->    _[4]=[0,y,z,-z2]
//->    _[5]=[0,0,y+z,0,0,-z2]
//->    _[6]=[1,0,1,y,-x,-z]
//->    _[7]=[0,1,-1,z,0,-x]
//->    _[8]=[0,-1,-1,0,z,-y+z]
//-> [3]:
//->    _[1]=[y,-z,0,x]
//->    _[2]=[0,0,y+z,-z,x,0,-z2]
//->    _[3]=[-1,0,-1,1,-1,z,-y+z,x]
```

Finally, let us minimize the computed resolution:

```
print(minres(Se));
//-> [1]:
//->    _[1]=xy+z2
//->    _[2]=y2+xz
//->    _[3]=yz+z2
//-> [2]:
//->    _[1]=[-2y2-xz-yz,2xy+xz-yz,-x2+y2+xz+yz]
//->    _[2]=[-xz-z2,-xz+z2,x2+xy-yz+z2]
//->    _[3]=[-yz-z2,yz+z2,xy-y2-xz+z2]
//->    _[4]=[yz+z2,yz+z2,-xy-y2-xz-z2]
//-> [3]:
//->    _[1]=[-z,-y,x+y,-y]
//->    _[2]=[0,y+z,-z,x]
```

Exercises

2.5.1. Find a standard basis with respect to the lexicographical ordering and a standard basis of the syzygies for $I = \langle x^2, y^2, xy + yz \rangle$. Compare Schreyer's method and the method of Lemma 2.5.3.

2.5.2. Let $I \subset K[x_1, \ldots, x_n]$ be a homogeneous ideal. Give an algorithm to compute a minimal resolution of $K[x_1, \ldots, x_n]/I$, modifying the algorithm RESOLUTION.

2.5.3. Compute a minimal resolution for $I = \langle x^2, y^2, xy + yz \rangle \subset K[x, y, z]$.

2.5.4. Give an algorithm to obtain a minimal resolution in the case of local rings from Schreyer's resolution.

2.5.5. Prove *Schanuel's Lemma*: Let R be a Noetherian ring and M a finitely generated R–module. Moreover, assume that the following sequences are exact:

$$0 \longrightarrow K_1 \longrightarrow R^{n_1} \xrightarrow{\pi_1} M \longrightarrow 0,$$

$$0 \longrightarrow K_2 \longrightarrow R^{n_2} \xrightarrow{\pi_2} M \longrightarrow 0.$$

Then $K_1 \oplus R^{n_2} \cong K_2 \oplus R^{n_1}$.
(Hint: Prove that both of them are isomorphic to $\mathrm{Ker}(R^{n_1} \oplus R^{n_2} \xrightarrow{\pi_1 + \pi_2} M)$.)

2.5.6. Let R be a Noetherian ring and $M = \langle f_1, \ldots, f_k \rangle = \langle g_1, \ldots, g_s \rangle \subset R^r$. Prove that $\mathrm{syz}(f_1, \ldots, f_k) \oplus R^s \cong \mathrm{syz}(g_1, \ldots, g_s) \oplus R^k$.

2.5.7. Let R be a local Noetherian ring, let M be a finitely generated R–module, and let $\{f_1, \ldots, f_k\}$, $\{g_1, \ldots, g_k\}$ be two minimal sets of generators. Prove that $\mathrm{syz}(f_1, \ldots, f_k) \cong \mathrm{syz}(g_1, \ldots, g_k)$, and conclude that the k–th syzygy module $\mathrm{syz}_k(M)$ is well–defined up to isomorphism.

2.5.8. Let R be a local Noetherian graded K–algebra, K a field, and let M be a finitely generated graded R–module. Show that, for a system of homogeneous generators $\{f_1, \ldots, f_k\}$ of M, the module of syzygies $\mathrm{syz}(f_1, \ldots, f_k)$ is a homogeneous submodule of R^k.

Moreover, show that for two minimal systems of homogeneous generators $\{f_1, \ldots, f_k\}$, $\{g_1, \ldots, g_k\}$ of M, the modules of syzygies are isomorphic as graded R–modules.

Conclude that the k–th syzygy module $\mathrm{syz}_k(M)$ is well–defined up to graded isomorphism and that the Betti numbers of M depend only on the graded isomorphism class of M.

2.5.9. Let $f, g \in K[x]$. Prove that $\gcd(f, g)$ and $\mathrm{lcm}(f, g)$ can be computed using the syzygies of f and g. Write a SINGULAR procedure to compute the gcd and lcm.

2.6 Modules over Principal Ideal Domains

In this section we shall study the structure of finitely generated modules over principal ideal domains. It will be proved that they can be decomposed in a unique way into a direct sum of cyclic modules with special properties. Examples are given for the case of a univariate polynomial ring over a field. We show how this decomposition can be computed by using standard bases (actually, we need only interreduction).

Theorem 2.6.1. *Let R be a principal ideal domain and M a finitely generated R–module, then M is a direct sum of cyclic modules.*

Proof. Let $R^m \to R^n \to M \to 0$ be a presentation of M given by the matrix $A = (a_{ij})$ with respect to the bases $B = \{e_1, \ldots, e_n\}$, $B' = \{f_1, \ldots, f_m\}$ of R^n, R^m, respectively. If A is the zero–matrix, then $M \cong R^n$, and we are done. Otherwise, we may assume that $a_{11} \neq 0$. We shall show that, for a suitable choice of the bases, the presentation matrix has *diagonal form*, that is, $a_{ij} = 0$ if $i \neq j$. For some $k > 1$ with $a_{k1} \neq 0$, let h be a generator of the ideal $\langle a_{11}, a_{k1} \rangle$, and let $a, b, c, d \in R$ be such that $h = aa_{11} + ba_{k1}$, $a_{11} = ch$, $a_{k1} = dh$ (we choose $a := 1$, $b := 0$, $c := 1$ if $\langle a_{11} \rangle = \langle a_{11}, a_{k1} \rangle$). Now we change the basis B to $\bar{B} = \{ce_1 + de_k, e_2, \ldots, e_{k-1}, -be_1 + ae_k, e_{k+1}, \ldots, e_n\}$. $\bar{B}$ is a basis because $\det \left(\begin{smallmatrix} c & -b \\ d & a \end{smallmatrix} \right) = 1$. Let $\bar{A} = (\bar{a}_{ij})$ be the presentation matrix with respect to this basis, then $\bar{a}_{11} = h$ and $\bar{a}_{k1} = 0$, while $\bar{a}_{i1} = a_{i1}$ for $i \neq 1, k$. Note that the first row of A and $\bar{A}$ are equal if and only if $\langle a_{11} \rangle = \langle a_{11}, a_{k1} \rangle$. Doing this with every $k > 1$, we may assume that $a_{k1} = 0$ for $k = 2, \ldots, n$.

Now, applying the same procedure to the transposed matrix ${}^t A$ (which corresponds to base changes in B'), we obtain a matrix ${}^t A_1$,

$$A_1 = \begin{pmatrix} a_{11}^{(1)} & 0 & \cdots & 0 \\ a_{21}^{(1)} & a_{22}^{(1)} & \cdots & a_{2m}^{(1)} \\ \vdots & & & \vdots \\ a_{n1}^{(1)} & a_{n2}^{(1)} & \cdots & a_{nm}^{(1)} \end{pmatrix},$$

with the property: $\langle a_{11} \rangle \subset \langle a_{11}^{(1)} \rangle$ and $a_{21}^{(1)} = \cdots = a_{n1}^{(1)} = 0$, if $\langle a_{11} \rangle = \langle a_{11}^{(1)} \rangle$.

Repeating this procedure, if $\langle a_{11} \rangle \subsetneq \langle a_{11}^{(1)} \rangle$, we obtain matrices $A_2, \ldots, A_\ell$ such that $\langle a_{11} \rangle \subset \langle a_{11}^{(1)} \rangle \subset \cdots \subset \langle a_{11}^{(\ell)} \rangle$. The ring R is Noetherian and, therefore, we find an ℓ such that $\langle a_{11}^{(\ell)} \rangle = \langle a_{11}^{(\ell+1)} \rangle$. This implies that, in the matrix $A_{\ell+1}$, $a_{1j}^{(\ell+1)} = 0$ for all j and $a_{j1}^{(\ell+1)} = 0$ for all j. After this step, we may assume that, for the matrix A, with respect to the bases B and B', $a_{11} \neq 0$, $a_{1j} = 0$ and $a_{j1} = 0$ for all $j > 1$.

Now we use induction to prove that, for suitable changes of $\{e_2, \ldots, e_n\}$ and $\{f_2, \ldots, f_m\}$, the presentation matrix A has diagonal form, that is, $a_{ij} = 0$ for $i \neq j$. Let M_i be the submodule of M generated by the image of the i-th basis element. Then $M = \bigoplus_{i=1}^{n} M_i$ and $M_i \cong R/\langle a_{ii}\rangle$ if $i \le m$, respectively $M_i \cong R$ if $i > m$. $\square$

From the proof of Theorem 2.6.1 we deduce the following algorithm to compute a diagonal form of a presentation matrix of a module over the polynomial ring $K[x]$, K a field.

Algorithm 2.6.2 (DIAGONALFORM).

Input: A matrix A with entries in $K[x]$.
Output: A matrix D in diagonal form such that $D = BAC$ for invertible
        matrices B, C with entries in $K[x]$.

- if A has no non–zero entry, return A;
- exchange rows and columns to obtain $a_{11} \neq 0$;
- while there exist $i > 1$ such that $a_{1i} \neq 0$ or $a_{i1} \neq 0$
    $A := \text{RowNF}(A)$;
    $A := \text{transpose}(\text{RowNF}(\text{transpose}(A)))$;
- let $A = \begin{pmatrix} a_{11} & 0 \\ 0 & A' \end{pmatrix}$, then return $\begin{pmatrix} a_{11} & 0 \\ 0 & \text{DIAGONALFORM}(A') \end{pmatrix}$.

We use the following procedure $\text{RowNF}(A)$ to obtain zeros in the first column of the matrix, except at the place $(1,1)$. Let $A = (a_{ij})$ be an $n \times m$–matrix with entries in $K[x]$ and assume that $a_{11} \neq 0$.

Input: $A = (a_{ij})$ an $n \times m$–matrix with entries in $K[x]$, $a_{11} \neq 0$.
Output: $\text{RowNF}(A)$, an $n \times m$–matrix, such that $\text{RowNF}(A) = C \cdot A$ for
        a suitable invertible matrix C, and the first column is of the form
        $^t(h, 0, \dots, 0)$ with $h \mid a_{11}$.

- For $i = 2, \dots, n$
    compute $h := \gcd(a_{11}, a_{i1})$;
    choose $a, b, c, d \in K[x]$, such that $h = aa_{11} + ba_{i1}$, $a_{11} = ch$, $a_{i1} = dh$
    (if a_{11} divides a_{i1} then choose $a := 1$, $b := 0$, $c := 1$);
    change A by multiplying the first row with a and add to it the b-th
    multiple of the i-th row;
    change A by subtracting from the i-th row the d-th multiple of the
    (new) first row;
- return A.

SINGULAR Example 2.6.3 (diagonal form).
In this example we shall use the SINGULAR command **interred** to diagonalize a matrix. This command replaces the procedure RowNF in Algorithm 2.6.2. In the ring $K[x]$ (with the two possible orderings **dp** and **ds**) consider, on $K[x]^n = \bigoplus_{i=1}^{n} K[x]e_i$, the ordering $> = (\text{C,dp})$, respectively (C,ds). Recall that $x^\alpha e_i < x^\beta e_j$ if $i < j$, or if $i = j$, $x^\alpha <_{dp} x^\beta$ (respectively $x^\alpha <_{ds} x^\beta$).
    The command **interred** applied to a matrix interreduces the columns of the matrix considered as elements of $K[x]^n$ (which is, in the case of one

variable, the same as to compute a standard basis). The result is an upper triangular matrix (the columns are ordered with respect to their leading terms, the first column with the smallest leading term).

```
proc diagonalForm(matrix M)
{
   int n=nrows(M);
   int m=ncols(M);
   matrix N,K;
   matrix L[n][m];
   while(N!=M)
   {
      N=M;
      M=L;
      K=transpose(interred(transpose(interred(N))));
      M[1..nrows(K),1..ncols(K)]=K;
   }
   return(N);
}
```

Here are two examples:

```
option(redSB);
ring R=0,(x),(C,dp);

matrix M[2][3]=(x2+1)^2,0,     0,
               0,        x3-x-1,0;
matrix N1[2][2]=1, 1,
                2,-2;
matrix N2[3][3]=1,2, 3,
                4,5, 6,
                7,8,-1;
M=N1*M*N2;
print(M);
//->x4+4x3+2x2-4x-3,  2x4+5x3+4x2-5x-3,   3x4+6x3+6x2-6x-3,
//->2x4-8x3+4x2+8x+10,4x4-10x3+8x2+10x+14,6x4-12x3+12x2+12x+18
```

Let us diagonalize M:

```
diagonalForm(M);
//-> _[1,1]=x7+x5-x4-x3-2x2-x-1
//-> _[1,2]=0
//-> _[1,3]=0
//-> _[2,1]=0
//-> _[2,2]=1
//-> _[2,3]=0
```

The second example:

```
matrix M0[5][5]=1, 1,0, 0,0,
                3,-1,0, 0,0,
                0, 0,1, 1,0,
                0, 0,3,-1,0,
                0, 0,0, 0,2;
matrix N[5][5]=1, 1, -1,  1,-1,
                2, 2,  1,  1, 0,
               -1, 2,  2,  1, 1,
               -2, 1,  1, -1, 0,
                1, 2, -2,  1, 1;
```

Now we want to compute $M = N^{-1} M_0 N - x E_5$ [11] and its normal form:

```
M=lift(N,freemodule(nrows(N)))*M0*N-x*freemodule(5);
print(M);
//-> -x+29/50,-71/50,  -143/50,1/25,    -71/50,
//-> 16/25,   -x+66/25,3/25,   58/25,    16/25,
//-> -24/25,  -24/25,  -x+8/25,-12/25,   -24/25,
//-> -12/25,  -12/25,  29/25,  -x-56/25,-12/25,
//-> -13/10,  -13/10,  -19/10, -7/5,     -x+7/10


print(diagonalForm(M));
//-> x2-4,0,  0,   0,0,
//-> 0,   x-2,0,   0,0,
//-> 0,   0,  x2-4,0,0,
//-> 0,   0,  0,   1,0,
//-> 0,   0,  0,   0,1
```

Corollary 2.6.4. *Let R be a principal ideal domain and M a finitely generated R–module. If M is torsion free, then M is free.*

In the following we further analyze the structure of modules over a principal ideal domain in order to obtain a unique decomposition.

Proposition 2.6.5. *Let R be a principal ideal domain and M a finitely generated R–module, then $M = F \oplus \mathrm{Tors}(M)$, F a free submodule of M. If $M \cong R^n \oplus T$, T a torsion module, then $R^n \cong F$ and $T \cong \mathrm{Tors}(M)$.*

Proof. $M/\mathrm{Tors}(M)$ is torsion free and, therefore, because of Corollary 2.6.4, free. Let $x_1, \ldots, x_s \in M$ be representatives of a basis of $M/\mathrm{Tors}(M)$, then $F := \langle x_1, \ldots, x_s \rangle$ is a free module and $F \cap \mathrm{Tors}(M) = \langle 0 \rangle$. This implies that $M = F \oplus \mathrm{Tors}(M)$.

To prove the second part, note that $\mathrm{Tors}(R^n \oplus T) = T$ and T is mapped via the isomorphism $M \cong R^n \oplus T$ to $\mathrm{Tors}(M)$.   □

---

[11] Here E_n denotes the $n \times n$ unit matrix.

Proposition 2.6.6. *Let R be a principal ideal domain and M a finitely generated torsion R–module. Let $\langle f \rangle = \mathrm{Ann}(M)$ and $f = p_1^{c_1} \cdot \ldots \cdot p_n^{c_n}$, with p_i prime and $\langle p_i, p_j \rangle = R$ for $i \neq j$.[12] Let $T_{p_i}(M) := \{x \in M \mid p_i^{c_i} x = 0\}$. Then*

$$M = \bigoplus_{i=1}^{n} T_{p_i}(M).$$

Proof. Let $x \in T_{p_i}(M) \cap \sum_{j \neq i} T_{p_j}(M)$. Then $p_i^{c_i} x = 0$ and

$$p_1^{c_1} \cdot \ldots \cdot p_{i-1}^{c_{i-1}} \cdot p_{i+1}^{c_{i+1}} \cdot \ldots \cdot p_n^{c_n} x = 0 \,.$$

But $\langle p_i^{c_i}, p_1^{c_1} \cdot \ldots \cdot p_{i-1}^{c_{i-1}} \cdot p_{i+1}^{c_{i+1}} \cdot \ldots \cdot p_n^{c_n} \rangle = R$, because $\langle p_i, p_j \rangle = R$ for $i \neq j$ and p_i is prime for all i (Exercise 2.6.7). This implies that $x = 0$, that is, $T_{p_i}(M) \cap \sum_{j \neq i} T_{p_j}(M) = \langle 0 \rangle$.

Let $x \in M$, and choose $a, b \in R$ such that $a p_n^{c_n} + b p_1^{c_1} \cdot \ldots \cdot p_{n-1}^{c_{n-1}} = 1$. We write $x = x' + x_n$ with $x' := a p_n^{c_n} x$ and $x_n := b p_1^{c_1} \cdot \ldots \cdot p_{n-1}^{c_{n-1}} x$, and obtain $x_n \in T_{p_n}(M)$ and $p_1^{c_1} \cdot \ldots \cdot p_{n-1}^{c_{n-1}} x' = 0$. Using induction, we can continue to decompose x' and obtain $x = x_1 + \cdots + x_n \in \sum_{i=1}^{n} T_{p_i}(M)$. $\square$

Proposition 2.6.7. *Let R be a principal ideal domain, M a torsion R–module and $\mathrm{Ann}(M) = \langle p^c \rangle$, p prime. Then $M = \bigoplus_{i=1}^{s} C_i$ with cyclic R–modules C_i such that $\mathrm{Ann}(C_i) = \langle p^{n_i} \rangle$, $n_1 \geq \cdots \geq n_s$. The numbers $n_1, \ldots, n_s$ are uniquely determined by M.*

Proof. Set $M_i := \{x \in M \mid p^i x = 0\}$, which are submodules of M satisfying $M_i \supset M_{i-1}$, $i = 1, \ldots, c$. The factor modules M_i/M_{i-1} are annihilated by p, hence, $R/\langle p \rangle$–vector spaces. Let $m_1 := \dim_{R/\langle p \rangle} M_c/M_{c-1}$ and choose $x_1, \ldots, x_{m_1} \in M$ representing a basis of M_c/M_{c-1}. Then $p x_1, \ldots, p x_{m_1}$ are linearly independent, considered as elements in M_{c-1}/M_{c-2}: assume that $\sum_{i=1}^{m_1} h_i p x_i \in M_{c-2}$ for some $h_1, \ldots, h_{m_1} \in R$, then $p^{c-2} \cdot \sum_{i=1}^{m_1} h_i p x_i = 0$. Therefore, $p^{c-1} \cdot \sum_{i=1}^{m_1} h_i x_i = 0$, that is, $\sum_{i=1}^{m_1} h_i x_i \in M_{c-1}$, which implies $h_1 = \cdots = h_{m_1} = 0$, due to the choice of $x_1, \ldots, x_{m_1}$.

Now, choose elements $x_{m_1+1}, \ldots, x_{m_2} \in M_{c-1}$ such that $p x_1, \ldots, p x_{m_1}$, $x_{m_1+1}, \ldots, x_{m_2}$ represent a basis of M_{c-1}/M_{c-2} ($m_1 = m_2$ is possible). Continuing like this, we obtain a sequence $x_1, \ldots, x_{m_c} \in M$ such that, for $\nu = 0, \ldots, c-1$, the set

$$\{p^\nu x_1, \ldots, p^\nu x_{m_1}, p^{\nu-1} x_{m_1+1}, \ldots, p^{\nu-1} x_{m_2}, \ldots, x_{m_\nu+1}, \ldots, x_{m_{\nu+1}}\}$$

induces an $R/\langle p \rangle$–basis of $M_{c-\nu}/M_{c-\nu-1}$ (with the convention $m_0 = 0$).

For $i = 1, \ldots, m_c$, define $C_i := \langle x_i \rangle$ and n_i by $\mathrm{Ann}(C_i) = \langle p^{n_i} \rangle$. Obviously, $\sum_{i=1}^{m_c} C_i = M$, and we have to show $C_i \cap \sum_{j \neq i} C_j = \langle 0 \rangle$: if $\sum_{i=1}^{m_1} h_i x_i = 0$ for some $h_i \in R$ then $\sum_{i=1}^{m_1} h_i p^{c-1} x_i = 0$ and, therefore, p divides h_i for

---

[12] Such a decomposition always exists and is uniquely determined up to permutations and multiplication with units. This is proved in the Exercises 1.3.4, 1.3.5 and later in Chapter 4.

$i = 1, \ldots, m_1$. This implies $p^{c-2} \left(\sum_{i=1}^{m_1} (h_i/p) \cdot p x_i + \sum_{i=m_1+1}^{m_2} h_i x_i \right) = 0$ and, therefore, p divides h_i/p for $i = 1, \ldots, m_1$ and h_i for $i = m_1 + 1, \ldots, m_2$. Continuing like this we obtain that $p^{c-\nu}$ divides h_i for $i = m_\nu + 1, \ldots, m_{\nu+1}$ and $\nu = 0, \ldots, c - 1$. This implies $h_i x_i = 0$ for all i, and we conclude that $C_i \cap \sum_{j \neq i} C_j = \langle 0 \rangle$.

It remains to prove that $n_1, \ldots, n_s$ are uniquely determined by M. By construction of the x_i and definition of n_i we have $n_{m_k+1} = \ldots = n_{m_{k+1}} = c - k$, $k = 0, \ldots, c - 1$. Therefore, $m_1, \ldots, m_c$ and $n_1, \ldots, n_s$ determine each other. But $m_1, \ldots, m_c$ are give by the equations $\sum_{i=1}^{k} m_i = \dim_{R/\langle p \rangle} M_{c-k+1}/M_{c-k}$, $k = 1, \ldots, c$. □

Summarizing the results obtained, we have the following theorem:

Theorem 2.6.8. *Let R be a principal ideal domain and M a finitely generated R–module, then M is a direct sum of cyclic modules, $M = \bigoplus_{i=1}^{s} C_i$. The cyclic modules C_i are free or $\mathrm{Ann}(C_i) = \langle p_i^{n_i} \rangle$, p_i prime. The number of the free cyclic modules, the prime ideals $\langle p_i \rangle$ and the numbers n_i are uniquely determined by M.*

$p_1^{n_1}, \ldots, p_s^{n_s} \in R$ are called the *elementary divisors* of M, respectively of the torsion submodule $\mathrm{Tors}(M)$.

Corollary 2.6.9. *Let R be a principal ideal domain and M be a finitely generated R–module, then M is a direct sum of cyclic modules, $M = \bigoplus_{i=1}^{r} D_i$ such that $\mathrm{Ann}(D_1) \subset \mathrm{Ann}(D_2) \subset \cdots \subset \mathrm{Ann}(D_s)$. The ideals $\mathrm{Ann}(D_i)$ are uniquely determined.*

Proof. We use Theorem 2.6.8 and write

$$M = C_1 \oplus \cdots \oplus C_t \oplus C_{1,1} \cdots \oplus C_{1,n_1} \oplus \cdots \oplus C_{r,1} \oplus \cdots \oplus C_{r,n_r}$$

such that $C_1, \ldots, C_t$ are free and $\mathrm{Ann}(C_{ij}) = \langle p_i^{m_{i,j}} \rangle$ and $m_{i,1} \leq \cdots \leq m_{i,n_i}$, p_i prime. Let $D_i := C_i$ for $i = 1, \ldots, t$. Define $D_{t+1} := \bigoplus_{i=1}^{r} C_{i,n_i}$ then

$$\langle 0 \rangle = \mathrm{Ann}(D_1) = \cdots = \mathrm{Ann}(D_t) \subset \mathrm{Ann}(D_{t+1}) = \langle p_1^{m_1,n_1} \cdot \ldots \cdot p_r^{m_r,n_r} \rangle .$$

We continue in this manner, defining $D_{t+k} = \bigoplus_{i=1}^{r} C_{i,n_i-k}$ with the convention $C_{i,n_i-k} = \langle 0 \rangle$ if $n_i - k \leq 0$. Then

$$\mathrm{Ann}(D_{t+k}) = \left\langle p_1^{m_1,n_1-k+1} \cdot \ldots \cdot p_r^{m_r,n_r-k+1} \right\rangle ,$$

with the convention $m_{i,n_i-k+1} = 0$ if $n_i - k + 1 \leq 0$. It remains to prove that the D_i are cyclic. This will be done in the following lemma. □

Lemma 2.6.10. *Let R be a ring and $f, g \in R$ such that $\langle f, g \rangle = R$, then $R/\langle f \rangle \oplus R/\langle g \rangle \cong R/\langle f \cdot g \rangle$.*

Proof. Consider the map $\pi : R \to R/\langle f \rangle \oplus R/\langle g \rangle$ defined by

$$\pi(h) = (h \ \mathrm{mod} \ \langle f \rangle, \ h \ \mathrm{mod} \ \langle g \rangle).$$

This map is surjective: Let $a, b \in R$ such that $af + bg = 1$. If

$$(h \ \mathrm{mod} \ \langle f \rangle, \ k \ \mathrm{mod} \ \langle g \rangle) \in R/\langle f \rangle \oplus R/\langle g \rangle$$

then $h - k = a(h - k)f + b(h - k)g$, that is,

$$h - a(k - h)f = k + b(h - k)g =: c$$

and $\pi(c) = (h \ \mathrm{mod} \ \langle f \rangle, \ k \ \mathrm{mod} \ \langle g \rangle)$.

Let $h \in \mathrm{Ker}(\pi) = \langle f \rangle \cap \langle g \rangle$, that is, $h = h_1 f = h_2 g$ for some $h_1, h_2 \in R$. But, by the above, $h_2 = ah_2 f + bh_2 g = (ah_2 + bh_1)f$, which implies $h \in \langle fg \rangle$. Finally, we obtain $\mathrm{Ker}(\pi) = \langle fg \rangle$, therefore, $R/\langle f \cdot g \rangle \cong R/\langle f \rangle \oplus R/\langle g \rangle$. $\square$

SINGULAR Example 2.6.11 (cyclic decomposition).
To obtain the complete decomposition as in Theorem 2.6.8, we have to diagonalize the presentation matrix of a given module and factorize the diagonal elements.

```
option(redSB);
ring R=0,(x),(C,dp);

matrix M0[5][5]=1, 1,0,0,0,
               -2,-1,0,0,0,
                0, 0,2,1,0,
                0, 0,0,2,0,
                0, 0,0,0,3;
matrix N[5][5]=1, 1, -1,  1,-1,
               2, 2,  1,  1, 0,
              -1, 2,  2,  1, 1,
              -2, 1,  1, -1, 0,
               1, 2, -2,  1, 1;
```

Now we compute the matrix $M = N^{-1}M_0 N - xE_5$:

```
matrix M=lift(N,freemodule(nrows(N)))*M0*N-x*freemodule(5);
print(M);
//-> -x-9/10,-183/50, -59/50,  -43/25,  29/50,
//-> -6/5,    -x+18/25,-11/25,  -19/25,  16/25,
//-> -11/5,   -52/25,  -x+54/25,-34/25,  1/25,
//-> 12/5,    99/25,   52/25,   -x+83/25,-12/25,
//-> -1/2,    1/10,    -17/10,  1/5,     -x+17/10

N=diagonalForm(M);
```

```
print(N);
//-> x5-7x4+17x3-19x2+16x-12,0,0,0,0,
//-> 0,                       1,0,0,0,
//-> 0,                       0,1,0,0,
//-> 0,                       0,0,1,0,
//-> 0,                       0,0,0,1
```

This shows that the module defined by the presentation matrix M is isomorphic to $\mathbb{Q}[x]/\langle x^5 - 7x^4 + 17x^3 - 19x^2 - 16x - 12\rangle$.

```
factorize(N[1,1]);
//->[1]:
//->    _[1]=1
//->    _[2]=x-3
//->    _[3]=x2+1
//->    _[4]=x-2
//->[2]:
//->    1,1,1,2
```

This shows that the decomposition of the module defined by the presentation matrix M is $\mathbb{Q}[x]/\langle x - 3\rangle \oplus \mathbb{Q}[x]/\langle x^2 + 1\rangle \oplus \mathbb{Q}[x]/\langle x - 2\rangle$.

Let us be given a finite presentation for a module N,

$$K[x]^m \xrightarrow{M} K[x]^n \xrightarrow{\pi} N \longrightarrow 0,$$

with presentation matrix M (with respect to the canonical basis). Algorithm 2.6.2 transforms M into a diagonal matrix, but it does not return the transformation matrices. The following procedure computes invertible matrices B and C with entries in $K[x]$ such that $MB = CD$ with D a matrix in diagonal form. Instead of M we consider the matrix

$$\begin{pmatrix} 0 & E_m \\ E_n & M \end{pmatrix} \in \mathrm{GL}(n + m, K[x]),$$

(E_n denoting the $n \times n$ unit matrix), and perform on this matrix appropriate row and column operations to obtain

$$\begin{pmatrix} 0 & B \\ F & D \end{pmatrix} \in \mathrm{GL}(n + m, K[x]),$$

with D an $n \times m$–matrix in diagonal form. Then, by construction, we obtain that $M \cdot B = C \cdot D$ with $C := F^{-1}$.

If $f_1, \ldots, f_n$ are the columns of C and if the matrix D has the entries $p_1, \ldots, p_k \in K[x]$, $k = \min\{n, m\}$, on the diagonal, then $N = \bigoplus_{i=1}^{n}\langle \pi(f_i)\rangle$ and $\langle \pi(f_i)\rangle \cong K[x]/\langle p_i\rangle$, for $i \leq k$, respectively $\langle \pi(f_i)\rangle \cong K[x]$, for $i > k$.

```
proc extendedDiagonalForm(matrix M)
{
   int n=nrows(M);
   int m=ncols(M);
   intvec v=1..n;
   intvec w=n+1..n+m;
   intvec u=1..m;
   intvec x=m+1..n+m;
   matrix E=unitmat(n);
   matrix B=unitmat(m);
   matrix N=M;                //to keep M for the test
   matrix D,K;

   while(D!=N)
   {
      D=N;
      K=transpose(interred(transpose(concat(E,D))));
      E=submat(K,v,v);
      N=submat(K,v,w);
      K=interred(transpose(concat(transpose(B),transpose(N))));
      K=simplify(K,1);     //here we normalize
      B=submat(K,u,u);
      N=submat(K,x,u);
   }
   matrix C=inverse(E);
   if(M*B!=C*D){ERROR("something went wrong");}  //test
   list Re=B,C,D;
   return(Re);
}
```

Let us apply the procedure to an example:

```
LIB"matrix.lib";
LIB"linalg.lib";
matrix M1[2][2]=x2+1,  0,
               0   ,x-1;
matrix N1[2][2]=1, 1,
               1, 2;
matrix N2[2][2]=0,-1,
               1, 1;
M=N1*M1*N2;
print(M);
//-> x-1, -x2+x-2,
//-> 2x-2,-x2+2x-3

list L=extendedDiagonalForm(M);
```

```
print(L[1]);
//-> 1,        0,
//-> 1/2x2-1/2,1/2

print(L[2]);
//-> -1/2x,     -1/2x2+1/2x-1,
//-> -1/2x+1/2,-1/2x2+x-3/2

print(L[3]);
//-> x3-x2+x-1,0,
//-> 0,        1
```

Proposition 2.6.12. *Let V be a finite dimensional K–vector space, and let $\varphi : V \to V$ be an endomorphism. Then V can be considered as $K[x]$–module via φ, setting $xv := \varphi(v)$ for $v \in V$. Let M be the matrix corresponding to φ with respect to a fixed basis $\{b_1, \ldots, b_n\}$, and define $\pi : K[x]^n \to V$ by $\pi(e_i) := b_i$, where $\{e_1, \ldots, e_n\}$ is the canonical basis of $K[x]^n$. Then*

$$K[x]^n \xrightarrow{M-xE} K[x]^n \xrightarrow{\pi} V \longrightarrow 0 \,,$$

is a presentation of the $K[x]$–module V. [13]

Remark 2.6.13. Any finitely generated torsion $K[x]$–module N can be obtained in the way described in Proposition 2.6.12: let $\langle f \rangle = \mathrm{Ann}(N)$, then $K[x]/\langle f \rangle$ is a finite dimensional K–vector space and N is a finitely generated $K[x]/\langle f \rangle$–module. This implies that N is a finite dimensional K–vector space and multiplication by x defines the above endomorphism $\varphi : N \to N$.

Proof of Proposition 2.6.12. By definition of the $K[x]$–module structure, we have $\pi \circ (M - xE) = 0$. Diagonalizing the matrix $M - xE$ as in Theorem 2.6.1 does not change its determinant. In particular, the product of the diagonal elements of the diagonal form of $M - xE$ is the characteristic polynomial of M, which has degree n. This implies that

$$\dim_K K[x]^n / \mathrm{Im}(M - xE) = n = \dim(V)$$

and, therefore, $V \cong K[x]^n / \mathrm{Im}(M - xE)$. $\qquad\square$

Remark 2.6.14. Let M be an $n \times n$–matrix with entries in K and consider, as in Proposition 2.6.12, K^n via M as $K[x]$–module. Let $B, C \in \mathrm{GL}(n, K[x])$ be invertible matrices such that

$$C^{-1} \cdot (M - xE) \cdot B = D = \begin{pmatrix} d_1 & & 0 \\ & \ddots & \\ 0 & & d_n \end{pmatrix} \in \mathrm{Mat}(n \times n, K[x])$$

---

[13] Note that $\det(M - xE)$ is the *characteristic polynomial* of M, E being the unit matrix.

is the diagonal form corresponding to $M - xE$ (as computed by extended-DiagonalForm(M-xE)). Let $f_1, \ldots, f_n$ be the columns of $C = (c_{ij})$, that is, a basis of $K[x]^n$ with $f_i = \sum_{j=1}^n c_{ij} e_j$, $\{e_1, \ldots, e_n\}$ being the canonical basis of $K[x]^n$. We use this notation also for the canonical basis of K^n.

Setting $v_i := \sum_{j=1}^n c_{ij}(M) \cdot e_j \in K^n$ (that is, replacing x by M in f_i), we obtain that $\langle v_i \rangle_{K[x]} \cong K[x]/\langle d_i \rangle$, and $K^n = \bigoplus_{i=1}^n \langle v_i \rangle_{K[x]}$ is a decomposition of K^n as $K[x]$–module into a direct sum of cyclic modules [14]. Finally, we can factorize the d_i and split the submodules $\langle v_i \rangle$ into direct sums, using Lemma 2.6.10. If $d_i = \prod_j d_{ji}^{\rho_{ji}}$ is a decomposition into irreducible polynomials, then $\langle v_i \rangle = \bigoplus_j V_{ij}$, where

$$V_{ij} := \left\{ w \in \langle v_i \rangle \mid d_{ji}^{\rho_{ji}} w = 0 \right\} = \left\langle \prod_{k \neq j} d_{ki}^{\rho_{ki}} \cdot v_i \right\rangle \cong K[x]/\langle d_{ji}^{\rho_{ji}} \rangle .$$

The $V_{ij} \subset K^n$ are *invariant subspaces* for the endomorphism $M : K^n \to K^n$, that is, $M \cdot V_{ij} \subset V_{ij}$. Moreover,

$$\det(M - xE) = \det(C) \cdot \det(B)^{-1} \cdot \prod_{i=1}^n d_i = \det(C) \cdot \det(B)^{-1} \cdot \prod_{i,j} d_{ji}^{\rho_{ji}} ,$$

that is, the characteristic polynomial of M is (up to multiplication by a non–zero constant) equal to $\prod_{i,j} d_{ji}^{\rho_{ji}}$.

This leads to a procedure for computing a decomposition of M into block matrices as shown in the example below.

If the characteristic polynomial splits into linear factors (for instance, if the field K is algebraically closed), we obtain the decomposition corresponding to the *Jordan normal form* of M. In the general case, a better decomposition is given by the *rational normal form*, treated in Exercise 2.6.3.

SINGULAR Example 2.6.15 (Jordan normal form).
We start with a matrix M whose characteristic polynomial does not split into linear factors, but which is already in the form described in Remark 2.6.14. We conjugate M by some invertible matrix N and, finally, compute the original form (up to normalization), according to the procedure described in Remark 2.6.14. The same method leads to the Jordan normal form of a matrix if its characteristic polynomial splits into linear factors.

```
ring R=0,(x),(C,dp);
matrix M[5][5]=1, 1,0,0,0,
              -2,-1,0,0,0,
               0, 0,2,1,0,
               0, 0,0,2,0,
               0, 0,0,0,2;
```

---

[14] Note that some of the v_i may be zero.

```
matrix N[5][5]=1,2, 2, 2,-1 ,           //an invertible
               1,1, 2, 1, 1,            //matrix over Q
              -1,1, 2,-1, 2,
              -1,1, 1,-1, 2,
               1,2,-1, 2, 1;
M=lift(N,freemodule(nrows(N)))*M*N;     //inverse(N)*M*N
print(M);
//-> -3/2,-21,-53/2,-8,  -14,
//-> 5/4, -6, -35/4,-1/2,-13/2,
//-> -1,  1,  3,     -1,  2,
//-> 3/4, 18, 87/4, 13/2,27/2,
//-> -3/2,2,  3/2,   -1,  4
```

We want to compute the normal form of M according to Remark 2.6.14:

```
matrix A = M-x*freemodule(5);           //the matrix M-xE
LIB"linalg.lib";
option(redSB);
list L = extendedDiagonalForm(A);       //A*L[1]=L[2]*L[3]

print(L[2]);                            //the new basis
//-> 0,          0,     -53/3250,-8/24375,        -14/24375
//-> 0,          -1/650,-7/1300,-1/48750,         -1/3750
//-> 168/1625,   1/325,-1/1625x+3/1625,-1/24375,  2/24375
//-> -3451/24375,-1/390,87/6500,-1/24375x+1/3750, 9/16250
//-> -2798/24375,-2/975,3/3250,-1/24375,-1/24375x+4/24375

print(L[3]);                            //the diagonal form
//-> x4-4x3+5x2-4x+4,0,  0,0,0,          //of M-xE
//-> 0,               x-2,0,0,0,
//-> 0,               0,  1,0,0,
//-> 0,               0,  0,1,0,
//-> 0,               0,  0,0,1
```

At this level we know that the vector space $\mathbb{Q}^5$ considered as $\mathbb{Q}[x]$–module, where x acts via the matrix M, is isomorphic to

$$\mathbb{Q}[x]/\langle x^4 - 4x^3 + 5x^2 - 4x + 4\rangle \oplus \mathbb{Q} =: V_1 \oplus V_2\,,$$

where V_1 and V_2 are invariant subspaces.

```
matrix V1[5][4]=concat(L[1][1],M*L[1][1],M*M*L[1][1],
                                M*M*M*L[1][1]);
matrix V2[5][1]=L[1][2];                //the 2 invariant
                                        //subspaces
list F=factorize(L[2][1,1]);
```

```
F;
//-> [1]:
//->    _[1]=1
//->    _[2]=x2+1
//->    _[3]=x-2
//-> [2]:
//->    1,1,2
```

The first diagonal element of L[3] does not split into linear factors over $\mathbb{Q}$.

We need a procedure to compute the matrix $p(B)$, where $p \in \mathbb{Q}[x]$ is a polynomial and B a matrix.

```
proc polyOfEndo(matrix B,poly p)
{
    int i;
    int d=nrows(B);
    matrix A=coeffs(p,var(1));
    matrix E[d][d]=freemodule(d);
    matrix C[d][d]=A[1,1]*E;
    for(i=2;i<=nrows(A);i++)
    {
        E=E*B;
        C=C+A[i,1]*E;
    }
    return(C);
}
```

Now we are able to compute bases for the invariant subspaces V_1, V_2. Since V_2 is already one–dimensional, we need only to consider V_1.

```
matrix S=polyOfEndo(M,F[1][3]^2); //the decomposition of V1
matrix V11=std(S*V1);
print(V11);
//-> -4,1,
//-> -1,-1,
//-> 0, 0,
//-> 3, 0,
//-> 0, 1

S=polyOfEndo(M,F[1][2]);
matrix V12=std(S*V1);
print(V12);
//-> -9776,13195,
//-> -7107,14214,
//-> 4888, -17258,
//-> 7107, 0,
//-> 0,    7107
```

We test whether we obtained bases, that is, whether $V_{11} \oplus V_{12} \oplus V_2 = \mathbb{Q}^5$, and whether the subspaces V_{11} and V_{12} are invariant.

```
matrix B=concat(V11,V12,V2);
det(B);                       //we obtained a basis
//-> -28428
reduce(M*V11,std(V11));       //subspaces are invariant
//-> _[1]=0
//-> _[2]=0
reduce(M*V12,std(V12));
//-> _[1]=0
//-> _[2]=0
reduce(M*V2,std(V2));
//-> _[1]=0
```

Compute the matrix with respect to the new bases, given by the invariant subspaces.

```
matrix C=lift(B,M*B);       //the matrix M with respect to
print(C);                   //the basis B
//-> -1/2,-5/4,0,          0,           0,
//-> 1,    1/2, 0,          0,           0,
//-> 0,    0,   11501/4738,-18225/9476,0,
//-> 0,    0,   225/2369,   7451/4738,  0,
//-> 0,    0,   0,          0,           2
```

We compute special bases to obtain the normal form.

```
matrix v[5][1]=V12[1];      //special basis for normal form
B=concat(V11,M*v-2*v,v,V2);
C=lift(B,M*B);              //the matrix M with respect to
print(C);                   //the basis B
//-> -1/2,-5/4,0,0,0,
//-> 1,    1/2, 0,0,0,
//-> 0,    0,   2,1,0,
//-> 0,    0,   0,2,0,
//-> 0,    0,   0,0,2
```

Hence, we obtain, up to normalization, the original matrix M we started with.

Exercises

2.6.1. Prove that Corollary 2.6.9 implies Theorem 2.6.8.

2.6.2. Write a SINGULAR procedure to compute the Jordan normal form of an endomorphism under the assumptions that the characteristic polynomial splits into linear factors over the ground field.

2.6.3. Write a SINGULAR procedure to compute the *rational normal form* of an endomorphism, that is, a block in the matrix corresponding to a cyclic submodule has the shape

$$\begin{pmatrix} 0 & 0 & \ldots & 0 & -a_0 \\ 1 & 0 & \ldots & 0 & -a_1 \\ 0 & 1 & \ldots & 0 & -a_2 \\ \vdots & & & & \vdots \\ 0 & 0 & \ldots & 1 & -a_{n-1} \end{pmatrix},$$

where $x^n + a_{n-1}x^{n-1} + \cdots + a_0$ is the characteristic polynomial.

(Hint: compute $V = \bigoplus_i \langle v_i \rangle_{K[x]}$, a decomposition of the K–vector space V into a direct sum of cyclic $K[x]$–modules as in SINGULAR Example 2.6.15. Then consider in $\langle v_i \rangle_{K[x]}$ the K–basis $v_i, xv_i, x^2v_i, \ldots$.)

2.6.4. Write a SINGULAR procedure using Algorithm 2.6.2 to diagonalize the presentation matrix for modules over the ring $K[x]$, respectively $K[x]_{\langle x \rangle}$.

2.6.5. Let A be a principal ideal domain and K its quotient field. Prove that K is not a free A–module and that K/A is not a direct sum of cyclic modules.

2.6.6. Let $A = K[x]$ be the polynomial ring in one variable, $a_1, \ldots, a_r \in A$, and $M := A/\langle a_1 \rangle \oplus \cdots \oplus A/\langle a_r \rangle$. Give an algorithm to compute the decomposition of M as in Corollary 2.6.9.

2.6.7. Let R be a principal ideal domain and $p_1, \ldots, p_n \in R$ prime elements such that $\langle p_i, p_j \rangle = R$ for $i \neq j$. Prove that $\langle p_i^{c_i}, p_1^{c_1} \cdot \ldots \cdot p_{i-1}^{c_{i-1}} p_{i+1}^{c_{i+1}} \cdot \ldots \cdot p_n^{c_n} \rangle = R$ for $c_1, \ldots, c_n \in \mathbb{N}$.

2.6.8. Use SINGULAR to compute the Jordan normal form of $\begin{pmatrix} 3 & -1 & 2 \\ 1 & 1 & 2 \\ 2 & -2 & 2 \end{pmatrix}$.

2.7 Tensor Product

Let A be a ring, and let M, N, and P be A–modules. Let $B(M, N; P)$ be the A–module of bilinear maps $M \times N \to P$. In this section we want to construct a module $M \otimes_A N$, the tensor product of M and N, together with a bilinear map $M \times N \to M \otimes_A N$, $(m, n) \mapsto m \otimes n$, such that this map induces a canonical isomorphism

$$B(M, N; P) \cong \mathrm{Hom}_A(M \otimes_A N, P)$$

of A–modules, and study its properties. The tensor product reduces the theory of bilinear maps to linear maps, for the price that the modules become more complicated.

Let $\sigma : M \times N \to P$ be a *bilinear map*, that is, for all $a \in A$, $m, m' \in M$, $n, n' \in N$,

(B1) $\sigma(am, n) = \sigma(m, an) = a\sigma(m, n)$,
(B2) $\sigma(m + m', n) = \sigma(m, n) + \sigma(m', n)$,
(B3) $\sigma(m, n + n') = \sigma(m, n) + \sigma(m, n')$.

To obtain the isomorphism above, the elements of type $m \otimes n$ of the module to construct have to satisfy the following properties:

(T1) $(am) \otimes n = m \otimes (an) = a(m \otimes n)$,
(T2) $(m + m') \otimes n = m \otimes n + m' \otimes n$,
(T3) $m \otimes (n + n') = m \otimes n + m \otimes n'$,

for all $a \in A$, $m, m' \in M$, $n, n' \in N$. The properties (T1)–(T3) imply the bilinearity of the map $(m, n) \mapsto m \otimes n$.

To obtain a "minimal" module with this property, the following is necessary: $M \otimes_A N$ is generated by $\{m \otimes n \mid m \in M, \ n \in N\}$ and

(T4)   all relations between the generators $\{m \otimes n\}$ are generated by relations of type (T1), (T2) and (T3).

This motivates the following definition:

Definition 2.7.1. Let T be the free A–module generated by the pairs (m, n), $m \in M$, $n \in N$, and let U be the submodule of T generated by the elements

$$(am, n) - a(m, n),$$
$$(m, an) - a(m, n),$$
$$(m + m', n) - (m, n) - (m', n),$$
$$(m, n + n') - (m, n) - (m, n'),$$

with $a \in A$, $m, m' \in M$ and $n, n' \in N$.

Then we define the *tensor product* $M \otimes_A N$ of M and N to be the A–module T/U, and denote by $m \otimes n$ the equivalence class of (m, n) in T/U.

Proposition 2.7.2. *There are canonical isomorphisms of A–modules*

(1) $B(M, N; P) \cong \mathrm{Hom}_A(M \otimes_A N, P)$,
(2) $B(M, N; P) \cong \mathrm{Hom}_A\big(M, \mathrm{Hom}_A(N, P)\big)$.

Proof. To prove (1), let $\varphi : M \otimes_A N \to P$ be a homomorphism. Then the properties (T1), (T2) and (T3) imply that $\phi(\varphi)(m, n) := \varphi(m \otimes n)$ defines a bilinear map $\phi(\varphi) : M \times N \to P$. Thus, we obtain a map

$$\phi : \mathrm{Hom}_A(M \otimes_A N, P) \longrightarrow B(M, N; P), \quad \varphi \longmapsto \phi(\varphi),$$

which is obviously A–linear.

If $\phi(\varphi) = 0$ then $\varphi(m \otimes n) = 0$ for all $m \in M$, $n \in N$. Because $M \otimes_A N$ is generated by the elements of the form $m \otimes n$, this implies that $\varphi = 0$.

To see that ϕ is surjective, let $\sigma : M \times N \to P$ be a bilinear map. Then we can define a linear map $\varphi : M \otimes_A N \to P$ by setting $\varphi(m \otimes n) := \sigma(m, n)$. This map is well–defined and linear by the properties (B1), (B2), (B3) of σ. Obviously, $\phi(\varphi) = \sigma$ and, therefore, ϕ is an isomorphism.

To prove (2), let $\varphi : M \to \operatorname{Hom}_A(N, P)$ be a homomorphism and define $\psi(\varphi)(m, n) := \varphi(m)(n)$. Thus,

$$\psi : \operatorname{Hom}_A\big(M, \operatorname{Hom}_A(N, P)\big) \longrightarrow B(M, N; P)$$

is a map which is obviously A–linear.

If $\psi(\varphi) = 0$ then $\varphi(m)(n) = 0$ for all $m \in M$, $n \in N$. This implies that $\varphi(m)$ is the zero map for all m and, therefore, $\varphi = 0$. Now let $\sigma : M \times N \to P$ be a bilinear map, then we can define a linear map $\varphi : M \to \operatorname{Hom}_A(N, P)$ by setting $\varphi(m)(n) := \sigma(m, n)$ and obtain $\psi(\varphi) = \sigma$. This implies that ψ is an isomorphism. $\qquad\square$

The following properties of the tensor product are easy to prove and, therefore, left as an exercise (Exercise 2.7.1).

Proposition 2.7.3. *Let M, M', N, N', and P be A–modules, let $S \subset A$ be a multiplicatively closed subset, and let $\varphi : M \to M'$ and $\psi : N \to N'$ be A–module homomorphisms. Then we have the following isomorphisms of A–modules, respectively $S^{-1}A$–modules,*

(1) $M \otimes_A N \cong N \otimes_A M$,
(2) $(M \otimes_A N) \otimes_A P \cong M \otimes_A (N \otimes_A P)$,
(3) $A \otimes_A M \cong M$,
(4) $(M \oplus N) \otimes_A P \cong (M \otimes_A P) \oplus (N \otimes_A P)$,
(5) $S^{-1}(M \otimes_A N) \cong S^{-1}M \otimes_{S^{-1}A} S^{-1}N$,

Moreover,

(6) $(\varphi \otimes \psi)(m \otimes n) := \varphi(m) \otimes \psi(n)$ defines a homomorphism

$$\varphi \otimes \psi : M \otimes_A N \to M' \otimes_A N'.$$

Example 2.7.4.

(1) $A^r \otimes_A A^s \cong A^{rs}$, and if $\{e_1, \ldots, e_r\}$, respectively $\{f_1, \ldots, f_r\}$, is a basis for A^r, respectively A^s, then $\{e_i \otimes f_j \mid i = 1, \ldots, r, \ j = 1, \ldots, s\}$ is a basis for $A^r \otimes_A A^s$.

(2) Let $\varphi : A^r \to A^s$ and $\psi : A^p \to A^q$ be linear maps, defined by the matrices $M = (m_{ij})_{i,j}$ (with respect to the bases $\{e_1, \ldots, e_r\}$ of A^r and $\{f_1, \ldots, f_s\}$ of A^s), respectively $N = (n_{ij})_{i,j}$ (with respect to the bases $\{g_1, \ldots, g_p\}$ of A^p and $\{h_1, \ldots, h_q\}$ of A^q). Then $\varphi \otimes \psi$ has the matrix $(m_{ca}n_{db})_{a,b;c,d}$ (with respect to the bases $\{e_1 \otimes g_1, e_1 \otimes g_2, \ldots, e_r \otimes g_p\}$

of $A^r \otimes_A A^p$ and $\{f_1 \otimes h_1, f_1 \otimes h_2, \ldots, f_s \otimes h_q\}$ of $A^s \otimes_A A^q$). More precisely, if

$$\varphi(e_a) = \sum_{c=1}^{s} m_{ca} f_c, \quad \psi(g_b) = \sum_{d=1}^{q} n_{db} h_d$$

then

$$(\varphi \otimes \psi)(e_a \otimes g_b) = \sum_{c=1}^{s} \sum_{d=1}^{q} m_{ca} n_{db} \cdot f_c \otimes h_d .$$

If $i = (c-1)q + d$ and $j = (a-1)p + b$ then the element in the i-th row and j-th column of the matrix of $\varphi \otimes \psi$ is $m_{ca} n_{db}$.

SINGULAR Example 2.7.5 (tensor product of maps).
Let M, N be matrices defining maps $\varphi : A^r \to A^s$, respectively $\psi : A^p \to A^q$. The matrix of $\varphi \otimes \psi$ can be computed as follows:

```
proc tensorMaps(matrix M, matrix N)
{
    int r=ncols(M);
    int s=nrows(M);
    int p=ncols(N);
    int q=nrows(N);
    int a,b,c,d;
    matrix R[s*q][r*p];
    for(b=1;b<=p;b++)
    {
        for(d=1;d<=q;d++)
        {
            for(a=1;a<=r;a++)
            {
                for(c=1;c<=s;c++)
                {
                    R[(c-1)*q+d,(a-1)*p+b]=M[c,a]*N[d,b];
                }
            }
        }
    }
    return(R);
}
```

Let us try an example.

```
ring A=0,(x,y,z),dp;
matrix M[3][3]=1,2,3,4,5,6,7,8,9;
matrix N[2][2]=x,y,0,z;
print(M);
```

```
//-> 1,2,3,
//-> 4,5,6,
//-> 7,8,9
print(N);
//-> x,y,
//-> 0,z


print(tensorMaps(M,N));
//->  x, y,2x,2y,3x,3y,
//->  0, z, 0,2z, 0,3z,
//-> 4x,4y,5x,5y,6x,6y,
//->  0,4z, 0,5z, 0,6z,
//-> 7x,7y,8x,8y,9x,9y,
//->  0,7z, 0,8z, 0,9z
```

The next theorem gives a very important property of the tensor product.

Theorem 2.7.6. *Let $M \xrightarrow{i} N \xrightarrow{\pi} P \to 0$ be an exact sequence of A–modules, and L an A–module, then*

$$M \otimes_A L \xrightarrow{i \otimes 1_L} N \otimes_A L \xrightarrow{\pi \otimes 1_L} P \otimes_A L \to 0$$

is exact (the tensor product is right exact).

Proof. We know from Section 2.4 that it is enough to prove that

$$0 \to \operatorname{Hom}_A(P \otimes_A L, S) \to \operatorname{Hom}_A(N \otimes_A L, S) \to \operatorname{Hom}_A(M \otimes_A L, S)$$

is exact for all A–modules S.

Using both isomorphisms of Proposition 2.7.2, we see that this is equivalent to the exactness of

$$0 \to \operatorname{Hom}_A\big(P, \operatorname{Hom}_A(L, S)\big) \to \operatorname{Hom}_A\big(N, \operatorname{Hom}_A(L, S)\big)$$
$$\to \operatorname{Hom}_A\big(M, \operatorname{Hom}_A(L, S)\big).$$

This is the left exactness of Hom already proved in Section 2.4. $\square$

Example 2.7.7. Let $A = \mathbb{Z}$ and consider the exact sequence $0 \to \mathbb{Z} \xrightarrow{i} \mathbb{Z} \xrightarrow{\pi} \mathbb{Z}/\langle 2 \rangle \to 0$, $i(x) = 2x$, then

$$\mathbb{Z} \otimes_{\mathbb{Z}} \mathbb{Z}/\langle 2 \rangle \xrightarrow{i \otimes 1} \mathbb{Z} \otimes_{\mathbb{Z}} \mathbb{Z}/\langle 2 \rangle \xrightarrow{\pi \otimes 1} \mathbb{Z}/\langle 2 \rangle \to 0$$

is exact but $i \otimes 1$ is not injective.

Namely, $(i \otimes 1)\big(a \otimes (b + \langle 2 \rangle)\big) = 2a \otimes (b + \langle 2 \rangle) = a \otimes \langle 2 \rangle = 0$. That is, $i \otimes 1$ is, in fact, the zero map.

Corollary 2.7.8. *Let* $A^r \xrightarrow{\varphi} A^s \xrightarrow{\pi} M \to 0$ *and* $A^p \xrightarrow{\psi} A^q \xrightarrow{\lambda} N \to 0$ *be presentations of the A–modules M and N, then*

$$A^{sp+rq} = (A^s \otimes_A A^p) \oplus (A^r \otimes_A A^q) \xrightarrow{\sigma} A^{sq} = A^s \otimes_A A^q \xrightarrow{\pi \otimes \lambda} M \otimes_A N \to 0$$

is a presentation of the tensor product $M \otimes_A N$, *where* σ *is the composition of the addition* $A^{sq} \oplus A^{sq} \xrightarrow{+} A^{sq}$ *and* $(1_{A^s} \otimes \varphi) \oplus (\varphi \otimes 1_{A^s})$.

Proof. Consider the following commutative diagram:

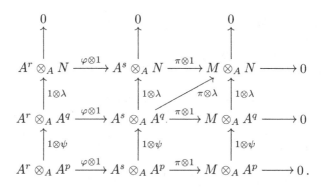

Because of Theorem 2.7.6 the rows and columns are exact. An easy diagram chase shows that $\pi \otimes \lambda$ is surjective and $\mathrm{Ker}(\pi \otimes \lambda) = \mathrm{Im}(1 \otimes \psi) + \mathrm{Im}(\varphi \otimes 1)$. $\square$

SINGULAR Example 2.7.9 (tensor product of modules).

Let φ and ψ be matrices describing presentations of the modules M, respectively N. We give a procedure for computing the presentation matrix of $M \otimes_A N$ as described in Corollary 2.7.8.

```
LIB"matrix.lib";

proc tensorMod(matrix Phi, matrix Psi)
{
    int s=nrows(Phi);
    int q=nrows(Psi);
    matrix A=tensorMaps(unitmat(s),Psi);   //I_s tensor Psi
    matrix B=tensorMaps(Phi,unitmat(q));   //Phi tensor I_q
    matrix R=concat(A,B);                  //sum of A and B
    return(R);
}
```

We consider an example:

```
ring A=0,(x,y,z),dp;
matrix M[3][3]=1,2,3,4,5,6,7,8,9;
matrix N[2][2]=x,y,0,z;
```

```
print(M);
//-> 1,2,3,
//-> 4,5,6,
//-> 7,8,9

print(N);
//-> x,y,
//-> 0,z

print(tensorMod(M,N));
//-> x,y,0,0,0,0,1,0,2,0,3,0,
//-> 0,z,0,0,0,0,0,1,0,2,0,3,
//-> 0,0,x,y,0,0,4,0,5,0,6,0,
//-> 0,0,0,z,0,0,0,4,0,5,0,6,
//-> 0,0,0,0,x,y,7,0,8,0,9,0,
//-> 0,0,0,0,0,z,0,7,0,8,0,9
```

For further applications we need a criterion for $\sum_i m_i \otimes n_i$ to be zero.

Proposition 2.7.10. *Let M and N be A–modules, $m_i \in M$ for $i \in I$, and $N = \langle n_i \mid i \in I \rangle$. Then $\sum_{i \in I} m_i \otimes n_i = 0$[15] if and only if there exist $a_{ij} \in A$ and $\bar{m}_j \in M$, for $i \in I$ and $j \in J$, such that $\sum_{j \in J} a_{ij} \bar{m}_j = m_i$ for all $i \in I$, and $\sum_{i \in I} a_{ij} n_i = 0$ for all $j \in J$.*

Proof. Suppose $\sum_{j \in J} a_{ij} \bar{m}_j = m_i$ and $\sum_{i \in I} a_{ij} n_i = 0$, then

$$\sum_{i \in I} m_i \otimes n_i = \sum_{i \in I} \left(\sum_{j \in J} a_{ij} \bar{m}_j \right) \otimes n_i = \sum_{j \in J} \left(\bar{m}_j \otimes \sum_{i \in I} a_{ij} n_i \right) = 0.$$

To prove the other direction, we consider first the special case that N is free and $\{n_i\}_{i \in I}$ is a basis of N. Using Proposition 2.7.3 (3), (4), we obtain that $\sum_{i \in I} m_i \otimes n_i \mapsto \sum_{i \in I} m_i$ induces an isomorphism $M \otimes_A N \cong \bigoplus_{i \in I} M$. This implies that $\sum_{i \in I} m_i \otimes n_i = 0$ if and only if $m_i = 0$ for all $i \in I$.

Now let N be arbitrary, and let $F_1 \xrightarrow{\lambda} F_0 \xrightarrow{\pi} N \to 0$ be a presentation of N such that there is a basis $\{e_i\}_{i \in I}$ of F_0 with $\pi(e_i) = n_i$ for all $i \in I$. Using Theorem 2.7.6, we obtain that the induced sequence

$$M \otimes_A F_1 \xrightarrow{1 \otimes \lambda} M \otimes_A F_0 \xrightarrow{1 \otimes \pi} M \otimes_A N \to 0$$

is exact, too. In these terms our assumption reads $(1 \otimes \pi)\left(\sum_{i \in I} m_i \otimes e_i \right) = 0$, which implies $\sum_{i \in I} m_i \otimes e_i = \sum_{j \in J} \bar{m}_j \otimes f_j$ for suitable $\bar{m}_j \in M$, $f_j \in \text{Im}(\lambda)$ (using the exactness of the induced sequence). Let $f_j =: \sum_{i \in I} a_{ij} e_i$, $j \in J$, then $\sum_{i \in I} m_i \otimes e_i - \sum_{i \in I} \left(\sum_{j \in J} a_{ij} \bar{m}_j \right) \otimes e_i = 0$. This is the situation of our special case and, therefore, $m_i = \sum_{j \in J} a_{ij} \bar{m}_j$ for all $i \in I$. On the other hand, $f_j \in \text{Im}(\lambda) = \text{Ker}(\pi)$ and, therefore, $\sum_{i \in I} a_{ij} n_i = 0$. $\quad\square$

---

[15] Of course, there are only finitely many indices $i \in I$ with $m_i \neq 0$ in such a sum.

Now we turn to the tensor product of algebras.

Proposition 2.7.11. *Let B, C be A–algebras, then $B \otimes_A C$ is an A–algebra having the following* universal *property: for any commutative diagram*

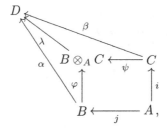

of A–algebras, there exists a unique A–algebra homomorphism $\lambda : B \otimes_A C \to D$ such that the diagram

$$
\begin{array}{c}
D
\end{array}
$$

with $\psi : c \mapsto 1 \otimes c$ and $\varphi : b \mapsto b \otimes 1$, commutes.

Proof. It is easy to see that $B \otimes_A C$ together with the multiplication defined by $(b \otimes c)(b' \otimes c') := (bb') \otimes (cc')$ is an A–algebra and φ and ψ are A–algebra homomorphisms such that $\varphi \circ j = \psi \circ i$.

To prove the second part of the proposition we set $\lambda(b \otimes c) := \alpha(b)\beta(c)$. If λ is well–defined, then it is obviously an A–algebra homomorphism, making the diagram commutative, and it is uniquely determined. That is, we have to prove that $\sum_{\ell \in L} b_\ell \otimes c_\ell = 0$ implies $\sum_{\ell \in L} \alpha(b_\ell)\beta(c_\ell) = 0$. Let $\{x_i\}_{i \in I}$ be a set of generators of C as an A–module, and let $c_\ell = \sum_{i \in I} c_{\ell i} x_i$ for some $c_{\ell i} \in A$. Then

$$
0 = \sum_{\ell \in L} b_\ell \otimes c_\ell = \sum_{i \in I} \left(\sum_{\ell \in L} c_{\ell i} b_\ell \right) \otimes x_i .
$$

Using Proposition 2.7.10, we obtain some $\bar{b}_j \in B$ and $a_{ij} \in A$, $j \in J$, such that $\sum_{j \in J} a_{ij}\bar{b}_j = \sum_{\ell \in L} c_{\ell i} b_\ell$ and $\sum_{i \in I} a_{ij} x_i = 0$ for all j. Now

$$
\sum_{\ell \in L} \alpha(b_\ell)\beta(c_\ell) = \sum_{\ell \in L} \alpha(b_\ell)\beta\left(\sum_{i \in I} c_{\ell i} x_i\right) = \sum_{i \in I} \alpha\left(\sum_{\ell \in L} c_{\ell i} b_\ell\right)\beta(x_i)
$$

$$
= \sum_{i \in I} \alpha\left(\sum_{j \in J} a_{ij}\bar{b}_j\right)\beta(x_i) = \sum_{j \in J} \alpha(\bar{b}_j)\beta\left(\sum_{i \in I} a_{ij}x_i\right) = 0 .
$$

$\square$

Corollary 2.7.12. *Let A, B, C be as in Proposition 2.7.11 and T be an A-algebra having the universal property described in 2.7.11, then there exists a unique isomorphism $T \cong B \otimes_A C$.*

Proof. We leave the proof as Exercise 2.7.3.   □

Corollary 2.7.13. *Let $B = A[x_1, \ldots, x_n]/I$ and $C = A[y_1, \ldots, y_m]/J$, then $B \otimes_A C = A[x_1, \ldots, x_n, y_1, \ldots, y_m]/\langle I, J \rangle$.*

Proof. We use Corollary 2.7.12. Let

$$
\begin{array}{ccc}
D & \xleftarrow{\ \beta\ } & C \\
{\scriptstyle\alpha}\big\uparrow & & \big\uparrow \\
B & \xleftarrow{\hspace{1cm}} & A
\end{array}
$$

be a commutative diagram of A–algebras. We define a ring map

$$\lambda : A[x_1, \ldots, x_n, y_1, \ldots, y_m]/\langle I, J \rangle \longrightarrow D$$

by setting $\lambda(x_i + \langle I, J \rangle) := \alpha(x_i + I)$ and $\lambda(y_i + \langle I, J \rangle) := \beta(y_i + J)$. If λ is well–defined then it obviously has the minimal property required for the tensor product. To prove that this is the case, let $F \in \langle I, J \rangle$, that is, we can write $F = \sum_i G_i g_i + \sum_j H_j h_j$ for suitable $g_i \in I$, $h_j \in J$, $G_i \in A[y_1, \ldots, y_m]$, $H_j \in A[x_1, \ldots, x_n]$. Then $\lambda(F) = \sum_i \beta(G_i)\alpha(g_i) + \sum_j \alpha(H_j)\beta(h_j)$. But we have $\alpha(g_i) = 0$, $\beta(h_j) = 0$, which implies that $\lambda(F) = 0$.   □

SINGULAR Example 2.7.14 (tensor product of rings).
In the following, we apply Corollary 2.7.12 to compute the tensor product of rings: let $A := \mathbb{Q}[a, b, c]/\langle ab - c^2 \rangle$, $B := \mathbb{Q}[x, y, z, a, b, c]/\langle x^2, y, ab - c^2 \rangle$ and $C := \mathbb{Q}[u, v, a, b, c]/\langle uv, ab - c^2 \rangle$, and let $A \to B$, $A \to C$ be the canonical maps.

```
ring A1=0,(a,b,c),dp;
ideal P=ab-c2;
qring A=std(P);              // A=A1/P
poly p=abc;


ring B1=0,(x,y,z,a,b,c),dp;
ideal I=x2,y,ab-c2;
qring B=std(I);              // B=B1/I
map ib=A,a,b,c;             // the canonical map A-->B


ring C1=0,(u,v,a,b,c),lp;
ideal J=uv,ab-c2;
qring C=std(J);              // C=C1/J
map ic=A,a,b,c;             // the canonical map A-->C
```

We compute the tensor product $T = B \otimes_A C$, together with the maps $B \to T$ and $C \to T$:

```
ring T1=0,(x,y,z,u,v,a,b,c),dp; // B1 tensor C1 over A1
ideal K=imap(C1,J)+imap(B1,I);
qring T=std(K);                 // B tensor C over A
map jb=B,x,y,z,a,b,c;           // the canonical map B-->T
map jc=C,u,v,a,b,c;             // the canonical map C-->T
```

Finally, we check that the tensor product diagram commutes:

```
map psi=jc(ic);
map phi=jb(ib);
psi(p);
//-> abc
phi(p);
//-> abc
```

Exercises

2.7.1. Prove Proposition 2.7.3.

2.7.2. Let A be a ring, B an A–algebra, and M an A–module. Prove that $M \otimes_A B$ has a canonical B–module structure.

2.7.3. Prove Corollary 2.7.12.

2.7.4. Prove that $\mathbb{Z}/\langle a \rangle \otimes_{\mathbb{Z}} \mathbb{Z}/\langle b \rangle = 0$ if a, b are coprime.

2.7.5. Let A be a ring, $I \subset A$ an ideal, and M an A–module. Prove that $M/IM \cong (A/I) \otimes_A M$.

2.7.6. Let A be a local ring and M, N finitely generated A–modules. Prove that $M \otimes_A N = 0$ implies $M = 0$ or $N = 0$.

2.7.7. Let $0 \to M \to N \to P \to 0$ be an exact sequence of A–modules and F a free A–module. Prove that

$$0 \to M \otimes_A F \to N \otimes_A F \to P \otimes_A F \to 0$$

is exact. (Hint: study the proof of Corollary 2.7.8.)

2.7.8. Let $\dots \to F_i \to F_{i-1} \to \dots \to F_0 \to 0$ be an exact sequence of A–modules and F a free A–module. Prove that

$$\dots \to F_i \otimes_A F \to F_{i-1} \otimes_A F \to \dots \to F_0 \otimes_A F \to 0$$

is exact.

2.7.9. Let A be a ring, let $a \in A$, and let M be an A–module. Prove that $\langle a \rangle \otimes_A M = \{a \otimes m \mid m \in M\}$.

2.7.10. Write a SINGULAR procedure to compute, for two ideals $I, J \subset K[x] = K[x_1, \ldots, x_n]$, the dimension of $K[x]/I \otimes_{K[x]} K[x]/J$.

2.7.11. Write a SINGULAR procedure to compute $\mathrm{Ker}(J \otimes_A M \to M)$, for $A = K[x_1, \ldots, x_n]/I$, $J \subset A$ an ideal and M a finitely generated A–module. Compute this kernel for $J = \langle x, y \rangle$, $M = A/J$ and $I = \langle x^2 + y^3 \rangle \subset \mathbb{Q}[x, y]$.

2.7.12. Let $x = (x_1, \ldots, x_n)$ and $y = (y_1, \ldots, y_m)$, let $>_1$ be a global monomial ordering on $\mathrm{Mon}(x)$ and $>_2$ be an arbitrary monomial ordering on $\mathrm{Mon}(y)$. Moreover, let $> = (>_1, >_2)$ be the corresponding product ordering on $\mathrm{Mon}(x, y)$. Show that there is an isomorphism of K–algebras

$$K[x, y]_> \xrightarrow{\cong} K[y]_{>_2} \otimes_K K[x] \,.$$

Show by an example that $K[x, y]_> \not\cong K[y]_{>_2} \otimes_K K[x]_{>_1}$ if $>_1$ is local.

2.8 Operations on Modules and Their Computation

Throughout the following, let K be a field, $>$ a monomial ordering on $K[x]$, $x = (x_1, \ldots, x_n)$, and $R = K[x]_>$. Moreover, let $>_m$ be any module ordering on $K[x]^r$.

2.8.1 Module Membership Problem

The module membership problem can be formulated as follows:

Problem: Given polynomial vectors $f, f_1, \ldots, f_k \in K[x]^r$, decide whether $f \in I := \langle f_1, \ldots, f_k \rangle \subset R^r$ or not.

Solution: Compute a standard basis $G = \{g_1, \ldots, g_s\}$ of I with respect to $>_m$ and choose any weak normal form NF on R^r. Then

$$f \in I \iff \mathrm{NF}(f \mid G) = 0 \,.$$

This is proved in Lemma 2.3.5.   □

Additional Problem: If $f \in I = \langle f_1, \ldots, f_r \rangle \subset R^r$ then express f as a linear combination $uf = \sum_{i=1}^k g_i f_i$ with $u, g_i \in K[x]$, u a unit in R.

If $\{f_1, \ldots, f_k\}$ is a standard basis then we could compute a standard representation for f by applying NFMORA. For an arbitrary set of generators this is not possible, and we have to use a more tricky

Solution: Compute a standard basis G of $\mathrm{syz}(f, f_1, \ldots, f_k) \subset R^{k+1}$ w.r.t. the ordering $(c, >)$. Now choose any vector $h = (u, -g_1, \ldots, -g_k) \in G$ whose first component u satisfies $\mathrm{LM}(u) = 1$. Then $uf = \sum_{i=1}^k g_i f_i$.

Proof. By definition, $h = (u, -g_1, \ldots, -g_k) \in \text{syz}(f, f_1, \ldots, f_k)$ if and only if $uf = \sum_{i=1}^{k} g_i f_i$. Moreover, for the chosen ordering $(c, >)$, $\text{LM}(h) = \text{LM}(u)\varepsilon_1$. Hence, $f \in I$ implies that we find a vector in G whose first component is a unit in R.   □

The built–in commands in SINGULAR for this computation are lift(I,f) (which returns $g_1, \ldots, g_k$), respectively division(f,I) (which returns, additionally, the unit u).

SINGULAR Example 2.8.1 (module membership).

```
ring R=0,(x,y,z),(c,dp);
module M=[-z,-y,x+y,x],[yz+z2,yz+z2,-xy-y2-xz-z2];
vector v=[-xz-z2,-xz+z2,x2+xy-yz+z2];
reduce(v,std(M));
//-> [0,xy-xz+yz+z2,-xz-2yz+z2,-x2-xz]   //v is not in M

v=M[1]-x5*M[2];
v;
//-> [-x5yz-x5z2-z,-x5yz-x5z2-y,x6y+x5y2+x6z+x5z2+x+y,x]
reduce(v,std(M));
//-> 0                                   //v is in M
```

Now we want to express v in terms of generators of M.

```
syz(v+M);
//-> _[1]=[1,-1,x5]
```

This shows that $v = M[1] - x^5 M[2]$. By the built–in command lift, we obtain

```
lift(M,v);
//-> _[1,1]=1
//-> _[2,1]=-x5
```

In the local case one should use the built–in command division (cf. SINGULAR Example 1.8.2).

```
ring S=0,(x,y),(c,ds);
vector v=[x2,xy];
module M=[x+x3+y2,y3+y],[y,-x2+y2];
list L=division(v,M);
L;
//-> [1]:
//->      _[1,1]=x
//->      _[2,1]=-xy
//-> [2]:
//->      _[1]=0
//-> [3]:
//->      _[1,1]=1+x2
```

From the output, we read that $(1 + x^2) \cdot v = x \cdot M[1] - xy \cdot M[2]$. The second entry of the list L is the remainder, which is 0.

2.8.2 Intersection with Free Submodules (Elimination of Module Components)

Let $R^r = \bigoplus_{i=1}^{r} Re_i$, where $\{e_1, \ldots, e_r\}$ denotes the canonical basis of R^r.

Problem: Given $f_1, \ldots, f_k \in K[x]^r$, $I = \langle f_1, \ldots, f_k \rangle \subset R^r$, find a (polynomial) system of generators for the submodule

$$I' := I \cap \bigoplus_{i=s+1}^{r} Re_i \, .$$

Elements of the submodule I' are said to be obtained from $f_1, \ldots, f_k$ *by eliminating* $e_1, \ldots, e_s$.

The following lemma is the basis for solving the elimination problem.

Lemma 2.8.2. *Let $>$ be any monomial ordering on* $\mathrm{Mon}(x_1, \ldots, x_n)$ *and* $R = K[x]_>$. *Moreover, let $I \subset R^r = \bigoplus_{i=1}^{r} Re_i$ be a submodule and S a standard basis of I w.r.t. the module ordering $>_m = (c, >)$ defined by*

$$x^\alpha e_i < x^\beta e_j \ :\Longleftrightarrow \ j < i \ \text{ or } \ (j = i \text{ and } x^\alpha < x^\beta) \, .$$

Then, for any $s = 0, \ldots, r - 1$, $S' := S \cap \bigoplus_{i=s+1}^{r} Re_i$ is a standard basis of $I' = I \cap \bigoplus_{i=s+1}^{r} Re_i$ w.r.t. $(c, >)$. In particular, S' generates I'.

Proof. Let $h \in I'$, then we have to prove that there exists $f \in S'$ such that $\mathrm{LM}(f) \mid \mathrm{LM}(h)$.

Because S is a standard basis of I there exists $f \in S$ such that $\mathrm{LM}(f)$ divides $\mathrm{LM}(h)$. In particular, $\mathrm{LM}(f) \in \bigoplus_{i=s+1}^{r} K[x]e_i$. Now, by definition of the ordering, we obtain $f \in \bigoplus_{i=s+1}^{r} Re_i$, in particular, $f \in S'$. □

Hence, we obtain

Solution: Compute a standard basis $G = \{g_1, \ldots, g_s\}$ of I w.r.t. $(c, >)$. Then

$$G' := \left\{ g \in G \ \middle| \ \mathrm{LM}(g) \in \bigoplus_{i=s+1}^{r} K[x]e_i \right\}$$

is a standard basis for I'. □

SINGULAR Example 2.8.3 (elimination of module components).

```
ring R=0,(x,y,z),(c,dp);
module T=[xy,1,0,0],[yz,0,1,0],[xz,0,0,1];
module N=std(T);
N;
//-> N[1]=[0,0,x,-y]
//-> N[2]=[0,z,0,-y]
//-> N[3]=[yz,0,1]
//-> N[4]=[xz,0,0,1]
//-> N[5]=[xy,1]
```

From the output, we read

$$N \cap \bigoplus_{i=2}^{4} Re_i = \left\langle \begin{pmatrix} 0 \\ z \\ 0 \\ -y \end{pmatrix}, \begin{pmatrix} 0 \\ 0 \\ x \\ -y \end{pmatrix} \right\rangle, \quad N \cap \bigoplus_{i=3}^{4} Re_i = \left\langle \begin{pmatrix} 0 \\ 0 \\ x \\ -y \end{pmatrix} \right\rangle.$$

2.8.3 Intersection of Submodules

Problem: Given $f_1, \ldots, f_k, h_1, \ldots, h_s \in K[x]^r$, let $I_1 = \langle f_1, \ldots, f_k \rangle R^r$ and $I_2 = \langle h_1, \ldots, h_s \rangle R^r$. We want to compute a (polynomial) system of generators for the intersection $I_1 \cap I_2$.

One solution would be to generalize the procedure described in Section 1.8.7 (cf. Exercise 2.8.3). Here we describe an alternative procedure, based on syzygies.

Lemma 2.8.4. *With the above assumptions, let $g \in K[x]^r$. Moreover, let $c_1, \ldots, c_{r+k+s} \in K[x]^{2r}$ be the columns of the $2r \times (r+k+s)$–matrix*

$$\begin{pmatrix} 1 & & 0 & & & & \\ & \ddots & & f_1 \ldots f_k & 0 \ldots 0 \\ 0 & & 1 & & & & \\ \hline 1 & & 0 & & & & \\ & \ddots & & 0 \ldots 0 & h_1 \ldots h_s \\ 0 & & 1 & & & & \end{pmatrix}.$$

Then $g \in I_1 \cap I_2 \subset R^r$ if and only if g appears as the first r components of some $g' \in \mathrm{syz}(c_1, \ldots, c_{r+k+s}) \subset R^{r+k+s}$.

The proof is easy and left as Exercise 2.8.2.

Solution: Let $c_1, \ldots, c_{r+k+s} \in K[x]^{2r}$ be as in Lemma 2.8.4 and compute a generating set $M = \{p_1, \ldots, p_\ell\}$ of $\mathrm{syz}(c_1, \ldots, c_{r+k+s})$. The projections of $p_1, \ldots, p_\ell$ to their first r components generate $I_1 \cap I_2$. □

The corresponding SINGULAR command is `intersect`(I_1, I_2).

SINGULAR Example 2.8.5 (intersection of submodules).

```
ring R=0,(x,y),(c,dp);
module I1=[x,y],[y,1];
module I2=[0,y-1],[x,1],[y,x];

intersect(I1,I2);
//-> _[1]=[y2-y,y-1]
//-> _[2]=[xy+y,x+1]
//-> _[3]=[x,y]
```

When using the procedure described before, we obtain a different set of generators:

```
vector c1=[1,0,1,0];
vector c2=[0,1,0,1];
vector c3=[x,y,0,0];
vector c4=[y,1,0,0];
vector c5=[0,0,0,y-1];
vector c6=[0,0,x,1];
vector c7=[0,0,y,x];
module M=c1,c2,c3,c4,c5,c6,c7;
syz(M);
//-> _[1]=[y,-y2+x+1,y,-x-1,y+1,0,-1]
//-> _[2]=[x,y,-1,0,-1,-1]
//-> _[3]=[y2-y,y-1,0,-y+1,x-1,0,-y+1]
```

From the output, we read that $I_1 \cap I_2$ is generated by the following vectors:

```
vector r1=[y,-y2+x+1];
vector r2=[x,y];
vector r3=[y2-y,y-1];
```

Both computations give the same module, for instance,

```
r1+y*r2;
//-> [xy+y,x+1]
```

2.8.4 Quotients of Submodules

Problem: Let I_1 and $I_2 \subset R^r$ be as in Section 2.8.3. Find a (polynomial) system of generators for the quotient

$$I_1 :_R I_2 = \{g \in R \mid gI_2 \subset I_1\}.$$

Note that $I_1 :_R I_2 = \mathrm{Ann}_R\big((I_1 + I_2)/I_1\big)$, in particular, if $I_1 \subset I_2$, then we have $I_1 :_R I_2 = \mathrm{Ann}_R(I_2/I_1)$.

We proceed as in the ideal case (cf. Section 1.8.8):

Solution 1: Compute generating sets G_i of $I_1 \cap \langle h_i \rangle$, $i = 1, \ldots, s$, according to Section 2.8.3, "divide" the generators by the vector h_i, getting the generating set $G_i' = \{g \in R \mid gh_i \in G_i\}$ for $I_1 :_R \langle h_i \rangle$. Finally, compute the intersection $\bigcap_i (I_1 :_R \langle h_i \rangle)$, again according to 2.8.3. $\qquad\square$

Note that there is a trick which can be used to "divide" by the vector h_i: let $G_i = \{v_1, \ldots, v_\ell\}$ and compute a generating system $\{w_1, \ldots, w_m\}$ for $\mathrm{syz}(v_1, \ldots, v_\ell, h_i)$. Then the last $(= (\ell+1)$-th$)$ components of $w_1, \ldots, w_m$ generate the R–module $I_1 :_R \langle h_i \rangle$.

Solution 2: Define $h := h_1 + t_1 h_2 + \ldots + t_{s-1} h_s \in K[t_1, \ldots, t_{s-1}, x_1, \ldots, x_n]^r$, and compute a generating system for $(I_1 \cdot R[t]) :_{R[t]} \langle h \rangle_{R[t]}$, as before. Finally, eliminate $t_1, \ldots, t_{s-1}$ from $(I_1 \cdot R[t]) :_{R[t]} \langle h \rangle_{R[t]}$ (cf. Section 1.8.2). $\qquad\square$

Let us consider the same problem for modules which are given by a presentation matrix.

Problem: Let $A = R/I$ for some ideal $I \subset R$, and let $\varphi : M_1 \to M_2$ be an A–module homomorphism, given by matrices B, B_1, B_2 with entries in R, such that the induced diagram

$$
\begin{array}{ccccccc}
A^r & \xrightarrow{B_1} & A^p & \longrightarrow & M_1 & \longrightarrow & 0 \\
 & & \downarrow{\scriptstyle B} & & \downarrow{\scriptstyle \varphi} & & \\
A^s & \xrightarrow{B_2} & A^q & \longrightarrow & M_2 & \longrightarrow & 0
\end{array}
$$

is commutative with exact rows. Compute generators for the ideal

$$
\varphi(M_1) :_A M_2 = \mathrm{Ann}_A\big(M_2 / \varphi(M_1)\big).
$$

Solution: Let $b_1, \ldots, b_p \in R^q$ and $g_1, \ldots, g_s \in R^q$ represent the columns of B and B_2, respectively, and let $\{e_1, \ldots, e_q\}$ denote the canonical basis of R^q. Then

$$
\begin{aligned}
\varphi(M_1) :_A M_2 &= \big((BA^p + \mathrm{Im}(B_2))/\mathrm{Im}(B_2)\big) :_A \big(A^q/\mathrm{Im}(B_2)\big) \\
&= \mathrm{Ann}_A\big(A^q/(BA^p + \mathrm{Im}(B_2))\big) \\
&= \langle g_1, \ldots, g_s, b_1, \ldots, b_p \rangle :_A \langle e_1, \ldots, e_q \rangle \\
&= \pi\big(\langle g_1, \ldots, g_s, b_1, \ldots, b_p \rangle + I \cdot R^q\big) :_R \langle e_1, \ldots, e_q \rangle,
\end{aligned}
$$

where $\pi : R \to A = R/I$ denotes the canonical projection. Hence, we can apply again the method from above. $\qquad\square$

The built-in command in SINGULAR for this computation is `quotient` (I_1, I_2), which also works over quotient rings of R.

SINGULAR Example 2.8.6 (quotient of submodules).

```
ring R=0,(x,y,z),(c,dp);
module I=[xy,xz],[yz,xy];
module J=[y,z],[z,y];
ideal K=quotient(I,J);
K;
//-> K[1]=x2y2-xyz2
reduce(K*J,std(I));          //test if KJ is contained in I
//-> _[1]=0
//-> _[2]=0
```

Since R has no zerodivisors, $\mathrm{Ann}_R(J) = \langle 0 \rangle :_R J = \langle 0 \rangle$, but the annihilator $\mathrm{Ann}_A(J)$ over the quotient ring $A = R/(I :_R J)$ is not trivial:

```
qring A=std(K);
module Null;
module J=[xy,xyz-x2y2],[xy2,y2x];
ideal ann=quotient(Null,J);      //annihilator of J
ann;
//-> ann[1]=xy-z2
```

Now let $M_1, M_2 \subset A^3$ be given by presentation matrices $B_1 \in \mathrm{Mat}(3 \times 2, A)$, respectively, $B_2 \in \mathrm{Mat}(3 \times 4, A)$, and let $\varphi : M_1 \to M_2$ be given by a matrix $B \in \mathrm{Mat}(3 \times 3, A)$. We compute $\varphi(M_1) :_A M_2 = \mathrm{Ann}_A(M_2/\varphi(M_1))$.

```
module B1 = [x2,xy,y2],[xy,xz,yz];   //presentation of M1
module B  = [x,zx,zy],[y,zy,xy],[x+y,zx+zy,zy+xy];
module B2 = B[1],B[2],B1[1],B1[2];   //presentation of M2

reduce(B*B1,std(B2)); //test if im(B*B1) contained in im(B2)
//-> _[1]=0
//-> _[2]=0

quotient(B2+B,freemodule(3));        //the annihilator
//-> _[1]=x3y-xy2z
```

2.8.5 Radical and Zerodivisors of Modules

Let $A = R/I$ for some ideal $I \subset R$ and let M, N be two A–modules with $N \subset M$. Define the *radical of N in M* as the ideal

$$\mathrm{rad}_M(N) := \sqrt[M]{N} := \left\{ g \in A \mid g^q M \subset N \text{ for some } q > 0 \right\}.$$

Problem: Solve the *radical membership problem* for modules, that is, decide whether $f \in A$ is contained in $\sqrt[M]{N}$, or not. [16]

---

[16] Cf. Exercises 4.1.13 − 4.1.15 to see how $\sqrt[M]{N}$ is related to a primary decomposition of M.

Solution: By Exercise 2.8.6, $\sqrt[M]{N} = \sqrt{\text{Ann}_A(M/N)}$. Hence, we can compute generators for $\text{Ann}_A(M/N) = N :_A M \subset A$ as in Section 2.8.4, and then we are reduced to solving the radical membership problem for ideals which was solved in Section 1.8.6. [17] $\square$

A slightly different problem is the zerodivisor test:

Problem: Decide whether a given $f \in A$ is a zerodivisor of M, that is, whether there exists some $m \in M \smallsetminus \{0\}$ such that $f \cdot m = 0$.

Solution: The element f defines an endomorphism $\varphi_f : M \to M, m \mapsto f \cdot m$, which is induced by $f \cdot E_r : A^r \to A^r$, with E_r the $r \times r$ unit matrix. It follows that f is a zerodivisor of M if and only if $\text{Ker}(\varphi_f) \neq 0$. The computation of $\text{Ker}(\varphi_f)$ is explained in Section 2.8.7. $\square$

Note that $\text{Ker}(\varphi_f)$ is a special case of the *quotient of a module by an ideal*

$$\text{Ker}(\varphi_f) = \langle 0 \rangle :_M \langle f \rangle = \{m \in M \mid f \cdot m = 0\}.$$

Hence, f is a zerodivisor of $M = A^q/N$ if and only if $N :_{A^q} \langle f \rangle \neq 0$.

SINGULAR Example 2.8.7 (radical, zerodivisors of modules).
We check first whether a polynomial f is in the radical of N in A^3, with $A = Q[x, y, z]/\langle xy(xy - z^2)\rangle$.

```
ring R    = 0,(x,y,z),(c,dp);
ideal I   = x2y2-xyz2;
qring A   = std(I);
poly f    = xy*(y-z)*(y-1);
module N  = [x,xz,y2],[y,yz,z2],[x2,xy,y2],[xy,xz,yz];
ideal ann= quotient(N,freemodule(3));   //annihilator of
                                        // Coker(N)
ring Rt   = 0,(t,x,y,z),dp;
ideal I   = imap(R,I);
ideal ann= imap(A,ann),I;
poly f    = imap(A,f);
ideal J   = ann,1-t*f;
eliminate(J,t);
//-> _[1]=1
```

Hence, f is contained in $\sqrt[A^3]{N}$, that is, some power of f maps A^3 to N. In particular, f should be a zerodivisor of A^3/N. Let us check this:

```
setring A;
size(quotient(N,f));
//-> 9
```

Hence, $N :_{A^3} \langle f \rangle \neq 0$ and f is a zerodivisor of A^3/N.

---

[17] If $M, N \subset A^r$ are given by generating systems $m_1, \ldots, m_p \in R^r, n_1, \ldots, n_q \in R^r$, then $\sqrt{N :_A M} = \sqrt{(\langle n_1, \ldots, n_q \rangle + I R^r) :_R \langle m_1, \ldots, m_p \rangle} \mod I$.

2.8.6 Annihilator and Support

Let $I = \langle f_1, \ldots, f_k \rangle \subset R$ be an ideal, and M an $A = R/I$–module. By Lemma 2.1.41, the *support of* M is the zero–set of the annihilator ideal of M,

$$\text{supp}(M) = V\big(\text{Ann}_A(M)\big),$$

where $\text{Ann}_A(M) = \{g \in A \mid gM = 0\} = \langle 0 \rangle :_A M$.

Problem: Compute a system of generators of some ideal $J \subset A$ satisfying

$$\text{supp}(M) = V(J) = \{P \subset A \text{ prime ideal } \mid P \supset J\}$$

There are two cases of interest:

(I) $M \subset A^r$ is given by a system of generators $m_1, \ldots, m_s \in R^r$,

(II) M has the presentation $A^p \xrightarrow{B} A^q \to M \to 0$, given in form of a matrix $B \in \text{Mat}(q \times p, R)$.

Note that $\text{Ann}_R(M) = 0$ for $M \subset R^r$, since $R = K[x]_>$ has no zerodivisors. Hence, the first case is only interesting if $I \neq 0$.

Solution 1: We compute a system of generators for $J = \text{Ann}_A(M)$.

In Case (I), this can be done by computing a system of generators for $\langle 0 \rangle :_A M$ as described in Section 2.8.4, with $A = R/I$ as basering. The latter means to compute a system of generators for the quotient $(I \cdot R^r) :_R \langle m_1, \ldots, m_r \rangle$, with R as basering and with $I \cdot R^r = \langle f_i e_j \mid 1 \leq i \leq k, \ 1 \leq j \leq r \rangle$, [18] and then to project modulo I.

In Case (II), we have $M \cong A^q / \text{Im}(B)$ where $\text{Im}(B) \subset A^q$ is the submodule generated by the columns of B. Hence, $\text{Ann}_A(M) = \text{Im}(B) :_A A^q$, and the generators of $\text{Im}(B) :_A A^q$ computed as in Section 2.8.4 with A as basering, generate $\text{Ann}_A(M)$. □

We shall see in Section 7.2 on Fitting ideals, that the 0–th *Fitting ideal* $F_0(M)$ satisfies also $\text{supp}(M) = V\big(F_0(M)\big)$ (cf. Exercise 7.2.5). This leads to the following

Solution 2: We compute a system of generators for $J = F_0(M)$.

In Case (I), this can be done by computing a system of generators $\{b_1, \ldots, b_p\}$ of the module of syzygies $\text{syz}^A(m_1, \ldots, m_s) \subset A^q$, cf. Remark 2.5.6. The b_i are the columns of a matrix B, defining a presentation $A^p \xrightarrow{B} A^q \to M \to 0$. Then the q–minors of B generate $F_0(M)$, cf. Definition 7.2.4.

In Case (II), it is clear that the q–minors of the presentation matrix B generate $F_0(M)$.

---

[18] Here $\{e_1, \ldots, e_r\}$ denotes the canonical basis of R^r.

SINGULAR Example 2.8.8 (annihilator and Fitting ideal).
We compute the annihilator and the Fitting ideal of a module $M = \mathrm{Coker}(B)$
given by a presentation matrix B over the quotient ring $A = R/I$, where
$R = \mathbb{Q}[x, y, z, u]$ and $I = \langle x^2y^2 - xyz^2 \rangle$.

```
ring R    = 0,(x,y,z,u),dp;
ideal I   = x2y2-xyz2;
qring A   = std(I);
module B = [x,xz,y2],[y,yz,z2],[x2,xy,y2];
ideal ann= quotient(B,freemodule(3));
ideal fit= minor(B,3);
ann;                              //annihilator of Coker(B)
//-> ann[1]=x3z3-xy2z3+xy4-x2yz2
//-> ann[2]=x4yz2-xy2z4+xy4z-x2yz3
//-> ann[3]=x5z2-x2yz4+xy4-x2yz2

fit;                             //Fitting ideal of Coker(B)
//-> fit[1]=-x2y3z+x3z3+xy4-x2yz2
```

We now check that $\mathtt{fit} \subset \mathtt{ann} \not\subset \mathtt{fit}$, but $\mathtt{ann}^2 \subset \mathtt{fit}$, by counting the number of non–zero generators after reduction.

```
size(reduce(fit,std(ann)));
//-> 0
size(reduce(ann,std(fit)));
//-> 2
size(reduce(ann*ann,std(fit)));
//-> 0
```

2.8.7 Kernel of a Module Homomorphism

Let $A = R/I$ for some ideal $I \subset R$, let $U \subset A^n$, $V = \langle v_1, \ldots, v_s \rangle \subset A^m$ be submodules, and let $\varphi : A^n/U \to A^m/V$ be an A–module homomorphism defined by the matrix $B = (b_1, \ldots, b_n)$, $b_i \in A^m$.

Problem: Compute a system of generators in $A^n = \bigoplus_{i=1}^n Ae_i$ for $\mathrm{Ker}(\varphi)$.

Note that $f = \sum_{i=1}^n f_i e_i \in \mathrm{Ker}(\varphi)$ if and only if there exist $y_1, \ldots, y_s \in A$ such that $\sum_{i=1}^n f_i b_i = \sum_{j=1}^s y_j v_j$, in particular, $(f_1, \ldots, f_n, -y_1, \ldots, -y_s)$ is a syzygy of $b_1, \ldots, b_n, v_1, \ldots, v_s$. Hence, we obtain the following

Solution: Compute a system of generators $\{h_1, \ldots, h_\ell\}$ for the module of syzygies $\mathrm{syz}(b_1, \ldots, b_n, v_1, \ldots, v_s) \subset A^{n+s}$ (cf. Remark 2.5.6). Let $h_i' \in A^n$ be the vector obtained from h_i when omitting the last s components, $i = 1, \ldots, \ell$. Then $\{h_1', \ldots, h_\ell'\}$ is a generating system for $\mathrm{Ker}(\varphi)$.   □

SINGULAR Example 2.8.9 (kernel of a module homomorphism).

```
ring R=0,(x,y,z),(c,dp);
ideal I=x2y2-xyz2;
qring A=std(I);                    // quotient ring A=R/I
module V=[x2,xy,y2],[xy,xz,yz];
matrix B[3][2]=x,y,zx,zy,y2,z2;
module N=B[1],B[2],V[1],V[2];
module Re=syz(N);                  // syzygy module of N
module Ker;
int i;
for(i=1;i<=size(Re);i++){Ker=Ker+Re[i][1..2];}
Ker;                               // kernel of B
//-> Ker[1]=[xy2-yz2]
//-> Ker[2]=[y3z-x2z2-xyz2+y2z2-xz3+yz3,xy3-x2yz]
//-> Ker[3]=[x2yz-xz3]
//-> Ker[4]=[x3z+x2z2-xyz2-y2z2+xz3-yz3,x3y-xy2z]
//-> Ker[5]=[x3y-x2z2]

reduce(B*Ker,std(V));              // test
//-> _[1]=0
//-> _[2]=0
//-> _[3]=0
//-> _[4]=0
//-> _[5]=0
```

We can use the built–in command `modulo` (cf. SINGULAR Example 2.1.26).

```
modulo(B,V);
```

gives the same result as `Ker;`.

2.8.8 Solving Systems of Linear Equations

Let $I = \langle f_1, \ldots, f_k \rangle \subset R$ be an ideal, with $f_i \in K[x]$ polynomials, and let $S = R/I$. Moreover, let

$$
\begin{aligned}
a_{11}Z_1 + \ldots + a_{1m}Z_m &= b_1 \\
\vdots \qquad\qquad \vdots \quad&\quad \vdots \\
a_{r1}Z_1 + \ldots + a_{rm}Z_m &= b_r
\end{aligned}
$$

be a system of linear equations, with $a_{ij}, b_i \in K[x]$ and indeterminates Z_j. We can write the above system as a matrix equation $AZ = b$, where

$$
A = \begin{pmatrix} a_{11} & \cdots & a_{1m} \\ \vdots & & \vdots \\ a_{r1} & \cdots & a_{rm} \end{pmatrix}, \quad Z = \begin{pmatrix} Z_1 \\ \vdots \\ Z_m \end{pmatrix}, \quad b = \begin{pmatrix} b_1 \\ \vdots \\ b_r \end{pmatrix}.
$$

A is called the *coefficient matrix* of the system and the concatenated matrix $(A \mid b)$ is called the *extended coefficient matrix*.

Problems:

(1) Check whether the system $AZ = b$ is solvable over S, and if so, then find a solution. That is, we are looking for a vector $z \in S^m$ such that the equation $Az = b$ holds in S^r. The set of all such solutions z is denoted by $\mathrm{Sol}_S(AZ = b)$.

(2) Consider the homogeneous system $AZ = 0$. The set of solutions is the kernel of the linear map $A : S^m \to S^r$ and, hence, a submodule of S^m. The problem is to find a set of generators of $\mathrm{Sol}_S(AZ = 0)$. Note that if $z \in \mathrm{Sol}_S(AZ = b)$ is a "special" solution of the inhomogeneous system, then $z + \mathrm{Sol}_S(AZ = 0) := \{z + w \mid w \in \mathrm{Sol}_S(AZ = 0)\}$ is the set of all solutions for $AZ = b$.

Solution 1: We assume that the a_{ij} and b_j are elements of a field F. [19] We create the ideal

$$
E = \left\langle \sum_{j=1}^{m} a_{1j} Z_j - b_1 , \, \ldots \, , \sum_{j=1}^{m} a_{rj} Z_j - b_r \right\rangle
$$

in $F[Z_1, \ldots, Z_m]$ and compute a reduced standard basis G of E with respect to a global monomial ordering. The system $AZ = b$ is solvable over F if and only if $G \neq \{1\}$ and then the solutions can be read from G.

In case $F = K(x_1, \ldots, x_n)$, the x_i are considered as parameters and if $AZ = b$ is not solvable over F this means that there is no vector $z(x)$ of rational functions, satisfying $AZ = b$. Hence $A(x) \cdot z(x) = b(x)$ has no solution for general x. Nevertheless, there may exist solutions for special x which satisfy some constraints. What we can do now is to create a ring with the x_i as variables, and then a standard basis of the ideal E in this ring, eventually, computes equations in the x_i such that $A(\overline{x}) \cdot Z = b(\overline{x})$ is solvable for all $\overline{x} = (\overline{x}_1, \ldots, \overline{x}_n) \in S^n$ satisfying these equations (see SINGULAR Example 2.8.10 below).

Solution 2: We consider now the general case.

(1) $AZ = b$ is solvable over S if and only if b is in the image of the module homomorphism $A : S^m \to S^r$. Hence, we have to check whether b is contained in the submodule $\mathrm{Im}(A) \subset S^r$, generated by the columns $a_1, \ldots, a_m$ of A. If the latter is given, then we have to find $z_1, \ldots, z_m \in S$, such that $b = \sum_{i=1}^{m} z_i a_i$. However, this problem was already solved in Section 2.8.1.

---

[19] For instance, consider a_{ij}, b_j as elements of the quotient field $K(x_1, \ldots, x_n)$.

(2) Finding generators for $\mathrm{Sol}_S(AZ = 0)$ means nothing else but finding generators for $\mathrm{Ker}(A : S^m \to S^r)$, and this problem was solved in Section 2.8.7.

Note that solving over $K[x]/I$ means that the solution $z(p)$ satisfies the linear equation $A(p) \cdot z(p) = b(p)$ for p in the zero–set of $f_1, \ldots, f_k$ (*solving with polynomial constraints*).[20]

SINGULAR Example 2.8.10 (solving linear equations).
Consider the system of linear equations in x, y, z, u,

$$
\begin{array}{rl}
3x + y + z - u = a \\
13x + 8y + 6z - 7u = b \\
14x + 10y + 6z - 7u = c \\
7x + 4y + 3z - 3u = d
\end{array}
\qquad (*)
$$

where a, b, c, d are parameters. Moreover, we also consider the system $(**)$, which has the additional equation

$$
x + y + z - u = 0.
$$

We want to solve these systems by expressing x, y, z, u as functions of the parameters a, b, c, d if possible, respectively find conditions for the parameters such that the system is solvable.

Following Solution 1, we can express the systems $(*)$ and $(**)$ as ideals **E** and **EE** in a ring with a, b, c, d as parameters:

```
ring R = (0,a,b,c,d),(x,y,z,u),(c,dp);
ideal E =   3x +   y +  z -  u - a,
           13x +  8y + 6z - 7u - b,
           14x + 10y + 6z - 7u - c,
            7x +  4y + 3z - 3u - d;


ideal EE =  E, x + y + z - u;
```

Computing a reduced standard basis performs a complete Gaussian elimination, showing that there is a unique solution to the system $(*)$:

---

[20] Hilbert's Nullstellensatz 3.5.2 implies that, for a radical ideal I, to solve the system $AZ = b$ over $K[x]/I$ means nothing else but to look for polynomial vectors $z = (z_1, \ldots, z_m) \in K[x]^m$ such that $A(p) \cdot z(p) = b(p)$, for all $p \in V_{\overline{K}}(I)$. Here

$$
V_{\overline{K}}(I) := \left\{ p = (p_1, \ldots, p_n) \in \overline{K}^n \mid f_1(p) = \ldots = f_k(p) = 0 \right\}
$$

denotes the zero–set of I over the algebraic closure $\overline{K}$ of K.

The same remark applies for solving over $K[x]_{\langle x \rangle}/I$, except that we are looking for solutions in some Zariski open neighbourhood of 0, subject to the constraints $f_1 = \ldots = f_k = 0$ (cf. Appendix A.2).

```
option(redSB);
simplify(std(E),1);                    //compute reduced SB
//-> _[1]=u+(6/5a+4/5b+1/5c-12/5d)
//-> _[2]=z+(16/5a-1/5b+6/5c-17/5d)
//-> _[3]=y+(3/5a+2/5b-2/5c-1/5d)
//-> _[4]=x+(-6/5a+1/5b-1/5c+2/5d)
```

We read the solution of (∗):

$$x = \tfrac{1}{5} \cdot \left(6a - b + c - 2d\right),$$
$$y = \tfrac{1}{5} \cdot \left(-3a - 2b + 2c + d\right),$$
$$z = \tfrac{1}{5} \cdot \left(-16a + b - 6c + 17d\right),$$
$$u = \tfrac{1}{5} \cdot \left(-6a - 4b - c + 12d\right).$$

Doing the same for EE gives 1:

```
std(EE);
//-> _[1]=1
```

This means that the system (∗∗) has no solution over the field $\mathbb{Q}(a, b, c, d)$, or, in other words, (∗∗) has no complex solutions for fixed *general* values of a, b, c, d. This is clear, since the extra equation for (∗∗), $x + y + z = u$ gives a linear condition in a, b, c, d for the solutions of (∗). Since general values of the parameters do not satisfy this equation, (∗∗) has no solution over the field of rational functions in a, b, c, d.

If we want to find conditions for a, b, c, d under which (∗∗) has a solution and then to solve it, we have to pass to a ring with a, b, c, d as variables. Since we have to solve for x, y, z, u, these variables have to come first.

```
ring R1 = 0,(x,y,z,u,a,b,c,d),(c,dp);
```

Now we compute a reduced standard basis (up to normalization) for the ideal generated by the rows of the system (∗∗).

```
ideal EE = imap(R,EE);
std(EE);
//-> _[1]=7a-2b+2c-4d
//-> _[2]=7u+8b-c-12d
//-> _[3]=7z+5b+2c-11d
//-> _[4]=7y+4b-4c+d
//-> _[5]=7x-b+c-2d
```

The first polynomial gives $7a - 2b + 2c - 4d = 0$, which must be satisfied by the parameters to have a solution for (∗∗). The remaining polynomials then give the solutions for x, y, z, u in terms of b, c, d (a has already been substituted).

Another method is to work directly on the (extended) coefficient matrix. The standard basis algorithm applied to a matrix operates on the module generated by the columns of the matrix. Hence, in order to solve the linear system, we have to transpose the matrix and to make sure that the module ordering gives priority to the columns (for example, (c,dp)).

First we have to write two procedures which return the coefficient matrix, respectively the extended coefficient matrix, of an ideal, describing the system of linear equations as above.

```
LIB "matrix.lib";
proc coeffMat (ideal i)
{
    int ii;
    int n = nvars(basering);
    int m = ncols(i);
    matrix C[m][n];
    for ( ii=1; ii<=n; ii++)
    {
        C[1..m,ii] = i/var(ii);
    }
    return(C);
}

proc coeffMatExt(ideal i)
{
    matrix C = coeffMat(i);
    C = concat(C,transpose(jet(i,0)));
    return(C);
}
```

Now we consider again the above example, and, of course, obtain the same answers as before:

```
setring R;
matrix CE = coeffMatExt(EE);
matrix C = coeffMat(E);

setring R1;
matrix CE = imap(R,CE);
std(transpose(CE));
//-> _[1]=[0,0,0,0,7a-2b+2c-4d]
//-> _[2]=[0,0,0,7,8b-c-12d]
//-> _[3]=[0,0,7,0,5b+2c-11d]
//-> _[4]=[0,7,0,0,4b-4c+d]
//-> _[5]=[7,0,0,0,-b+c-2d]
```

Finally let us proceed as in Solution 2 with the above system (**), already knowing that $7a - 2b + 2c - 4d = 0$ is the condition for solvability. Hence, we can work over the corresponding quotient ring.

```
ring R2 = 0,(x,y,z,u,a,b,c,d),(c,dp);
ideal p = 7a-2b+2c-4d;
qring qR= std(p);
matrix C = imap(R,C);
matrix CE= imap(R,CE);
matrix b[5][1]=CE[1..5,5];          //the r.h.s. of (**)
lift(C,b);
//-> __[1,1]=-1/7b+1/7c-2/7d
//-> __[2,1]=4/7b-4/7c+1/7d
//-> __[3,1]=5/7b+2/7c-11/7d
//-> __[4,1]=8/7b-1/7c-12/7d
```

Exercises

2.8.1. Let $>$ be any monomial ordering on $\mathrm{Mon}(x_1, \ldots, x_n)$ and $R = K[x]_>$, $x = (x_1, \ldots, x_n)$. Moreover, let $I \subset J \subset R^r$ be submodules and $f \in R^r/I$. Find a procedure to decide whether $f \in J/I$ or not.

2.8.2. Prove Lemma 2.8.4.

2.8.3. Let $>$ be any monomial ordering on $\mathrm{Mon}(x_1, \ldots, x_n)$ and $R = K[x]_>$. Show that the procedure of Section 1.8.7 can be generalized to a procedure computing the intersection $I \cap J$ of two submodules $I, J \subset R^r$.

2.8.4. Use the `modulo` command to write a SINGULAR procedure to compute the quotient of two modules.

2.8.5. Let $R = \mathbb{Q}[x, y, z]/\langle x^2 + y^2 + z^2 \rangle$, $M = R^3/\langle(x, xy, xz)\rangle$, and let $N = R^2/\langle(1, y)\rangle$. Moreover, let $\varphi = \varphi_A : M \to N$ be the R-module homomorphism given by the matrix

$$A = \begin{pmatrix} x^2+1 & y & z \\ yz & 1 & -y \end{pmatrix} .$$

(1) Compute $\mathrm{Ker}(\varphi)$.
(2) Test whether $(x^2, y^2) \in \mathrm{Im}(\varphi)$, or not.
(3) Compute $\mathrm{Im}(\varphi) \cap \{f \in N \mid f \equiv (h, 0) \mod \langle(x, 1)\rangle \text{ for some } h \in R\}$.
(4) Compute $\mathrm{Ann}_R(\mathrm{Im}(\varphi))$.

2.8.6. Let A be a ring, M an A–module and $N \subset M$ a submodule. Then

$$\sqrt[M]{N} := \{g \in A \mid g^q M \subset N \text{ for some } q > 0\},$$

is called the *radical of N in M*. Prove the following statements:

(1) $\sqrt[M]{N} = \sqrt{\mathrm{Ann}(M/N)} = \sqrt{N :_A M}$.
(2) $\sqrt[M]{N} = A$ if and only if $M = N$.
(3) $\sqrt[M]{N \cap N'} = \sqrt[M]{N} \cap \sqrt[M]{N'}$, for each submodule $N' \subset M$.
(4) $\sqrt[M]{N + N'} = \sqrt[M]{N} + \sqrt[M]{N'}$, for each submodule $N' \subset M$.

3. Noether Normalization and Applications

Integral extension of a ring means adjoining roots of monic polynomials over the ring. This is an important tool for studying affine rings, and it is used in many places, for example, in dimension theory, ring normalization and primary decomposition. Integral extensions are closely related to finite maps which, geometrically, can be thought of as projections with finite fibres plus some algebraic conditions. We shall give a constructive introduction with explicit algorithms to these subjects.

3.1 Finite and Integral Extensions

This section contains the basic algebraic theory of finite and algebraic extensions and their relationship. Moreover, important criteria for integral dependence (Proposition 3.1.3) and finiteness (Proposition 3.1.5) are proven.

Definition 3.1.1. Let $A \subset B$ be rings.

(1) $b \in B$ is called *integral* over A if there is a monic polynomial $f \in A[x]$ satisfying $f(b) = 0$, that is, b satisfies a relation of degree p,

$$b^p + a_1 b^{p-1} + \cdots + a_p = 0, \quad a_i \in A ,$$

for some $p > 0$.
(2) B is called *integral over* A or an *integral extension* of A if every $b \in B$ is integral over A.
(3) B is called a *finite extension* of A if B is a finitely generated A–module.
(4) If $\varphi : A \to B$ is a ring map then φ is called an *integral,* respectively *finite, extension* if this holds for the subring $\varphi(A) \subset B$.

If there is no doubt about φ, we say also, in this situation, that B is integral, respectively finite, over A. Often we omit φ in the notation, for example we write IM instead of $\varphi(I)M$ if $I \subset A$ is an ideal and M a B–module.

Proposition 3.1.2. *Let A, B be rings.*

(1) If $\varphi : A \to B$ is a finite extension, then it is integral. More generally, if $I \subset A$ is an ideal and M a finitely generated B–module then any $b \in B$ with $bM \subset IM$ satisfies a relation

$$b^p + a_1 b^{p-1} + \cdots + a_p = 0 , \quad a_i \in I^i \subset A .$$

(2) If B is a finitely generated A–algebra of the form $B = A[b_1, \ldots, b_n]$ with $b_i \in B$ integral over A then B is finite over A.

Proof. (1) Replacing A by the image of A, we may assume that $A \subset B$. Any $b \in B$ defines an endomorphism of the finitely generated A–module B. The characteristic polynomial of this endomorphism defines an integral relation for b, by the *Cayley–Hamilton Theorem* (this is sometimes called the "determinantal trick").

In concrete terms, let $b_1, \ldots, b_k$ be a system of generators for B as A–module, then $b \cdot b_i = \sum_{j=1}^{k} a_{ij} b_j$, $1 \leq i \leq k$, for suitable $a_{ij} \in A$. This implies

$$\left(b \cdot E_k - (a_{ij}) \right) \begin{pmatrix} b_1 \\ \vdots \\ b_k \end{pmatrix} = 0 ,$$

therefore, by Cramer's rule, $\det\left(b \cdot E_k - (a_{ij}) \right) \cdot b_i = 0$ for $i = 1, \ldots, k$.[1] But, since $1 = \sum_i e_i b_i \in B$ for suitable $e_1, \ldots, e_k$, we obtain $\det\left(b E_k - (a_{ij}) \right) = 0$, which is the required integral relation for b.

In the general case, let $b_1, \ldots, b_k$ be a system of generators of M as A–module. Then we can choose the a_{ij} from I and it follows that the coefficient of b^{k-i} in $\det(b E_k - (a_{ij}))$ is a sum of $i \times i$–minors of (a_{ij}) and, therefore, contained in I^i.

(2) We proceed by induction on n. If b_1 satisfies an integral relation of degree p, then b_1^p and hence, any power b_1^q, $q \geq p$, can be expressed as an A–linear combination of $b_1^0, \ldots, b_1^{p-1}$. That is, the A–module $B = A[b_1]$ is generated by $b_1^0, \ldots, b_1^{p-1}$, in particular, it is finite over A.

For $n > 1$ we may assume, by induction, that $A[b_1, \ldots, b_{n-1}]$ is finite over A. Since taking finite extension is clearly transitive, $(A[b_1, \ldots b_{n-1}])[b_n]$ is finite over A. $\square$

Let K be a field, $I \subset K[x] := K[x_1, \ldots, x_n]$ an ideal and $f_1, \ldots, f_k \in K[x]$. The residue classes $\bar{f}_i = f_i \bmod I$ generate a subring

$$A := K[\bar{f}_1, \ldots, \bar{f}_k] \subset B := K[x]/I .$$

We want to check whether a given $b \in K[x]$ is integral over $K[f_1, \ldots, f_k]$ mod I, that is, whether $\bar{b}$ is integral over A.

---

[1] Here E_n denotes the $n \times n$ unit matrix.

The following two results are the basis for an algorithm to check for integral dependence respectively finiteness.

Proposition 3.1.3 (Criterion for integral dependence).
Let $b, f_1, \ldots, f_k \in K[x]$, $I = \langle g_1, \ldots, g_s \rangle \subset K[x]$ an ideal and $t, y_1, \ldots, y_k$ new variables. Consider the ideal

$$M = \langle t - b, \, y_1 - f_1, \ldots, \, y_k - f_k, \, g_1, \ldots, g_s \rangle \subset K[x_1, \ldots, x_n, t, y_1, \ldots, y_k] \,.$$

Let $>$ be an ordering on $K[x, t, y]$ with $x \gg t \gg y$,[2] and let G be a standard basis of M with respect to this ordering.
  Then b is integral over $K[f] = K[f_1, \ldots, f_k] \bmod I$ if and only if G contains an element g with leading monomial $\mathrm{LM}(g) = t^p$ for some $p > 0$. Moreover, any such g defines an integral relation for b over $K[f] \bmod I$.

Proof. If $\mathrm{LM}(g) = t^p$ then g must have the form

$$g(t, y) = a_0 t^p + a_1(y) t^{p-1} + \cdots + a_p(y) \in K[t, y], \quad a_0 \in K \smallsetminus \{0\} \,.$$

We may assume that $a_0 = 1$. Since $g \in M$ we have $g(b, f) \in I$. Thus, g defines an integral relation for b over $K[f] \bmod I$.
  Conversely, if b is integral, then there exists a $g \in K[t, y]$ as above. By Taylor's formula, $g(t, y) = g(b, f) + b_0 \cdot (t - b) + \sum_{i=1}^{k} b_i \cdot (y_i - f_i)$ for some $b_i \in K[t, y]$, $i = 0, \ldots, k$. Hence, $g \in M$ and, therefore, $t^p = \mathrm{LM}(g) \in L(M)$. Since G is a standard basis, t^p is divisible by the leading monomial of some element of G which implies the result. □

SINGULAR Example 3.1.4 (integral elements).
Let $K = \mathbb{Q}$, the field of rational numbers, $I = \langle x_1^2 - x_2^3 \rangle \subset A = K[x_1, \ldots, x_4]$, let $f_1 = x_3^2 - 1$ and $f_2 = x_1^2 x_2$. We want to check whether the elements $b = x_3$ (respectively x_4) are integral over $K[f_1, f_2] \bmod I$.

```
ring A = 0,(x(1..4),t,y(1..2)),lp;
//For complicated examples the ordering (dp(n),dp(1),dp(k))
//is preferable.

ideal I    =x(1)^2-x(2)^3;
poly f1,f2=x(3)^2-1,x(1)^2*x(2);
poly b     =x(3);
ideal M    =t-b,y(1)-f1,y(2)-f2,I;

groebner(M);
//-> _[1]=t^2-y(1)-1           _[2]=x(3)-t
//-> _[3]=x(2)^4-y(2)          _[4]=x(1)^2-x(2)^3
```

---

[2] Recall that $x \gg y$ refers to a block ordering where terms in $x = (x_1, \ldots, x_n)$ are always greater than terms in $y = (y_1, \ldots, y_k)$.

```
b =x(4);
M =t-b,y(1)-f1,y(2)-f2,I;

groebner(M);
//-> _[1]=x(4)-t               _[2]=x(3)^2-y(1)-1
//-> _[3]=x(2)^4-y(2)          _[4]=x(1)^2-x(2)^3
```

We see that in the first case t^2 is one of the leading monomials of the standard basis of M and, therefore, x_3 is integral over $K[\bar{f}_1, \bar{f}_2]$ with integral relation $x_3^2 - \bar{f}_1 - 1$. In the second case we see that x_4 is not integral over $K[\bar{f}_1, \bar{f}_2]$.

Proposition 3.1.5 (Criterion for finiteness). *Let K be a field, and let $x = (x_1, \ldots, x_n)$, $y = (y_1, \ldots, y_m)$ be two sets of variables. Moreover, let $I \subset K[x]$, $J = \langle h_1, \ldots, h_s \rangle \subset K[y]$ be ideals and $\varphi : K[x]/I \to K[y]/J$ a morphism, defined by $\varphi(x_i) := f_i$. Set*

$$M := \langle x_1 - f_1, \ldots, x_n - f_n, h_1, \ldots, h_s \rangle \subset K[x, y],$$

and let $>$ be a block ordering on $K[x, y]$ such that $>$ is the lexicographical ordering for y, $y_1 > \cdots > y_m$, and $y \gg x$. Let $G = \{g_1, \ldots, g_t\}$ be a standard basis of M with respect to this ordering.

   Then φ is finite if and only if for each $j \in \{1, \ldots, m\}$ there exists some $g \in G$ such that $\mathrm{LM}(g) = y_j^{\nu_j}$ for some $\nu_j > 0$.

Proof. If $g_{s_j} = y_j^{\nu_j} + \sum_{\nu=0}^{\nu_j - 1} a_{j\nu}(x, y_{j+1}, \ldots, y_m) \cdot y_j^\nu \in M$ then

$$g_{s_j}\big|_{x=f} := g_{s_j}\big(f_1(y), \ldots, f_n(y), y_{j+1}, \ldots, y_m\big) \in J$$

for $j = 1, \ldots, m$. Therefore, $y_m \bmod J$ is integral over $K[x]/I$. Using induction and the transitivity of integrality, we obtain that $y_j \bmod J$ is integral over $K[x]/I$, hence $K[y]/J$ is finite over $K[x]/I$ by Proposition 3.1.2 (2).

   Conversely, the finiteness of φ guarantees, again by 3.1.2, an integral relation $y_j^{\nu_j} + \sum_{\nu=0}^{\nu_j - 1} a_{j\nu}\big(f_1(y), \ldots, f_n(y)\big) \cdot y_j^\nu \in J$ for suitable $a_{j\nu} \in K[x]$. Using Taylor's formula, as in the proof of Proposition 3.1.3, we obtain

$$y_j^{\nu_j} + \sum_{\nu=0}^{\nu_j - 1} a_{j\nu}(x_1, \ldots, x_n) \cdot y_j^\nu \in M,$$

and, therefore, its leading monomial, $y_j^{\nu_j}$, is an element of $L(M)$.   □

SINGULAR Example 3.1.6 (finite maps).
Let $\varphi : K[a, b, c] \to K[x, y, z]/\langle xy \rangle$ be given by $a \mapsto (xy)^3 + x^2 + z$, $b \mapsto y^2 - 1$, $c \mapsto z^3$. To check whether φ is finite we have to compute a standard basis of the ideal

$$M := \langle a - (xy)^3 - x^2 - z, \, b - y^2 + 1, \, c - z^3, \, xy \rangle \subset K[a, b, c, x, y, z]$$

with respect to a block ordering $x \gg y \gg z \gg a, b, c$. We choose the lexico-graphical ordering $x > y > z > a > b > c$.

```
ring A  =0,(x,y,z,a,b,c),lp;
ideal M =a-(xy)^3-x2-z,b-y2+1,c-z3,xy;
ideal SM=std(M);
lead(SM);                    //get leading terms of SM
//-> -[1]=a3b      _[2]=zb       _[3]=z3
//-> _[4]=ya3      _[5]=yz       _[6]=y2
//-> _[7]=xb       _[8]=xy       _[9]=x2
kill A;
```

We see that the map is finite because z^3, y^2, x^2 appear as leading terms in the standard basis. We could also have used the built–in procedure `mapIsFinite`, which checks for finiteness (cf. below).

Remark 3.1.7. Usually the above method is not the fastest. In most cases it appears to be faster, first to eliminate the x_i from M (notations from Proposition 3.1.5) and then to compute a standard basis of $M \cap K[t, y]$ for an ordering with $t \gg y_i$, see also Exercise 3.1.3.

Remark 3.1.8. For a finite map $\varphi : A \to B$ and $M \subset A$ a maximal ideal, B/MB is a finite dimensional (A/M)–vector space. This implies that the fibres of closed points of the induced map $\phi : \operatorname{Max} B \to \operatorname{Max} A$ (cf. Appendix A) are finite sets. To be specific, let $A = K[x]/I$ and $B = K[y]/J$ (K an algebraically closed field), and let

$$\mathbb{A}^m \supset V(J) \xrightarrow{\phi} V(I) \subset \mathbb{A}^n$$

be the induced map. If $M = \langle x_1 - p_1, \dots, x_n - p_n \rangle \subset K[x]$ is the maximal ideal of the point $p = (p_1, \dots, p_n) \in V(I)$ then $MB = (J + N)/J$ with $N := \langle \varphi(x_1) - p_1, \dots, \varphi(x_n) - p_n \rangle \subset K[y]$. $V(J + N) = \phi^{-1}(p)$ is the fibre of ϕ over p, which is a finite set, since $\dim_K(K[y]/(J + N)) < \infty$.

The converse, however, is not true, not even for local rings (cf. Exercise 3.1.7). But, if $\varphi : A \to B$ is a map between local analytic K–algebras, then φ is finite if and only if $\dim_K B/\varphi(\mathfrak{m}_A)B < \infty$ (cf. Corollary 6.2.14).

We illustrate a finite and a non–finite map of varieties by a picture (cf. Figure 3.1), which is created by the following SINGULAR session:

```
ring B  = 0,(x,y,z),dp;
ideal I = x-zy;
LIB"surf.lib";
plot(I);       // cf. Fig. 3.1
```

Fig. 3.1. The "blown up" (x, y)-plane.

We see that the projection ϕ_1 to the (x, y)-plane cannot be finite, since the preimage of 0 is a line. However, all fibres of the projection ϕ_2 to the (y, z)-plane consist of just one point, $\phi_2^{-1}(b, c) = (bc, b, c)$. Indeed, we can check that ϕ_2 is finite by using `mapIsFinite` from the library `algebra.lib`:

```
LIB"algebra.lib";
ring A = 0,(u,v),dp;
setring B;
map phi1 = A,x,y;        //projection to (x,y)-plane
mapIsFinite(phi1,A,I);
// -> 0
map phi2 = A,y,z;        //projection to (y,z)-plane
mapIsFinite(phi2,A,I);
// -> 1
```

Lemma 3.1.9. *Let* $\varphi : A \to B$ *be a ring map.*

(1) If $P \subset B$ *is a prime ideal, then* $\varphi^{-1}(P) \subset A$ *is a prime ideal.*

(2) If φ *is an integral extension, and if* $\varphi(x)$ *is a unit in* B, *then* $\varphi(x)$ *is a unit in the ring* $\varphi(A)$, *too.*

(3) Let φ *be an integral extension,* B *an integral domain. Then* B *is a field if and only if* $A/\mathrm{Ker}(\varphi)$ *is a field.*

(4) If φ *is an integral extension and* $M \subset B$ *a maximal ideal, then* $\varphi^{-1}(M)$ *is a maximal ideal in* A.

For a ring map $\varphi : A \to B$ and an ideal $I \subset B$ the ideal $\varphi^{-1}(I) \subset A$ is called the *contraction* of I; for $A \subset B$ the contraction of I is $I \cap A$.

Proof. (1) is obvious. To prove (2) let $\varphi(x) \cdot y = 1$ for some $y \in B$. Since B is integral over A, we can choose $a_0, \dots, a_{n-1} \in A$ such that

$$y^n + \varphi(a_{n-1})y^{n-1} + \cdots + \varphi(a_0) = 0.$$

Multiplication with $\varphi(x)^{n-1}$ gives

$$y = y^n \varphi(x)^{n-1} = -\varphi(a_{n-1} + a_{n-2}x + \cdots + a_0 x^{n-1}) \in \varphi(A).$$

(3) is a consequence of (2). For the if–direction, choose an integral relation as in (2) of minimal degree and use that B is integral.

Finally, (4) is a consequence of (3) because $A/\varphi^{-1}(M) \subset B/M$ is again an integral extension. $\square$

Proposition 3.1.10 (lying over, going up). *Let* $\varphi : A \to B$ *be an integral extension.*

(1) *If* $P \subset A$ *is a prime ideal, then there is a prime ideal* $Q \subset B$ *such that* $\varphi^{-1}(Q) = P$ *(lying over–property)* .

(2) *Let* $P \subset P' \subset A$ *and* $Q \subset B$ *be prime ideals, with* $\varphi^{-1}(Q) = P$. *Then there exists a prime ideal* $Q' \subset B$ *such that* $Q \subset Q'$ *and* $\varphi^{-1}(Q') = P'$ *(going up–property)*.

Proof. Let $S = A \smallsetminus P$ and consider $\varphi_P : S^{-1}A = A_P \to S^{-1}B := \varphi(S)^{-1}B$. A_P is a local ring and, therefore, for any maximal ideal $M \subset \varphi(S)^{-1}B$ we have $\varphi_P^{-1}(M) = PA_P$ (Lemma 3.1.9 (4), Exercise 3.1.2 (3)).

Now $P = \varphi_A^{-1}(M) \cap A = \varphi^{-1}(M \cap B)$ and $Q = M \cap B$ is prime (Lemma 3.1.9 (1), Exercise 3.1.2 (2)). This proves (1).

To prove (2) consider the integral extension $A/P =: \bar{A} \subset B/Q =: \bar{B}$. We apply (1) to this extension and the prime ideal $\bar{P}' \subset \bar{A}$ to obtain a prime ideal $\bar{Q}' \subset \bar{B}$ such that $\bar{Q}' \cap \bar{A} = \bar{P}'$. We set $Q' := \{q \in B \mid \bar{q} \in \bar{Q}'\}$. Then $Q' \subset B$ is a prime ideal which has the required properties. $\square$

Remark 3.1.11. The meaning of "lying over" and "going up" is best explained geometrically. Let $\varphi^\# : \operatorname{Spec} B \to \operatorname{Spec} A$ denote the induced map (cf. A.3). Then lying over just means that $\varphi^\#$ is surjective, that is, over each point of $\operatorname{Spec} A$ lies a point of $\operatorname{Spec} B$.

Going up means that for any point $P' \in V(P)$ and any $Q \in (\varphi^\#)^{-1}(P)$ there exists a point $Q' \in V(Q)$ such that $\varphi^\#(Q') = P'$, that is, the induced map $\varphi^\# : V(P) \to V(Q)$ is surjective, and we can "go up" from $V(Q)$ to $V(P)$.

Exercises

3.1.1. Let $A \subset B$ be rings. Show that $C := \{b \in B \mid b$ is integral over $A\}$, is a subring of B.
(Hint: consider $A[b_1, b_2]$ to show that $b_1 - b_2$, $b_1 b_2 \in C$.)

3.1.2. Check the following properties of integral dependence. Let $A \subset B \subset C$ be rings.

(1) (Transitivity)   If B is integral over A and C integral over B, then C is integral over A.

(2) (Compatibility with passing to quotient rings)   If $I \subset B$ is an ideal and B integral over A, then B/I is integral over $A/(I \cap A)$.

(3) (Compatibility with localization) If S is a multiplicatively closed set in A and B is integral over A, then $S^{-1}B$ is integral over $S^{-1}A$.

(4) Let $A \subset B$ be integral, $N \subset B$ a maximal ideal and $M = N \cap A$. Is B_N integral over A_M? Study the case $A = K[x^2 - 1]$, $B = K[x]$ and $N = \langle x - 1 \rangle$.

3.1.3. Prove that the method for checking finiteness proposed in Remark 3.1.7 is correct. Implement both methods (that of the Proposition and of the Remark) and compare their performance.

3.1.4. (1) Let $f = x^3 - y^6$, $g = x^5 + y^3 \in K[x, y]$. Show that $K[x, y]$ is finite over $K[f, g]$ (hence, $F = (f, g) : \mathbb{A}^2 \to \mathbb{A}^2$ is a finite morphism of varieties).

(2) To find the integral relations for x and y in (1) is already difficult without a computer. Compute the first three terms of an integral relation of x over $K[f, g]$ in Example (1) by hand.

(3) Use SINGULAR to find the integral relations for x and y in (2).

3.1.5. Let $\varphi : A \to B$ be an integral extension, and let $\psi : A \to K$ be a homomorphism to an algebraically closed field K. Prove that there exists an extension $\lambda : B \to K$ such that $\lambda \circ \varphi = \psi$.

3.1.6. Let $A \subset B_i$ be integral extensions of rings, $i = 1, \dots, s$. Prove that $A \subset \bigoplus_{i=1}^{s} B_i$ is integral.

3.1.7. Let $\varphi : A \to B$ be a ring map of Noetherian rings and $\varphi^{\#} : \mathrm{Spec}(B) \to \mathrm{Spec}(A)$ the induced map.

(1) Prove that for φ finite, $\varphi^{\#}$ has finite fibres, that is, $(\varphi^*)^{-1}(P)$ is a finite set for each prime ideal $P \subset A$.

(2) Show that the converse of (1) is not true in general, not even if A and B are local (consider the hyperbola and $A = K[x]_{\langle x \rangle}$ where x is one variable).

3.1.8. Let K be a field and $f = y^2 + 2y - x^2 \in K[x, y]$. Prove that

(1) the canonical map $K[x] \to K[x, y]/\langle f \rangle$ is injective and finite,

(2) the induced map between local rings $K[x]_{\langle x \rangle} \to K[x, y]_{\langle x, y \rangle}/\langle f \rangle$ is injective but not finite.

(Hint: $R = K[x]_{\langle x \rangle}[y]/\langle f \rangle$ is a semi–local ring with maximal ideals $\langle x, y \rangle$ and $\langle x, y + 2 \rangle$. Show that $R \subset R_{y+2}$ is not finite, that is, $\frac{1}{y+2}$ is not integral over R.)

3.2 The Integral Closure

We explain the notion of integral closure by an example. Assume we have a *parametrization* of an affine plane curve which is given by a polynomial map $\mathbb{A}^1 \to \mathbb{A}^2$, $t \mapsto \big(x(t), y(t)\big)$ such that t is contained in the quotient field $K\big(x(t), y(t)\big)$ of $A = K[x(t), y(t)]$. Let $A \subset B = K[t]$ denote the corresponding ring map, then t is integral over A and $A[t] = B$. We shall see that $K[t]$ is integrally closed in the quotient field $Q(A) = Q(K[t]) = K(t)$, and the "smallest ring" with this property containing A (Exercise 3.6.5). For example, $K[t^2, t^3] \subset K[t]$ corresponds to the parametrization of the cuspidal cubic (cf. Figure 3.2).

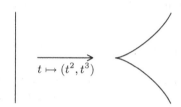

Fig. 3.2. The parametrization of the cuspidal cubic.

For arbitrary reduced affine curves with coordinate ring $A = K[x]/I$ the normalization of A, that is, the integral closure of A in $Q(A)$, is the affine ring of a "desingularization" of the curve. For higher dimensional varieties, the normalization of the coordinate ring will not necessarily be a desingularization, but an improvement of the singularities, for example, the codimension of the singular locus will be ≥ 2. Here we shall treat only some algebraic properties of the normalization.

More generally, we shall study the process associating to a ring extension $A \subset B$ the smallest subring $\widetilde{A} \subset B$ containing all elements of B which are integral over A.

Definition 3.2.1. Let $A \subset B$ be a ring extension and $I \subset A$ be an ideal (the case $I = A$ is not excluded). An element $b \in B$ which satisfies a relation

$$b^n + a_1 b^{n-1} + \cdots + a_n = 0, \quad a_i \in I$$

is called *integral over I*. We denote by

$$C(I, B) = \{b \in B \mid b \text{ integral over } I\}$$

the *(weak) integral closure* of I in B. If, moreover, $a_i \in I^i$, we say that b is *strongly integral over I* and call

$$C_s(I, B) = \{b \in B \mid b \text{ strongly integral over } I\}$$

the *strong integral closure* of I in B.

Proposition 3.2.2. *Let A be Noetherian, $A \subset B$ a ring extension and $I \subset A$ an ideal. Moreover, let $S \subset A$ be multiplicatively closed. Then*

(1) $C(A, B)$ is a subring of B containing A and $C(I, B)$ is a $C(A, B)$–ideal.
(2) $S^{-1}C(A, B) = C(S^{-1}A, S^{-1}B)$.
(3) $IC(A, B) \subset C_s(I, B) \subset C(I, B) = \sqrt{IC(A, B)}$.

Proof. (1) Let $x, y \in B$ be integral over A, then $A[x, y] \supset A$ is finite and, therefore, $A[x + y]$ and $A[xy]$ are finite A–modules (A is Noetherian). This implies that $x + y$ and $x \cdot y$ are integral over A by Proposition 3.1.2. The second statement follows from the expression of $C(I, B)$ in (3).

(2) Let $\frac{b}{s} \in S^{-1}B$ be integral over $S^{-1}A$ with a relation

$$\left(\frac{b}{s}\right)^n + \frac{a_{n-1}}{s_{n-1}}\left(\frac{b}{s}\right)^{n-1} + \ldots + \frac{a_0}{s_0} = 0, \quad s_i \in S, \ a_i \in A.$$

Let $t := \prod_i s_i$ and multiply the above equation with $(ts)^n$. We obtain that $bt \in B$ is integral over A, that is, $bt \in C(A, B)$ and, therefore, $b \in S^{-1}C(A, B)$. The other inclusion is obvious.

(3) Let $x \in IC(A, B)$, then Proposition 3.1.2, applied to $A \subset C(A, B)$ implies $x \in C_s(I, B)$, hence $IC(A, B) \subset C_s(I, B)$. If $x \in \sqrt{IC(A, B)}$, then $x^n \in IC(A, B)$ for some n, hence $x^n \in C_s(I, B)$ which implies obviously $x \in C(I, B)$. Conversely, let $x \in C(I, B)$ and $x^n + a_{n-1}x^{n-1} + \cdots + a_0 = 0$, $a_i \in I$. Then $x \in C(A, B)$ and, therefore, $x^n = -\sum_{\nu=0}^{n-1} a_\nu x^\nu \in IC(A, B)$. This implies that $C(I, B) \subset \sqrt{IC(A, B)}$. $\qquad\square$

The proposition shows that the $C(A, B)$–ideal $C(I, B)$ can be computed if $C(A, B)$ is computable, since the radical is computable as we shall see in Section 4.3. This is the case if $B = Q(A)$ is the total ring of fractions of A (see Example 3.2.3).

The strong integral closure $C_s(I, B)$ is mainly of interest for $B = A$. In this case we have $I \subset C_s(I, A) \subset C(I, A) = \sqrt{I}$. In particular, if $I = \sqrt{I}$ then $I = C_s(I, A)$. We shall see in Exercise 3.6.4 that $C_s(I, A)$ is an ideal.

SINGULAR Example 3.2.3 (integral closure of an ideal).

We consider $A = \mathbb{Q}[x, y, z]/\langle zy^2 - zx^3 - x^6 \rangle$ and $I = \langle y \rangle$. We want to compute $C\big(I, Q(A)\big)$, the integral closure of I in the quotient field of A. Using Proposition 3.2.2 we have to compute $\sqrt{IR}$, R the integral closure of A in $Q(A)$. We use the SINGULAR procedure **normal** from **normal.lib** to compute the integral closure as $R = \mathbb{Q}[T_1, T_2, T_3, T_4]/\text{norid}$. Using the map **normap**: $A \to R$, we can compute $\sqrt{IR}$.

```
LIB"normal.lib";
ring A   =0,(x,y,z),wp(2,3,6);
ideal I =y;
ideal J =zy2-zx3-x6;
list nor=normal(J);        //compute the normalization
def R   =nor[1];
setring R;
```

```
norid;
//-> norid[1]=T(2)*T(3)-T(1)*T(4)
//-> norid[2]=T(1)^5+T(1)^2*T(3)-T(2)*T(4)
//-> norid[3]=T(1)^4*T(3)+T(1)*T(3)^2-T(4)^2

normap;
//-> normap[1]=T(1)   normap[2]=T(2)   normap[3]=T(3)

map phi=A,normap;
ideal I=phi(I)+norid;

std(I);
//-> _[1]=T(2)    _[2]=T(1)*T(4)    _[3]=T(1)^5+T(1)^2*T(3)
//-> _[4]=T(1)^4*T(3)+T(1)*T(3)^2-T(4)^2    _[5]=T(4)^3

radical(I);
//-> _[1]=T(2)    _[2]=T(4)    _[3]=T(1)^4+T(1)*T(3)

kill A;
```

Hence, $C\big(I, Q(A)\big)$ is generated by T_2, T_4 and $T_1^4 + T_1 T_3$ in the integral closure $R = \mathbb{Q}[T_1, T_2, T_3, T_4]/\mathtt{norid}$.

Definition 3.2.4. Let A be a reduced ring. The *integral closure* $\overline{A}$ of A is the integral closure of A in the total ring of fractions $Q(A)$, that is,

$$\overline{A} = C\big(A, Q(A)\big).$$

$\overline{A}$ is also called the *normalization* of A. A is called *integrally closed*, or *normal*, if $A = \overline{A}$.

Proposition 3.2.5. *Let A be a Noetherian ring, the following conditions are equivalent:*

(1) A is normal;
(2) A_P is normal for all prime ideals $P \subset A$;
(3) A_M is normal for all maximal ideals $M \subset A$.

Proof. By Proposition 3.2.2, (1) implies (2). That (2) implies (3) is obvious. In order to prove that (3) implies (1), let $C := C(A, Q(A))$. By assumption and Proposition 3.2.2, we have $A_M = C_M$ for all maximal ideals $M \subset A$.

Assume $C \supsetneq A$, and let $c \in C \setminus A$. Define $I := \{s \in A \mid cs \in A\}$ which is a proper ideal in A. There exists a maximal ideal $M \subset A$ such that $I \subset M$. But $A_M = C_M$ and, therefore, there exists an $s \in A \setminus M$ such that $sc \in A$. This is a contradiction to the definition of I and the choice of M. $\square$

Example 3.2.6. Let K be a field, then the polynomial ring $K[x_1, \ldots, x_n]$ is normal.

Proof. This is proved by induction on n. Let $n = 1$ and $f, g \in K[x_1]$ such that $\gcd(f, g) = 1$. Consider an integral relation for f/g,

$$\left(\frac{f}{g}\right)^m + a_{m-1}\left(\frac{f}{g}\right)^{m-1} + \cdots + a_0 = 0, \quad a_i \in K[x_1].$$

This implies $g \mid f^m$ which contradicts $\gcd(f, g) = 1$. For the induction step we use the following lemma. □

Lemma 3.2.7. *Let A be Noetherian and $A \subset B$ be integrally closed, then $A[x] \subset B[x]$ is integrally closed.*

Assume that Lemma 3.2.7 is proved. We obtain from the induction hypothesis that $K[x_1, \ldots, x_{n-1}] \subset K(x_1, \ldots, x_{n-1})$ is integrally closed and, therefore, by the lemma, $K[x_1, \ldots, x_n] \subset K(x_1, \ldots, x_{n-1})[x_n]$ is integrally closed. The case $n = 1$ shows that $K(x_1, \ldots, x_{n-1})[x_n] \subset K(x_1, \ldots, x_n)$ is integrally closed. The result follows from transitivity (Exercise 3.1.2).

Proof of Lemma 3.2.7. Let $f \in B[x]$ be integral over $A[x]$. We use induction over $n := \deg(f)$ to prove that $f \in A[x]$. First of all, $M := A[x][f]$ is a finitely generated $A[x]$–module. Let $M_0 \subset B$ be the A–module generated by all coefficients of the powers of x of all elements of M, considered as elements of $B[x]$. M_0 is a finitely generated A–module (generated by the coefficients of $1, f, \ldots, f^{m-1}$ if $f^m + \sum_{\nu=0}^{m-1} a_\nu f^\nu = 0$, $a_\nu \in A[x]$).
If $f = \sum_{\nu=0}^{n} b_\nu x^\nu$, $b_n \neq 0$, then $A[b_n] \subset M_0$ is a finite A–module. This implies that b_n is integral over A, therefore $b_n \in A$. Now $g := f - b_n x^n$ is integral over $A[x]$. If $n = 1$ this implies that b_0 is integral and, by assumption, $b_0 \in A$. In this case, we obtain $f \in A[x]$. If $n > 1$ then $g \in A[x]$ by the induction hypothesis and, again, $f \in A[x]$. □

Example 3.2.8. $A := K[x, y]/\langle x^4 + 6x^2 y - y^3 \rangle \cong K[-t^3 + 6t, t^4 - 6t^2]$ is not normal. Namely, $-t = y/x \in Q(A) \setminus A$, but $(y/x)^3 - 6(y/x) - x = 0$ is an integral relation for t.

An important example for normalization is the parametrization of an affine curve C, $\mathbb{A}^1 \ni t \mapsto (x_1(t), \ldots, x_n(t)) \in C \subset \mathbb{A}^n$, such that $K(x_1(t), \ldots, x_n(t)) = K(t)$. Then $K[x_1(t), \ldots, x_n(t)] \hookrightarrow K[t]$ is the normalization (cf. Exercise 3.6.5). Given the parametrization, we obtain generators for an ideal I satisfying $V(I) = C$ by eliminating t from $x_1 - x_1(t), \ldots, x_n - x_n(t)$. This process is called *implicitization*, the inverse process to parametrization (which is not always possible). Let us look at an example:

```
LIB"surf.lib";
ring  A=0,(t,x,y),dp;
ideal J=t3-6t+x,-t4+6t2+y;
ideal I=eliminate(J,t);        //implicitization
I;
//-> I[1]=x4+6x2y-y3
plot(I);
```

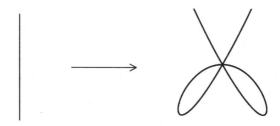

Fig. 3.3. The parametrization of a deformed E_6–singularity.

The picture produced by the `plot(I)` (cf. Figure 3.3) shows the normalization (parametrization) $t \mapsto (-t^3 + 6t,\ t^4 - 6t^2)$ of a deformed E_6–singularity.

Theorem 3.2.9 (going down). *Let $A \subset B$ be a ring extension of Noetherian integral domains, A normal and B integral over A. Let $Q \subset B$ be a prime ideal, $P = Q \cap A$ and $P' \subset P$ a prime ideal in A. Then there exists a prime ideal $Q' \subset Q$ in B such that $Q' \cap A = P'$.*

Lemma 3.2.10. *With the hypothesis of the theorem, let $x \in B$ be integral over the ideal $I \subset A$ and $f = t^n + \sum_{\nu=0}^{n-1} a_\nu t^\nu \in Q(A)[t]$, be the minimal polynomial of x. Then $a_\nu \in \sqrt{I}$ for $\nu = 0, \ldots, n-1$.*

Proof. Let L be the splitting field of f and $x_1 := x$, $x_2, \ldots, x_n \in L$ be the roots of f. Then all the x_i are integral over I because f divides the polynomial which defines the integral relation of x over I.

The elementary symmetric polynomials in $x_1, \ldots, x_n$ are the coefficients of f and, therefore, the coefficients are integral over I, that is, in $C\big(I, Q(A)\big)$. But A is integrally closed, hence equal to $C\big(A, Q(A)\big)$ and, therefore, by Proposition 3.2.2 these coefficients are in $\sqrt{I}$.   □

Proof of Theorem 3.2.9. It is sufficient to prove that $P'B_Q \cap A = P'$. Namely, let $S' = A \smallsetminus P'$, then $P'S'^{-1}B_Q$ is a proper ideal, since $P'S'^{-1}B_Q = S'^{-1}B_Q$ would imply $s' \in P'B_Q$ for some $s' \in S'$, which contradicts $P'B_Q \cap A = P'$. Let $M \supset P'S'^{-1}B_Q$ be a maximal ideal of $S'^{-1}B_Q$ then $P' = M \cap A$, because $P' \subsetneq M \cap A$ would imply $s' \in M$ for some $s' \in S'$, and M would be the whole ring. Defining $Q' = M \cap B$, then $Q' \subset Q$ and $Q' \cap A = P'$.

We have to show the inclusion $P'B_Q \cap A \subset P'$. Let $x \in P'B_Q \cap A$ then $x = y/s$, for some $s \in B \smallsetminus Q$ and $y \in P'B \subset \sqrt{P'B} = C(P', B)$. This implies that y is integral over P' and, using Lemma 3.2.10, that the coefficients a_ν of the (monic) minimal polynomial $f = t^n + \sum_{\nu=0}^{n-1} a_\nu t^\nu$ of y over $Q(A)$ are already in P'. Then $f(y) = 0$ implies

$$\left(\frac{y}{x}\right)^n + \frac{a_{n-1}}{x}\left(\frac{y}{x}\right)^{n-1} + \cdots + \frac{a_0}{x^n} = 0\,.$$

The polynomial $g := t^n + \sum_{\nu=0}^{n-1}(a_\nu/x^{n-\nu}) \cdot t^\nu$ is a polynomial in $Q(A)[t]$, because $x \in A$ by assumption. Since $y/x = s$, we have $g(s) = 0$.

But then g is the minimal polynomial of s, since otherwise f would not be the minimal polynomial of y. This implies, using Lemma 3.2.10 with $I = A$, that $a_\nu / x^{n-\nu} \in A$ for $\nu = 0, \dots, n - 1$.

Now $(a_\nu / x^{n-\nu}) \cdot x^{n-\nu} = a_\nu \in P'$. If $x \notin P'$ then we obtain $a_\nu / x^{n-\nu} \in P'$ for all ν, which implies $s^n \in P'B \subset PB \subset Q$. This is a contradiction to the choice of s and, hence, proves $x \in P'$. $\square$

Example 3.2.11. Let $A := K[y, z] \subset B := K[x, y, z]/(\langle x, y \rangle \cap \langle x + z \rangle)$. Then the extension $A \subset B$ is integral and A is normal, but the going down theorem fails for $Q = \langle x, y \rangle$ and $P' = \langle 0 \rangle$, because B is not an integral domain (see Figure 3.4).

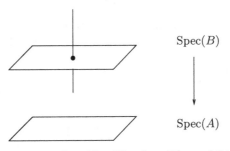

Spec(B)

Spec(A)

Fig. 3.4. The projection induced by $K[y, z] \hookrightarrow K[x, y, z]/(\langle x, y \rangle \cap \langle x + z \rangle)$.

Exercises

3.2.1. Let A be a *unique factorization domain* (that is, A is a domain and every non–unit of A can be written as a product of irreducible elements so that the factors are up to multiplication with units uniquely determined). Prove that A is a normal ring.

3.2.2. Prove that principal ideal domains are normal.
(Hint: Use Exercise 3.2.1.)

3.2.3. Let K be a field.

(1) Prove that $K[x, y, z]/\langle z^2 - xy \rangle$ is normal.
(2) Prove that $K[x, y]/\langle xy \rangle$ is not normal.

3.2.4. Use SINGULAR to test the examples of Exercise 3.2.3.
(Hint: Type ? `normal.lib`;.)

3.2.5. Check Example 3.2.11.

3.2.6. Let $A \subset B$ be integral and $C = C(A, B)$. Prove that C is integrally closed in B.

3.2.7. Let $A \subset B$ be integral and $B \setminus A$ be multiplicatively closed in B. Prove that A is integrally closed in B.

3.2.8. Let $A \subset B$ be integral domains and $C = C(A, B)$. Let $f, g \in B[x]$ be monic polynomials such that $f \cdot g \in C[x]$. Prove that $f, g \in C[x]$.

3.2.9. Let $A = \bigoplus_{i=0}^{\infty} A_i$ be a Noetherian graded ring, $d \geq 1$ an integer, and let $A^{(d)} = \bigoplus_{i=0}^{\infty} A_{id}$. Prove that A is integral over $A^{(d)}$.

3.2.10. Prove that a normal local ring is an integral domain.

3.3 Dimension

In this section we shall use chains of prime ideals to define the dimension of a ring. This is one possibility to define the dimension. It is called the *Krull dimension*. We shall show that the dimension of the polynomial ring $K[x_1, \ldots, x_n]$ in the variables $x_1, \ldots, x_n$ over a field K equals n, given by the chain $\langle 0 \rangle \subset \langle x_1 \rangle \subset \langle x_1, x_2 \rangle \subset \cdots \subset \langle x_1, \ldots, x_n \rangle$.

Definition 3.3.1. Let A be a ring

(1) Let $\mathcal{C}(A)$ denote the set of all *chains of prime ideals* in A, that is,

$$\mathcal{C}(A) := \left\{ \wp = (P_0 \subsetneq \cdots \subsetneq P_m \subsetneq A) \mid P_i \text{ prime ideal} \right\}.$$

(2) If $\wp = (P_0 \subsetneq \cdots \subsetneq P_m \subsetneq A) \in \mathcal{C}(A)$ then $length(\wp) := m$.
(3) The *dimension* of A is defined as $\dim(A) = \sup\{length(\wp) \mid \wp \in \mathcal{C}(A)\}$.
(4) For $P \subset A$ a prime ideal, let

$$\mathcal{C}(A, P) = \left\{ \wp = (P_0 \subsetneq \cdots \subsetneq P_m) \in \mathcal{C}(A) \mid P_m = P \right\}$$

denote the set of prime ideal chains ending in P. We define the *height* of P as $ht(P) = \sup\{length(\wp) \mid \wp \in \mathcal{C}(A, P)\}$.
(5) For an arbitrary ideal $I \subset A$, $ht(I) = \inf\{ht(P) \mid P \supset I \text{ prime}\}$ is called the *height* of I and $\dim(I) := \dim(A/I)$ is called the *dimension* of I.

Example 3.3.2.

(1) We shall see in Section 3.5 that in the polynomial ring $K[x_1, \ldots, x_n]$ over a field K all maximal chains of prime ideals have the same length n. Here a chain of prime ideals is called *maximal* if it cannot be refined.
(2) Let $A = K[x]_{(x)}[y]$ then $(0) \subset (xy - 1)$ and $(0) \subset (x) \subset (x, y)$ are two maximal chains of different length.
(3) Let $A = K[x, y, z]/\langle xz, yz \rangle$, then $\dim(A) = 2$. Let $P = \langle x, y, z - 1 \rangle$, then $\dim(A_P) = 1$ (cf. Figure 3.5 on page 227).

Corollary 3.3.3. *Let $A \subset B$ be an integral extension, then $Q \mapsto Q \cap A$ defines a surjection $\mathcal{C}(B) \to \mathcal{C}(A)$ preserving the length of chains, in particular, $\dim(A) = \dim(B)$.*

Proof. Using Proposition 3.1.10 we see that the map is surjective. We have to prove that the length is also preserved. Let $Q \subsetneqq Q'$ be prime ideals in B, assume $Q \cap A = Q' \cap A = P$.

Now $A_P \subset B_P$ is an integral extension, and A_P is local with maximal ideal PA_P. Moreover, $QB_P \subset Q'B_P$ are prime ideals in B_P, with the property $QB_P \cap A_P = Q'B_P \cap A_P = PA_P$. Because of Lemma 3.1.9 (3), QB_P and $Q'B_P$ are maximal and, therefore, $QB_P = Q'B_P$. This implies $Q = Q'$. □

Definition 3.3.4. Let A be a ring and $I \subset A$ an ideal. A prime ideal P with $I \subset P$ is called *minimal associated prime ideal* of I, if, for any prime ideal $Q \subset A$ with $I \subset Q \subset P$ we have $Q = P$. The set of minimal associated prime ideals of I is denoted by $\min\mathrm{Ass}(I)$.

Proposition 3.3.5. *Let A be a Noetherian ring and $I \subset A$ be an ideal. Then $\min\mathrm{Ass}(I) = \{P_1, \dots, P_n\}$ is finite and*

$$\sqrt{I} = P_1 \cap \cdots \cap P_n \,.$$

In particular, $\sqrt{I}$ is the intersection of all prime ideals containing I.[3]

Proof. Obviously we have $\min\mathrm{Ass}(I) = \min\mathrm{Ass}(\sqrt{I})$ and, therefore, we may assume that $I = \sqrt{I}$.

If I is prime, the statement is trivial. Hence, we assume that there exist $a, b \notin I$ with $ab \in I$. We show that $\sqrt{I : \langle a \rangle} = I : \langle a \rangle = I : \langle a^2 \rangle \supsetneqq I$. Namely, $f \in \sqrt{I : \langle a \rangle}$ implies $f^\rho \in I : \langle a \rangle$ for a suitable ρ. Therefore, $af^\rho \in I$ and $(af)^\rho \in I$, which implies $af \in \sqrt{I} = I$, that is, $f \in I : \langle a \rangle$. On the other hand, $f \in I : \langle a^2 \rangle$ implies $a^2 f \in I$ and $(af)^2 \in I$, that is, $af \in \sqrt{I} = I$ and, therefore, $f \in I : \langle a \rangle$. Finally, $b \in I : \langle a \rangle$ but $b \notin I$. Now, because of Lemma 3.3.6 below, we obtain $I = (I : \langle a \rangle) \cap \langle I, a \rangle$. In particular, we obtain

$$I = \sqrt{I} = \sqrt{(I : \langle a \rangle)} \cap \sqrt{\langle I, a \rangle} = (I : \langle a \rangle) \cap \sqrt{\langle I, a \rangle} \,.$$

If $I : \langle a \rangle$ or $\sqrt{\langle I, a \rangle}$ are not prime, we can continue with these ideals as we did with I. This process has to stop because A is Noetherian and, finally, we obtain $I = \bigcap_{i=1}^n P_i$ with P_i prime. We may assume that $P_i \not\subset P_j$ for $i \neq j$ by deleting unnecessary primes. In this case we have $\min\mathrm{Ass}(I) = \{P_1, \dots, P_n\}$. For, if $P \supset I$ is a prime ideal, then $P \supset \bigcap_{i=1}^n P_i$, and, therefore, there exist j such that $P \supset P_j$ by Lemma 1.3.12. This proves the proposition. □

Lemma 3.3.6. (Splitting tool) Let A be a ring, $I \subset A$ an ideal, and let $I : \langle a \rangle = I : \langle a^2 \rangle$ for some $a \in A$. Then $I = (I : \langle a \rangle) \cap \langle I, a \rangle$.

Proof. Let $f \in (I : \langle a \rangle) \cap \langle I, a \rangle$, and let $f = g + xa$ for some $g \in I$. Then $af = ag + xa^2 \in I$ and, therefore, $xa^2 \in I$. That is, $x \in I : \langle a^2 \rangle = I : \langle a \rangle$, which implies $xa \in I$ and, consequently, $f \in I$. □

---

[3] The latter statement is also true for not necessarily Noetherian rings, see also Exercise 3.3.1.

Example 3.3.7.

(1) Let $I = \langle wx, wy, wz, vx, vy, vz, ux, uy, uz, y^3 - x^2 \rangle \subset K[v, w, x, y, z]$. Then
$I = \langle x, y, z \rangle \cap \langle u, v, w, x^2 - y^3 \rangle$, $\mathrm{minAss}(I) = \{\langle x, y, z \rangle, \langle u, v, w, x^2 - y^3 \rangle\}$.
(2) Let $I = \langle x^2, xy \rangle \subset K[x, y]$ then $I = \langle x \rangle \cap \langle x^2, y \rangle$, $\sqrt{I} = \langle x \rangle$ and, hence,
$\mathrm{minAss}(I) = \{\langle x \rangle\}$.

The minimal associated primes can be computed (with two different algorithms) using the SINGULAR library `primdec.lib`:

SINGULAR Example 3.3.8 (minimal associated primes).

```
ring A=0,(u,v,w,x,y,z),dp;
ideal I=wx,wy,wz,vx,vy,vz,ux,uy,uz,y3-x2;
LIB"primdec.lib";
minAssGTZ(I);
//-> [1]:                    [2]:
//->    _[1]=z                  _[1]=-y3+x2
//->    _[2]=y                  _[2]=w
//->    _[3]=x                  _[3]=v
//->                            _[4]=u
ring B=0,(x,y,z),dp;
ideal I=zx,zy;
minAssChar(I);
//-> [1]:                    [2]:
//->    _[1]=y                  _[1]=z
//->    _[2]=x
```

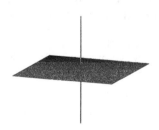

Fig. 3.5. The variety $V(xz, yz)$.

The minimal associated primes of $\langle zx, zy \rangle$ are $\langle z \rangle$ and $\langle x, y \rangle$ which correspond to two components of dimension 2, respectively 1, that is, to the plane, respectively the line, in Figure 3.5.

The following lemma is easy to prove and left as an exercise.

Lemma 3.3.9. *Let A be a ring and $A_{\mathrm{red}} := A/\sqrt{\langle 0 \rangle}$ the reduction of A, then*

$$\dim(A) = \dim(A_{\mathrm{red}}) = \max_{P \in \mathrm{minAss}(\langle 0 \rangle)} \{\dim(A/P)\}.$$

Remark 3.3.10. It is possible for a Noetherian integral domain to have infinite dimension: let K be a field and let $A = k[x_1, x_2, \dots]$ be a polynomial ring in countably many indeterminates. Let $(\nu_j)_{j \geq 1}$ be a strictly increasing sequence of positive integers such that $(\nu_{j+1} - \nu_j)_{j \geq 1}$ is also strictly increasing. Let $P_i := \langle x_{\nu_i+1}, \dots, x_{\nu_{i+1}} \rangle$ and $S = A \setminus \bigcup_i P_i$. Then $S^{-1}A$ is a Noetherian integral domain (using Exercise 3.3.3). But $\mathrm{ht}(S^{-1}P_i) = \nu_{i+1} - \nu_i$ implies $\dim(S^{-1}A) = \infty$.

Remark 3.3.11. Notice that the ring in the previous remark is not local. We shall see in Chapter 5 that local Noetherian rings have finite dimension. In particular, this implies (using localization) that in a Noetherian ring the height of an ideal is always finite.

Remark 3.3.12. For graded rings, we shall obtain in Chapter 5 another description of the dimension as the degree of the Hilbert polynomial. This will be the basis to compute the dimension due to the fact that for an ideal $I \subset K[x_1, \dots, x_n]$

$$\dim(K[x_1, \dots, x_n]/I) = \dim(K[x_1, \dots, x_n]/L(I)),$$

where $L(I)$ is the leading ideal of I (cf. Corollary 5.3.14).

Thus, after a Gröbner basis computation, the computation of the dimension is reduced to a pure combinatorial problem.

SINGULAR Example 3.3.13 (computation of the dimension).
Let I be the ideal of Example 3.3.7 (1). We want to compute the dimension:

```
ring A=0,(u,v,w,x,y,z),dp;
ideal I=wx,wy,wz,vx,vy,vz,ux,uy,uz,y3-x2;
I=std(I);
dim(I);
//-> 3
```

The next lemmas prepare applications of the Noether normalization theorem (see Section 3.4).

Lemma 3.3.14. *Let A be a ring such that for each prime ideal $P \subset A$ there exists a normal Noetherian integral domain $C \subset A$ with $C \subset A/P$ being finite. Then the following holds:*

If $A \subset B$ is a finite ring extension then the map $\mathcal{C}(B) \to \mathcal{C}(A)$ induced by the contraction $P \mapsto P \cap A$ maps maximal chains to maximal chains.

Proof. Let $\langle 0 \rangle = Q_0 \subset Q_1 \subset \cdots \subset Q_n$ be a maximal chain of prime ideals in B and consider $\langle 0 \rangle = Q_0 \cap A \subset Q_1 \cap A \subset \cdots \subset Q_n \cap A$. We have to prove that this chain is maximal. Assume $Q \subset Q' \subset B$ are two prime ideals and there exists a prime ideal $P \subset A$ such that $Q \cap A \subsetneqq P \subsetneqq Q' \cap A$. We choose for $Q \cap A$ an integrally closed, Noetherian integral domain A' such that $A' \subset A/(Q \cap A)$ is finite. Note that the ideals $\langle 0 \rangle$, $P/(Q \cap A) \cap A'$, $Q'/(Q \cap A) \cap A'$ are pairwise different as $A' \subset A/(Q \cap A)$ is finite (Corollary 3.3.3). Moreover, B/Q is also finite over A', and we can apply the going down theorem to find a prime ideal $\bar{P}' \neq \langle 0 \rangle$ in B/Q, $\bar{P}' \subset Q'/Q$ such that $\bar{P}' \cap A' = P \cap A'$. Therefore, $\bar{P}' \subsetneqq Q'/Q$. This implies the existence of a prime ideal P', $Q \subsetneqq P' \subsetneqq Q'$ and proves the lemma, because the case $Q' \cap A \subsetneqq P$, can be handled similarly. $\qquad\square$

Remark 3.3.15. It is a consequence of the Noether normalization theorem (cf. Section 3.4) that all rings of finite type over a field K have the property required in the assumption of the lemma.

Lemma 3.3.16. *Let A, B satisfy the assumptions of the going down theorem (Theorem 3.2.9). If $Q \subset B$ is a prime ideal then $\mathrm{ht}(Q) = \mathrm{ht}(Q \cap A)$.*

Proof. Due to Corollary 3.3.3, the map $\mathcal{C}(B) \to \mathcal{C}(A)$ induced by $P \mapsto P \cap A$ induces a map $\mathcal{C}(B, Q) \to \mathcal{C}(A, Q \cap A)$, preserving the length of prime ideal chains. To see that this map is surjective, let $Q \cap A = P_s \supsetneqq P_{s-1} \supsetneqq \cdots \supsetneqq P_0$ be a chain of prime ideals in A. Starting with P_{s-1}, and using s times the going down theorem, we obtain a chain $Q = Q_s \supsetneqq Q_{s-1} \supsetneqq \cdots \supsetneqq Q_0$ of prime ideals in B. This proves the lemma. $\qquad\square$

Exercises

3.3.1. Let A be a ring and $I \subsetneqq A$ a proper ideal. Prove that $\sqrt{I}$ is the intersection of all prime ideals containing I.

(Hint: reduce the statement to the case $I = \langle 0 \rangle$ and consider, for $f \in A$ not nilpotent, the set of all ideals not containing any power of f. Show that this set contains a prime ideal by using Zorn's lemma.)

3.3.2. Prove Lemma 3.3.9.

3.3.3. Let A be a ring such that

(1) for each maximal ideal M of A, the localization A_M is Noetherian;
(2) for each $x \neq 0$ in A the set of maximal ideals of A which contain x is finite.

Prove that A is Noetherian.

3.3.4. Check the statement of Example 3.3.2 (2), (3). Moreover, draw the zero–set in $\mathbb{R}^3$ of the ideal

$$I := \langle x \cdot (x^2 + y^2 + z^2 - 1),\ y \cdot (x^2 + y^2 + z^2 - 1) \rangle \subset \mathbb{R}[x, y, z]$$

to see the phenomena occurring in (3).

3.3.5. Use SINGULAR to compute

(1) the minimal associated primes of the ideal

$$I = \langle t - b - d,\ x + y + z + t - a - c - d,\ xz + yz + xt + zt - ac - ad - cd,$$
$$xzt - acd \rangle \subset \mathbb{Q}[a, b, c, d, t, x, y, z]\,,$$

(2) the intersection of the minimal associated primes, and
(3) the radical of I,

and verify the statement of Proposition 3.3.5.

3.3.6. Let $P \subset K[x_1, \ldots, x_s]$ be a homogeneous prime ideal of height r. Prove that there exists a chain $P_0 \subsetneq P_1 \subsetneq \cdots \subsetneq P_r = P$ of homogeneous prime ideals.

3.3.7. Prove that a principal ideal domain has dimension at most 1.

3.3.8. Let A be a Noetherian ring. Prove that $\dim(A[x]) = \dim(A) + 1$.

3.3.9. Let A be a Noetherian ring. Use the previous Exercise 3.3.8 to prove that $\dim(A[x, x^{-1}]) = \dim(A) + 1$.

3.3.10. Let A be a Noetherian ring, and let $P \in \mathrm{minAss}(\langle 0 \rangle)$. Prove that $A_P = Q(A)_{PQ(A)}$.

3.3.11. Let K be a field and $A = K[x, y]_{\langle x, y \rangle}/\langle x^2, xy \rangle$. Prove that A is equal to its total ring of fractions, $A = Q(A)$, and $\dim(A) = 1$.

3.3.12. Show that $A := \mathbb{Q}[x, y, z]/\langle xy, xz, yz, (x-y)(x+1), z^3 \rangle$ has dimension 0. Compute the $\mathbb{Q}$–vector space dimension of A and compare it to the $\mathbb{Q}$–vector space dimension of the localization of A in $\langle x, y, z \rangle$.

3.3.13. Let K be a field and $A = K[x, y]$. Prove that Lemma 3.3.14 does not hold for $B = K[x, y, z]/\langle xz, z^2 - yz \rangle$.

3.4 Noether Normalization

Let K be a field, $A = K[x_1, \ldots, x_n]$ be the polynomial ring and $I \subset A$ an ideal.

Noether normalization is a basic tool in the theory of affine K–algebras, that is, algebras of type A/I. It is the basis for many applications of the theorems of the previous chapters, because it provides us with a polynomial ring $K[x_{s+1}, \ldots, x_n] \subset A/I$ such that the extension is finite.

Theorem 3.4.1 (Noether normalization). *Let K be a field, and let $I \subset K[x_1, \ldots, x_n]$ be an ideal. Then there exist an integer $s \leq n$ and an isomorphism*

$$\varphi : K[x_1, \ldots, x_n] \to A := K[y_1, \ldots, y_n],$$

such that:

(1) the canonical map $K[y_{s+1}, \ldots, y_n] \to A/\varphi(I)$, $y_i \mapsto y_i \bmod \varphi(I)$ is injective and finite.

(2) Moreover, φ can be chosen such that, for $j = 1, \ldots, s$, there exist polynomials

$$g_j = y_j^{e_j} + \sum_{k=0}^{e_j - 1} \xi_{j,k}(y_{j+1}, \ldots, y_n) \cdot y_j^k \in \varphi(I)$$

satisfying $e_j \geq \deg(\xi_{j,k}) + k$ for $k = 0, \ldots, e_j - 1$.

(3) If I is homogeneous then the g_j can be chosen to be homogeneous, too. If I is a prime ideal, the g_j can be chosen to be irreducible.

(4) If K is perfect and if I is prime, then the morphism φ can be chosen such that, additionally, $Q\big(A/\varphi(I)\big) \supset Q(K[y_{s+1}, \ldots, y_n])$ is a separable field extension and, moreover, if K is infinite then

$$Q\big(A/\varphi(I)\big) = Q(K[y_{s+1}, \ldots, y_n])[y_s]/\langle g_s \rangle.$$

(5) If K is infinite then φ can be chosen to be linear, $\varphi(x_i) = \sum_j m_{ij} y_j$ with $M = (m_{ij}) \in \mathrm{GL}(n, K)$.

Definition 3.4.2. Let $I \subset A = K[y_1, \ldots, y_n]$ be an ideal. A finite and injective map $K[y_{s+1}, \ldots, y_n] \to A/I$ is called a *Noether normalization* of A/I. If, moreover, I contains $g_1, \ldots, g_s$ as in Theorem 3.4.1 (2), then it is called a *general Noether normalization*.

Example 3.4.3. $K[x] \subset K[x, y]/\langle x^3 - y^2 \rangle$ is a Noether normalization, but not a general Noether normalization, while $K[y] \subset K[x, y]/\langle x^3 - y^2 \rangle$ is a general Noether normalization.

Proof of Theorem 3.4.1. We prove the theorem for infinite fields, while the proof for finite fields is left as Exercise 3.4.1. The case $I = \langle 0 \rangle$ being trivial, we can suppose $I \neq \langle 0 \rangle$. We proceed by induction on n. Let $n = 1$, and let $I = \langle f \rangle$, f a polynomial of degree d. Then $K[x_1]/I = K + x_1 K + \cdots + x_1^{d-1} K$ is a finite dimensional K–vector space, and the theorem holds with $s = 1$.

Assume now that the theorem is proved for $n - 1 \geq 1$, and let $f \in I$ be a polynomial of degree $d \geq 1$. If I is homogeneous we choose f to be homogeneous. Let $f = \sum_{\nu=0}^d f_\nu$ be the decomposition of f into homogeneous parts f_ν of degree ν. To keep notations short in the following construction of the morphism φ, we identify the x_i (resp. the y_j) with their images in $K[y_1, \ldots, y_n]$ (resp. in $K[z_2, \ldots, z_n]$). Let $M_1 = (m_{ij}) \in \mathrm{GL}(n, K)$,

$$M_1 \cdot \begin{pmatrix} y_1 \\ \vdots \\ y_n \end{pmatrix} = \begin{pmatrix} x_1 \\ \vdots \\ x_n \end{pmatrix}.$$

Then we obtain

$$f_d(x_1, \ldots, x_n) = f_d \left(\sum_{j=1}^{n} m_{1j} y_j, \ldots, \sum_{j=1}^{n} m_{nj} y_j \right)$$
$$= f_d(m_{11}, \ldots, m_{n1}) \cdot y_1^d + \text{ lower terms in } y_1.$$

Now the condition for M_1 becomes $f_d(m_{11}, \ldots, m_{n1}) \neq 0$, which can be satisfied as K is infinite. Then, obviously, $K[y_2, \ldots, y_n] \to A/\langle f \rangle$ is injective and finite by Proposition 3.1.2, since y_1 satisfies an integral relation, and $\widetilde{g}_1 := f(M_1 y)$ has, after normalizing, the property required in (2).

Note that $K[y_2, \ldots, y_n]/(I \cap K[y_2, \ldots, y_n]) \to A/I$ is injective and still finite (we write I instead of $\varphi(I)$). If $I \cap K[y_2, \ldots, y_n] = \langle 0 \rangle$ then there is nothing to show.

Otherwise, let $I_0 := I \cap K[y_2, \ldots, y_n]$. By the induction hypothesis there is some matrix $M_0 \in \mathrm{GL}(n-1, K)$ such that, for

$$M_0 \cdot \begin{pmatrix} z_2 \\ \vdots \\ z_n \end{pmatrix} = \begin{pmatrix} y_2 \\ \vdots \\ y_n \end{pmatrix}$$

and some $s \leq n$, the map $K[z_{s+1}, \ldots, z_n] \to K[y_2, \ldots, y_n]/I_0$ is injective and finite. Moreover, for $j = 2, \ldots, s$ there exist polynomials

$$g_j = z_j^{e_j} + \sum_{k=0}^{e_j - 1} \xi_{j,k}(z_{j+1}, \ldots, z_n) \cdot z_j^k \in I$$

such that $e_j \geq \deg(\xi_{j,k}) + k$ for $k = 0, \ldots, e_j - 1$. Again, the g_j can be chosen to be homogeneous if I is homogeneous.

This implies that $K[z_{s+1}, \ldots, z_n] \to A/I$ is injective and finite. The theorem is proved for

$$M = M_1 \cdot \begin{pmatrix} 1 & 0 \ldots 0 \\ 0 & \\ \vdots & M_0 \\ 0 & \end{pmatrix}$$

and $g_1 := \widetilde{g}_1(y_1, M_0 z)$, $g_2, \ldots, g_s \in K[y_1, z_2, \ldots, z_n]$.

If I is prime and if a g_j splits into irreducible factors then already one of the factors must be in I. We take this factor which has the desired shape.

The proof of (4) in the case of characteristic zero is left as Exercise 3.4.4. For the general case, we refer to [66, 159]. (5) was already proved in (1). $\square$

Remark 3.4.4. The proof of Theorem 3.4.1 shows that

(1) the theorem holds for M arbitrarily chosen in

- some open dense subset $U \subset \mathrm{GL}(n, K)$, respectively
- some open dense subset U' in the set of all lower triangular matrices with entries 1 on the diagonal;

(2) the theorem holds also for finite fields if the characteristic is large;
(3) for a finite field of small characteristic the theorem also holds, when replacing the linear coordinate change $M \cdot y = x$ by a coordinate change of type $x_i = y_i + h_i(y)$, $\deg(h_i) \geq 2$, see Exercises 3.4.1 and 3.4.2.

The general Noether normalization is necessary in the theory of Hilbert functions, as we shall see in Chapter 5.

Analyzing the proof we obtain the following algorithm to compute a Noether normalization, which is correct and works well for characteristic 0 and large characteristic:

Algorithm 3.4.5 (NOETHERNORMALIZATION(I)).

Input: $I := \langle f_1, \ldots, f_k \rangle \subset K[x]$, $x = (x_1, \ldots, x_n)$.
Output: A set of variables $\{x_{s+1}, \ldots, x_n\}$ and a map $\varphi : K[x] \to K[x]$ such that $K[x_{s+1}, \ldots, x_n] \subset K[x]/\varphi(I)$ is a Noether normalization.

- perform a random lower triangular linear coordinate change

$$\varphi(x) = \begin{pmatrix} 1 & & 0 \\ & \ddots & \\ * & & 1 \end{pmatrix} \cdot \begin{pmatrix} x_1 \\ \vdots \\ x_n \end{pmatrix} ;$$

- compute a reduced standard basis $\{f_1, \ldots, f_r\}$ of $\varphi(I)$ with respect to the lexicographical ordering with $x_1 > \cdots > x_n$, and order the f_i such that $\mathrm{LM}(f_r) > \cdots > \mathrm{LM}(f_1)$;
- choose s maximal such that $\{f_1, \ldots, f_r\} \cap K[x_{s+1}, \ldots, x_n] = \emptyset$;
- for each $i = 1, \ldots, s$, test whether $\{f_1, \ldots, f_r\}$ contains polynomials with leading monomial $x_i^{\rho_i}$ for some ρ_i;
- if the test is true for all i then return φ and $x_{s+1}, \ldots, x_n$ (note that in this case $K[x_{s+1}, \ldots, x_n] \subset K[x_1, \ldots, x_n]/\varphi(I)$ is finite);
- return NOETHERNORMALIZATION(I).

Let us try an example:

SINGULAR Example 3.4.6 (Noether normalization).

```
LIB"random.lib";
ring R=0,(x,y,z),lp;
ideal I=xy,xz;
dim(std(I));
//-> 2
ideal M=ideal(sparsetriag(3,3,0,100)
                *transpose(maxideal(1)));
```

```
M;
//-> M[1]=x    M[2]=65x+y    M[3]=85x+82y+z

map phi=R,M;
ideal J=phi(I);    //the random coordinate change
J;
//-> J[1]=65x2+xy    J[2]=85x2+82xy+xz

//dim(I)=2 implies R/J <--Q[y,z] is a Noether normalization
```

The algorithm in the SINGULAR programming language can be found in Section 3.7 at the end of this chapter.

Exercises

3.4.1. (*Noether normalization over finite fields*).
Let K be a finite field, and let $f \in K[x_1, \ldots, x_n] \setminus K$. Prove that there exist $y_2, \ldots, y_n \in K[x_1, \ldots, x_n]$ such that $K[y_2, \ldots, y_n] \subset K[x_1, \ldots, x_n]/\langle f \rangle$ is finite. For any sufficiently large e one can choose $y_i = x_i - x_1^{e^i}$.
    If f is homogeneous then $y_2, \ldots, y_n$ can be chosen to be homogeneous.

3.4.2. Use Exercise 3.4.1 to prove the Noether normalization theorem 3.4.1 in the case of a finite field, replacing the linear coordinate change $M \cdot y = x$ by a coordinate change of the form $x_i = y_i + h_i(y_1, \ldots, y_{i-1})$, $i = 1, \ldots, n$, $h_1 = 0$.

3.4.3. Write a SINGULAR procedure to compute a Noether normalization over finite fields.

3.4.4. With the notations and assumptions of Theorem 3.4.1 prove that if I is a prime ideal and the characteristic of K is zero, then g_s can be chosen such that $Q(A/I) = Q(K[y_{s+1}, \ldots, y_n])[y_s]/\langle g_s \rangle$.

3.4.5. Compute a general Noether normalization for the ideal

$$I = \langle x^3 + xy - z, \, y^3 - t + z, \, x^2y + xy^2 - u \rangle \subset \mathbb{Q}[x, y, z, t, u].$$

Prove that I is a prime ideal. Check this using SINGULAR. Check whether your Noether normalization has the properties of Exercise 3.4.4.

3.4.6. (*Noether normalization for local rings*).
Let K be a field. Then $f \in K[x_1, \ldots, x_n]_{\langle x_1, \ldots, x_n \rangle}$ is called a *Weierstraß polynomial* of degree s with respect to x_n if $f = x_n^s + a_{s-1}x_n^{s-1} + \cdots + a_0$, with $a_i \in \langle x_1, \ldots, x_{n-1} \rangle \cdot K[x_1, \ldots, x_{n-1}]_{\langle x_1, \ldots, x_{n-1} \rangle}$. Prove that

$$K[x_1, \ldots, x_{n-1}]_{\langle x_1, \ldots, x_{n-1} \rangle} \subset K[x_1, \ldots, x_n]_{\langle x_1, \ldots, x_n \rangle}/\langle g \rangle$$

is finite and injective for $g \in K[x_1, \dots, x_n]_{\langle x_1, \dots, x_n \rangle}$ if and only if $u \cdot g$ is a Weierstraß polynomial of degree ≥ 1 with respect to x_n for a suitable unit u.

(Hint: prove that the extension above is finite if and only if the localization $S^{-1}K[x_1, \dots, x_n]/\langle g \rangle$ w.r.t. $S := \{g \in K[x_1, \dots, x_{n-1}] \mid g(0) \neq 0\}$ is a local ring.)

3.4.7. Let $f = x^4 + y^4 + x^3 + y^3 + x^2 + y^2 + x + y \in K[x, y]$. Prove that there exists no linear automorphism $\varphi : K[x, y]_{\langle x, y \rangle} \to K[x, y]_{\langle x, y \rangle}$ such that $\varphi(f)$ is a product of a unit and a Weierstraß polynomial (cf. Exercise 3.4.6). This proves that, in general, Noether normalization as in Theorem 3.4.1 does not hold for localizations of polynomial rings.

(Hint: use the fact that the polynomial ring is a unique factorization domain and that f is irreducible to prove that $\varphi(f) = ug$, u a unit, g a Weierstraß polynomial, implies that $\varphi(f)$ is already a Weierstraß polynomial.)

3.4.8. Formulate and prove Theorem 3.4.1 for ideals generated by homogeneous polynomials in $K[x_1, \dots, x_n]_{\langle x_1, \dots, x_n \rangle}$.

3.5 Applications

In this section we shall use the Noether normalization to develop the dimension theory for the polynomial ring $K[x_1, \dots, x_n]$ and, more generally, affine algebras $K[x_1, \dots, x_n]/I$. We shall prove Hilbert's Nullstellensatz and give an algorithm to compute the dimension of an affine algebra. Finally, we prove that the normalization of an affine algebra R, being an integral domain, is finite over R and, therefore, again an affine algebra.

Theorem 3.5.1. *Let K be a field and $A = K[x]$, $x = \{x_1, \dots, x_n\}$. Then*

(1) $\dim(A) = n$, *moreover, all maximal chains in $\mathcal{C}(A)$ have length n.*

(2) *If $f \in A$, $\deg(f) \geq 1$, then $\dim(A/\langle f \rangle) = n - 1$ (Krull's principal ideal theorem).*

(3) *If $P \subset A$ is a prime ideal then $\operatorname{ht}(P) + \dim(A/P) = \dim(A) = n$.*

(4) *If $P \subset A$ is a prime ideal then $\dim(A/P) = \operatorname{trdeg}_K Q(A/P)$, the transcendence degree of the field extension $K \subset Q(A/P)$. Moreover, all maximal chains in $\mathcal{C}(A/P)$ have the length $\dim(A/P)$.*

(5) *If $M \subset A$ is a maximal ideal, then $A/M \supset K$ is finite (Hilbert's Nullstellensatz)[4].*

(6) *Let $I \subset A$ be an ideal and $u \subset x$ be a subset such that $I \cap K[u] = 0$, then $\dim(A/I) \geq \#u$. Furthermore, there exists some $u \subset x$ with $I \cap K[u] = 0$ and $\dim(A/I) = \#u$.[5]*

---

[4] This is a weak form of Hilbert's Nullstellensatz. For K algebraically closed we obtain $A/M = K$, hence, the maximal ideals are of type $\langle x_1 - a_1, \dots, x_n - a_n \rangle$.

[5] Note that u is allowed to be empty, that is, $\#u = 0$.

(7) *Let $I \subsetneq A$ be an ideal, and let S be a standard basis of I with respect to any global ordering $>$ on $\mathrm{Mon}(x_1, \ldots, x_n)$. Then $\dim(I) = 0$ if and only if $L(I)$ contains suitable powers of each variable x_i, $i = 1, \ldots, n$. This is the case if and only if S contains, for each variable x_i, an element whose leading monomial is $x_i^{a_i}$ for some a_i.*

Proof. We use induction on n, the case $n = 0$ being trivial. To prove (1), let $\langle 0 \rangle = P_0 \subsetneq \cdots \subsetneq P_m \subsetneq A$ be a maximal chain in $C(A)$. Choose an irreducible $f \in P_1$ and coordinates $y_1, \ldots, y_{n-1}$ (Theorem 3.4.1 and Exercise 3.4.2) such that $K[x_1, \ldots, x_n]/\langle f \rangle \supset K[y_1, \ldots, y_{n-1}]$ is finite. Then, clearly, the chain

$$\langle 0 \rangle = P_1/\langle f \rangle \subsetneq P_1/\langle f \rangle \subsetneq \cdots \subsetneq P_m/\langle f \rangle$$

is maximal, too. Using Lemma 3.3.14 and the Noether normalization theorem, this chain induces a maximal chain in $K[y_1, \ldots, y_{n-1}]$ which, due to the induction hypothesis, has length $n - 1$.

(3), (2) are immediate consequences of (1), respectively its proof. To prove (4), we may assume, again, that $A/P \supset K[y_1, \ldots, y_s]$ is finite. Then Corollary 3.3.3 implies $\dim(A/P) = \dim(K[y_1, \ldots, y_s])$, which equals s due to (1). On the other hand, $\mathrm{trdeg}_K Q(A/P) = \mathrm{trdeg}_K Q(K[y_1, \ldots, y_s]) = s$.

Moreover, each maximal chain $\bar{P}_m \supsetneq \cdots \supsetneq \bar{P}_0 = \langle 0 \rangle$ of primes in A/P lifts to a maximal chain $P_m \supsetneq \cdots \supsetneq P_0 = P \supsetneq \cdots \supsetneq \langle 0 \rangle$ of prime ideals in A.

(5) is a consequence of Theorem 3.4.1 and the fact that M is maximal: if $K[y_1, \ldots, y_s] \subset A/M$ is finite, then, because A/M is a field and by Lemma 3.1.9, $K[y_1, \ldots, y_s]$ is also a field, hence, $s = 0$.

(6) Let $u \subset x$ be a subset such that $I \cap K[u] = \langle 0 \rangle$. In particular, we have $\sqrt{I} \cap K[u] = \langle 0 \rangle$, hence, $\bigcap_{P \in \mathrm{minAss}(I)} (P \cap K[u]) = \langle 0 \rangle$ (Proposition 3.3.5). This implies $P \cap K[u] = \langle 0 \rangle$ for some $P \in \mathrm{minAss}(I)$ (Lemma 1.3.12) and, therefore, $K(u) \subset Q(K[x]/P)$. We obtain

$$\dim(K[x]/I) \geq \dim(K[x]/P) = \mathrm{trdeg}_K Q(K[x]/P) \geq \#u.$$

Now let $P \in \mathrm{minAss}(I)$ with $d = \dim(K[x]/I) = \dim(K[x]/P)$. Then, due to (4), we may choose $x_{i_1}, \ldots, x_{i_d}$ being algebraically independent modulo P. Then $P \cap K[u] = \langle 0 \rangle$ for $u := \{x_{i_1}, \ldots, x_{i_d}\}$ and, therefore, $I \cap K[u] = \langle 0 \rangle$.

(7) We use (6) to see that $I \cap K[x_i] \neq \langle 0 \rangle$ for all i because I is zero–dimensional. Let $f \in I \cap K[x_i]$, $f \neq 0$, then $\mathrm{LM}(f) = x_i^{a_i}$ for a suitable $a_i > 0$ (f is not constant and $>$ is a well–ordering). By definition of a standard basis there exist $g \in S$ and $\mathrm{LM}(g) \mid \mathrm{LM}(f)$. This proves the "only if"–direction.

For the "if"–direction, we show that under our assumption on S, $K[x]/I$ is, indeed, a finite dimensional K–vector space. Let $p \in K[x]$ be any polynomial, and consider $\mathrm{NF}(p \mid S)$, the reduced normal form of p with respect to S. Then, clearly $\mathrm{NF}(p \mid S) = \sum_\beta c_\beta x^\beta$, where $c_\beta \neq 0$ implies $\beta_i < a_i$ for all i. In particular, the images of the monomials x^β with $\beta_i < a_i$ for all i generate $K[x]/I$ as K–vector space. □

Theorem 3.5.2 (Hilbert's Nullstellensatz). *Assume that $K = \overline{K}$ is an algebraically closed field. Let $I \subset K[x] = K[x_1, \ldots, x_n]$ be an ideal, and let*

$$V(I) = \left\{ x \in K^n \mid f(x) = 0 \text{ for all } f \in I \right\}.$$

If, for some $g \in K[x]$, $g(x) = 0$ for all $x \in V(I)$ then $g \in \sqrt{I}$.

Proof. We consider the ideal $J := IK[x, t] + \langle 1 - tg \rangle$ in the polynomial ring $K[x, t] = K[x_1, \ldots, x_n, t]$.

If $J = K[x, t]$ then there exist $g_1, \ldots, g_s \in I$ and $h, h_1, \ldots, h_s \in K[x, t]$ such that $1 = \sum_{i=1}^s g_i h_i + h(1 - tg)$. Setting $t := \frac{1}{g} \in K[x]_g$, this implies

$$1 = \sum_{i=1}^{s} g_i \cdot h_i \left(x, \frac{1}{g} \right) \in K[x]_g.$$

Clearing denominators, we obtain $g^\rho = \sum_i g_i h_i'$ for some $\rho > 0$, $h_i' \in K[x]$, and, therefore, $g \in \sqrt{I}$.

Now assume that $J \subsetneq K[x, t]$. We choose a maximal ideal $M \subset K[x, t]$ such that $J \subset M$. Using Theorem 3.5.1 (5) we know (since K is algebraically closed) that $K[x, t]/M \cong K$, and, hence, $M = \langle x_1 - a_1, \ldots, x_n - a_n, t - a \rangle$, for some $a_i, a \in K$. Now $J \subset M$ implies $(a_1, \ldots, a_n, a) \in V(J)$.

If $(a_1, \ldots, a_n) \in V(I)$ then $g(a_1, \ldots, a_n) = 0$. Hence, $1 - tg \in J$ does not vanish at $(a_1, \ldots, a_n)$, contradicting the assumption $(a_1, \ldots, a_n, a) \in V(J)$. If $(a_1, \ldots, a_n) \notin V(I)$ then there is some $h \in I$ such that $h(a_1, \ldots, a_n) \neq 0$, in particular (as h does not depend on t) $h(a_1, \ldots, a_n, a) \neq 0$ and, therefore, $(a_1, \ldots, a_n, a) \notin V(J)$, again contradicting our assumption. □

Definition 3.5.3. Let $I \subset K[x_1, \ldots, x_n]$ be an ideal. Then a subset

$$u \subset x = \{x_1, \ldots, x_n\}$$

is called an *independent set* (with respect to I) if $I \cap K[u] = 0$. An independent set $u \subset x$ (with respect to I) is called *maximal* if $\dim(K[x]/I) = \#u$.

Example 3.5.4. Let $I = \langle xz, yz \rangle \subset K[x, y, z]$, then $\{x, y\} \subset \{x, y, z\}$ is a maximal independent set. Notice that $\{z\} \subset \{x, y, z\}$ is independent and non–extendable (that is, cannot be enlarged) but it is not a maximal independent set.

Note that all maximal (resp. all non–extendable) independent sets of the leading ideal $L(I)$ are computed by the SINGULAR commands `indepSet(std(I))` (respectively by `indepSet(std(I),1)`). Thus, using these commands, we obtain independent sets of I but maybe not all. Exercises 3.5.1 and 3.5.2 show how to compute independent sets.

SINGULAR Example 3.5.5 (independent set).

```
ring R=0,(x,y,z),dp;
ideal I=yz,xz;
indepSet(std(I));
//-> 1,1,0
```

This means, $\{x, y\}$ is a maximal independent set for I.

```
indepSet(std(I),1);
//-> [1]:         [2]:
          1,1,0        0,0,1
```

This means, the only independent sets which cannot be enlarged are $\{x, y\}$ and $\{z\}$.

The geometrical meaning of $u \subset x$ being an independent set for I is that the projection of $V(I)$ to the affine space of the variables in u is surjective, since $V(I \cap K[u]) = V(\langle 0 \rangle)$ is the whole affine space.

```
ring A  = 0,(x,y,t),dp;
ideal I = y2-x3-3t2x2;
indepSet(std(I),1);
//-> [1]:         [2]:
          1,1,0        0,1,1
```

Hence, $\{x, y\}$ and $\{y, t\}$ are the only non–extendable independent sets for the leading ideal $L(I) = \langle t^2 x^2 \rangle$. However, the ideal I itself has, additionally, $\{x, t\}$ as a non–extendable independent set. The difference is seen in the pictures in Figure 3.6, which are generated by the following SINGULAR session:

```
LIB"surf.lib";
plot(lead(I),"clip=cube;");
plot(I,"rot_x=1.4; rot_y=3.0; rot_z=1.44;"); //see Fig. 3.6
```

The first surface is $V(L(I))$ and the second $V(I)$. The projection of $V(L(I))$ to the $\{x, t\}$–plane is not dominant, but the projection of $V(I)$ is.

Next we want to compute the dimension of monomial ideals.

Definition 3.5.6. Let $I = \langle m_1, \ldots, m_s \rangle \subset K[x] = K[x_1, \ldots, x_n]$ be a monomial ideal (with $m_i \in \mathrm{Mon}(x_1, \ldots, x_n)$ for $i = 1, \ldots, s$). Then we define an integer $d(I, K[x])$ by the recursive formula: $d(\langle 0 \rangle, K[x]) := n$ and

$$d(I, K[x]) := \max \left\{ d\left(I\big|_{x_i=0}, K[x \smallsetminus x_i]\right) \,\middle|\, x_i \text{ divides } m_1 \right\},$$

where $x \smallsetminus x_i = (x_1, \ldots, x_{i-1}, x_{i+1}, \ldots, x_n)$.

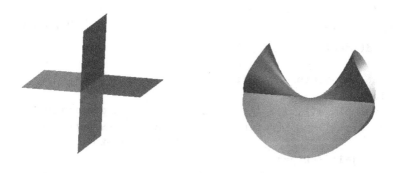

Fig. 3.6. The zero–sets of t^2x^2, respectively $y^2 - x^3 - 3t^2x^2$.

Example 3.5.7. Let $I = \langle xz, yz \rangle \subset K[x, y, z]$ then

$$d(I, K[x, y, z]) = \max\Big\{\underbrace{d(\langle yz \rangle, K[y, z])}_{=1}, \underbrace{d(\langle 0 \rangle, K[x, y])}_{=2}\Big\} = 2.$$

Proposition 3.5.8. *Let $I = \langle m_1, \ldots, m_s \rangle \subset K[x]$ be a monomial ideal, then*

$$\dim(K[x]/I) = d(I, K[x]).$$

Proof. Let $P \supset I$ be a prime ideal, then for all i one factor of m_i has to be in P. In particular, for every $P \in \mathrm{minAss}(I)$ there exists some ρ, such that $x_\rho \in P$ divides m_1. In particular, we have $I|_{x_\rho=0} \subset P|_{x_\rho=0} \subset K[x \smallsetminus x_\rho]$. Using the induction hypothesis we may assume that

$$d\Big(I\big|_{x_\rho=0}, K[x \smallsetminus x_\rho]\Big) = \dim\Big(K[x \smallsetminus x_\rho]/(I\big|_{x_\rho=0})\Big)$$

$$\geq \dim\Big(K[x \smallsetminus x_\rho]/(P\big|_{x_\rho=0})\Big) = \dim(K[x]/P).$$

This implies that $d(I, K[x]) \geq \max_{P \in \mathrm{minAss}(I)} \dim(K[x]/P) = \dim(K[x]/I)$. Let us assume that $d(I, K[x]) > \dim(K[x]/I)$. Then there would exist some i such that x_i divides m_1 and

$$\dim(K[x]/I) < d\Big(I\big|_{x_i=0}, K[x \smallsetminus x_i]\Big) = \dim\Big(K[x \smallsetminus x_i]/(I\big|_{x_i=0})\Big),$$

the latter equality being implied by the induction hypothesis. But

$$\dim\Big(K[x \smallsetminus x_i]/(I\big|_{x_i=0})\Big) = \dim(K[x]/\langle I, x_i \rangle) \leq \dim(K[x]/I),$$

whence a contradiction. $\qquad\square$

SINGULAR Example 3.5.9 (computation of $d(I, K[x])$).
We give a procedure to compute the function $d(I, K[x])$ of Definition 3.5.6:

```
proc d(ideal I)
{
    int n=nvars(basering);
    int j,b,a;
    I=simplify(I,2);  //cancels zeros in the generators of I
    if(size(I)==0) {return(n);}  //size counts generators
                                 //not equal to 0
    for(j=1;j<=n;j++)
    {
        if(I[1]/var(j)!=0)
        {
            a=d(subst(I,var(j),0))-1;
            //we need -1 here because we stay in the basering
            if(a>b) {b=a;}
        }
    }
    return(b);
}
```

Let us test the procedure:

```
ring R=0,(x,y,z),dp;
ideal I=yz,xz;

d(I);
//-> 2
dim(std(I));
//-> 2
```

We shall prove later that for any ideal $I \subset K[x]$,

$$\dim(K[x]/I) = \dim\big(K[x]/L(I)\big) .$$

Hence, $\dim(K[x]/I) = d\big(L(I), K[x]\big)$ is very easy to compute once we know generators for $L(I)$, which are the leading terms of a standard basis of I.

Now we prove the finiteness of the normalization.

Theorem 3.5.10 (E. Noether). *Let $P \subset K[x_1, \ldots, x_n]$ be a prime ideal, and let $A = K[x_1, \ldots, x_n]/P$, then the normalization $\overline{A} \supset A$ is a finite A–module.*

Remark 3.5.11. In general, that is, for an arbitrary Noetherian integral domain, Theorem 3.5.10 is incorrect, as discovered by Nagata [183, Ex. 5, p. 207]. The polynomial ring $K[x_1, \ldots, x_n]$ and, more generally, each affine algebra $R = K[x_1, \ldots, x_n]/I$ satisfy the following stronger[6] condition: for each

---

[6] Strictly speaking, this is only a stronger condition if K has characteristic $p > 0$.

prime ideal $P \subset R$ and for each finite extension field L of $Q(R/P)$, the integral closure of R/P in L is a finite R/P–module. A Noetherian ring with this property is called *universally Japanese* (in honour of Nagata).

To prove Theorem 3.5.10 we need an additional lemma. We shall give a proof for the case that K is a perfect field (for example, $\mathrm{char}(K) = 0$, or K is finite, or K is algebraically closed, cf. Exercise 1.1.6). For a proof in the general case, see [66, Corollary 13.15].

Lemma 3.5.12. *Let A be a normal Noetherian integral domain, $L \supset Q(A)$ a finite separable field extension and B the integral closure of A in L. Let $\alpha \in B$ be a primitive element of the field extension, F the minimal polynomial of α and Δ the discriminant of F.[7] Then $B \subset \frac{1}{\Delta} A[\alpha]$. In particular, B is a finite A–module.*

Proof of Theorem 3.5.10. We use the Noether normalization theorem and choose $y_1, \ldots, y_s$ such that $K[y_1, \ldots, y_s] \hookrightarrow K[x_1, \ldots, x_n]/P$ is finite. We obtain a commutative diagram

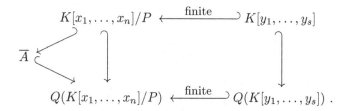

Notice that $\overline{A}$ is also the integral closure of $K[y_1, \ldots, y_s]$ in the quotient field $Q(K[x_1, \ldots, x_n]/P)$, which is a finite separable extension of $Q(K[y_1, \ldots, y_s])$. Since $K[y_1, \ldots, y_s]$ is a normal Noetherian integral domain, we obtain the assumption of Lemma 3.5.12, which proves the theorem. $\square$

Proof of Lemma 3.5.12. Let L_0 be the splitting field of F, and let $\alpha = \alpha_1$, $\alpha_2, \ldots, \alpha_n \in L_0$ be the roots of F. Further, let B_0 be the integral closure of A in L_0. Then $\alpha_1, \ldots, \alpha_n \in B_0$ (since F is monic), and we have the following diagram

$$
\begin{array}{ccccc}
A & \subset & B & \subset & B_0 \\
\cap & & \cap & & \cap \\
Q(A) & \subset & L & \subset & L_0,
\end{array}
$$

where $L_0 \supset Q(A)$ is Galois. We consider the matrix

---

[7] The *discriminant* of a univariate polynomial $F \in K[x]$ is defined to be the resultant of F and its derivative F'. If $\alpha_1, \ldots, \alpha_n$ are the roots of F in the algebraic closure $\overline{K}$ of K then the discriminant equals $\prod_{i \neq j}(\alpha_i - \alpha_j)$, see, e.g., [162].

$$M := \begin{pmatrix} 1 & \alpha_1 & \dots & \alpha_1^{n-1} \\ \vdots & \vdots & & \vdots \\ 1 & \alpha_n & \dots & \alpha_n^{n-1} \end{pmatrix} \in \mathrm{Mat}(n \times n, B_0).$$

For $d := \det(M) = \pm \prod_{i<j}(\alpha_i - \alpha_j) \neq 0$ we have $d^2 \in Q(A)$, because d^2 is the discriminant of F.

Let $\mathrm{Gal}(L_0 \mid Q(A)) = \{\sigma_1, \dots, \sigma_n\}$ with $\sigma_i(\alpha_1) = \alpha_i$. Let $b \in B$, then there exist $c_i \in Q(A)$ such that $b = \sum_{i=0}^{n-1} c_i \alpha_1^i$. We obtain

$$M \cdot \begin{pmatrix} c_0 \\ \vdots \\ c_{n-1} \end{pmatrix} = \begin{pmatrix} \sum_i c_i \alpha_1^i \\ \vdots \\ \sum_i c_i \alpha_n^i \end{pmatrix} = \begin{pmatrix} \sigma_1(b) \\ \vdots \\ \sigma_n(b) \end{pmatrix} \in B_0^n$$

and, consequently,

$$dM^{-1}M \cdot \begin{pmatrix} c_0 \\ \vdots \\ c_{n-1} \end{pmatrix} = \begin{pmatrix} dc_0 \\ \vdots \\ dc_{n-1} \end{pmatrix} \in B_0^n,$$

which yields $B \subset (1/d^2) \cdot \sum_{i=0}^{n-1} A\alpha^i$, because $d^2 c_i \in B_0 \cap Q(A) = A$. This implies, in particular, that B is a finite A–module. $\qquad\square$

Lemma 3.5.13. *Let A and B be rings of finite type over a field K. Assume $A \subset B$ is finite, and let $P \subset A$ be a prime ideal. Moreover, let $PB \subset Q$ be a prime ideal with $\dim B/Q = \dim B/PB$, then $Q \cap A = P$.*

Proof. We may assume $P = \langle 0 \rangle$ and use Corollary 3.3.3. $\qquad\square$

Note that the assumption $\dim B/Q = \dim B/PB$ is important as the following example shows: $B := K[x, y, z]/\langle x^2, y \rangle \cap \langle x + z \rangle^2 \subset A := K[y, z]$ and $P = \langle 0 \rangle$.

Remark 3.5.14. Lemma 3.5.13 is a basis for computing the prime ideals occurring in the going up, lying over and going down theorems:

- If $P \subset A$ is prime, compute the minimal associated primes of PB and choose one with the correct dimension to obtain the required prime ideal of the lying over theorem.
- For the going up theorem we do the same: for given $Q \subset B$, $P = Q \cap A$ and $P \subset P'$ we choose a $Q' \in \mathrm{minAss}(P'B)$ such that $\dim B/Q' = \dim B/P'B$ and $Q \subset Q'$.
- For the going down theorem we choose for given $Q \subset B$, $P = Q \cap A$ and $P' \subset P$ a $Q' \in \mathrm{minAss}(P'B)$ such that $\dim B/Q' = \dim B/P'B$ and $Q' \subset Q$.

SINGULAR Example 3.5.15 (lying over theorem).
Let $A = \mathbb{Q}[x, y]$, $P = \langle x \rangle$ and $B = \mathbb{Q}[x, y, z]/\langle z^2 - xz - 1 \rangle$. We want to find
a prime ideal $Q \supset PB$ such that $Q \cap A = P$.

```
LIB"primdec.lib";
ring B=0,(x,y,z),dp;
ideal PB=x,z2-xz-1;
list pr=minAssGTZ(PB);
pr;
[1]:            [2]:
  _[1]=z-1       _[1]=z+1
  _[2]=x         _[2]=x
```

Both prime ideals $\langle z - 1, x \rangle$, $\langle z + 1, x \rangle$ in B give as intersection with A the
ideal P.

Exercises

3.5.1. Let $I \subset K[x_1, \ldots, x_n]$ be an ideal and $>$ an ordering of the set of
monomials $\mathrm{Mon}(x_1, \ldots, x_n)$. Let $u \subset x = \{x_1, \ldots, x_n\}$ be an independent set
for $L(I)$. Prove that u is an independent set for I. Use Remark 3.3.12 to see
that a maximal independent set for $L(I)$ is also a maximal independent set
for I.

3.5.2. Modify the procedure of Example 3.5.9 to compute a maximal independent set for I.

3.5.3. Use similar methods as in Example 3.5.15 to test the going down theorem in the following case: $A = Q[x, y]$, $B = \mathbb{Q}[x, y, z]/\langle z^2 - xy \rangle$, $Q = \langle x, y, z \rangle$,
$P = \langle x, y \rangle$, $P' = \langle y^2 - x^3 \rangle$.

3.5.4. Let K be a field, and let $P \subset K[x_1, \ldots, x_n]$ be a prime ideal. Moreover,
let $u \subset x = \{x_1, \ldots, x_n\}$ be a maximal independent set for P. Prove that
$K[x]_P = K(u)[x \smallsetminus u]_{PK(u)[x \smallsetminus u]}$.

3.5.5. Let K be a field, and let $P \subset K[x_1, \ldots, x_n]$ be a prime ideal. Moreover, let $u \subset x = \{x_1, \ldots, x_n\}$ be a maximal independent set for P. Prove
that $K(u)[x \smallsetminus u]/PK(u)[x \smallsetminus u] = Q(K[x_1, \ldots, x_n]/P)$. Note that, because
of SINGULAR Example 1.3.13 (4), we can use SINGULAR to compute in this
field. Use Exercise 3.4.4 to prove that, after a generic linear coordinate change
φ (in case that K is of characteristic 0) for a suitable $x_i \in x \smallsetminus u$ and $f \in \varphi(P)$,
$Q(K[x_1, \ldots, x_n]/(P)) = K(u)[x_i]/\langle f \rangle$.

3.5.6. Let A be a normal integral domain, K its quotient field, and $L \supset K$ be
a field extension. Let $x \in L$ be integral over A and $f \in K[t]$ be the minimal
polynomial of x. Then $f \in A[t]$.

3.6 An Algorithm to Compute the Normalization

At the end of this chapter we give a criterion for normality which is the basis for an algorithm to compute the normalization and the non–normal locus for affine K–algebras.

In this section let A be a reduced Noetherian ring and $\overline{A}$ its normalization, that is, the integral closure of A in $Q(A)$, the total ring of fractions of A. A is normal if and only if $A = \overline{A}$. The idea to compute $\overline{A}$ is to enlarge the ring A by an endomorphism ring $A' = \operatorname{Hom}_A(J, J)$ for a suitable ideal J such that $A' \subset \overline{A}$. Repeating the procedure with A' instead of A, we continue until the normality criterion allows us to stop.

Lemma 3.6.1. Let A be a reduced Noetherian ring and $J \subset A$ an ideal containing a non–zerodivisor x of A. Then there are natural inclusions of rings

$$A \subset \operatorname{Hom}_A(J, J) \cong \frac{1}{x} \cdot (xJ : J) \subset \overline{A}.$$

Proof. For $a \in A$, let $m_a : J \to J$ denote the multiplication with a. If $m_a = 0$, then $m_a(x) = ax = 0$ and, hence, $a = 0$, since x is a non–zerodivisor. Thus, $a \mapsto m_a$ defines an inclusion $A \subset \operatorname{Hom}_A(J, J)$.

It is easy to see that for $\varphi \in \operatorname{Hom}_A(J, J)$ the element $\varphi(x)/x \in Q(A)$ is independent of x: for any $a \in J$ we have $x \cdot \varphi(a) = \varphi(xa) = a \cdot \varphi(x)$, since φ is A–linear.

Hence, $\varphi \mapsto \varphi(x)/x$ defines an inclusion $\operatorname{Hom}_A(J, J) \subset Q(A)$ mapping $x \cdot \operatorname{Hom}_A(J, J)$ into $xJ : J = \{b \in A \mid bJ \subset xJ\}$. The latter map is also surjective, since any $b \in xJ : J$ defines, via multiplication with b/x, an element $\varphi \in \operatorname{Hom}_A(J, J)$ with $\varphi(x) = b$. Since x is a non–zerodivisor, we obtain the isomorphism $\operatorname{Hom}_A(J, J) \cong (1/x) \cdot (xJ : J)$.

It follows from Proposition 3.1.2 that any $b \in xJ : J$ satisfies an integral relation $b^p + a_1 b^{p-1} + \cdots + a_0 = 0$ with $a_i \in \langle x^i \rangle$. Hence, b/x is integral over A, showing $(1/x) \cdot (xJ : J) \subset \overline{A}$. $\square$

By Proposition 3.2.5, A is normal if and only if A_P is normal for each prime ideal $P \subset A$.

Definition 3.6.2. The *non–normal locus* of A is defined as

$$N(A) = \{P \in \operatorname{Spec} A \mid A_P \text{ is not normal}\}.$$

Lemma 3.6.3. Let $\overline{A}$ be finitely generated over A, and let $C = \operatorname{Ann}_A(\overline{A}/A) = \{a \in A \mid a\overline{A} \subset A\}$ be the conductor of A in $\overline{A}$. Then

$$N(A) = V(C) = \{P \in \operatorname{Spec} A \mid P \supset C\}.$$

In particular, $N(A)$ is closed[8] in $\operatorname{Spec} A$.

---

[8] In general, if $\overline{A}$ is not finite over A, the proof of Lemma 3.6.3 shows that $\bigcup_{h \in \overline{A}} V(C_h) \subset N(A) \subset V(C)$. If $\overline{A}$ is finite over A, generated by $h_1, \ldots, h_s$, then $\bigcup_{h \in \overline{A}} V(C_h) = \bigcup_{i=1}^s V(C_{h_i}) = V(C)$.

Proof. If $P \in N(A)$ then $A_P \neq \overline{A}_P$, hence, $C_P = \mathrm{Ann}_{A_P}(\overline{A}_P/A_P) \subsetneq A_P$ and, therefore, $P \supset C$. To show the converse inclusion, note that $C = \bigcap_{h \in \overline{A}} C_h$, where $C_h := \{a \in A \mid ah \in A\}$. Thus, we only need to show $V(C_h) \subset N(A)$: let $P \notin N(A)$, then $A_P = \overline{A}_P$ and, hence, $h = p/q$ for suitable $p, q \in A$, $q \notin P$. This implies $qh \in A$, that is, $q \in C_h$. Therefore, $C_h \not\subset P$ and $P \notin V(C_h)$. $\square$

Lemma 3.6.4. *Let $J \subset A$ be an ideal containing a non–zerodivisor of A.*

(1) There are natural inclusions of A–modules

$$\mathrm{Hom}_A(J, J) \subset \mathrm{Hom}_A(J, A) \cap \overline{A} \subset \mathrm{Hom}_A(J, \sqrt{J}).$$

(2) If $N(A) \subset V(J)$ then $J^d \overline{A} \subset A$ for some d.

Proof. (1) The embedding of $\mathrm{Hom}_A(J, A)$ in $Q(A)$ is given by $\varphi \mapsto \varphi(x)/x$, where x is a non–zerodivisor of J (cf. proof of Lemma 3.6.1). With this identification we obtain

$$\mathrm{Hom}_A(J, A) = A :_{Q(A)} J = \{h \in Q(A) \mid hJ \subset A\}$$

and $\mathrm{Hom}_A(J, J)$, respectively $\mathrm{Hom}_A(J, \sqrt{J})$, is identified with those $h \in Q(A)$ such that $hJ \subset J$, respectively $hJ \subset \sqrt{J}$. Then the first inclusion follows from Lemma 3.6.1.

For the second inclusion let $h \in \overline{A}$ satisfy $hJ \subset A$. Consider an integral relation $h^n + a_1 h^{n-1} + \cdots + a_n = 0$ with $a_i \in A$. Let $g \in J$ and multiply the above equation with g^n. Then

$$(hg)^n + ga_1(hg)^{n-1} + \cdots + g^n a_n = 0.$$

Since $g \in J$, $hg \in A$ and, therefore, $(hg)^n \in J$ and $hg \in \sqrt{J}$. This shows the second inclusion.

(2) By assumption, and by Lemma 3.6.3, we have $V(C) \subset V(J)$ and, hence, $J \subset \sqrt{C}$ by Theorem A.3.4, that is, $J^d \subset C$ for some d which implies the claim. $\square$

The following criterion for normality is due to Grauert and Remmert [99].

Proposition 3.6.5 (Criterion for normality). *Let A be a Noetherian reduced ring and $J \subset A$ an ideal satisfying*

(1) J contains a non–zerodivisor of A,
(2) J is a radical ideal,
(3) $N(A) \subset V(J)$.

Then A is normal if and only if $A = \mathrm{Hom}_A(J, J)$.

Proof. If $A = \overline{A}$ then $\mathrm{Hom}_A(J, J) = A$, by Lemma 3.6.1. To see the converse, we choose $d \geq 0$ minimal such that $J^d \overline{A} \subset A$ (Lemma 3.6.4 (2)). If $d > 0$ then there exists some $a \in J^{d-1}$ and $h \in \overline{A}$ such that $ah \notin A$. But $ah \in \overline{A}$ and $ah \cdot J \subset hJ^d \subset A$, that is, $ah \in \mathrm{Hom}_A(J, A) \cap \overline{A}$, which is equal to $\mathrm{Hom}_A(J, J)$ by Lemma 3.6.4, since $J = \sqrt{J}$. By assumption $\mathrm{Hom}_A(J, J) = A$ and, hence, $ah \in A$, which is a contradiction. We conclude that $d = 0$ and $A = \overline{A}$. $\qquad\square$

Remark 3.6.6.

(1) An ideal J, as in Proposition 3.6.5, is called a *test ideal for normality*. It is not difficult to see that the conductor ideal C is a test ideal. But C cannot be computed as long as we do not know $\overline{A}$. We shall see that the ideal of the singular locus is, indeed, a test ideal which can be computed.

(2) Let $J \subset A$ be a test ideal, and let $x \in J$ be any non–zero element. Then $N(A) \subset V(x)$. If x is a non–zerodivisor of A then $\sqrt{\langle x \rangle}$ is a test ideal which is smaller than J and, hence, preferable for computations. Since $\mathrm{Hom}_A(\langle x \rangle, \langle x \rangle) = A$, it also follows that A is normal if $\langle x \rangle = \sqrt{\langle x \rangle}$.

If x is a zerodivisor, then A cannot be normal (Exercise 3.6.7), and we consider the annihilator of x, $\mathrm{Ann}_A(x) = \langle 0 \rangle : \langle x \rangle$. By Exercise 3.6.8, we obtain that $A \subset A/\langle x \rangle \oplus A/\mathrm{Ann}_A(x)$ is a finite extension, and we can continue with the rings $A/\langle x \rangle$ and $A/\mathrm{Ann}_A(x)$.

In case that A is not normal, $A \subsetneq \mathrm{Hom}_A(J, J) =: A'$, and we can continue with A' instead of A. To do this, we need to present A' as an A–algebra of finite type.

The following lemma describes the A–algebra structure of $\mathrm{Hom}_A(J, J)$:

Lemma 3.6.7. *Let A be a reduced Noetherian ring, let $J \subset A$ be an ideal and $x \in J$ a non–zerodivisor. Then*

(1) $A = \mathrm{Hom}_A(J, J)$ if and only if $xJ : J = \langle x \rangle$.

Moreover, let $\{u_0 = x, u_1, \ldots, u_s\}$ be a system of generators for the A–module $xJ : J$. Then we can write

(2) $u_i \cdot u_j = \displaystyle\sum_{k=0}^{s} x\xi_k^{ij} u_k$ with suitable $\xi_k^{ij} \in A$, $1 \leq i \leq j \leq s$.

Let $(\eta_0^{(k)}, \ldots, \eta_s^{(k)}) \in A^{s+1}$, $k = 1, \ldots, m$, generate $\mathrm{syz}(u_0, \ldots, u_s)$, and let $I \subset A[t_1, \ldots, t_s]$ be the ideal

$$I := \left\langle \left\{ t_i t_j - \sum_{k=0}^{s} \xi_k^{ij} t_k \;\middle|\; 1 \leq i \leq j \leq s \right\}, \left\{ \sum_{\nu=0}^{s} \eta_\nu^{(k)} t_\nu \;\middle|\; 1 \leq k \leq m \right\} \right\rangle,$$

where $t_0 := 1$. Then

(3) $t_i \mapsto u_i/x$, $i = 1, \ldots, s$, *defines an isomorphism*

$$A[t_1, \ldots, t_s]/I \xrightarrow{\cong} \mathrm{Hom}_A(J, J) \cong \frac{1}{x} \cdot (xJ : J).$$

Proof. (1) follows immediately from Lemma 3.6.1.

To prove (2), note that $\mathrm{Hom}_A(J, J) = (1/x) \cdot (xJ : J)$ is a ring, which is generated as A–module by $u_0/x, \ldots, u_s/x$. Therefore, there exist $\xi_k^{ij} \in A$ such that $(u_i/x) \cdot (u_j/x) = \sum_{k=0}^{s} \xi_k^{ij} \cdot (u_k/x)$.

(3) Obviously, $I \subset \mathrm{Ker}(\phi)$, where $\phi : A[t_1, \ldots, t_s] \to (1/x) \cdot (xJ : J)$ is the ring map defined by $t_i \mapsto u_i/x$, $i = 1, \ldots, s$. On the other hand, let $h \in \mathrm{Ker}(\phi)$. Then, using the relations $t_i t_j - \sum_{k=0}^{s} \xi_k^{ij} t_k$, $1 \le i \le j \le s$, we can write $h \equiv h_0 + \sum_{i=1}^{s} h_i t_i \bmod I$, for some $h_0, h_1, \ldots, h_s \in A$.

Now $\phi(h) = 0$ implies $h_0 + \sum_{i=1}^{s} h_i \cdot (u_i/x) = 0$, hence, $(h_0, \ldots, h_s)$ is a syzygy of $u_0 = x, u_1, \ldots, u_s$ and, therefore, $h \in I$. □

Example 3.6.8. Let $A := K[x, y]/\langle x^2 - y^3 \rangle$ and $J := \langle x, y \rangle \subset A$. Then $x \in J$ is a non–zerodivisor in A with $xJ : J = x\langle x, y \rangle : \langle x, y \rangle = \langle x, y^2 \rangle$, therefore, $\mathrm{Hom}_A(J, J) = \langle 1, y^2/x \rangle$ (using Lemma 3.6.1). Setting $u_0 := x$, $u_1 := y^2$, we obtain $u_1^2 = y^4 = x^2 y$, that is, $\xi_0^{11} = y$. Hence, we obtain an isomorphism

$$A[t]/\langle t^2 - y, \, xt - y^2, \, yt - x \rangle \xrightarrow{\cong} \mathrm{Hom}_A(J, J).$$

of A–algebras. Note that $A[t]/\langle t^2 - y, \, xt - y^2, \, yt - x \rangle \simeq K[t]$.

Now, using Proposition 3.6.5 and Lemma 3.6.7 we obtain an algorithm to compute the integral closure. We describe the algorithm for the case that $A = K[x_1, \ldots, x_n]/I$ is an integral domain over a field K of characteristic 0, that is, especially I is prime. Let $I = \langle f_1, \ldots, f_m \rangle$ and $r = \dim(A)$. In order to apply Proposition 3.6.5, we need to find an ideal J such that $V(J)$ contains the non–normal locus. We shall see later (cf. Chapter 5) that we can use the *Jacobian ideal* of I, that is, $J = I + $ the ideal generated by the $(n-r)$–minors of the Jacobian matrix $\left(\frac{\partial f_i}{\partial x_j} \right)$.

Now we are prepared to give the normalization algorithm. We restrict ourselves to the case of affine *integral domains* $A = K[x_1, \ldots, x_n]/I$.

Algorithm 3.6.9 (NORMALIZATION(I)).

Input: $I := \langle f_1, \ldots, f_k \rangle \subset K[x]$ a prime ideal, $x = (x_1, \ldots, x_n)$.
Output: A polynomial ring $K[t]$, $t = (t_1, \ldots, t_N)$, a prime ideal $P \subset K[t]$ and
$\pi : K[x] \to K[t]$ such that the induced map $\pi : K[x]/I \to K[t]/P$ is the normalization of $K[x]/I$.

- if $I = \langle 0 \rangle$ then return $(K[x], \langle 0 \rangle, \mathrm{id}_{K[x]})$;
- compute $r := \dim(I)$;

- if we know that the singular locus of I is $V(x_1, \ldots, x_n)$[9]
    $$J := \langle x_1, \ldots, x_n \rangle;$$
  else
    compute $J :=$ the ideal of the $(n-r)$–minors of the Jacobian matrix I;
- $J := \text{RADICAL}(I + J)$;
- choose $a \in J \smallsetminus \{0\}$;
- if $aJ : J = \langle a \rangle$ return $(K[x], I, \text{id}_{K[x]})$;
- compute a generating system $u_0 = a, u_1, \ldots, u_s$ for $aJ : J$;
- compute a generating system $\left\{ (\eta_0^{(1)}, \ldots, \eta_s^{(1)}), \ldots, (\eta_0^{(m)}, \ldots, \eta_s^{(m)}) \right\}$ for the module of syzygies $\text{syz}(u_0, \ldots, u_s) \subset (K[x]/I)^{s+1}$;
- compute ξ_k^{ij} such that $u_i \cdot u_j = \sum_{k=0}^{s} a \cdot \xi_k^{ij} u_k$, $i, j = 1, \ldots s$;
- change ring to $K[x_1, \ldots, x_n, t_1, \ldots, t_s]$, and set (with $t_0 := 1$)
    $$I_1 := \left\langle \{ t_i t_j - \textstyle\sum_{k=0}^{s} \xi_k^{ij} t_k \}_{1 \leq i \leq j \leq s}, \{ \textstyle\sum_{\nu=0}^{s} \eta_\nu^{(k)} t_\nu \}_{1 \leq k \leq m} \right\rangle + IK[x, t];$$
- return $\text{NORMALIZATION}(I_1)$.

Note that I_1 is again a prime ideal, since

$$K[x_1, \ldots, x_n, t_1, \ldots, t_s]/I_1 \cong \text{Hom}_A(J, J) \subset Q(A)$$

is an integral domain.

Correctness of the algorithm follows from Proposition 3.6.5, and termination follows from Theorem 3.5.10. An implementation in the programming language of SINGULAR can be found in Section 3.7.

SINGULAR Example 3.6.10 (normalization).
Let us illustrate the normalization with Whitney's umbrella

```
ring A = 0,(x,y,z),dp;
ideal I = y2-zx2;
LIB "surf.lib";
plot(I,"rot_x=1.45;rot_y=1.36;rot_z=4.5;");

list nor = normal (I);
def R = nor[1]; setring R;
norid;
//-> norid[1]=0
normap;
//-> normap[1]=T(1)   normap[2]=T(1)*T(2)   normap[3]=T(2)^2
```

Hence, the normalization of A/I is $K[T_1, T_2]$ with normalization map $x \mapsto T_1$, $y \mapsto -T_2^2$, $z \mapsto -T_1 T_2$.

Proposition 3.6.5 and Lemma 3.6.7 also give us the possibility to compute the non–normal locus:

---

[9] This is useful information because, in this case, we can avoid computing the minors of the Jacobian matrix and the radical (which can be expensive). The property of being an isolated singularity is kept during the normalization loops.

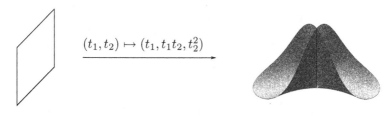

$$(t_1, t_2) \mapsto (t_1, t_1 t_2, t_2^2)$$

Fig. 3.7. The normalization of Whitney's umbrella.

Corollary 3.6.11. *Let A and J be as in Proposition 3.6.5, then the ideal* $\mathrm{Ann}_A\big(\mathrm{Hom}_A(J,J)/A\big) \subset A$ *defines the non–normal locus. Moreover,*

$$\mathrm{Ann}_A\big(\mathrm{Hom}_A(J,J)/A\big) = \langle x \rangle : (xJ : J)$$

for any non–zerodivisor $x \in J$.

We can take the first part of the normalization algorithm to compute the non–normal locus:

Algorithm 3.6.12 (NON–NORMALLOCUS(I)).

Input: $I := \langle f_1, \ldots, f_k \rangle \subset K[x]$ a prime ideal, $x = (x_1, \ldots, x_n)$.
Output: An ideal $NN \subset K[x]$, defining the non–normal locus in $V(I)$.

- If $I = \langle 0 \rangle$ then return $(K[x])$;
- compute $r = \dim(I)$;
- compute J the ideal of the $(n-r)$–minors of the Jacobian matrix of I;
- $J = $ RADICAL $(I + J)$;
- choose $a \in J \smallsetminus \{0\}$;
- return $\big(\langle a \rangle : (aJ : J)\big)$.

We can test this procedure in SINGULAR:

SINGULAR Example 3.6.13 (non–normal locus).
We compute the non–normal locus of $A := K[x,y,z]/\langle zy^2 - zx^3 - x^6 \rangle$.

```
LIB"primdec.lib";
ring A = 0,(x,y,z),dp;
ideal I = zy2-zx3-x6;
ideal sing = I+jacob(I);
ideal J = radical(sing);
qring R = std(I);
ideal J = fetch(A,J);
ideal a = J[1];
ideal re = quotient(a,quotient(a*J,J));
re;
//-> re[1]=y
//-> re[2]=x
```

From the output, we read that the non–normal locus is the z–axis (the zero–set of $\langle x, y \rangle$).

Exercises

3.6.1. Compute the normalization of $\mathbb{Q}[x, y, z]/\langle z(y^3 - x^5) + x^{10} \rangle$.

3.6.2. Let $A \subset B$ be rings with A Noetherian. Use Lemma 3.2.7 to show $C(A, B)[x] = C(A[x], B[x])$. Conclude that $\overline{A}[x] = \overline{A[x]}$; in particular, if A is normal then $A[x]$ is normal.

3.6.3. Prove that the normalization of a graded integral domain is graded.

3.6.4. Let A be a Noetherian ring, $I \subset A$ an ideal, and

$$A[tI] := A + tI + t^2 I^2 + \ldots \subset A[t]$$

the *Rees–Algebra* of I. Show that

$$C(A[tI], A[t]) = \bigoplus_{\nu=0}^{\infty} C_s(I^\nu, A) \cdot t^\nu .$$

In particular, $C_s(I, A)$ is an ideal.

Prove this along the following lines:

- If $f = \sum_{i=0}^{k} f_i t^i$ is integral over $A[tI]$ then each summand $f_i t^i$ is integral over $A[tI]$, too.
- If $C(A[tI]), A[t]) = \bigoplus_{\nu=0}^{\infty} I_\nu t^\nu$ then $I_\nu = C_s(I^\nu, A)$.

The ring $R = A[tI]$ is also called the *blow–up algebra* of I. The arguments above show that the normalization of R is

$$\overline{R} = \overline{A} + t \cdot \overline{I} + t^2 \cdot \overline{I^2} + \ldots$$

where $\overline{I^\nu}$ is the strong integral closure of I^ν in A.

3.6.5. Let $x_1(t), \ldots, x_n(t) \in K[t]$ such that $A = K[x_1(t), \ldots, x_n(t)] \subset K[t]$ is finite and $K\big(x_1(t), \ldots, x_n(t)\big) = K(t)$. Prove that $K[t]$ is the normalization of A.

(Hint: prove that there exists an N such that $t^N K[t] \subset A$, and use this to prove that t is integral over A.)

3.6.6. For $n \in \mathbb{Z}_{\geq 0}$ compute the normalization of $\mathbb{Z}[\sqrt{n}]$.

3.6.7. Let A be a reduced Noetherian ring, and let $x \in A$ a zerodivisor such that $N(A) \subset V(x)$. Prove that A is not normal.

3.6.8. Let A be a reduced Noetherian ring, and let $x \in A$ be a zerodivisor, $I = \langle 0 \rangle : \langle x \rangle$. Prove that $A \subset A/\langle x \rangle \oplus A/I$ is a finite extension.

3.6.9 (Vasconcelos). Let $K \subset L$ be a field extension and consider the ring $A := K[x,y] + \langle x,y \rangle^n L[x,y]$. Prove that the normalization of A is $L[x,y]$. (Hint: prove that $A_x = L[x,y]_x$, $A_y = L[x,y]_y$ and, therefore, the maximal ideal $M = \langle x,y \rangle K[x,y] + \langle x,y \rangle^n L[x,y]$ defines the non–normal locus. Then prove that $\mathrm{Hom}_A(M,M) = K[x,y] + \langle x,y \rangle^{n-1} L[x,y]$.)

3.7 Procedures

The following procedures are in the same style as the SINGULAR procedures in the distributed libraries. Hence, they may be used as a pattern for the reader's own procedures. They are fully working procedures, however, all specialties which are incorporated in the distributed libraries are missing, and, therefore, the performance cannot be expected to be very good.

3.7.1. We start with the Noether normalization using finitenessTest as a sub–procedure. The procedure in the distributed libraries is noetherNormal.

proc NoetherNormal(ideal id)

```
"USAGE:   NoetherNormal(id);  id ideal
 RETURN:   two ideals i1,i2, where i2 is given by a subset of
           the variables and i1 defines a map:
           map phi=basering,i1 such that
           k[i2] --> k[var(1),...,var(n)]/phi(id)
           is a Noether normalization
 "
 {
    def r=basering;
    int n=nvars(r);
 //----- change to lexicographical ordering ------------
    //a procedure from ring.lib changing the order to lp
    //creating a new basering s
    changeord("s","lp");
 //----- make a random coordinate change ---------------
    //creating lower triangular random generators for the
    //maximal ideal a procedure form random.lib
    ideal m=
    ideal(sparsetriag(n,n,0,100)*transpose(maxideal(1)));

    map phi=r,m;
    ideal i=std(phi(id));
 //---------- check finiteness -------------------------
    //from theoretical point of view Noether normalization
    //should be o.k. but we need a test whether the
```

```
//coordinate change was random enough
list l=finitenessTest(i);

setring r;
list l=imap(s,l);

if(l[1]==1)
{
   //the good case, coordinate change was random enough
   return(list(fetch(s,m),l[2]));
}
kill s;
//-------- the bad case, try again --------------------
return(NoetherNormal(i));
}
```

Example 3.7.1.

```
LIB"ring.lib";
LIB"random.lib";
ring R=0,(x,y,z),dp;
ideal I = xy,xz;
NoetherNormal(I);
//-> [1]:                    [2]:
        _[1]=x                  _[1]=0
        _[2]=14x+y              _[2]=y
        _[3]=64x+100y+z         _[3]=z
```

3.7.2. The following procedure uses Proposition 3.1.5 for testing finiteness. Since the input is assumed to be a reduced lexicographical standard basis, this test is a simple inspection of the leading exponents.

proc finitenessTest(ideal I)

```
"USAGE:  finitenessTest(ideal I)
RETURN: A list l, l[1] is 1 or 0 and l[2] is an ideal gener-
        ated by a subset of the variables. l[1]=1 if the map
        basering/I <-- K[l[2]] is finite and 0 else.
NOTE:   It is assumed that I is a reduced standard basis
        with respect to the lexicographical ordering lp,
        sorted w.r.t. increasing leading terms.
"
{
   intvec w=leadexp(I[1]);
   int j,t;
   int s=1;
   ideal k;
```

```
// --------- check leading exponents ---------------------
//compute s such that lead(I[1]) depends only on
//var(s),...,var(n) by inspection of the leading exponents
while (w[s]==0) {s++;}
for (j=1; j<= size(I); j++)
{
  w=leadexp(I[j]);
  if (size(ideal(w))==1) {t++;}
}
//-------------check finiteness -----------------------
//t is the number of elements of the standard basis which
//have pure powers in the variables var(1),...,var(s) as
//leading term. The map is finite iff s=t.
if(s!=t) {return(list(0,k));}
for (j=s+1; j<= nvars(basering);j++)
{
  k[j]=var(j);
}
return (list(1,k));
}
```

Example 3.7.2.

```
ring R=0,(x,y,z),lp;
ideal I = y2+z3,x3+xyz;
finitenessTest(I);
//-> [1]:     [2]:
//->    1        _[1]=0
//->             _[2]=0
//->             _[3]=z
```

3.7.3. We provide a procedure for computing the normalization in the most simple case of a prime ideal. The general procedure in the distributed libraries is called `normal`.

proc normalization(ideal i)

```
"USAGE:  list L=normalization(i); i prime ideal
 RETURN:  a list L of one ring L[1]=R; R contains the ideal
          norid such that R/norid is the normalization of
          basering/i.
 NOTE:    to use the ring type def S=L[1];setring S;norid;
 "
{
   def BAS=basering;
   list result;
```

```
ideal rf;
int   ds = -1;
int isIso;

if (typeof(attrib(i,"isIsolatedSingularity"))=="int")
{
  if(attrib(i,"isIsolatedSingularity")==1){isIso=1;}
}

if(size(i)!=0)
{
  list SM=mstd(i);
  i=SM[2];
  ideal SBi=SM[1];

  int n=nvars(BAS);
  int d=dim(SBi);
//----------------- the singular locus ---------------
  if(isIso)
  {
    list singM=maxideal(1), maxideal(1);
    ds=0;
  }
  else
  {
    ideal sing=minor(jacob(i),n-d)+i;
    list singM=mstd(sing);
    ds=dim(singM[1]);
   }
   if(ds!=-1)
   {
//---------------- computation of the radical ---------

  if (isIso)
  {
    ideal J=maxideal(1);
  }
  else
  {
    ideal J=radical(singM[2]);
  }
//----------------- go to quotient ring --------------
    qring R=SBi;
    ideal J=fetch(BAS,J);
```

```
      ideal i=fetch(BAS,i);
      poly p=J[1];
//-------- computation of p*Hom(J,J) as R-ideal---------
      ideal f=quotient(p*J,J);
      ideal rf = interred(reduce(f,std(p)));
// represents p*Hom(J,J)/p*R = Hom(J,J)/R
    }
  }
//-------- Test: Hom(J,J) == R ?, if yes, go home ------
  if ( size(rf) == 0 )
  {
    execute("ring newR="+charstr(basering)+",
      ("+varstr(basering)+"),("+ordstr(basering)+");");
    ideal norid=fetch(BAS,i);
    export norid;
    result=newR;
    setring BAS;
    return(result);
  }
//------------------ Case: Hom(J,J)!= R ------------------
// create new ring and map form old ring, the ring
// newR/SBi+syzf will be isomorphic to Hom(J,J) as R-module

  f=p,rf;
  //generates pJ:J mod(p), i.e. p*Hom(J,J)/p*R as R-module
  int q=size(f);
  module syzf=syz(f);

  ring newR1 = char(R),(X(1..nvars(R)),T(1..q)),dp;
  map psi1 = BAS,maxideal(1);
  ideal SBi = psi1(SBi);
  attrib(SBi,"isSB",1);

  qring newRq = SBi;
  map psi = R,ideal(X(1..nvars(R)));
  ideal i = psi(i);
  ideal f = psi(f);
  module syzf = psi(syzf);

//------- computation of Hom(J,J) as ring ----------------
// determine kernel of:
// R[T1,...,Tq] -> J:J >-> R[1/p]=R[t]/(t*p-1),
// Ti -> fi/p -> t*fi (p=f1=f[1]), to get ring structure.
// This is of course the same as the kernel of
```

```
//  R[T1,...,Tq] -> pJ:J >-> R, Ti -> fi.
// It is a fact, that the kernel is generated by the linear
// and the quadratic relations

    ideal pf=f[1]*f;
    matrix T=matrix(ideal(T(1..q)),1,q);
    ideal Lin = ideal(T*syzf); //the linear relations

    int ii,jj;
    matrix A;
    ideal Q;

    for (ii=2; ii<=q; ii++ )
    {
       for ( jj=2; jj<=ii; jj++ )
       {
          A = lift(pf,f[ii]*f[jj]);
          Q = Q, ideal(T(jj)*T(ii) - T*A);
          //the quadratic relations
       }
    }
    Q = Lin+Q;
    Q = subst(Q,T(1),1);
    Q = interred(reduce(Q,std(0)));

    ring newR = char(R),(X(1..nvars(R)),T(2..q)),dp;
    ideal k=imap(newRq,Q)+imap(newRq,i);
    if(isIso)
    {
       attrib(k,"isIsolatedSingularity",1);
    }
    result=normalization(k);
    setring BAS;
    return(result);
}
```

Example 3.7.3.

```
LIB"primdec.lib";
ring R=0, (x,y,z),dp;
ideal I=zy2-zx3-x6;
list nor=normalization(I);
def S=nor[1];
setring S;
norid;
```

```
//-> norid[1]=X(2)*X(3)-X(1)*T(2)
//-> norid[2]=X(1)^4*X(3)+X(1)*X(3)^2-T(2)^2
//-> norid[3]=X(1)^5+X(1)^2*X(3)-X(2)*T(2)
```

Remark 3.7.4. The procedure `normal` from `normal.lib` does also compute the normalization map which is omitted in the above procedure; `normal` works also for arbitrary radical ideals which are not necessarily prime.

In the above procedure, we use a feature of SINGULAR, namely to attach *attributes* to an object, which is extremely useful since it allows the use of mathematical knowledge. For instance, if the ideal I has an isolated singularity at the origin and no other singularities, the radical of the singular locus is known to be the maximal ideal and need not be computed. Moreover, this property is preserved during the normalization loop.[10]

Attributes can be defined and used freely. For instance, to check whether I defines a complete intersection can be done without extra cost, since a standard basis of I, together with a minimal generating set in the homogeneous case, will be computed anyway. Since a complete intersection (more generally, a Cohen–Macaulay ideal) with singular locus of codimension ≥ 2 is automatically normal, such an attribute, checked and set during the normalization loop, can speed up the algorithm considerably. The distributed libraries make use of several such attributes.

---

[10] In the above situation, we may attach the attribute `isIsolatedSingularity`: `attrib(I,"isIsolatedSingularity",1);` then, after asking for the attribute, we obtain the following answer: `attrib(I,"isIsolatedSingularity"); //-> 1.`

4. Primary Decomposition and Related Topics

4.1 The Theory of Primary Decomposition

It is well–known that every integer is a product of prime numbers, for instance $10 = 2 \cdot 5$. This equation can also be written as an equality of ideals, $\langle 10 \rangle = \langle 2 \rangle \cap \langle 5 \rangle$ in the ring $\mathbb{Z}$. The aim of this section is to generalize this fact to ideals in arbitrary Noetherian rings.

Ideals generated by prime elements are prime ideals. Therefore, $\langle 10 \rangle$ is the intersection of finitely many prime ideals. In Proposition 3.3.5 this is generalized to radical ideals: in a Noetherian ring every radical ideal I, that is, $I = \sqrt{I}$, is the intersection of finitely many prime ideals. However, what can we expect if the ideal is not radical? For example, $20 = 2^2 \cdot 5$, respectively $\langle 20 \rangle = \langle 2 \rangle^2 \cap \langle 5 \rangle$; in the ring of integers $\mathbb{Z}$ every ideal is the intersection of finitely many ideals which are powers of prime ideals. This is, for arbitrary Noetherian rings, no longer true. A generalization of the powers of prime ideals are the so–called primary ideals. We shall prove in this section that, in a Noetherian ring, every ideal is the intersection of finitely many primary ideals.

Definition 4.1.1. Let A be a Noetherian ring, and let $I \subsetneq A$ be an ideal.

(1) The set of *associated primes* of I, denoted by $\mathrm{Ass}(I)$, is defined as

$$\mathrm{Ass}(I) = \left\{ P \subset A \mid P \text{ prime}, P = I : \langle b \rangle \text{ for some } b \in A \right\}.$$

Elements of $\mathrm{Ass}(\langle 0 \rangle)$ are also called *associated primes* of A.

(2) Let $P, Q \in \mathrm{Ass}(I)$ and $Q \subsetneq P$, then P is called an *embedded prime ideal* of I. We define $\mathrm{Ass}(I, P) := \{Q \mid Q \in \mathrm{Ass}(I), Q \subset P\}$.

(3) I is called *equidimensional* or *pure dimensional* if all associated primes of I have the same dimension.

(4) I is a *primary ideal* if, for any $a, b \in A$, $ab \in I$ and $a \notin I$ imply $b \in \sqrt{I}$. Let P be a prime ideal, then a primary ideal I is called *P–primary* if $P = \sqrt{I}$.

(5) A *primary decomposition* of I, that is, a decomposition $I = Q_1 \cap \cdots \cap Q_s$ with Q_i primary ideals, is called *irredundant* if no Q_i can be omitted in the decomposition and if $\sqrt{Q_i} \neq \sqrt{Q_j}$ for all $i \neq j$.

Example 4.1.2.

(1) Let A be a ring, and let $I \subset A$ be an ideal such that $\sqrt{I}$ is a maximal ideal, then I is primary (cf. Exercise 4.1.4).

(2) Let $A = K[x,y]$ and $I = \langle x^2, xy \rangle = \langle x \rangle \cap \langle x, y \rangle^2 = \langle x \rangle \cap \langle x^2, y \rangle$. Then $\langle x \rangle$, $\langle x, y \rangle^2$, $\langle x^2, y \rangle$ are primary ideals, and $\mathrm{Ass}(I) = \{\langle x \rangle, \langle x, y \rangle\}$. In particular, $\langle x, y \rangle$ is an embedded prime of I with $\mathrm{Ass}(I, \langle x, y \rangle) = \{\langle x \rangle, \langle x, y \rangle\}$, while $\mathrm{Ass}(I, \langle x \rangle) = \{\langle x \rangle\}$. Note that both decompositions are irredundant primary decompositions of I, which shows that an irredundant primary decomposition might be *not unique*.

(3) $\mathrm{minAss}(I) \subset \mathrm{Ass}(I)$ and $\mathrm{minAss}(I) = \mathrm{Ass}(I)$ if and only if I has no embedded primes (Exercise 4.1.5), showing that $\mathrm{minAss}(I)$ is the set of minimal elements (with respect to inclusion) of $\mathrm{Ass}(I)$.

The following lemma collects the properties of primary ideals needed for the primary decomposition.

Lemma 4.1.3. *Let A be a Noetherian ring and $Q \subset A$ a P–primary ideal.*

(1) The radical of a primary ideal is a prime ideal.

(2) Let Q' be a P–primary ideal, then $Q \cap Q'$ is a P–primary ideal.

(3) Let $b \in A$, $b \notin Q$, then $Q : \langle b \rangle$ is P–primary. $b \in P$ if and only if $Q \subsetneq Q : \langle b \rangle$.

(4) Let $P' \supset Q$ be a prime ideal, then $QA_{P'} \cap A = Q$.

(5) There exists $d \in A$ such that $P = Q : \langle d \rangle$. Especially, $P \in \mathrm{Ass}(Q)$.

Proof. (1) and (2) are left as exercises. To prove (3), let $b \in A$, $b \notin Q$. If $b \notin P$, then $Q : \langle b \rangle = Q$ because $ab \in Q$, $a \notin Q$ implies $b \in P$ by definition of a primary ideal. If $b \in P$ then $b^n \in Q$ for a suitable n. We may assume $n \geq 2$ and $b^{n-1} \notin Q$. Then $b^{n-1} \in Q : \langle b \rangle$ and, therefore, $Q \subsetneq Q : \langle b \rangle$. Let $xy \in Q : \langle b \rangle$ and $x \notin Q : \langle b \rangle$. This implies $bxy \in Q$ and $bx \notin Q$. By definition of a primary ideal, we obtain $y^n \in Q$ for a suitable n. This implies that $Q : \langle b \rangle$ is a primary ideal. Finally, $\sqrt{Q : \langle b \rangle} \supset \sqrt{Q} = P$. Let $x \in \sqrt{Q : \langle b \rangle}$, that is, $bx^n \in Q$ for some n but $b \notin Q$ and, therefore, $x^n \in P$. Now P is prime and we obtain $x \in P$ which proves $\sqrt{Q : \langle b \rangle} = P$.

To prove (4), let $x \in QA_{P'} \cap A$. This means that $sx \in Q$ for a suitable $s \notin P'$. If $x \notin Q$, then, by definition of a primary ideal, $s \in \sqrt{Q} \subset P'$ in contradiction to the choice of s. We obtain $QA_{P'} \cap A \subset Q$. The other inclusion is trivial.

To prove (5), we consider first the case $Q = P$. In this case, we can use $d = 1$ and are finished. If $Q \subsetneq P$ we choose $g_1 \in P \smallsetminus Q$ and obtain, using (3), that $Q : \langle g_1 \rangle \supsetneq Q$ is P–primary and $\sqrt{Q : \langle g_1 \rangle} = P$. Again, if $Q : \langle g_1 \rangle \subsetneq P$ we can choose $g_2 \in P \smallsetminus (Q : \langle g_1 \rangle)$ such that $(Q : \langle g_1 \rangle) : \langle g_2 \rangle \supsetneq Q : \langle g_1 \rangle$. Now $(Q : \langle g_1 \rangle) : \langle g_2 \rangle = Q : \langle g_1 g_2 \rangle$ (Exercise 4.1.2), and continuing in this way we obtain an increasing chain of ideals $Q \subsetneq Q : \langle g_1 \rangle \subsetneq Q : \langle g_1 g_2 \rangle \subsetneq \dots$. The ring A is Noetherian and, therefore, this chain has to stop, that is, we find n and $g_1, \dots, g_n \in P$ such that $Q : \langle g_1 \cdots g_n \rangle = P$. $\quad\square$

Theorem 4.1.4. *Let A be a Noetherian ring and $I \subsetneq A$ be an ideal, then there exists an irredundant decomposition $I = Q_1 \cap \cdots \cap Q_r$ of I as intersection of primary ideals $Q_1, \ldots, Q_r$.*

Proof. Because of Lemma 4.1.3 (2) it is enough to prove that every ideal is the intersection of finitely many primary ideals. Suppose this is not true, and let $\mathfrak{M}$ be the set of ideals which are not an intersection of finitely many primary ideals. The ring A is Noetherian and, by Proposition 1.3.6, $\mathfrak{M}$ has a maximal element with respect to the inclusion. Let $I \in \mathfrak{M}$ be maximal. Since I is not primary, there exist $a, b \in A$, $a \notin I$, $b^n \notin I$ for all n and $ab \in I$. Now consider the chain $I : \langle b \rangle \subset I : \langle b^2 \rangle \subset \cdots$. As A is Noetherian, there exists an n with $I : \langle b^n \rangle = I : \langle b^{n+1} \rangle = \cdots$. Using Lemma 3.3.6, we obtain $I = (I : \langle b^n \rangle) \cap \langle I, b^n \rangle$. Since $b^n \notin I$ we have $I \subsetneq \langle I, b^n \rangle$. Since $a \notin I$ and $ab^n \in I$ we have $I \subsetneq I : \langle b^n \rangle$. As I is maximal in $\mathfrak{M}$, $I : \langle b^n \rangle$ and $\langle I, b^n \rangle$ are not in $\mathfrak{M}$. This implies that both ideals are intersections of finitely many primary ideals and, therefore, I is an intersection of finitely many primary ideals, too, in contradiction to the assumption. $\qquad\square$

Theorem 4.1.5. *Let A be a ring and $I \subset A$ be an ideal with irredundant primary decomposition $I = Q_1 \cap \cdots \cap Q_r$. Then $r = \# \operatorname{Ass}(I)$,*

$$\operatorname{Ass}(I) = \{ \sqrt{Q_1}, \ldots, \sqrt{Q_r} \},$$

and if $\{ \sqrt{Q_{i_1}}, \ldots, \sqrt{Q_{i_s}} \} = \operatorname{Ass}(I, P)$ for $P \in \operatorname{Ass}(I)$ then $Q_{i_1} \cap \cdots \cap Q_{i_s}$ is independent of the decomposition.

Proof. Let $I = Q_1 \cap \cdots \cap Q_r$ be an irredundant primary decomposition. If $P \in \operatorname{Ass}(I)$, $P = I : \langle b \rangle$ for a suitable b, then $P = (Q_1 : \langle b \rangle) \cap \cdots \cap (Q_r : \langle b \rangle)$ (Exercise 4.1.3). In particular, $\bigcap_{i=1}^r (Q_i : \langle b \rangle) \subset P$, hence, $Q_j : \langle b \rangle \subset P$ for a suitable j (Lemma 1.3.12). On the other hand, since $P = I : \langle b \rangle \subset Q_j : \langle b \rangle$, we obtain $P = Q_j : \langle b \rangle$. Now $Q_j : \langle b \rangle \subset \sqrt{Q_j}$ (Lemma 4.1.3 (3)), which implies $P = \sqrt{Q_j}$. This proves that $\{ \sqrt{Q_1}, \ldots, \sqrt{Q_r} \} \supset \operatorname{Ass}(I)$.

It remains to prove that $\sqrt{Q_i} = I : \langle b_i \rangle$ for a suitable b_i. But this is a consequence of Lemma 4.1.3 (5): let $J = Q_1 \cap \cdots \cap Q_{i-1} \cap Q_{i+1} \cap \cdots \cap Q_r$, then $J \not\subset Q_i$, since the decomposition is irredundant. We can choose $d \in J \setminus Q_i$ and obtain, using Exercise 4.1.3, $I : \langle d \rangle = Q_i : \langle d \rangle$. By Lemma 4.1.3 (3), (5), respectively Exercise 4.1.2, $\sqrt{Q_i} = \sqrt{Q_i : \langle d \rangle} = (Q_i : \langle d \rangle) : \langle g \rangle = I : \langle dg \rangle$ for a suitable g. We obtain $\operatorname{Ass}(I) = \{ \sqrt{Q_1}, \ldots, \sqrt{Q_r} \}$.

Now let $\operatorname{Ass}(I, P) = \{ \sqrt{Q_{i_1}}, \ldots, \sqrt{Q_{i_s}} \}$, then Lemma 4.1.3 (4) gives that $Q_{i_\nu} A_P \cap A = Q_{i_\nu}$. If $j \notin \{i_1, \ldots, i_s\}$ then $Q_j \not\subset P$, therefore, $Q_j A_P = A_P$. This implies that $I A_P \cap A = \bigcap_{j=1}^r (Q_j A_P \cap A) = Q_{i_1} \cap \cdots \cap Q_{i_s}$ is independent of the decomposition, since $\operatorname{Ass}(I, P)$ is. $\qquad\square$

Example 4.1.6.

(1) If $I = \langle f \rangle \subset K[x_1, \ldots, x_n]$ is a principal ideal and $f = f_1^{n_1} \cdots f_s^{n_s}$ is the factorization of f into irreducible factors, then

$$I = \langle f_1^{n_1} \rangle \cap \cdots \cap \langle f_r^{n_r} \rangle$$

is the primary decomposition, and the $\langle f_i \rangle$ are the associated prime ideals which are all minimal.

(2) Let $I = \langle xy, xz, yz \rangle = \langle x, y \rangle \cap \langle x, z \rangle \cap \langle y, z \rangle \subset K[x, y, z]$. Then the zero–set $V(I)$ (cf. A.1) is the union of the coordinate axes (cf. Figure 4.1).

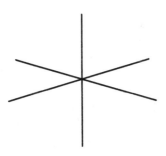

Fig. 4.1. The zero–set of $\langle xy, xz, yz \rangle$.

(3) Let $I = \langle (y^2 - xz) \cdot (z^2 - x^2 y), (y^2 - xz) \cdot z \rangle \subset K[x, y, z]$. Then we obtain the irredundant primary decomposition $I = \langle y^2 - xz \rangle \cap \langle x^2, z \rangle \cap \langle y, z^2 \rangle$, $\mathrm{Ass}(I) = \{\langle y^2 - xz \rangle, \langle x, z \rangle, \langle y, z \rangle\}$ and $\mathrm{minAss}(I) = \{\langle y^2 - xz \rangle, \langle x, z \rangle\}$. $\langle y, z \rangle$ is an embedded prime with $\mathrm{Ass}(I, \langle y, z \rangle) = \{\langle y^2 - xz \rangle, \langle y, z \rangle\}$. The zero–set of I (cf. A.1) is displayed in Figure 4.2.

Fig. 4.2. The zero–set of $I = \langle y^2 - xz \rangle \cap \langle x^2, z \rangle \cap \langle y, z^2 \rangle$.

Remark 4.1.7.

(1) Primary decomposition does not hold, in general, in non–Noetherian rings, even if we allow infinite intersections.
(2) There exists a concept of primary decomposition for finitely generated modules over Noetherian rings (Exercise 4.1.13). Primary decomposition of modules has been implemented in the SINGULAR library `mprimdec.lib`.

Exercises

For these exercises let A be a Noetherian ring, K a field and I, J ideals in A.

4.1.1. Prove that $\sqrt{I}$ is prime if I is primary.

4.1.2. Prove that, for $a, b \in A$, $(I : \langle a \rangle) : \langle b \rangle = I : \langle ab \rangle$.

4.1.3. Prove that, for any $b \in A$, $(I \cap J) : \langle b \rangle = (I : \langle b \rangle) \cap (J : \langle b \rangle)$.

4.1.4. Prove that I is primary if $\sqrt{I}$ is a maximal ideal.

4.1.5. Prove that $\mathrm{minAss}(I) \subset \mathrm{Ass}(I)$ with equality if and only if I has no embedded primes.

4.1.6. Let $P \subset A$ be a prime ideal, and let $Q_1, Q_2 \subset A$ be P–primary. Prove that $Q_1 \cap Q_2$ is a P–primary ideal.

4.1.7. Let $f_1, f_2 \in A$ such that $f = f_1 \cdot f_2 \in I$ and $\langle f_1, f_2 \rangle = A$. Prove that $I = \langle I, f_1 \rangle \cap \langle I, f_2 \rangle$.

4.1.8. Let $I \subset K[x_1, \ldots, x_n]$ be a *homogeneous* ideal (that is, generated by homogeneous polynomials). Prove that the ideals in $\mathrm{Ass}(I)$ are homogeneous.

4.1.9. Let $w = (w_1, \ldots, w_n) \in \mathbb{Z}^n$, $w_i \neq 0$ for all i, and let $I \subset K[x_1, \ldots, x_n]$ be an ideal. Moreover, let $I^h \subset K[x_1, \ldots, x_n, t]$ be the ideal generated by the weighted homogenizations of the elements of I with respect to t (see Exercise 1.7.5). Prove the following statements:

(1) I^h is primary (prime) if and only if I is primary (prime).
(2) Let $I = Q_1 \cap \ldots \cap Q_r$ be an irredundant primary decomposition, then $I^h = Q_1^h \cap \ldots \cap Q_r^h$ is an irredundant primary decomposition, too.

(Hint: to show (1), first prove the analogue of Exercise 2.2.5 for primary instead of prime ideals. For (2), prove that $(I_1 \cap I_2)^h = I_1^h \cap I_2^h$.)

4.1.10. Let $\mathrm{Ass}(\langle 0 \rangle) = \{P_1, \ldots, P_s\}$. Prove that $\bigcup_{i=1}^{s} P_i$ is the set of zerodivisors of A.

4.1.11. Let $I = Q_1 \cap \cdots \cap Q_m$ be an irredundant primary decomposition, and let $J := Q_2 \cap \cdots \cap Q_m$. Prove that $\dim\big(A/(Q_1 + J)\big) < \dim(A/J)$.

4.1.12. Use SINGULAR to show the following equality of ideals in $K[x, y, z]$:

$$\langle y^2 - xz \rangle \cap \langle x^2, z \rangle \cap \langle y, z^2 \rangle = \left\langle (y^2 - xz)(z^2 - x^2 y), \, (y^2 - xz) \cdot z \right\rangle.$$

4.1.13. Let M be a finitely generated A–module and $N \subset M$ a submodule. Then N is called *primary* in M if $N \neq M$ and for every zerodivisor x of M/N there exists ρ such that $x^\rho \in \mathrm{Ann}(M/N)$. Prove the following statements:

(1) If $N \subset M$ is primary then $N : M$ is a primary ideal ($\sqrt{N : M}$ is called the *associated prime* to N).

(2) N has an irredundant primary decomposition and the associated primes are uniquely determined.

(3) If P is an associated prime of N, then $P = N : \langle m \rangle$ for some $m \in M$.

(4) Let $P_1, \ldots, P_s$ be the set of associated primes of N, then the zerodivisors of M/N are $\bigcup_{i=1}^s P_i$.

(Hint: recall that $\sqrt{N : M} = \sqrt{\mathrm{Ann}(M/N)} = \sqrt[M]{N}$, see Exercise 2.8.6.)

4.1.14. Let M be a finitely generated A–module. Let $\mathrm{Ass}(M)$ be the set of *associated prime ideals* to $\langle 0 \rangle \subset M$ in the sense of Exercise 4.1.13, that is,

$$\mathrm{Ass}(M) := \left\{ P \subset A \text{ prime} \mid P = \mathrm{Ann}(m), \, m \in M \smallsetminus \{0\} \right\}.$$

Let $\mathcal{M} := \{ \mathrm{Ann}(m) \mid 0 \neq m \in M \}$. Prove that the maximal elements in $\mathcal{M}$ are associated prime ideals.

4.1.15. Let A be a Noetherian ring and $M \neq \langle 0 \rangle$ a finitely generated A–module. Prove that there exists a chain $M = M_0 \supset M_1 \supset \cdots \supset M_n = \langle 0 \rangle$ of submodules of M such that $M_i/M_{i+1} \cong A/P_i$ for a suitable prime ideal $P_i \subset A$, $i = 0, \ldots, n - 1$.

(Hint: choose an associated prime $P_1 \in \mathrm{Ass}(M)$, and let $P_1 = \mathrm{Ann}(m_1)$. If $M = \langle m_1 \rangle$ then $M \cong A/P_1$, otherwise continue with $M/\langle m_1 \rangle$.)

4.2 Zero–dimensional Primary Decomposition

In this section we shall give an algorithm to compute a primary decomposition for zero–dimensional ideals in a polynomial ring over a field of characteristic 0. This algorithm was published by Gianni, Trager, and Zacharias ([90]). Let K be a field of characteristic 0. In the case of one variable x, any ideal $I \subset K[x]$ is a principal ideal and the primary decomposition is given by the factorization of a generator of I: let $I = \langle f \rangle$, $f = f_1^{n_1} \ldots f_r^{n_r}$ with f_i irreducible and $\langle f_i, f_j \rangle = K[x]$ for $i \neq j$, then $I = \langle f_1 \rangle^{n_1} \cap \cdots \cap \langle f_r \rangle^{n_r}$ is the primary decomposition of I. In the case of n variables, the univariate polynomial factorization is also an essential ingredient. We shall see that, after a generic coordinate change, the factorization of a polynomial in one variable leads to a primary decomposition. By definition, all associated prime ideals of a zero–dimensional ideal are maximal. We need the concept for an ideal in general position.

Definition 4.2.1.

(1) A maximal ideal $M \subset K[x_1, \ldots, x_n]$ is called in *general position* with respect to the lexicographical ordering with $x_1 > \cdots > x_n$, if there exist $g_1, \ldots, g_n \in K[x_n]$ with $M = \langle x_1 + g_1(x_n), \ldots, x_{n-1} + g_{n-1}(x_n), g_n(x_n) \rangle$.

(2) A zero–dimensional ideal $I \subset K[x_1, \ldots, x_n]$ is called in *general position* with respect to the lexicographical ordering with $x_1 > \cdots > x_n$, if all associated primes $P_1, \ldots, P_k$ are in general position and if $P_i \cap K[x_n] \neq P_j \cap K[x_n]$ for $i \neq j$.

Proposition 4.2.2. *Let K be a field of characteristic 0, and let $I \subset K[x]$, $x = (x_1, \ldots, x_n)$, be a zero–dimensional ideal. Then there exists a non–empty, Zariski open subset $U \subset K^{n-1}$ such that for all $\underline{a} = (a_1, \ldots, a_{n-1}) \in U$, the coordinate change $\varphi_{\underline{a}} : K[x] \to K[x]$ defined by $\varphi_{\underline{a}}(x_i) = x_i$ if $i < n$, and*

$$\varphi_{\underline{a}}(x_n) = x_n + \sum_{i=1}^{n-1} a_i x_i$$

has the property that $\varphi_{\underline{a}}(I)$ is in general position with respect to the lexicographical ordering defined by $x_1 > \cdots > x_n$.

Proof. We consider first the case that $I \subset K[x_1, \ldots, x_n]$ is a maximal ideal. The field $K[x_1, \ldots, x_n]/I$ is a finite extension of K (Theorem 3.5.1), and there exists a dense, Zariski open subset $U \subset K^{n-1}$ such that for $\underline{a} \in U$ the element $z = x_n + \sum_{i=1}^{n-1} a_i x_i$ is a primitive element for the field extension (Primitive Element Theorem, cf. [238], here it is necessary that K is a perfect, infinite field).

Since $\varphi_{\underline{a}+\underline{b}} = \varphi_{\underline{b}} \circ \varphi_{\underline{a}}$, we may assume that $\underline{0} \in U$, that is,

$$K[x_1, \ldots, x_n]/I \cong K[x_n]/\langle f_n(x_n) \rangle$$

for some irreducible polynomial $f_n(x_n)$. Via this isomorphism $x_i \bmod I$ corresponds to some $f_i(x_n) \bmod \langle f_n(x_n) \rangle$ and we obtain

$$\langle x_1 - f_1(x_n), \ldots, x_{n-1} - f_{n-1}(x_n), f_n(x_n) \rangle = I \, .$$

The set of these generators is obviously a Gröbner basis with the required properties.

Now let I be an arbitrary zero–dimensional ideal and let $P_1, \ldots, P_s$ be the associated primes of I, then $\varphi_{\underline{a}}(P_j)$ are in general position with respect to the lexicographical ordering $x_1 > \cdots > x_n$ for almost all $\underline{a} \in K^{n-1}$. It remains to prove that $\varphi_{\underline{a}}(P_i) \cap K[x_n] \neq \varphi_{\underline{a}}(P_j) \cap K[x_n]$ for $i \neq j$ and almost all $\underline{a}$. We may assume that the P_i's are already in general position with respect to the lexicographical ordering $x_1 > \cdots > x_n$. We study the behaviour of a maximal ideal $P = \langle x_1 - g_1(x_n), \ldots, x_{n-1} - g_{n-1}(x_n), g_n(x_n) \rangle$ under the automorphism $\varphi_{\underline{a}}$.

If $\varphi_{\underline{a}}(P)$ is again in general position with respect to the lexicographical ordering $x_1 > \cdots > x_n$, then $\varphi_{\underline{a}}(P) \cap K[x_n] = \langle h^{(\underline{a})} \rangle$ for a monic polynomial $h^{(\underline{a})}$ of degree

$$r := \dim_K K[x_n]/\langle h^{(\underline{a})} \rangle = \dim_K K[x]/\varphi_{\underline{a}}(P) = \dim_K K[x]/P = \deg(g_n).$$

To compute $h^{(\underline{a})}$, we consider the algebraic closure $\overline{K}$ of K. Let $\alpha_1, \ldots, \alpha_r \in \overline{K}$ be the roots of $g_n(x_n)$. Because of Exercise 4.2.1 (b), $g_n(x_n)$ is squarefree in $\overline{K}[x_n]$. Then $g_n(x_n) = c(x_n - \alpha_1) \cdot \ldots \cdot (x_n - \alpha_r)$, $c \in K$ and, because of Exercise 4.1.7,

$$P\overline{K}[x] = \bigcap_{i=1}^{r} \langle x_1 - g_1(\alpha_i), \ldots, x_{n-1} - g_{n-1}(\alpha_i), x_n - \alpha_i \rangle.$$

Now

$$\varphi_{\underline{a}}\big(\langle x_1 - g_1(\alpha_i), \ldots, x_{n-1} - g_{n-1}(\alpha_i), x_n - \alpha_i \rangle\big)$$
$$= \Big\langle x_1 - g_1(\alpha_i), \ldots, x_{n-1} - g_{n-1}(\alpha_i), x_n - \alpha_i + \sum_{\nu=1}^{n-1} a_\nu g_\nu(\alpha_i) \Big\rangle.$$

This implies that $\varphi_{\underline{a}}(P\overline{K}[x]) \cap \overline{K}[x_n] \supset \langle \prod_{i=1}^{r}(x_n - \alpha_i + \sum_{\nu=1}^{n-1} a_\nu g_\nu(\alpha_i)) \rangle$.

Since $\langle h^{(\underline{a})} \rangle = \varphi_{\underline{a}}(P) \cap K[x_n] = \varphi_{\underline{a}}(P\overline{K}[x]) \cap K[x_n]$ (Exercise 4.2.1 (a)), and since $h^{(\underline{a})}$, as well as $\prod_{i=1}^{r}(x_n - \alpha_i + \sum_{\nu=1}^{n-1} a_\nu g_\nu(\alpha_i))$, are monic polynomials in $K[x_n]^1$ of degree r, it follows that

$$h^{(\underline{a})} = \prod_{i=1}^{r}\left(x_n - \alpha_i + \sum_{\nu=1}^{n-1} a_\nu g_\nu(\alpha_i) \right).$$

Now let $\varphi_{\underline{a}}(P_1) \cap K[x_n] = \langle h_1^{(\underline{a})} \rangle$, $\ldots$, $\varphi_{\underline{a}}(P_s) \cap K[x_n] = \langle h_s^{(\underline{a})} \rangle$ with monic polynomials $h_i^{(\underline{a})} \in K[x_n]$, and assume that the prime ideals $\varphi_{\underline{a}}(P_i)$ are in general position with respect to the lexicographical ordering $x_1 > \cdots > x_n$. The condition $\varphi_{\underline{a}}(P_i) \cap K[x_n] = \varphi_{\underline{a}}(P_j) \cap K[x_n]$, that is, $h_i^{(\underline{a})} = h_j^{(\underline{a})}$ leads, because of $P_i \neq P_j$, to a non-trivial polynomial system of equations for $\underline{a}$. This implies that for almost all $\underline{a}$, $\varphi_{\underline{a}}(P_i) \cap K[x_n] \neq \varphi_{\underline{a}}(P_j) \cap K[x_n]$ if $i \neq j$. $\qquad\square$

Proposition 4.2.3. Let $I \subset K[x_1, \ldots, x_n]$ be a zero–dimensional ideal. Let $\langle g \rangle = I \cap K[x_n]$, $g = g_1^{\nu_1} \ldots g_s^{\nu_s}$, g_i monic and prime and $g_i \neq g_j$ for $i \neq j$. Then

---

[1] $\prod_{i=1}^{r}(x_n - \alpha_i + \sum_{\nu=1}^{n-1} a_\nu g_\nu(\alpha_i)) \in K[x_n]$ is a consequence of Galois theory, since the product is invariant under the action of the Galois group (the K–automorphisms of $K(\alpha_1, \ldots, \alpha_r)$ are given by permutations of the roots $\alpha_1, \ldots, \alpha_r$).

(1) $I = \bigcap_{i=1}^{s} \langle I, g_i^{\nu_i} \rangle$.

If I is in general position with respect to the lexicographical ordering with $x_1 > \cdots > x_n$, then

(2) $\langle I, g_i^{\nu_i} \rangle$ is a primary ideal for all i.

Proof. To prove (1) note that, obviously, $I \subset \bigcap_{i=1}^{s} \langle I, g_i^{\nu_i} \rangle$. To prove the other inclusion let $g^{(i)} := g/g_i^{\nu_i}$ for $i = 1, \ldots, s$. Then the univariate polynomials $g^{(1)}, \ldots, g^{(s)} \in K[x_n]$ have the greatest common divisor 1. Hence, we can find $a_1, \ldots, a_s \in K[x_n]$ with $\sum_{i=1}^{s} a_i g^{(i)} = 1$. Now let $f \in \bigcap_{i=1}^{s} \langle I, g_i^{\nu_i} \rangle$, in particular, there exist $f_i \in I$, $\xi_i \in K[x]$ such that $f = f_i + \xi_i g_i^{\nu_i}$, $i = 1, \ldots, s$. Hence,

$$ f = \sum_{i=1}^{s} a_i g^{(i)}(f_i + \xi_i g_i^{\nu_i}) = \sum_{i=1}^{s} (a_i g^{(i)} f_i + a_i \xi_i g) \in I \,, $$

which proves (1).

(2) First note that $\langle I, g_i^{\nu_i} \rangle \subsetneq K[x]$ and $\mathrm{Ass}(\langle I, g_i^{\nu_i} \rangle) \subset \mathrm{Ass}(I)$. This can be seen as follows: if we could write $1 = f + a g_i^{\nu_i}$ for some $f \in I$, $a \in K[x]$, then $g/g_i^{\nu_i} \in \langle f, g \rangle \subset I$, contradicting the assumption $I \cap K[x_n] = \langle g \rangle$. Moreover, $I \subset \langle I, g_i^{\nu_i} \rangle$ and the uniqueness of associated primes implies that each associated prime of $\langle I, g_i^{\nu_i} \rangle$ has to contain some associated prime of I. But, since I is zero–dimensional, its associated primes are maximal ideals.

Now, let $P_1, \ldots, P_\ell$ be the associated primes of I and let $P_i \cap K[x_n] = \langle p_i \rangle$. Then, by assumption, the polynomials $p_1, \ldots, p_\ell$ are pairwise coprime and, therefore, $\bigcap_{i=1}^{\ell}(P_i \cap K[x_n]) = \bigcap_{i=1}^{\ell} \langle p_i \rangle = \langle \prod_{i=1}^{\ell} p_i \rangle$. On the other hand, we have $\bigcap_{i=1}^{\ell}(P_i \cap K[x_n]) = (\bigcap_{i=1}^{\ell} P_i) \cap K[x_n] = \sqrt{I} \cap K[x_n]$. Hence, the assumption $I \cap K[x_n] = \langle g \rangle$ implies that $\prod_{i=1}^{\ell} p_i$ divides g and g divides a power of $\prod_{i=1}^{\ell} p_i$. The latter implies $\ell = s$, and we may assume $g_i = p_i$ for $i = 1, \ldots, s$. It follows that P_i is the unique associated prime of I containing $g_i^{\nu_i}$, and, by the above, we can conclude that $\mathrm{Ass}(\langle I, g_i^{\nu_i} \rangle) = \{P_i\}$. Hence, $\langle I, g_i^{\nu_i} \rangle$ is a primary ideal. $\qquad\square$

Proposition 4.2.3 shows how to obtain a primary decomposition of a zero–dimensional ideal in general position by using the factorization of g. In the algorithm for the zero–dimensional decomposition we try to put I in general position via a map $\varphi_{\underline{a}}$, $\underline{a} \in K^{n-1}$ chosen randomly. But we cannot be sure, in practice, that for a random choice of $\underline{a}$ made by the computer, $\varphi_{\underline{a}}(I)$ is in general position. We need a test to decide whether $\langle I, g_i^{\nu_i} \rangle$ is primary and in general position. Using Definition 4.2.1 we obtain the following criterion:

Criterion 4.2.4. *Let $I \subset K[x_1, \ldots, x_n]$ be a proper ideal. Then the following conditions are equivalent:*

(1) I is zero–dimensional, primary and in general position with respect to the lexicographical ordering with $x_1 > \cdots > x_n$.

(2) There exist $g_1, \ldots, g_n \in K[x_n]$ and positive integers $\nu_1, \ldots, \nu_n$ such that

    a) $I \cap K[x_n] = \langle g_n^{\nu_n} \rangle$, g_n irreducible;

    b) for each $j < n$, I contains the element $(x_j + g_j)^{\nu_j}$.

(3) Let S be a reduced Gröbner basis of I with respect to the lexicographical ordering with $x_1 > \ldots > x_n$. Then there exist $g_1, \ldots, g_n \in K[x_n]$ and positive integers $\nu_1, \ldots, \nu_n$ such that

    a) $g_n^{\nu_n} \in S$ and g_n is irreducible;

    b) $(x_j + g_j)^{\nu_j}$ is congruent to an element in $S \cap K[x_j, \ldots, x_n]$ modulo $\langle g_n, x_{n-1} + g_{n-1}, \ldots, x_{j+1} + g_{j+1} \rangle \subset K[x]$ for $j = 1, \ldots, n-1$.

Proof. To prove $(3) \Rightarrow (2)$, let $M := \sqrt{I}$. Then $g_n \in M$, and, inductively, we obtain $x_j + g_j \in M$ for all j. This implies

$$M = \langle x_1 + g_1, \ldots, x_{n-1} + g_{n-1}, g_n \rangle,$$

because g_n is irreducible and, therefore, $\langle x_1 + g_1, \ldots, x_{n-1} + g_{n-1}, g_n \rangle \subset K[x]$ is a maximal ideal. Finally, $M = \sqrt{I}$ implies now a) and b) in (2).

$(2) \Rightarrow (1)$ is clear because $M = \langle x_1 + g_1, \ldots, x_{n-1} + g_{n-1}, g_n \rangle \subset \sqrt{I}$ is a maximal ideal and, by definition, in general position with respect to the lexicographical ordering with $x_1 > \cdots > x_n$.

To prove $(1) \Rightarrow (3)$, let $M := \sqrt{I}$. Since I is in general position and primary, $M = \langle x_1 + g_1, \ldots, x_{n-1} + g_{n-1}, g_n \rangle$ with $g_n \in K[x_n]$ irreducible and $g_1, \ldots, g_{n-1} \in K[x_n]$. We may assume that g_n is monic. Now, let S be a reduced Gröbner basis of I (in particular, all elements are supposed to be monic, too). Then, due to the elimination property of $>_{lp}$, $S \cap K[x_n] = \{g\}$ generates $I \cap K[x_n]$, which is a primary ideal with $\sqrt{I \cap K[x_n]} = \langle g_n \rangle$. This implies $g = g_n^{\nu_n}$ for a suitable ν_n.

Now let $j \in \{1, \ldots, n-1\}$. Since I is zero-dimensional and S is a reduced Gröbner basis of I, there exists a unique $h \in S$ such that $\mathrm{LM}(h)$ is a power of x_j, $\mathrm{LM}(h) = x_j^m$ (Theorem 3.5.1 (7)). Note that the latter implies, in particular, that $h \in K[x_j, \ldots, x_n]$ (again due to the elimination property of $>_{lp}$). We set $M' := M \cap K[x_{j+1}, \ldots, x_n]$, $K' := K[x_{j+1}, \ldots, x_n]/M' \cong K[x_n]/\langle g_n \rangle$, and consider the canonical projection

$$\Phi : K[x_1, \ldots, x_n] = (K[x_{j+1}, \ldots, x_n])[x_1, \ldots, x_j] \longrightarrow K'[x_1, \ldots, x_j].$$

Step 1. We show $\Phi(S \cap K[x_j, \ldots, x_n]) = \{\Phi(h), 0\}$. Since $S \cap K[x_j, \ldots, x_n]$ is a standard basis (w.r.t. $>_{lp}$) of $I \cap K[x_j, \ldots, x_n]$, this implies

$$I \cap K[x_j, \ldots, x_n] \equiv \langle h \rangle_{K[x_j, \ldots, x_n]} \mod M' \cdot K[x_j, \ldots, x_n].$$

Let $K[x'] := K[x_{j+1}, \ldots, x_n]$ and consider

$$L := \left\langle f_s \in K[x'] \,\middle|\, \exists\, f_0, \ldots, f_{s-1} \in K[x'], s < m, \text{ such that } \sum_{i=0}^{s} f_i x_j^i \in I \right\rangle.$$

Then, clearly, $I \cap K[x'] \subset L \subsetneq K[x']$. Since $I \cap K[x']$ is primary and zero-dimensional, $\sqrt{I \cap K[x']}$ is the unique associated prime of $I \cap K[x']$ (Theorem 4.1.5) and a maximal ideal in $K[x']$. Hence, $L \subset \sqrt{I \cap K[x']} \subsetneq K[x']$.

Now, let $f \in S \cap K[x_j, \dots, x_n] \subset I$, $f \neq h$. We write $f = \sum_{i=0}^{s} f_i x_j^i$, with $f_i \in K[x']$. Since S is reduced and $\mathrm{LM}(h) = x_j^m$, we have $s < m$, hence $f_s \in L$. Moreover, $f' := x_j^{m-s} f - f_s h \in I$, and, writing $f' = \sum_{i=0}^{m-1} f_i' x_j^i$, we obtain $f_{m-1}' \in L$ and $f_i' \equiv f_{i+s-m} \bmod L$, $i = m - s, \dots, m - 1$. Therefore, $f_{s-1} \in L$, and proceeding inductively we obtain $f_i \in L$, $i = 0, \dots, s$.

The above implies now that $f_i \in \sqrt{I \cap K[x']} = M'$ for $i = 0, \dots, s$. Thus, $\Phi(f) = 0$.

Step 2. On the other hand, $\sqrt{\Phi(I)} = \Phi\big(\sqrt{I + \mathrm{Ker}(\Phi)}\big) = \Phi(M)$. It follows that $\sqrt{\Phi(I)} \cap K'[x_j] = \langle x_j + \overline{g}_j \rangle_{K'[x_j]}$, where $\overline{g}_j := g_j \bmod M'$, and we conclude that $\Phi(I \cap K[x_j, \dots, x_n]) \cap K'[x_j] = \Phi(I) \cap K'[x_j] = \langle (x_j + \overline{g}_j)^{\ell} \rangle_{K'[x_j]}$ for a positive integer ℓ. Together with the result of Step 1, this implies that $h \equiv (x_j + g_j)^{\ell} \bmod M' \cdot K[x_j, \dots, x_n]$, in particular, $\ell = m =: \nu_j$. $\qquad\square$

Criterion 4.2.4 is the basis of the following algorithm to test whether a zero–dimensional ideal is primary and in general position.

Algorithm 4.2.5 (PRIMARYTEST(I)).

Input: A zero–dimensional ideal $I := \langle f_1, \dots, f_k \rangle \subset K[x]$, $x = (x_1, \dots, x_n)$.
Output: $\langle 0 \rangle$ if I is either not primary or not in general position, or $\sqrt{I}$ if I is primary and in general position.

- compute a reduced Gröbner basis S of I with respect to the lexicographical ordering with $x_1 > \cdots > x_n$;
- factorize $g \in S$, the element with smallest leading monomial;
- if ($g = g_n^{\nu_n}$ with g_n irreducible)
      prim := $\langle g_n \rangle$
  else
      return $\langle 0 \rangle$.
- $i := n$;
  while ($i > 1$)
      $i := i - 1$;
      choose $f \in S$ with $\mathrm{LM}(f) = x_i^m$;
      $b :=$ the coefficient of x_i^{m-1} in f considered as polynomial in x_i;
      $q := x_i + b/m$;
      if ($q^m \equiv f \bmod \mathrm{prim}$)
          prim := prim $+ \langle q \rangle$;
      else
          return $\langle 0 \rangle$;
- return prim.

SINGULAR Example 4.2.6 (primary test).

```
option(redSB);
ring R=0,(x,y),lp;
ideal I=y4-4y3-10y2+28y+49,x3-6x2y+3x2+12xy2-12xy+3x-8y3
+13y2-8y-6;
//the generators are a Groebner basis
```

We want to check whether the ideal I is primary and in general position.

```
factorize(I[1]);    //to test if Criterion 4.2.4 (3) a) holds
//-> [1]:
//->    _[1]=1
//->    _[2]=y2-2y-7
//-> [2]:
//->    1,2 //I[1] is the square of an irreducible element
ideal prim=std(y2-2y-7);
poly q=3x-6y+3;
poly f2=I[2];
reduce(q^3-27*f2,prim);

//-> 0
```

The ideal is primary and in general position and $\langle y^2 - 2y - 7,\ x - 2y + 1 \rangle$ is the associated prime ideal.

Now we are ready to give the procedure for the zero–dimensional decomposition. We describe first the main steps:

Algorithm 4.2.7 (ZERODECOMP(I)).

Input: a zero-dimensional ideal $I := \langle f_1, \ldots, f_k \rangle \subset K[x]$, $x = (x_1, \ldots, x_n)$.
Output: a set of pairs (Q_i, P_i) of ideals in $K[x]$, $i = 1, \ldots, r$, such that
- $I = Q_1 \cap \cdots \cap Q_r$ is a primary decomposition of I, and
- $P_i = \sqrt{Q_i}$, $i = 1, \ldots, r$.

- result $:= \emptyset$;
- choose a random $\underline{a} \in K^{n-1}$, and apply the coordinate change $I' := \varphi_{\underline{a}}(I)$ (cf. Proposition 4.2.2);
- compute a Gröbner basis G of I' with respect to the lexicographical ordering with $x_1 > \cdots > x_n$, and let $g \in G$ be the element with smallest leading monomial.
- factorize $g = g_1^{\nu_1} \cdot \ldots \cdot g_s^{\nu_s} \in K[x_n]$;
- for $i = 1$ to s do
    set $Q_i' := \langle I', g_i^{\nu_i} \rangle$ and $Q_i := \langle I, \varphi_{\underline{a}}^{-1}(g_i)^{\nu_i} \rangle$;
    set $P_i' := \text{PRIMARYTEST}(Q_i')$;
    if $P_i' \neq \langle 0 \rangle$

      set $P_i := \varphi_{\underline{a}}^{-1}(P_i')$;
      result := result $\cup \{(Q_i, P_i)\}$;
    else
      result := result $\cup$ ZERODECOMP (Q_i);
• return result.

In the programming language of SINGULAR the procedure can be found in Section 4.6.

SINGULAR Example 4.2.8 (zero–dim primary decomposition).
We give an example for a zero-dimensional primary decomposition.

```
option(redSB);
ring R=0,(x,y),lp;
ideal I=(y2-1)^2,x2-(y+1)^3;
```

The ideal I is not in general position with respect to lp, since the minimal associated prime $\langle x^2 - 8, y - 1 \rangle$ is not.

```
map phi=R,x,x+y;       //we choose a generic coordinate change
map psi=R,x,-x+y;      //and the inverse map
I=std(phi(I));
I;
//-> I[1]=y7-y6-19y5-13y4+99y3+221y2+175y+49
//-> I[2]=112xy+112x-27y6+64y5+431y4-264y3-2277y2-2520y-847
//-> I[3]=56x2+65y6-159y5-1014y4+662y3+5505y2+6153y+2100
factorize(I[1]);
//-> [1]:
//->    _[1]=1
//->    _[2]=y2-2y-7
//->    _[3]=y+1
//-> [2]:
//->    1,2,3

ideal Q1=std(I,(y2-2y-7)^2); //the candidates for the
                             //primary ideals
ideal Q2=std(I,(y+1)^3);     //in general position
Q1; Q2;

//-> Q1[1]=y4-4y3-10y2+28y+49   Q2[1]=y3+3y2+3y+1
//-> Q1[2]=56x+y3-9y2+63y-7     Q2[2]=2xy+2x+y2+2y+1
                               Q2[3]=x2

factorize(Q1[1]);   //primary and general position test
                    //for Q1
//-> [1]:
```

```
//->     _[1]=1
//->     _[2]=y2-2y-7
//-> [2]:
//->     1,2
```

```
factorize(Q2[1]);    //primary and general position test
                     //for Q2
//-> [1]:
//->     _[1]=1
//->     _[2]=y+1
//-> [2]:
//->     1,3
```

Both ideals are primary and in general position.

```
Q1=std(psi(Q1));     //the inverse coordinate change
Q2=std(psi(Q2));     //the result
Q1; Q2;
```

```
//-> Q1[1]=y2-2y+1     Q2[1]=y2+2y+1
//-> Q1[2]=x2-12y+4    Q2[2]=x2
```

We obtain that I is the intersection of the primary ideals Q_1 and Q_2 with associated prime ideals $\langle y - 1, x^2 - 8 \rangle$ and $\langle y + 1, x \rangle$.

Exercises

4.2.1. Let K be a field of characteristic 0, $\overline{K}$ the algebraic closure of K and $I \subset K[x]$ an ideal. Prove that

(1) $I\overline{K}[x] \cap K[x] = I$;
(2) if $f \in K[x]$ is squarefree, then $f \in \overline{K}[x]$ is squarefree.

Condition (1) says that $\overline{K}[x]$ is a flat $K[x]$–module (cf. Chapter 7).

4.2.2. Let $I \subset K[x] = K[x_1, \ldots, x_n]$ be a zero–dimensional, and $J \subset K[x]$ a homogeneous ideal with $I \subset J \subset \sqrt{I}$. Prove that $\sqrt{I} = \langle x_1, \ldots, x_n \rangle$.

4.2.3. Let $I \subset K[x_1, \ldots, x_n]$ be a zero–dimensional ideal, and let $f \in K[x_n]$ be irreducible such that $I \cap K[x_n] = \langle f \rangle$. Let $\dim_K K[x_1, \ldots, x_n]/I = \deg(f)$. Prove that I is a prime ideal in general position with respect to the lexicographical ordering with $x_1 > \cdots > x_n$.

4.2.4. Compute a primary decomposition of $\langle x^2 + 1, y^2 + 1 \rangle \subset \mathbb{Q}[x, y]$, by following Algorithm 4.2.7 (without using SINGULAR).

4.2.5. Let K be a field of characteristic 0 and $M \subset K[x_1, \ldots, x_n]$ a maximal ideal. Prove that $K[x_1, \ldots, x_n]_M \cong K[x_1, \ldots, x_n]_{\langle x_1, \ldots, x_{n-1}, f \rangle}$ for a suitable $f \in K[x_n]$.

4.2.6. Give an example for a zero–dimensional ideal in $\mathbb{F}_2[x, y]$ which is not in general position with respect to the lexicographical ordering with $x > y$.

4.3 Higher Dimensional Primary Decomposition

In this section we show how to reduce the primary decomposition of an arbitrary ideal in $K[x]$ to the zero–dimensional case. We use the following idea:

Let K be a field and $I \subset K[x]$ an ideal. Let $u \subset x = \{x_1, \ldots, x_n\}$ be a maximal independent set with respect to the ideal I (cf. Definition 3.5.3) then $\emptyset \subset x \smallsetminus u$ is a maximal independent set with respect to $IK(u)[x \smallsetminus u]$ and, therefore, $IK(u)[x \smallsetminus u] \subset K(u)[x \smallsetminus u]$ is a zero–dimensional ideal (Theorem 3.5.1 (6)). Now, let $Q_1 \cap \cdots \cap Q_s = IK(u)[x \smallsetminus u]$ be an irredundant primary decomposition (which we can compute as we are in the zero–dimensional case), then also $IK(u)[x \smallsetminus u] \cap K[x] = (Q_1 \cap K[x]) \cap \cdots \cap (Q_s \cap K[x])$ is an irredundant primary decomposition. It turns out that $IK(u)[x \smallsetminus u] \cap K[x]$ is equal to the saturation $I : \langle h^\infty \rangle = \bigcup_{m>0} I : \langle h^m \rangle$ for some $h \in K[u]$ which can be read from an appropriate Gröbner basis of $IK(u)[x \smallsetminus u]$. Assume that $I : \langle h^\infty \rangle = I : \langle h^m \rangle$ for a suitable m (the ring is Noetherian). Then, using Lemma 3.3.6, we have $I = (I : \langle h^m \rangle) \cap \langle I, h^m \rangle$. Because we computed already the primary decomposition for $I : \langle h^m \rangle$ (an equidimensional ideal of dimension $\dim(I)$) we can use induction, that is, apply the procedure again to $\langle I, h^m \rangle$.

This approach terminates because either $\dim(\langle I, h^m \rangle) < \dim(I)$ or the number of maximal independent sets with respect to $\langle I, h^m \rangle$ is smaller than the number of maximal independent sets with respect to I (since u is not an independent set with respect to $\langle I, h^m \rangle$). The basis of this reduction procedure to the zero–dimensional case is the following proposition:

Proposition 4.3.1. *Let $I \subset K[x]$ be an ideal and $u \subset x = \{x_1, \ldots, x_n\}$ be a maximal independent set of variables with respect to I.*

(1) $IK(u)[x \smallsetminus u] \subset K(u)[x \smallsetminus u]$ is a zero–dimensional ideal.
(2) Let $S = \{g_1, \ldots, g_s\} \subset I \subset K[x]$ be a Gröbner basis of $IK(u)[x \smallsetminus u]$, and let $h := \mathrm{lcm}\big(\mathrm{LC}(g_1), \ldots, \mathrm{LC}(g_s)\big) \in K[u]$, then

$$IK(u)[x \smallsetminus u] \cap K[x] = I : \langle h^\infty \rangle ,$$

and this ideal is equidimensional of dimension $\dim(I)$.
(3) Let $IK(u)[x \smallsetminus u] = Q_1 \cap \cdots \cap Q_s$ be an irredundant primary decomposition, then also $IK(u)[x \smallsetminus u] \cap K[x] = (Q_1 \cap K[x]) \cap \cdots \cap (Q_s \cap K[x])$ is an irredundant primary decomposition.

Proof. (1) is obvious by definition of u and Theorem 3.5.1 (6).

(2) Obviously, $I : \langle h^\infty \rangle \subset IK(u)[x \smallsetminus u]$. To prove the inverse inclusion, let $f \in IK(u)[x \smallsetminus u] \cap K[x]$. S being a Gröbner basis, we obtain $\mathrm{NF}(f \mid S) = 0$,

where NF denotes the Buchberger normal form in $K(u)[x \smallsetminus u]$. But the Buchberger normal form algorithm requires only to divide by the leading coefficients $\mathrm{LC}(g_i)$ of the g_i, $i = 1, \ldots, s$. Hence, we obtain a standard representation $f = \sum_{i=1}^{s} \xi_i g_i$ with $\xi_i \in K[x]_h$. Therefore, $h^N f \in K[x]$ for some N. This proves $IK(u)[x \smallsetminus u] \cap K[x] \subset I : \langle h^\infty \rangle$.

To show that $I : \langle h^\infty \rangle \subset K[x]$ is an equidimensional ideal, suppose that $I = Q_1 \cap \cdots \cap Q_r$ is a primary decomposition of I with $Q_i \cap K[u] = \langle 0 \rangle$ for $i = 1, \ldots, s$ and $Q_i \cap K[u] \neq \langle 0 \rangle$ for $i = s+1, \ldots, r$. Then $IK(u)[x \smallsetminus u] = \bigcap_{i=1}^{s} Q_i K(u)[x \smallsetminus u]$ is a primary decomposition (Exercise 4.3.3). Since u is an independent set w.r.t. the ideals $\sqrt{Q_i K(u)[x \smallsetminus u]}$, $i = 1, \ldots, s$, it follows that all associated primes of $IK(u)[x \smallsetminus u]$ have at least dimension $\dim(I) = \#u$ (cf. Theorem 3.5.1 (6)).

(3) Obviously $Q_i \cap K[x]$ is primary and $\sqrt{Q_i \cap K[x]} \neq \sqrt{Q_j \cap K[x]}$ for $i \neq j$. Namely, $f \in \sqrt{Q_i}$ implies $f^m \in Q_i$ for a suitable m. It follows that $hf^m \in Q_i \cap K[x]$ for a suitable $h \in K[u]$, in particular, $(hf)^m \in Q_i \cap K[x]$. This implies $hf \in \sqrt{Q_i \cap K[x]}$. Assuming $\sqrt{Q_i \cap K[x]} = \sqrt{Q_j \cap K[x]}$, we would obtain $(hf)^\ell \in Q_j \cap K[x]$ for a suitable ℓ, that is, $f \in \sqrt{Q_j}$. This, together with the same reasoning applied to (j, i) in place of (i, j), would give $\sqrt{Q_i} = \sqrt{Q_j}$, contradicting the irredundance assumption. Similarly, we obtain a contradiction if we assume that $Q_i \cap K[x]$ can be omitted in the decomposition. □

Now we are prepared to give the algorithms. We start with a "universal" algorithm to compute all the ingredients we need for the reduction to the zero–dimensional case, as described above. We need this procedure for the primary decomposition and also for the computation of the equidimensional decomposition and the radical.

Algorithm 4.3.2 (REDUCTIONTOZERO(I)).

Input: $I := \langle f_1, \ldots, f_k \rangle \subset K[x]$, $x = (x_1, \ldots, x_n)$.
Output: A list (u, G, h), where
  – $u \subset x$ is a maximal independent set with respect to I,
  – $G = \{g_1, \ldots, g_s\} \subset I$ is a Gröbner basis of $IK(u)[x \smallsetminus u]$,
  – $h \in K[u]$ such that $IK(u)[x \smallsetminus u] \cap K[x] = I : \langle h \rangle = I : \langle h^\infty \rangle$.

- compute a maximal independent set $u \subset x$ with respect to I; [2]
- compute a Gröbner basis $G = \{g_1, \ldots, g_s\}$ of I with respect to the lexicographical ordering with $x \smallsetminus u > u$;
- $h := \prod_{i=1}^{s} \mathrm{LC}(g_i) \in K[u]$, where the g_i are considered as polynomials in $x \smallsetminus u$ with coefficients in $K(u)$;
- compute m such that $\langle g_1, \ldots, g_s \rangle : \langle h^m \rangle = \langle g_1, \ldots, g_s \rangle : \langle h^{m+1} \rangle$; [3]
- return $u, \{g_1, \ldots, g_s\}, h^m$.

---

[2] For the computation of a maximal independent set, cf. Exercises 3.5.1 and 3.5.2.
[3] For the computation of the saturation exponent m, cf. Section 1.8.9.

Note that G is, indeed, a Gröbner basis of $IK(u)[x \smallsetminus u]$ (with respect to the induced lexicographical ordering), since, for each $f \in IK(u)[x \smallsetminus u]$, we obtain $\mathrm{LM}(f) \in L(I) \cdot K(u)$.

SINGULAR Example 4.3.3 (reduction to zero–dimensional case).

```
option(redSB);
ring R=0,(x,y),lp;
ideal a1=x;                   //preparation of the example
ideal a2=y2+2y+1,x-1;
ideal a3=y2-2y+1,x-1;
ideal I=intersect(a1,a2,a3);
I;
//-> I[1]=xy4-2xy2+x
//-> I[2]=x2-x

ideal G=std(I);
indepSet(G);
//-> 0,1                     //the independent set is u={y}

ring S=(0,y),(x),lp;        //the ring K(u)[x\u]
ideal G=imap(R,G);
G;
//-> G[1]=(y4-2y2+1)*x
//-> G[2]=x2-x
```

This ideal in $K(y)[x]$ is obviously the prime ideal generated by x.

```
setring R;
poly h=y4-2y2+1;   //the lcm of the leading coefficients

ideal I1=quotient(I,h);
I1;
//-> I1[1]=x
```

Therefore, we obtain $I : \langle h \rangle = I : \langle h^\infty \rangle = G \cap K[x,y] = \langle x \rangle$, as predicted by Proposition 4.3.1 (2).

Combining everything so far, we obtain the following algorithm to compute a higher dimensional primary decomposition:

Algorithm 4.3.4 (DECOMP(I)).

Input: $I := \langle f_1, \ldots, f_k \rangle \subset K[x]$, $x = (x_1, \ldots, x_n)$.
Output: a set of pairs (Q_i, P_i) of ideals in $K[x]$, $i = 1, \ldots, r$, such that
  $- I = Q_1 \cap \cdots \cap Q_r$ is a primary decomposition of I, and
  $- P_i = \sqrt{(Q_i)}$, $i = 1, \ldots, r$.

- $(u, G, h) :=$ REDUCTIONTOZERO (I);
- change ring to $K(u)[x \smallsetminus u]$ and compute
    qprimary := ZERODECOMP $(\langle G \rangle_{K(u)[x \smallsetminus u]})$;
- change ring to $K[x]$ and compute
    primary := $\{(Q' \cap K[x], P' \cap K[x]) \mid (Q', P') \in$ qprimary$\}$;
- primary := primary $\cup$ DECOMP $(\langle I, h \rangle)$;
- return primary.

The intersection $Q' \cap K[x]$ may be computed by saturation: let Q' be given by a Gröbner basis $\{g'_1, \ldots, g'_m\} \subset K[x]$, and let $g' := \prod_{i=1}^{m} \mathrm{LC}(g'_i) \in K[u]$, then $Q' \cap K[x] = \langle g'_1, \ldots, g'_m \rangle : \langle g'^\infty \rangle \subset K[x]$ (Exercise 4.3.4).

The procedure in the SINGULAR programming language can be found in Section 4.6.[4]

SINGULAR Example 4.3.5 (primary decomposition).
Use the results of Example 4.3.3.

```
ideal I2=std(I+ideal(h));
//we compute now the decomposition of I2
indepSet(I2);
//-> 0,0        // we are in the zero-dimensional case now

list fac=factorize(I2[1]);
fac;
//-> [1]:
//->    _[1]=1
//->    _[2]=y+1
//->    _[3]=y-1
//-> [2]:
//->    1,2,2

ideal J1=std(I2,(y+1)^2);     // the two candidates
ideal J2=std(I2,(y-1)^2);     // for primary ideals

J1; J2;
//-> J1[1]=y2+2y+1    J2[1]=y2-2y+1
//-> J1[2]=x2-x       J2[2]=x2-x
```

J1 and J2 are not in general position with respect to 1p. We choose a generic coordinate change.

```
map phi=R,x,x+y;     // coordinate change
map psi=R,x,-x+y;    // and the inverse map
```

---

[4] Note that the algorithm described above computes a primary decomposition which is not necessarily irredundant. Check this using Example 4.1.6 (3).

```
ideal K1=std(phi(J1));
ideal K2=std(phi(J2));
factorize(K1[1]);
//-> [1]:
//->    _[1]=1
//->    _[2]=y+2
//->    _[3]=y+1
//-> [2]:
//->    1,2,2

ideal K11=std(K1,(y+1)^2);        // the new candidates
                                  // for primary ideals
ideal K12=std(K1,(y+2)^2);        // coming from K1
factorize(K2[1]);
//-> [1]:
//->    _[1]=1
//->    _[2]=y
//->    _[3]=y-1
//-> [2]:
//->    1,2,2

ideal K21=std(K2,(y-1)^2);        // the new candidates
                                  // for primary ideals
ideal K22=std(K2,y2);             // coming from K2
K11=std(psi(K11));                // the inverse coordinate
                                  // transformation
K12=std(psi(K12));
K21=std(psi(K21));
K22=std(psi(K22));

K11; K12; K21; K22;               // the result
//-> K11[1]=y2+2y+1    K12[1]=y2+2y+1
//-> K11[2]=x          K12[2]=x-1

//-> K21[1]=y2-2y+1    K22[1]=y2-2y+1
//-> K21[2]=x          K22[2]=x-1
```

$K_{11}, \ldots, K_{22}$ are now primary and in general position with respect to 1p. K_{11} and K_{21} are redundant, because they contain I_1. We obtain $a_1 = I_1$, $a_2 = K_{12}$, $a_3 = K_{22}$ for the primary decomposition of I, as it should be, from the definition of I in Example 4.3.3.

Exercises

4.3.1. Compute the primary decomposition of the ideals $\langle xy, xz \rangle$ and $\langle x^2, xy \rangle$ in $\mathbb{Q}[x, y]$ using the algorithm `decomp`.

4.3.2. Let $I \subset K[x_1, \ldots, x_n]$ be an ideal, and let $u \subset x = \{x_1, \ldots, x_n\}$ be an independent set with respect to I. Prove that $IK(u)[x \smallsetminus u]$ is primary (respectively prime) if I is primary (respectively prime).

4.3.3. Let $I \subset K[x_1, \ldots, x_n]$ be an ideal, and let $I = Q_1 \cap \cdots \cap Q_r$ be an irredundant primary decomposition. Moreover, let $u \subset x = \{x_1, \ldots, x_n\}$ be an independent set with respect to I. Assume that $Q_i \cap K[u] = \langle 0 \rangle$ for $i = 1, \ldots, s$ and $Q_i \cap K[u] \neq \langle 0 \rangle$ for $i = s + 1, \ldots, r$.

Prove that $IK(u)[x \smallsetminus u] = \bigcap_{i=1}^{s} Q_i K(u)[x \smallsetminus u]$ is an irredundant primary decomposition.

4.3.4. Let $u \subset x = \{x_1, \ldots, x_n\}$ be a subset, $J \subset K(u)[x \smallsetminus u]$ an ideal, and let $\{g_1, \ldots, g_s\} \subset K[x_1, \ldots, x_n]$ be a Gröbner basis of J with respect to any global monomial ordering on $K(u)[x \smallsetminus u]$. Let $\widetilde{h} \in K[u]$ be the least common multiple of the leading coefficients of the g_i and h the squarefree part of $\widetilde{h}$. Prove that $J \cap K[x] = \langle g_1, \ldots, g_s \rangle : \langle h^\infty \rangle$.

4.3.5. Follow Examples 4.3.3 and 4.3.5 to compute an irredundant primary decomposition of the intersection of the Clebsch cubic (Figure A.1) and the Cayley cubic (Figure A.2).

4.4 The Equidimensional Part of an Ideal

In this section we shall compute the equidimensional part of an ideal and an equidimensional decomposition.

Definition 4.4.1. Let A be a Noetherian ring, let $I \subset A$ be an ideal, and let $I = Q_1 \cap \cdots \cap Q_s$ be an irredundant primary decomposition. The *equidimensional part* $E(I)$ is the intersection of all primary ideals Q_i with $\dim(Q_i) = \dim(I)$.[5] The ideal I (respectively the ring A/I) is called *equidimensional* or *pure dimensional* if $E(I) = I$. In particular, the ring A is called *equidimensional* if $E(\langle 0 \rangle) = \langle 0 \rangle$.

Example 4.4.2.

(1) Let $I = \langle x^2, xy \rangle = \langle x \rangle \cap \langle x, y \rangle^2 \subset K[x, y]$, K any field. Then $E(I) = \langle x \rangle$.
(2) Let $A = K[x, y, z]$ and $I = \langle xy, xz \rangle = \langle x \rangle \cap \langle y, z \rangle$ then $E(I) = \langle x \rangle$. The zero–set of I is shown in Figure 4.3, the plane being the zero–set of the equidimensional part.

---

[5] Note that because of Theorem 4.1.5 the definition is independent of the choice of the irredundant primary decomposition.

Fig. 4.3. The zero–set of $\langle xy, xz \rangle \subset K[x, y, z]$.

Using Proposition 4.3.1 (2) we obtain the following algorithm to compute the equidimensional part of an ideal:

Algorithm 4.4.3 (EQUIDIMENSIONAL(I)).

Input: $I := \langle f_1, \ldots, f_k \rangle \subset K[x]$, $x = (x_1, \ldots, x_n)$.
Output: $E(I) \subset K[x]$, the equidimensional part of I.

- set $(u, G, h) :=$ REDUCTIONToZERO (I);
- if $(\dim(\langle I, h \rangle) < \dim(I))$
      return $(\langle G \rangle : \langle h \rangle)$;
  else
      return $((\langle G \rangle : \langle h \rangle) \cap$ EQUIDIMENSIONAL $(\langle I, h \rangle))$.

SINGULAR Example 4.4.4 (equidimensional part).
We compute $E(I)$ for $I = \langle xy^4 - 2xy^2 + x, \ x^2 - x \rangle \subset K[x, y]$ (cf. SINGULAR Example 4.3.3). As seen above, REDUCTIONToZERO(I) returns $u = \{y\}$, $G = \{xy^4 - 2xy^2 + x, \ x^2 - x\}$ and $h = y^4 - 2y^2 + 1$. Using the results of Example 4.3.3, we compute the dimension of $\langle I, h \rangle$:

```
dim(std(I+ideal(h)));
//-> 0
```

Since $\dim(I) = \#u = 1$ and $\dim(\langle I, h \rangle) = 0$ as computed, we can stop here. The equidimensional part is $I_1 = \langle x \rangle$.

A little more advanced algorithm, returning the equidimensional part $E(I)$ and an ideal $J \subset K[x]$ with $I = E(I) \cap J$, written in the SINGULAR programming language, can be found in Section 4.6.
    We should just like to mention another method to compute the equidimensional part of an ideal (cf. [67]). Let $A = K[x_1, \ldots, x_n]$, K a field, and $I \subset A$ be an ideal. Then

$$E(I) = \text{Ann}\big(\text{Ext}_A^{n-d}(A/I, A)\big), \quad d = \dim(A/I)$$

(for the definition of Ext see Chapter 7).

Definition 4.4.5. Let A be a Noetherian ring, and let $I \subset A$ be an ideal without embedded prime ideals. Moreover, let $I = \bigcap_{i=1}^s Q_i$ be an irredundant primary decomposition. For $\nu \leq d = \dim(I)$ we define the ν-th equidimensional part $E_\nu(I)$ to be the intersection of all Q_i with $\dim(Q_i) = \nu$.[6]

Example 4.4.6. Let $I = \langle xy, xz \rangle = \langle x \rangle \cap \langle y, z \rangle \subset K[x, y, z]$, then $E_2(I) = \langle x \rangle$ and $E_1(I) = \langle y, z \rangle$.

Lemma 4.4.7. *Let A be a Noetherian ring and $I \subset A$ be an ideal without embedded prime ideals. Let $I = \bigcap_{i=1}^s Q_i$ be an irredundant primary decomposition such that $E(I) = \bigcap_{i=1}^k Q_i$. Then*

$$I : E(I) = \bigcap_{i=k+1}^s Q_i \, .$$

In particular, $I = E(I) \cap \big(I : E(I)\big)$.

Proof. $I : E(I) = \bigcap_{i=1}^s (Q_i : E(I)) = \bigcap_{i=k+1}^s (Q_i : E(I))$. Now $E(I) \not\subset \sqrt{Q_i}$ for $i = k+1, \ldots, s$, because the primary decomposition is irredundant and all associated primes are minimal by assumption. This implies $Q_i : E(I) = Q_i$, since otherwise $E(I) \subset Q_i : \langle f \rangle$ for some $f \notin Q_i$ and, by Lemma 4.1.3 (3), $E(I) \subset \sqrt{Q_i : \langle f \rangle} = \sqrt{Q_i}$. $\qquad\square$

Remark 4.4.8. Let A be a Noetherian ring, let $I \subset A$ be an ideal, and let $I = \bigcap_{i=1}^s Q_i$ be an irredundant primary decomposition with $E(I) = \bigcap_{i=1}^k Q_i$. Then $I : E(I) = \bigcap_{i=k+1}^s \widetilde{Q}_i$ for some primary ideals $\widetilde{Q}_i$ with $Q_i \subset \widetilde{Q}_i \subset \sqrt{Q_i}$, but $I = E(I) \cap \big(I : E(I)\big)$ need not be true. Just consider the following example: $I = \langle x^2, xy \rangle = \langle x \rangle \cap \langle x^2, y \rangle$, $E(I) = \langle x \rangle$ and $I : E(I) = \langle x, y \rangle$.

The following algorithm, based on Lemma 4.4.7, computes, for a given ideal I without embedded primes, all equidimensional parts.[7]

Algorithm 4.4.9 (EQUIDIMENSIONALDECOMP(I)).

Input: $I := \langle f_1, \ldots, f_k \rangle \subset K[x]$, $x = (x_1, \ldots, x_n)$.
Output: A list of ideals $I_1, \ldots, I_n \subset K[x]$ such that $I_1 = E(I)$, $I_2 = E(I : I_1)$, $\ldots$, $I_n = E(I_{n-2} : I_{n-1})$, and $\sqrt{I} = \bigcap_{j=1}^n \sqrt{I_j}$. If I is radical then the I_j are radical, too. for all j.

---

[6] The $E_\nu(I)$ are well-defined, because, under the above assumptions, the primary decomposition is uniquely determined (Theorem 4.1.5).

[7] If we apply the algorithm to an arbitrary ideal then we obtain a set of equidimensional ideals such that the intersection of their radicals is the radical of the given ideal. In case of $\langle x^2, xy \rangle$ we obtain $\langle x \rangle, \langle x, y \rangle$.

- $E :=$ EQUIDIMENSIONAL (I);
- return $\{E\} \cup$ EQUIDIMENSIONALDECOMP $(I : E)$.

SINGULAR Example 4.4.10 (equidimensional decomposition).
We use the results of SINGULAR Example 4.3.3:

```
ideal I2=quotient(I,I1);
I2;
//-> I2[1]=y4-2y2+1
//-> I2[2]=x-1
```

$I_2 = E(I_2)$, because I_2 is zero–dimensional (SINGULAR Example 4.4.4). Therefore, we obtain $E_1(I) = I_1 = \langle x \rangle$ and $E_0(I) = I_2 = \langle y^4 - 2y^2 + 1, x - 1 \rangle$ as the equidimensional components of I.

Exercises

4.4.1. Write a SINGULAR procedure to compute the equidimensional decomposition using Procedure 4.8.6.

4.4.2. Use the algorithm EQUIDIMENSIONAL to compute the equidimensional part of $\langle xy, xz \rangle \subset K[x, y, z]$.

4.4.3. Let $I \subset K[x_1, \dots, x_n]$ be an ideal and assume that $K[x_1, \dots, x_r] \subset K[x_1, \dots, x_n]/I$ is a Noether normalization. Prove that I is equidimensional if and only if every non–zero $f \in K[x_1, \dots, x_r]$ is a non–zerodivisor in $K[x_1, \dots, x_n]/I$.

4.4.4. Use Exercise 4.4.3 to check whether $\langle x^2 + xy, xz \rangle$ is equidimensional.

4.4.5. Follow the SINGULAR Examples of this section to compute an equidimensional decomposition of the ideal

$$\langle x^3 + x^2 y + x^2 z - x^2 - xz - yz - z^2 + z,\ x^2 xz + x^2 y - yz^2 - yz,$$
$$x^2 y^2 - x^2 y - y^2 z + yz \rangle\,,$$

and verify it by using the procedure EQUIDIMENSIONAL.

4.5 The Radical

In this section we describe the algorithm of Krick and Logar (cf. [139]) to compute the radical of an ideal. Similarly to the algorithm for primary decomposition, using maximal independent sets, the computation of the radical is reduced to the zero–dimensional case.

Proposition 4.5.1. *Let $I \subset K[x_1, \ldots, x_n]$ be a zero–dimensional ideal and $I \cap K[x_i] = \langle f_i \rangle$ for $i = 1, \ldots, n$. Moreover, let g_i be the squarefree part of f_i, then $\sqrt{I} = I + \langle g_1, \ldots, g_n \rangle$.*

Proof. Obviously, $I \subset I + \langle g_1, \ldots, g_n \rangle \subset \sqrt{I}$. Hence, it remains to show that $a^k \in I$ implies that $a \in I + \langle g_1, \ldots, g_n \rangle$. Let $\overline{K}$ be the algebraic closure of K. Using Exercise 4.2.1 we see that each g_i is the product of different linear factors of $\overline{K}[x_i]$. Due to Exercise 4.1.7, these linear factors of the g_i induce a splitting of the ideal $(I + \langle g_1, \ldots, g_n \rangle)\overline{K}[x]$ into an intersection of maximal ideals. Hence, $(I + \langle g_1, \ldots, g_n \rangle)\overline{K}[x]$ is radical (Exercise 4.5.7). Now consider $a \in K[x]$ with $a^k \in I + \langle g_1, \ldots, g_n \rangle$. Using Exercise 4.2.1 again, we obtain $a \in (I + \langle g_1, \ldots, g_n \rangle)\overline{K}[x] \cap K[x] = I + \langle g_1, \ldots, g_n \rangle$. $\square$

This leads to the following algorithm:

Algorithm 4.5.2 (ZERORADICAL(I)).

Input: a zero–dimensional ideal $I := \langle f_1, \ldots, f_k \rangle \subset K[x]$, $x = (x_1, \ldots, x_n)$.
Output: $\sqrt{I} \subset K[x]$, the radical of I.

- for $i = 1, \ldots, n$, compute $f_i \in K[x_i]$ such that $I \cap K[x_i] = \langle f_i \rangle$;
- return $I + \langle \text{SQUAREFREE}(f_1), \ldots, \text{SQUAREFREE}(f_n) \rangle$.

To reduce the computation of the radical for an arbitrary ideal to the zero–dimensional case we proceed as in Section 4.3. Let $u \subset x$ be a maximal independent set for the ideal $I \subset K[x]$, $x = (x_1, \ldots, x_n)$, and let $h \in K[u]$ satisfy

$$IK(u)[x \smallsetminus u] \cap K[x] = I : \langle h \rangle = I : \langle h^\infty \rangle$$

(cf. Proposition 4.3.1 (2)). Then $I = (I : \langle h \rangle) \cap \langle I, h \rangle$ (Lemma 3.3.6), which implies that $\sqrt{I} = \sqrt{I : \langle h \rangle} \cap \sqrt{\langle I, h \rangle}$ (Exercise 4.5.7). Now $IK(u)[x \smallsetminus u]$ is a zero–dimensional ideal (Theorem 3.5.1 (6)), hence, we may compute its radical by applying ZERORADICAL. Clearly,

$$\sqrt{IK(u)[x \smallsetminus u]} \cap K[x] = \sqrt{IK(u)[x \smallsetminus u] \cap K[x]} = \sqrt{I : \langle h \rangle},$$

and it remains to compute the radical of the ideal $\langle I, h \rangle \subset K[x]$. This inductive approach terminates similarly to the corresponding approach for the primary decomposition.

We obtain the following algorithm for computing the radical of an arbitrary ideal:

Algorithm 4.5.3 (RADICAL(I)).

Input: $I := \langle f_1, \ldots, f_k \rangle \subset K[x]$, $x = (x_1, \ldots, x_n)$.
Output: $\sqrt{I} \subset K[x]$, the radical of I.

- $(u, G, h) := \text{REDUCTIONTOZERO}(I)$;

- change ring to $K(u)[x \setminus u]$ and compute $J :=$ ZERORADICAL $(\langle G \rangle)$;
- compute a Gröbner basis $\{g_1, \ldots, g_\ell\} \subset K[x]$ of J;
- set $p := \prod_{i=1}^{\ell} \mathrm{LC}(g_i) \in K[u]$;
- change ring to $K[x]$ and compute $J \cap K[x] = \langle g_1, \ldots, g_\ell \rangle : \langle p^\infty \rangle$;
- return $(J \cap K[x]) \cap$ RADICAL $(\langle I, h \rangle)$.

SINGULAR Example 4.5.4 (radical).
Use the results of Example 4.3.3.

```
ideal rad=I1;
ideal I2=std(I+ideal(h));
dim(I2);
//-> 0          //we are in the zero-dimensional case now

ideal u=finduni(I2);    //finds univariate polynomials
                        //in each variable in I2
u;
//-> u[1]=x2-x
//-> u[2]=y4-2y2+1

I2=I2,x2-1,y2-1;        //the squarefree parts of
                        //u[1],u[2] are added to I2
rad=intersect(rad,I2);
rad;
//-> rad[1]=xy2-x
//-> rad[2]=x2-x
```

From the output, we read $\sqrt{I} = \langle xy^2 - x, x^2 - x \rangle$.

Exercises

4.5.1. Let K be a field of characteristic 0, $\overline{K}$ its algebraic closure and $P \subset K[x_1, \ldots, x_n]$ a maximal ideal. Prove that $P\overline{K}[x_1, \ldots, x_n]$ is a radical ideal.

4.5.2. Let K be a field of characteristic 0, let $\overline{K}$ be its algebraic closure, and let $I \subset K[x_1, \ldots, x_n]$ be a zero–dimensional radical ideal. Prove that $\dim_K K[x_1, \ldots, x_n]/I$ is equal to the number of associated prime ideals of $I\overline{K}[x_1, \ldots, x_n]$. This means, geometrically, that the number of points of the zero–set $V(I) \subset \overline{K}^n$ is equal to the dimension of the factor ring.

4.5.3. Let A be a ring, $I \subset A$ an ideal. Prove that

(1) $\sqrt{\langle I, fg \rangle} = \sqrt{\langle I, f \rangle} \cap \sqrt{\langle I, g \rangle}$,

(2) $\sqrt{I} = \sqrt{IA_f \cap A} \cap \sqrt{\langle I, f \rangle}$.

4.5.4 (Factorizing Gröbner basis algorithm). The idea of the *factorizing Gröbner basis algorithm* is to factorize, during Algorithm 1.7.1, a new polynomial when it occurs and then split the computations. A simple version is described in the following algorithm (we use the notations of Chapter 1).

Algorithm (FACSTD(G,NF)).
Let $>$ be a well–ordering.

Input: $G \in \mathcal{G}$, NF an algorithm returning a weak normal form.
Output: $S_1, \ldots, S_r \in \mathcal{G}$ such that $\sqrt{\langle S_1 \rangle} \cap \cdots \cap \sqrt{\langle S_r \rangle} = \sqrt{\langle G \rangle}$ and S_i is a
          standard basis of $\langle S_i \rangle$.

- $S := G$;
- if there exist non–constant polynomials g_1, g_2 with $g_1 g_2 \in S$
        return FACSTD$(S \cup \{g_1\}, \text{NF}) \cup$ FACSTD$(S \cup \{g_2\}, \text{NF})$;
- $P := \{(f, g) \mid f, g \in S, \ f \neq g\}$, the pair–set;
- while $(P \neq \emptyset)$
        choose $(f, g) \in P$;
        $P := P \smallsetminus \{(f, g)\}$;
        $h := \text{NF}(\text{spoly}(f, g) \mid S)$;
        if $(h \neq 0)$
            if $(h = h_1 h_2$ with non–constant polynomials $h_1, h_2)$
                return FACSTD$(S \cup \{h_1\}, \text{NF}) \cup$ FACSTD$(S \cup \{h_2\}, \text{NF})$;
            $P := P \cup \{(h, f) \mid f \in S\}$;
            $S := S \cup \{h\}$;
- return S.

Prove that the output of FACSTD has the required properties. Moreover, use the command `facstd` of SINGULAR to compute a decomposition of the ideal I of Example 4.3.3.

Note that FACSTD can be used for the computation of the radical.

4.5.5. Let I_1 be primary and $I_2 \not\subset \sqrt{I_1}$. Prove that $\sqrt{I_1 : I_2^i} = \sqrt{I_1}$ for $i \geq 1$.

4.5.6. Let I be a radical ideal. Prove that, for every $h \notin I$, the ideal quotient $I : \langle h \rangle$ is a radical ideal.

4.5.7. Prove that $\sqrt{I \cap J} = \sqrt{I} \cap \sqrt{J}$.

4.5.8. *(Shape Lemma)* Let K be a field of characteristic 0, and let $I \subset K[x], x = (x_1, \ldots, x_n)$, be a zero–dimensional radical ideal. Prove that for almost all changes of coordinates $I = \langle x_1 + g_1(x_n), \ldots, x_{n-1} + g_{n-1}(x_n), g_n(x_n) \rangle$ for suitable $g_1, \ldots, g_n \in K[x_n]$.

4.6 Characteristic Sets

In this chapter we introduce characteristic sets and develop another method to compute the minimal associated primes of an ideal. The concept of characteristic sets goes back to Ritt and Wu (cf. [195], [236]).

Let R be an integral domain and $f, g \in R[x]$, the univariate polynomial ring over R. For $f = \sum_{\nu=0}^{m} f_\nu x^\nu$ of degree $\deg(f) = m$ with $f_m \neq 0$ we call $f_m =: \text{In}(f, x)$ the *initial form* of f (with respect to x). Here and in the following discussion we mention x explicitly since in our application, $R[x]$ will be a polynomial ring in several variables $x_1, \ldots, x_n$ where x will be one of the variables x_i.

Proposition 4.6.1. *For $f \in R[x] \setminus \{0\}$ and $g \in R[x]$ there exist uniquely determined $q, r \in R[x]$ with the following properties:*

(1) $\text{In}(f, x)^\alpha \cdot g = qf + r$, $\alpha = \max\{0, \deg(g) - \deg(f) + 1\}$,
(2) $r = 0$ or $\deg(r) < \deg(f)$.

Definition 4.6.2. The element $q =: \text{pquot}(g \mid f, x)$, is called the *pseudo quotient* of g with respect to f (and the variable x) and $r =: \text{prem}(g \mid f, x)$ the *pseudo remainder* of g with respect to f.

Proof. We use induction on α. If $\alpha = 0$ then $\deg(g) < \deg(f) = m$ and (1) holds with $q = 0$ and $r = g$.

Let $\alpha \geq 1$ and $g = \sum_{\nu=0}^{s} g_\nu x^\nu, g_s \neq 0$, then $s \geq m$ and

$$f_m^{s-m+1} g - f_m^{s-m} g_s f x^{s-m} = f_m^{s-m} \sum_{\nu=0}^{s-1} (f_m g_\nu - g_s f_{\nu-s+m}) x^\nu$$

(here we use the convention $f_\nu = 0$ if $\nu < 0$).

Now, using the induction hypothesis, we obtain

$$f_m^{s-m} \sum_{\nu=0}^{s-1} (f_m g_\nu - g_s f_{\nu-s+m}) x^\nu = q'f + r$$

and $r = 0$ or $\deg(r) < m$.

Then for $q = q' + f_m^{s-m} g_s x^{s-m}$ we have

$$f_m^{s-m+1} g = qf + r.$$

To see uniqueness assume $qf + r = q'f + r'$ which implies $(q - q')f = r' - r$. If $r' - r \neq 0$ then $\deg(r' - r) < m = \deg(f)$. But this is impossible since R is an integral domain. Hence, $r = r'$ and $q = q'$ since $R[x]$ is an integral domain too. $\qquad\square$

Now we extend this concept to several variables:

Let K be a field and $x_1, \ldots, x_n$ be variables[8]. Let $f_1, \ldots, f_r \in K[x_1, \ldots, x_n]$ be given with the property that for $1 \leq i_1 < i_2 < \cdots < i_r \leq n$, $f_k \in K[x_1, \ldots, x_{i_k}] \setminus K[x_1, \ldots, x_{i_k-1}]$ does not depend on the variables $> x_{i_k}$ but x_{i_k} really appears in f_k for $k = 1, \ldots, r$. The variable x_{i_k} is called the *principal variable* of f_k. We additionally allow that $f_1 \in K$ with the principal variable x_1.

For $g \in K[x_1, \ldots, x_n]$ define a sequence of pseudo–remainders (with respect to $1 \leq i_1 < i_2 < \cdots < i_r \leq n$ and $f_1, \ldots, f_r$ as above) inductively as follows:

$$R_r := g \text{ and for } 0 \leq k < r,$$
$$R_k := \operatorname{prem}(R_{k+1}|f_{k+1}, x_{i_{k+1}}),$$

to be understood in the polynomial ring $(K[x_1, \ldots, \widehat{x}_{i_{k+1}}, \ldots, x_n])[x_{i_{k+1}}]$[9].

Applying 4.6.1 successively we get $\operatorname{In}(f_r, x_{i_r})^{\alpha_r} R_r = q_r f_r + R_{r-1}$, $\operatorname{In}(f_{r-1}, x_{i_{r-1}})^{\alpha_{r-1}} R_{r-1} = q_{r-1} f_{r-1} + R_{r-2}$ and so on. Substituting, we finally obtain

Lemma 4.6.3. *With the notations above we have:*

(1) $\operatorname{In}(f_1, x_{i_1})^{\alpha_1} \cdot \ldots \cdot \operatorname{In}(f_r, x_{i_r})^{\alpha_r} g = \sum\limits_{\nu=1}^{r} \operatorname{pquot}(R_\nu|f_\nu, x_{i_\nu}) f_\nu + R_0,$

$\alpha_k = \max\{0, \deg_{x_{i_k}}(R_k) - \deg_{x_{i_k}}(f_k) + 1\},$

(2) $R_0 = 0$ *or* $\deg_{x_{i_k}}(R_0) < \deg_{x_{i_k}}(f_k).$

Definition 4.6.4. Keeping the above notations we define for given $f_1, \ldots, f_r$:

(1) $\operatorname{In}(f_\nu) := \operatorname{In}(f_\nu, x_{i_\nu})$ the *initial form* of f_ν (w.r.t. the principal variable).
(2) $R_0 =: \operatorname{prem}(g|\{f_1, \ldots, f_r\})$, the *pseudo remainder* of g (w.r.t. $\{f_1, \ldots, f_r\}$).
(3) If $g = \operatorname{prem}(g|\{f_1, \ldots, f_r\})$ we say that g is *reduced* with respect to $\{f_1, \ldots, f_r\}$.
(4) $\{f_1, \ldots, f_r\}$ is called an *ascending set*[10] if f_i is reduced with respect to $\{f_1, \ldots, f_{i-1}\}$ for $i = 2, \ldots, r$.
(5) Let $t_\nu := \begin{cases} \deg_{x_{i_k}}(f_k) & \text{if } \nu = i_k \text{ for some } k \\ \infty & \text{else} \end{cases}$

then $\operatorname{type}(\{f_1, \ldots, f_r\}) := (t_1, \ldots, t_n)$ is called the *type* of $\{f_1, \ldots, f_r\}$.
(6) The type of f_ν is the type of $\{f_\nu\}$.

Hence, if x_ν is a principal variable of some f_k then $t_\nu = \deg_{x_\nu}(f_k)$, otherwise $t_\nu = \infty$.

---

[8] In this chapter we fix an ordering of the variables such that $x_1 < x_2 < \ldots < x_n$. The definitions and constructions will depend on this ordering.

[9] ^ means that the variable below ^ is omitted

[10] We consider ascending sets (and later characteristic sets) as ordered sets but keep the notation $\{f_1, \ldots, f_r\}$.

Example 4.6.5. Let $f_1 = (x_1 + 1)^2$, $f_2 = (x_1 + 1)x_2^2 + x_1$, $g = (x_2^2 + 1)^2 f_1$, and $h = f_2(f_2 + 2)$. Then we have

(1) $\{f_1, f_2\}$ is an ascending set of type $(2, 2)$,
(2) $0 = \mathrm{prem}(g | \{f_1, f_2\})$,
(3) $0 = \mathrm{prem}(h | \{f_1, f_2\})$,
(4) $1 = g - h$, that is, $\mathrm{prem}(g - h | \{f_1, f_2\}) = 1$.

To see this just check that $(x_1 + 1)^3 = \mathrm{prem}(g \mid f_2, x_2)$.

The example shows that $\mathrm{prem}(- \mid \{f_1, f_2\})$ is not additive, hence not a good notion in general. Especially the set $\{h | \mathrm{prem}(h | \{f_1, f_2\}) = 0\}$ is not an ideal. In good cases, however, this set is a (prime) ideal and an important object as we shall see below (cf. Proposition 4.6.16 and 4.6.18).

The aim now is to prove the following proposition:

Proposition 4.6.6. *Let K be a field and let $I = \langle h_1, \ldots, h_s \rangle \subsetneq K[x_1, \ldots, x_n]$ be an ideal. There exists an ascending set $T = \{g_1, \ldots, g_r\}$ with the following properties:*

(1) $g_i \in I$, $i = 1, \ldots, r$.
(2) $\mathrm{prem}(h_j | T) = 0$, $j = 1, \ldots, s$.

To prove the proposition we need the possibility to compare ascending sets.

Definition 4.6.7. Let $T = \{f_1, \ldots, f_r\}$, $T' = \{g_1, \ldots, g_s\}$ be ascending sets. We define $T < T'$ if $\mathrm{type}(T) < \mathrm{type}(T')$ with respect to the lexicographical ordering.

Example 4.6.8. Let $f_1 = (x_1 + 1)^2$, $f_2 = (x_1 + 1)x_2^2 + x_1$ then $\mathrm{type}(f_1) = (1, \infty)$, $\mathrm{type}(f_2) = (\infty, 2)$, $\mathrm{type}(\{x_1, f_2\}) = (1, 2)$, $\mathrm{type}(\{f_1, f_2\}) = (2, 2)$, $\mathrm{type}(x_1^3) = (3, \infty)$. Hence, $\{x_1, f_2\} < \{f_1, f_2\} < \{x_1^3\}$.

Lemma 4.6.9. *Let $T = \{f_1, \ldots, f_r\}$ be an ascending set in $K[x_1, \ldots, x_n]$ and assume $g \neq 0$ is reduced with respect to T. Then $T \cup \{g\}$ contains an ascending subset T' such that $T' < T$.*

Proof. Since g is reduced with respect to T we have, for each i, either $\{g\} < \{f_i\}$ or $\{f_i\} < \{g\}$.
If $\{g\} < \{f_1\}$ then $T' := \{g\} < T$.
If $\{f_r\} < \{g\}$ then $T' := \{f_1, \ldots, f_r, g\} < T$.
If $\{f_i\} < \{g\} < \{f_{i+1}\}$ then $T' := \{f_1, \ldots, f_i, g\} < T$.   □

Lemma 4.6.10. *Let $\mathfrak{M}$ be a set of ascending subsets of $K[x_1, \ldots, x_n]$, then $\mathfrak{M}$ has a minimal element ($\mathfrak{M}$ is partially well–ordered with respect to $<$).*

Proof. Let $\tau = \{\mathrm{type}(T) | T \in \mathfrak{M}\} \subset (\mathbb{N} \cup \{\infty\})^n$. The lexicographical ordering is a well–ordering and, therefore, τ has a minimal element. By definition, the corresponding element in $\mathfrak{M}$ is minimal.   □

Proof of Proposition 4.6.6. Let $F^{(0)} := \{h_1, \ldots, h_s\}$ and $T^{(0)}$ be minimal among the ascending sets contained in $F^{(0)}$ and let $R^{(0)} = F^{(0)} \smallsetminus T^{(0)}$. Assume $T^{(i-1)}, F^{(i-1)}$ and $R^{(i-1)}$ are already defined. If $R^{(i-1)} \neq \emptyset$ and

$$P := \left\{ \mathrm{prem}(g | T^{(i-1)}) | g \in R^{(i-1)} \right\} \neq \{0\}$$

then let $T^{(i)}$ be a minimal ascending set in $F^{(i)} := F^{(i-1)} \cup P$. In this case we have (Lemma 4.6.9) that $T^{(i)} < T^{(i-1)}$ and we define $R^{(i)} := F^{(i)} \smallsetminus T^{(i)}$.

If $R^{(i)} = \emptyset$ or $P = \{0\}$ we are done with $T = T^{(i)}$ because $F^{(0)} \subseteq F^{(i)}$. Moreover, due to Lemma 4.6.10 this case occurs after finitely many steps. □

Definition 4.6.11. Let $F = \{f_1, \ldots, f_s\}$ be a subset of $K[x_1, \ldots, x_n]$ and let $I = \langle f_1, \ldots, f_s \rangle$. An ascending set T with the properties (1) and (2) of Proposition 4.6.6 is called a *characteristic set* for F.

The proof of Proposition 4.6.6 provides the following algorithm to compute a characteristic set for F:

Algorithm 4.6.12 (CHARACTERISTIC(F)).

Input: $F = \{f_1, \ldots, f_s\}$
Output: a characteristic set for F

- Rest = F; $G = F$;

- While Rest $\neq \emptyset$
     Result = minAscending(G)
     If Result = $\{f\}$ with $f \in K$
        Rest = $\emptyset$
     else
        Rest = $\{\mathrm{prem}(g|\mathrm{Result}) \neq 0 \mid g \in G \smallsetminus \mathrm{Result}\}$
     $G = G \cup \mathrm{Rest}$

- return Result

Note that the proof of Proposition 4.6.1 provides an algorithm to compute the pseudo remainder (and the pseudo quotient). Moreover, we used in Algorithm 4.6.12 the algorithm minAscending(G):

Algorithm 4.6.13 (MINASCENDING(G)).

Input: $G = \{g_1, \ldots, g_s\}$
Output: a minimal ascending subset of G

- Result = $\emptyset$; Rest = G;

- While Rest $\neq \emptyset$

  Choose $f \in$ Rest of minimal type

  Result = Result $\cup \{f\}$

  If $f \in K$

  Rest $= \emptyset$

  else

  Rest $= \{g \in$ Rest $|g$ reduced with respect to $f\}$
- return Result

Example 4.6.14. Let $F = \{f_1, f_2, f_3\}$ with $f_1 = x_1x_4 + x_3 - x_1x_2$, $f_2 = x_3x_4 - 2x_2^2 - x_1x_2 - 1$, $f_3 = (x_1 + 1)x_4^2 - x_2(x_1 + 1)x_4 + x_1x_2 + 3x_2$.

We follow Algorithms 4.6.12 and 4.6.13 to compute a characteristic set for F. We start with Rest $= F$ and $G = F$:

(1) Result = MINASCENDING $(G) = \{f_1\}$

   Rest $= \{\text{prem}(f_2|\{f_1\}) =: f_4,\ \text{prem}(f_3|\{f_1\}) =: f_5\}$

   $f_4 = -x_3^2 + x_1x_2x_3 - 2x_1x_2^2 - x_1^2x_2 - x_1$

   $f_5 = (x_1 + 1)x_3^2 - x_1(x_1 + 1)x_2x_3 + x_1^3x_2 + 3x_1^2x_2$

   $G = G \cup$ Rest $= \{f_1, \ldots, f_5\}$.

(2) Result = MINASCENDING $(G) = \{f_4, f_1\}$

   Rest $= \{\text{prem}(f_3|\{f_4, f_1\}) =: f_6,\ \text{prem}(f_5|\{f_4, f_1\}) =: f_7\}$

   $f_6 = -2x_1(x_1 + 1)x_2^2 + 2x_1^2x_2 - x_1^2 - x_1$

   $f_7 = f_6$

   $G = G \cup$ Rest $= \{f_1, \ldots, f_7\}$.

(3) Result = MINASCENDING $(G) = \{f_6, f_1\}$

   Rest $= \{\text{prem}(f_4|\{f_6, f_1\} =: f_8\}$.

   $f_8 = 2x_1(x_1 + 1)x_3^2 - 2x_1^2(x_1 + 1)x_2x_3 + 2x_1^3(x_1 + 3)x_2$

   $G = G \cup$ Rest $= \{f_1, \ldots, f_8\}$.

(4) Result = MINASCENDING $(G) = \{f_6, f_8, f_1\}$

   Rest $= \emptyset$.

We obtain $T = \{f_6, f_8, f_1\}$ as characteristic set for F.

Remark 4.6.15. Different choices in the above algorithms give different characteristic sets. We illustrate this with two examples. In step (2) we could have chosen $\{f_5, f_1\}$ as minimal ascending set. This would result in $\{f_6, f_5, f_1\}$ as characteristic set for F.

In step (1) we could have chosen $\{f_2\}$ as minimal ascending set. This would give $\{\bar{f}_6, \bar{f}_5, f_2\}$ as minimal ascending set with

$$\bar{f}_5 = (-2x_1x_2^3 + 2x_1x_2^2 - x_1x_2 - 2x_2^3 - x_2)x_3$$
$$+2x_1^2x_2^3 - 2x_1^2x_2^2 + x_1^2x_2 + 4x_1x_2^4 - 2x_1x_2^3 + 4x_1x_2^2 - x_1x_2 + x_1 +$$
$$4x_2^4 + 4x_2^2 + 1$$

$$\bar{f}_6 = (-16x_1^2 - 32x_1 - 16)x_2^8$$
$$+(-16x_1^3 + 16x_1)x_2^7$$
$$+(-4x_1^4 + 24x_1^3 - 20x_1^2 - 64x_1 - 32)x_2^6$$
$$+(8x_1^4 - 32x_1^3 + 24x_1)x_2^5$$
$$+(-8x_1^4 + 24x_1^3 - 12x_1^2 - 48x_1 - 24)x_2^4$$
$$+(4x_1^4 - 16x_1^3 + 12x_1)x_2^3$$
$$+(-x_1^4 + 6x_1^3 - 5x_1^2 - 16x_1 - 8)x_2^2$$
$$+(-2x_1^3 + 2x_1)x_2$$
$$+(-x_1^2 - 2x_1 - 1).$$

The following proposition shows that, in general, a characteristic set G of a set of generators of an ideal does not generate the ideal. The difference is, however, controlled by products of initial forms of G.

Proposition 4.6.16. *Let* $I = \langle f_1, \dots, f_r \rangle \subseteq K[x_1, \dots, x_n]$ *be an ideal and* $G = \{g_1, \dots, g_s\}$ *a characteristic set for* $\{f_1, \dots, f_r\}$. *Let* $J := \langle G \rangle$ *and* $S := \{\text{In}(g_1)^{\alpha_1} \cdot \dots \cdot \text{In}(g_s)^{\alpha_s} \mid \alpha_1, \dots, \alpha_s \in \mathbb{N}\}$. *Let* H *be the ideal generated by all polynomials* h *with* $\text{prem}(h|G) = 0$. *Then we have the following inclusion of ideals:*

$$J \subseteq I \subseteq H \subseteq J : S.$$

Proof. We have $J \subseteq I \subseteq H$ by definition of G being a characteristic set for $\{f_1, \dots, f_r\}$. Now let h be a polynomial with $\text{prem}(h|G) = 0$. Then, by Lemma 4.6.3, there exist $g \in S$ such that $gh \in J$, that is, $h \in J : S$. This implies $H \subseteq J : S$. $\qquad \square$

Let us now explain how characteristic sets are related to primary decomposition.

Definition 4.6.17. *Let* $F = \{f_1, \dots, f_r\} \subset K[x_1, \dots, x_n]$ *be an ascending set and assume that* $f_\nu \in K[x_1, \dots, x_{i_\nu}] \setminus K[x_1, \dots, x_{i_\nu-1}]$ *for all* ν, $1 \leq i_1 < \cdots < i_r \leq n$. *Define inductively*
$$K_1 := K(x_1, \dots, x_{i_1-1}) \text{ and}$$
$$K_\nu = (K_{\nu-1}[x_{i_\nu-1}]/\langle f_{\nu-1}\rangle)(x_{i_{\nu-1}+1}, \dots, x_{i_\nu-1})$$
for $\nu = 2, \dots, r$.
F *is called an* irreducible ascending *set if each* f_ν *is irreducible in* $K_\nu[x_{i_\nu}]$.

Note that K_ν is a field if $f_{\nu-1}$ is irreducible. Hence, if F is an irreducible ascending set then all rings K_ν, $\nu = 1, \dots, r$ are field extensions of K.

Proposition 4.6.18. *Let* $F = \{f_1, \dots, f_r\} \subset K[x_1, \dots, x_n]$ *be an irreducible ascending set, then the set*

$$P = \{g \in K[x_1, \dots, x_n] \mid \text{prem}(g|F) = 0\}$$

is a prime ideal.

More precisely, let $1 \leq i_1 < \cdots < i_r \leq n$, $f_\nu \in K[x_1, \ldots, x_{i_\nu}] \smallsetminus K[x_1, \ldots, x_{i_\nu - 1}]$ *for* $\nu = 1, \ldots, r$, *and* $L := \{1, \ldots, n\} \smallsetminus \{i_1, \ldots, i_r\}$. *Then* F *generates a maximal ideal in* $K(\{x_\nu\}_{\nu \in L})[x_{i_1}, \ldots, x_{i_r}]$ *and*

$$P = (\langle F \rangle K(\{x_\nu\}_{\nu \in L})[x_{i_1}, \ldots, x_{i_r}]) \cap K[x_1, \ldots, x_n].$$

We call P the *prime ideal associated to the irreducible ascending set* F.

Proof. After a suitable coordinate change we may assume that $(i_1, \ldots, i_r) = (n-r+1, \ldots, n)$ and hence that $f_i \in K[x_1, \ldots, x_{n-r+i}] \smallsetminus K[x_1, \ldots, x_{n-r+i-1}]$. With the notations of Proposition 4.6.16 we have

$$J := \langle F \rangle \subseteq H = \langle P \rangle \subseteq J : S.$$

Now, by definition of an irreducible ascending set, we have that

$$K_\nu[x_{n-r+\nu}]/\langle f_\nu \rangle = K(x_1, \ldots, x_{n-r})[x_{n-r+1}, \ldots, x_{n-r+\nu}]/\langle f_1, \ldots, f_\nu \rangle$$

and hence $K(x_1, \ldots, x_{n-r})[x_{n-r+1}, \ldots, x_n]/J$ is a field. Therefore, $JK(x_1, \ldots, x_{n-r})[x_{n-r+1}, \ldots, x_n]$ is a maximal ideal.

Let $I := K[x_1, \ldots, x_n] \cap JK(x_1, \ldots, x_{n-r})[x_{n-r+1}, \ldots, x_n]$, then, by Lemma 4.6.3, $P \subseteq I$. We claim that $\mathrm{prem}(g|F) = 0$ for all $g \in I$, that is, $I \subseteq P$. Let $g \in I$ and choose $a \in K[x_1, \ldots, x_{n-r}]$ such that $ag \in J$. Now $a \cdot \mathrm{prem}(g|F) = \mathrm{prem}(Ag|F)$ and, since F generates J, we may assume that $g \in J$ and $g = \mathrm{prem}(g|F)$. We have to prove that $g = 0$. Assume $g \neq 0$ then $g \in K[x_1, \ldots, x_s] \smallsetminus K[x_1, \ldots, x_{s-1}]$ for some $s > n - r$ because $J \cap K[x_1, \ldots, x_{n-r}] = 0$.

On the other hand, since g is reduced, g satisfies the inequalities $0 < \deg_{x_s}(g) < \deg_{x_s}(f_{s+r-n})$. But this is impossible, because g is in the ideal generated by f_{s+r-n} which is irreducible in the ring

$$\big(K(x_1, \ldots, x_{n-r})[x_{n-r+1}, \ldots, x_{s-1}]/\langle f_1, \ldots, f_{s+r-n-1}\rangle\big)[x_s].$$

We proved $I = P$ and, therefore, P is a prime ideal. $\square$

Remark: The condition that F is irreducible is used to prove that P is an ideal which is wrong in general (cf. Example 4.6.5).

Let us now show the converse of Proposition 4.6.18

Proposition 4.6.19. *Let* $F = \{f_1, \ldots, f_r\} \subset K[x_1, \ldots, x_n]$ *be an ascending set. If* $P = \{g \in K[x_1, \ldots, x_n] \mid \mathrm{prem}(g|F) = 0\}$ *is a prime ideal then* F *is irreducible.*

Proof. Assume that F is not irreducible and assume as in the proof of Proposition 4.6.18 that $f_i \in K[x_1, \ldots, x_{n-r+i}] \smallsetminus [x_1, \ldots, x_{n-r+i-1}]$. Choose i minimal such that $\{f_1, \ldots, f_i\}$ is not reducible and assume $f_i = \bar{g} \cdot \bar{h}$ in $K_i[x_{n-r+i}]$ with $\deg_{x_{n-r+i}}(\bar{g}) > 0$, $\deg_{x_{n-r+i}}(\bar{h}) > 0$, where

$$K_i := K(x_1, \ldots, x_{n-r})[x_{n-r+1}, \ldots, x_{n-r+i-1}]/\langle f_1, \ldots, f_{i-1} \rangle.$$

This implies that $a f_i - g \cdot h = \sum_{\nu=1}^{i-1} g_\nu f_\nu$ for suitable $a \in K[x_1, \ldots, x_{n-r}]$, $g, h, g_\nu \in K[x_1, \ldots, x_{n-r+i}]$, $\deg_{x_{n-r+i}}(g) > 0$ and $\deg_{x_{n-r+i}}(h) > 0$ and $\deg_{x_{n-r+i}}(g) + \deg_{x_{n-r+i}}(h) = \deg_{x_{n-r+i}}(f_i)$.

Now $g \cdot h \in P$ by Proposition 4.6.16 and P is a prime ideal by assumption. Therefore, we may assume that $g \in P$. This implies $0 = \mathrm{prem}(g|F) = \mathrm{prem}(g|\{f_1, \ldots, f_i\}) = \mathrm{prem}(g|\{f_1, \ldots, f_{i-1}\})$ because of $\deg_{x_{n-r+i}}(g) < \deg_{x_{n-r+i}}(f_i)$. Hence, $\mathrm{In}(f_1)^{\alpha_1} \cdot \ldots \cdot \mathrm{In}(f_{i-1})^{\alpha_{i-1}} g \in \langle f_1, \ldots, f_{i-1} \rangle$ and, therefore, g is zero in $K_i[x_{n-r+i}]$ which implies f_i is zero in $K_i[x_{n-r+i}]$ and this gives a contradiction. $\square$

Let $I = \langle f_1, \ldots, f_r \rangle$ be an ideal and $\sqrt{I} = P_1 \cap \cdots \cap P_s$, P_i the minimal associated primes of I. We want to give an algorithm to compute irreducible ascending sets $G^{(1)}, \ldots, G^{(s)}$ such that $P_i = \{h \mid \mathrm{prem}(h \mid G^{(i)}) = 0\}$.

The algorithm is based on the following lemma.

Lemma 4.6.20. *Let $I = \langle f_1, \ldots, f_r \rangle$ be an ideal and $G = \{g_1, \ldots, g_s\}$ be a characteristic set for $\{f_1, \ldots, f_r\}$. Suppose that G is irreducible and let P be the prime ideal associated to G, then*

$$\sqrt{I} = P \cap \sqrt{\langle F_1 \rangle} \cap \cdots \cap \sqrt{\langle F_s \rangle}$$

with $F_i = \{f_1, \ldots, f_r, \mathrm{In}(g_i)\}$.

Proof. The lemma is a special consequence of Proposition 4.6.16 and left as an exercise (cf. proof of Proposition 3.3.5). $\square$

Lemma 4.6.21. *Let $F = \{f_1, \ldots, f_r\}$ be an ascending set. Assume that $\{f_1, \ldots, f_{k-1}\}$ is irreducible and F is not. With the notations of Proposition 4.6.18 there exist $a, b \in K[\{x_\nu\}_{\nu \in L}]$, irreducible polynomials $h_1, \ldots, h_s \in K[x_1, \ldots, x_{i_k}] \setminus K[x_1, \ldots, x_{i_k-1}]$ and $\rho_1, \ldots, \rho_s, \alpha_1, \ldots, \alpha_{k-1} \in \mathbb{N}$ such that*

(1) $a f_k = b h_1^{\rho_1} \cdot \ldots \cdot h_s^{\rho_s}$ in $K_k[x_{i_k}]$,
(2) $\mathrm{In}(f_1)^{\alpha_1} \cdot \ldots \cdot \mathrm{In}(f_{k-1})^{\alpha_{k-1}} b a h_1^{\rho_1} \cdot \ldots \cdot h_s^{\rho_s} \in \langle F \rangle$.

Here $K_k = K(\{x_\nu\}_{\nu \in L}) [x_{i_1}, \ldots, x_{i_{k-1}}]/\langle f_1, \ldots, f_{k-1} \rangle$

Proof. Let $f_k = \bar{h}_1^{\rho_1} \cdot \ldots \cdot \bar{h}_s^{\rho_s}$ be the factorisation of f_k in $K_k[x_{i_k}]$ into irreducible factors. Then we can write $\bar{h}_i = \frac{b_i h_i}{a_i}$, $h_i \in K[x_1, \ldots, x_n]$ irreducible, $b_i, a_i \in K[\{x_\nu\}_{\nu \in L}]$. For $a = a_1^{\rho_1} \cdot \ldots \cdot a_s^{\rho_s}$ and $b = b_1^{\rho_1} \cdot \ldots \cdot b_s^{\rho_s}$ we obtain $a f_k = b h_1^{\rho_1} \cdot \ldots \cdot h_s^{\rho_s}$ in $K_k[x_{i_k}]$ and hence (1). Let $g := a f_k - b h_1^{\rho_1} \cdot \ldots \cdot h_s^{\rho_s}$. Since the class of g in $K_k[x_{i_k}]$, and hence in K_k is zero, $g \in \langle f_1, \ldots, f_{k-1} \rangle$. Then $\mathrm{prem}(g|\{f_1, \ldots, f_{k-1}\}) = 0$ by Proposition 4.6.18 because $\{f_1, \ldots, f_{k-1}\}$ is irreducible. This implies (2). $\square$

If we combine now Exercise 4.5.3 Lemma 4.6.20 and Lemma 4.6.21 we obtain an algorithm to compute irreducible ascending sets for the associated prime ideals of an ideal:

(1) We try to find an element in I which factors as $f \cdot g$ and apply Exercise 4.5.3 in order to reduce the problem to the consideration of $\langle I, f \rangle$ and $\langle I, g \rangle$ separately. Indeed, we try to factor the generators of I.

(2) If $I = \langle g_1, \ldots, g_m \rangle$ with $g_1, \ldots, g_m$ irreducible, we compute an ascending set $T = \{f_1, \ldots, f_r\}$ for $\{g_1, \ldots, g_m\}$.

(3) If $T = \{f_1\}$ and $f_1 \in K$ then $I = \langle 1 \rangle$.

(4) If T is irreducible we obtain an associated prime P of I, where $P = \{h | \operatorname{prem}(h|T) = 0\}$, and use Lemma 4.6.20 for a decomposition

$$\sqrt{I} = P \cap \text{ Rest}$$

and continue with Rest.

(5) If T is not irreducible, we use Lemma 4.6.21 to obtain $b, h_1, \ldots, h_s$ such that

$$\operatorname{In}(f_1) \cdot \ldots \cdot \operatorname{In}(f_{k-1}) \cdot b \cdot h_1 \cdot \ldots \cdot h_s \in \sqrt{I}.$$

Then we use Exercise 4.5.3 to obtain

$$\sqrt{I} = \sqrt{\langle I, \operatorname{In}(f_1) \rangle} \cap \cdots \cap \sqrt{\langle I, \operatorname{In}(f_{k-1}) \rangle} \cap \sqrt{\langle I, b \rangle} \cap \sqrt{\langle I, h_1 \rangle} \cap \cdots \cap \sqrt{\langle I, h_s \rangle}$$

and continue with the $\langle I, \operatorname{In}(f_j) \rangle$, $\langle I, b \rangle$ and $\langle I, h_j \rangle$. If we know the factors of b we may split $\langle I, b \rangle$ again.

Altogether we obtain the following algorithm which computes irreducible ascending sets of a given set of generators of I such that the associated prime ideals (Proposition 4.6.18) are the minimal prime ideals of I.

Algorithm 4.6.22 (IRRASCENDING(F)).

Input: a set of polynomials $F \subseteq K[x_1, \ldots, x_n]$
Output: a set of irreducible ascending sets of the minimal prime ideals of $\langle F \rangle$

- Result $= \emptyset$;   Rest $= \{F\}$;
- While Rest $\neq \emptyset$
  Choose $X \in$ Rest;
  Rest $=$ Rest $\setminus \{X\}$;
  $T = $ CHARACTERISTIC(X);
  If $T = \{f\}$ with $f \in K$
       If Rest $= \emptyset$ and Result $= \emptyset$
       return $\{\{1\}\}$
  else
     If T is irreducible
       If Result $\neq \emptyset$ and $\operatorname{prem}(S \mid T) \neq 0$ for all $S \in$ Result
         Result $=$ Result $\cup \{T\}$
         Rest $=$ Rest $\cup \Big(\bigcup_{f \in I, \deg\left(\operatorname{In}(f)\right) > 0} \{T \cup X \cup \{\operatorname{In}(f)\}\} \Big)$
     else

choose $f_1, \ldots, f_{k-1} \in T$, $b, h_1, \ldots, h_s$ as in Lemma 4.6.21

$$\text{Rest} = \text{Rest} \cup \left(\bigcup_{j=1}^{s} \{T \cup X \cup \{h_j\}\} \right) \cup \{T \cup X \cup \{b\}\}$$

$$\cup \left(\bigcup_{j=1,\ldots,k-1,\deg\left(\text{In}(f_j)\right)>0} \{T \cup X \cup \{\text{In}(f_j)\}\} \right)$$

$\text{Rest} = \text{CLEAR (Result)}$

- return Result.

We used the following procedure:

Algorithm 4.6.23 (CLEAR(R)).

Input: R, a set of polynomials
Output: S, a subset of the input R with the following properties:
      (1) $X, Y \in S$ implies $\text{prem}(X|Y) \neq 0$,
      (2) given $X \in R$ there exists $Y \in S$ such that $\text{prem}(X|Y) = 0$

- $S = R$
- $t = 0$;
- while $t = 0$;
    If exist $X, Y \in S$ such that $\text{prem}(X|Y) = 0$
    $S = S \setminus \{X\}$;
    else
        $t = 1$;
- return S

One possibility to refine the algorithm is to use a splitting with the following procedure before the computation of the ascending sets.

Algorithm 4.6.24 (SPLIT(X)).

Input: X, a set of polynomials
Output: Result=$\{W_1, \ldots, W_k\}$, W_i set of irreducible polynomials such that
    $\bigcap_i \sqrt{\langle W_i \rangle} = \sqrt{\langle X \rangle}$

- $\text{Rest} = \{X\}$; $\text{Result} = \emptyset$;
- While $\text{Rest} \neq \emptyset$
    Choose $X \in \text{Rest}$;
    $\text{Rest} = \text{Rest} \setminus X$;
    If all elements of X are irreducible
        $\text{Result} = \text{Result} \cup \{X\}$;
    else
        Choose $f = g \cdot h \in X$, g, h nontrivial factors of f;
        $X = X \setminus \{f\}$;
        $\text{Rest} = \text{Rest} \cup \{X \cup \{g\}, X \cup \{h\}\}$;
- return Result

Example 4.6.25. Let $F = \{f_1, f_2, f_3\}$ be the set of polynomials of Example 4.6.14 and let $I = \langle F \rangle$. Let us compute a primary decomposition of $\sqrt{I}$.

The initialisation of the algorithm gives

(0)   Result $= \emptyset$, Rest $= \{F\}$.

(1) $X = F$, Rest $= \emptyset$.

As a result of Example 4.6.14 we obtain

$T = \text{CHARACTERISTIC } (X) = \{f_6, f_8, f_1\}$.

T is not irreducible:

$f_8 = 2x_1(x_1 + 1)(x_3 - 2x_1 x_2 + x_1)(x_3 + x_1 x_2 - x_1) - 2x_1^2 f_6$

and we obtain

$$
\begin{aligned}
h_1 &= x_3 - 2x_1 x_2 + x_1 \\
h_2 &= x_3 + x_1 x_2 - x_1 \\
b &= 2x_1(x_1 + 1) \\
\text{Rest} &= \{Y_1, Y_2, Y_3, Y_4\} \\
Y_1 &= F \cup T \cup \{h_1\} \\
Y_2 &= F \cup T \cup \{h_2\} \\
Y_3 &= F \cup T \cup \{b\} \\
Y_4 &= F \cup T \cup \{\text{In}(f_6)\}
\end{aligned}
$$

(2) $\begin{aligned}[t]
X &= Y_1, \quad \text{Rest} = \{Y_2, Y_3, Y_4\} \\
T &= \text{CHARACTERISTIC}(X) = \{f_6, h_1, \bar{f}_1\} \\
\bar{f}_1 &= x_1 x_4 + x_1 x_2 - x_1
\end{aligned}$

T is irreducible

$\begin{aligned}
\text{Result} &= \text{Result} \cup \{T\} = \{\{f_6, h_1, \bar{f}_1\}\} \\
\text{Rest} &= \text{Rest} \cup \{Y_5, Y_6\} && (\text{In}(h_1) = 1) \\
Y_5 &= Y_1 \cup T \cup \{-2x_1(x_1 + 1)\} && (\text{In}(f_6) = -2x_1(x_1 + 1)) \\
Y_6 &= Y_1 \cup T \cup \{x_1\} && (\text{In}(\bar{f}_1) = x_1)
\end{aligned}$

(3) $\begin{aligned}[t]
X &= Y_2, \quad \text{Rest} = \{Y_3, \ldots, Y_6\} \\
T &= \text{CHARACTERISTIC}(X) = \{f_6, h_2, \bar{f}_2\} \\
\bar{f}_2 &= x_1 x_4 - x_1 x_2 + x_1
\end{aligned}$

T is irreducible

$\text{prem}(\{f_6, h_1, \bar{f}_1\} | T) \neq 0$

$\begin{aligned}
\text{Result} &= \text{Result} \cup \{T\} = \{\{f_6, h_1, \bar{f}_1\}, \{f_6, h_2, \bar{f}_2\}\} \\
\text{Rest} &= \text{Rest} \cup \{Y_7, Y_8\} \\
Y_7 &= Y_2 \cup T \cup \{-2x_1(x_1 + 1)\} \\
Y_8 &= Y_2 \cup T \cup \{x_1\}
\end{aligned}$

Now we continue as before, leaving the details to the reader. If we end with an irreducible ascending set T in the algorithm, we always have $\text{prem}(\{f_6, h_1, \bar{f}_1\} | T) = 0$ or $\text{prem}(\{f_6, h_2, \bar{f}_2\} | T) = 0$.

No further ascending set is added to the result. We obtain as a result two irreducible ascending sets

$$T_1 := \{f_6, h_1, \bar{f}_1\}, \quad T_2 := \{f_6, h_2, \bar{f}_2\},$$

and as associated prime ideals

$$P_1 = \langle T_1 \rangle K(x_1)[x_2, x_3, x_4] \cap K[x_1, x_2, x_3, x_4]$$
$$= \langle -2(x_1 + 1)x_2^2 + 2x_1x_2 - x_1 - 1, x_3 - 2x_1x_2 + x_1, x_4 + x_2 - 1 \rangle,$$
$$P_2 = \langle T_2 \rangle K(x_1)[x_2, x_3, x_4] \cap K[x_1, x_2, x_3, x_4]$$
$$= \langle -2(x_1 + 1)x_2^2 + 2x_1x_21 - x_1 - 1, x_3 + x_1x_2 - x_1, x_4 - x_2 + 1 \rangle.$$

Finally, we have $\sqrt{I} = P_1 \cap P_2$.

SINGULAR Example 4.6.26 (irreducible ascending set).
We compute Example 4.6.25 by using the command `char_series`. This command computes the irreducible ascending sets associated to the generators of a given ideal using some (internally chosen) heuristic ordering of the variables. In this example internally the ordering $x_1 > x_2 > x_3 > x_4$ is chosen.

```
ring    R=0,x(4..1),dp;
ideal   I=-x(1)*x(2)+x(1)*x(4)+x(3),
          -x(1)*x(2)-2*x(2)^2+x(3)*x(4)-1,
          -x(1)*x(2)*x(4)+x(1)*x(4)^2+x(1)*x(2)-x(2)*x(4)
          +x(4)^2+3*x(2);
matrix M=char_series(I);
ring    S=(0,x(4)),x(1..3),dp;//to see the result with re-
matrix M=imap(R,M);          //spect to the choosen ordering
M;
//-> M[1,1]=(-2*x(4)^2+2*x(4)-1)*x(3)+(4*x(4)^3-10*x(4)^2
            +10*x(4)-3)
//-> M[1,2]=x(2)+(x(4)-1)
//-> M[1,3]=(2*x(4)^2-2*x(4)+1)*x(1)+(2*x(4)^2-4*x(4)+3)
//-> M[2,1]=(-2*x(4)^2-2)*x(3)+(x(4)^3+x(4)^2+x(4)-3)
//-> M[2,2]=2*x(2)+(-x(4)-1)
//-> M[2,3]=(x(4)^2+1)*x(1)+(x(4)^2+2*x(4)+3)
```

So far, we developed characteristic sets for a given set of polynomials. This is sufficient for practical computations and for implementations. In the remaining part of this section we take an invariant point of view by considering the corresponding concept for ideals without a specific set of generators. This is mainly of theoretical interest.

Definition 4.6.27. Let $I \subseteq K[x_1, \ldots, x_n]$ be an ideal and $G \subseteq I$ an ascending set. G is called a *characteristic set of I* if $\mathrm{prem}(h|G) = 0$ for all $h \in I$.

Example 4.6.28. Let G be an irreducible ascending set. Then the set $P = \{h | \mathrm{prem}(h|G) = 0\}$ is a (prime) ideal and G is a characteristic set of P.

Example 4.6.29. Let $I = \langle x_1^2, x_1 x_2, x_2^2 \rangle \subseteq K[x_1, x_2]$ then I is zero-dimensional and $F = \{x_1^2, x_1 x_2\}$ is a characteristic set of I. It is not difficult to see that even $\langle x_1^2, x_2 \rangle \subseteq \{h \mid \operatorname{prem}(h \mid F) = 0\}$. On the other hand, also $\operatorname{prem}(x_2^2 + 1 \mid F) = 0$. This implies that $\{h \mid \operatorname{prem}(h \mid F) = 0\}$ is not an ideal.

Below, we show that characteristic sets of an ideal I can be computed with the help of Gröbner basis. The examples below show that this is not completely obvious: a lexicographical Gröbner basis needs not be a characteristic set of I and even if we apply the algorithm CHARACTERISTIC to a lexicographical Gröbner basis, we need not get a characteristic set of I.

Example 4.6.30. Let $P = \langle f_1, f_2 \rangle \subseteq K[x_1, x_2, x_3, x_4]$, $f_1 = x_2 x_3^2 + x_1$, $f_2 = x_4^2 + x_3^3$, then P is a prime ideal and $\{f_1, f_2\}$ is a reduced Gröbner basis of P with respect to the lexicographical ordering $x_1 < \cdots < x_4$ but not an ascending set because $\operatorname{prem}(f_2 \mid \{f_1\}) = x_2^2 x_4^2 - x_1 x_2 =: g_2$. However, $\{f_1, g_2\}$ is a characteristic set of P and $\langle f_1, g_2 \rangle : x_1 x_2 = P$.

Example 4.6.31. Let $I = \langle x_1 x_2^3, x_2^3 x_3 \rangle \subseteq K[x_1, x_2, x_3]$. Then $\{x_1 x_2^3, x_2^3 x_3\}$ is a reduced Gröbner basis with respect to the lexicographical ordering and $\operatorname{prem}(x_2^3 x_3 \mid \{x_1 x_2^3\}) = 0$. We get that $\{x_1 x_2^3\}$ is a characteristic set of I but $\{h \mid \operatorname{prem}(h \mid \{x_1 x_2^3\} = 0\} = \langle x_2^3 \rangle$ is strictly bigger than I.

Example 4.6.32. Let $I = \langle x_1^2, x_1 x_2^2, x_2^5, x_3^3 - x_2^3 \rangle \subseteq K[x_1, x_2, x_3]$ then I is zero-dimensional and $F := \{x_1^2, x_1 x_2^2, x_2^5, x_3^3 - x_2^3\}$ is a reduced Gröbner basis of I with respect to the lexicographical ordering. The algorithm CHARACTERISTIC gives CHARACTERISTIC $(F) = \{x_1^2, x_1 x_2^2\}$ because $\operatorname{prem}(x_2^5 \mid \{x_1^2, x_1 x_2^2\}) = 0$ and $\operatorname{prem}(x_3^3 - x_2^3 \mid \{x_1^2, x_1 x_2^2\}) = 0$. This means that $T := \{x_1^2, x_1 x_2^2\}$ is a characteristic set for F. But $x_1 x_2^3 \in I$ and $\operatorname{prem}(x_1 x_2^3 \mid T) = x_1 x_2^3$. This implies that $I \not\subseteq \{h \mid \operatorname{prem}(h \mid T) = 0\}$ even though we started with a reduced Gröbner basis. Notice that $\{x_1^2, x_1 x_2^2, x_1 x_2^3\}$ is a characteristic set of I.

Proposition 4.6.33. *Let $I \subseteq K[x_1, \ldots, x_n]$ be an ideal and $G \subseteq I$ an ascending set. Then G is a characteristic set of I if and only if G is a minimal ascending set of I (w.r.t. the ordering of Definition 4.6.7).*

Proof. Suppose G is a minimal ascending set of I, that is, if $G' \subseteq I$ is ascending then $G' < G$ is not possible. Let $h \in I$ and $h = \operatorname{prem}(h \mid G)$. If $h \neq 0$, then, using Lemma 4.6.9, $G \cup \{h\}$ contains is an ascending subset G' and $G' < G$. This is a contradiction and hence G is a characteristic set of I.

Now assume that G is the a characteristic set of I and that $G' < G$ with $G' \subset I$ is an ascending set. Let $g' \in G'$ be responsible for $G' < G$, that is, if $G = \{g_1, \ldots, g_s\}$ then one of the following conditions is satisfied:

(1) $\{g_1, \ldots, g_s, g'\} \subseteq G'$,
(2) $\{g'\} < \{g_1\}$,
(3) $\{g_1, \ldots, g_{i-1}, g'\} \subseteq G'$ and $\{g'\} < \{g_i\}$.

This implies that g' is reduced with respect to G in all cases. Therefore, $g' = \text{prem}(g'|G) \neq 0$ which is a contradiction to $g' \in I$ and G being a characteristic set of I. $\square$

At the end of this chapter we shall give the idea of an algorithm to compute a characteristic set for an ideal. It will not be used later on.

First of all we treat the zero–dimensional case.

Proposition 4.6.34. *Let $I \subseteq K[x_1, \ldots, x_n]$ be a zero–dimensional ideal and $T = \{g_1, \ldots, g_s\}$ a characteristic set of I. Then $s = n$, $g_i \in K[x_1, \ldots, x_i] \smallsetminus K[x_1, \ldots, x_{i-1}]$ and $\deg_{x_i}(g_i) \leq \deg_{x_i}(h_i)$ where $\langle h_i \rangle = I \cap K[x_i]$.*

Proof. Assume $s < n$, then there exist $j < n$ such that $g_j \in K[x_1, \ldots, x_j] \smallsetminus K[x_1, \ldots, x_{j-1}]$ and, if $j < s$, $g_{j+1} \notin K[x_1, \ldots, x_{j+1}]$.

But $I \cap K[x_{j+1}] = \langle h_{j+1} \rangle \neq 0$ and, therefore, $\{g_1, \ldots, g_j, h_{j+1}, \ldots, h_n\}$ is an ascending set of smaller type than T, which is a contradiction to the minimality of T by Proposition 4.6.33. Obviously, $H := \{h_1, \ldots, h_n\}$ is an ascending set. If $\deg_{x_j}(g_j) > \deg_{x_j}(h_j)$ for some j then we have $\{g_1, \ldots, g_{j-1}, h_j, \ldots, h_n\} < T$ which is again a contradiction to the minimality of T. $\square$

Remark 4.6.35. With the notations of Proposition 4.6.34 the h_i can be computed from some given Gröbner basis (with respect to any given ordering) of I using linear algebra. Therefore, especially their degrees can be computed. This gives us an estimate for the degrees of the polynomials in a characteristic set.

We obtain the following algorithm:

Algorithm 4.6.36 (ZeroCharsets (F)).

Input: A set F of polynomials such that $\langle F \rangle$ is zero–dimensional
Output: A characteristic set T for $\langle F \rangle$

- Choose a (global) monomial ordering and compute a Gröbner basis G of $\langle F \rangle$;
- $d = \dim(\langle G \rangle)$;
- $i = 0$;
- while $i < n$; $i = i + 1$;
  $M = \{1, x_i, \ldots, x_i^d\}$;
  $h_i = \text{MINREL}(M, G)$;   (computes $\langle h_i \rangle = \langle F \rangle \cap K[x_i]$)
  $d_i = \deg(h_i)$;
- $T = \{h_1\}$;
- $I = \{(\alpha_1, \alpha_2, 0, \ldots, 0) \mid \alpha_1 < d_1, \alpha_2 \leq d_2)\}$; $i = 1$;

- while $i < n$

    $i = i + 1;$

    $M = \{x^\alpha | \alpha \in I\};$

    $f = \mathrm{MINREL}(M, G);$

    $d = \deg_{x_i}(f)$

    $T = T \cup \{f\}$

    $I = I \smallsetminus \{(\beta_1, \ldots, \beta_i, 0, \ldots, 0) \in I \mid \beta_i \geq d\}$

    $I = I \cup \{(\alpha_1, \ldots, \alpha_{i+1}, 0 \ldots, 0) \mid (\alpha_1, \ldots, \alpha_i, 0, \ldots, 0) \in I, \alpha_{i+1} \leq d_i\}$

- return T

In the algorithm above we used the algorithm MINREL:

Algorithm 4.6.37 (MINREL (M, G)).

Input: $M = \{m_1, \ldots, m_s\}$ a set of monomials, ordered with respect to the lexicographical ordering $x_1 < \cdots < x_n$, G a Gröbner basis with respect to a given ordering such that $\dim(\langle G \rangle) = 0$.

Output: 0 if M is linearly independent modulo $\langle G \rangle$ or a polynomial

$$h = \sum_{i=1}^{k} c_i m_i, \ c_k = 1, \ k \leq s \text{ minimal such that } h \in \langle G \rangle.$$

- $i = 0;$
- $d = \dim_K K[x_1, \ldots, x_n]/\langle G \rangle$
- while $i < s$

    $i = i + 1;$

    $f = \mathrm{NF}(m_i | G) = \sum_{j=1}^{d} c_{ji} x^{\alpha j}$

    $A = (c_{ab})_{a \leq d, b \leq i}$

    If $A \begin{pmatrix} y_1 \\ \vdots \\ y_i \end{pmatrix} = 0$ has a solution $(y_1, \ldots, y_{i-1}, 1)$

    return $m_i + \sum_{j=1}^{i-1} y_j m_j$

- return 0

If the ideal I is not zero–dimensional we can reduce the computation of a characteristic set to the zero–dimensional case using the following lemma:

Lemma 4.6.38. *Let $I \subseteq K[x_1, \ldots, x_n]$ be an ideal, $u \subseteq \{x_1, \ldots, x_n\}$ a maximal independent set of variables for I.*

Let $T = \{h_1, \ldots, h_s\} \subseteq K[x]$ be a characteristic set for $IK(u)[x \smallsetminus u]$ and assume that $IK(u)[x \smallsetminus u] \cap K[x] = I : h$ for $h \in K[u]$. Then $T' = \{hh_1, \ldots, hh_s\}$ is a characteristic set for I.

Proof. Let $f \in I$ then $\mathrm{prem}(f|T) = 0$. This implies $\mathrm{prem}(f|T') = 0$. On the other hand, by definition of h, we have $T' \subseteq I$. This proves the lemma. $\square$

Example 4.6.39. Let $F = \{f_1, f_2, f_3\}$ be as in Example 4.6.14. Then $u = \{x_1\}$ is a maximal independent set of variables for $\langle F \rangle$.

In $K(x_1)[x_2, x_3, x_4]$ we obtain

$$\text{ZeroCharsets}(F) = \left\{ -\frac{1}{x_1} f_6, -(x_1 + 1)f_4 + f_6, f_1 \right\}$$

(with the notations of Example 4.6.14).

Since $-\frac{1}{x_1} f_6 \in \langle F \rangle$ we obtain $\{-\frac{1}{x_1} f_6, -(x_1 + 1)f_4 + f_6, f_1\}$ as characteristic set for $\langle F \rangle$.

Exercises

4.6.1. Compute Example 4.6.25 with respect to the ordering $x_1 > x_2 > x_3 > x_4$ and compare the result with 4.6.26.

4.6.2. Let $>$ be the lexicographical ordering with $x_1 < \ldots < x_n$ and $f_1, \ldots, f_n \in K[x_1, \ldots, x_n]$. Assume that $\text{LM}(f_i) = x_i^{m_i}$ for $i = 1, \ldots, n$ (such a set of polynomials is called a triangular set and will be studied in the next chapter). Prove that $\{f_1, \ldots, f_n\}$ is a Gröbner basis. Assume furthermore that $\text{NF}(f_i \mid \{f_1, \ldots, f_{i-1}\}) = f_i$ and prove that $\{f_1, \ldots, f_n\}$ is a characteristic set for $I = \langle f_1, \ldots, f_n \rangle$.

4.6.3. Prove that $\{x_1^2 + 1, x_1 x_2 + 1\}$ is the characteristic set of a prime ideal in $\mathbb{Q}[x_1, x_2]$. Note that it is not a Gröbner basis with respect to any well-ordering, especially it is not a triangular set (cf. Exercise 4.6.2).

4.6.4. With the notations of Proposition 4.6.18 assume that F is a Gröbner basis of $\langle F \rangle K(\{x_\nu\}_{\nu \in L})[x_{i_1}, \ldots, x_{i_r}]$. Prove that $P = \langle F \rangle : h^\infty$ where $h = \prod_{\nu=1}^{r} In(f_\nu)$.

4.6.5. Prove Lemma 4.6.20.

4.7 Triangular Sets

In this chapter we introduce another method, triangular sets, in order to show how to decompose a zero–dimensional ideal in $K[x_1, \ldots, x_n]$ into so–called triangular ideals, ideals generated by a lexicographical Gröbner basis of n elements. This is a basic tool for symbolic pre-processing to solve zero–dimensional systems of polynomial equations.

In this chapter we fix the lexicographical ordering lp.

Definition 4.7.1. A set of polynomials $F = \{f_1, \ldots, f_n\} \subset K[x_1, \ldots, x_n]$ is called a *triangular set* if for each i

(1) $f_i \in K[x_{n-i+1}, \ldots, x_n]$,

(2) $\mathrm{LM}(f_i) = x_{n-i+1}^{m_i}$, for some $m_i > 0$.

Hence, f_1 depends only on x_n, f_2 on x_{n-1}, x_n and so on, until f_n which depends on all variables.

A list of triangular sets $F_1, \ldots, F_s$ is called a *triangular decomposition* of the zero–dimensional ideal I if

$$\sqrt{I} = \sqrt{\langle F_1 \rangle} \cap \ldots \cap \sqrt{\langle F_s \rangle}.$$

Remark 4.7.2. If F is a triangular set then Exercise 1.7.1 implies that F is a Gröbner basis of $\langle F \rangle$.

Proposition 4.7.3. *Let* $M \subset K[x_1, \ldots, x_n]$ *be a maximal ideal and* $G = \{g_1, \ldots, g_r\}$ *a minimal Gröbner basis of* M *such that* $\mathrm{LM}(g_1) < \ldots < \mathrm{LM}(g_r)$. *Then* G *is a triangular set, in particular* $r = n$.

Proof. We use induction on the number of variables, the case $n = 1$ being trivial. Since $M \cap K[x_2, \ldots, x_n]$ is maximal we may assume by Lemma 1.8.3 that $G \cap K[x_2, \ldots, x_n] = \{g_1, \ldots, g_{n-1}\}$ is a triangular set. In particular $r \geq n$, since M is a maximal ideal. Consider the ideal $\overline{M}$ induced by M in $(K[x_2, \ldots, x_n]/M \cap K[x_2, \ldots, x_n])[x_1]$. $\overline{M}$ is generated by the elements induced by $g_n, \ldots, g_r$. Because $\mathrm{LM}(g_1) < \ldots < \mathrm{LM}(g_r)$ and since G is a minimal lexicographical Gröbner basis we have

$$\deg_{x_1}(g_n) \leq \ldots \leq \deg_{x_1}(g_r).$$

$\overline{M}$ is a principal ideal as $K[x_2, \ldots, x_n]/M \cap K[x_2, \ldots, x_n]$ is a field. Using Euclid's algorithm we deduce that g_n induces a generator of $\overline{M}$, i.e. $M = \langle g_1, \ldots, g_n \rangle$.

We have still to prove that $r = n$.

Assume $r > n$. M being 0–dimensional and $\mathrm{LM}(g_r)$ maximal with respect to lp implies that $\mathrm{LM}(g_r) = x_1^m$ for some integer $m \geq 1$. By assumption we have $1 \leq \deg_{x_1}(g_n) < m$. Let $k \geq n$ be defined by $\deg_{x_1}(g_n) = \ldots = \deg_{x_1}(g_k) < \deg_{x_1}(g_{k+1})$. We claim that $G' = \{g_1, \ldots, g_k\}$ is a Gröbner basis of M. Since G is a Gröbner basis we have $\mathrm{NF}(\mathrm{spoly}(g_i, g_j) \mid G) = 0$ for $i, j \leq k$. But

$$\deg_{x_1}(\mathrm{spoly}(g_i, g_j)) \leq \deg_{x_1}(g_n) \text{ for all } i, j \leq k$$

implies that $\mathrm{LM}(\mathrm{spoly}(g_i, g_j)) < \mathrm{LM}(g_l)$ if $i, j \leq k$ and $l > k$. This shows that in the reduction process to compute the normal form of the s–polynomials the elements $g_{k+1}, \ldots, g_r$ are not used. Therefore $\mathrm{NF}(\mathrm{spoly}(g_i, g_j) \mid G') = 0$ for $i, j \leq k$, i.e. G' is a Gröbner basis of M. This implies $\mathrm{LM}(g_k) = x_1^s$ for some s because M is zero–dimensional. However, this contradicts the minimality of G. We proved $r = n$ and therefore the proposition. $\square$

Since a zero–dimensional prime ideal is maximal, we can apply Proposition 4.7.3 to a primary decomposition of $\sqrt{I}$ and get the following existence of a triangular decomposition of I.

Corollary 4.7.4. *If I is a zero–dimensional ideal then there exist triangular sets $F_1, \ldots, F_s$ such that*

(1) $\sqrt{I} = \sqrt{\langle F_1 \rangle} \cap \ldots \cap \sqrt{\langle F_s \rangle}$
(2) $\langle F_i \rangle + \langle F_j \rangle = K[x_1, \ldots, x_n]$ *for* $i \neq j$.

Using a primary decomposition is not satisfactory for practical computation. Our aim is to find triangular decompositions of a zero–dimensional ideal with less effort than the computation of the minimal associated primes. The following lemma is the basis for the algorithm by Möller (see [122], [169]) which avoids this computation.

Lemma 4.7.5. *Let $G = \{g_1, \ldots, g_r\}$ be a reduced (lexicographical) Gröbner basis for the zero–dimensional ideal $I \subset K[x_1, \ldots, x_n]$ and assume $\mathrm{LM}(g_1) < \ldots < \mathrm{LM}(g_r)$.*

Let $g_i = \sum\limits_{j=0}^{n_i} h_j^{(i)} x_1^j$, $h_j^{(i)} \in K[x_2, \ldots, x_n]$, $h_{n_i}^{(i)} \neq 0$ and $F = \left\{ h_{n_1}^{(1)}, \ldots, h_{n_{r-1}}^{(r-1)} \right\}$.

Then the following holds:
(1) F is a Gröbner basis for $\langle g_1, \ldots, g_{r-1} \rangle : g_r$ and
(2) $\sqrt{\langle F, g_r \rangle} = \sqrt{\langle F, G \rangle}$.

Proof. First we claim that $\{g_1, \ldots, g_{r-1}\}$ is a Gröbner basis of $\langle g_1, \ldots, g_{r-1} \rangle$. We have $\mathrm{NF}(\mathrm{spoly}(g_i, g_j) \mid G) = 0$ using Buchberger's criterion (Theorem 2.5.9). But if $i, j \leq r-1$ then g_r is not used in the reduction of the $\mathrm{spoly}(g_i, g_j)$ because $\mathrm{LM}(g_r) = x_1^m$ for some $m \in \mathbb{N}$ and $\mathrm{LM}(\mathrm{spoly}(g_i, g_j)) < x_1^m$. This implies $\mathrm{NF}(\mathrm{spoly}(g_i, g_j | G \smallsetminus \{g_r\}) = 0$ for $i, j \leq r-1$ and therefore $\{g_1, \ldots, g_{r-1}\}$ is a Gröbner basis again by Buchberger's criterion. If we set $h(g_i, g_r) := h_{n_i}^{(i)} \cdot g_r - x_1^{m-n_i} \cdot g_i$ then $\mathrm{NF}(h(g_i, g_r) | G) = 0$ and, as before, g_r is not used in the reduction, i.e. $\mathrm{NF}(h(g_i, g_r) | G \smallsetminus \{g_r\}) = 0$.

This implies that $h(g_i, g_r) \in \langle g_1, \ldots, g_{r-1} \rangle$ and, by definition of $h(g_i, g_r)$, that $h_{n_i}^{(i)} \cdot g_r \in \langle g_1, \ldots, g_{r-1} \rangle$. This implies that $h_{n_i}^{(i)} \in \langle g_1, \ldots, g_{r-1} \rangle : g_r$, i.e. $F \subseteq \langle g_1, \ldots, g_{r-1} \rangle : g_r$.

Conversely, let $f \in \langle g_1, \ldots, g_{r-1} \rangle : g_r$, i.e. $fg_r \in \langle g_1, \ldots, g_{r-1} \rangle$. There exists an i such that $\mathrm{LM}(g_i) \mid \mathrm{LM}(fg_r) = \mathrm{LM}(f) \cdot x_1^m$. However, this implies $\mathrm{LM}(h_{n_i}^{(i)}) \mid \mathrm{LM}(f)$ because $\mathrm{LM}(g_i) = \mathrm{LM}(h_{n_i}^{(i)}) x_1^{n_i}$ and $n_i < m$ and therefore F is a Gröbner basis of $\langle g_1, \ldots, g_{r-1} \rangle : g_r$.

The proof of (2) is left as Exercise 4.7.3. This proves the lemma. $\qquad \square$

Example 4.7.6. Let $I = \langle z^2 - 2, y^2 + 2y - 1, (y + z + 1)x + yz + z + 2,$ $x^2 + x + y - 1 \rangle \subset \mathbb{Q}[x, y, z]$. Then $I = P_1 \cap P_2 \cap P_3$ with the prime ideals
$$P_1 = \langle z^2 - 2, y - z + 1, x + z \rangle,$$
$$P_2 = \langle z^2 - 2, y + z + 1, x - z \rangle,$$
$$P_3 = \langle z^2 - 2, y + z + 1, x + z + 1 \rangle,$$
which are generated by triangular sets.

There is another triangular decomposition of I, namely

$$I = \langle I, y + z + 1 \rangle \cap (I : (y + z + 1)),$$

with
$$\langle I, y + z + 1 \rangle = \langle z^2 - 2, y + z + 1, x^2 + x + y - 1 \rangle$$
$$I : (y + z + 1) = \langle z^2 - 1, y - z + 1, x + z \rangle.$$

SINGULAR Example 4.7.7 (triangular decomposition).
We consider again Example 4.7.6 and compute the minimal associated primes, a triangular decomposition by using the command `triangMH`, and by applying the method of Lemma 4.7.5.

```
LIB"primdec.lib";
ring R=0,(x,y,z),lp;
ideal I=z2-2, y2+2y-1, (y+z+1)*x+yz+z+2, x2+x+y-1;
minAssGTZ(I);

//-> [1]:              [2]:              [3]:
//->    _[1]=z2-2         _[1]=z2-2         _[1]=z2-2
//->    _[2]=x+z          _[2]=x-z          _[2]=x2+x+y-1
//->    _[3]=x2+x+y-1     _[3]=x2+x+y-1     _[3]=x+z+1

option(redSB);       //a reduced lex Groebner basis is needed
I=std(I);            //as input for triangMH (algorithm
triangMH(I,2);       //of Moeller, Hillebrand)

//-> [1]:              [2]:
//->    _[1]=z2-2         _[1]=z2-2
//->    _[2]=y+z+1        _[2]=y-z+1
//->    _[3]=x2+x-z-2     _[3]=x+z

std(quotient(I,y+z+1));  //the second triangular set

//-> _[1]=z2-2
//-> _[2]=y-z+1
//-> _[3]=x+z

std(I,y+z+1);            //the first triangular set
                        //(recall the meaning of std(I, f))
//-> _[1]=z2-2
//-> _[2]=y+z+1
//-> _[3]=x2+x-z-2
```

We will now describe an algorithm to compute a triangular decomposition of a zero–dimensional ideal I

$$\sqrt{I} = \sqrt{\langle F_1 \rangle} \cap \ldots \cap \sqrt{\langle F_s \rangle}$$

with triangular sets F_i satisfying

$$\langle F_i \rangle + \langle F_j \rangle = K[x_1 \ldots x_n] \text{ for } i \neq j.$$

The algorithm is based on Lemma 4.7.5 and Exercise 4.7.1. We use the notations of Lemma 4.7.5. By applying first Exercise 4.7.1 and then 4.7.5 we obtain the following (defining additionally $h_{n_r}^{(r)} = 1$):

$$\sqrt{I} = \bigcap_{i=0}^{r-1} \sqrt{\langle G, h_{n_1}^{(1)}, \ldots, h_{n_i}^{(i)} \rangle : h_{n_{i+1}}^{(i+1)\infty}}$$

$$= \sqrt{\langle g_r, F \rangle} \cap \left(\bigcap_{i=0}^{r-2} \sqrt{\langle G, h_{n_1}^{(1)}, \ldots, h_{n_i}^{(i)} \rangle : h_{n_{i+1}}^{(i+1)\infty}} \right)$$

Now $\sqrt{\langle g_r, F \rangle}$ is nicely prepared for induction since $F \subseteq K[x_2, \ldots, x_n]$ and $\mathrm{LM}(g_r) = x_1^m$ for some $m \in \mathbb{N}$. This implies that a triangular set $T' \subset K[x_2, \ldots, x_n], \langle F \rangle \subseteq \sqrt{\langle T' \rangle}$, leads to a triangular set $T = T' \cup \{g_r\}$, $\sqrt{I} \subseteq \sqrt{\langle g_r, F \rangle} \subseteq \sqrt{\langle T \rangle}$. Therefore the decomposition above gives the possibility to compute a triangular decomposition inductively. This leads to the following recursive algorithm.

Algorithm 4.7.8 (TRIANGDECOMP (I)).

Input: a zero-dimensional ideal $I := \langle f_1, \ldots, f_m \rangle$

Output: A list of triangular sets $F_1, \ldots, F_s$ such that $\sqrt{I} = \bigcap_{i=1}^{s} \sqrt{\langle F_i \rangle}$ and

$$\langle F_i \rangle + \langle F_j \rangle = K[x_1, \ldots, x_n] \text{ for } i \neq j$$

- Compute $G = \{g_1, \ldots, g_r\}$ a reduced Gröbner basis for $\langle f_1, \ldots, f_m \rangle$ with respect to $>_{\mathrm{lp}}$ such that $\mathrm{LM}(g_1) < \ldots < \mathrm{LM}(g_r)$.
- Compute $G' = \{h_1, \ldots, h_{r-1}\} \subset K[x_2, \ldots, x_n]$, with h_i the leading coefficient of g_i considered as polynomial in x_1.
- $L' =$ TRIANGDECOMP($\langle G' \rangle$)
- $L = \{T' \cup \{g_r\} \mid T' \in L'\}$
- $i = 0$
- while $(i < r - 1)$
    $i = i + 1$
    If $h_i \notin G$
        $L = L \cup$ TRIANGDECOMP($\langle G \rangle : h_i^\infty$)
        $G = G \cup \{h_i\}$
- return L

Exercises

4.7.1. Let I be an ideal in a Noetherian ring R and $a_1, \ldots a_r \in R, a_r = 1$. Prove that $\sqrt{I} = \bigcap_{s=0}^{r-1} \sqrt{\langle I, a_1, \ldots, a_s \rangle : a_{s+1}^\infty}$.

Hint: Use Lemma 3.3.6 and Exercise 4.5.7.

4.7.2. Compute a triangular decomposition for Example 4.7.6 considered as an ideal in $\mathbb{Q}[z, y, x]$ (permutation of variables) and compare it with the result from Example 4.7.6.

4.7.3. Prove (2) of Lemma 4.7.5.

4.7.4. Consider the following system of equations over $\mathbb{Q}(a)[x, y, z]$:

$$
\begin{aligned}
ax^2 + 2y + a + 1 &= 0 \\
y^2 z + xy &= 0 \\
ayz^2 + z - a^2 + 1 &= 0
\end{aligned}
$$

Use the procedure **triangL** of the library **triang.lib** to compute the solutions depending on the parameter a (you may assume that a is generic).

4.7.5. Use the procedure **solve** of the library **solve.lib** to compute the solutions of the system of equations of 4.7.4 numerically for several specified parameters a (including $a = 1$).

4.7.6. Substitute in 4.7.4 special values for a (including $a = 1$) and recompute the triangular set. Substitute the same values for a in the result of the computation in Exercise 4.7.4 and compare the results.

4.8 Procedures

We collect the main procedures of this section as fully functioning SINGULAR procedures. However, since they are in no way optimized, one cannot expect them to be very fast. Each procedure has a small example to test it. This section demonstrates that it is not too difficult to implement a full primary decomposition, the equidimensional part and the radical.

4.8.1. We begin with a procedure to test whether a zero–dimensional ideal is primary and in general position.

proc primaryTest (ideal i, poly p)

```
"USAGE:   primaryTest(i,p); i standard basis with respect to
          lp, p irreducible polynomial in K[var(n)],
          p^a=i[1] for some a;
ASSUME:   i is a zero-dimensional ideal.
RETURN:   an ideal, the radical of i if i is primary and in
          general position with respect to lp,
          the zero ideal else.
"
{
    int m,e;
    int n=nvars(basering);
```

```
    poly t;
    ideal prm=p;

    for(m=2;m<=size(i);m++)
    {
      if(size(ideal(leadexp(i[m]))))==1)
      {
        n--;
//---------------i[m] has a power of var(n) as leading term
        attrib(prm,"isSB",1);
//--- ?? i[m]=(c*var(n)+h)^e modulo prm for h
//      in K[var(n+1),...], c in K ??
        e=deg(lead(i[m]));
        t=leadcoef(i[m])*e*var(n)+(i[m]-lead(i[m]))
        /var(n)^(e-1);
        i[m]=poly(e)^e*leadcoef(i[m])^(e-1)*i[m];
//---if not (0) is returned, else c*var(n)+h is added to prm
        if (reduce(i[m]-t^e,prm,1) !=0)
        {
          return(ideal(0));
        }
        prm = prm,cleardenom(simplify(t,1));
      }
    }
    return(prm);
}


ring s=(0,x),(d,e,f,g),lp;
ideal i=g^5,(x*f-g)^3,5*e-g^2,x*d^3;
primaryTest(i,g);
```

4.8.2. The next procedure computes the primary decomposition of a zero-dimensional ideal.

proc zeroDecomp (ideal i)

```
    "USAGE:   zeroDecomp(i); i zero-dimensional ideal
    RETURN:   list l of lists of two ideals such that the
              intersection(l[j][1], j=1..)=i, the l[i][1] are
              primary and the l[i][2] their radicals
    NOTE:     algorithm of Gianni/Trager/Zacharias
    "
    {
      def  BAS = basering;
//----the ordering is changed to the lexicographical one
      changeord("R","lp");
```

```
    ideal i=fetch(BAS,i);
    int n=nvars(R);
    int k;
    list result,rest;
    ideal primary,prim;
    option(redSB);

//------the random coordinate change and its inverse
    ideal m=maxideal(1);
    m[n]=0;
    poly p=(random(100,1,n)*transpose(m))[1,1]+var(n);
    m[n]=p;
    map phi=R,m;
    m[n]=2*var(n)-p;
    map invphi=R,m;
    ideal j=groebner(phi(i));
//-------------factorization of the first element in i
    list fac=factorize(j[1],2);
//-------------computation of the primaries and primes
    for(k=1;k<=size(fac[1]);k++)

      p=fac[1][k]^fac[2][k];
      primary=groebner(j+p);
      prim=primaryTest(primary,fac[1][k]);
//---test whether all ideals were primary and in general
//    position
      if(prim==0)
      {
        rest[size(rest)+1]=i+invphi(p);
      }
      else
      {
        result[size(result)+1]=
          list(std(i+invphi(p)),std(invphi(prim)));
      }
    }
//------treat the bad cases collected in the rest again
    for(k=1;k<=size(rest);k++)
    {
      result=result+zeroDecomp(rest[k]);
    }
    option(noredSB);
    setring BAS;
    list result=imap(R,result);
```

```
        kill R;
        return(result);
}

ring  r = 32003,(x,y,z),dp;
poly  p = z2+1;
poly  q = z4+2;
ideal i = p^2*q^3,(y-z3)^3,(x-yz+z4)^4;
list pr = zeroDecomp(i);
pr;
```

4.8.3. Procedure to define for an independent set $u \subset x$ the ring $K(u)[x \setminus u]$.

proc prepareQuotientring(ideal i)

```
    "USAGE:    prepareQuotientring(i); i standard basis
     RETURN:    a list l of two strings:
                l[1] to define K[x\u,u ], u a maximal independent
                set for i
                l[2] to define K(u)[x\u ], u a maximal independent
                set for i
                both rings with lexicographical ordering
     "
{
    string va,pa;
//v describes the independent set u: var(j) is in
//u iff v[j]!=0
    intvec v=indepSet(i);
    int k;

    for(k=1;k<=size(v);k++)
    {
        if(v[k]!=0)
        {
            pa=pa+"var("+string(k)+"),";
        }
        else
        {
            va=va+"var("+string(k)+"),";
        }
    }

    pa=pa[1..size(pa)-1];
    va=va[1..size(va)-1];

    string newring="
```

```
ring nring=("+charstr(basering)+"),("+va+","+pa+"),lp;";
string quotring="
ring quring=("+charstr(basering)+","+pa+"),("+va+"),lp;";
return(newring,quotring);
}


ring s=(0,x),(a,b,c,d,e,f,g),dp;
ideal i=x*b*c,d^2,f-g;
i=std(i);
def Q=basering;
list l= prepareQuotientring(i);
l;
execute (l[1]);
basering;
execute (l[2]);
basering;
setring Q;
```

4.8.4. A procedure to collect the leading coefficients of a standard basis of an ideal in $K(u)[x \setminus u]$. They are needed to compute $IK(u)[x \setminus u] \cap K[x]$ via saturation.

proc prepareSat(ideal i)

```
{
    int k;
    poly p=leadcoef(i[1]);
    for(k=2;k<=size(i);k++)
    {
      p=p*leadcoef(i[k]);
    }
    return(p);
}
```

4.8.5. Using the above procedures, we can now present our procedure to compute a primary decomposition of an ideal.

proc decomp (ideal i)

```
"USAGE:  decomp(i); i ideal
RETURN:  list l of lists of two ideals such that the
         intersection(l[j][1], j=1..)=i, the l[i][1] are
         primary and the l[i][2] their radicals
NOTE:    algorithm of Gianni/Trager/Zacharias
"
{
    if(i==0)
```

```
  {
    return(list(i,i));
  }
  def  BAS = basering;
  ideal j;
  int n=nvars(BAS);
  int k;

  ideal SBi=std(i);
  int d=dim(SBi);
//---the trivial case and the zero-dimensional case
  if ((d==0)||(d==-1))
  {
     return(zeroDecomp(i));
  }
//---prepare the quotient ring with respect to a maximal
//   independent set
  list quotring=prepareQuotientring(SBi);
  execute (quotring[1]);
//---used to compute a standard basis of i*quring
//   which is in i
  ideal i=std(imap(BAS,i));
//---pass to the quotient ring with respect to a maximal
//   independent set
  execute (quotring[2]);
  ideal i=imap(nring,i);
  kill nring;
//---computation of the zero-dimensional decomposition
  list ra=zeroDecomp(i);
//---preparation for saturation
  list p;
  for(k=1;k<=size(ra);k++)
  {
    p[k]=list(prepareSat(ra[k][1]),prepareSat(ra[k][2]));
  }
  poly q=prepareSat(i);
//---back to the original ring
  setring BAS;
  list p=imap(quring,p);
  list ra=imap(quring,ra);
  poly q=imap(quring,q);
  kill quring;
//---compute the intersection of ra with BAS
  for(k=1;k<=size(ra);k++)
```

```
   {
      ra[k]=list(sat(ra[k][1],p[k][1])[1],
                 sat(ra[k][2],p[k][2])[1]);
   }
   q=q^sat(i,q)[2];
//---i=intersection((i:q),(i,q)) and ra is the primary
//   decomposition of i:q
   if(deg(q)>0)
   {
      ra=ra+decomp(i+q);
   }
   return(ra);
}


ring  r = 0,(x,y,z),dp;
ideal i = intersect(ideal(x,y,z)^3,ideal(x-y-z)^2,
          ideal(x-y,x-z)^2);
list pr = decomp(i);
pr;
```

4.8.6. We pass to the computation of the equidimensional part of an ideal.

proc equidimensional (ideal i)

```
"USAGE:   equidimensional(i); i ideal
RETURN:   list l of two ideals such that intersection(l[1],
          l[2])=i if there are no embedded primes
          l[1] is equidimensional and dim(l[1])>dim(l[2])
"
{
   def  BAS = basering;

   ideal SBi=std(i);
   int d=dim(SBi);
   int n=nvars(BAS);
   int k;
   list result;

//----the trivial cases
   if ((d==-1)||(n==d)||(n==1)||(d==0))
   {
      result=i,ideal(1);
      return(result);
   }
//----prepare the quotient ring with respect to a maximal
//       independent set
```

```
    list quotring=prepareQuotientring(SBi);
    execute (quotring[1]);
//----we use this ring to compute a standard basis of
//     i*quring which is in i
    ideal eq=std(imap(BAS,i));
//----pass to the quotient ring with respect to a maximal
//     independent set
    execute (quotring[2]);
    ideal eq=imap(nring,eq);
    kill nring;
//----preparation for saturation
    poly p=prepareSat(eq);
//----back to the original ring
    setring BAS;
    poly p=imap(quring,p);
    ideal eq=imap(quring,eq);
    kill quring;
//----compute the intersection of eq with BAS
    eq=sat(eq,p)[1];
    SBi=std(quotient(i,eq));

    if(d>dim(SBi))
    {
      result=eq,SBi;
      return(result);
    }
    result=equidimensional(i);
    result=intersect(result[1],eq),result[2];
    return(result);
}

ring  r = 0,(x,y,z),dp;
ideal i = intersect(ideal(x,y,z)^3,ideal(x-y-z)^2,
          ideal(x-y,x-z)^2);
list pr = equidimensional(i); pr;
dim(std(pr[1]));
dim(std(pr[2]));
option(redSB);
std(i);
std(intersect(pr[1],pr[2]));
```

4.8.7. Compute the squarefree part of a univariate polynomial f over a field of characteristic 0, depending on the i–th variable.

proc squarefree (poly f, int i)

```
{
    poly h=gcd(f,diff(f,var(i)));
    poly g=lift(h,f)[1][1];
    return(g);
}
```

4.8.8. Finally, a procedure to compute the radical of an ideal.

proc radical(ideal i)

```
"USAGE:   radical(i); i ideal
RETURN:   ideal = the radical of i
NOTE:     algorithm of Krick/Logar
"
{
    def  BAS = basering;
    ideal j;
    int n=nvars(BAS);
    int k;

    option(redSB);
    ideal SBi=std(i);
    option(noredSB);
    int d=dim(SBi);

//-----the trivial cases
    if ((d==-1)||(n==d)||(n==1))
    {
        return(ideal(squarefree(SBi[1],1)));
    }
//-----the zero-dimensional case
    if (d==0)
    {
        j=finduni(SBi);
        for(k=1;k<=size(j);k++)
        {
            i=i,squarefree(cleardenom(j[k]),k);
        }
        return(std(i));
    }
//-----prepare the quotientring with respect to a maximal
//      independent set
    list quotring=prepareQuotientring(SBi);
    execute (quotring[1]);
//-----we use this ring to compute a standardbasis of
```

```
//      i*quring which is in i
   ideal i=std(imap(BAS,i));
//-----pass to the quotientring with respect to a maximal
//      independent set
   execute( quotring[2]);
   ideal i=imap(nring,i);
   kill nring;
//-----computation of the zerodimensional radical
   ideal ra=radical(i);
//-----preparation for saturation
   poly p=prepareSat(ra);
   poly q=prepareSat(i);
//-----back to the original ring
   setring BAS;
   poly p=imap(quring,p);
   poly q=imap(quring,q);
   ideal ra=imap(quring,ra);
   kill quring;
//-----compute the intersection of ra with BAS
   ra=sat(ra,p)[1];
//----now we have radical(i)=intersection(ra,radical((i,q)))
   return(intersect(ra,radical(i+q)));
}


ring  r = 0,(x,y,z),dp;
ideal i =
intersect(ideal(x,y,z)^3,ideal(x-y-z)^2,ideal(x-y,x-z)^2);
ideal pr= radical(i);
pr;
```

The algorithms and, hence, the procedures work in characteristic 0. However, by our experience, the procedures in the library primdec.lib, distributed with SINGULAR, do also work for prime fields of finite characteristic provided that it is not too small. In fact, the procedures, although designed for characteristic 0, give a correct result for finite prime field whenever they terminate.

5. Hilbert Function and Dimension

5.1 The Hilbert Function and the Hilbert Polynomial

The Hilbert function of a graded module associates to an integer n the dimension of the n–th graded part of the given module. For sufficiently large n, the values of this function are given by a polynomial, the Hilbert polynomial. To show this, we use the Hilbert–Poincaré series, a formal power series in t with coefficients being the values of the Hilbert function. This power series turns out to be a rational function.

Let K be a field.

Definition 5.1.1. Let $A = \bigoplus_{\nu \geq 0} A_\nu$ be a Noetherian graded K–algebra (cf. Definition 2.2.1), and let $M = \bigoplus_{\nu \in \mathbb{Z}} M_\nu$ be a finitely generated graded A–module. The *Hilbert function* $H_M : \mathbb{Z} \to \mathbb{Z}$ of M is defined by

$$H_M(n) := \dim_K(M_n),$$

and the *Hilbert–Poincaré series* HP_M of M is defined by

$$\mathrm{HP}_M(t) := \sum_{\nu \in \mathbb{Z}} H_M(\nu) \cdot t^\nu \in \mathbb{Z}[[t]][t^{-1}].$$

By definition, H_M (and, hence, HP_M) depend only on the graded structure of M. Hence, if $\varphi : B \to A$ is a graded K–algebra map, then it does not matter whether we consider M as A–module or as B–module. In particular, since $A/\operatorname{Ann}_A(M)$ is a graded A–algebra (cf. Exercise 2.2.3), we may always consider M as $A/\operatorname{Ann}_A(M)$–module when computing the Hilbert function (or Hilbert–Poincaré series).

Using the results of Section 2.2, we obtain the first elementary properties of H_M and HP_M (note that $\dim_K(M_n) < \infty$ by Lemma 2.2.14):

Lemma 5.1.2. *Let $A = \bigoplus_{\nu \geq 0} A_\nu$ be a Noetherian graded K–algebra, and let M be a finitely generated graded A–module.*

(1) Let $N \subset M$ be a graded submodule, then

$$H_M(n) = H_N(n) + H_{M/N}(n)$$

for all n, in particular, $\mathrm{HP}_M(t) = \mathrm{HP}_N(t) + \mathrm{HP}_{M/N}(t)$.

(2) Let d be an integer, then

$$H_{M(d)}(n) = H_M(n + d)$$

for all n, in particular, $\mathrm{HP}_{M(d)}(t) = t^{-d}\,\mathrm{HP}_M(t)$.
(3) Let d be a non–negative integer, let $f \in A_d$, and let $\varphi : M(-d) \to M$ be defined by $\varphi(m) := f \cdot m$, then $\mathrm{Ker}(\varphi)$ and $\mathrm{Coker}(\varphi)$ are graded $A/\langle f \rangle$–modules with the induced gradings and

$$H_M(n) - H_M(n - d) = H_{\mathrm{Coker}(\varphi)}(n) - H_{\mathrm{Ker}(\varphi)}(n - d),$$

in particular, $\mathrm{HP}_M(t) - t^d\,\mathrm{HP}_M(t) = \mathrm{HP}_{\mathrm{Coker}(\varphi)}(t) - t^d\,\mathrm{HP}_{\mathrm{Ker}(\varphi)}(t)$.

Proof. (1) holds, because $N_\nu = N \cap M_\nu$ and $(M/N)_\nu = M_\nu/N_\nu$. (2) is an immediate consequence of the definition of $M(d)$, and (3) is a consequence of (1) and (2). $\square$

Theorem 5.1.3. *Let $A = \bigoplus_{\nu \geq 0} A_\nu$ be a graded K–algebra, and assume that A is generated, as K–algebra, by $x_1, \dots, x_r \in A_1$. Then, for any finitely generated (positively) graded A–module $M = \bigoplus_{\nu \geq 0} M_\nu$,*

$$\mathrm{HP}_M(t) = \frac{Q(t)}{(1 - t)^r} \quad \text{for some } Q(t) \in \mathbb{Z}[t].$$

Proof. We prove the theorem using induction on r. In the case $r = 0$, M is a finite dimensional K–vector space, and, therefore, there exists an integer n such that $M_\nu = \langle 0 \rangle$ for $\nu \geq n$. This implies $\mathrm{HP}_M(t) \in \mathbb{Z}[t]$.

Assume that $r > 0$, and consider the map $\varphi : M(-1) \to M$ defined by $\varphi(m) := x_1 \cdot m$. Using Lemma 5.1.2 (3), we obtain

$$(1 - t) \cdot \mathrm{HP}_M(t) = \mathrm{HP}_{\mathrm{Coker}(\varphi)}(t) - t\,\mathrm{HP}_{\mathrm{Ker}(\varphi)}(t).$$

Now both $\mathrm{Ker}(\varphi)$ and $\mathrm{Coker}(\varphi)$ are graded $A/\langle x_1 \rangle \cong A_0[\bar{x}_2, \dots, \bar{x}_r]$–modules, where $\bar{x}_i := x_i \bmod \langle x_1 \rangle$, $i = 2, \dots, r$. Using the induction hypothesis we obtain $\mathrm{HP}_{\mathrm{Coker}(\varphi)}(t) = Q_1(t)/(1 - t)^{r-1}$ and $\mathrm{HP}_{\mathrm{Ker}(\varphi)}(t) = Q_2(t)/(1 - t)^{r-1}$ for some $Q_1, Q_2 \in \mathbb{Z}[t]$. This implies $\mathrm{HP}_M(t) = \big(Q_1(t) - Q_2(t)\big)/(1 - t)^r$. $\square$

With the notations of Theorem 5.1.3, we cancel all common factors in the numerator and denominator of $\mathrm{HP}_M(t) = Q(t)/(1 - t)^r$, and we obtain

$$\mathrm{HP}_M(t) = \frac{G(t)}{(1 - t)^s}, \quad 0 \leq s \leq r, \quad G(t) = \sum_{\nu=0}^{d} g_\nu t^\nu \in \mathbb{Z}[t],$$

such that $g_d \neq 0$ and $G(1) \neq 0$, that is, s is the pole order of $\mathrm{HP}_M(t)$ at $t = 1$.[1]

---

[1] We set $G(t) := 0$, $s := 0$, if $M = 0$. In this chapter the zero–polynomial has degree -1, and we set $\binom{n}{-1} := 0$ if $n \geq 0$ and $\binom{-1}{-1} := 1$.

Definition 5.1.4. Let $A = \bigoplus_{\nu \geq 0} A_\nu$ be a Noetherian graded K–algebra, and let $M = \bigoplus_{\nu \geq 0} M_\nu$ be a finitely generated (positively) graded A–module.

(1) The polynomial $Q(t)$, respectively $G(t)$, defined above, is called the *first Hilbert series*, respectively the *second Hilbert series*, of M.
(2) Let d be the degree of the second Hilbert series $G(t)$, and let s be the pole order of the Hilbert–Poincaré series $\mathrm{HP}_M(t)$ at $t = 1$, then

$$P_M := \sum_{\nu=0}^{d} g_\nu \cdot \binom{s-1+n-\nu}{s-1} \in \mathbb{Q}[n]$$

is called the *Hilbert polynomial* of M (with $\binom{n}{k} = 0$ for $k < 0$).

Corollary 5.1.5. *With the above assumptions, P_M is a polynomial in n with rational coefficients, of degree $s - 1$, and satisfies $P_M(n) = H_M(n)$ for $n \geq d$. Moreover, there exist $a_\nu \in \mathbb{Z}$ such that*

$$P_M = \sum_{\nu=0}^{s-1} a_\nu \cdot \binom{n}{\nu} = \frac{a_{s-1}}{(s-1)!} \cdot n^{s-1} + \text{lower terms in } n,$$

where $a_{s-1} = G(1) > 0$.

Proof. The equality $1/(1-t)^s = \sum_{\nu=0}^{\infty} \binom{s-1+\nu}{s-1} \cdot t^\nu$ implies

$$\sum_{\nu=0}^{\infty} H_M(\nu) t^\nu = \mathrm{HP}_M(t) = \left(\sum_{\nu=0}^{d} g_\nu t^\nu \right) \cdot \sum_{\mu=0}^{\infty} \binom{s-1+\mu}{s-1} t^\mu.$$

Therefore, for $n \geq d$, we obtain

$$H_M(n) = \sum_{\nu=0}^{d} g_\nu \cdot \binom{s-1+n-\nu}{s-1} = P_M(n).$$

It is easy to see that the leading term of $P_M \in \mathbb{Q}[n]$ is $\sum_{\nu=0}^{d} g_\nu n^{s-1}/(s-1)!$ which equals $G(1) \cdot n^{s-1}/(s-1)!$. In particular, we obtain $\deg P_M = s - 1$.

Finally, we have to prove that $P_M = \sum_{\nu=0}^{s-1} a_\nu \binom{n}{\nu}$ for suitable $a_\nu \in \mathbb{Z}$ and $a_{s-1} > 0$. Suppose that we can find such $a_\nu \in \mathbb{Z}$. Then

$$P_M = \frac{a_{s-1}}{(s-1)!} \cdot n^{s-1} + \text{lower terms in } n.$$

Now, $P_M(n) = H_M(n) > 0$ for n sufficiently large implies $a_{s-1} > 0$. Finally, the existence of suitable integer coefficients a_ν is a consequence of the following general lemma. $\square$

Lemma 5.1.6. *Let $f \in \mathbb{Q}[t]$ be a polynomial of degree m and $n_0 \in \mathbb{N}$ such that $f(n) \in \mathbb{Z}$ for all $n \geq n_0$. Then $f(n) = \sum_{\nu=0}^{m} a_\nu \binom{n}{\nu}$ for suitable $a_\nu \in \mathbb{Z}$.*

Proof. Let $g(n) := f(n+1) - f(n)$. Then g is a polynomial of degree $m-1$ and $g(n) \in \mathbb{Z}$ for $n \geq n_0$. By induction on m we may assume that there exist $b_\nu \in \mathbb{Z}$ such that $g(n) = \sum_{\nu=0}^{m-1} b_\nu \binom{n}{\nu}$. Now consider the function

$$h(n) := f(n) - \sum_{\nu=1}^{m} b_{\nu-1} \binom{n}{\nu}.$$

Then

$$h(n+1) - h(n) = g(n) - \sum_{\nu=1}^{m} b_{\nu-1} \left(\binom{n+1}{\nu} - \binom{n}{\nu} \right)$$

$$= g(n) - \sum_{\nu=1}^{m} b_{\nu-1} \binom{n}{\nu-1} = 0.$$

It follows that $h(n) = h(0)$ for all $n \in \mathbb{N}$, hence $f(n) = h(0) + \sum_{\nu=1}^{m} b_{\nu-1} \binom{n}{\nu}$. This implies $f(n) = \sum_{\nu=0}^{m} a_\nu \binom{n}{\nu}$ with $a_0 = h(0)$ and $a_\nu = b_{\nu-1}$ for $\nu \geq 1$. $\square$

Exercises

5.1.1. Let $A = \bigoplus_{\nu \geq 0} A_\nu$ be a Noetherian K–algebra with K a field. Assume $\dim_K A_0 < \infty$, $A_i A_j \subset A_{i+j}$ and that A is generated by $x_1, \ldots, x_r$, which are homogeneous of degrees $d_1, \ldots, d_r$, respectively. Let $M = \bigoplus_{\nu \geq 0} M_\nu$ be a finitely generated (positively) graded A–module. Prove that

$$\mathrm{HP}_M(t) = \frac{Q_M(t)}{\prod_{i=1}^{r}(1 - t^{d_i})},$$

for some $Q_M(t) \in \mathbb{Z}[t]$, which is called the *weighted first Hilbert series* of M. This generalizes Theorem 5.1.3 which is the case $d_1 = \cdots = d_r = 1$.

Note that, in a ring with a weighted monomial ordering, the SINGULAR–command `hilb` computes the weighted first Hilbert series.

5.1.2. Let $A = K[x, y, z]/\langle x^3 + y^3 + z^3 \rangle$, considered as a graded K–algebra with the canonical grading. Compute the Hilbert–Poincaré series and the Hilbert polynomial of A.

5.1.3. Let $A = K[x, y, z]/\langle x^3 + y^4 + z^5 \rangle$, considered as a graded K–algebra with the grading defined by $\deg(x) := 20$, $\deg(y) := 15$ and $\deg(z) := 12$. Use SINGULAR to compute the (weighted) Hilbert-Poincaré series of A.

5.1.4. Let $x^\alpha, x^\beta \in \mathrm{Mon}(x_1, \ldots, x_r)$ be monomials of the same degree. Prove that $\mathrm{HP}_{K[x]/\langle x^\alpha \rangle} = \mathrm{HP}_{K[x]/\langle x^\beta \rangle}$.

5.1.5. Let $I \subseteq \mathbb{Q}[x]$ be a homogeneous ideal and $I_0 = I \cap \mathbb{Z}[x]$, $x = (x_1, \ldots, x_n)$. Let p be a prime number and $I_p = I_0 \mathbb{Z}/p[x]$. Prove that $H_{\mathbb{Q}[x]/I}^{(n)} \leq H_{\mathbb{Z}/p[x]/I_p}^{(n)}$ for all n .

Hint: Use the fact that $I_0 \cap \mathbb{Z}[x]_n = \{ f \in I_0 | f \text{ homogeneous of degree } n \} \cup \{0\}$ is a finitely generated free $\mathbb{Z}$–module of rank $\dim_{\mathbb{Q}}(I \cap \mathbb{Q}[x]_n)$.

5.1.6. Let A be a graded K–algebra and M and M' be finitely generated graded A–modules with first Hilbert series $Q_M = \sum_{v=0}^{d} g_v t^v$ and $Q_{M'} = \sum_{v=0}^{d'} g'_v t^v$.

Assume that $g_v = g'_v$ for $v = 1, \ldots, k-1$ and $g_k < g'_k$. Prove that $H_M(v) = H_{M'}(v)$ for $v = 1, \ldots, k-1$ and $H_M(k) < H_{M'}(k)$.

5.1.7. Let $H : \mathbb{N} \to \mathbb{Z}$ be any function and $d > 0$. Show that the following are equivalent for sufficiently large n:

(i) $H(n)$ is a polynomial of degree d,
(ii) $D(n) := H(n+1) - (H(n)$ is a polynomial of degree $d-1$,
(iii) $S(n) := \sum_{v=0}^{n-1} H(v)$ is a polynomial of degree $d+1$.

Moreover, if $\frac{a_d}{d!}$ is the leading coefficient of H then D resp. S have leading coefficient $\frac{a_d}{(d-1)!}$ resp. $\frac{a_d}{(d+1)!}$.

5.2 Computation of the Hilbert–Poincaré Series

The main result in this section is that, for a homogeneous ideal I and any monomial ordering, the K–algebras $K[x_1, \ldots, x_r]/I$ and $K[x_1, \ldots, x_r]/L(I)$ have the same Hilbert polynomial.

Recall that the Hilbert polynomial and the Hilbert–Poincaré series determine each other (cf. Definition 5.1.4). Hence, it suffices to study and compute the Hilbert–Poincaré series.

Example 5.2.1. Let $K[x] := K[x_1, \ldots, x_r]$ be the polynomial ring in r indeterminates, considered as graded K–algebra (as in Example 2.2.3). Then $H_{K[x]}(n) = P_{K[x]}(n) = \binom{n+r-1}{r-1}$ and, therefore,

$$\mathrm{HP}_{K[x]}(t) = \sum_{v=0}^{\infty} \binom{r-1+v}{r-1} t^v = \frac{1}{(1-t)^r} \,.$$

Lemma 5.2.2. *Let $I \subset K[x] := K[x_1, \ldots, x_r]$ be a homogeneous ideal, and let $f \in K[x]$ be a homogeneous polynomial of degree d then*

$$\mathrm{HP}_{K[x]/I}(t) = \mathrm{HP}_{K[x]/\langle I, f \rangle}(t) + t^d \, \mathrm{HP}_{K[x]/(I:\langle f \rangle)}(t) \,.$$

Proof. We consider the following exact sequence

$$0 \longrightarrow \left(K[x]/(I : \langle f \rangle) \right)(-d) \xrightarrow{\cdot f} K[x]/I \longrightarrow K[x]/\langle I, f \rangle \longrightarrow 0$$

and use Lemma 5.1.2. $\qquad\qquad\qquad\qquad\qquad\qquad\qquad\qquad\qquad\qquad\qquad$ $\square$

Example 5.2.3. Let $I := \langle xz, yz \rangle \subset K[x, y, z]$. Using Lemma 5.2.2 for $f := z$
we obtain

$$\mathrm{HP}_{K[x,y,z]/\langle xz,yz \rangle}(t) = \mathrm{HP}_{K[x,y]}(t) + t \cdot \mathrm{HP}_{K[z]}(t) = \frac{-t^2 + t + 1}{(1-t)^2},$$

and, therefore,

$$P_{K[x,y,z]/\langle xz,yz \rangle}(n) = \binom{n+1}{1} + \binom{n}{1} - \binom{n-1}{1} = n + 2.$$

Using Example 5.2.1 and Lemma 5.2.2, we obtain the following algorithm to
compute the Hilbert–Poincaré series for a monomial ideal, more precisely, it
computes the polynomial $Q(t) \in \mathbb{Z}[t]$ (notation of Theorem 5.1.3).

Algorithm 5.2.4 (MONOMIALHILBERTPOINCARE(I)).

Input: $I := \langle m_1, \ldots, m_k \rangle \subset K[x]$, m_i monomials in $x = (x_1, \ldots, x_r)$.
Output: A polynomial $Q(t) \in \mathbb{Z}[t]$ such that $Q(t)/(1-t)^r$ is the Hilbert–
         Poincaré series of $K[x]/I$.

- choose $S = \{x^{\alpha_1}, \ldots, x^{\alpha_s}\} \subset \{m_1, \ldots, m_k\}$ to be the minimal set of monomial generators of I;
- if $S = \{0\}$ then return 1;
- if $S = \{1\}$ then return 0;
- if all elements of S have degree 1 then return $(1-t)^s$;
- choose $1 \leq i \leq s$ such that $\deg(x^{\alpha_i}) > 1$ and $1 \leq k \leq r$ such that $x_k \mid x^{\alpha_i}$;
- return $\big($MONOMIALHILBERTPOINCARE $(\langle I, x_k \rangle)$
        $+ t \cdot$ MONOMIALHILBERTPOINCARE $(I : \langle x_k \rangle)\big)$.

In the following example we present an implementation of this procedure in
the programming language of SINGULAR .

SINGULAR Example 5.2.5 (Hilbert–Poincaré series).
We compute the polynomial $Q(t) \in \mathbb{Z}[t]$ such that $Q(t)/(1-t)^r$ is the Hilbert–
Poincaré series of $\mathbb{Q}[x_1, \ldots, x_r]/I$ for an ideal I given by monomial generators.
For simplicity we compute in the ring $\mathbb{Q}[t, x_1, \ldots, x_r]$, where the first vari-
able t is reserved for the polynomial $Q(t) \in \mathbb{Z}[t]$. We use the degree reverse
lexicographical ordering dp.

```
proc MonomialHilbertPoincare(ideal I)
{
  I=interred(I);    //computes a minimal set of generators
  int s=size(I);    //of the monomial ideal I

  if(I[1]==0){return(1);}                    //I=<0>
  if(I[1]==1){return(0);}                    //I=<1>
  if(deg(I[s])==1){return((1-var(1))^s);}//I is generated by
```

```
                                          //s of the {var(j)}
   int j=1;
   while(leadexp(I[s])[j]==0){j++;}        //I[s]=var(j)*m
   return(MonomialHilbertPoincare(I+var(j))
        +var(1)*MonomialHilbertPoincare(quotient(I,var(j)))));
}


ring A=0,(t,x,y,z),dp;
ideal I=x5y2,x3,y3,xy4,xy7;

MonomialHilbertPoincare(I);
//-> t6-2t3+1
```

Note that SINGULAR has a command which computes the numerator $Q(t)$ for the Hilbert–Poincaré series:

```
intvec v = hilb(std(I),1);
v;
//-> 1,0,0,-2,0,0,1,0
```

The latter output has to be interpreted as follows: if $v = (v_0, \ldots, v_d, 0)$ then $Q(t) = \sum_{i=0}^{d} v_i t^i$. [2]

The following theorem is fundamental for the computation of the Hilbert–Poincaré series for arbitrary graded K–algebras:

Theorem 5.2.6. *Let $>$ be any monomial ordering on $K[x] := K[x_1, \ldots, x_r]$, and let $I \subset K[x]$ be a homogeneous ideal. Then*

$$\mathrm{HP}_{K[x]/I}(t) = \mathrm{HP}_{K[x]/L(I)}(t),$$

where $L(I)$ is the leading ideal of I with respect to $>$.

Note that the theorem holds for any monomial ordering. However, in a ring with a weighted monomial ordering, the SINGULAR–command `hilb` computes the weighted first Hilbert series of the leading ideal.

Proof. We have to show that $H_{K[x]/I}(n) = H_{K[x]/L(I)}(n)$, or, equivalently, $\dim_K K[x]_n/I_n = \dim_K K[x]_n/L(I)_n$ for all n.

Let $S := \{x^\alpha \notin L(I) \mid \deg(x^\alpha) = n\}$. We shall prove that S represents a K–basis in $K[x]_n/I_n$ and $K[x]_n/L(I)_n$. To do so, choose a standard basis G of I, let $f \in K[x]_n$, and let NF denote the Mora normal form (which preserves homogeneity), then we obtain that both, $\mathrm{NF}(f \mid G)$ and $\mathrm{NF}(f \mid L(G))$, are elements of $K[x]_n$. Iterating this process by computing the Mora normal form

---

[2] Since $Q(t) \in \mathbb{Z}[t]$, SINGULAR cannot return $Q(t)$ as a polynomial in the current basering (which may have a positive characteristic). Therefore, SINGULAR returns an integer vector.

of the tail of $\mathrm{NF}(f \mid G)$, respectively $\mathrm{NF}\big(f \mid L(G)\big)$, hence we can assume that $\mathrm{NF}(f \mid G)$, $\mathrm{NF}\big(f \mid L(G)\big) \in \sum_{x^\alpha \in S} K \cdot x^\alpha$.

Since $\mathrm{NF}(f \mid G) = 0$ (respectively $\mathrm{NF}\big(f \mid L(G)\big) = 0$) if and only if $f \in I$ (respectively $f \in L(I)$), the latter implies that S represents a K–basis of $K[x]_n/I_n$ and $K[x]_n/L(I)_n$. This proves the theorem. $\qquad\square$

Corollary 5.2.7. *With the notation of Theorem 5.2.6 we have*

$$\dim_K\big(K[x]/I\big) = \dim_K\big(K[x]/L(I)\big).$$

In Section 7.5, we shall show that this result holds also if I is not homogeneous.

Proof. For each graded $K[x]$–module M, $\dim_K(M) = \mathrm{HP}_M(1)$, where M may be finite dimensional or not. $\qquad\square$

For any homogeneous ideal $I \subset K[x] := K[x_1,\dots,x_r]$, the following algorithm computes the Hilbert–Poincaré series of $K[x]/I$:

Algorithm 5.2.8 (HILBERTPOINCARE(I)).

Input: $I := \langle f_1,\dots,f_k \rangle \subset K[x]$ a homogeneous ideal, $x = (x_1,\dots,x_r)$.
Output: A polynomial $Q(t) \in \mathbb{Z}[t]$ such that $Q(t)/(1-t)^r$ is the Hilbert–Poincaré series of $K[x]/I$.

- compute a standard basis $\{g_1,\dots,g_s\}$ of I w.r.t. any monomial ordering;
- return MONOMIALHILBERTPOINCARE($\langle \mathrm{LM}(g_1),\dots,\mathrm{LM}(g_s)\rangle$).

Remark 5.2.9 (Hilbert–driven Buchberger Algorithm). We can use Theorem 5.2.6 to speed up the standard basis algorithm using the Hilbert function (for more details see [219]). This can be useful if we need to compute a standard basis with respect to an "expensive" ordering (as for instance lp): We first compute a standard basis of the ideal I with respect to the ordering dp (which is often sufficiently fast) and use it to compute the Hilbert function. Then we start the computation for the "expensive" ordering using the Hilbert function as a bound in the following way: The pair-set P is ordered by increasing degree of the s–polynomials of the pairs. Whenever the standard basis algorithm finds a new polynomial, say of degree d, which is added to the set S (the set collecting the elements of the standard basis), the Hilbert function is used to check whether $\dim_K(K[x]/I)_d = \dim_K(K[x]/\langle L(S)\rangle)_d$. If this condition holds, the remaining elements of degree d in P are cancelled.

In SINGULAR the command std directly computes a standard basis while the command groebner uses the Hilbert–driven approach for "expensive" orderings. If option(prot) is set, each cancellation by the Hilbert function is shown by printing the letter h on the screen.

Remark 5.2.10 (Modular Gröbner basis computations.).

It is well known that the computation of standard bases in a polynomial ring over the rational number $\mathbb{Q}$ is much more difficult than in a polynomial ring over a finite field $\mathbb{F}_p = \mathbb{Z}/p$. The reason is the enormous growth of the coefficients during the computation even if the result may have relatively small coefficients. To avoid these problems one can try to compute the standard bases over $\mathbb{F}_p$ for one suitable prime p (resp. several suitable primes) and use Hensel lifting as proposed in [234] (resp. Chinese remainder theorem) as proposed in [200] to lift the coefficients to $\mathbb{Z}$. Then Farey fractions (cf. [118], [134]) can be used to obtain the "correct" coefficients over $\mathbb{Q}$. This approach has been discussed since a long time (cf. [4],[64],[96],[185],[200],[234]). Here we follow the approach using Chinese remainder theorem. After the lifting there are two problems to be solved. It has to be checked whether the lifting of the standard bases to characteristic zero remains a standard basis and that it generates the ideal we started with. For the case of Gröbner bases (the monomial ordering is a global, i.e. a well–ordering) and homogeneous ideals a reasonable solution can be found for instance in the paper of Arnold (cf. [4]). It turns out that this method can also be used for standard basis with respect to local orderings. The case of mixed orderings or global orderings and non–homogeneous ideals is more complicated. With the same methods as in the homogeneous resp. local case one just obtains a standard basis generating an ideal containing the ideal we started with. For experiments this is already interesting, for proofs this is not enough.

In SINGULAR the command modStd of the Library modstd.lib computes standard basis using the modular methods described above.

Exercises

5.2.1. Write a SINGULAR procedure to compute the Hilbert polynomial of $K[x_1, \ldots, x_r]/I$ for a given homogeneous ideal I.

5.2.2. State and prove Theorem 5.2.6 for the case of graded modules.

5.2.3. Use SINGULAR to compute the Hilbert–Poincaré series of the rational quartic curve $C \subset \mathbb{P}^3$ defined by the equations

$$z^3 - yt^2 = 0, \quad yz - xt = 0, \quad y^3 - x^2 z = 0, \quad xz^2 - y^2 t = 0.$$

5.2.4. Use SINGULAR to compute a standard basis for the following ideal with respect to the ordering lp, using a direct computation by the command std and a Hilbert-driven approach by groebner:

```
ring  R = 32003,(t,x,y,z,w),lp;
ideal I = 4t2zw+3txz2+6tz3+3zw3,
          5t7xyw+5t3x2yz4+3tw9,
          6t2yzw+6ty2z2+2xw4+2yw4;
```

Compare the two approaches. (Use `option(prot)` to obtain more information about the progress of the computations.)

5.2.5. (*Hilbert polynomial of a hypersurface*) Let $f \in K[x] = K[x_1, \ldots, x_r]$ be homogeneous of degree d. Prove that

$$P_{K[x]/\langle f \rangle}(n) = \binom{n+r-1}{r-1} - \binom{n-d+r-1}{r-1}$$

$$= \frac{d}{(r-2)!} \cdot n^{r-2} + \text{terms of lower degree}.$$

(Hint: Use Theorem 5.2.6 and Example 5.2.1.)

5.2.6. Let $I \subseteq \mathbb{Q}[x]$ be a homogeneous ideal and $I_0 = I \cap \mathbb{Z}[x]$, $x = (x_1, \ldots, x_n)$. Let p be a prime and $I_p = I_0\mathbb{Z}/p[x]$. Let $g_1, \ldots, g_s \in \mathbb{Z}[x]$ and $J = \langle g_1, \ldots, g_s \rangle \mathbb{Q}[x]$. Let $\bar{g}_i := g_i \mod p\mathbb{Z}[x]$.
   Assume that

(1) $\{g_1, \ldots, g_s\}$ is a minimal Gröbner basis of J.
(2) $\{\bar{g}_1, \ldots, \bar{g}_s\}$ is a minimal Gröbner basis of I_p.
(3) $I \subseteq J$.
   Prove that $I = J$.
   Hint: Use Exerccise 5.1.5 and Theorem 5.2.6.

5.3 Properties of the Hilbert Polynomial

In this section we prove that, for a graded K–algebra $A = K[x_1, \ldots, x_r]/I$, $\dim(A) - 1$ is equal to the degree of the Hilbert polynomial P_A. This implies, in particular, using the main result of the previous section (Theorem 5.2.6), that

$$\dim(K[x_1, \ldots, x_r]/I) = \dim\big(K[x_1, \ldots, x_r]/L(I)\big).$$

Definition 5.3.1. Let $A = \bigoplus_{\nu \geq 0} A_\nu$ be a Noetherian graded K–algebra, and let $M = \bigoplus_{\nu \in \mathbb{Z}} M_\nu$ be a finitely generated, (not necessarily positively) graded A–module. Then we introduce

$$M^{(0)} := \bigoplus_{\nu \geq 0} M_\nu,$$

and define the *Hilbert polynomial* of M to be the Hilbert polynomial of $M^{(0)}$, that is, $P_M := P_{M^{(0)}}$.

Example 5.3.2. Let $A = \bigoplus_{\nu \geq 0} A_\nu$ be a Noetherian graded K–algebra, then $P_{A(d)}(n) = P_A(n+d) = P_A(n) +$ terms of lower degree in n.

Definition 5.3.3. Let A be a Noetherian graded K–algebra and M a finitely generated graded A–module, and let $P_M = \sum_{\nu=0}^{d} a_\nu n^\nu$, $a_d \neq 0$, be the Hilbert polynomial of M. Then we set

$$d(M) := \deg(P_M) = d\,,$$

and we define the *degree of* M as

$$\deg(M) := d! \cdot a_d$$

if $M \neq 0$, and $\deg(M) := 0$ if $M = 0$.

Remark 5.3.4. If M is positively graded and $\mathrm{HP}_M(t) = G(t)/(1-t)^s$ with $G(1) \neq 0$, then $d(M) = s - 1$ and $\deg(M) = G(1)$ (cf. Corollary 5.1.5). In particular, $d(K[x_1,\ldots,x_r]) = r - 1$ and $\deg(K[x_1,\ldots,x_r]) = 1$ (cf. Example 5.2.1).

Lemma 5.3.5. *Let $f \in K[x_1,\ldots,x_r] \setminus \{0\}$ be a homogeneous polynomial of degree m, then $d(K[x]/\langle f \rangle) = r - 2$ and $\deg(K[x]/\langle f \rangle) = m$.*

Proof. The statement is an immediate consequence of Exercise 5.2.5. $\square$

Proposition 5.3.6. *Let A be a Noetherian graded K–algebra, and let M, N be finitely generated graded A–modules.*

(1) If there is a surjective graded morphism $\varphi : M \to N$ then $d(M) \geq d(N)$.
(2) $d(M) \leq d(A)$.
(3) If there is a homogeneous element $m \in M$ such that $\mathrm{Ann}_A(\langle m \rangle) = \langle 0 \rangle$ then $d(M) = d(A)$.
(4) Let $x \in A_d$ be a homogeneous non–zerodivisor for M, then

$$d(M/xM) = d(M) - 1\,, \quad \deg(M/xM) = d \cdot \deg(M)\,.$$

Proof. (1) Let $\varphi : M \to N$ be a graded and surjective homomorphism of A–modules. Then, for all n, the restriction to M_n, $\varphi|_{M_n} : M_n \to N_n$ is surjective, too. This implies $H_M(n) = \dim_K(M_n) \geq \dim_K(N_n) = H_N(n)$.

Hence, $P_M(n) \geq P_N(n)$ for all n sufficiently large, which is only possible if $\deg(P_M) \geq \deg(P_N)$, since the leading coefficients are positive.

(2) Since M is finitely generated, we may choose homogeneous generators $m_1,\ldots,m_k$ of degree $d_1,\ldots,d_k$. Now consider the map

$$\varphi : \bigoplus_{i=1}^{k} A(-d_i) \longrightarrow M$$

defined by $\varphi(a_1,\ldots,a_k) = \sum_{i=1}^{k} a_i m_i$. Obviously, φ is graded and surjective. Using (1) we obtain $d\big(\bigoplus_{i=1}^{k} A(-d_i)\big) \geq d(M)$. On the other hand, for n sufficiently large

$$P_{\bigoplus_{i=1}^{k} A(-d_i)}(n) = \sum_{i=1}^{k} P_{A(-d_i)}(n) = \sum_{i=1}^{k} P_A(n - d_i)$$

$$= k \cdot P_A(n) + \text{ terms of lower degree in } n,$$

which implies $d\big(\bigoplus_{i=1}^{k} A(-d_i)\big) = d(A)$.

(3) Let $m \in M_d$ such that $\mathrm{Ann}_A(\langle m \rangle) = \langle 0 \rangle$. Then $\varphi : A(-d) \to M$ defined by $\varphi(a) := am$ is graded and injective. This implies that, for n sufficiently large, $P_A(n - d) = P_{A(-d)}(n) \leq P_M(n)$, which is only possible if $\deg(P_M) \geq \deg(P_A)$. Together with (2) this implies $d(M) = d(A)$.

(4) Using the exact sequence $0 \to M(-d) \xrightarrow{x} M \to M/xM \to 0$ of graded A–modules, we obtain, by Lemma 5.1.2 (3), $(1 - t^d) \, \mathrm{HP}_M(t) = \mathrm{HP}_{M/xM}(t)$. If $\mathrm{HP}_M(t) = G(t)/(1 - t)^{d(M)+1}$ with $G(1) \neq 0$, then

$$\mathrm{HP}_{M/xM} = \frac{G(t)(1 - t^d)}{(1 - t)^{d(M)}(1 - t)} = \frac{G(t) \cdot \sum_{\nu=0}^{d-1} t^\nu}{(1 - t)^{d(M)}}.$$

Then $\mathrm{HP}_{M/xM}$ has pole order $d(M)$ at $t = 1$, hence, $d(M/xM) = d(M) - 1$ and $\deg(M/xM) = \big(G(t) \cdot \sum_{\nu=0}^{d-1} t^\nu\big)\big|_{t=1} = G(1) \cdot d = \deg(M) \cdot d$.   □

Note that Proposition 5.3.6 (4) generalizes Lemma 5.3.5.

Theorem 5.3.7. *Let $I \subset K[x_1, \ldots, x_r]$ be a homogeneous ideal, then*

$$\dim(K[x_1, \ldots, x_r]/I) = d(K[x_1, \ldots, x_r]/I) + 1.$$

Proof. Using Noether normalization, $K[x_1, \ldots, x_r]/I$ can be considered as finitely generated graded $K[y_1, \ldots, y_s]$–module. The assumptions of Proposition 5.3.6 (3) are satisfied and, therefore,

$$\deg(P_{K[x_1, \ldots, x_r]/I}) = \deg(P_{K[y_1, \ldots, y_s]}) = s - 1 = \dim(K[x_1, \ldots, x_r]/I) - 1.$$

□

For a more general statement see Exercise 5.3.5.

Corollary 5.3.8. *Let $I \subset K[x] = K[x_1, \ldots, x_r]$ be a homogeneous complete intersection ideal (that is, $I = \langle f_1, \ldots, f_k \rangle$ with f_i homogeneous and a non-zerodivisor in $K[x]/\langle f_1, \ldots, f_{i-1} \rangle$, $i = 1, \ldots, k$). Then*

$$\dim(K[x]/I) = r - k, \quad \deg(K[x]/I) = \deg(f_1) \cdot \ldots \cdot \deg(f_k).$$

Proof. Apply Proposition 5.3.6 (4) and Theorem 5.3.7.   □

Corollary 5.3.9. *Let $>$ be any monomial ordering on $K[x] = K[x_1, \ldots, x_r]$, and let $I \subset K[x]$ be a homogeneous ideal, then*

$$\dim(K[x]/I) = \dim\big(K[x]/L(I)\big).$$

Proof. We just combine Theorems 5.3.7 and 5.2.6.   □

Proposition 5.3.10. *Let* $A = K[x_1,\ldots,x_r]/P$, P *a homogeneous prime ideal, and let* $K[x_1,\ldots,x_s] \subset A$ *be a Noether normalization such that the field extension* $Q(K[x_1,\ldots,x_s]) \subset Q(A)$ *is separable. Then*

$$\deg(A) = \left[Q(A) : Q(K[x_1,\ldots,x_s])\right].$$

Proof. We prove the proposition for the case that K is an infinite field. Using the Primitive Element Theorem ([238]), we know that the field extension can be generated by $\alpha_{s+1} x_{s+1} + \cdots + \alpha_r x_r \bmod P$ for generic $\alpha_i \in K$. We may, therefore, assume that the extension is generated by $x_{s+1} \bmod P$. Let $f \in K[x_1,\ldots,x_{s+1}] \cap P$ be a monic (with respect to x_{s+1}) homogeneous irreducible polynomial. It exists, because P is prime and homogeneous and $A \supset K[x_1,\ldots,x_s]$ is finite. By the Gauß lemma (see, e.g., [128], Section 1.4), f is also irreducible in $Q(K[x_1,\ldots,x_s])[x_{s+1}]$. Therefore, f is the minimal polynomial of the field extension. In particular, we have

$$\deg_{x_{s+1}}(f) = \left[Q(A) : Q(K[x_1,\ldots,x_s])\right].$$

Let $\Delta \in K[x_1,\ldots,x_s]$ be the discriminant of f and $B := K[x_1,\ldots,x_{s+1}]/\langle f \rangle$. Since $K[x_1,\ldots,x_s]$ is a Noetherian normal integral domain, $K[x_1,\ldots,x_s] \subset A$ an integral extension in $Q(B) = Q(A)$ and $Q(K[x_1,\ldots,x_s]) \subset Q(B)$ a separable field extension, Lemma 3.5.12 implies, in particular, that $A \subset \frac{1}{\Delta}B$. Now consider the exact sequence $0 \to B \to A \to A/B \to 0$ of finitely generated $K[x_1,\ldots,x_s]$–modules.

Since $A \subset \frac{1}{\Delta}B$, we obtain $\Delta A/B = 0$. This implies that A/B is already a $K[x_1,\ldots,x_s]/\Delta$–module and, therefore,

$$\deg(P_{A/B}) = d(A/B) \le d(K[x_1,\ldots,x_s]) - 1 = s - 2.$$

B is a free $K[x_1,\ldots,x_s]$–module of finite rank, that is, $\deg(P_B) = s - 1$. From the exact sequence we obtain $P_A - P_B = P_{A/B}$.

This implies $\deg(P_B) = \deg(P_A) = s - 1$, and the leading coefficients of P_B and P_A are the same, namely $\deg(A) = \deg(B)$. Finally, Lemma 5.3.5 implies that $\deg(B) = \deg_{x_{s+1}}(f)$.   □

Lemma 5.3.11. *Let* $K[x] := K[x_1,\ldots,x_r]$, *and let* $I \subset K[x]$ *be a homogeneous ideal. Moreover, let* $I = Q_1 \cap \cdots \cap Q_m$ *be an irredundant homogeneous primary decomposition and assume that*

$$\dim(K[x]/Q_1) = \cdots = \dim(K[x]/Q_s) = \dim(K[x]/I)$$

and $\dim(K[x]/Q_j) < \dim(K[x]/I)$ *for* $j > s$. *Then*

$$\deg(K[x]/I) = \sum_{i=1}^{s} \deg(K[x]/Q_i).$$

Proof. We use induction on m. If $J := Q_2 \cap \cdots \cap Q_m$ then, because of Exercise 5.3.3, we have an exact sequence

$$0 \longrightarrow K[x]/I \longrightarrow K[x]/J \oplus K[x]/Q_1 \longrightarrow K[x]/(J+Q_1) \longrightarrow 0.$$

This implies $P_{K[x]/I} + P_{K[x]/J+Q_1} = P_{K[x]/J} + P_{K[x]/Q_1}$. The primary decomposition being irredundant, we have $\dim\big(K[x]/(J+Q_1)\big) < \dim(K[x]/J)$ (cf. Exercise 4.1.11), which implies $\deg\big(P_{K[x]/(J+Q_1)}\big) < \deg(P_{K[x]/I})$.

If $s = 1$ then, by definition, $\dim(K[x]/J) < \dim(K[x]/I)$ and, therefore, $\deg(P_{K[x]/J}) < \deg(P_{K[x]/I})$. This implies $\deg(K[x]/I) = \deg(K[x]/Q_1)$. If $s > 1$ then we obtain $\deg(K[x]/I) = \deg(K[x]/Q_1) + \deg(K[x]/J)$, and we can use induction to obtain the lemma. $\square$

SINGULAR Example 5.3.12 (dimension, degree and Hilbert function of a homogeneous ideal).

We compute the dimension, degree and Hilbert function of an ideal arising from a geometrical problem (cf. Appendix A, SINGULAR Exercise A.8.12). Let φ be a morphism given by

$$\mathbb{P}^1 \xrightarrow{\varphi} \mathbb{P}^d$$
$$(s : t) \longmapsto (t^d : t^{d-1}s : \cdots : s^d).$$

It defines the so-called *rational normal curve* of degree d in $\mathbb{P}^d$ and has the parametrization $x_0 = t^d$, $x_1 = t^{d-1}s$, ..., $x_d = s^d$. First, we compute the ideal defining the rational normal curve, by eliminating s and t from the parametrization (cf. Section 1.8.2):

```
ring R=0,(s,t,x(0..4)),dp;
ideal I=x(0)-t4,x(1)-t3s,x(2)-t2s2,x(3)-ts3,x(4)-s4;
ideal J=eliminate(I,st);
J;
//-> J[1]=x(3)^2-x(2)*x(4)
//-> J[2]=x(2)*x(3)-x(1)*x(4)
//-> J[3]=x(1)*x(3)-x(0)*x(4)
//-> J[4]=x(2)^2-x(0)*x(4)
//-> J[5]=x(1)*x(2)-x(0)*x(3)
//-> J[6]=x(1)^2-x(0)*x(2)
```

We now compute the Hilbert–Poincaré series of the ideal J.

```
ring S=0,(x(0..4)),dp;
ideal J=imap(R,J);
J=std(J);
hilb(J);
//->        1 t^0
//->       -6 t^2
```

```
//->        8 t^3
//->       -3 t^4
//->
//->        1 t^0
//->        3 t^1
//->
//-> // dimension  = 1
//-> // degree     = 4
```

We obtain $Q(t) = -3t^4 + 8t^3 - 6t^2 + 1$ (first Hilbert series) and $G(t) = 3t + 1$ (second Hilbert series). The dimension of the projective variety defined by J is 1 — it is a curve; the degree is 4. Let us check this by computing the Hilbert polynomial.

```
LIB"poly.lib";
hilbPoly(J);
//-> 1,4
```

The Hilbert polynomial is $4t + 1$ which leads to the dimension and degree we already computed using the Hilbert–Poincaré series. Let us now use the SINGULAR commands `degree` and `dim` for the same computations:

```
degree(J);
//-> 4
dim(J);
//-> 2
```

Here we obtain the dimension of the affine variety given by J, that is, the affine cone over the rational normal curve, which is 2.

Up to this point, we have only considered homogeneous ideals. In the general case (that is, without assuming homogeneity) studying the Hilbert function of the homogenization leads to results about the relation between an ideal and its leading ideal as well.

Lemma 5.3.13. *Let* $I \subset K[x] := K[x_1, \ldots, x_r]$ *be an ideal and* I^h *its homogenization with respect to* t *in* $K[t, x]$, *then*

$$\dim(K[x]/I) + 1 = \dim(K[t, x]/I^h).$$

Proof. We shall prove the lemma for the case that the characteristic of K is zero. We choose coordinates $M \cdot y = x$, $y = (y_1, \ldots, y_r)$, for a Noether normalization such that $I \cap K[y_{s+1}, \ldots, y_r] = 0$ and, for $j = 1, \ldots, s$, there exist polynomials

$$g_j = y_j^{e_j} + \sum_{k=0}^{e_j-1} \xi_{jk}(y_{j+1}, \ldots, y_r) \cdot y_j^k \in I \text{ with } e_j \geq \deg(\xi_{j,k}) + k$$

(cf. Theorem 3.4.1). Then the homogenizations of g_j with respect to t are

$$g_j^h = y_j^{e_j} + \sum_{k=0}^{e_j-1} t^{e_j-\deg(\xi_{jk})-k} \xi_{jk}^h \cdot y_j^k \in I^h$$

and, therefore, the map $K[t, y_1, \ldots, y_r]/I^h \leftarrow K[t, y_{s+1}, \ldots, y_r]$ is finite. Moreover, $I^h \cap K[t, y_{s+1}, \ldots, y_r] = \langle 0 \rangle$, because $f \in I^h$ implies $f|_{t=1} \in I$. Therefore, $K[t, y_1, \ldots, y_r]/I^h \leftarrow K[t, y_{s+1}, \ldots, y_r]$ is a Noether normalization, which implies

$$\dim(K[t, y_1, \ldots, y_r]/I^h) = r - s + 1 = \dim(K[x_1, \ldots, x_r]/I) + 1. \qquad \square$$

Corollary 5.3.14. *Let $>$ be a degree ordering on $K[x] := K[x_1, \ldots, x_r]$ (that is, $\deg(x^\alpha) > \deg(x^\beta)$ implies $x^\alpha > x^\beta$) and $I \subset K[x]$ any ideal, then*

$$\dim(K[x]/I) = \dim(K[x]/L(I)).$$

Proof. Let $I^h = \langle f^h \mid f \in I \rangle$ be the homogenization of I with respect to t in $K[t, x] = K[t, x_1, \ldots, x_r]$. We extend $>$ to a monomial ordering $>_h$ on $K[t, x]$ by setting

$$t^{\alpha_0} x^\alpha >_h t^{\beta_0} x^\beta :\Longleftrightarrow \alpha_0 + |\alpha| > \beta_0 + |\beta|$$
$$\text{or } \big(\alpha_0 + |\alpha| = \beta_0 + |\beta| \text{ and } x^\alpha > x^\beta\big)$$

(see page 56). Recall that this ordering has the following property:

$$L(I^h) = L(I) \cdot K[t, x] = L(I)^h.$$

Now, using Theorem 5.2.6 we obtain

$$P_{K[t,x]/I^h} = P_{K[t,x]/L(I^h)} = P_{K[t,x]/L(I)^h}.$$

The corollary follows now from Lemma 5.3.13. $\qquad \square$

Definition 5.3.15. For $I \subset K[x] := K[x_1, \ldots, x_r]$ an ideal we define the *affine Hilbert function* as follows: let $K[x]_{\leq s} = \{f \in K[x] \mid \deg(f) \leq s\}$ and $I_{\leq s} = I \cap K[x]_{\leq s}$. Then the affine Hilbert function AH_I of I is given by

$$\text{AH}_I(n) := \dim_K\big(K[x]_{\leq n}/I_{\leq n}\big).$$

Remark 5.3.16. Let $I \subset K[x] := K[x_1, \ldots, x_r]$ be an ideal, and $I^h \subset K[t, x]$ the homogenization of I with respect to t. Then

$$\text{AH}_I(n) = H_{K[t,x]/I^h}(n),$$

because $\varphi : K[x]_{\leq s} \longrightarrow K[t, x]_s$, $f \mapsto t^s f(x_1/t, \ldots, x_r/t)$, the homogenization map, defines a K–isomorphism with $\varphi(I_{\leq s}) = I_s^h$. Therefore, we obtain $\text{AH}_I(n) = \text{AH}_{L(I)}(n)$ for any degree ordering.

On the other hand, $\text{AH}_I(n) = P_{K[t,x]/I^h}(n)$ is, for n sufficiently large, polynomial in n, of degree $\dim(K[x]/I)$.

Corollary 5.3.17. *Let $I \subset K[x] := K[x_1, \ldots, x_r]$ be an ideal.*

(1) $\dim(K[x]/I) = 0$ *if and only if* $\dim_K(K[x]/I) < \infty$.
(2) *Let $>$ be a degree ordering, then* $\dim_K(K[x]/I) = \dim_K(K[x]/L(I))$.

Proof. To prove (1), choose any degree ordering on $K[x]$. If $\dim(K[x]/I) = 0$ then $\mathrm{AH}_I(n) = \dim_K(K[x]/I)$ for large n. Therefore, $\dim_K(K[x]/I) < \infty$.

On the other hand, assume that $\dim_K(K[x]/I) < \infty$. Then $\mathrm{AH}_I(n)$ is bounded, but, for large n, it is a polynomial of degree equal to the dimension of $K[x]/I$. Hence, $\dim(K[x]/I) = 0$.

To prove (2) we may assume that $\dim(K[x]/I) = 0$, because, otherwise, both dimensions are infinite. This implies that for large n, $\mathrm{AH}_I(n)$ is a polynomial of degree zero, that is, constant. Hence, for large n,

$$\dim_K(K[x]/I) = \mathrm{AH}_I(n) = \mathrm{AH}_{L(I)}(n) = \dim_K(K[x]/L(I)). \qquad \square$$

Remark 5.3.18. Once we have the notion of flatness, we shall show that Corollaries 5.3.9, 5.3.14 and 5.3.17 can be generalized to ideals $I \subset K[x_1, \ldots, x_n]_>$, where $>$ is any monomial ordering (see Section 7.5).

Exercises

5.3.1. Write a SINGULAR procedure to compute the affine Hilbert function.

5.3.2. Let $I = \langle x^3 + y^2 + z \rangle$. Compute the affine Hilbert function AH_I.

5.3.3. Let A be a ring and $I, J \subset A$ ideals. Prove that the sequence

$$0 \longrightarrow A/(I \cap J) \longrightarrow A/I \oplus A/J \longrightarrow A/(I + J) \longrightarrow 0$$

is exact, the first map being given by $a \bmod (I \cap J) \mapsto (a \bmod I, a \bmod J)$ and the second by $(a \bmod I, b \bmod J) \mapsto (a - b) \bmod (I + J)$.

5.3.4. Let $A = K[x_1, \ldots, x_5]$ and let I be the ideal generated by the 3–minors of the matrix

$$M = \begin{pmatrix} x_0 & x_1 & x_2 & x_3 \\ x_1 & x_2 & x_3 & x_4 \\ x_2 & x_3 & x_4 & x_5 \end{pmatrix}.$$

Compute the Hilbert–Poincaré series and the Hilbert polynomial of A/I.

5.3.5. Let A be a graded K–algebra and M a (positively) graded, finitely generated A–module. Show that $\deg(P_M) = \dim(M) - 1$ (cf. Definition 5.6.4).

5.4 Filtrations and the Lemma of Artin–Rees

In the next section, we shall study modules over local rings and the filtration induced by the powers of the maximal ideal. In this section, we provide the technical prerequisites.

Let A be a Noetherian ring and $Q \subset A$ be an ideal.

Definition 5.4.1. A set $\{M_n\}_{n \geq 0}$ of submodules of an A–module M is called Q–*filtration* of M if

(1) $M = M_0 \supset M_1 \supset M_2 \supset \ldots$.
(2) $QM_n \subset M_{n+1}$ for all $n \geq 0$.

A Q–filtration $\{M_n\}_{n \geq 0}$ of M is called *stable* if $QM_n = M_{n+1}$ for all sufficiently large n.

Example 5.4.2. Let M be an A–module and $M_n := Q^n M$ for $n \geq 0$. Then $\{M_n\}_{n \geq 0}$ is a stable Q–filtration of M.

Lemma 5.4.3. *Let $\{M_n\}_{n \geq 0}$ and $\{N_n\}_{n \geq 0}$ be two stable Q–filtrations of M. Then there exists some non–negative integer n_0 such that $M_{n+n_0} \subset N_n$ and $N_{n+n_0} \subset M_n$ for all $n \geq 0$.*

Proof. We may assume that one of the two filtrations is the one of Example 5.4.2, say $N_n := Q^n M$. Now, $\{M_n\}_{n \geq 0}$ being stable implies that there exists some non–negative integer n_0 such that $M_{n_0+n} = Q^n M_{n_0}$ for all $n \geq 0$. On the other hand, $Q^n M_{n_0} \subset Q^n M = N_n$ implies $M_{n_0+n} \subset N_n$.

As $\{M_n\}_{n \geq 0}$ is a Q–filtration, we have $QM_n \subset M_{n+1}$ for all $n \geq 0$, which implies, in particular, $N_n = Q^n M = Q^n M_0 \subset M_n$. But $N_{n_0+n} \subset N_n$ which proves the claim. $\qquad\square$

Lemma 5.4.4. *Let $\{M_n\}_{n \geq 0}$ be a Q–filtration of M, and let N be a submodule of M. Then $\{N \cap M_n\}_{n \geq 0}$ is a Q–filtration of N.*

Proof. $QM_n \subset M_{n+1}$ implies $Q(M_n \cap N) \subset QM_n \cap QN \subset M_{n+1} \cap N$ for all $n \geq 0$. $\qquad\square$

Lemma 5.4.5 (Artin–Rees). *Let $\{M_n\}_{n \geq 0}$ be a stable Q–filtration of the finitely generated A–module M and $N \subset M$ a submodule, then $\{M_n \cap N\}_{n \geq 0}$ is a stable Q–filtration of N.*

To prove the lemma, we need a criterion for stability. Let M be a finitely generated A–module and $\{M_n\}_{n \geq 0}$ be a Q–filtration of M. Let

$$\overline{A} := \bigoplus_{n=0}^{\infty} Q^n, \qquad \overline{M} := \bigoplus_{n=0}^{\infty} M_n.$$

Lemma 5.4.6 (Criterion for stability). $\overline{M}$ *is a finitely generated $\overline{A}$–module if and only if $\{M_n\}_{n\geq 0}$ is Q–stable.*

Proof. Since A is Noetherian and M is finitely generated, if follows that the submodules M_n, $n \geq 0$, are finitely generated. Let

$$\overline{M}_n := M_0 \oplus \cdots \oplus M_{n-1} \oplus \overline{A}M_n \,,$$

then $\overline{M}_n$ is a finitely generated $\overline{A}$–module, because $\bigoplus_{\nu=0}^{n} M_\nu$ is a finitely generated A–module. Moreover, $\overline{M}_n \subset \overline{M}_{n+1}$ for all $n \geq 0$, and $\bigcup_{n=0}^{\infty} \overline{M}_n = \overline{M}$. $\overline{A}$ is Noetherian, because $\overline{A} = A[f_1,\ldots,f_r]$ if $Q = \langle f_1,\ldots,f_r\rangle$. This implies (finitely generated modules over Noetherian rings being Noetherian) that $\overline{M}$ is a finitely generated $\overline{A}$–module if and only if there exists a non–negative integer n_0 such that $\overline{M}_{n_0} = \overline{M}$. This is the case if and only if $M_{n_0+r} = Q^r M_{n_0}$ for all $r \geq 0$. $\qquad\square$

Proof of Lemma 5.4.5. Using Lemma 5.4.4 we obtain that $\{M_n \cap N\}_{n\geq 0}$ is a Q–filtration of N. We know by assumption and Lemma 5.4.6 that $\overline{M}$ is a finitely generated $\overline{A}$–module. Moreover, $\overline{N} := \bigoplus_{n=0}^{\infty}(M_n \cap N)$ is an $\overline{A}$–submodule of $\overline{M}$ and, therefore, also finitely generated. Using Lemma 5.4.6 again, we obtain that $\{M_n \cap N\}_{n\geq 0}$ is a stable Q–filtration of N. $\qquad\square$

Corollary 5.4.7. *Let M be a finitely generated A–module and $N \subset M$ a submodule, then there exists a non–negative integer n_0 such that, for all $n \geq n_0$,*

$$(Q^n M) \cap N = Q^{n-n_0}\big((Q^{n_0}M) \cap N\big).$$

Proof. Apply Lemma 5.4.5 for the filtration given by $M_n := Q^n M$, $n \geq 0$. $\quad\square$

Exercises

5.4.1. Consider $M := K[x,y]/\langle y^2 + y + x\rangle$ as $A := K[x]$–module with the following $Q = \langle x\rangle$–filtration: $M_n = \langle x,y\rangle^n$. Decide whether this is a stable Q–filtration.

5.4.2. Let A be a Noetherian ring, M a finitely generated A–module and $N \subset M$ a submodule. Let $Q \subset A$ be an ideal and $\{M_n\}_{n\geq 0}$ a stable Q–filtration of M. Prove that there exist stable Q–filtrations $\{N_n\}_{n\geq 0}$ of N and $\{\widetilde{M}_n\}_{n\geq 0}$ of M/N such that $0 \to N_n \to M_n \to \widetilde{M}_n \to 0$ is exact.

5.4.3. Give an example for an A–module M, a submodule $N \subset M$ and an ideal $Q \subset A$, such that the filtrations $\{Q^n N\}_{n\geq 0}$ and $\{Q^n M \cap N\}_{n\geq 0}$ of N are different.

5.5 The Hilbert–Samuel Function

The Hilbert–Samuel function is the counterpart to the Hilbert function in the local case. To a module M over the local ring $(A, \mathfrak{m})$ and to an integer n, it associates the dimension of $M/\mathfrak{m}^n M$. Similarly to the homogeneous case, this function is a polynomial for large n, the Hilbert–Samuel polynomial. By passing to the associated graded module $\mathrm{Gr}_\mathfrak{m}(M) = \bigoplus_{\nu=0}^\infty \mathfrak{m}^\nu M/\mathfrak{m}^{\nu+1} M$, the results from the homogeneous case can be used.

Let A be a local Noetherian ring with maximal ideal $\mathfrak{m}$. We assume (just for simplicity) that $K = A/\mathfrak{m} \subset A$. Moreover, let Q be an $\mathfrak{m}$–primary ideal and M a finitely generated A–module. Recall that the associated graded ring to $Q \subset A$ is defined as $\mathrm{Gr}_Q(A) = \bigoplus_{\nu=0}^\infty (Q^\nu/Q^{\nu+1})$ (cf. Example 2.2.3).

Lemma 5.5.1. *Let $\{M_n\}_{n\geq 0}$ be a stable Q–filtration of M, and let*

$$\mathrm{HS}_{\{M_n\}_{n\geq 0}}(k) := \dim_K(M/M_k)\,.$$

Moreover, suppose that Q is generated by r elements. Then

(1) $\mathrm{HS}_{\{M_n\}_{n\geq 0}}(k) < \infty$ for all $k \geq 0$;

(2) there exists a polynomial $\mathrm{HSP}_{\{M_n\}_{n\geq 0}}(t) \in \mathbb{Q}[t]$ of degree at most r such that $\mathrm{HS}_{\{M_n\}_{n\geq 0}}(k) = \mathrm{HSP}_{\{M_n\}_{n\geq 0}}(k)$ for all sufficiently large k;

(3) the degree of $\mathrm{HSP}_{\{M_n\}_{n\geq 0}}$ and its leading coefficient do not depend on the choice of the stable Q–filtration $\{M_n\}_{n\geq 0}$.

Proof. $\mathrm{Gr}_Q(A) = \bigoplus_{\nu\geq 0} Q^\nu/Q^{\nu+1}$ is a graded K–algebra (Example 2.2.3), which is generated by r elements of degree 1. Now, let

$$\mathrm{Gr}_{\{M_n\}}(M) := \bigoplus_{\nu\geq 0} M_\nu/M_{\nu+1}\,.$$

Since $\{M_n\}_{n\geq 0}$ is a stable Q–filtration, $\mathrm{Gr}_{\{M_n\}}(M)$ is a finitely generated, graded $\mathrm{Gr}_Q(A)$–module.

Now, as $QM_\nu \subset M_{\nu+1}$, the quotients $M_\nu/M_{\nu+1}$, $\nu \geq 0$, are annihilated by Q and, therefore, are finitely generated A/Q–modules. But A/Q is a finite dimensional K–vector space (Q is $\mathfrak{m}$–primary), hence, $\dim_K(M_\nu/M_{\nu+1}) < \infty$. This implies $\dim_K(M/M_n) = \sum_{\nu=1}^n \dim_K(M_{\nu-1}/M_\nu) < \infty$, which proves (1).

To prove (2) notice that $H_{\mathrm{Gr}_{\{M_n\}}(M)}(k) = \dim_K(M_k/M_{k+1})$. For sufficiently large k, $H_{\mathrm{Gr}_{\{M_n\}}(M)}(k) = P_{\mathrm{Gr}_{\{M_n\}}(M)}(k)$, and $P_{\mathrm{Gr}_{\{M_n\}}(M)}$ is a polynomial of degree at most $r - 1$ (cf. Theorem 5.1.3 and Corollary 5.1.5). Let

$$P_{\mathrm{Gr}_{\{M_n\}}(M)}(k) = \sum_{\nu=0}^{r-1} a_\nu \binom{k}{\nu}\,,$$

then we have

$$\mathrm{HS}_{\{M_n\}_{n\geq0}}(k+1) - \mathrm{HS}_{\{M_n\}_{n\geq0}}(k) = \dim_K(M/M_{k+1}) - \dim_K(M/M_k)$$
$$= \dim_K(M_k/M_{k+1}) = H_{\mathrm{Gr}_{\{M_n\}}(M)}(k) = P_{\mathrm{Gr}_{\{M_n\}}(M)}(k),$$

for sufficiently large k. On the other hand,

$$\sum_{\nu=1}^{r} a_{\nu-1}\binom{k+1}{\nu} - \sum_{\nu=1}^{r} a_{\nu-1}\binom{k}{\nu} = \sum_{\nu=0}^{r-1} a_\nu\binom{k}{\nu}$$
$$= \mathrm{HS}_{\{M_n\}_{n\geq0}}(k+1) - \mathrm{HS}_{\{M_n\}_{n\geq0}}(k).$$

Hence $\mathrm{HS}_{\{M_n\}_{n\geq0}}(k) - \sum_{\nu=1}^{r} a_{\nu-1}\binom{k}{\nu}$ is constant if k is sufficiently large. Let C be this constant and set $\mathrm{HSP}_{\{M_n\}_{n\geq0}}(k) := \sum_{\nu=1}^{r} a_{\nu-1}\binom{k}{\nu} + C$. Then $\mathrm{HS}_{\{M_n\}_{n\geq0}}(k) = \mathrm{HSP}_{\{M_n\}_{n\geq0}}(k)$, a polynomial of degree at most r, for sufficiently large k.

To prove (3) let $\{M'_n\}_{n\geq0}$ be another stable Q–filtration of M, and choose, using Lemma 5.4.3, k_0 such that $M_{k+k_0} \subset M'_k$ and $M'_{k+k_0} \subset M_k$ for all $k \geq 0$. This implies the inequalities $\mathrm{HS}_{\{M_n\}_{n\geq0}}(k) \leq \mathrm{HS}_{\{M'_n\}_{n\geq0}}(k+k_0)$ and $\mathrm{HS}_{\{M'_n\}_{n\geq0}}(k) \leq \mathrm{HS}_{\{M_n\}_{n\geq0}}(k+k_0)$ and, therefore,

$$1 = \lim_{k\to\infty} \frac{\mathrm{HS}_{\{M_n\}}(k)}{\mathrm{HS}_{\{M'_n\}}(k)} = \lim_{k\to\infty} \frac{\mathrm{HSP}_{\{M_n\}}(k)}{\mathrm{HSP}_{\{M'_n\}}(k)},$$

which proves (3). □

Definition 5.5.2. With the notation of Lemma 5.5.1 we define:

(1) $\mathrm{HS}_{M,Q} := \mathrm{HS}_{\{Q^n M\}_{n\geq0}}$ is called the *Hilbert–Samuel function* of M with respect to Q;

(2) $\mathrm{HSP}_{M,Q} := \mathrm{HSP}_{\{Q^n M\}_{n\geq0}}$ is called the *Hilbert–Samuel polynomial* of M with respect to Q;

(3) let $\mathrm{HSP}_{M,Q}(k) = \sum_{\nu=0}^{d} a_\nu k^\nu$ with $a_d \neq 0$. Then $\mathrm{mult}(M,Q) := d! \cdot a_d$ is called the *(Hilbert–Samuel) multiplicity* of M with respect to Q;

(4) $\mathrm{mult}(M) := \mathrm{mult}(M,\mathfrak{m})$ is the *(Hilbert–Samuel) multiplicity* of M.

Remark 5.5.3. Let $0 \to N \to M \to M/N \to 0$ be an exact sequence of finitely generated A–modules, and Q an $\mathfrak{m}$–primary ideal. Then

$$0 \to N/(Q^n M \cap N) \to M/Q^n M \to (M/N)/Q^n(M/N) \to 0$$

is exact and, hence, $\mathrm{HSP}_{M,Q} = \mathrm{HSP}_{M/N,Q} + \mathrm{HSP}_{\{Q^n M \cap N\}}$. The proof of Lemma 5.5.1 shows that, indeed,

$$\mathrm{HSP}_{M,Q} = \mathrm{HSP}_{M/N,Q} + \mathrm{HSP}_{N,Q} - R,$$

where R is a polynomial of degree strictly smaller than that of $\mathrm{HSP}_{N,Q}$.[3] More precisely, $\deg R = \dim(\mathrm{Ker}(\mathrm{Gr}_Q M/\mathrm{Gr}_Q N \twoheadrightarrow \mathrm{Gr}_Q(M/N))) - 1$. Moreover, if $R \neq 0$, then its leading coefficient is positive. We shall see in Proposition 5.6.3 that $\deg(\mathrm{HSP}_{M,Q}) = \deg(\mathrm{HSP}_{M,\mathfrak{m}})$.

---

[3] Note the difference to the Hilbert polynomial of graded modules where no remainder R appears, cf. Lemma 5.1.2.

Example 5.5.4. Let $A = K[x_1, \ldots, x_r]_{\langle x_1, \ldots, x_r \rangle}$ then $\mathrm{HSP}_{A,\mathfrak{m}}(k) = \binom{k+r-1}{r}$. In particular, $\mathrm{mult}(A) = 1$.

In the proof of Lemma 5.5.1, we actually proved more than just the claim. Using the notation of Definition 5.5.2, we can summarize the additional results as a comparison between the Hilbert–Samuel polynomial of M with respect to Q and the Hilbert polynomial of the graded $\mathrm{Gr}_Q(A)$-module $\mathrm{Gr}_Q(M)$ $\cong \oplus_{\nu=0}^{\infty}(Q^\nu M/Q^{\nu+1}M)$.

Corollary 5.5.5. *Let $(A, \mathfrak{m})$ be a Noetherian local ring, $Q \subset A$ an $\mathfrak{m}$-primary ideal and M a finitely generated A-module.*

(1) $\mathrm{HSP}_{M,Q}(k+1) - \mathrm{HSP}_{M,Q}(k) = P_{\mathrm{Gr}_Q(M)}(k)$.
(2) If $P_{\mathrm{Gr}_Q(M)}(k) = \sum_{\nu=0}^{s-1} a_\nu \binom{k}{\nu}$ *then*

$$\mathrm{HSP}_{M,Q}(k) = \sum_{\nu=1}^{s} a_{\nu-1}\binom{k}{\nu} + c$$

with $c = \dim_K(M/Q^\ell M) - \sum_{\nu=1}^{s} a_{\nu-1}\binom{\ell}{\nu}$ for any sufficiently large ℓ. In particular, we obtain $\mathrm{mult}(M, Q) = \deg(\mathrm{Gr}_Q(M))$ and

$$\deg(\mathrm{HSP}_{M,Q}) = \deg(P_{\mathrm{Gr}_Q(M)}) + 1.$$

Example 5.5.6. Let $I \subset K[x] := K[x_1, \ldots, x_r]$ be a monomial ideal, and let $A = K[x]_{\langle x \rangle}/I$.[4] Then, because of $K[x]_{\langle x \rangle}/(I + \langle x \rangle^k) \cong K[x]/(I + \langle x \rangle^k)$, we obtain $\mathrm{Gr}_{\langle x \rangle}(A) = K[x]/I$, and Corollary 5.5.5 implies

$$\mathrm{HSP}_{A,\langle x \rangle}(k) - \mathrm{HSP}_{A,\langle x \rangle}(0) = \sum_{\nu=0}^{k-1} P_{K[x]/I}(\nu).$$

The following proposition enables the use of Example 5.5.6 for the computation of the Hilbert–Samuel polynomial:

Proposition 5.5.7. *Let $>$ be a local degree ordering on $K[x] = K[x_1, \ldots, x_r]$ (that is, $\deg(x^\alpha) > \deg(x^\beta)$ implies $x^\alpha < x^\beta$). Let $I \subset \langle x \rangle = \langle x_1, \ldots, x_r \rangle \subset K[x]$ be an ideal and $A = K[x]_{\langle x \rangle}/I$, then*

$$\mathrm{HS}_{A,\mathfrak{m}} = \mathrm{HS}_{K[x]_{\langle x \rangle}/L(I), \langle x \rangle}.$$

In particular, $K[x]_{\langle x \rangle}/I$ and $K[x]_{\langle x \rangle}/L(I)$ have the same multiplicity with respect to $\langle x \rangle$.

---

[4] Throughout this chapter, we denote the ideal generated by $I \subset K[x]$ in the localization $K[x]_{\langle x \rangle}$ by I, too, in order to keep the notations short.

Proof. We have to prove that

$$\dim_K K[x]_{\langle x \rangle}/(I + \langle x \rangle^k) = \dim_K K[x]_{\langle x \rangle}/(L(I) + \langle x \rangle^k).$$

Clearly, for each $k \geq 0$, the set $S := \{x^\alpha \notin L(I) \mid \deg(x^\alpha) < k\}$ represents a K–basis of $K[x]_{\langle x \rangle}/(L(I) + \langle x \rangle^k) \cong K[x]/(L(I) + \langle x \rangle^k)$. On the other hand (using reduction by a standard basis of I), we can write each $f \in K[x]$ as

$$f = g + \sum_{x^\alpha \in S} c_\alpha x^\alpha \mod \langle x_1, \ldots, x_r \rangle^k$$

for some $g \in I$ and uniquely determined $c_\alpha \in K$. This is possible without multiplying f by a unit, because we are working modulo $\langle x_1, \ldots, x_r \rangle^k$. Therefore, S also represents a K–basis of $K[x]/(I + \langle x \rangle^k) \cong K[x]_{\langle x \rangle}/(I + \langle x \rangle^k)$, which proves the proposition. $\square$

Definition 5.5.8. For any power series $f \in K[x]_{\langle x \rangle} \smallsetminus \{0\}$, $x = (x_1, \ldots, x_r)$, we can find $u \in K[x]$, $u(0) \neq 0$, such that $uf = \sum_\alpha a_\alpha x^\alpha \in K[x]$. Then

$$\mathrm{ord}(f) := \min\{|\alpha| \mid a_\alpha \neq 0\}$$

is called the *order* of f. Note that $\mathrm{ord}(f)$ is well–defined (Exercise 5.5.6).

Corollary 5.5.9. *Let $f \in K[x] \smallsetminus \{0\}$, $x = (x_1, \ldots, x_r)$. Then*

$$\mathrm{mult}(K[x]_{\langle x \rangle}/\langle f \rangle) = \mathrm{ord}(f).$$

Proof. Proposition 5.5.7 implies that, for a local degree ordering, we have $\mathrm{HSP}_{K[x]_{\langle x \rangle}/\langle f \rangle, \langle x \rangle} = \mathrm{HSP}_{K[x]_{\langle x \rangle}/\langle \mathrm{LM}(f) \rangle}$. Using Corollary 5.5.5, we conclude that $\mathrm{mult}(K[x]_{\langle x \rangle}/\langle \mathrm{LM}(f) \rangle) = \deg(K[x]/\langle \mathrm{LM}(f) \rangle) = \mathrm{ord}(f)$. $\square$

Next we shall describe how to compute $\mathrm{Gr}_\mathfrak{m}(A)$, the so–called tangent cone of A.

Definition 5.5.10.

(1) For $(A, \mathfrak{m})$ a local ring, the graded ring $\mathrm{Gr}_\mathfrak{m}(A)$ is called the *tangent cone* of A.

(2) For $f \in K[x]_{\langle x \rangle} \smallsetminus \{0\}$, $x = (x_1, \ldots, x_r)$, choose $u \in K[x]$, $u(0) = 1$, such that $uf = \sum_\alpha a_\alpha x^\alpha \in K[x]$. Let $d := \mathrm{ord}(f)$, and call

$$\mathrm{In}(f) := \sum_{\deg(x^\alpha) = d} a_\alpha x^\alpha$$

the *initial form* of f. Note that $\mathrm{In}(f)$ is well–defined (Exercise 5.5.6).

(3) Let $I \subset K[x]_{\langle x \rangle}$ be an ideal. The ideal

$$\mathrm{In}(I) := \langle \mathrm{In}(f) \mid f \in I \smallsetminus \{0\} \rangle \subset K[x]$$

is called the *initial ideal* of I.

Lemma 5.5.11. *Let $I \subset \langle x \rangle \subset K[x]$, $x = (x_1, \ldots, x_r)$, be an ideal, and let $\{f_1, \ldots, f_s\}$ be a standard basis of I with respect to a local degree ordering $>$. Then $\mathrm{In}(I) = \langle \mathrm{In}(f_1), \ldots, \mathrm{In}(f_s) \rangle$.*

Proof. Let $f \in I$, then, using the property of being a standard basis (Theorem 1.7.3 (3)), there exist $u, g_1, \ldots, g_s \in K[x]$, $u \in S_<$, such that $uf = \sum_{i=1}^s g_i f_i$ and $\mathrm{LM}(uf) \geq \mathrm{LM}(g_i f_i)$ for all i, especially, $\deg(\mathrm{LM}(uf)) \leq \deg(\mathrm{LM}(g_i f_i))$. Now, let

$$J := \{1 \leq i \leq s \mid \deg(LM(uf)) = \deg(LM(g_i f_i))\}.$$

Then $\mathrm{In}(f) = \sum_{i \in J} \mathrm{In}(g_i) \mathrm{In}(f_i)$. $\qquad\square$

Proposition 5.5.12. *Let $I \subset \langle x \rangle \subset K[x]$, $x = (x_1, \ldots, x_r)$, be an ideal, let $A := K[x]_{\langle x \rangle}/I$, and let $\mathfrak{m}$ be the maximal ideal of A. Then*

$$\mathrm{Gr}_{\mathfrak{m}}(A) \cong K[x]/\mathrm{In}(I).$$

Proof. Clearly, $\mathfrak{m}^\nu/\mathfrak{m}^{\nu+1} \cong \langle x \rangle^\nu/(\langle x \rangle^{\nu+1} + (I \cap \langle x \rangle^\nu))$. Moreover, it is easy to see that the kernel of the canonical surjection

$$K[x]_\nu \longrightarrow \langle x \rangle^\nu/(\langle x \rangle^{\nu+1} + (I \cap \langle x \rangle^\nu))$$

is $\mathrm{In}(I)_\nu$. $\qquad\square$

Proposition 5.5.12 and Lemma 5.5.11 are the basis for determining $\mathrm{Gr}_{\mathfrak{m}} A$ by computing the initial ideal of I as the ideal generated by the initial forms of a standard basis with respect to a local degree ordering.

SINGULAR Example 5.5.13 (initial ideal, Poincaré series, Hilbert polynomial).
Let $I := \langle yz + z^2 + x^3, y^2 + xz + y^4 \rangle \subset K[x, y, z]$.

```
ring A=0,(x,y,z),ds;
ideal I=yz+z2+x3,y2+xz+y4;
ideal J=std(I);
J;
//-> J[1]=y2+xz+y4   J[2]=yz+z2+x3   J[3]=xz2-yz2-x3y+y4z

int k;
ideal In=jet(J[1],ord(J[1]));
for(k=2;k<=size(J);k++){ In=In,jet(J[k],ord(J[k]));}

In;
//-> In[1]=y2+xz   In[2]=yz+z2   In[3]=xz2-yz2
```

From the output, we read the initial ideal $\mathrm{In}(I) = \mathrm{In}$ of I and know, therefore, the tangent cone $K[x, y, z]/\mathrm{In}(I)$ of $A = K[x, y, z]_{\langle x, y, z \rangle}/I$. Now let us compute the Poincaré series and Hilbert polynomial of $K[x, y, z]/\mathrm{In}(I)$:

```
intvec v=hilb(J,1);
v;
//-> 1,0,-2,0,1,0
```

Hence,

$$\mathrm{HP}_{K[x,y,z]/\operatorname{In}(I)}(t) = \frac{t^4 - 2t^2 + 1}{(1-t)^3} = \frac{(t+1)^2}{1-t} = 1 + 3t + \sum_{i=1}^{\infty} 4t^i,$$

and, therefore, $P_{K[x,y,z]/\operatorname{In}(I)}(n) = 4$. This implies $\mathrm{HSP}_{A,\mathfrak{m}}(n) = 4 \cdot \binom{n}{1} + C$. To compute the constant C we compute a value of the Hilbert–Samuel function, for a sufficiently large n. We have seen that $H_{K[x,y,z]/\operatorname{In}(I)}(n)$ coincides with $P_{K[x,y,z]/\operatorname{In}(I)}(n)$ for all $n \geq 3$ and that the highest monomial in the standard basis of I is of degree 3 as well. Therefore, we know that the choice $n = 5$ is sufficiently large. [5]

```
ideal I1=I+maxideal(5);
vdim(std(I1));
//-> 16
```

This implies $C = \dim_K(A/\mathfrak{m}^5) - 4 \cdot \binom{5}{1} = -4$. The Hilbert polynomial shows that the multiplicity of A is 4. We can check this directly:

```
mult(J);
//-> 4
```

Exercises

5.5.1. Let I be the ideal of the 2–minors of the matrix $\left(\begin{smallmatrix} x & y & z \\ y & z & w \end{smallmatrix}\right)$ in $K[x, y, z, w]$. Let $A = K[x, y, z, w]_{\langle x,y,z,w \rangle}/I$. Compute $\mathrm{HSP}_{A,\langle x,w \rangle}$.

5.5.2. Formulate and prove Proposition 5.5.12 for modules. Try examples with SINGULAR.

5.5.3. Write a SINGULAR procedure to compute the Hilbert–Samuel polynomial of $K[x]_{\langle x \rangle}/I$ for an ideal $I \subset K[x]$ using Propositions 5.5.5 and 5.5.7.

5.5.4. Let $f \in K[x_1, \ldots, x_r]$ be a polynomial with $\operatorname{ord}(f) = m$. Prove that

$$\mathrm{HSP}_{K[x]_{\langle x \rangle}/\langle f \rangle}(k) = \sum_{\nu=1}^{m} \binom{r + k - \nu - 1}{r - 1}.$$

---

[5] Just checking, whether the difference $\dim_K(A/Q^\ell) - \sum_{\nu=1}^{s} a_{\nu-1}\binom{\ell}{\nu}$ is constant, is not sufficient.

5.5.5. Let $I \subset \mathfrak{m} = \langle x \rangle \subset K[x_1, \ldots, x_r]$ be an ideal and $A := K[x]/I$. The *normal cone* is defined to be the graded ring $\mathrm{Gr}_\mathfrak{m}(A)$. Let $>$ be a degree ordering on $K[x]$ (that is, $\deg(x^\alpha) < \deg(x^\beta)$ implies $x^\alpha < x^\beta$). For $f \in K[x]$, $f = \sum_\alpha a_\alpha x^\alpha$ and $d = \deg(f)$ let $\mathrm{In}(f) = \sum_{\deg(x^\alpha)=d} a_\alpha x^\alpha$ be the *initial form* of f. Similarly, the *initial ideal* of I is $\mathrm{In}(I) = \langle \mathrm{In}(f) \mid f \in I \smallsetminus \{0\} \rangle$.

Prove analogues of Lemma 5.5.11 and Proposition 5.5.12, under these assumptions.

5.5.6. Show that $\mathrm{ord}(f)$ and $\mathrm{In}(f)$ for $f \in K[x]_{\langle x \rangle}$ are well–defined, that is, there exist units u as required and the definition is independent of the choice of u.

5.5.7. Compute the tangent cone for the local ring $A = \mathbb{Q}[x, y, z]_{\langle x,y,z \rangle}/I$, where $I := \langle x^2 + y^3 + z^4, xy + xz + z^3 \rangle$.

5.5.8. Compute the normal cone for $A = Q[x, y, z]/I$, I the ideal in Exercise 5.5.7.

5.5.9. Write a SINGULAR procedure to compute the tangent cone.

5.5.10. Let A be a Noetherian local ring and M a finitely generated A–module such that $\mathrm{Ann}_A(M) = \langle 0 \rangle$. Prove that $\deg(\mathrm{HSP}_{M,\mathfrak{m}}) = \deg(\mathrm{HSP}_{A,\mathfrak{m}})$.

5.5.11. Let A be a local Noetherian ring, and let $I \subset A$ be an ideal. Moreover, let $I = Q_1 \cap \cdots \cap Q_m$ be an irredundant primary decomposition and assume that $\dim(A/I) = \dim(A/Q_i)$, $i = 1, \ldots, s$, and $\dim(A/I) > \dim(A/Q_i)$, $i = s+1, \ldots, m$. Prove that $\mathrm{mult}(A/I) = \sum_{i=1}^s \mathrm{mult}(A/Q_i)$.

(Hint: compare with Lemma 5.3.11.)

5.5.12. Compute the multiplicity of $K[x, y, z]_{\langle x,y,z \rangle}/\langle z^2 - z, yz - y \rangle$. Compute the degree of $K[x, y, z, t]/\langle z^2 - zt, yz - yt \rangle$.

5.5.13. Formulate and prove the corresponding statements to Exercise 5.1.5 and 5.2.6.

5.6 Characterization of the Dimension of Local Rings

Let A be a Noetherian local ring, $\mathfrak{m}$ its maximal ideal and assume, as before, for simplicity that $K = A/\mathfrak{m} \subset A$.

In this section, we shall prove that the dimension of a local ring is equal to the degree of the Hilbert–Samuel polynomial and equal to the least number of generators of an $\mathfrak{m}$–primary ideal. In particular, we shall define and study regular local rings.

Definition 5.6.1. We introduce the following non–negative integers:

- $\delta(A) :=$ the minimal number of generators of an $\mathfrak{m}$–primary ideal of A,
- $d(A) := \deg(\mathrm{HSP}_{A,\mathfrak{m}})$,

- edim(A) := the *embedding dimension* of A, defined as minimal number of generators for $\mathfrak{m}$. Hence, edim(A) = $\dim_K(\mathfrak{m}/\mathfrak{m}^2)$, by Nakayama's Lemma.

Theorem 5.6.2. *Let $(A, \mathfrak{m})$ be a Noetherian local ring, then, with the above notation, $\delta(A) = d(A) = \dim(A)$.*

We first prove the following proposition:

Proposition 5.6.3. *Let $(A, \mathfrak{m})$ be a Noetherian local ring, let M be a finitely generated A-module, and let Q an $\mathfrak{m}$-primary ideal. Then*

(1) $\deg(\mathrm{HSP}_{M,Q}) = \deg(\mathrm{HSP}_{M,\mathfrak{m}});$

Moreover, let $x \in A$ be a non-zerodivisor for M, then

(2) $\deg(\mathrm{HSP}_{M/xM,Q}) \leq \deg(\mathrm{HSP}_{M,Q}) - 1.$

Proof. For (1) we choose s such that $\mathfrak{m} \supset Q \supset \mathfrak{m}^s$. Then $\mathfrak{m}^k \supset Q^k \supset \mathfrak{m}^{sk}$ for all k implies $\mathrm{HSP}_{M,\mathfrak{m}}(k) \leq \mathrm{HSP}_{M,Q}(k) \leq \mathrm{HSP}_{M,\mathfrak{m}}(sk)$ for sufficiently large k. But this is only possible if $\deg(\mathrm{HSP}_{M,Q}) = \deg(\mathrm{HSP}_{M,\mathfrak{m}})$. To prove (2), we apply Remark 5.5.3 to the exact sequence

$$0 \longrightarrow M \overset{\cdot x}{\longrightarrow} M \longrightarrow M/xM \longrightarrow 0,$$

and conclude that $\deg(\mathrm{HSP}_{M/xM,Q}) \leq \deg(\mathrm{HSP}_{M,Q}) - 1$. □

Definition 5.6.4. Let R be a ring and M an R-module. The *dimension* of M, $\dim(M)$, is defined by $\dim(M) = \dim\big(R/\mathrm{Ann}(M)\big)$.

SINGULAR Example 5.6.5 (dimension of a module).

```
LIB "primdec.lib";
ring R=0,(x,y,z),ds;
module I=[x2,0,0],[0,xz,0],[0,0,x2+zx3];

Ann(I);          //the annihilator of M=R^3/I
//-> _[1]=x2z

dim(std(Ann(I)));
//-> 2
```

This shows that $\dim(R^3/I) = 2$.

Corollary 5.6.6. *Let A be a Noetherian local ring and M a finitely generated A-module. Then $\dim(M) = \deg(\mathrm{HSP}_{M,\mathfrak{m}})$.*

Proof. The corollary is an immediate consequence of Exercise 5.5.10 and Theorem 5.6.2. □

Proof of Theorem 5.6.2. We shall prove that

(1) $\delta(A) \geq d(A)$;
(2) $d(A) \geq \dim(A)$;
(3) $\dim(A) \geq \delta(A)$.

(1) holds because of Lemma 5.5.1 (2) and Proposition 5.6.3 (1).

To prove (2) we use induction on $d = d(A)$. If $d = 0$ then $\dim_K A/\mathfrak{m}^n$ is constant for sufficiently large n. This implies $\mathfrak{m}^n = \mathfrak{m}^{n+1}$ for sufficiently large n and, therefore, by Nakayama's lemma $\mathfrak{m}^n = \langle 0 \rangle$. But then $\dim(A) = 0$ because $\mathfrak{m}$ is the only prime ideal in A.

Assume $d > 0$, and let $P_0 \subset \cdots \subset P_s = \mathfrak{m}$ be a maximal chain of prime ideals in A, $s = \dim(A)$. Let $\bar{A} := A/P_0$, then $\dim(\bar{A}) = s$. On the other hand, the obvious map $A/\mathfrak{m}^n \twoheadrightarrow \bar{A}/\bar{\mathfrak{m}}^n$, $\bar{\mathfrak{m}}$ the maximal ideal of $\bar{A}$, is surjective and, therefore, $\dim_K(A/\mathfrak{m}^n) \geq \dim_K(\bar{A}/\bar{\mathfrak{m}}^n)$. This implies $d(A) \geq d(\bar{A})$, and we may assume that $A = \bar{A}$ is an integral domain. If $s = 0$ then (2) is proved. If $s > 0$ then we choose a non–zerodivisor $x \in P_1$. Using Proposition 5.6.3 (2), we obtain $d(A/\langle x \rangle) \leq d(A) - 1$.

Using the induction hypothesis, we have $d(A/\langle x \rangle) \geq \dim(A/\langle x \rangle)$. But $x \in P_1$ and hence we have $\dim(A/\langle x \rangle) \geq s - 1$. Combining these inequalities, we obtain $d - 1 \geq d(A/\langle x \rangle) \geq \dim(A/\langle x \rangle) \geq s - 1$, which implies that $d(A) = d \geq s = \dim(A)$.

To prove (3) we have to construct an $\mathfrak{m}$–primary ideal $Q = \langle x_1, \ldots, x_d \rangle$ such that $d \leq \dim(A)$.

Suppose $i > 0$ and $x_1, \ldots, x_{i-1}$ are constructed such that every prime ideal P containing $\langle x_1, \ldots, x_{i-1} \rangle$ has height $\geq i - 1$. Let P_j, $1 \leq j \leq s$, be the minimal prime ideals of $\langle x_1, \ldots, x_{i-1} \rangle$ which have height $i - 1$. If $i - 1 = \dim(A)$ then (3) is proved.

If $i - 1 < \dim(A) = \mathrm{ht}(\mathfrak{m})$ then $\mathfrak{m} \neq P_j$ for $j = 1, \ldots, s$. This implies $\mathfrak{m} \neq \bigcup_{j=1}^s P_j$ (cf. Lemma 1.3.12, prime avoidance). Thus we can choose $x_i \in \mathfrak{m}$, $x_i \notin \bigcup_{j=1}^s P_j$. Let P be a prime ideal, $\langle x_1, \ldots, x_i \rangle \subset P$. Then P contains a minimal prime ideal Q of $\langle x_1, \ldots, x_{i-1} \rangle$.

If $Q = P_j$ for some j, then we have $P \supsetneq Q$ ($x \in P \smallsetminus P_j$ by the choice of x) and, therefore, $\mathrm{ht}(P) \geq i$. If $Q \neq P_j$ for $j = 1, \ldots, s$ then $\mathrm{ht}(Q) \geq i$ and, moreover, $\mathrm{ht}(P) \geq i$. We conclude that every prime ideal containing $\langle x_1, \ldots, x_i \rangle$ has height at least i.

Using induction, we obtain an ideal $\langle x_1, \ldots, x_d \rangle$, $d = \dim(A) = \mathrm{ht}(\mathfrak{m})$, such that all primes containing $\langle x_1, \ldots, x_d \rangle$ have height at least d. This implies that $\langle x_1, \ldots, x_d \rangle$ is $\mathfrak{m}$–primary and proves (3). $\quad\square$

Remark 5.6.7. Notice that the theorem implies that $\dim(A) < \infty$ for a Noetherian local ring, which is not true in the non–local case (cf. Remark 3.3.10).

Theorem 5.6.8 (Krull's Principal Ideal Theorem). *Let A be a Noetherian ring, $f \in A$ a non–zerodivisor and $P \in \mathrm{minAss}(\langle f \rangle)$ then $\mathrm{ht}(P) = 1$.*

Proof. Let $\langle f \rangle = Q_1 \cap \cdots \cap Q_r$ be an irredundant primary decomposition. We may assume that $P = \sqrt{Q_1}$. As $P \in \mathrm{minAss}(\langle f \rangle)$ we have $\sqrt{Q_i} \not\subset P$ for $i > 1$. Especially $\langle f \rangle A_P = Q_1 A_P$ is a PR_P–primary ideal. Theorem 5.6.2 implies that $\dim(A_P) \leq 1$. This implies that $\mathrm{ht}(P) \leq 1$. If $\mathrm{ht}(P) = 0$ then $P \in \mathrm{minAss}(\langle 0 \rangle)$ and, therefore, f is a zerodivisor (Exercise 4.1.10). This is a contradiction to the assumption and proves the theorem. $\qquad\square$

Definition 5.6.9. Let $d = \dim(A)$, $\{x_1, \ldots, x_d\}$ is called a *system of parameters* of A, if $\langle x_1, \ldots, x_d \rangle$ is $\mathfrak{m}$–primary. Let $\{x_1, \ldots, x_d\}$ be a system of parameters. If $\langle x_1, \ldots, x_d \rangle = \mathfrak{m}$ it is called a *regular system of parameters*.

Corollary 5.6.10.

(1) $\dim(A) \leq \mathrm{edim}(A)$.
(2) If $x \in \mathfrak{m}$ then $\dim(A/\langle x \rangle) \geq \dim(A) - 1$, with equality if x is a non–zerodivisor.

Proof. Let $\mathfrak{m} = \langle x_1, \ldots, x_s \rangle$ be minimally generated by s elements. Using Nakayama's lemma, we obtain $s = \dim_K(\mathfrak{m}/\mathfrak{m}^2)$. Theorem 5.6.2 implies $\dim(A) = \delta(A) \leq s$, which proves (1). To prove (2), we use Proposition 5.6.3 (2) and obtain $\dim(A/\langle x \rangle) \leq \dim(A) - 1$ if x is a non–zerodivisor. On the other hand, let $x_1, \ldots, x_d$ be elements of $\mathfrak{m}$ whose images in $A/\langle x \rangle$ generate an $\mathfrak{m}/\langle x \rangle$–primary ideal. Then $\langle x, x_1, \ldots, x_d \rangle$ is $\mathfrak{m}$–primary and hence $d + 1 \geq \dim(A)$. Using the theorem again, we obtain (2). $\qquad\square$

Remark 5.6.11. The restriction made at the beginning of this chapter, that A contains its residue field, is not necessary. It is made just for better understanding. In particular, Theorem 5.6.2 holds for any Noetherian local ring.

Using this in full generality, we obtain the following consequences: let $(A, \mathfrak{m})$ be a Noetherian local ring, then

(1) every minimal prime ideal associated to $\langle x_1, \ldots, x_r \rangle \subset \mathfrak{m}$ has height $\leq r$;
(2) if $x \in A$ is neither a zerodivisor nor a unit then every minimal prime ideal associated to $\langle x \rangle$ has height 1 (Krull's Principal Ideal Theorem).

Next we give a characterization of the embedding dimension, using the Jacobian matrix.

Theorem 5.6.12. *Let* $A = K[x_1, \ldots, x_n]_{\langle x_1, \ldots, x_n \rangle}/\langle f_1, \ldots, f_m \rangle$, *then*

$$\mathrm{edim}(A) = n - \mathrm{rank}\left(\frac{\partial f_i}{\partial x_j}(0)\right).$$

Proof. By definition we have $\mathrm{edim}(A) = \dim_K(\mathfrak{m}/\mathfrak{m}^2)$. Let $\mathfrak{n} = \langle x_1, \ldots, x_n \rangle \subset K[x_1, \ldots, x_n]_{\langle x_1, \ldots, x_n \rangle}$ then

$$\dim_K(\mathfrak{m}/\mathfrak{m}^2) = \dim_K(\mathfrak{n}/\mathfrak{n}^2 + \langle f_1, \ldots, f_m \rangle)$$
$$= \dim_K(\mathfrak{n}/\mathfrak{n}^2) - \dim_K(\mathfrak{n}^2 + \langle f_1, \ldots, f_m \rangle/\mathfrak{n}^2)$$
$$= n - \dim_K(\mathfrak{n}^2 + \langle f_1, \ldots, f_m \rangle/\mathfrak{n}^2).$$

The last dimension is equal to the number of linearly independent linear forms among the $f_1 \bmod \mathfrak{n}^2, \ldots, f_m \bmod \mathfrak{n}^2$. This is equal to rank $\left(\frac{\partial f_i}{\partial x_j}(0)\right)$.   □

Definition 5.6.13. A Noetherian local ring A is called a *regular local ring* if its dimension and embedding dimension coincide, that is, $\dim(A) = \operatorname{edim}(A)$.

Corollary 5.6.14 (Jacobian criterion).
Let K be an algebraically closed field, $A = K[x_1, \ldots, x_n] / \langle f_1, \ldots, f_m \rangle$ and $\mathfrak{m} \subset A$ a maximal ideal.[6] Then $A_\mathfrak{m}$ is regular if and only if

$$\operatorname{rank}\left(\tfrac{\partial f_i}{\partial x_j} \bmod \mathfrak{m}\right) = n - \dim(A_\mathfrak{m}).$$

SINGULAR Example 5.6.15 (Jacobian criterion, regular system of parameters, embedding dimension).
We want to study the local ring $A = \mathbb{Q}[x, y, z]_{\langle xyz \rangle}/I$ where I is generated by the two polynomials $x + y^2 + z^3$ and $x + y + xyz$. More precisely, we want to find out whether A is regular. To this end, we use the Jacobian criterion.

```
ring R=0,(x,y,z),ds;
ideal I=x+y2+z3,x+y+xyz;
matrix J = jacob(I);

print(J);
//-> 1,   2y,  3z2,
//-> 1+yz,1+xz,xy

print(subst(J,x,0,y,0,z,0));
//-> 1,0,0,
//-> 1,1,0
```

The rank of the Jacobian matrix at the point $(0, 0, 0)$ is 2. This implies that $\operatorname{edim}(A) = 1$.

```
dim(std(I));
//-> 1
```

Dimension and embedding dimension coincide and, hence, A is regular.

```
ideal K=std(I+ideal(z));
K;
//-> K[1]=x
//-> K[2]=y
//-> K[3]=z
```

---

[6] The Jacobian criterion holds more generally for perfect fields and prime ideals $P \subset A$, as we shall see in the next section.

This implies that the maximal ideal of A is generated by z, that is, $\{z\}$ is a regular system of parameters.

Theorem 5.6.16. *Let $(A, \mathfrak{m})$ be a Noetherian local ring and $d = \dim A$. Then the following conditions are equivalent:*

(1) $\mathrm{Gr}_{\mathfrak{m}}(A) = K[t_1, \ldots, t_d]$ is a polynomial ring in d variables.
(2) A is regular of dimension d.
(3) $\mathfrak{m}$ can be generated by d elements.

Proof. (1) clearly implies (2). That (2) implies (3), immediately follows from the lemma of Nakayama. To prove that (3) implies (1), let $\mathfrak{m} = \langle x_1, \ldots, x_d \rangle$, and let $\varphi : K[y_1, \ldots, y_d] \to \mathrm{Gr}_{\mathfrak{m}}(A) = K[x_1, \ldots, x_d]$ be the surjective morphism of graded rings defined by $\varphi(y_i) := x_i$. Assume that $\mathrm{Ker}\, \varphi \neq \langle 0 \rangle$, and let $f \in \mathrm{Ker}\, \varphi$ be homogeneous of degree n. Using Proposition 5.3.6 (4), we obtain

$$d(K[y_1, \ldots, y_d]/\langle f \rangle) = d(K[y_1, \ldots, y_d]) - 1 = d - 2.$$

On the other hand, $\mathrm{Gr}_{\mathfrak{m}}(A)$ is a finitely generated graded $K[y_1, \ldots, y_d]/\langle f \rangle$–module (via φ). This implies, using Proposition 5.3.6 (2) that

$$d\big(\mathrm{Gr}_{\mathfrak{m}}(A)\big) \leq d(K[y_1, \ldots, y_d]/\langle f \rangle) = d - 2.$$

But $d\big(\mathrm{Gr}_{\mathfrak{m}}(A)\big) = d - 1$, because of Proposition 5.5.5 (2) and Theorem 5.6.2. This is a contradiction, and the theorem is proved. $\qquad\square$

Proposition 5.6.17. *Let A be an r–dimensional regular local ring with maximal ideal $\mathfrak{m}$ (containing its residue field) and $x_1, \ldots, x_i \in \mathfrak{m}$. The following conditions are equivalent:*

(1) $\{x_1, \ldots, x_i\}$ is a subset of a regular system of parameters of A;
(2) the images in $\mathfrak{m}/\mathfrak{m}^2$ of $x_1, \ldots, x_i$ are linearly independent over $A/\mathfrak{m} = K$;
(3) $A/\langle x_1, \ldots, x_i \rangle$ is an $(r - i)$–dimensional regular local ring.

Proof. (1) implies (2): let $\{x_1, \ldots, x_r\}$ be a regular system of parameters of A, then their images generate $\mathfrak{m}/\mathfrak{m}^2$ as a K–vector space. Since $\dim_K(\mathfrak{m}/\mathfrak{m}^2) = r$ they must be linearly independent.

(2) implies (3): we choose $x_{i+1}, \ldots, x_r$ such that the images of $x_1, \ldots, x_r$ in $\mathfrak{m}/\mathfrak{m}^2$ are a K–basis. Nakayama's Lemma implies that $\{x_1, \ldots, x_r\}$ is a regular system of parameters. Now x_1 is not a zerodivisor of A because A is an integral domain (Exercise 5.6.2). Using Corollary 5.6.10 (2) we obtain $\dim(A/\langle x_1 \rangle) = \dim(A) - 1$. On the other hand, the images of $x_2, \ldots, x_r$ generate the maximal ideal in $A/\langle x_1 \rangle$. This implies that $A/\langle x_1 \rangle$ is regular and $x_2, \ldots, x_r \bmod \langle x_1 \rangle$ is a regular system of parameters. This implies inductively that $A/\langle x_1, \ldots, x_i \rangle$ is regular.

To prove (3) implies (1), let $\mathfrak{m}/\langle x_1, \ldots, x_i \rangle$ be generated by the images of $y_1, \ldots, y_{r-i} \in \mathfrak{m}$. Then $\mathfrak{m}$ is generated by $x_1, \ldots, x_i, y_1, \ldots, y_{r-i}$. $\qquad\square$

Corollary 5.6.18. *If $x_1, \ldots, x_r$ is a regular system of parameters in A, then it is a regular A–sequence, that is, for all $1 \leq j \leq r$, x_j is not a zerodivisor in $A/\langle x_1, \ldots, x_{j-1} \rangle$.*

Proof. By Proposition 5.6.17 $A/\langle x_1, \ldots, x_{j-1} \rangle$ is regular and, therefore, by Exercise 5.6.2 an integral domain. $\square$

Exercises

5.6.1. Let $x = (x_1, \ldots, x_r)$, and let $f \in \langle x \rangle K[x]$ be an irreducible polynomial. Prove that $A := K[x]_{\langle x \rangle}/\langle f \rangle$ is a regular local ring if and only if

$$\langle \tfrac{\partial f}{\partial x_1}, \ldots, \tfrac{\partial f}{\partial x_r} \rangle \not\subset \langle x \rangle .$$

5.6.2. Let A be a ring, and let $I \subset A$ be an ideal such that $\bigcap_{n \geq 0} I^n = \langle 0 \rangle$. Suppose that $\mathrm{Gr}_I(A)$ is an integral domain. Then A is an integral domain. In particular, regular local rings are integral domains.

5.6.3. Let K be a field and $A := K[x, y]_{\langle x, y \rangle}/\langle y^2 - x^3 \rangle$. Then A is an integral domain, but $\mathrm{Gr}_{\mathfrak{m}}(A)$ is not an integral domain.

5.6.4. Let A be a Noetherian ring. Prove that $\dim(A[x]) = \dim(A) + 1$.

5.6.5. Let A be a Noetherian local ring with maximal ideal $\mathfrak{m}$. For simplicity we assume $K = A/\mathfrak{m} \subset A$. Let M be a finitely generated A–module, and let $\delta(M)$ be the smallest value of n such that there exist $x_1, \ldots, x_n \in \mathfrak{m}$, for which $\dim_K(M/x_1 M + \cdots + x_n M) < \infty$. Generalize Theorem 5.6.2 and prove that $\dim(M) = d(M) = \delta(M)$.

5.6.6. Let A and M be as in Exercise 5.6.5. Prove that $\dim(M) = \dim(\mathrm{Gr}_Q(M))$ for any $\mathfrak{m}$–primary ideal Q.

5.6.7. Let A be a regular local ring of dimension 1. Prove that A is a principal ideal domain.

5.6.8. Write a SINGULAR procedure to compute the embedding dimension.

5.7 Singular Locus

The aim of this section is to describe the singular locus and prove that the non–normal locus is contained in the singular locus. This means that regular local rings are normal. The proof of this result is, in general, difficult and uses the following result of Serre: the localization of a regular local ring in a prime ideal is again regular. In this section, we shall prove this result for rings of type $K[x_1, \ldots, x_n]_P/\langle f_1, \ldots, f_m \rangle$, P a prime ideal, using a generalization of the Jacobian criterion. A proof for the general case is given in Chapter 7.

Another way to prove that regular rings are normal is used in [66] proving that regular rings are factorial.

Theorem 5.7.1 (General Jacobian criterion). *Let* $I = \langle f_1, \ldots, f_m \rangle \subset K[x_1, \ldots, x_n]$ *be an ideal and* P *an associated prime ideal of* I. *Moreover, let* $Q \supset P$ *be a prime ideal such that the quotient field of* $K[x_1, \ldots, x_n]/Q$ *is separable over* K. *Then*

$$\mathrm{rank}\left(\tfrac{\partial f_i}{\partial x_j} \bmod Q\right) \leq \mathrm{ht}(P),$$

and $K[x_1, \ldots, x_n]_Q/I_Q$ *is a regular local ring if and only if*

$$\mathrm{rank}\left(\tfrac{\partial f_i}{\partial x_j} \bmod Q\right) = \mathrm{ht}(P).$$

Remark 5.7.2. If $Q = \mathfrak{m}$ is a maximal ideal and K is algebraically closed, then we obtain the Jacobian criterion proved in the previous section.

Before proving Theorem 5.7.1, we give some consequences:

Theorem 5.7.3. *Let* A *be a regular local ring. Then* A_Q *is regular for every prime ideal* $Q \subset A$.

In the following, we prove Theorem 5.7.3 only for a special case, namely, we assume that $A = K[x_1, \ldots, x_n]_M/\langle f_1, \ldots, f_m \rangle$, M a prime ideal and K a field of characteristic 0. A proof for the general case uses homological methods and is given in Chapter 7, Section 7.9.

Proof of Theorem 5.7.3. Theorem 5.7.1 implies $\mathrm{rank}\left(\tfrac{\partial f_i}{\partial x_j} \bmod M\right) = \mathrm{ht}(P)$, P an associated prime of $\langle f_1, \ldots, f_m \rangle$ such that $P \subset M$. Let Q' be a prime ideal in $K[x_1, \ldots, x_n]$ such that $P \subset Q' \subset M$ and $Q' \bmod \langle f_1, \ldots, f_m \rangle = Q$. Then $\mathrm{rank}\left(\tfrac{\partial f_i}{\partial x_j} \bmod Q'\right) = \mathrm{ht}(P)$. This implies, again using Theorem 5.7.1, that A_Q is regular. $\qquad\square$

Definition 5.7.4. For any ring A, the set

$$\mathrm{Sing}(A) := \{P \in \mathrm{Spec}(A) \mid A_P \text{ is not regular}\}$$

is called the *singular locus* of A.

The above theorem implies that for affine rings over a perfect field, the singular locus is closed and can be computed.

Corollary 5.7.5. *Let* K *be a perfect field,* $A = K[x_1, \ldots, x_n]/\langle f_1, \ldots, f_m \rangle$ *be equidimensional and let* $J \subset A$ *be the ideal generated by the* $(n - \dim(A))-$ *minors of the Jacobian matrix* $\left(\tfrac{\partial f_i}{\partial x_j}\right)$. *Then*

$$\mathrm{Sing}(A) = V(J) = \{Q \subset A \text{ prime} \mid J \subset Q\}.$$

Proof. For the proof, we assume that K is a field of characteristic 0. A prime ideal $Q \subset A$ contains J if and only if rank $\left(\frac{\partial f_i}{\partial x_j} \bmod Q\right) < n - \dim(A)$. Let $P' \subset K[x_1, \ldots, x_n]$ be an associated prime ideal of $\langle f_1, \ldots, f_m \rangle$, then by assumption $\operatorname{ht}(P') = n - \dim(A)$. Let $Q' \subset K[x_1, \ldots, x_n]$ be a prime ideal such that $Q = Q' \bmod \langle f_1, \ldots, f_m \rangle$, and assume $P' \subset Q'$. Then Theorem 5.7.1 implies that $K[x_1, \ldots, x_n]_{Q'}/\langle f_1, \ldots, f_m \rangle_{Q'} = A_Q$ is not regular. □

We obtain the following algorithm to compute the singular locus of an equidimensional ideal $I = \langle f_1, \ldots, f_m \rangle \subset K[x_1, \ldots, x_n]$.

Algorithm 5.7.6 (SINGULARLOCUSEQUI(I)).

Input: $I := \langle f_1, \ldots, f_k \rangle \subset K[x]$ an equidimensional ideal, $x = (x_1, \ldots, x_n)$.
Output: An ideal $J \subset K[x]$ such that $V(J) = \operatorname{Sing}(K[x]/I)$.

- $d := \dim(I)$;
- compute J, the ideal generated by the $(n - d)$–minors of the Jacobian matrix $\left(\frac{\partial f_i}{\partial x_j}\right)$;
- return($I + J$).

The following lemma is the basis for the description of the singular locus in general.

Lemma 5.7.7. *Let A be a Noetherian ring and $\langle 0 \rangle = I \cap I'$, I, I' non–zero ideals. Let $P \subset A$ be a prime ideal. Then A_P is singular if and only if one of the following three conditions holds:*

(a) A_P/I_P is singular
(b) A_P/I'_P is singular
(c) $P \supset I + I'$.

Proof. If $P \not\supset I$ then $A_P = A_P/I'_P$. If $P \not\supset I'$ then $A_P = A_P/I_P$. Moreover, if $P \supset I + I'$ then A_P is not an integral domain. Using Exercise 5.6.2 this implies that A_P is not regular. □

We obtain the following algorithm to compute the singular locus:

Algorithm 5.7.8 (SINGULARLOCUS(I)).

Input: $I := \langle f_1, \ldots, f_k \rangle \subset K[x]$, $x = (x_1, \ldots, x_n)$.
Output: An ideal $J \subset K[x]$ such that $V(J) = \operatorname{Sing}(K[x]/I)$.

- $D :=$ EQUIDIMENSIONAL(I);
- $S := \bigcap_{J \in D}$ SINGULARLOCUSEQUI(J);
- $R := \bigcap_{\substack{J, J' \in D \\ J \neq J'}} (J + J')$;
- return($S \cap R$).

SINGULAR Example 5.7.9 (singular locus).

```
ring R=0,(u,v,w,x,y,z),dp;
ideal I=wx,wy,wz,vx,vy,vz,ux,uy,uz,y3-x2;
LIB"primdec.lib";
radical(I);
//-> _[1]=wz    _[2]=vz    _[3]=uz    _[4]=wy    _[5]=vy
//-> _[6]=uy    _[7]=wx    _[8]=vx    _[9]=ux    _[10]=y3-x2
```

The ideal I is radical.

```
list l=minAssGTZ(I);
l;
//-> [1]:              [2]:
//->    _[1]=z             _[1]=-y3+x2
//->    _[2]=y             _[2]=w
//->    _[3]=x             _[3]=v
//->                       _[4]=u
```

I is the intersection of the two primes l[1], l[2].

```
ideal J=l[1]+l[2];
std(J);
//-> _[2]=z    _[2]=y    _[3]=x
//-> _[4]=w    _[5]=v    _[6]=u
```

The intersection of the two irreducible components defined by l[1] and l[2] is the point 0.

```
ideal sing=l[2]+minor(jacob(l[2]),4);
std(sing);
//-> _[2]=x    _[2]=w    _[3]=v    _[4]=u    _[5]=y2
```

The singular locus of the components defined by l[2] is a line defined by the ideal **sing** above which is also the singular locus of $V(I)$.

We try to visualize $V(I)$ by the picture in Figure 5.1, showing a 3–dimensional and a 2–dimensional component, both meeting transversally in $\mathbb{A}^6$.

To prove the general Jacobian criterion, we need some preparations.

Proposition 5.7.10. *Let* $P \subset K[x_1, \ldots, x_n]$ *be a prime ideal, then the localization* $K[x_1, \ldots, x_n]_P$ *is regular.*

Proof. We prove the proposition for the case that K is a field of characteristic 0. A proof for the general case will be given in Chapter 7, Section 7.9. Using Theorem 3.4.1 (respectively Exercise 3.4.4), we may assume that $K[x_1, \ldots, x_k] \subset K[x_1, \ldots, x_n]/P$ is a Noether normalization such that the field extension $Q(K[x_1, \ldots, x_n]/P) \supset Q(K[x_1, \ldots, x_k])$ is generated by the class $\bar{x}_{k+1} := x_{k+1} \bmod P$, and such that the minimal

Fig. 5.1. A visualization of $V(wx, wy, wz, vx, vy, vz, ux, uy, uz, y^3 - x^2) \subset \mathbb{A}^6$.

polynomial of $\bar{x}_{k+1}$, $g = x_{k+1}^s + \sum_{i=1}^s g_{s-i}(x_1, \ldots, x_k)x_{k+1}^{s-i}$, is in P. Now $\mathrm{ht}(P) = n - \dim(P) = n - k$.

Using Theorem 5.6.16, we have to prove that $PK[x_1, \ldots, x_n]_P$ is generated by $n - k$ elements. Let Δ be the discriminant of g (with respect to x_{k+1}). $\Delta \in K[x_1, \ldots, x_k]$ and, therefore, $\Delta \notin P$. Now we can apply Lemma 3.5.12 and obtain $K[x_1, \ldots, x_n]/P \subset \frac{1}{\Delta} \cdot K[x_1, \ldots, x_{k+1}]/\langle g \rangle$. Therefore, we can find elements $q_{k+2}, \ldots, q_n \in K[x_1, \ldots, x_{k+1}]$ such that $Q_i := x_i\Delta - q_i \in P$ for all $i \geq k + 2$. But, $\Delta \notin P$ implies that $x_{k+2} - q_{k+2}/\Delta$, $\ldots$, $x_n - q_n/\Delta$ are elements of $PK[x_1, \ldots, x_n]_P$. Finally, we conclude that $g, Q_{k+2}, \ldots, Q_n$ generate $PK[x_1, \ldots, x_n]_P$. $\qquad\square$

Proposition 5.7.10 leads to the following definition:

Definition 5.7.11. Let A be a Noetherian ring. A is called a *regular ring*, if the localization A_P is a regular local ring, for every prime ideal P.

Note that, because of Theorem 5.7.3, this is equivalent to the property that A_P is regular for every maximal ideal P.

Corollary 5.7.12. *Let K be a field, and let $>$ be any monomial ordering on $K[x_1, \ldots, x_n]$, then $K[x_1, \ldots, x_n]_>$ is a regular ring.*

Proof. Let $P \subset K[x_1, \ldots, x_n]_>$ be a prime ideal and $Q = P \cap K[x_1, \ldots, x_n]$, then Q is a prime ideal and $K[x_1, \ldots, x_n]_Q = (K[x_1, \ldots, x_n]_>)_P$ (Exercise 1.4.5). Because of Proposition 5.7.10 we obtain that $K[x_1, \ldots, x_n]_>$ is a regular ring. $\qquad\square$

Proof of Theorem 5.7.1. We prove the theorem for the case that K is a field of characteristic 0; the proof of the first part of the statement, that is, of the inequality, is left to the reader. For the second statement, let us first consider the case $I = P = Q$. In this case $K[x_1, \ldots, x_n]_P/P_P$ is a field, hence, regular. We have to prove that $rank\left(\frac{\partial f_i}{\partial x_j} \bmod P\right) = \mathrm{ht}(P)$. As in the proof of Proposition 5.7.10, we may assume that $PK[x_1,\ldots,x_n]_P$ is generated by $g, x_{k+2}\Delta - q_{k+2},\ldots,x_n\Delta - q_n$, where $g, q_{k+2},\ldots,q_n \in K[x_1,\ldots,x_{k+1}]$, Δ and $\frac{\partial g}{\partial x_{k+1}} \notin P$ and $\mathrm{ht}(P) = n - k$.

By Exercise 5.7.1 the rank of the Jacobian matrix is independent of the choice of the generators of $PK[x_1,\ldots,x_n]_P$. As $PK[x_1,\ldots,x_n]_P$ is generated by $n-k$ elements, it follows that the rank of the Jacobian matrix modulo P is, at most, $n - k$. The last $n - k$ columns of the Jacobian matrix corresponding to $g, \Delta x_{k+2} - q_{k+2},\ldots,\Delta x_n - q_n$ are

$$
M := \begin{pmatrix}
\frac{\partial g}{\partial x_{k+1}} & 0 \cdots 0 \\
\hline
-\frac{\partial q_{k+2}}{\partial x_{k+1}} & \Delta \quad 0 \\
\vdots & \quad\ddots \\
-\frac{\partial q_n}{\partial x_{k+1}} & 0 \quad \Delta
\end{pmatrix}.
$$

The determinant of this matrix is $\Delta^{n-k-1} \cdot \frac{\partial g}{\partial x_{k+1}} \notin P$. Therefore, we have

$$
\mathrm{rank}\left(\frac{\partial f_i}{\partial x_j} \bmod P\right) = \mathrm{rank}\,(M \bmod P) = n - k = \mathrm{ht}(P).
$$

To prove the general case, assume that $K[x_1,\ldots,x_n]_Q/I_Q$ is a regular local ring. Using Exercise 5.6.2, we obtain that it is an integral domain. This implies that I_Q is a prime ideal and, therefore, $I_Q = P_Q$. As a consequence, we may assume that $I = P$. Let $a = \mathrm{ht}(P)$ and $b = \mathrm{ht}(Q)$. Using Exercise 5.7.3, we can find $w_1,\ldots,w_a \in Q \setminus Q^2$ generating P_Q. Using Proposition 5.6.17, we can extend this sequence to a regular sequence $w_1,\ldots,w_b \in Q \setminus Q^2$ generating $QK[x_1,\ldots,x_n]_Q$. From the first part of the proof we obtain that

$$
\mathrm{rank}\left(\frac{\partial w_i}{\partial x_j} \bmod Q\right) = \mathrm{ht}(Q) = b,
$$

which is the maximal possible rank. Deleting the columns corresponding to $w_{a+1},\ldots,w_b$ we obtain $\mathrm{rank}\left(\frac{\partial w_i}{\partial x_j} \bmod Q\right)_{i \leq a, j \leq n} = a = \mathrm{ht}(P)$, which is again the maximal possible rank. This implies, using Exercise 5.7.1, that $\mathrm{rank}\left(\frac{\partial f_i}{\partial x_j} \bmod Q\right) = a = \mathrm{ht}(P)$.

To prove the other direction, assume that $\mathrm{rank}\left(\frac{\partial f_i}{\partial x_j} \bmod Q\right) = \mathrm{ht}(P) = a$. We may even assume that $\mathrm{rank}\left(\frac{\partial f_i}{\partial x_j} \bmod Q\right)_{i,j \leq a} = a$. We shall show that $f_1,\ldots,f_a$ is a subset of a regular system of parameters of $K[x_1,\ldots,x_n]_Q$. Indeed, if a linear combination satisfies $\sum_{i=1}^{a} c_i f_i \in Q^2$, then $\sum_{i=1}^{a} c_i \frac{\partial f_i}{\partial x_j} \in Q$, and $\mathrm{rank}\left(\frac{\partial f_i}{\partial x_j} \bmod Q\right)_{i,j \leq a} = a$ implies $c_i = 0$ for all i.

Using Proposition 5.6.17 we obtain that $K[x_1,\ldots,x_n]_Q/\langle f_1,\ldots,f_a\rangle$ is a regular local ring. Therefore, $\langle f_1,\ldots,f_a\rangle K[x_1,\ldots,x_n]_Q$ is a prime ideal. On the other hand, $\langle f_1,\ldots,f_a\rangle_Q \subset I_Q \subset P_Q$ and $\mathrm{ht}(\langle f_1,\ldots,f_a\rangle_Q) = \mathrm{ht}(P_Q)$. This implies $\langle f_1,\ldots,f_a\rangle_Q = I_Q$ and proves the theorem. $\qquad\square$

Now we want to prove that regular local rings are normal. We need a criterion for checking normality, different from the criterion in Proposition 3.6.5, which needs an ideal describing the non–normal locus.

Theorem 5.7.13 (Serre). *Let A be a reduced Noetherian ring. Then A is normal, if and only if the following conditions are satisfied:*

(R1) A_P is a regular local ring for every prime ideal P of height one.
(S2) Let $f \in A$ be a non–zerodivisor, then $\mathrm{minAss}(\langle f \rangle) = \mathrm{Ass}(\langle f \rangle)$.

The above conditions (R1) and (S2) are called *Serre's conditions*, see also Exercise 7.7.5.

Proof. Assume that A is normal, then A_P is normal for every prime ideal P (Proposition 3.2.5). If P has height one, then $\dim(A_P) = 1$. A normal local ring is an integral domain (Exercise 3.2.10). Assume that A_P is not regular and let $\mathfrak{m} := PA_P$. Consider the A_P–module $\mathfrak{m}^* = \{x \in Q(A_P) \mid x\mathfrak{m} \subset A_P\}$. We shall see that $A_P \subset \mathfrak{m}^*$ is a strict inclusion. Namely, let $a \in \mathfrak{m}$, $a \neq 0$ then $\sqrt{\langle a \rangle} = \mathfrak{m}$. We can choose an integer s such that $\mathfrak{m}^s \subset \langle a \rangle$ and $\mathfrak{m}^{s-1} \not\subset \langle a \rangle$. $s > 1$ because A_P is assumed not to be regular. We choose any element $b \in \mathfrak{m}^{s-1} \setminus \langle a \rangle$. Then $\frac{b}{a} \notin A_P$ but $\frac{b}{a}\mathfrak{m} \subset A_P$, that is, $\frac{b}{a} \in \mathfrak{m}^* \setminus A_P$. Now, by definition of $\mathfrak{m}^*$, we have $\mathfrak{m} \subset \mathfrak{m}\mathfrak{m}^* \subset A_P$. If $\mathfrak{m}\mathfrak{m}^* = \mathfrak{m}$, then $\mathfrak{m}^*$ is an $A_P[\frac{b}{a}]$–module, finitely generated as A_P–module. This implies that $\frac{b}{a}$ is integral over A_P (Cayley–Hamilton Theorem). But A_P is normal and $\frac{a}{b} \notin A_P$ gives a contradiction. Therefore, we obtain $\mathfrak{m}\mathfrak{m}^* = A_P$. Let $t \in \mathfrak{m} \setminus \mathfrak{m}^2$ (note that $\mathfrak{m} = \mathfrak{m}^2$ implies $\mathfrak{m} = \langle 0 \rangle$ by Nakayama's Lemma), then $t\mathfrak{m}^* \subset A_P$, $t\mathfrak{m}^* \not\subset \mathfrak{m}$ because $t \notin \mathfrak{m}^2$. This implies $t\mathfrak{m}^* = A_P$ and, therefore, $\mathfrak{m} = \langle t \rangle$. This is, again, a contradiction to the assumption and proves that A_P is a regular local ring. Therefore, (R1) holds for normal rings.

To prove (S2) assume that P is an embedded prime ideal for some $\langle f \rangle$. By Definition 4.1.1 we can find $b \in A$ such that $P = \langle f \rangle : \langle b \rangle$. Let $\mathfrak{m} := PA_P$ and consider, as before, $\mathfrak{m}^* = \{x \in Q(A_P) \mid x\mathfrak{m} \subset A_P\}$. Then $\frac{b}{f} \in \mathfrak{m}^*$. As before, we can deduce that $\mathfrak{m}\mathfrak{m}^* = A_P$ implies that $\mathfrak{m} = \langle t \rangle$ for a suitable t. This implies that $\dim(A_P) \leq 1$. But this is a contradiction to the fact that P is not a minimal associated prime ideal of $\langle f \rangle$. We obtain $\mathfrak{m}\mathfrak{m}^* = \mathfrak{m}$. As before, we see that $\frac{b}{f} = \frac{a}{s}$, $a, s \in A$, $s \notin P$. Then, because A_P is an integral domain, we have $PA_P = \langle f \rangle : \langle b \rangle = \langle fa \rangle : \langle ba \rangle = \langle bs \rangle : \langle ba \rangle = \langle s \rangle : \langle a \rangle$.

But this implies that $s \in P$ and is a contradiction to the choice of s. This proves that $\langle f \rangle$ cannot have embedded primes and, therefore, (S2) holds for normal rings.

To prove the other implication, assume that (R1) and (S2) hold and let $\frac{a}{b}$ be integral over A. Let $\langle b \rangle = Q_1 \cap \cdots \cap Q_s$ be an irredundant primary decomposition. By (S2) all $P_i := \sqrt{Q_i}$ are minimal associated prime ideals to $\langle b \rangle$. By Krull's principal ideal theorem (Theorem 5.6.8) all the P_i have height one. By (R1) we know that A_{P_i} is regular for all i. But then the A_{P_i} are principal ideal domains because $\dim(A_{P_i}) = 1$ (Exercise 5.6.7). But principal ideal domains are normal because they are unique factorization domains (Exercise 3.2.10). It follows that $\frac{a}{b} \in A_{P_i}$ for all i. Therefore, $a \in \langle b \rangle A_{P_i}$ for all i. Using Lemma 4.1.3 (4), we have $Q_i = A \cap \langle b \rangle A_{P_i}$ and, therefore, $a \in Q_i$ for all i. Hence, $a \in Q_1 \cap \cdots \cap Q_s = \langle b \rangle$ and, therefore, $\frac{a}{b} \in A$. This implies that A is normal. $\qquad \square$

Theorem 5.7.14. *Let A be a regular local ring, then A is normal.*

Proof. According to the previous theorem, we have to prove that (R1) and (S2) are satisfied. (R1) is a consequence of Theorem 5.7.3. (S2) is a consequence of Corollary 7.7.11 and Example 7.7.2 (which uses Corollary 5.6.18). $\qquad \square$

Exercises

5.7.1. Let $I = \langle f_1, \ldots, f_m \rangle \subset K[x_1, \ldots, x_n]$ be an ideal and $Q \supset I$ be a prime ideal. Let $g_1, \ldots, g_s \in I$ such that $I_Q = \langle g_1, \ldots, g_s \rangle K[x_1, \ldots, x_n]_Q$. Prove that $\mathrm{rank}\left(\frac{\partial f_i}{\partial x_j} \bmod Q\right) = \mathrm{rank}\left(\frac{\partial g_i}{\partial x_j} \bmod Q\right)$.

5.7.2. Let $(A, \mathfrak{m})$ be a local ring and $P \subset \mathfrak{m}^2$ be a prime ideal. Prove that A/P is not regular.

5.7.3. Let $(A, \mathfrak{m})$ be a regular local ring and $P \subset A$ a prime ideal. Suppose that A/P is a regular local ring. Prove that P can be generated by a regular sequence $x_1, \ldots, x_r$ from $\mathfrak{m} \setminus \mathfrak{m}^2$.

5.7.4. Compute the dimension of the singular locus of

$$A := Q[x, y, z]/\langle x^5 + y^{11} + xy^9 + z^4 \rangle.$$

5.7.5. Compute the singular locus of

$$A := Q[x, y, z]/\langle x^6 - x^3y^2 - x^3z^2 + 2x^3z + x^3 + y^2z^2 - 2y^2z - y^2 \rangle.$$

5.7.6. Let K be a field and $f \in K[x]$ irreducible. Prove (without using Proposition 5.7.10) that $K[x]_{\langle f \rangle}$ is regular.

5.7.7. Prove the first statement of Theorem 5.7.1 under the assumption that K is a field of characteristic zero.

(Hint: Use Propositions 5.6.10 and 5.7.10).

5.7.8. Let K be a field of characteristic 0. Use Exercise 5.7.6 and Exercise 3.5.4 to give a proof of Proposition 5.7.10 without using Lemma 3.5.12.

5.7.9. Prove that a regular ring is normal.

5.7.10. Prove that a Noetherian ring A is regular if and only if A_P is regular for all maximal ideals P.

5.7.11. Let K be a field and $A = K[x_1, \ldots, x_n]/I$. Let $P \subset A$ be a prime ideal such that $Q(A/P)$ is separable over K. Prove that A_P is regular if and only if $(\Omega_{A|K})_P$ is free of rank $n - \dim(A_P)$. Here $\Omega_{A|K}$, the *module of differentials*[7] is defined by the exact sequence

$$I/I^2 \xrightarrow{\varphi} A^n \to \Omega_{A|K} \to 0$$

with $\varphi(f) = \left(\frac{\partial f}{\partial x_1}, \ldots, \frac{\partial f}{\partial x_n} \right)$.

(Hint: use the Jacobian criterion and Theorem 7.2.7.)

---

[7] The module of differentials can be defined in a more general situation. Let A be a ring and B an A–algebra. Let $m : B \otimes_A B \to B$ be defined by $m(b \otimes c) = bc$ and $J = \mathrm{Ker}(m)$. Then $\Omega_{B|A} = J/J^2$. For more details see [66] and [160].

6. Complete Local Rings

For certain applications the local rings $K[x]_{\langle x \rangle}$, $x = (x_1, \ldots, x_n)$, are not "sufficiently local". As explained in Appendix A, Sections A.8 and A.9, the latter rings contain informations about arbitrary small Zariski neighbourhoods of $0 \in K^n$. Such neighbourhoods turn out to be still quite large, for instance, if $n = 1$ then they consist of K minus a finite number of points. If we are working over the field $K = \mathbb{C}$, respectively $K = \mathbb{R}$, we can use the convergent power series ring $K\{x\}$ which contains information about arbitrary small Euclidean neighbourhoods of 0, and this is what we are usually interested in. For arbitrary fields, however, we have to consider the formal power series ring $K[[x]]$ instead.

6.1 Formal Power Series Rings

Let K be a field, $x = (x_1, \ldots, x_n)$ variables, $\alpha = (\alpha_1, \ldots, \alpha_n) \in \mathbb{N}^n$. As before, we write $x^\alpha = x_1^{\alpha_1} \cdot \ldots \cdot x_n^{\alpha_n}$ and $\deg(x^\alpha) = |\alpha| = \sum_{i=1}^{n} \alpha_i$.

Let $w = (w_1, \ldots, w_n)$, $w_i \in \mathbb{Z}$, $w_i > 0$, be a so-called *weight–vector*. We define the *weighted degree* w–$\deg(x^\alpha) = |\alpha|_w = \sum_{i=1}^{n} w_i \alpha_i$, especially if $w = (1, \ldots, 1)$ then w–$\deg(x^\alpha) = \deg(x^\alpha)$.

Definition 6.1.1.

(1) An expression $\sum_{\alpha \in \mathbb{N}^n} a_\alpha x^\alpha$, $a_\alpha \in K$, is called a *formal power series*. We also use the notation $\sum_{|\alpha|=0}^{\infty} a_\alpha x^\alpha$ or $\sum a_\alpha x^\alpha$.

(2) Let $f = \sum_{\alpha \in \mathbb{N}^n} a_\alpha x^\alpha$ be a formal power series, then we set

$$\mathrm{ord}(f) := \min\{|\alpha| \mid a_\alpha \neq 0\} \text{ and } \text{w–}\mathrm{ord}(f) := \min\{|\alpha|_w \mid a_\alpha \neq 0\}$$

and call it the *order*, respectively *weighted order*, of f.

(3) $K[[x]] := \{\sum_{\alpha \in \mathbb{N}^n} a_\alpha x^\alpha \mid a_\alpha \in K, \; \alpha \in \mathbb{N}^n\}$ denotes the ring of formal power series, with addition and multiplication given by

$$\sum_{\alpha \in \mathbb{N}^n} a_\alpha x^\alpha + \sum_{\alpha \in \mathbb{N}^n} b_\alpha x^\alpha := \sum_{\alpha \in \mathbb{N}^n} (a_\alpha + b_\alpha) x^\alpha .$$

$$\sum_{\alpha \in \mathbb{N}^n} a_\alpha x^\alpha \cdot \sum_{\alpha \in \mathbb{N}^n} b_\alpha x^\alpha := \sum_{\gamma \in \mathbb{N}^n} \left(\sum_{\alpha + \beta = \gamma} a_\alpha b_\beta \right) x^\gamma .$$

(4) We also write recursively $f = \sum_{\nu=0}^{\infty} b_\nu(x_1, \dots, x_{n-1}) x_n^\nu = \sum_{\nu=0}^{\infty} b_\nu x_n^\nu$ with

$$b_\nu = \sum_{\substack{\alpha \in \mathbb{N}^n \\ \alpha_n = \nu}} a_\alpha x_1^{\alpha_1} \cdot \dots \cdot x_{n-1}^{\alpha_{n-1}} \in K[[x_1, \dots, x_{n-1}]] ,$$

for $f = \sum_{\alpha \in \mathbb{N}^n} a_\alpha x^\alpha \in K[[x]]$.

(5) Let $f = \sum a_\alpha x^\alpha \in K[[x]]$ and $k \in \mathbb{N}$ then the k–jet of f is defined by $j_k(f) := \sum_{|\alpha| \le k} a_\alpha x^\alpha$, the sum of terms of order $\le k$.

Lemma 6.1.2. $K[[x]]$ *is, with the operations defined in (3), a (commutative) local ring. The canonical injection $K[x]_{\langle x \rangle} \to K[[x]]$ is a homomorphism of local rings and induces isomorphisms*

$$K[x]/\langle x \rangle^\nu \cong K[x]_{\langle x \rangle}/\langle x \rangle^\nu \cong K[[x]]/\langle x \rangle^\nu.$$

Proof. It is left to the reader to prove that $K[[x]]$ is a commutative ring and the canonical injection is a homomorphism of local rings inducing the isomorphisms above. To prove that $K[[x]]$ is a local ring, we show that $\langle x \rangle$ is the unique maximal ideal. We have to prove that $f = \sum_\alpha a_\alpha x^\alpha \in K[[x]]$ with $a_0 \ne 0$ has an inverse in $K[[x]]$. We use the recursive description of f and apply induction on n. We may assume $f = \sum_{\nu=0}^{\infty} b_\nu x_n^\nu$, $b_\nu \in K[[x_1, \dots, x_{n-1}]]$ and b_0 a unit in $K[[x_1, \dots, x_{n-1}]]$, that is, there exists $c_0 \in K[[x_1, \dots, x_{n-1}]]$ such that $b_0 c_0 = 1$. We want to determine $c_\nu \in K[[x_1, \dots, x_{n-1}]]$, $\nu \ge 1$, such that $g := \sum_{\nu=0}^{\infty} c_\nu x_n^\nu$ satisfies $fg = 1$. We already have $b_0 c_0 = 1$ and we have to choose the c_ν in such a way that $\sum_{i=0}^{\nu} b_i c_{\nu-i} = 0$ for all $\nu > 0$. Assume we have already found $c_0, \dots, c_{\nu-1}$ with this property. We define $c_\nu := -c_0 \sum_{i=1}^{\nu} b_i c_{\nu-i}$. □

In particular, if $f \in K[x] \smallsetminus \langle x \rangle$, then we can express the inverse $1/f \in K[x]_{\langle x \rangle}$ as a (unique) power series, that is, we have an injection $K[x]_{\langle x \rangle} \hookrightarrow K[[x]]$.

SINGULAR Example 6.1.3 (inverse of a power series).
The following procedure computes the inverse of a polynomial p with non–vanishing constant term (in the current basering, which is assumed to be equipped with a local ordering) as a power series up to a given order k:

```
proc invers(poly p, int k)
{
    poly q=1/p[1];            //assume that p[1]<>0
    poly re=q;
```

```
p=q*(p[1]-jet(p,k));
poly s=p;
while(p!=0)
{
    re=re+q*p;
    p=jet(p*s,k);
}
return(re);
}

ring R=0,(x,y),ds;
poly p=2+x+y2;
poly q=invers(p,4);
q;
//->1/2-1/4x+1/8x2-1/4y2-1/16x3+1/4xy2+1/32x4-3/16x2y2+1/8y4

jet(p*q,4);
//-> 1
```

Lemma 6.1.4. *Let* $f, g \in K[[x]]$ *then* $\mathrm{ord}(f + g) \geq \min\{\mathrm{ord}(f), \mathrm{ord}(g)\}$ *and* $\mathrm{ord}(fg) = \mathrm{ord}(f) + \mathrm{ord}(g)$. *A similar statement holds for* w–ord.

Proof. The proof is left as an exercise. ☐

Lemma 6.1.5. $\bigcap_{\nu=1}^{\infty} \langle x \rangle^{\nu} = \langle 0 \rangle$.

Proof. $f \in \langle x \rangle^{\nu}$ implies $\mathrm{ord}(f) \geq \nu$. If $f = \sum_{\alpha} a_{\alpha} x^{\alpha}$ then $a_{\alpha} = 0$ if $|\alpha| < \nu$.
If $f \in \langle x \rangle^{\nu}$ for all ν then $a_{\alpha} = 0$ for all α and, therefore, $f = 0$. ☐

Definition 6.1.6.

(1) Consider $K[[x]]$ together with the set $F := \{\langle x \rangle^{\nu} \mid \nu \in \mathbb{N}\}$ as a topological space, where F is a fundamental system of neighbourhoods of 0. This topology is called $\langle x \rangle$–*adic topology*.

(2) A sequence $\{f_{\nu}\} = \{f_{\nu}\}_{\nu \in \mathbb{N}}$, $f_{\nu} \in K[[x]]$, is called *Cauchy sequence* if, for every $k \in \mathbb{N}$, there exists $\ell \in \mathbb{N}$ such that

$$f_{\nu} - f_m \in \langle x \rangle^k \text{ for all } \nu, m \geq \ell.$$

(3) A sequence $\{f_{\nu}\}_{\nu \in \mathbb{N}}$, $f_{\nu} \in K[[x]]$, is called *convergent* then if there exists a power series $f \in K[[x]]$ such that for every $k \in \mathbb{N}$ there exists some $\ell \in \mathbb{N}$ satisfying $f - f_{\nu} \in \langle x \rangle^k$ for all $\nu \geq \ell$. Then f is uniquely determined, and we write as usual $f =: \lim_{\nu \to \infty} f_{\nu}$.

(4) If $\{f_{\nu}\}$ is a convergent sequence with limit $\lim_{\nu \to \infty} f_{\nu} = 0$, then the sequence of partial sums, $\left\{\sum_{\nu=0}^{m} f_{\nu}\right\}_{\nu \in \mathbb{N}}$, converges[1], and we define

---

[1] Do not confuse convergence of the sequence of partial sums in the $\langle x \rangle$–adic topology with the notion of convergent power series if K is either $\mathbb{C}$ or $\mathbb{R}$.

$$\sum_{\nu=0}^{\infty} f_{\nu} := \lim_{m \to \infty} \left(\sum_{\nu=0}^{m} f_{\nu} \right).$$

Remark 6.1.7. It follows in the usual way that if $\{f_{\nu}\}, \{g_{\nu}\}$ are convergent sequences

$$\lim_{\nu \to \infty} (f_{\nu} + g_{\nu}) = \lim_{\nu \to \infty} f_{\nu} + \lim_{\nu \to \infty} g_{\nu}, \quad \lim_{\nu \to \infty} (f_{\nu} g_{\nu}) = \lim_{\nu \to \infty} f_{\nu} \cdot \lim_{\nu \to \infty} g_{\nu},$$

and if $\lim_{\nu \to \infty} f_{\nu} = \lim_{\nu \to \infty} g_{\nu} = 0$ then

$$\sum_{\nu=0}^{\infty} f_{\nu} + \sum_{\nu=0}^{\infty} g_{\nu} = \sum_{\nu=0}^{\infty} (f_{\nu} + g_{\nu}),$$

$$\sum_{\nu=0}^{\infty} f_{\nu} \cdot \sum_{\nu=0}^{\infty} g_{\nu} = \sum_{\nu=0}^{\infty} \sum_{i=0}^{\nu} f_{\nu-i} g_i.$$

Theorem 6.1.8. $K[[x]]$ *is complete, that is, every Cauchy sequence in* $K[[x]]$ *is convergent, and Hausdorff.*

Proof. The Hausdorff property is an immediate consequence of Lemma 6.1.5.

Let $\{f_{\nu}\}$ be a Cauchy sequence in $K[[x]]$. We have to prove that f_{ν} has a limit in $K[[x]]$, that is, there exists some power series $f \in K[[x]]$, such that, for all $k \in \mathbb{N}$, there exists some $\ell \in \mathbb{N}$ satisfying $f - f_{\nu} \in \langle x \rangle^k$ if $\nu \geq \ell$.

Let $f_{\nu} = \sum a_{\alpha}^{(\nu)} x^{\alpha}$, then $f_{\nu} - f_{\mu} \in \langle x \rangle^k$ implies $a_{\alpha}^{(\nu)} = a_{\alpha}^{(\mu)}$ for all α with $|\alpha| < k$. Now define $f = \sum_{\alpha} b_{\alpha} x^{\alpha}$ as follows. Let $\alpha \in \mathbb{N}^n$ with $|\alpha| = s$, and choose ℓ such that $f_{\nu} - f_{\mu} \in \langle x \rangle^{s+1}$ if $\nu, \mu \geq \ell$, then we define $b_{\alpha} := a_{\alpha}^{(\ell)}$. Obviously, $f = \lim_{\nu \to \infty} f_{\nu}$. $\square$

Definition 6.1.9. Let $f, g_1, \dots, g_n \in K[[x]]$ and assume that $\operatorname{ord}(g_i) \geq 1$ for all i, then we define the *substitution*

$$f(g_1, \dots, g_n) = \lim_{m \to \infty} j_m(f)(j_m(g_1), \dots, j_m(g_n)).$$

We leave it as an exercise to prove that the sequence used in the definition is a Cauchy sequence.

Corollary 6.1.10. *Let* $y = (y_1, \dots, y_m)$, *and let* $\varphi : K[[x]] \to K[[y]]$ *be a continuous K-algebra homomorphism with* $f_i := \varphi(x_i)$, $i = 1, \dots, n$. *Then* $\varphi(g) = g(f_1, \dots, f_n)$ *for all* $g \in K[[x]]$.

Proof. For each m we have $\varphi(j_m(g)) = j_m(g)(f_1, \dots, f_n)$ because φ is a K-algebra homomorphism. Now φ being continuous implies $\varphi(g) = g(f_1, \dots, f_n)$. $\square$

Remark 6.1.11.

(1) Any K–algebra homomorphism $\varphi : K[[x]] \to K[[y]]$ is automatically *local* (that is, maps the maximal ideal $\langle x \rangle$ to $\langle y \rangle$). To see this, let $f \in \langle x \rangle$ and $\varphi(f) = g + c$ with $g \in \langle y \rangle$, $c \in K$, and assume $c \neq 0$. Clearly, $f - c$ is a unit in $K[[x]]$, hence $\varphi(f - c) = g$ is a unit, too, contradicting the assumption.

(2) In particular, any K–algebra homomorphism $\varphi : K[[x]] \to K[[y]]$ is continuous, and Corollary 6.1.10 shows that φ is uniquely determined by the images $f_i := \varphi(x_i)$, $i = 1, \ldots, n$, where the f_i are power series with $f_1, \ldots, f_n \in \langle x \rangle$.

Conversely, any such collection of power series $f_1, \ldots, f_n \in \langle x \rangle$ defines a unique (continuous) morphism by setting

$$\varphi(g) := \varphi \left(\sum_{\alpha \in \mathbb{N}^n} a_\alpha x^\alpha \right) = \sum_{\alpha \in \mathbb{N}^n} a_\alpha f_1^{\alpha_1} \cdot \ldots \cdot f_n^{\alpha_n} = g(f_1, \ldots, f_n) \,.$$

Exercises

6.1.1. Prove Remark 6.1.7.

6.1.2. Prove that the sequence defined in Definition 6.1.9 is a Cauchy sequence.

6.1.3. Prove that the procedure of SINGULAR Example 6.1.3 is correct.

6.1.4. Compute the power series expansion of the inverse of $1 + x^2 + y^2 + z^2$ in $K[x, y, z]_{\langle x,y,z \rangle}$ up to order 10.

6.1.5. Show that every formal power series $f = 1 + \sum_{|\alpha| \geq 1} a_\alpha x^\alpha \in K[[x]]$ has a square root $g \in K[[x]]$ (that is, $g^2 = f$), provided that K is a field of characteristic $\neq 2$. What about k–th roots for $k \geq 3$?

6.1.6. Write a SINGULAR procedure computing the square root of f as in Exercise 6.1.5 up to a given order.

6.2 Weierstraß Preparation Theorem

Definition 6.2.1. $f \in K[[x]]$ is called x_n–*general of order* m (or x_n–*regular of order* m) if

$$f(0, \ldots, 0, x_n) = x_n^m \cdot g(x_n), \quad g(0) \neq 0 \,.$$

Lemma 6.2.2. *Let $f \in K[[x]]$ and $f = \sum_{\nu \geq m} f_\nu$ with f_ν homogeneous polynomials of degree ν, $f_m \neq 0$. Let $(a_1, \ldots, a_{n-1}) \in K^{n-1}$, such that*

$$f_m(a_1, \ldots, a_{n-1}, 1) \neq 0.$$

Then $f(x_1 + a_1 x_n, \ldots, x_{n-1} + a_{n-1} x_n, x_n)$ is x_n-general of order m.

Proof. $f_m(x_1 + a_1 x_n, \ldots, x_{n-1} + a_{n-1} x_n, x_n) = f_m(a_1, \ldots, a_{n-1}, 1)x_n^m +$ terms of lower degree with respect to x_n because of Taylor's formula. On the other hand, $f_\nu(x_1 + a_1 x_n, \ldots, x_{n-1} + a_{n-1} x_n, x_n)$ are homogeneous polynomials of degree ν. This implies that $f(x_1 + a_1 x_n, \ldots, x_{n-1} + a_{n-1} x_n, x_n)$ is x_n-general of order m. □

Lemma 6.2.3. *Let K be an infinite field and $f \in K[x]$ be a homogeneous polynomial of degree $m > 0$, then there exist $(a_1, \ldots, a_{n-1}) \in K^{n-1}$ such that $f(a_1, \ldots, a_{n-1}, 1) \neq 0$, that is, f is x_n-general of some order.*

The proof is left as an exercise.

We can test the statement of the lemma in SINGULAR:

SINGULAR Example 6.2.4 (z–general power series).

```
ring R=0,(x,y,z),ls;
poly p=xyz+x2yz+xy2z;
ideal m=x+random(-5,5)*z,y+random(-5,5)*z,z;
map phi=R,m;
phi(p);
//-> z3+2z4+yz2+3yz3+y2z2+xz2+3xz3+xyz+4xyz2+xy2z+x2z2+x2yz
```

Remark 6.2.5. If K is finite and $f_m(a_1, \ldots, a_{n-1}, 1) = 0$ (with the notations of Lemma 6.2.2) for all $(a_1, \ldots, a_{n-1}) \in K^{n-1}$ then one can use the transformation $x_i \mapsto x_i + x_n^{\alpha_i}$, $x_n \mapsto x_n$ for suitable $\alpha_1, \ldots, \alpha_{n-1}$ to obtain a x_n-general power series $f(x_1 + x_n^{\alpha_1}, \ldots, x_{n-1} + x_n^{\alpha_{n-1}}, x_n)$.

Theorem 6.2.6 (Weierstraß Division Theorem). *Let $f \in K[[x]]$ be x_n-general of order m, $g \in K[[x]]$, then there exist uniquely determined $q \in K[[x]]$ and $r_0, \ldots, r_{m-1} \in K[[x_1, \ldots, x_{n-1}]]$ such that*

$$g = qf + r, \quad \text{with } r = \sum_{\nu=0}^{m-1} r_\nu x_n^\nu.$$

Proof. We define two $K[[x_1, \ldots, x_{n-1}]]$–linear maps h, r of $K[[x_1, \ldots, x_n]]$. Let $p = \sum_{\nu=0}^{\infty} a_\nu x_n^\nu$, $a_\nu \in K[[x_1, \ldots, x_{n-1}]]$, then $r(p) = \sum_{\nu=0}^{m-1} a_\nu x_n^\nu$ and $h(p) = (p - r(p))/x_n^m$. This means $p = h(p)x_n^m + r(p)$.

To prove the theorem we have to find $q \in K[[x]]$ such that $h(g) = h(qf)$. Now $qf = qr(f) + qh(f) \cdot x_n^m$ and, therefore, we are looking for some q satisfying

$$h(g) = h\big(qr(f) + qh(f)x_n^m\big) = h\big(qr(f)\big) + qh(f)\,.$$

Let $v := qh(f)$, $w = -h(f)^{-1}r(f)$, ($h(f)$ is a unit because f is x_n-general of order m) and $u = h(g)$, then it is sufficient to find v such that

$$v = u + h(w \cdot v)\,.$$

Let $H(y) := h(wy)$. Again, by assumption, $w \in \langle x_1, \dots, x_{n-1}\rangle \subset K[[x]]$ and, therefore, $H(y) \in \langle x_1, \dots, x_{n-1}\rangle^{i+1}$ if $y \in \langle x_1, \dots, x_{n-1}\rangle^i$. Now we can iterate $v = u + H(v) = u + H\big(u + H(v)\big) = u + H(u) + H^2(v)$ and obtain successively

$$v = u + H(u) + H^2(u) + \dots + H^s(u) + H^{s+1}(v)\,.$$

We just saw that $H^i(u) \in \langle x_1, \dots, x_{n-1}\rangle^i$ and $H^{s+1}(v) \in \langle x_1, \dots, x_{n-1}\rangle^{s+1}$. Therefore, $v := \sum_{i=0}^{\infty} H^i(u)$ converges and satisfies the equation $v = u + H(v)$ and is, therefore, uniquely determined by this property. $\qquad\square$

Definition 6.2.7. $p = x_n^m + \sum_{\nu=0}^{m-1} a_\nu x_n^\nu \in K[[x_1, \dots, x_{n-1}]][x_n]$, a polynomial in x_n with coefficients $a_\nu \in K[[x_1, \dots, x_{n-1}]]$, is called a *Weierstraß polynomial* with respect to x_n if $a_\nu \in \langle x_1, \dots, x_{n-1}\rangle$ for all ν.

Corollary 6.2.8 (Weierstraß Preparation Theorem). *Let $f \in K[[x]]$ be x_n-general of order m, then there exists a unit $u \in K[[x]]$ and a Weierstraß polynomial p of degree m with respect to x_n such that $f = u \cdot p$. Here, u and p are uniquely determined.*

Proof. Apply Theorem 6.2.6 to $g = x_n^m$ and define $p = x_n^m - r$, $u = q^{-1}$. $\quad\square$

Corollary 6.2.9. *Let $f \in K[[x]]$ be x_n-general of order m then $K[[x]]/\langle f\rangle$ is a free $K[[x_1, \dots, x_{n-1}]]$-module of rank m.*

Proof. The Weierstraß Division Theorem implies that $K[[x]]/\langle f\rangle$ is a finitely generated $K[[x_1, \dots, x_{n-1}]]$-module, being generated by $1, x_n, \dots, x_n^{m-1}$. The uniqueness of the division implies that this is also a basis. $\qquad\square$

SINGULAR Example 6.2.10 (Weierstraß polynomial).
The following procedure follows the proof of the Weierstraß preparation theorem to compute a Weierstraß polynomial up to a given order k for an x_n-general power series $f \in K[[x]][x_n]$, $x = (x_1, \dots, x_{n-1})$, of order m. The second input is the polynomial $g = x_n^m$.

```
proc Weierstrass(poly f,poly g,int k)
{
    int i;
    int n=nvars(basering);
    poly p=f;
    for(i=1;i<=n-1;i++)
    {
```

```
      p=subst(p,var(i),0);
   }
   if(p==0)
   {
      "the polynomial is not regular";
      return(0);
   }
   int m=ord(p);
   poly hf=f/var(n)^m;
   poly rf=f-var(n)^m*hf;
   poly invhf=invers(f/var(n)^m,k);
   poly w=-invhf*rf;
   poly u=g/var(n)^m;
   poly v=u;
   poly H=jet((w*u)/var(n)^m,k);
   while(H!=0)
   {
      v=v+H;
      H=jet((w*H)/var(n)^m,k);
   }
   poly q=v*invhf;
   return(q);
}


ring R=0,(x,y),ds;
poly f=y4+xy+x2y6+x7;
poly g=y4;
poly q=Weierstrass(f,g,10);
poly w=jet(q*f,10); //the Weierstrass polynomial
ring S=(0,x),y,ds;
poly w=imap(R,w);
w;
//-> (x7)+(-x9+x)*y+(-x6)+y2+(-x3)*y3+y4

setring R;
q=Weierstrass(f,g,15);
w=jet(q*f,15);   //the Weierstrass polynomial
setring S;
w=imap(R,w);
w;
//-> (-2x15+x7)+(-2x12-2x9+x)*y+(-x9-x6)*y2+(4x11-x3)*y3+y4
```

See the library weierstr.lib for further procedures in connection with the Weierstraß preparation theorem.

Corollary 6.2.11. *Let K be a field, then $K[[x_1, \ldots, x_n]]$ is Noetherian.*

Proof. We prove the corollary using induction on n. If $I \subset K[[x_1]]$ is a non–zero ideal then we choose $f \in I$, $f \neq 0$ such that $m = \operatorname{ord}(f)$ is minimal. Then $f = x_1^m \cdot u$, u a unit. Obviously, $I = \langle x_1^m \rangle$. Assume $K[[x_1, \ldots, x_{n-1}]]$ is Noetherian. Let $I \subset K[[x_1, \ldots, x_n]]$ be a non–zero ideal and $f \in I$, $f \neq 0$. We may assume, using 6.2.2 or 6.2.5 that f is x_n–general. By the induction assumption $K[[x_1, \ldots, x_{n-1}]]$ and, hence, $K[[x_1, \ldots, x_{n-1}]][x_n]$ (Hilbert's basis theorem) are Noetherian. Thus, $I \cap K[[x_1, \ldots, x_{n-1}]][x_n]$ is finitely generated, say by $f_1, \ldots, f_m$. Now we claim that $I = \langle f, f_1, \ldots, f_m \rangle$. Let $g \in I$, then we use the Weierstraß Division Theorem to write $g = qf + r$ with $r \in I \cap K[[x_1, \ldots, x_{n-1}]][x_n]$, hence, for suitable $\xi_i \in K[[x_1, \ldots, x_{n-1}]][x_n]$ we have $r = \sum_{i=1}^m \xi_i f_i$. This proves the claim. $\qquad\square$

Corollary 6.2.12. *Let* $y = (y_1, \ldots, y_m)$, *and let* M *be a finitely generated* $K[[x, y]]$*–module. If* $\dim_K(M/\langle x \rangle M) < \infty$ *then* M *is a finitely generated* $K[[x]]$*–module.*

Proof. We use induction on m. The difficult part is the case $m = 1$. Consider the map $M/\langle x \rangle M \to M/\langle x \rangle M$ defined by multiplication with y_1. As $M/\langle x \rangle M$ is a finite dimensional K–vector space it follows from the Cayley–Hamilton theorem that, for suitable $c_i \in K$,

$$f := y_1^q + c_1 y_1^{q-1} + \cdots + c_q \in \operatorname{Ann}(M/\langle x \rangle M).$$

This implies that $fM \subset \langle x \rangle M$. Applying again the Cayley–Hamilton theorem to the $K[[x, y_1]]$–linear map $M \to M$ defined by multiplication with f, we obtain

$$g := f^r + h_1 f^{r-1} + \cdots + h_r \in \operatorname{Ann}(M)$$

for suitable $h_i \in \langle x, y_1 \rangle$. Therefore, M is a finitely generated $K[[x, y_1]]/\langle g \rangle$–module.

By construction, g is y_1–general. Using Corollary 6.2.9 we obtain that $K[[x, y_1]]/\langle g \rangle$ is finite over $K[[x]]$. This implies, indeed, that M is finitely generated as $K[[x]]$–module. $\qquad\square$

Definition 6.2.13. An *analytic* K*–algebra* A is a factor ring of a formal power series ring, $A = K[[x_1, \ldots, x_n]]/I$.

Corollary 6.2.14. *Let* A, B *be analytic* K*–algebras and* $\varphi : A \longrightarrow B$ *a morphism of local rings. Then* φ *is finite if and only if* $\dim_K B/\varphi(\mathfrak{m}_A)B < \infty$.

Proof. One implication is trivial; for the second, let $\dim_K B/\varphi(\mathfrak{m}_A)B < \infty$. Let $A = K[[x]]/I$ and $B = K[[y]]/J$, $y = (y_1, \ldots, y_m)$. Then B is a finitely generated $K[[x, y]]$–module, where the module structure is defined via φ: let $b \in B$, then $x_i b = \varphi(x_i \bmod I) \cdot b$, $y_i b = (y_i \bmod J) \cdot b$. Furthermore, $B/\langle x \rangle B = B/\varphi(\mathfrak{m}_A)B$. Now the statement is a consequence of Corollary 6.2.12. $\qquad\square$

We are now able to test with SINGULAR whether a map is finite. This is much simpler than for the polynomial case.

SINGULAR Example 6.2.15 (finiteness test).

```
proc mapIsFinite(R,map phi,ideal I)
{
    def S=basering;
    setring R;
    ideal ma=maxideal(1);
    setring S;
    ideal ma=phi(ma);
    ma=std(ma+I);
    if(dim(ma)==0)
    {
        return(1);
    }
    return(0);
}
```

Let us try an example.

```
ring A=0,(x,y),ds;
ring B=0,(x,y,z),ds;
map phi=A,x,y;
ideal I=z2-x2y;
mapIsFinite(A,phi,I);
//-> 1
```

Note that we want to compute $\dim(K[[x]]/IK[[x]])$, for some ideal $I \subset K[x]$, but what we actually compute is the dimension $\dim(K[x]_{\langle x \rangle}/IK[x]_{\langle x \rangle})$. We shall see soon that both dimensions are equal.

Theorem 6.2.16 (Noether Normalization). *Let K be an infinite field, $A = K[[x_1, \ldots, x_n]]$ and $I \subset A$ an ideal. Then there exists an integer $s \leq n$ and a matrix $M \in \mathrm{GL}(n, K)$ such that for*

$$\begin{pmatrix} y_1 \\ \vdots \\ y_n \end{pmatrix} = M^{-1} \cdot \begin{pmatrix} x_1 \\ \vdots \\ x_n \end{pmatrix}$$

the canonical map $K[[y_{s+1}, \ldots, y_n]] \rightarrow A/I$, defined by $y_i \rightarrow y_i \bmod I$, is injective and finite.

Proof. The proof is similar to the proof of Theorem 3.4.1, therefore, we just supply the idea. Because of Lemma 6.2.2 we may assume that there is some x_n–general $f \in I \setminus \{0\}$. Now, Corollary 6.2.9 implies that $K[[x]]/\langle f \rangle$ is a finitely generated $K[[x_1, \ldots, x_{n-1}]]$–module. Therefore, $K[[x]]/I$ is a finitely generated $K[[x_1, \ldots, x_{n-1}]]$–module. Let $I' = I \cap K[[x_1, \ldots, x_{n-1}]]$. If $I' = \langle 0 \rangle$, we are done. If $I' \neq \langle 0 \rangle$, we use the induction hypothesis for $I' \subset K[[x_1, \ldots, x_{n-1}]]$. $\qquad\square$

Theorem 6.2.17 (Implicit Function Theorem). *Let K be a field and $F \in K[[x_1, \ldots, x_n, y]]$ such that*

$$F(x_1, \ldots, x_n, 0) \in \langle x_1, \ldots, x_n \rangle \text{ and } \frac{\partial F}{\partial y}(x_1, \ldots, x_n, 0) \notin \langle x_1, \ldots, x_n \rangle,$$

then there exists a unique $y(x_1, \ldots, x_n) \in \langle x_1, \ldots, x_n \rangle K[[x_1, \ldots, x_n]]$ such that $F\big(x_1, \ldots, x_n, y(x_1, \ldots, x_n)\big) = 0$.

Proof. The conditions on F imply $F(0, \ldots, 0, y) = c \cdot y+$ higher terms in y, $c \neq 0$, that is, F is y–general of order 1. The Weierstraß Preparation Theorem (6.2.8) implies that $F = u \cdot \big(y - y(x_1, \ldots, x_n)\big)$, where $y(x_1, \ldots, x_n) \in K[[x_1, \ldots, x_n]]$ and $u \in K[[x_1, \ldots, x_n, y]]$ a unit. $\qquad\square$

Theorem 6.2.18 (Inverse Function Theorem). *Let K be a field, and let $f_1, \ldots, f_n \in K[[x_1, \ldots, x_n]]$, satisfying $f_1(0) = \cdots = f_n(0) = 0$. Then the K–algebra homomorphism*

$$\varphi : K[[x_1, \ldots, x_n]] \to K[[x_1, \ldots, x_n]],$$

defined by $\varphi(x_i) = f_i$, is an isomorphism if and only if $\det\big(\frac{\partial f_i}{\partial x_j}(0)\big) \neq 0$.

Proof. Assume that φ is an isomorphism, let ψ be its inverse and $g_i := \psi(x_i)$. Then we have $x_i = \psi \circ \varphi(x_i) = \psi(f_i) = f_i(g_1, \ldots, g_n)$ for all i which implies

$$\delta_{ij} = \sum_{k=1}^{n} \frac{\partial f_i}{\partial x_k}(\psi) \frac{\partial g_k}{\partial x_j}.$$

We obtain $(\delta_{ij}) = \big(\frac{\partial f_i}{\partial x_k}(0)\big)\big(\frac{\partial g_k}{\partial x_j}(0)\big)$ and, therefore, $\det\big(\frac{\partial f_i}{\partial x_j}(0)\big) \neq 0$.

Now assume that $\det\big(\frac{\partial f_i}{\partial x_j}(0)\big) \neq 0$. Then the matrix $\big(\frac{\partial f_i}{\partial x_j}(0)\big)$ is invertible and we may replace $f_1, \ldots, f_n$ by suitable linear combinations and assume that $\big(\frac{\partial f_i}{\partial x_j}(0)\big) = (\delta_{ij})$ is the unit matrix. We construct the inverse ψ of φ by applying the Implicit Function Theorem successively to

$$F_1 := f_1(y_1, \ldots, y_n) - x_1, \ldots, F_n := f_n(y_1, \ldots, y_n) - x_n.$$

In the first step we obtain $g \in \langle y_2, \ldots, y_n, x_1 \rangle \cdot K[[y_2, \ldots, y_n, x_1]]$ such that $f_1(g, y_2, \ldots, y_n) = x_1$. Now we consider

$$\widetilde{F}_2 = f_2(g, y_2, \ldots, y_n) - x_2, \ldots, \widetilde{F}_n = f_n(g, y_2, \ldots, y_n) - x_n.$$

An easy computation shows that, again, $\big(\frac{\partial \widetilde{F}_i}{\partial y_j}(0)\big) = (\delta_{ij})$. Hence, we can use induction and can assume that after applying the Implicit Function Theorem $n - 1$ times, we would find $g_2, \ldots, g_n \in \langle x_1, \ldots, x_n \rangle K[[x_1, \ldots, x_n]]$ with $\widetilde{F}_i(g_2, \ldots, g_n) = 0$ for $i = 2, \ldots, n$.

Let $g_1 := g(g_2, \ldots, g_n, x_1)$, and let $\psi : K[[x]] \to K[[x]]$ be the K–algebra homomorphism defined by $\psi(x_i) := g_i$, then

$$\psi \circ \varphi(x_i) = \psi(f_i) = f_i(g_1, \ldots, g_n) = x_i$$

for $i = 1, \ldots, n$. It follows that φ is injective. On the other hand, the same consideration as above shows that $\det\left(\frac{\partial g_i}{\partial x_j}(0)\right) \neq 0$. Hence, we can conclude that ψ is, indeed, an isomorphism, which completes the proof. $\square$

Exercises

6.2.1. Let K be a field and $I \subset K[[x_1, \ldots, x_n]]$ an ideal. Prove that there is a Noether normalization $K[[x_1, \ldots, x_r]] \subset K[[x_1, \ldots, x_n]]/I$ with the following property: for $i = r + 1, \ldots, n$ there exist Weierstraß polynomials

$$p_i = x_i^{n_i} + \sum_{\nu=0}^{n_i-1} p_{i,\nu}(x_1, \ldots, x_{i-1}) x_i^{\nu} \in I \,,$$

$p_{i,\nu} \in K[[x_1, \ldots, x_{i-1}]]$ and $\mathrm{ord}(p_{i,\nu}) \geq n_i - \nu$. We call such a Noether normalization a *general Noether normalization*.

In the case that I is a prime ideal, prove that p_{r+1} can be chosen such that $Q(K[[x_1, \ldots, x_n]]/I) = Q(K[[x_1, \ldots, x_r]])[x_{r+1}]/\langle p_{r+1}\rangle$.

6.2.2. Prove Remark 6.2.5.

6.2.3. Write a SINGULAR procedure for the Weierstraß division theorem up to a given order (cf. SINGULAR Example 6.2.10).

6.2.4. Write a SINGULAR procedure to compute the Noether normalization of an ideal $I \subset K[[x_1, \ldots, x_n]]$, I being generated by polynomials.

6.2.5. Prove *Newton's Lemma*: there exists $\bar{y} \in K[[x_1, \ldots, x_n]]$ such that $f(\bar{y}) = 0$ and $a \equiv \bar{y} \bmod \left(\frac{\partial f}{\partial y}(a)\right)(x_1, \ldots, x_n)^c$ if $f \in K[[x_1, \ldots, x_n, y]]$ and $a \in K[[x_1, \ldots, x_n]]$ such that $f(a) \equiv 0 \bmod \left(\frac{\partial f}{\partial y}(a)\right)^2 (x_1, \ldots, x_n)^c$.

6.2.6. Prove *Hensel's Lemma*: let $F \in K[[x_1, \ldots, x_n]][y]$ be a monic polynomial with respect to y and assume that $F(0, \ldots, 0, y) = g_1 \cdot g_2$ for monic polynomials g_1, g_2 such that $\langle g_1, g_2 \rangle = K[y]$. Then there exist monic polynomials $G_1, G_2 \in K[[x_1, \ldots, x_n]][y]$ such that

(1) $F = G_1 G_2$,
(2) $G_i(0, \ldots, 0, y) = g_i$, $i = 1, 2$.

6.3 Completions

Let A be a Noetherian local ring with maximal ideal $\mathfrak{m}$. Similar to Definition 6.1.6 we define Cauchy sequences in A with respect to the $\mathfrak{m}$–adic topology.

Definition 6.3.1. A is called a *complete local ring*, if every Cauchy sequence in A has a limit in A.

We shall now show another characterization of complete local rings.

Definition 6.3.2. Let M be an A–module and $\{M_n\}$ be a stable $\mathfrak{m}$–filtration of M, then the module

$$\widehat{M} := \left\{ (m_1, m_2, \ldots) \in \prod_{i=1}^{\infty} M/M_i \;\middle|\; m_i \equiv m_j \bmod M_i \text{ if } j > i \right\}$$

is called the *completion* of M.

Lemma 6.3.3. *With the notations of Definition 6.3.2, $\widehat{M}$ is an A–module and does not depend on the choice of the stable filtration.*

Proof. Obviously $\widehat{M}$ is an A–module, but we have to prove that $\widehat{M}$ does not depend on the filtration. Let $\left\{ \widetilde{M}_n \right\}$ be another stable $\mathfrak{m}$–filtration of M, and choose, using Lemma 5.4.3, k_0 such that $M_{k+k_0} \subset \widetilde{M}_k$ and $\widetilde{M}_{k+k_0} \subset M_k$ for all k.

We have canonical maps

$$M/M_{k+2k_0} \xrightarrow{\varphi_k} M/\widetilde{M}_{k+k_0} \xrightarrow{\Psi_k} M/M_k$$

for all k and obtain, therefore, maps

$$\widehat{M}_{\{M_n\}} \xrightarrow{\widehat{\varphi}} \widehat{M}_{\{\widetilde{M}_n\}} \xrightarrow{\widehat{\Psi}} \widehat{M}_{\{M_n\}}$$

(to distinguish we use here the notation $\widehat{M}_{\{M_n\}}$, respectively $\widehat{M}_{\{\widetilde{M}_n\}}$, for the completion of M with respect to the filtration $\{M_n\}$, respectively $\{\widetilde{M}_n\}$).

Let $\overline{m} = (m_1, \ldots) \in \widehat{M}_{\{M_n\}}$, then $\widehat{\varphi}(\overline{m}) = (n_1, \ldots) =: \overline{n}$ satisfies $n_{k_0+k} = \varphi_k(m_{2k_0+k})$ and $n_i = n_{k_0+1} \bmod \widetilde{M}_i$ if $i \leq k_0$. Now $\widehat{\Psi}(\overline{n}) = (\widetilde{m}_1, \ldots)$ with the property $\widetilde{m}_k = \Psi_k(n_{k_0+k})$. But $\widetilde{m}_k = m_{2k_0+k} \bmod M_k = m_k$ and, therefore, $\widehat{\Psi} \circ \widehat{\varphi} = \text{id}$. Similarly, we can see that $\widehat{\varphi} \circ \widehat{\Psi} = \text{id}$. $\square$

Example 6.3.4. The completion $\widehat{A}$ of a Noetherian local ring A is given by

$$\widehat{A} = \left\{ (a_1, a_2, \ldots) \in \prod_{i=1}^{\infty} A/\mathfrak{m}^i \;\middle|\; a_i \equiv a_j \bmod \mathfrak{m}^i \text{ if } j > i \right\}.$$

Note that $\widehat{A}$ has a natural ring structure, given by component wise addition and multiplication,

$$(a_1, a_2, \ldots) \cdot (a_1', a_2', \ldots) := (a_1 a_1', a_2 a_2', \ldots).$$

Theorem 6.3.5. *Let A be a Noetherian local ring. Then $\widehat{A}$ is a complete local ring with maximal ideal*

$$\widehat{\mathfrak{m}} := \left\{ (a_1, a_2, \ldots) \in \widehat{A} \,\middle|\, a_1 = 0 \right\},$$

and $\widehat{A}/\widehat{\mathfrak{m}}^\nu \cong A/\mathfrak{m}^\nu$. The canonical map $A \to \widehat{A}$ is an isomorphism if and only if A is complete.

Proof. Any $(a_1, a_2, \ldots) \in \widehat{\mathfrak{m}}$ satisfies, in particular, $a_i \equiv 0 \bmod \mathfrak{m}$, that is, $a_i \in \mathfrak{m}$ for all i. Now it is easy to see that

$$\widehat{\mathfrak{m}}^i = \left\{ (a_1, a_2, \ldots) \in \widehat{A} \,\middle|\, a_j = 0 \text{ for all } j \leq i \right\}.$$

It follows that the canonical map $\widehat{A} \to A/\mathfrak{m}^i$, $(a_1, a_2, \ldots) \mapsto a_i$ induces an isomorphism $\widehat{A}/\widehat{\mathfrak{m}}^i \cong A/\mathfrak{m}^i$. On the other hand, consider the canonical map $\varphi : A \to \widehat{A}$ defined by

$$\varphi(a) := (a \bmod \mathfrak{m}, \, a \bmod \mathfrak{m}^2, \ldots).$$

Note that the above consideration shows that φ is, indeed, a continuous map.

If A is complete and $(a_1, a_2, \ldots) \in \widehat{A}$, then $\{a_i\}$ is a Cauchy sequence because, for $j > i$, $a_j - a_i \in \mathfrak{m}^i$. This sequence has a limit $a \in A$ which satisfies $\varphi(a) = (a_1, a_2, \ldots)$. Thus, φ is surjective. Because $\bigcap_i \mathfrak{m}^i = \langle 0 \rangle$ we have that φ is always injective.

To prove the other direction it suffices to prove that $\widehat{A}$ is complete. Let $\{\bar{a}_n\}$ be a Cauchy sequence in $\widehat{A}$, $\bar{a}_n = \left(a_1^{(n)}, a_2^{(n)}, \ldots \right)$. This means that for all given N there exists $i(N)$ such that $\bar{a}_n - \bar{a}_m \in \widehat{\mathfrak{m}}^N$ if $n, m \geq i(N)$. Now $\bar{a}_n - \bar{a}_m \in \widehat{\mathfrak{m}}^N$ if and only if $a_i^{(n)} - a_i^{(m)} \in \mathfrak{m}^N$ for all i. Define

$$\bar{b} := (b_1, b_2, \ldots) \in \prod_{i=1}^{\infty} A/\mathfrak{m}^i, \qquad b_n := a_n^{(i(n))}.$$

Then

(1) $b_n - a_n^{(j)} \in \mathfrak{m}^n$ if $j \geq i(n)$;
(2) $b_m - b_n \in \mathfrak{m}^n$ if $m \geq n$.

This implies $\bar{b} \in \widehat{A}$ and $\bar{b} - \bar{a}_j \in \widehat{\mathfrak{m}}^n$ for $j \geq i(n)$. $\square$

Corollary 6.3.6. *Let A be a Noetherian local ring, and let $\widehat{A}$ be the completion. Then the following holds:*

(1) $\mathrm{HS}_{\widehat{A},\widehat{\mathfrak{m}}} = \mathrm{HS}_{A,\mathfrak{m}}$, *especially* $\dim(A) = \dim(\widehat{A})$, *and* $\mathrm{mult}(A) = \mathrm{mult}(\widehat{A})$;
(2) A *is regular if and only if* $\widehat{A}$ *is regular.*

Proposition 6.3.7. *Let A be a Noetherian local ring, and let*

$$0 \to N \xrightarrow{j} M \xrightarrow{\pi} P \to 0$$

be an exact sequence of finitely generated A–modules, then the induced sequence of $\widehat{A}$–modules

$$0 \longrightarrow \widehat{N} \xrightarrow{\widehat{j}} \widehat{M} \xrightarrow{\widehat{\pi}} \widehat{P} \longrightarrow 0$$

is exact.

Proof. Artin–Rees' Lemma 5.4.5 implies that $\{\mathfrak{m}^i M \cap N\}$ is an $\mathfrak{m}$–stable filtration of N, therefore, $\widehat{N} = \widehat{N}_{\{\mathfrak{m}^i M \cap N\}}$. Now, since for all i the sequence

$$0 \longrightarrow N/\mathfrak{m}^i M \cap N \longrightarrow M/\mathfrak{m}^i M \longrightarrow P/\mathfrak{m}^i P \longrightarrow 0$$

is exact, we obtain that the induced sequence

$$0 \longrightarrow \widehat{N} \xrightarrow{\widehat{j}} \widehat{M} \xrightarrow{\widehat{\pi}} \widehat{P}$$

is exact. It remains to prove that $\widehat{\pi}$ is surjective.

Let $p = (p_1, p_2, \ldots) \in \widehat{P}$. Choose $m_1, \overline{m}_2 \in M$ such that $\pi(m_1) + \mathfrak{m}P = p_1$ and $\pi(\overline{m}_2) + \mathfrak{m}^2 P = p_2$. Now $p_1 \equiv p_2 \mod \mathfrak{m}P$. This implies that there exist $t \in \mathfrak{m}M$, $n \in N$ such that $m_1 - \overline{m}_2 = n + t$. Define $m_2 := \overline{m}_2 + n$, then $\pi(m_2) + \mathfrak{m}^2 P = p_2$ and $m_1 - m_2 \in \mathfrak{m}M$. Continuing in this way, we define a sequence $m = (m_1 + \mathfrak{m}M, m_2 + \mathfrak{m}^2 M, \ldots) \in \widehat{M}$ with $\widehat{\pi}(m) = p$. $\square$

Corollary 6.3.8. *Let A be a Noetherian local ring and M a finitely generated A–module, then $M \otimes_A \widehat{A} \cong \widehat{M}$. Especially, if A is complete, then M is complete, that is, $M = \widehat{M}$. In particular, A/I is complete if A is complete.*

Proof. Let $A^n \to A^m \to M \to 0$ be a presentation of M, then Proposition 6.3.7 and Theorem 2.7.6 give a commutative diagram

$$
\begin{array}{ccccccc}
\widehat{A}^n & \longrightarrow & \widehat{A}^m & \longrightarrow & \widehat{M} & \longrightarrow & 0 \\
\uparrow & & \uparrow & & \uparrow & & \\
A^n \otimes_A \widehat{A} & \longrightarrow & A^m \otimes_A \widehat{A} & \longrightarrow & M \otimes_A \widehat{A} & \longrightarrow & 0
\end{array}
$$

with exact rows. The vertical arrows are the canonical maps. The two left–hand arrows are isomorphisms, since completion commutes with direct sums. This implies that the canonical map $M \otimes_A \widehat{A} \to \widehat{M}$ is an isomorphism. $\square$

Example 6.3.9. Let K be a field, and let $I \subset K[[x_1, \ldots, x_n]]$ be an ideal, then $A = K[[x_1, \ldots, x_n]]/I$ is complete.

We shall see now that, in a sense, the converse is also true.

Theorem 6.3.10 (Cohen). *Let A be a Noetherian complete local ring with maximal ideal $\mathfrak{m}$. Suppose that A contains a field and let $K := A/\mathfrak{m}$. Then $A \cong K[[x_1, \ldots, x_n]]/I$ for some ideal I.*

Proof. Here we shall give the proof for the special case that $K \subset A$. Let $\mathfrak{m} = \langle m_1, \ldots, m_n \rangle$, and consider the map $\varphi : K[x_1, \ldots, x_n]_{\langle x_1, \ldots, x_n \rangle} \to A$ defined by $\varphi(x_i) := m_i$. Now a Cauchy sequence $\{a_\nu\}$ in $K[x_1, \ldots, x_n]_{\langle x_1, \ldots, x_n \rangle}$ with respect to $\langle x_1, \ldots, x_n \rangle$ is mapped to a Cauchy sequence $\{\varphi(a_\nu)\}$ in A with respect to $\mathfrak{m}$.

Because A is complete, φ can be extended to $\widehat{\varphi} : K[[x_1, \ldots, x_n]] \to A$. Let $a \in A$, then $a = a_0 + m$, $a_0 \in K$, $m \in \mathfrak{m}$, $m = \sum_i c_i m_i$. Again we decompose $c_i = c_0^i + c_1^i$, $c_1^i \in \mathfrak{m}$, $c_0^i \in K$. Continuing like that we obtain, for all j,

$$a = \sum_{\alpha_1 + \cdots + \alpha_n \leq j} a_{\alpha_1, \ldots, \alpha_n} m_1^{\alpha_1} \cdot \ldots \cdot m_n^{\alpha_n} + h_j, \quad h_j \in \mathfrak{m}^{j+1}.$$

Because A is complete, the sequence $\left\{ \sum_{|\alpha| \leq j} a_{\alpha_1, \ldots, \alpha_n} m_1^{\alpha_1} \cdot \ldots \cdot m_n^{\alpha_n} \right\}_{j \in \mathbb{N}}$ converges to a and $a = \widehat{\varphi}(\sum_\alpha a_{\alpha_1, \ldots, \alpha_n} x_1^{\alpha_1} \cdot \ldots \cdot x_n^{\alpha_n})$. Hence, $\widehat{\varphi}$ is surjective, and we set $I := \mathrm{Ker}\, \widehat{\varphi}$. □

As an application we want to compute the completion of the local ring $K[x_1, \ldots, x_n]_P$, $P \subset K[x_1, \ldots, x_n]$ a prime ideal. Because of Exercise 3.5.4,

$$K[x_1, \ldots, x_n]_P = K(u)[x \smallsetminus u]_{PK(u)[x \smallsetminus u]},$$

$u \subset x = \{x_1, \ldots, x_n\}$ a maximal independent set. Since $PK(u)[x \smallsetminus u]$ is a maximal ideal in $K(u)[x \smallsetminus u]$, it is enough to describe the computation of the completion in the case that P is a maximal ideal. We shall study here the case $P = \langle x_1, \ldots, x_{n-1}, f \rangle$, $f \in K[x_n]$ irreducible and separable. If K is a field of characteristic 0, we can always find an automorphism φ of $K[x_1, \ldots, x_n]$ mapping a given maximal ideal P to an ideal $\varphi(P) = \langle x_1, \ldots, x_{n-1}, f \rangle$ with $f \in K(x_n)$ irreducible (Proposition 4.2.2).

Theorem 6.3.11. *Let K be a field, $f \in K[x_n]$ an irreducible and separable polynomial and α a root of f. Then there is a canonical isomorphism*

$$\varphi : K[x_1, \ldots, x_n]\widehat{}_{\langle x_1, \ldots, x_{n-1}, f \rangle} \xrightarrow{\cong} K(\alpha)[[y_1, \ldots, y_n]]$$

defined by $\varphi(x_i) = y_i$, $i = 1, \ldots, n-1$, and $\varphi(f) = y_n$.

Proof. Let $f = \sum_{\nu=0}^{m} a_\nu x_n^\nu$ and consider

$$F(T) := \sum_{\nu=0}^{m} a_\nu (T + \alpha)^\nu - y_n \in K(\alpha)[[y_n]][T].$$

We have $F(0) = -y_n$ and $F'(0) = f'(\alpha)$ a unit. Using the Implicit Function Theorem we obtain a unique $t(y_n) \in \langle y_n \rangle \cdot K(\alpha)[[y_n]]$ such that $F\big(t(y_n)\big) = 0$. Now we define $\varphi_0 : K[x_1, \ldots, x_n] \to K(\alpha)[[y_1, \ldots, y_n]]$, setting $\varphi_0(x_i) := y_i$, $i = 1, \ldots, n-1$, and $\varphi_0(x_n) := t(y_n) + \alpha$.

We obtain $\varphi_0(f) = y_n$, in particular, $\varphi_0(\langle x_1, \ldots, x_{n-1}, f \rangle) \subset \langle y_1, \ldots, y_n \rangle$. If $h \in K[x_1, \ldots, x_n]$, $h \notin \langle x_1, \ldots, x_{n-1}, f \rangle$, then

$$ah + \sum_{i=1}^{n-1} b_i x_i + cf = 1$$

for suitable $a, b_1, \ldots, b_{n-1}, c \in K[x_1, \ldots, x_n]$ because $\langle x_1, \ldots, x_{n-1}, f \rangle$ is a maximal ideal. This implies $\varphi_0(a)\varphi_0(h) + \sum_{i=1}^{n-1} \varphi_0(b_i)y_i + \varphi_0(c)y_n = 1$ and, therefore, $\varphi_0(h) \notin \langle y_1, \ldots, y_n \rangle$. Hence, φ_0 extends to a map

$$\varphi_0 : K[x_1, \ldots, x_n]_{\langle x_1, \ldots, x_{n-1}, f \rangle} \longrightarrow K(\alpha)[[y_1, \ldots, y_n]].$$

Denote by $\varphi_0^{(k)}$ the induced maps

$$\varphi_0^{(k)} : K[x_1, \ldots, x_n]_{\langle x_1, \ldots, x_{n-1}, f \rangle} / \langle x_1, \ldots, x_{n-1}, f \rangle^k$$
$$\longrightarrow K(\alpha)[[y_1, \ldots, y_n]] / \langle y_1, \ldots, y_n \rangle^k.$$

It remains to prove that $\varphi_0^{(k)}$ is an isomorphism for all k.

Let $h \in K[x_1, \ldots, x_n]_{\langle x_1, \ldots, x_{n-1}, f \rangle}$ such that $\varphi_0(h) \in \langle y_1, \ldots, y_n \rangle^k$ and assume $h \notin \langle x_1, \ldots, x_{n-1}, f \rangle^k$. We may assume that $h \in \langle x_1, \ldots, x_{n-1}, f \rangle^{k-1}$ and

$$h = \sum_{|\alpha| = k-1} a_{\alpha_1, \ldots, \alpha_n} x_1^{\alpha_1} \cdots x_{n-1}^{\alpha_{n-1}} f^{\alpha_n},$$

with coefficients $a_{\alpha_1, \ldots, \alpha_n} \in K[x_1, \ldots, x_n]_{\langle x_1, \ldots, x_{n-1}, f \rangle}$ among which there is (at least one) coefficient $a_{\beta_1, \ldots, \beta_n} \notin \langle x_1, \ldots, x_{n-1}, f \rangle$. Then

$$\varphi_0(h) = \sum_{|\alpha| = k-1} \varphi_0(a_{\alpha_1, \ldots, \alpha_n}) \cdot y_1^{\alpha_1} \cdots y_n^{\alpha_n},$$

and $\varphi_0(a_{\beta_1, \ldots, \beta_n}) \notin \langle y_1, \ldots, y_n \rangle$, which contradicts $\varphi_0(h) \in \langle y_1, \ldots, y_n \rangle^k$. This proves the injectivity of $\varphi_0^{(k)}$.

To prove the surjectivity of $\varphi_0^{(k)}$, it is sufficient to show that there exists some $h \in K[x_1, \ldots, x_n]_{\langle x_1, \ldots, x_{n-1}, f \rangle}$ such that $\varphi_0(h) - \alpha \in \langle y_1, \ldots, y_n \rangle^k$. We know already that $\varphi_0(x_n) - \alpha \in \langle y_n \rangle \cdot K(\alpha)[[y_n]]$. Assume we have found $\overline{h}$ such that $\varphi_0(\overline{h}) - \alpha \in \langle y_n \rangle^{k-1} K(\alpha)[[y_n]]$, $\varphi_0(\overline{h}) - \alpha = y_n^{k-1}(q(\alpha) + y_n p(\alpha))$, $q(\alpha) \in K(\alpha) = K[x_n]/\langle f \rangle$. Now $\varphi_0(q(x_n)) - q(\alpha) \in \langle y_n \rangle K(\alpha)[[y_n]]$ implies that $\varphi_0(\overline{h} - f^{k-1} q(x_n)) - \alpha \in \langle y_n \rangle^k K(\alpha)[[y_n]]$. $\qquad \square$

Corollary 6.3.12. *Let K be a field, $f \in K[x_n]$ an irreducible and separable polynomial and α a root of f. Then there is a canonical isomorphism of the associated graded rings*

$$\psi : \mathrm{Gr}_{\langle x_1,\dots,x_{n-1},f\rangle}\big(K[x_1,\dots,x_n]_{\langle x_1,\dots,x_{n-1},f\rangle}\big) \xrightarrow{\cong} K(\alpha)[y_1,\dots,y_n],$$

defined by $\psi(x_i) = y_i$, $i = 1,\dots,n-1$, and $\psi(f) = y_n$.[2]

Exercises

6.3.1. Let A be a local ring and M an A–module. Compute the kernel of the canonical map $M \to \widehat{M}$.

6.3.2. Let A be a Noetherian local ring and $i : A \to \widehat{A}$ the canonical map. Prove that $i(x)$ is not a zerodivisor in $\widehat{A}$ if x is not a zerodivisor in A.

6.3.3. Let K be a field and $I \subset K[[x_1,\dots,x_n]]$ an ideal. Let $K[[x_1,\dots,x_r]] \subset A = K[[x_1,\dots,x_n]]/I$ be a general Noether normalization (cf. Exercise 6.2.1). Moreover, let $\mathfrak{m}_0 = \langle x_1,\dots,x_r\rangle$ be the maximal ideal in $K[[x_1,\dots,x_r]]$, $\mathfrak{m}$ the maximal ideal of A. Prove that $\{\mathfrak{m}^i\}$ is a stable $\mathfrak{m}_0$–filtration of A.

6.3.4. Use Exercise 6.3.3 to give another proof that $K[[x_1,\dots,x_n]]/I$ is complete.

6.3.5. Use Exercise 6.3.3 to prove the following: let $f \in K[[x_1,\dots,x_n]]$, $\mathrm{ord}(f) = m$ then $\mathrm{mult}(K[[x_1,\dots,x_n]]/\langle f\rangle) = m$.

6.3.6. Let K be an infinite field, let $P \subset K[[x_1,\dots,x_n]]$ be a prime ideal, and let $K[[x_1,\dots,x_r]] \subset A = K[[x_1,\dots,x_n]]/P$ be a general Noether normalization. Prove that $\mathrm{mult}(A) = [Q(A) : Q(K[[x_1,\dots,x_r]])]$

6.3.7. Compute the dimension and the multiplicity of $A = K[[x,y,z]]/I$, $I = \langle x^2 - xy, z^3 - xyz\rangle$.

6.3.8. Use Theorem 6.3.11 to give another proof of Proposition 5.7.10.

6.3.9. Let K be a field, and let $I = \langle f_1,\dots,f_m\rangle$ be an ideal in $K[[x_1,\dots,x_n]]$. Prove (similarly to Theorem 5.6.12) that

$$\mathrm{edim}(K[[x_1,\dots,x_n]]/I) = n - \mathrm{rank}\big(\tfrac{\partial f_i}{\partial x_j}(0)\big).$$

6.3.10. Let K be a field, and $I = \langle f_1,\dots,f_m\rangle$ an ideal in $K[[x_1,\dots,x_n]]$. Prove (using Exercise 6.3.9) the *Jacobian Criterion*: $K[[x_1,\dots,x_n]]$ is regular if and only if $\mathrm{rank}\big(\tfrac{\partial f_i}{\partial x_j}(0)\big) = n - \dim(K[[x_1,\dots,x_n]]/I)$.

---

[2] If K is one of the ground fields in SINGULAR, we can compute in $K(\alpha)[y_1,\dots,y_n]$ and, hence, in the associated graded ring of the localization of the polynomial ring $K[x_1,\dots x_n]$ in a prime ideal.

6.3.11. Let K be a field, and $I = \langle f_1, \ldots, f_m \rangle$ an ideal in $K[[x_1, \ldots, x_n]]$. Use Theorem 6.2.18 and Exercise 6.3.10 to show that $K[[x_1, \ldots, x_n]]/I$ is regular of dimension s if and only if $K[[x_1, \ldots, x_n]]/I \cong K[[y_1, \ldots, y_s]]$.

6.3.12. Let A be a Noetherian local ring. Prove that $\widehat{A^n} = \widehat{A}^n$.

6.4 Standard Bases

In this section we shall introduce standard bases for ideals in formal power series rings. The main result is that they can be computed, if the ideal is generated by polynomials. This is the basis for computations in local analytic geometry. The theory of standard bases in power series rings goes back to Hironaka (cf. [123]) and Grauert (cf. [98]).

Let K be a field and $x = (x_1, \ldots, x_n)$. Throughout this section, we fix a local degree ordering $>$ on $\mathrm{Mon}(x_1, \ldots, x_n)$, that is, $x^\alpha > x^\beta$ implies that $\mathrm{w\text{-}deg}(x^\alpha) \leq \mathrm{w\text{-}deg}(x^\beta)$ for a suitable weight vector $w = (w_1, \ldots, w_n)$ with $w_i > 0$. Such orderings are compatible with the $\langle x \rangle$–adic topology, which allows us to compare standard bases in $K[x]_{\langle x \rangle}$ and $K[[x]]$.

A non–zero element $f \in K[[x]]$ can be written as $\sum_{\nu=0}^{\infty} a_\nu x^{\alpha(\nu)}, a_\nu \in K$, $a_0 \neq 0$ and $x^{\alpha(\nu)} > x^{\alpha(\nu+1)}$ for all ν.

As in Definition 1.2.2, we define $\mathrm{LM}(f)$, $\mathrm{LE}(f)$, $\mathrm{LT}(f)$, $\mathrm{LC}(f)$ and $\mathrm{tail}(f)$. As in Definition 1.6.1, we define a *standard basis* of an ideal $I \subset K[[x]]$. Standard bases exist in $K[[x]]$ because the leading ideal is finitely generated (by Corollary 6.2.11). Finally, as in Definition 1.6.2, we define *minimal* and *completely reduced standard bases*.

We shall see in the exercises that completely reduced standard bases exist and are uniquely determined. To prove the existence of a reduced normal form, we give the formal version of Grauert's Division Theorem ([98]), a generalization of the Weierstraß Division Theorem.

Theorem 6.4.1 (Division Theorem). *Let* $f, f_1, \ldots, f_m \in K[[x_1, \ldots, x_n]]$ *then there exist* $q_j, r \in K[[x_1, \ldots, x_n]]$ *such that*

$$f = \sum_{j=1}^{m} q_j f_j + r$$

and, for all $j = 1, \ldots, m$,

(1) no monomial of r is divisible by $\mathrm{LM}(f_j)$;
(2) $\mathrm{LM}(q_j f_j) \leq \mathrm{LM}(f)$.

Proof. We may assume that $\mathrm{LC}(f_i) = 1$, $i = 1, \ldots, m$. Let $\mathrm{LM}(f_i) = x^{\alpha(i)}$, $i = 1, \ldots, m$, and let

$$\Gamma := \langle \alpha(1), \ldots, \alpha(m) \rangle := \{ \alpha \in \mathbb{N}^n \mid \alpha - \alpha(i) \in \mathbb{N}^n \text{ for some } i \}$$

be the semi–module in $\mathbb{N}^n$ generated by $\alpha(1), \dots, \alpha(m)$. We define inductively

$$\Gamma_1 := \langle \alpha(1) \rangle, \ \dots \ , \Gamma_i := \langle \alpha(i) \rangle \cap (\Gamma \smallsetminus \langle \alpha(1), \dots, \alpha(i-1) \rangle).$$

Let $w = \sum w_\alpha x^\alpha \in K[[x]]$, then we define

$$r(w) := \sum_{\alpha \notin \Gamma} w_\alpha x^\alpha, \quad q_j(w) := \frac{1}{x^{\alpha(j)}} \sum_{\alpha \in \Gamma_j} w_\alpha x^\alpha$$

and obtain

$$w = \sum_{j=1}^m q_j(w) x^{\alpha(j)} + r(w). \tag{$*$}$$

Now define a sequence $\{w_i\}_{i \in \mathbb{N}}$ by $w_0 := f$,

$$w_{i+1} := w_i - \sum_{j=1}^m q_j(w_i) f_j - r(w_i).$$

We claim that $\sum_{j=0}^\infty w_j =: w$ converges in the $\langle x \rangle$–adic topology and that we obtain the wanted decomposition as $f = \sum_{j=1}^m q_j(w) f_j + r(w)$. To see the convergence note that, due to $\mathrm{LT}(f_j) = x^{\alpha(j)}$ and $(*)$, $\mathrm{LM}(w_{i+1}) < \mathrm{LM}(w_i)$ and, by assumption, the ordering is compatible with the $\langle x \rangle$–adic filtration.

Now $\sum_{j=0}^\infty w_j = w$ implies $r(w) = \sum_{j=0}^\infty r(w_j)$ and $q_i(w) = \sum_{j=0}^\infty q_i(w_j)$. To prove that $f = \sum_{j=1}^m q_j(w) f_j + r(w)$, we write

$$f = \sum_{j=0}^\infty (w_j - w_{j+1}) = \sum_{j=0}^\infty \left(w_j - w_j + \sum_{k=1}^m q_k(w_j) f_k + r(w_j) \right)$$

$$= \sum_{k=1}^m \left(\sum_{j=0}^\infty q_k(w_j) \right) f_k + \sum_{j=0}^\infty r(w_j) = \sum_{k=1}^m q_k(w) f_k + r(w).$$

By construction, no monomial of $r(w)$ is divisible by $\mathrm{LM}(f_j) = x^{\alpha(j)}$ for all j. On the other hand, $\mathrm{LM}(q_k(w) f_k) = \mathrm{LM}(\sum_{j=0}^\infty q_k(w_j) f_k) \leq \mathrm{LM}(q_k(w_0) f_k)$ because $\mathrm{LM}(w_{k+1}) < \mathrm{LM}(w_k)$ for all k. But $w_0 = f$ and, therefore, we obtain $\mathrm{LM}(q_k(w) f_k) \leq \mathrm{LM}(f)$. $\square$

Definition 6.4.2. With the notation of Theorem 6.4.1, we define

$$\mathrm{NF}(f \mid \{f_1, \dots, f_m\}) := r$$

and obtain in this way a *reduced normal form*.

The existence of a reduced normal form is the basis to obtain, in the formal power series ring, all properties of standard bases already proved for $K[x]_>$:

- If S, S' are two standard bases of the ideal I, then $\mathrm{NF}(f \mid S) = \mathrm{NF}(f \mid S')$.

- Buchberger's criterion (Theorem 1.7.3).
- A reduced standard basis is uniquely determined.

For computations in local analytic geometry, the following theorem is important.

Theorem 6.4.3. *Let $K[x] \subset K[[x]]$ be equipped with compatible local degree orderings. Let I be an ideal in $K[x]$ and S a standard basis of I, then S is a standard basis of $IK[[x]]$.*

Proof. Let $\{f_1, \dots, f_m\}$ be a standard basis of $I \subset K[x]$ and

$$\overline{f} = \sum_{j=1}^{m} \overline{h}_j f_j \in IK[[x]]$$

be a non–zero element. We choose $c \in \mathbb{N}$ such that $\mathrm{LM}(\overline{f}) \notin \langle x \rangle^c$ and every monomial in $\langle x \rangle^c$ is smaller than $\mathrm{LM}(\overline{f})$. Then we choose $h_j \in K[x]$ such that $\overline{h}_j - h_j \in \langle x \rangle^c$. Let $f := \sum_{j=1}^{m} h_j f_j$, then $f \in I$ and $f - \overline{f} \in \langle x \rangle^c$.

This implies that $\mathrm{LM}(f) = \mathrm{LM}(\overline{f})$, because every monomial in $\langle x \rangle^c$ is smaller than $\mathrm{LM}(\overline{f})$. But $\mathrm{LM}(f) \in L(I) = \langle \mathrm{LM}(f_1), \dots, \mathrm{LM}(f_m) \rangle$. $\qquad\square$

Exercises

6.4.1. Prove Theorem 6.4.3 for a local ordering, satisfying the following property: given x^β and an infinite decreasing sequence $x^{\alpha(1)} > \cdots > x^{\alpha(k)} > \dots$ there exists j such that $x^\beta > x^{\alpha(j)}$.

6.4.2. Prove Buchberger's criterion for ideals in formal power series rings.

6.4.3. Prove that a reduced standard basis of an ideal in a formal power series ring is uniquely determined.

6.4.4. Let S, S' be two standard bases of the ideal $I \subset K[[x]]$. Prove that $\mathrm{NF}(f \mid S) = \mathrm{NF}(f \mid S')$.

6.4.5. Use SINGULAR to compute a standard basis for the ideal

$$\langle x^{10} + x^9 y^2, \ y^8 - x^2 y^7 \rangle \subset \mathbb{Q}[[x, y]]$$

with respect to the negative weighted degree lexicographical ordering with weight vector $w = (-2, -7)$.

6.4.6. Let $f = xy + z^4$ and $g = xz + y^5 + yz^2$. Use SINGULAR to compute

(1) $\dim_{\mathbb{Q}} \mathbb{Q}[[x, y, z]] / \langle f, g, M_1, M_2, M_3 \rangle$, where M_1, M_2, M_3 are the 2–minors of the Jacobian matrix of f and g;

(2) $\dim_{\mathbb{Q}} \mathbb{Q}[[x, y, z]]^3 / M$, where $M \subset \mathbb{Q}[[x, y, z]]^3$ is the submodule generated by $f \cdot \mathbb{Q}[[x, y, z]]^3$, $g \cdot \mathbb{Q}[[x, y, z]]^3$, and the vectors $\left(\frac{\partial f}{\partial y}, \frac{\partial f}{\partial z}, 0\right)$, $\left(\frac{\partial f}{\partial x}, 0, -\frac{\partial f}{\partial z}\right)$, $\left(0, \frac{\partial f}{\partial x}, \frac{\partial f}{\partial y}\right)$, $\left(\frac{\partial g}{\partial y}, \frac{\partial g}{\partial z}, 0\right)$, $\left(\frac{\partial g}{\partial x}, 0, -\frac{\partial g}{\partial z}\right)$, $\left(0, \frac{\partial g}{\partial x}, \frac{\partial g}{\partial y}\right)$.

The latter quotient is isomorphic to $\Omega^2_{A|\mathbb{Q}}$, where $A := \mathbb{Q}[[x, y, z]]/\langle f, g\rangle$. The dimension computed in (1) is the so–called Tjurina number of A.[3]

---

[3] This is one counterexample showing that the exactness of the Poincaré complex of a complete intersection curve does not imply that the curve is quasihomogeneous. To see that $V(f, g)$ is not quasihomogeneous, it suffices to compare the Tjurina and Milnor number, which are different [109]. To see that the Poincaré complex is exact, it suffices to show that the Milnor number is equal to the difference of the dimensions of $\Omega^2_{A|\mathbb{Q}}$ and $\Omega^3_{A|\mathbb{Q}}$ [190].

7. Homological Algebra

7.1 Tor and Exactness

In Section 2.7, we saw that the tensor product is right exact, but in general not exact. We shall establish criteria for the exactness in terms of homology.

Let $\ldots \longrightarrow F_{i+1} \xrightarrow{\varphi_{i+1}} F_i \xrightarrow{\varphi_i} \ldots F_0 \xrightarrow{\varphi_0} N \longrightarrow 0$ be a free resolution of the A–module N, that is, the sequence is exact and the F_i are free A–modules. Then, for any A–module M, the induced sequence of A–modules and homomorphisms

$$\ldots \longrightarrow M \otimes_A F_{i+1} \xrightarrow{1_M \otimes \varphi_{i+1}} M \otimes_A F_i \xrightarrow{1_M \otimes \varphi_i} \ldots \longrightarrow M \otimes_A F_0 \longrightarrow 0$$

defines a complex $M \otimes_A F_\bullet$.

Definition 7.1.1. We introduce the A–modules $\operatorname{Tor}_i^A(M, N)$, which are called *Tor–modules*:

(1) $\operatorname{Tor}_0^A(M, N) := M \otimes_A N$;

(2) $\operatorname{Tor}_i^A(M, N) := \operatorname{Ker}(1_M \otimes \varphi_i)/\operatorname{Im}(1_M \otimes \varphi_{i+1})$ for $i \geq 1$.

This definition is independent of the chosen free resolution of N, and, for all i, $\operatorname{Tor}_i^A(M, N) \cong \operatorname{Tor}_i^A(N, M)$, see Exercises 7.1.2 and 7.1.1.

Proposition 7.1.2. *Let $0 \longrightarrow M \xrightarrow{i} N \xrightarrow{\pi} P \longrightarrow 0$ be an exact sequence of A–modules and L an A–module. Then, with the canonical induced maps, the sequence*

$$\ldots \longrightarrow \operatorname{Tor}_2^A(P, L) \longrightarrow \operatorname{Tor}_1^A(M, L) \longrightarrow \operatorname{Tor}_1^A(N, L) \longrightarrow \operatorname{Tor}_1^A(P, L) \longrightarrow$$
$$\longrightarrow M \otimes_A L \longrightarrow N \otimes_A L \longrightarrow P \otimes_A L \longrightarrow 0$$

is exact.

This sequence is called the *long exact Tor–sequence*.

Proof. Consider a free resolution

$$\ldots \longrightarrow F_{i+1} \xrightarrow{\varphi_{i+1}} F_i \xrightarrow{\varphi_i} \ldots \longrightarrow F_0 \xrightarrow{\varphi_0} L \longrightarrow 0.$$

Tensoring the latter with $0 \to M \to N \to P \to 0$, we obtain the commutative diagram:

$$
\begin{array}{ccccccccc}
& & 0 & & 0 & & 0 & & \\
& & \downarrow & & \downarrow & & \downarrow & & \\
\cdots \to & M \otimes_A F_{i+1} & \xrightarrow{1_M \otimes \varphi_{i+1}} & M \otimes_A F_i & \to \cdots \to & M \otimes_A F_0 & \to & M \otimes_A L & \to 0 \\
& & & \downarrow i \otimes 1_{F_i} & & \downarrow & & \downarrow & \\
\cdots \to & N \otimes_A F_{i+1} & \xrightarrow{1_N \otimes \varphi_{i+1}} & N \otimes_A F_i & \to \cdots \to & N \otimes_A F_0 & \to & N \otimes_A L & \to 0 \\
& & & \downarrow \pi \otimes 1_{F_i} & & \downarrow & & \downarrow & \\
\cdots \to & P \otimes_A F_{i+1} & \xrightarrow{1_P \otimes \varphi_{i+1}} & P \otimes_A F_i & \to \cdots \to & P \otimes_A F_0 & \to & P \otimes_A L & \to 0 \\
& & \downarrow & & \downarrow & & \downarrow & & \downarrow \\
& & 0 & & 0 & & 0 & & 0 .
\end{array}
$$

The rows of the diagram are complexes and, because of Theorem 2.7.6 and Exercise 2.7.8, the columns are exact. From the definition of Tor, we obtain for each i a sequence

$$\operatorname{Tor}_i^A(M, L) \longrightarrow \operatorname{Tor}_i^A(N, L) \longrightarrow \operatorname{Tor}_i^A(P, L) ,$$

which is, indeed, exact, as an easy diagram chase (left as an exercise) shows. Now the key point is to find maps $\gamma_i : \operatorname{Tor}_{i+1}^A(P, L) \to \operatorname{Tor}_i^A(M, L)$ such that

$$\operatorname{Tor}_{i+1}^A(N, L) \longrightarrow \operatorname{Tor}_{i+1}^A(P, L) \xrightarrow{\gamma_i} \operatorname{Tor}_i^A(M, L) \longrightarrow \operatorname{Tor}_i^A(N, L)$$

is exact.

Let $x \in \operatorname{Ker}(1_P \otimes \varphi_{i+1})$ and choose any preimage $y \in N \otimes_A F_{i+1}$ of x, that is, $(\pi \otimes 1_{F_{i+1}})(y) = x$. Then $(\pi \otimes 1_{F_i}) \circ (1_N \otimes \varphi_{i+1})(y) = 0$. Therefore, there exists $z \in M \otimes_A F_i$ such that $(1_N \otimes \varphi_{i+1})(y) = (i \otimes 1_{F_i})(z)$. Now

$$
\begin{aligned}
(i \otimes 1_{F_{i-1}}) \circ (1_M \otimes \varphi_i)(z) &= (1_N \otimes \varphi_i) \circ (i \otimes 1_{F_i})(z) \\
&= (1_N \otimes \varphi_i) \circ (1_N \otimes \varphi_{i+1})(y) = 0 .
\end{aligned}
$$

This implies that $z \in \operatorname{Ker}(1_M \otimes \varphi_i)$. We define the map γ_i by

$$\gamma_i\big(x + \operatorname{Im}(1_P \otimes \varphi_{i+2})\big) := z + \operatorname{Im}(1_M \otimes \varphi_{i+1}) .$$

It is left as an exercise to prove that γ_i is well–defined, that is, independent of the choice of x and z, and A–linear.

Finally, we have to prove the exactness of the above sequence. Let $\gamma_i\big(x + \operatorname{Im}(1_P \otimes \varphi_{i+2})\big) = 0$ that is, $z \in \operatorname{Im}(1_M \otimes \varphi_{i+1})$. Let $w \in M \otimes_A F_{i+1}$ be any preimage of z then

$$
\begin{aligned}
(1_N \otimes \varphi_{i+1}) \circ (i \otimes 1_{F_{i+1}})(w) &= (i \otimes 1_{F_i}) \circ (1_M \otimes \varphi_{i+1})(w) = (i \otimes 1_{F_i})(z) \\
&= (1_N \otimes \varphi_{i+1})(y) .
\end{aligned}
$$

This implies that $y - (i \otimes 1_{F_{i+1}})(w) \in \mathrm{Ker}(1_N \otimes \varphi_{i+1})$. On the other hand, by construction, $(\pi \otimes 1_{F_{i+1}})\big(y - (i \otimes 1_{F_{i+1}})(w)\big) = x$. So $y - (i \otimes 1_{F_{i+1}})(w)$ represents an element in $\mathrm{Tor}_{i+1}^A(N, L)$ which is mapped to $x + \mathrm{Im}(1_P \otimes \varphi_{i+2})$. This proves the exactness of the above sequence at $\mathrm{Tor}_{i+1}^A(P, L)$. The proof of the exactness at $\mathrm{Tor}_i^A(M, L)$ is similar. $\qquad\square$

The following proposition is the basis for computing $\mathrm{Tor}_i^A(M, N)$. Let

$$\cdots \longrightarrow F_{i+1} \xrightarrow{\varphi_{i+1}} F_i \xrightarrow{\varphi_i} \cdots \xrightarrow{\varphi_1} F_0 \xrightarrow{\varphi_0} N \longrightarrow 0$$

be a free resolution of the A–module N and

$$G_1 \xrightarrow{\psi} G_0 \xrightarrow{\pi} M \longrightarrow 0$$

a presentation of the A–module M. Then we obtain the following commutative diagram:

$$
\begin{array}{ccccccc}
& 0 & & 0 & & 0 & \\
& \uparrow & & \uparrow & & \uparrow & \\
\cdots \longrightarrow & F_{i+1} \otimes_A M & \longrightarrow & F_i \otimes_A M & \longrightarrow & F_{i-1} \otimes_A M & \longrightarrow \cdots \\
& \uparrow & & \uparrow & & \uparrow & \\
\cdots \longrightarrow & F_{i+1} \otimes_A G_0 & \xrightarrow{\varphi_{i+1} \otimes 1_{G_0}} & F_i \otimes_A G_0 & \xrightarrow{\varphi_i \otimes 1_{G_0}} & F_{i-1} \otimes_A G_0 & \longrightarrow \cdots \\
& \uparrow & & \uparrow{\scriptstyle 1_{F_i} \otimes \psi} & & \uparrow{\scriptstyle 1_{F_{i-1}} \otimes \psi} & \\
\cdots \longrightarrow & F_{i+1} \otimes_A G_1 & \longrightarrow & F_i \otimes_A G_1 & \longrightarrow & F_{i-1} \otimes_A G_1 & \longrightarrow \cdots
\end{array}
$$

Proposition 7.1.3. *With the above notations*

$$\mathrm{Tor}_i^A(M, N) = (\varphi_i \otimes 1_{G_0})^{-1} \mathrm{Im}(1_{F_{i-1}} \otimes \psi) \big/ \big(\mathrm{Im}(1_{F_i} \otimes \psi) + \mathrm{Im}(\varphi_{i+1} \otimes 1_{G_0})\big).$$

Proof. The columns, the second and third row of the above diagram are exact (Theorem 2.7.6 and Exercise 2.7.8). By definition,

$$\mathrm{Tor}_i^A(M, N) = \mathrm{Ker}(\varphi_i \otimes 1_M) / \mathrm{Im}(\varphi_{i+1} \otimes 1_M).$$

Now $1_{F_i} \otimes \pi$ maps $(\varphi_i \otimes 1_{G_0})^{-1}\big(\mathrm{Im}(1_{F_{i-1}} \otimes \psi)\big)$ surjectively to $\mathrm{Ker}(\varphi_i \otimes 1_M)$. Therefore, we have a surjection

$$(\varphi_i \otimes 1_{G_0})^{-1}\big(\mathrm{Im}(1_{F_{i-1}} \otimes \psi)\big) \longrightarrow \mathrm{Tor}_i^A(M, N).$$

Obviously, $\mathrm{Im}(1_{F_i} \otimes \psi) + \mathrm{Im}(\varphi_{i+1} \otimes 1_{G_0})$ is contained in the kernel of this surjection. An easy diagram chase shows that this is already the kernel. $\quad\square$

Example 7.1.4. Let $x \in A$ be a non–zerodivisor and M an A–module. Then

$$\operatorname{Tor}_0^A(A/\langle x \rangle, M) \cong M/xM \,,$$
$$\operatorname{Tor}_1^A(A/\langle x \rangle, M) = \{m \in M \mid xm = 0\} \,,$$
$$\operatorname{Tor}_i^A(A/\langle x \rangle, M) = 0 \text{ if } i \geq 2 \,.$$

In particular, $\operatorname{Tor}_1^{\mathbb{Z}}(\mathbb{Z}/\langle 2 \rangle, \mathbb{Z}/\langle 2 \rangle) = \mathbb{Z}/\langle 2 \rangle$.

Proof. $0 \to A \xrightarrow{x} A \longrightarrow A/\langle x \rangle \to 0$ is a free resolution of $A/\langle x \rangle$. Tensoring with M we obtain the complex

$$M \otimes_A F_\bullet : \qquad \ldots \longrightarrow 0 \longrightarrow M \xrightarrow{x} M \,,$$

hence, $\operatorname{Tor}_i^A(A/\langle x \rangle, M)$, $i \geq 1$, is as above. Moreover, by Exercise 2.7.5 we obtain $\operatorname{Tor}_0^A(A/\langle x \rangle, M) = A/\langle x \rangle \otimes_A M \cong M/xM$. $\square$

SINGULAR Example 7.1.5 (computation of Tor).

Using Proposition 7.1.3 we write a procedure `Tor` to compute $\operatorname{Tor}_i^A(M, N)$ for finitely generated A–modules M and N, presented by the matrices `Ps` and `Ph`. We compute $\operatorname{Im} := \operatorname{Im}(1_{F_{i-1}} \otimes \psi)$, $\mathtt{f} := \operatorname{Im}(\varphi_i \otimes 1_{G_0})$, $\operatorname{Im1} := \operatorname{Im}(1_{F_i} \otimes \psi)$, $\operatorname{Im2} := \operatorname{Im}(\varphi_{i+1} \otimes 1_{G_0})$, and obtain

$$\operatorname{Tor}_i^A(M, N) = \operatorname{Ker}\left(F_i \otimes_A G_0 \xrightarrow{\overline{\varphi_i \otimes 1_{G_0}}} (F_{i-1} \otimes_A G_0)/\operatorname{Im} \right) \Big/ \left(\operatorname{Im1} + \operatorname{Im2}\right)$$
$$= \mathtt{modulo(f,Im)}/(\operatorname{Im1} + \operatorname{Im2}) \,.$$

For the applied SINGULAR commands `modulo` and `prune`, we refer to the SINGULAR Examples 2.1.26 and 2.1.34.

```
proc Tor(int i, matrix Ps, matrix Ph)
{
  if(i==0){return(module(tensorMod(Ps,Ph)));}
                            // the tensor product
  list Phi  =mres(Ph,i+1);       // a resolution of Ph
  module Im =tensorMaps(unitmat(nrows(Phi[i])),Ps);
  module f  =tensorMaps(matrix(Phi[i]),unitmat(nrows(Ps)));
  module Im1=tensorMaps(unitmat(ncols(Phi[i])),Ps);
  module Im2=tensorMaps(matrix(Phi[i+1]),unitmat(nrows(Ps)));
  module ker=modulo(f,Im);
  module tor=modulo(ker,Im1+Im2);
  tor       =prune(tor);
  return(tor);
}
```

As an example, we want to check the formula (cf. Exercise 7.1.7)

$$\operatorname{Tor}_1^A(A/I, A/J) = (I \cap J)/(I \cdot J)$$

for $A = \mathbb{Q}[x,y]$, $I = \langle x^2, y \rangle$, $J = \langle x \rangle$.

```
ring A=0,(x,y),dp;
matrix Ps[1][2]=x2,y;
matrix Ph[1][1]=x;

Tor(1,Ps,Ph);
//-> _[1]=y*gen(1)
//-> _[2]=x*gen(1)
```

Hence, $\mathrm{Tor}_1^A(A/\langle x^2, y\rangle, A/\langle x\rangle) = A/\langle x, y\rangle \cong \mathbb{Q}$. Now, let's compute the right–hand side of the above formula:

```
ideal I1=intersect(ideal(Ps),ideal(Ph));
ideal I2=ideal(Ps)*ideal(Ph);

modulo(I1,I2);
//-> _[1]=gen(1)
//-> _[2]=y*gen(2)
//-> _[3]=x*gen(2)
```

From the output, we read $(I \cap J)/(I \cdot J) \cong A^2/(Ae_1 + \langle x, y\rangle e_2) \cong A/\langle x, y\rangle$. To obtain, immediately, a minimal embedding of the quotient, one should use the command **prune**.

```
prune(modulo(I1,I2));       // I1/I2 minimally embedded
//-> _[1]=y*gen(1)
//-> _[2]=x*gen(1)
```

Finally, we compute $\mathrm{Tor}_2^A(A/\langle x^2, y\rangle, A/\langle x\rangle)$ and $\mathrm{Tor}_0^A(A/\langle x^2, y\rangle, A/\langle x\rangle)$:

```
Tor(2,Ps,Ph);
//-> _[1]=0

Tor(0,Ps,Ph);
//-> _[1]=x*gen(1)
//-> _[2]=x2*gen(1)
//-> _[3]=y*gen(1)
```

Hence, $\mathrm{Tor}_2^A(A/\langle x^2, y\rangle, A/\langle x\rangle) = 0$ and $\mathrm{Tor}_0^A(A/\langle x^2, y\rangle, A/\langle x\rangle) \cong \mathbb{Q}$.

Remark 7.1.6. Using Exercise 7.1.8 we can use the previous procedure to compute $\mathrm{Tor}_i^A(M, B)$ with $A = K[x_1, \ldots, x_n]/I$, M a finitely generated A–module and $B = A[y_1, \ldots, y_m]/J$.

Exercises

7.1.1. Prove that $\mathrm{Tor}_i^A(M, N) \cong \mathrm{Tor}_i^A(N, M)$.

7.1.2. Prove that the definition of $\operatorname{Tor}_i^A(M,N)$ is independent of the chosen free resolution of N.

7.1.3. Let $\ldots \longrightarrow F_{i+1} \xrightarrow{\varphi_{i+1}} F_i \xrightarrow{\varphi_i} \ldots \longrightarrow F_0 \xrightarrow{\varphi_0} N \longrightarrow 0$ be a free resolution of the A–module N, and let M be an A–module. Consider the complex

$$0 \to \operatorname{Hom}_A(N,M) \to \operatorname{Hom}_A(F_0,M) \to \ldots \to \operatorname{Hom}_A(F_i,M) \to \ldots .$$

Similarly to the Tor–modules define the *Ext–modules*,

$$\operatorname{Ext}_A^0(N,M) = \operatorname{Hom}_A(N,M);$$
$$\operatorname{Ext}_A^i(N,M) = \operatorname{Ker}\bigl(\operatorname{Hom}(\varphi_{i+1},1_M)\bigr)/\operatorname{Im}\bigl(\operatorname{Hom}(\varphi_i,1_M)\bigr) \text{ for } i \geq 1.$$

(1) Prove that the definition of $\operatorname{Ext}_A^i(N,M)$ is independent of the chosen free resolution of N. Show by an example that $\operatorname{Ext}_A^i(N,M) \not\cong \operatorname{Ext}_A^i(M,N)$ in general.

(2) Let $0 \to M \to N \to P \to 0$ be an exact sequence of A–modules and L an A–module, then, with the canonical induced maps, the sequence

$$0 \to \operatorname{Hom}(P,L) \to \operatorname{Hom}(N,L) \to \operatorname{Hom}(M,L) \to \operatorname{Ext}_A^1(P,L) \to \cdots$$

is exact. This sequence is called the *long exact Ext–sequence*.

(2') With the assumptions of (2) the sequence

$$0 \to \operatorname{Hom}(L,M) \to \operatorname{Hom}(L,N) \to \operatorname{Hom}(L,P) \to \operatorname{Ext}_A^1(L,M) \to \cdots$$

is exact.

(3) Prove a result similar to Proposition 7.1.3.
(4) Prove that, for $i \geq 1$, $\operatorname{Ext}_A^i(N,M) = 0$ if N is free.
(5) Prove that $\operatorname{Ann}(N) \subset \operatorname{Ann}\bigl(\operatorname{Ext}_A^i(N,M)\bigr)$.

7.1.4. Write a SINGULAR programme to compute Ext (similar to Example 7.1.5).

7.1.5. Let $S \subset A$ be a multiplicatively closed set. Prove that

$$S^{-1}\operatorname{Tor}_i^A(M,N) = \operatorname{Tor}_i^{S^{-1}A}(S^{-1}M, S^{-1}N).$$

7.1.6. Let M be an A–module and $x \in A$ a non–zerodivisor for M and A. Moreover, let N be an $A/\langle x \rangle$–module. Prove that

$$\operatorname{Tor}_i^{A/\langle x \rangle}(N, M/xM) \cong \operatorname{Tor}_i^A(N,M).$$

7.1.7. Let $I, J \subset A$ be ideals. Prove that $\operatorname{Tor}_1^A(A/I, A/J) = (I \cap J)/(I \cdot J)$.

7.1.8. Let M be an A–module and $B = A[x_1, \ldots, x_n]/J$. Prove that

$$\operatorname{Tor}_i^A(M,B) \cong \operatorname{Tor}_i^{A[x_1,\ldots,x_n]}(M \otimes_A A[x_1,\ldots,x_n], B).$$

7.1.9. Let $(A,\mathfrak{m})$ be a local ring, $k = A/\mathfrak{m}$ the residue field and

$$\ldots \to F_n \to \ldots \to F_0 \to M \to 0$$

be a minimal free resolution of M. Prove that $\dim_k \operatorname{Tor}_i^A(M,k) = \operatorname{rank}(F_i)$ for all i; $\operatorname{rank}(F_i)$ is called the i–th *Betti number* of M (see Section 2.4).

7.2 Fitting Ideals

In this section we develop the theory of Fitting ideals as a tool for checking locally the freeness of a module. If a finitely generated A–module M is given by a finite presentation, then the k–th *Fitting ideal* $F_k(M)$ is the ideal generated by the $(n-k)$–minors of the presentation matrix, n being the number of rows of the matrix. The Fitting ideals do not depend on the presentation of the module. Moreover, they are compatible with base change. We shall prove that M is locally free of constant rank r if and only if $F_r(M) = A$ and $F_{r-1}(M) = 0$.

First of all we should like to generalize the notion of rank to arbitrary modules.

Definition 7.2.1. Let A be a ring and M be a finitely generated A–module. We say that M has *rank* r if $M \otimes_A Q(A)$ is a free $Q(A)$–module of rank r, where $Q(A)$ denotes the total quotient ring of A.

Remark 7.2.2. If A is an integral domain, then $Q(A)$ is a field and, hence, every finitely generated A–module M has a rank, namely

$$\mathrm{rank}(M) = \dim_{Q(A)} M \otimes_A Q(A).$$

For arbitrary rings A, not every finitely generated A–module needs to have a rank (cf. Exercise 7.2.4).

Lemma 7.2.3. *Let A be a Noetherian ring and M be a finitely generated A–module. M has rank r if and only if M_P is a free A_P–module of rank r for all prime ideals $P \in \mathrm{Ass}(\langle 0 \rangle)$.*

Proof. Using Exercise 3.3.10 we may assume $A = Q(A)$. If M is free of rank r then M_P is free of rank r for all prime ideals P.

Now assume that M_P is free of rank r for all prime ideals $P \in \mathrm{Ass}(\langle 0 \rangle)$. Using Proposition 2.1.38 we may assume that $r > 0$ and choose $x \in M$ such that $x \notin PM_P$ for all P (Exercise 2.1.22). Now x is an element of a minimal system of generators of M_P for all P. Using Nakayama's Lemma we obtain that x is an element of a basis of the free module M_P for all P. This implies that $(M/xA)_P$ is free of rank $r-1$ for all P. Using induction we may assume that M/xA is free of rank $r-1$. This implies $M \cong xA \oplus M/xA$ is free of rank r (Exercise 2.4.1). $\qquad\square$

Definition 7.2.4. Let A be a ring and M be an A–module with presentation $A^m \xrightarrow{\varphi} A^n \to M \to 0$. Assume that φ is defined by the matrix S for some choice of bases in A^m and A^n. For all k let $F_k(M) = F_k^A(M) \subset A$ be the ideal generated by the $(n-k)$–minors[1] of the matrix S, the k-th *Fitting ideal* of M. We use the convention that $F_k(M) = 0$ if $n - k > \min\{n, m\}$ and $F_k(M) = A$ if $k \geq n$.

---

[1] If S is a matrix with entries in A, then by a k–*minor* we denote the determinant of a $k \times k$–submatrix of S.

In particular, we obtain $F_r(M) = 0$ if $r < 0$, $F_0(M)$ is generated by the n–minors of S, and $F_{n-1}(M)$ is generated by all entries of S.

The following lemma justifies the definition:

Lemma 7.2.5. *With the notations of Definition 7.2.4, the following holds:*

(1) $F_k(M)$ is independent of the choice of the bases in A^m and A^n;
(2) $F_k(M)$ is independent of the presentation of M;
(3) let B be an A–algebra, then $F_k^B(M \otimes_A B) = F_k^A(M) \cdot B$.

Property (3) is usually expressed by saying that *Fitting ideals are compatible with base change*. Note that this implies that Fitting ideals are compatible with localization, that is, $F_k^{A_P}(M_P) = \left(F_k^A(M)\right)_P$ for any prime ideal $P \subset A$.

Proof. (1) It is enough to prove that for two matrices S and T with the property that T is obtained from S by adding the i–th row to the j–th row, the ideal generated by the r–minors of S is equal to the ideal generated by the r–minors of T. But this is easy to see, using the additivity of the determinant, and is, therefore, left as an exercise.

(2) Let $A^{m_1} \xrightarrow{\varphi_1} A^{n_1} \to M \to 0$ and $A^{m_2} \xrightarrow{\varphi_2} A^{n_2} \to M \to 0$ be two presentations. We have to prove that the ideal generated by the $(n_1 - k)$–minors of φ_1 is equal to the ideal generated by the $(n_2 - k)$–minors of φ_2. It is enough to prove this for the localizations in the prime ideals $P \subset A$ (Exercise 2.1.23). Therefore, we may assume that $(A, \mathfrak{m})$ is local. We first treat the case that $\mathrm{Ker}(\varphi_j) \subset \mathfrak{m} A^{m_j}$ and $\mathrm{Im}(\varphi_j) \subset \mathfrak{m} A^{n_j}$. Let $\{e_i\}_{1 \le i \le n_1}$ be a basis of A^{n_1}, and assume that $\{e_{i_1}, \ldots, e_{i_k}\}$ induces a basis of the $A/\mathfrak{m}$–vector space $M/\mathfrak{m}M$, then $\langle e_{i_1}, \ldots, e_{i_k}\rangle + \mathrm{Im}(\varphi_1) = A^{n_1}$ (Lemma of Nakayama). But $\mathrm{Im}(\varphi_1) \subset \mathfrak{m} A^{n_1}$ implies that $\langle e_{i_1}, \ldots, e_{i_k}\rangle = A^{n_1}$, again by the Lemma of Nakayama. So $\{e_i\}_{1 \le i \le n_1}$ induces a basis of $M/\mathfrak{m}M$. Choose now elements $\{a_i\}_{1 \le i \le n_1}$ in A^{n_2} inducing in $M/\mathfrak{m}M$ the same elements as $\{e_i\}_{1 \le i \le n_1}$. As before, we obtain $A^{n_2} = \langle a_1, \ldots, a_{n_1}\rangle$. This implies that $n_1 = n_2$ and that there exists an isomorphism ϕ making the diagram

$$
\begin{array}{ccc}
A^{n_1} & \longrightarrow & M \\
{\scriptstyle \cong}\big\downarrow{\scriptstyle \phi} & & \big\| \\
A^{n_2} & \longrightarrow & M
\end{array}
$$

commutative.

Similarly we prove $m_1 = m_2$ and the existence of an isomorphism ψ making the diagram

$$
\begin{array}{ccccccc}
A^{m_1} & \xrightarrow{\varphi_1} & A^{n_1} & \longrightarrow & M & \longrightarrow & 0 \\
{\scriptstyle \psi}\big\downarrow{\scriptstyle \cong} & & {\scriptstyle \cong}\big\downarrow{\scriptstyle \phi} & & \big\| & & \\
A^{m_2} & \xrightarrow{\varphi_2} & A^{n_2} & \longrightarrow & M & \longrightarrow & 0
\end{array}
$$

commutative. Now we use (1) and obtain that the ideal of the $(n_1 - k)$–minors of φ_1 is equal to the ideal of the $(n_2 - k)$–minors of φ_2. To prove the general case it is enough to prove that, given a presentation $A^{m_1} \xrightarrow{\varphi_1} A^{n_1} \to M \to 0$, there exists a presentation $A^{m_2} \xrightarrow{\varphi_2} A^{n_2} \to M \to 0$ such that

(1) $\mathrm{Ker}(\varphi_2) \subset \mathfrak{m}A^{m_2}$ and $\mathrm{Im}(\varphi_2) \subset \mathfrak{m}A^{n_2}$;
(2) the ideal generated by the $(n_1 - k)$–minors of φ_1 is equal to the ideal generated by the $(n_2 - k)$–minors of φ_2.

It is enough to prove that, for a suitable choice of the bases in A^{n_1} and A^{m_1}, the matrix corresponding to φ_1 is of type $\left(\begin{smallmatrix} L & 0 & 0 \\ 0 & E_s & 0 \end{smallmatrix} \right)$ such that the entries of L are in $\mathfrak{m}$, and E_s is the s–unit matrix. Then define the presentation $A^{m_2} \xrightarrow{\varphi_2} A^{n_2} \to M \to 0$ by the matrix L. Then, obviously, the $(n_2 - k)$–minors of L generate the same ideal as the $(n_1 - k)$–minors of $\left(\begin{smallmatrix} L & 0 & 0 \\ 0 & E_s & 0 \end{smallmatrix} \right)$.

If $\mathrm{Im}(\varphi_1) \not\subset \mathfrak{m}A^{n_1}$, then the matrix corresponding to φ_1 contains a unit. Using suitable bases, we obtain a matrix of type $\left(\begin{smallmatrix} L' & 0 \\ 0 & 1 \end{smallmatrix} \right)$. If $\mathrm{Ker}(\varphi_1) \not\subset \mathfrak{m}A^{m_1}$ we can find a new basis of A^{m_1} such that the matrix is of type $(L'' \ 0)$. Combining both and using induction we obtain the result claimed above.

(3) It is enough to prove the following: let $A^m \xrightarrow{\varphi} A^n \to M \to 0$ be a presentation of M, then $B^m \xrightarrow{\varphi \otimes 1_B} B^n \to M \otimes_A B \to 0$ is a presentation of $M \otimes_A B$. But this is a consequence of Theorem 2.7.6. □

SINGULAR Example 7.2.6 (Fitting ideals).

In this example we give a procedure to compute the Fitting ideals of a module M.

```
proc fitting(matrix M, int n)
{
  n=nrows(M)-n;
  if(n<=0){return(ideal(1));}
  if((n>nrows(M))||(n>ncols(M))){return(ideal(0));}
  return(std(minor(M,n)));
}
```

The procedure returns, indeed, a standard basis of $F_n(M)$. This is, of course, not necessary, but often a standard basis has less elements than the number of minors of the given size. We consider a concrete example:

```
ring R=0,x(0..4),dp;
matrix M[2][4]=x(0),x(1),x(2),x(3),x(1),x(2),x(3),x(4);
print(M);
//-> x(0),x(1),x(2),x(3),
//-> x(1),x(2),x(3),x(4)

fitting(M,-1);
//-> _[1]=0
```

```
fitting(M,0);
//-> _[1]=x(3)^2-x(2)*x(4)      _[2]=x(2)*x(3)-x(1)*x(4)
//-> _[3]=x(1)*x(3)-x(0)*x(4)   _[4]=x(2)^2-x(0)*x(4)
//-> _[5]=x(1)*x(2)-x(0)*x(3)   _[6]=x(1)^2-x(0)*x(2)

fitting(M,1);
//-> _[1]=x(4) _[2]=x(3) _[3]=x(2) _[4]=x(1) _[5]=x(0)

fitting(M,2);
//-> _[1]=1
```

Let us proceed to the main result of this section.

Theorem 7.2.7. *Let A be a local ring and M be an A-module of finite presentation. The following conditions are equivalent:*

(1) M is a free module of rank r;
(2) $F_r(M) = A$ and $F_{r-1}(M) = 0$.

Proof. That (1) implies (2) is a trivial consequence of Definition 7.2.4. To prove that (2) implies (1), let $F_r(M) = A$ and $F_{r-1}(M) = 0$, and choose a presentation $A^m \to A^n \to M \to 0$ with presentation matrix S (with respect to some bases of A^m and A^n). Then either $n = r$ and S is the zero matrix, or $n > r$, one $(n - r)$–minor of S is a unit (A is a local ring) and all $(n-r+1)$–minors of S vanish. If $n = r$ and S is the zero matrix, then, obviously, M is free of rank r. In the second case, one $(n - r)$–minor is a unit, so we can choose new bases of A^m and A^n such that the presentation matrix is of type $\left(\begin{smallmatrix} E_{n-r} & 0 \\ 0 & C \end{smallmatrix} \right)$, E_{n-r} the $(n - r)$–unit matrix. Because all $(n-r+1)$–minors are zero, we obtain, indeed, $C = 0$. This implies that M is free and isomorphic to the submodule of A^n generated by the vectors $e_{n-r+1}, \ldots, e_n$, $(e_j = (0, \ldots, 1, \ldots, 0)$, with 1 at the j–th place). $\square$

Corollary 7.2.8. *Let A be a ring and M an A-module of finite presentation. The following conditions are equivalent:*

(1) M is locally free[2] of constant rank r;
(2) $F_r(M) = A$ and $F_{r-1}(M) = 0$.

Proof. The result is an immediate consequence of Theorem 7.2.7 and Lemma 7.2.5 (3). $\square$

Corollary 7.2.9. *Let $\varphi : A^m \to A^N$ be a homomorphism, and let S be a matrix of φ with respect to some bases of A^m and A^n. Then φ is surjective if and only if there exists an n–minor of S which is a unit in A.*

---

[2] By definition, an A–module M is *locally free* if the localization M_P of M at every maximal ideal (or, equivalently, at every prime ideal) $P \subset A$ is a free A_P–module.

Proof. Apply Corollary 7.2.8 to $\mathrm{Coker}(\varphi)$.   □

SINGULAR Example 7.2.10 (test for local freeness).

The following procedure tests whether an A–module M, given by a presentation matrix $S \in \mathrm{Mat}(n \times m, A)$ (A the current basering), is locally free of constant rank r. The procedure uses that `fitting` returns a standard basis for the respective Fitting ideal, and that SINGULAR returns 1 if this standard basis contains a unit.

```
proc isLocallyFree(matrix S, int r)
{
    ideal F=fitting(S,r);
    ideal G=fitting(S,r-1);
    if((deg(F[1])==0)&&(size(G)==0)){return(1);}
    return(0);
}
```

The procedure returns 1 if the module M is locally free of rank r, and 0 otherwise. We apply this procedure to $A = \mathbb{Q}[x, y, z]$ and the A–module M given by the presentation

$$\mathbb{Q}[x,y,z]^3 \xrightarrow{\left(\begin{smallmatrix} x-1 & y-1 & z \\ y-1 & x-2 & x \end{smallmatrix}\right)} \mathbb{Q}[x,y,z]^2 \longrightarrow M \longrightarrow 0 \,.$$

```
ring R=0,(x,y,z),dp;
matrix S[2][3];              //the presentation matrix
S=x-1,y-1,z,y-1,x-2,x;
ideal I=fitting(S,0);
```

By Exercise 7.2.5, M has also the structure of a $Q := \mathbb{Q}[x, y, z]/F_0(M)$–module. In the following, we change the basering to Q and consider M as a Q–module:

```
qring Q=I;
matrix S=fetch(R,S);
isLocallyFree(S,1);
//-> 1
```

Hence, as Q–module, M is locally free of rank 1. Just to check, let us show that $M \neq 0$:

```
isLocallyFree(S,0);
//-> 0
```

Exercises

7.2.1. Let A be a ring, let M be an A–module of finite presentation, and assume that M can be generated by n elements. Show that

$$0 = F_{-1}(M) \subset F_0(M) \subset F_1(M) \subset F_2(M) \subset \ldots \subset F_n(M) = A \,.$$

7.2.2. Let M be a matrix with rows $m_1, \ldots, m_k$ and N be the matrix with rows $m_1, \ldots, m_{j-1}, m_j + m_i, m_{j+1}, \ldots, m_k$ for $i \neq j$. Prove that the ideal generated by the r–minors of M is equal to the ideal generated by the r–minors of N.

7.2.3. Let M, respectively N, be the $K[x,y]$–modules defined by the presentation matrix $\begin{pmatrix} x & y \\ 0 & y^2 \end{pmatrix}$, respectively $\begin{pmatrix} x & 0 \\ y & y^2 \end{pmatrix}$. Prove that M and N are not isomorphic but have the same Fitting ideals.

7.2.4. Let $A = K[x,y]/\langle xy \rangle$ and $M = \langle x \rangle$. Prove that M does not have a rank.

7.2.5. Let A be a ring and M an A–module of finite presentation. Prove that $F_0(M) \subset \mathrm{Ann}(M)$ and $\sqrt{F_0(M)} = \sqrt{\mathrm{Ann}(M)}$. More precisely, if M can be generated by n elements, then show that $(\mathrm{Ann}(M))^n \subset F_0(M)$. Conclude that $\mathrm{supp}(M) = V(F_0(M))$ (cf. Definition 2.1.40 and Lemma 2.1.41).

7.2.6. Let A be a ring, and let $0 \to M' \to M \to M'' \to 0$ be an exact sequence of A–modules of finite presentation. Show that, for each $k, r \geq 0$,

$$F_k(M') \cdot F_r(M'') \subset F_{k+r}(M) \,.$$

7.2.7. Let A be a ring, and let M, N be A–modules of finite presentation. Show that, for each k,

$$F_k(M \oplus N) = \sum_{r+s=k} F_r(M) \cdot F_s(N) \,.$$

7.2.8. Let A be a ring, and let $I_1, \ldots, I_s \subset A$ be ideals. Prove that

$$F_0(A/I_1 \oplus \ldots \oplus A/I_s) = I_1 \cdot \ldots \cdot I_s \,.$$

7.3 Flatness

Definition 7.3.1. Let A be a ring. An A–module M is called *flat* if, for every injective homomorphism $i : N \to L$, the induced map $N \otimes_A M \to L \otimes_A M$ is again injective. An A–algebra B is called flat if it is flat as A–module.

Example 7.3.2. Free modules are flat (Exercise 2.7.7).

The following theorem gives a characterization of flatness using Tor.

Theorem 7.3.3. *Let A be a ring and M an A–module. Then M is flat if and only if $\mathrm{Tor}_1^A(A/I, M) = 0$ for all finitely generated ideals $I \subset A$.*

Proof. Consider the following exact sequence

$$0 \longrightarrow I \longrightarrow A \longrightarrow A/I \longrightarrow 0 \,.$$

We obtain a long exact sequence using Proposition 7.1.2:

$$0 = \operatorname{Tor}_1^A(A, M) \longrightarrow \operatorname{Tor}_1^A(A/I, M) \longrightarrow I \otimes_A M \longrightarrow A \otimes_A M = M \,.$$

If M is flat then $I \otimes_A M \to M$ is injective and, therefore, $\operatorname{Tor}_1^A(A/I, M) = 0$.

Now assume $\operatorname{Tor}_1^A(A/I, M) = 0$ for all finitely generated ideals $I \subset A$. We have to prove that, for any injection $i : N \to L$, the induced map $N \otimes_A M \to L \otimes_A M$ is injective.

We consider first the case $N = I$, $L = A$ and $I \subset A$ an ideal (but not necessarily finitely generated). If $I \otimes_A M \to M$ is not injective, then there exists $\sum_{\nu=1}^r a_\nu \otimes m_\nu \in I \otimes_A M$ different from zero with $\sum_{\nu=1}^r a_\nu m_\nu = 0$. Let $I_0 := \langle a_1, \ldots, a_r \rangle$ then $\sum_{\nu=1}^r a_\nu \otimes m_\nu \in I_0 \otimes_A M$ and, therefore, by assumption, it has to be zero. In particular, its image in $I \otimes_A M$ has to be zero, too. This implies that $I \otimes_A M \to M$ is injective for all ideals $I \subset A$.

Similarly, we may assume that L is finitely generated (the statement that an element from $N \otimes_A M$ is mapped to zero in $L \otimes_A M$ involves only finitely many elements in L). So we can find a chain

$$N = N_0 \subset N_1 \subset \cdots \subset N_r = L$$

of A–modules such that each quotient N_{i+1}/N_i is generated by one element. Since it is enough to prove that, for all i, $N_i \otimes_A M \to N_{i+1} \otimes_A M$ is injective, we have reduced the statement to the case that $L/N \cong A/I$. Now, consider the exact sequence

$$\operatorname{Tor}_1^A(A/I, M) = \operatorname{Tor}_1^A(L/N, M) \longrightarrow N \otimes_A M \longrightarrow L \otimes_A M \,.$$

By assumption, $\operatorname{Tor}_1^A(A/I, M) = 0$, which implies that $N \otimes_A M \to L \otimes_A M$ is injective. $\qquad\square$

Now we use Proposition 2.7.10 to give another criterion for flatness, in terms of equations in M.

Proposition 7.3.4. *Let A be a ring, and M an A–module. Then M is flat if and only if the following condition is satisfied: let $\sum_{i=1}^r a_i m_i = 0$, $a_i \in A$, $m_i \in M$, then there exist $a_{ij} \in A$, $\tilde{m}_j \in M$ such that*

(1) $\sum_{j=1}^s a_{ij} \tilde{m}_j = m_i$ *for all $i = 1, \ldots, r$,*

(2) $\sum_{i=1}^r a_{ij} a_i = 0$ *for all $j = 1, \ldots s$.*

Proof. Assume that M is flat. Let $\sum_{i=1}^r a_i m_i = 0$, $a_i \in A$, $m_i \in M$ and set $I := \langle a_1, \ldots, a_r \rangle$. Since M is flat, the map induced by $I \subset A$, $I \otimes_A M \to M$, is injective. This implies $\sum_{i=1}^r a_i \otimes m_i = 0$, and the result follows by Proposition 2.7.10.

On the other hand, assume that the condition above is satisfied, and let $I \subset A$ be a finitely generated ideal. By Theorem 7.3.3, it suffices to prove that

$\mathrm{Tor}_1^A(A/I, M) = 0$, or, equivalently, that the induced map $I \otimes_A M \to M$ is injective. Let $\sum_{i=1}^r a_i \otimes m_i \in I \otimes_A M$ and $\sum_i a_i m_i = 0$. Using the condition above, we obtain $a_{ij} \in A$, $\tilde{m}_i \in M$ such that $\sum_{j=1}^s a_{ij}\tilde{m}_j = m_i$ for all i and $\sum_{i=1}^r a_{ij}a_i = 0$ for all j. Thus, $\sum_i a_i \otimes m_i = \sum_j (\sum_i a_{ij}a_i) \otimes \tilde{m}_j = 0$.   □

Let $I = \langle a \rangle \subset A$ be a principal ideal. Then the preceding proof shows that the induced map $\langle a \rangle \otimes_A M \to M$ is injective if and only if the following holds: $am = 0$ for $m \in M$ implies that there exist $a_1, \ldots, a_s \in A$, $\tilde{m}_1, \ldots, \tilde{m}_s \in M$ such that $m = \sum_{i=1}^s a_i \tilde{m}_i$ and $aa_i = 0$ for all i.

In other words, $\langle a \rangle \otimes_A M \to M$ is injective if and only if

$$\mathrm{Ann}_M(a) := \{ m \in M \mid am = 0 \} \subset \mathrm{Ann}_A(a) \cdot M.$$

Since the other inclusion is obvious, we have shown

Corollary 7.3.5. *Let A be a principal ideal domain. Then an A–module M is flat if and only if, for every $a \in A$,*

$$\mathrm{Ann}_M(a) = \mathrm{Ann}_A(a) \cdot M.$$

If, moreover, A is integral, then M is flat if and only if it is torsion free.

If we apply Corollary 7.3.5 to the ring $A = K[\varepsilon] := K[t]/\langle t^2 \rangle$, where K is a field and ε the class of t mod $\langle t^2 \rangle$, then we obtain

Corollary 7.3.6. *A $K[\varepsilon]$–module M is flat if and only if $\mathrm{Ann}_M(\varepsilon) = \varepsilon M$, that is, the multiplication by ε induces an isomorphism $M/\varepsilon M \cong \varepsilon M$.*

The above considerations imply also

Corollary 7.3.7. *Let $a \in A$ be a non–zerodivisor of A, and let M be a flat A–module, then a is a non–zerodivisor for M.*

Corollary 7.3.8. *Let A be a local ring with maximal ideal $\mathfrak{m}$ and M a flat A–module. Moreover, let $m_1, \ldots, m_k \in M$, such that their classes $\overline{m}_1, \ldots, \overline{m}_k$ in $M/\mathfrak{m}M$ are linearly independent. Then $m_1, \ldots, m_k$ are linearly independent.*

In particular, a finitely generated A–module is flat if and only if it is free.

Proof. We use induction on k. Let $k = 1$ and assume $am_1 = 0$ for some $a \in A$. Using Proposition 7.3.4 we obtain $\tilde{m}_j \in M$, $a_j \in A$ such that $\sum_j a_j \tilde{m}_j = m_1$ and $aa_j = 0$ for all j. But $m_1 \notin \mathfrak{m}M$ implies $a_j \notin \mathfrak{m}$ for some j and, therefore, $a = 0$.

Assume the corollary is proved for $k - 1$. Let $\sum_{j=1}^k a_j m_j = 0$. We use Proposition 7.3.4 again and obtain $\tilde{m}_j \in M$, $a_{ij} \in A$ such that $m_i = \sum_j a_{ij}\tilde{m}_j$ and $\sum_i a_{ij}a_i = 0$ for all $i = 1, \ldots, k$, respectively $j = 1, \ldots, k$.

Because $m_k \notin \mathfrak{m}M$, we have $a_{kj} \notin \mathfrak{m}$ (and, hence, a unit) for some j. This implies that a_k is a linear combination of $a_1, \ldots, a_{k-1}$, $a_k = \sum_{\nu=1}^{k-1} h_\nu a_\nu$ for $h_\nu := -a_{\nu j}/a_{kj}$. Now

$$\sum_{\nu=1}^{k-1} a_\nu(m_\nu + h_\nu m_k) = a_k m_k + \sum_{\nu=1}^{k-1} a_\nu m_\nu = 0.$$

The induction hypothesis implies that $a_1 = \ldots = a_{k-1} = 0$ and, therefore, $a_k = 0$. □

Proposition 7.3.9. *Let A be a Noetherian ring and M an A–module of finite presentation. Let $F_i(M)$ be the i–th Fitting ideal of M. M is flat and has rank r if and only if $F_r(M) = A$ and $F_{r-1}(M) = 0$.*

Proof. We use the fact that M is flat if and only if it is locally flat, which will be proved in Proposition 7.4.1, below. Then the result is an immediate consequence of Corollaries 7.2.8, 7.3.8 and Lemma 7.2.3. □

Corollary 7.3.10. *Let A be a ring, M an A–module of finite presentation and $\rho : A \to B$ an A–algebra. $M \otimes_A B$ is a flat B–module of rank r if and only if $F_{r-1}(M) \subset \mathrm{Ker}(\rho)$ (that is, B is already an $A/F_{r-1}(M)$–algebra) and $F_r(M) \cdot B = B$.*

In particular, for any prime ideal $P \subset A$, such that $P \supset F_{r-1}(M)$ and $P \not\supset F_r(M)$, the module $M_P/(F_{r-1}(M)_P \cdot M_P)$ is a flat $A_P/F_{r-1}(M)_P$–module of rank r.

Proof. The corollary is an immediate consequence of Proposition 7.3.9 and the fact that the Fitting ideals are compatible with base change, see Lemma 7.2.5. □

Definition 7.3.11. For $r \geq 0$, the locally closed [3] subset of $\mathrm{Spec}(A)$,

$$\mathrm{Flat}_r(M) := \{P \subset A \text{ prime ideal} \mid P \supset F_{r-1}(M) \text{ and } P \not\supset F_r(M)\},$$

considered as an open subset in $\mathrm{Spec}(A/F_{r-1}(M))$, is called the *flattening stratum* of rank r for M. The collection $\{\mathrm{Flat}_r(M) \mid r \geq 0\}$ is called the *flattening stratification* of M.

To summarize the results of the discussion so far, we use the geometric language of Appendix A. Let $A^m \to A^n \to M \to 0$ be a presentation of M, and let $F_k = F_k(M)$ be the k–th Fitting ideal of M, $k \geq -1$. Then we have the inclusions (cf. Exercise 7.2.1)

$$\mathrm{Spec}(A) = V(F_{-1}) \supset V(F_0) \supset V(F_1) \supset V(F_2) \supset \ldots \supset V(F_n) = \emptyset$$

of closed subvarieties, respectively subschemes, of $\mathrm{Spec}(A)$. Moreover,

$$\mathrm{Flat}_r(M) = V(F_{r-1}) \smallsetminus V(F_r)$$

---

[3] A subset of $\mathrm{Spec}(A)$ is called *locally closed* if it is the intersection of a Zariski open and a Zariski closed subset. For the definition of the Zariski topology on $\mathrm{Spec}(A)$, see Appendix A, Definition A.3.1.

is locally closed in $\mathrm{Spec}(A)$, and $\mathrm{Spec}(A)$ is the disjoint union of the $\mathrm{Flat}_i(M)$, $i = 0, \ldots, n$.

By Exercise 7.2.5 and Lemma 2.1.40, $V(F_0)$ is the support of the module M, and $\mathrm{Flat}_0(M)$ is the open complement of the support. If we restrict M to $V(F_0)$, that is, if we pass to the A/F_0–module $M \otimes_A A/F_0$, then $\mathrm{Flat}_1(M)$ is exactly the set of all $P \in V(F_0)$ such that $(M \otimes_A A/F_0)_P$ is a flat (that is, free, by Corollary 7.3.8) $(A/F_0)_P$–module of rank 1.

Again, $\mathrm{Flat}_1(M)$ is an open subset of $V(F_0)$. Now we continue, restricting M to $V(F_1)$, and obtain that $\mathrm{Flat}_2(M)$ is the set of all $P \in V(F_1)$ such that $(M \otimes_A A/F_1)_P$ is a flat $(A/F_1)_P$–module of rank 2, which is an open subset of $V(F_1)$, etc.. In general, we have

$$P \in \mathrm{Flat}_r(M) \iff (M \otimes_A A/F_{r-1})_P \text{ is } (A/F_{r-1})_P\text{–flat of rank } r,$$

whence the name "flattening stratification".

Note that $P \in \mathrm{Flat}_r(M)$ does *not* mean that the localization M_P is A_P–flat (of rank r). The set of all P such that M_P is A_P–flat, the so–called *flat locus* of M, will be described in Corollary 7.3.13, below.

SINGULAR Example 7.3.12 (flattening stratification).

In this example we give a procedure which computes the non–trivial Fitting ideals of a module given by the presentation matrix M. The result is a list L of ideals $I_j = L[j]$, $j = 1, \ldots, r$, with the following property: set $I_0 := \langle 0 \rangle$, $I_{r+1} := \langle 1 \rangle$, then the flattening stratification of $\mathrm{Coker}(M)$ is given by the open sets defined as the complement of $V(I_j)$ in $\mathrm{Spec}(A/I_{j-1})$ for $j = 1, \ldots, r + 1$. Note that $\mathrm{Coker}(M)$ is flat if L is the empty list.

```
proc flatteningStrat(matrix M)
{
    list l;
    int v,w;
    ideal F;
    while(1)
    {
        F=interred(fitting(M,w));
        if(F[1]==1){return(l);}
        if(size(F)!=0){v++;l[v]=F;}
        w++;
    }
    return(l);
}
```

We apply this procedure to $A = \mathbb{Q}[x_0, \ldots, x_4]$ and the module N given by the presentation

$$A^4 \xrightarrow{\begin{pmatrix} x_0 & x_1 & x_2 & x_3 \\ x_1 & x_2 & x_3 & x_4 \end{pmatrix}} A^2 \longrightarrow N \longrightarrow 0.$$

```
ring A = 0,x(0..4),dp;
// the presentation matrix of N:
matrix M[2][4] = x(0),x(1),x(2),x(3),x(1),x(2),x(3),x(4);

flatteningStrat(M);
//-> [1]:                                    [2]:
//->     _[1]=x(3)^2-x(2)*x(4)                   _[1]=x(4)
//->     _[2]=x(2)*x(3)-x(1)*x(4)                _[2]=x(3)
//->     _[3]=x(1)*x(3)-x(0)*x(4)                _[3]=x(2)
//->     _[4]=x(2)^2-x(0)*x(4)                   _[4]=x(1)
//->     _[5]=x(1)*x(2)-x(0)*x(3)                _[5]=x(0)
//->     _[6]=x(1)^2-x(0)*x(2)
```

From the output, we read that the flattening stratification of N is given as follows: let

$$I := \langle x_3^2 - x_2 x_4,\ x_2 x_3 - x_1 x_4,\ x_1 x_3 - x_0 x_4,$$
$$x_2^2 - x_0 x_4,\ x_1 x_2 - x_0 x_3,\ x_1^2 - x_0 x_2 \rangle.$$

Then we have

$$\mathrm{Flat}_0(N) = \{P \subset A \text{ prime ideals} \mid P \not\supset I\} = \mathbb{A}^5 \smallsetminus V(I),$$
$$\mathrm{Flat}_1(N) = \{P \subset A/I \text{ prime ideals} \mid P \neq \langle x_0, \dots, x_4 \rangle\} = V(I) \smallsetminus \{0\},$$
$$\mathrm{Flat}_2(N) = \{\langle 0 \rangle \subset A/\langle x_0, \dots, x_4 \rangle = \mathbb{Q}\} = \{0\},$$

with $\{0\} = V(x_0, \dots, x_4) \subset \mathbb{A}^5$.

Corollary 7.3.13 (flat locus is open). *Let A be a ring, M an A–module of finite presentation and*

$$F(M) := \sum_\nu \big((0 : F_{\nu-1}(M)) \cap F_\nu(M) \big).$$

Then

(1) M is flat if and only if $F(M) = A$;
(2) let $P \subset A$ be a prime ideal, then M_P is flat if and only if $P \not\supset F(M)$.

Proof. Due to Proposition 7.4.1, below, (1) is a special case of (2). To prove (2), let $A^\ell \xrightarrow{\varphi} A^s \to M \to 0$ be a presentation of M. Then $F_\nu(M)$ is the ideal generated by the $(s - \nu)$–minors of φ with the convention $F_\nu(M) = A$ if $\nu \geq s$ and $F_\nu(M) = 0$ if $\nu < 0$. We have $F(M) = \sum_{\nu=0}^s \big((0 : F_{\nu-1}(M)) \cap F_\nu(M) \big)$. Let $P \subset A$ be a prime ideal. Because of Proposition 7.3.9, M_P is a free A_P–module of rank r if and only if $F_r(M)_P = A_P$ and $F_{r-1}(M)_P = 0$. Now $F_{r-1}(M)_P = 0$ if and only if there exists $h \notin P$ such that $hF_{r-1}(M) = 0$, that is, $P \not\supset \big(0 : F_{r-1}(M) \big)$. This implies that M_P is a free A_P–module of rank

r if and only if $P \not\supset (0 : F_{r-1}(M)) \cap F_r(M)$. So M_P is a free A_P–module implies $P \not\supset F(M)$. Now assume $P \not\supset F(M)$, then there exists r such that $(0 : F_{r-1}(M)) \cap F_r(M) \not\subset P$. This implies that M_P is a free A_P–module. $\square$

The geometric interpretation of Corollary 7.3.13 and its proof is the following: set $F_r := F_r(M)$, then

$$\mathrm{Spec}(A) \smallsetminus V((0 : F_{r-1}) \cap F_r) = \{P \in \mathrm{Spec}(A) \mid M_P \text{ is } A_P\text{–flat of rank } r\}$$

is the open subset of $\mathrm{Spec}(A)$, where M is flat of rank r, and

$$\mathrm{Flat}(M) := \bigcup_r \Big(\mathrm{Spec}(A) \smallsetminus V((0 : F_{r-1}) \cap F_r) \Big) = \mathrm{Spec}(A) \smallsetminus V(F(M))$$

is the flat locus (or locally free locus) of M.

SINGULAR Example 7.3.14 (test for flatness).

We use Proposition 7.3.9 to write a procedure to test the flatness of a finitely generated module defined by a presentation matrix M.

```
proc isFlat(matrix M)
{
   if (size(ideal(M))==0) {return(1);}
   int w;
   ideal F=fitting(M,0);
   while(size(F)==0)
   {
      w++;
      F=fitting(M,w);
   }
   if (deg(std(F)[1])==0) {return(1);}
   return(0);
}
```

The procedure returns 1 if $\mathrm{Coker}(M)$ is flat, and 0 otherwise. Let us try an example:

```
ring A=0,(x,y),dp;
matrix M[3][3]=x-1,y,x,x,x+1,y,x2,xy+x+1,x2+y;
print(M);
//-> x-1,y,     x,
//-> x,  x+1,   y,
//-> x2, xy+x+1,x2+y
isFlat(M);
//-> 0                      // coker(M) is not flat over A=Q[x,y]

qring B=std(x2+x-y);   // the ring B=Q[x,y]/<x2+x-y>
```

```
matrix M=fetch(A,M);
isFlat(M);
//-> 1                    // coker(M) is flat over B

setring A;
qring C=std(x2+x+y);    // the ring C=Q[x,y]/<x2+x+y>
matrix M=fetch(A,M);
isFlat(M);
//-> 0                    // coker(M) is not flat over C
```

Similarly, we can compute the flat locus. The procedure flatLocus returns an ideal I such that the complement of $V(I)$ is the flat locus of $\mathrm{Coker}(M)$.

```
proc flatLocus(matrix M)
{
    if (size(ideal(M))==0) {return(ideal(1));}
    int v,w;
    ideal F=fitting(M,0);
    while(size(F)==0)
    {
        w++;
        F=fitting(M,w);
    }
    if(typeof(basering)=="qring")
    {
        for(v=w+1;v<=nrows(M);v++)
        {
            F=F+intersect(fitting(M,v),quotient(ideal(0),
            fitting(M,v-1)));
        }
    }
    return(interred(F));
}
```

As an example, let us compute the flat locus of the $\mathbb{Q}[x,y,z]/\langle xyz\rangle$–module N given by the presentation

$$(\mathbb{Q}[x,y,z]/\langle xyz\rangle)^3 \xrightarrow{\left(\begin{smallmatrix} x & y & z \\ 0 & x^3 & z^3 \end{smallmatrix}\right)} (\mathbb{Q}[x,y,z]/\langle xyz\rangle)^2 \longrightarrow N \longrightarrow 0\,.$$

```
ring R=0,(x,y,z),dp;
qring S=std(xyz);
matrix M[2][3]=x,y,z,0,x3,z3;
flatLocus(M);
//-> _[1]=xz3  _[2]=x3z-yz3  _[3]=x4  _[4]=y2z3  _[5]=yz5
```

Hence, $N = \mathrm{Coker}(M)$ is flat outside $V(x,yz)$.

There is another possibility to describe the flat locus.

Proposition 7.3.15. *Let A be a Noetherian ring and M a finitely generated A–module. Assume that $0 \to K \to A^p \xrightarrow{\pi} M \to 0$ is exact, that is, $K = \mathrm{syz}_1(M)$. Then M is flat if and only if $\mathrm{Ext}_A^1(M, K) = 0$.*

Proof. Because of Proposition 7.4.1, below, and the fact that Ext is compatible with localizations, we may assume that A is local. If M is flat, then M is free and, therefore, $\mathrm{Ext}_A^1(M, K) = 0$. Now assume that $\mathrm{Ext}_A^1(M, K) = 0$. We have the following exact sequence

$$0 \to \mathrm{Hom}_A(M, K) \to \mathrm{Hom}_A(M, A^p) \xrightarrow{\pi \circ -} \mathrm{Hom}_A(M, M) \to \underbrace{\mathrm{Ext}_A^1(M, K)}_{=0}\,.$$

In particular, $\mathrm{id}_M \in \mathrm{Hom}_A(M, M)$ has a preimage $i \in \mathrm{Hom}_A(M, A^p)$, that is, $\pi : A^p \to M$ has a section $i : M \to A^p$ satisfying $\pi \circ i = \mathrm{id}_M$. Therefore, M is, as a direct summand of A^p, a free module (Exercise 2.1.18). $\square$

Corollary 7.3.16. *Let A be a Noetherian ring and M a finitely generated A–module given by the exact sequence $0 \to K \to A^P \to M \to 0$. Then*

$$\{P \subset A \text{ prime} \mid M_P \text{ is } A_P\text{–flat}\} = \{P \subset A \text{ prime} \mid P \not\supset \mathrm{Ann}\big(\mathrm{Ext}_A^1(M, K)\big)\},$$

that is, $\mathrm{Flat}(M) = \mathrm{Spec}(A) \smallsetminus V\big(\mathrm{Ann}\big(\mathrm{Ext}_A^1(M, K)\big)\big).$

SINGULAR Example 7.3.17 (flat locus).

We use Corollary 7.3.16 for a procedure `flatLocus1`, returning an ideal such that the complement of its zero–set is the flat locus of $\mathrm{Coker}(M)$.

```
LIB"homolog.lib";
LIB"primdec.lib";
proc flatLocus1(matrix M)
{
   list l=mres(M,2);
   return(Ann(Ext(1,M,l[2])));
}
ring R=0,(x,y,z),dp;
qring S=std(xyz);
matrix M[2][3]=x,y,z,0,x3,z3;
print(M);
//-> x,y , z,
//-> 0,x3,z3

flatLocus1(M);
//-> _[1]=xz3  _[2]=x3z-yz3  _[3]=x4  _[4]=y2z3  _[5]=yz5
```

Exercises

7.3.1. Let M be a flat A–module and $S \subset A$ a multiplicatively closed subset. Prove that $M \otimes_A S^{-1}A \cong S^{-1}M$ via $m \otimes \frac{a}{b} \mapsto \frac{ma}{b}$ and that $S^{-1}M$ is a flat $S^{-1}A$–module. Prove that $S^{-1}A$ is a flat A–module.

7.3.2. Prove that $M \oplus N$ is a flat A–module if and only if M, N are flat. More generally, let $\{M_i\}_{i \in I}$ be a family of flat A–modules. Prove that $\oplus_{i \in I} M_i$ is flat; especially $A[x_1, \ldots, x_n]$ is a flat A–algebra.

7.3.3. An A–module M is called *projective* if, for every surjective homomorphism $\pi : N \to L$, the induced map $\mathrm{Hom}_A(M, N) \to \mathrm{Hom}_A(M, L)$ is again surjective. Prove that M is projective if and only if M is a direct summand of a free module. Prove that a projective module is flat.

7.3.4. Prove that the ideal $\langle 2, 1 + \sqrt{-5} \rangle \subset \mathbb{Z}[\sqrt{-5}]$ is a projective module, but not free.

7.3.5. Let $\varphi : A \to B$ be a ring homomorphism and M a flat A–module. Prove that $M \otimes_A B$ is a flat B–module.

7.3.6. Let $0 \to M \to N \to P \to 0$ be an exact sequence of A–modules, and suppose that M and P are flat. Prove that N is flat. Show that P is not necesarily flat if M and N are flat (but cf. Exercise 7.3.8).

7.3.7. Let $0 \to M \to N \to P \to 0$ be an exact sequence of A–modules, and assume that N and P are flat. Prove that M is flat.

7.3.8. Let $0 \to M \to N \to P \to 0$ be an exact sequence of $K[\varepsilon]$–modules with M flat. Show that P is flat iff N is flat. [Hint: use Corollary 7.3.6]

7.3.9. Prove that M is flat if and only if $\mathrm{Tor}_i^A(N, M) = 0$ for all A–modules N and $i > 0$.

7.3.10. Prove that $\mathbb{C}[x, y]/\langle 1 + xy \rangle$ is a flat $\mathbb{C}[x]$–module but not a free $\mathbb{C}[x]$–module.

7.3.11. Let M, N be A–modules and B a flat A–algebra. Prove that

$$\mathrm{Tor}_i^A(M, N) \otimes_A B \cong \mathrm{Tor}_i^B(M \otimes_A B, N \otimes_A B).$$

7.3.12. Let A be Noetherian and $\{M_i\}_{i \in I}$ a family of flat A–modules. Then $\prod_{i \in I} M_i$ is a flat A–module. In particular, $A[[x_1, \ldots, x_n]]$ is a flat A–algebra.

7.3.13. An A–module M is called *faithfully flat* if every complex

$$\cdots \longrightarrow N_i \longrightarrow N_{i-1} \longrightarrow N_{i-2} \longrightarrow \cdots$$

of A–modules is exact if and only if the induced complex

$$\cdots \longrightarrow N_i \otimes_A M \longrightarrow N_{i-1} \otimes_A M \longrightarrow N_{i-2} \otimes_A M \longrightarrow \cdots$$

is exact. Prove that M is faithfully flat if and only if M is flat and $\mathfrak{m}M \neq M$ for every maximal ideal $\mathfrak{m}$ in A.

7.3.14. Let $K \subset L$ be a field extension. Prove that L is a faithfully flat K–module.

7.3.15. Let A be a ring and B a faithfully flat A–module. Prove that an A–module M is faithfully flat if and only if $M \otimes_A B$ is a faithfully flat B–module.

7.3.16. Let A be a local ring and M a finitely generated flat A–module. Prove that M is faithfully flat.

7.3.17. Let $\varphi : A \to B$ be a faithfully flat ring homomorphism, that is, B is, via φ, a faithfully flat A–module. Prove that, for any A–module M, the map $M \to M \otimes_A B$, defined by $m \mapsto m \otimes 1$, is injective. Use this result to prove that $IB \cap A = I$ for any ideal $I \subset A$.

7.3.18. Let K be a field and $I \subset K[x_1, \ldots, x_n, t]$ be a homogeneous ideal. Prove that $\operatorname{Tor}_1^{K[t]}(K, K[x_1, \ldots, x_n, t]/I) = (I : \langle t \rangle)/I$. Use this to prove that $K[x_1, \ldots, x_n, t]/I$ is flat over $K[t]$ if and only if $I : \langle t \rangle = I$.

7.3.19. Let K be a field, and let $I \subset K[x_1, \ldots, x_n]$ be an ideal. Moreover, let $w = (w_1, \ldots, w_n) \in (\mathbb{Z} \setminus \{0\})^n$, and let $I^h \subset K[x_1, \ldots, x_n, t]$ be the weighted homogenization of I with respect to t (see Exercise 1.7.5). Prove that $K[x_1, \ldots, x_n, t]/I^h$ is a flat $K[t]$–module.

7.3.20. Give an example of a homogeneous ideal $J \subset K[x_1, \ldots, x_n, t]$ such that $K[x_1, \ldots, x_n, t]/J$ is not a flat $K[t]$–module.

7.3.21. Let A be an integral domain and $I \subset A[x_1, \ldots, x_n]$ an ideal. Prove that $A[x_1, \ldots, x_n]/I$ is a torsion–free A–module if and only if

$$IQ(A)[x_1, \ldots, x_n] \cap A[x_1, \ldots, x_n] = I.$$

7.3.22. Use Exercise 7.3.21 and the results from Chapter 4 (Proposition 4.3.1) to write a SINGULAR procedure to test torsion–freeness.

7.3.23. Let K be a field. Prove that the $K[x]$–module $K[x][[y]]/\langle y^2 - xy \rangle$ is flat but not finitely generated.

The following two exercises provide further algorithms to compute the singular locus of a ring A, without using an equidimensional decomposition as in Algorithm 5.7.8.

7.3.24. Let K be a perfect field and $A = K[x_1, \ldots, x_n]/\langle f_1, \ldots, f_m \rangle$. Let $J \subset A^n$ be the submodule generated by the columns of the Jacobian matrix $\left(\frac{\partial f_i}{\partial x_j}\right)$. Let $I_0 := \operatorname{Ann}\left(\operatorname{Ext}_A^1(\Omega_{A|K}, J)\right)$. Prove that

$$\operatorname{Sing}(A) = V(I_0) = \{Q \subset A \text{ prime} \mid I_0 \subset Q\}.$$

(Hint: use Proposition 7.3.15 and Exercise 5.7.11.)

7.3.25. Use SINGULAR Examples 7.3.14 and 7.3.17 and write a procedure to compute the singular locus of $A = K[x_1, \ldots, x_n]/\langle f_1, \ldots, f_m \rangle$, using the fact that the flat locus of $\Omega_{A|K}$ is the regular locus of A.

7.4 Local Criteria for Flatness

In this section we give criteria for flatness over local rings, which can be checked with a computer. We shall weaken the condition $\operatorname{Tor}_1^A(A/I, M) = 0$ for all $I \subset A$ to $\operatorname{Tor}_1^A(A/\mathfrak{m}, M) = 0$, $\mathfrak{m}$ the maximal ideal. We shall see that we can compute $\operatorname{Tor}_1^A(A/\mathfrak{m}, M)$ and, therefore, check flatness.

Proposition 7.4.1. *Let M be an A–module. The following conditions are equivalent:*

(1) M is a flat A–module.
(2) M_P is a flat A_P–module for all prime ideals P.
(3) M_P is a flat A_P–module for all maximal ideals P.

Proof. (1) implies (2), by Exercise 7.3.1. That (2) implies (3) is trivial. Finally, to prove that (3) implies (1), let $I \subset A$ be an ideal. Then

$$(I \otimes_A M)_P = I_P \otimes_{A_P} M_P \longrightarrow M_P$$

is injective by assumption. This is true for all maximal ideals and, therefore, $I \otimes_A M \to M$ is injective by Corollary 2.1.39. $\qquad\qquad\square$

The following theorem can easily be proved if M is a finitely generated A–module, by using Nakayama's lemma. However, its importance is just the fact that we need only a much weaker finiteness assumption, which turns out to be extremely useful in applications.

Theorem 7.4.2. *Let $(A, \mathfrak{m})$ and $(R, \mathfrak{n})$ be Noetherian local rings, R an A–algebra and $\mathfrak{m}R \subset \mathfrak{n}$. Let M be a finitely generated R–module. Then M is flat as an A–module if and only if $\operatorname{Tor}_1^A(A/\mathfrak{m}, M) = 0$.*

Proof. If M is flat as an A–module then $\operatorname{Tor}_1^A(A/\mathfrak{m}, M) = 0$, because of Theorem 7.3.3.

Now assume that $\operatorname{Tor}_1^A(A/\mathfrak{m}, M) = 0$. Let $I \subset A$ be an ideal. We have to prove that $I \otimes_A M \to M$ is injective.

We first claim that $\bigcap_{n=0}^{\infty} \mathfrak{m}^n \cdot (I \otimes_A M) = 0$. To see this, we consider $I \otimes_A M$ as an R–module via the R–module structure of M. It is finitely generated as an R–module and, therefore, by Krull's Intersection Theorem $\bigcap_{n=0}^{\infty} \mathfrak{n}^n \cdot (I \otimes_A M) = 0$. But $\mathfrak{m}R \subset \mathfrak{n}$ implies the claim.

Let $x \in \operatorname{Ker}(I \otimes_A M \to M)$ then we will show that $x \in \mathfrak{m}^n \cdot (I \otimes_A M)$ for all n. To prove this, we consider the map

$$(\mathfrak{m}^n I) \otimes_A M \longrightarrow I \otimes_A M \,.$$

The image of this map is $\mathfrak{m}^n \cdot (I \otimes_A M)$. Using the lemma of Artin-Rees (Lemma 5.4.5), we obtain an integer s such that $\mathfrak{m}^s \cap I \subset \mathfrak{m}^n I$. Therefore, it is enough to prove that x is in the image of $(\mathfrak{m}^n \cap I) \otimes_A M \to I \otimes_A M$ for all n. From the exact sequence

$$(\mathfrak{m}^n \cap I) \otimes_A M \longrightarrow I \otimes_A M \longrightarrow (I/\mathfrak{m}^n \cap I) \otimes_A M \longrightarrow 0,$$

we deduce that it is sufficient to see that x maps to 0 in $(I/\mathfrak{m}^n \cap I) \otimes_A M$. Consider the following commutative diagram:

$$
\begin{array}{ccc}
I \otimes_A M & \xrightarrow{\ \gamma\ } & (I/\mathfrak{m}^n \cap I) \otimes_A M \\
\alpha \downarrow & & \downarrow \pi \\
M & \xrightarrow{\ \beta\ } & (A/\mathfrak{m}^n) \otimes_A M .
\end{array}
$$

We know that $\alpha(x) = 0$. Therefore, $\pi \circ \gamma(x) = 0$, and it is sufficient to prove that π is injective. To prove this, consider the following exact sequence

$$0 \longrightarrow I/(\mathfrak{m}^n \cap I) \longrightarrow A/\mathfrak{m}^n \longrightarrow A/(I + \mathfrak{m}^n) \longrightarrow 0$$

which induces an exact sequence

$$\mathrm{Tor}_1^A\big(A/(I+\mathfrak{m}^n), M\big) \to (I/I \cap \mathfrak{m}^n) \otimes_A M \xrightarrow{\ \pi\ } (A/\mathfrak{m}^n) \otimes_A M .$$

We see that, finally, it suffices to prove that $\mathrm{Tor}_1^A\big(A/(I+\mathfrak{m}^n), M\big) = 0$. But $A/(I+\mathfrak{m}^n)$ is an A–module of finite length. Therefore, the following lemma proves the theorem. $\square$

Lemma 7.4.3. *Let $(A, \mathfrak{m})$ be a local ring and M an A–module such that $\mathrm{Tor}_1^A(A/\mathfrak{m}, M) = 0$. Then $\mathrm{Tor}_1^A(P, M) = 0$ for all A–modules P of finite length.*

Proof. We use induction on the length. The case $\mathrm{length}(P) = 1$ is clear because it implies $P = A/\mathfrak{m}$. Let $N \subset P$ be a proper submodule, then we obtain the exact sequence

$$\mathrm{Tor}_1^A(N, M) \longrightarrow \mathrm{Tor}_1^A(P, M) \longrightarrow \mathrm{Tor}_1^A(P/N, M) .$$

By the induction hypothesis, $\mathrm{Tor}_1^A(N, M) = 0$ and $\mathrm{Tor}_1^A(P/N, M) = 0$. This implies $\mathrm{Tor}_1^A(P, M) = 0$. $\square$

Note that $\mathrm{Tor}_1^A(A/\mathfrak{m}, M) = 0$ if and only if the homomorphism $\mathfrak{m} \otimes_A M \to M$, $a \otimes m \mapsto am$, is injective. Hence, Theorem 7.4.2 shows that, for modules over local rings A, it suffices to consider a single injective homomorphism, $i : \mathfrak{m} \hookrightarrow A$, when studying flatness (cf. Definition 7.3.1).

Corollary 7.4.4. *Let $(A, \mathfrak{m})$ and $(R, \mathfrak{n})$ be Noetherian local rings, R an A–algebra and $\mathfrak{m}R \subset \mathfrak{n}$. Let M be a finitely generated R–module. Then M is flat as an A–module if and only if $M/\mathfrak{m}^n M$ is a flat $A/\mathfrak{m}^n$–module for all n.*

Proof. Let M be a flat A–module then $\mathfrak{m}/\mathfrak{m}^n \hookrightarrow A/\mathfrak{m}^n$ induces an injective (multiplication) map

$$(\mathfrak{m}/\mathfrak{m}^n) \otimes_{A/\mathfrak{m}^n} (M/\mathfrak{m}^n M) \cong (\mathfrak{m}/\mathfrak{m}^n) \otimes_A M \hookrightarrow M/\mathfrak{m}^n M .$$

Hence, by the above, $M/\mathfrak{m}^n M$ is a flat $A/\mathfrak{m}^n$–module.

For the other implication, we have to prove that $\mathfrak{m} \otimes_A M \to M$ is injective. As in the proof of Theorem 7.4.2, we know that $\mathfrak{m} \otimes_A M$ is a finitely generated R–module and, therefore, $\bigcap_{n=0}^\infty \mathfrak{m}^n \cdot (\mathfrak{m} \otimes_A M) = 0$.

Let $x \in \mathrm{Ker}(\mathfrak{m} \otimes_A M \to M)$ then we have to prove that $x \in \mathfrak{m}^n \cdot (\mathfrak{m} \otimes_A M)$ for all n. As before, it is enough to see that x maps to 0 in $(\mathfrak{m}/\mathfrak{m}^n) \otimes_A M$. But $(\mathfrak{m}/\mathfrak{m}^n) \otimes_A M \cong (\mathfrak{m}/\mathfrak{m}^n) \otimes_{A/\mathfrak{m}^n} (M/\mathfrak{m}^n M) \to M/\mathfrak{m}^n M$ is injective by assumption. $\qquad\square$

Corollary 7.4.5. *Let $(A, \mathfrak{m})$ be a Noetherian local ring and $\widehat{A}$ the $\mathfrak{m}$–adic completion, then $\widehat{A}$ is a faithfully flat A–algebra.*

Proof. The corollary follows from Corollary 7.4.4 and Exercise 7.3.11. $\qquad\square$

Corollary 7.4.6. *Let $>$ be any monomial ordering on $\mathrm{Mon}(x_1, \dots, x_n)$ and $I \subset K[x]_>$, $x = (x_1, \dots, x_n)$, be an ideal. Then the canonical inclusions*

$$K[x]_> / I \subset K[x]_{\langle x \rangle} / I \cdot K[x]_{\langle x \rangle} \subset K[[x]] / I K[[x]]$$

are flat, and the second inclusion is faithfully flat. Moreover, if $I \subset K[x]$, then the inclusion $K[x]/I \subset K[x]_>/I$ is flat, too.

Corollary 7.4.7. *Let $(A, \mathfrak{m})$ be a local ring with residue field K. Moreover, let C be $A[x]_{\langle \mathfrak{m}, x \rangle}$, respectively $A[[x]]$, $x = (x_1, \dots, x_n)$, and let $\overline{C}$ be $K[x]_{\langle x \rangle}$, respectively $K[[x]]$, the reduction of C mod $\mathfrak{m}$. Then, for any $F_1, \dots, F_m \in C$, the following are equivalent:*

(1) $C/\langle F_1, \dots, F_m \rangle$ is flat over A;
(2) every relation in $\overline{C}$, $\sum_{i=1}^m H_i F_i \equiv 0 \bmod \mathfrak{m}$, can be lifted to a relation $\sum_{i=1}^m \widetilde{H}_i F_i = 0$ in C.

Proof. By Theorem 7.4.2, the A–module $B := C/\langle F_1, \dots, F_m \rangle$ is flat if and only if $\mathrm{Tor}_1^A(K, B) = 0$. Now, consider the exact sequence

$$0 \longrightarrow \langle F_1, \dots, F_m \rangle \longrightarrow C \longrightarrow B \longrightarrow 0$$

and tensor with K. We obtain an exact sequence

$$0 \longrightarrow \mathrm{Tor}_1^A(K, B) \longrightarrow \langle F_1, \dots, F_m \rangle \otimes_A K \longrightarrow \overline{C} \longrightarrow \overline{B} \longrightarrow 0 ,$$

where $\overline{B} := \overline{C}/\langle f_1, \dots, f_m \rangle = B/\mathfrak{m}B$, $f_i := F_i \bmod \mathfrak{m}$. This implies that B is A–flat if and only if the map $\langle F_1, \dots, F_m \rangle \otimes_A K \longrightarrow \overline{C}$ is injective. We have

$$\langle F_1, \ldots, F_m \rangle \otimes_A K = \left\{ \left(\sum_{i=1}^{m} H_i F_i \right) \otimes 1 \;\middle|\; H_i \in C \right\}.$$

By Proposition 2.7.10, $\sum_{i=1}^{m} H_i F_i \otimes 1 = 0$ in $\langle F_1, \ldots, F_m \rangle \otimes_A K$ if and only if $\sum_{i=1}^{m} H_i F_i = \sum_{j=1}^{r} b_j m_j$ for suitable $b_j \in A$, $m_j \in \langle F_1, \ldots, F_m \rangle$, satisfying $b_j \cdot 1 = 0$ in K, that is, $b_j \in \mathfrak{m}$; hence, if and only if $\sum_{i=1}^{m} H_i F_i = \sum_{i=1}^{m} K_i F_i$ for suitable $K_i \in \mathfrak{m}C$. Setting $\widetilde{H}_i := H_i - K_i$, the statement follows. $\square$

Corollary 7.4.8. *Let $(A, \mathfrak{m})$ and $(R, \mathfrak{n})$ be Noetherian local rings, R an A-algebra and $\mathfrak{m}R \subset \mathfrak{n}$. Let $\pi : M \to N$ be a homomorphism of finitely generated R-modules with N being A-flat. Then the following conditions are equivalent.*

(1) π is injective and $L := M/\pi(N)$ is flat over A.
(2) $\pi \otimes 1_{A/\mathfrak{m}} : M \otimes_A A/\mathfrak{m} \to N \otimes_A A/\mathfrak{m}$ is injective.

Proof. (1) $\Rightarrow$ (2): assume that π is injective, then

$$\mathrm{Tor}_1^A(A/\mathfrak{m}, L) \longrightarrow M \otimes_A A/\mathfrak{m} \xrightarrow{\pi \otimes 1_{A/\mathfrak{m}}} N \otimes_A A/\mathfrak{m}$$

is exact. But L is flat over A, hence $\mathrm{Tor}_1^A(A/\mathfrak{m}, L) = 0$ and, therefore, $\pi \otimes 1_{A/\mathfrak{m}}$ is injective.

(2) $\Rightarrow$ (1): let $x \in M$ and $\pi(x) = 0$. Then by assumption, $x \in \mathfrak{m}M$. Assume that we have already proved that $x \in \mathfrak{m}^n M$ and let $m_1, \ldots, m_r$ be a minimal system of generators of $\mathfrak{m}^n$ (as an A-module). Let $x = \sum_{i=1}^{r} y_i m_i$ for suitable $y_i \in M$. Then $0 = \pi(x) = \sum_{i=1}^{r} \pi(y_i) m_i$. Now N is a flat A-module and, by Proposition 7.3.4, there exist $a_{ij} \in A$ and $z_j \in N$ such that $\sum_{i=1}^{r} m_i a_{ij} = 0$ and $\pi(y_i) = \sum_j a_{ij} z_j$. By the choice of $m_1, \ldots, m_r$ (to be a minimal system of generators) we obtain $a_{ij} \in \mathfrak{m}$. This implies $\pi(y_i) \in \mathfrak{m}N$ and, by the injectivity of $\pi \otimes 1_{A/\mathfrak{m}}$ we obtain $y_i \in \mathfrak{m}M$ and, therefore, $x \in \mathfrak{m}^{n+1}M$. Using induction, we obtain $x \in \bigcap_{n=0}^{\infty} \mathfrak{m}^n M = \langle 0 \rangle$. This implies that π is injective.

Now the sequence

$$0 \to M \xrightarrow{\pi} N \to L \to 0$$

is exact. This implies that

$$\mathrm{Tor}_1^A(A/\mathfrak{m}, N) \longrightarrow \mathrm{Tor}_1^A(A/\mathfrak{m}, L) \longrightarrow M/\mathfrak{m}M \xrightarrow{\pi \otimes 1_{A/\mathfrak{m}}} N/\mathfrak{m}N$$

is exact. As N is A-flat, $\mathrm{Tor}_1^A(A/\mathfrak{m}, N) = 0$. Because $\pi \otimes 1_{A/\mathfrak{m}}$ is injective, we obtain $\mathrm{Tor}_1^A(A/\mathfrak{m}, L) = 0$. This implies that L is A-flat by Theorem 7.4.2. $\square$

Exercises

7.4.1. Let $(A, \mathfrak{m})$, $(R, \mathfrak{n})$ be Noetherian local rings, R an A-algebra such that $\mathfrak{m}R \subset \mathfrak{n}$. Let $x \in \mathfrak{m}$ be a non-zerodivisor and M a finitely generated R-module. If x is a non-zerodivisor for M, then M is a flat A-module if and only if M/xM is a flat $A/\langle x \rangle$-module.

7.4.2. Let $(A, \mathfrak{m})$, $(R, \mathfrak{n})$ be Noetherian local rings and $\widehat{A}$, respectively $\widehat{R}$, their completions. Let R be an A–algebra such that $\mathfrak{m}R \subset \mathfrak{n}$. Let M be a finitely generated R–module and $\widehat{M} = M \otimes_R \widehat{R}$.

Prove that M is flat over A if and only if $\widehat{M}$ is flat over $\widehat{A}$.

7.4.3. Let A be a ring and B a flat A–algebra. Let $Q \subset B$ be a prime ideal and $P = Q \cap A$. Prove that the localization B_Q is a faithfully flat A_P–algebra.

7.4.4. Let A be a ring and B a faithfully flat A–algebra. Let $P \subset A$ be a prime ideal. Prove that there exists a prime ideal $Q \subset B$, such that $Q \cap A = P$. In the language of Appendix A, this means that the induced map $\mathrm{Spec}(B) \to \mathrm{Spec}(A)$ is surjective.

7.4.5. Let A be a ring and B a flat A–algebra. Prove that the going down theorem holds for $A \subset B$. More precisely, let $P \subset P'$ be prime ideals in A, and let $Q' \subset B$ be a prime ideal such that $Q' \cap A = P'$. Prove that there exists a prime ideal $Q \subset Q' \subset B$, such that $Q \cap A = P$.

(Hint: use Exercises 7.4.3, 7.4.4.)

7.4.6. Let $(A, \mathfrak{m})$, $(R, \mathfrak{n})$ be Noetherian local rings, and assume that R is an A–algebra with $\mathfrak{m}R \subset \mathfrak{n}$. Prove that

$$\dim(R) \leq \dim(A) + \dim(R/\mathfrak{m}R),$$

with equality if R is flat over A.

(Hint: use Exercise 7.4.5 to prove that $\dim(R) \geq \dim(A) + \dim(R/\mathfrak{m}R)$. To prove the other inequality, choose a system of parameters $x_1, \ldots, x_r \in A$, and choose $y_1, \ldots, y_s \in R$, inducing a system of parameters in $R/\mathfrak{m}R$. Prove that $\mathfrak{n}^k \subset \langle x_1, \ldots, x_r, y_1, \ldots, y_s \rangle_R$ for a suitable k.)

7.4.7. Let A be a Noetherian ring and B a Noetherian A–algebra. Moreover, let $Q \subset B$ be a prime ideal and $P := Q \cap A$. Prove that

$$\mathrm{ht}(Q) \leq \mathrm{ht}(P) + \dim B_Q/PB_Q,$$

with equality if B is flat over A.

7.4.8. (*Dimension of fibres in a faithfully flat family need not be constant.*) Let $A := K[y] \subset K[x, y, z]/(\langle xy - 1 \rangle \cap \langle x, z \rangle) =: B$. Prove that B is a faithfully flat A–algebra, and compute $\dim(B/\mathfrak{m}B)$ for $\mathfrak{m} = \langle y \rangle$ and $\mathfrak{m} = \langle y - 1 \rangle$.

7.4.9. Modify the example of Exercise 7.4.8 to obtain a faithfully flat A–algebra B such that, for all maximal ideals $\mathfrak{m} \subset A$, the fibres $B/\mathfrak{m}B$ are finite dimensional K–vector spaces, but B is not a finitely generated A–module.

7.4.10. Let A be a Noetherian ring and M a finitely generated A–module. Prove that the following conditions are equivalent:

(1) M is projective.
(2) M_P is free for all prime ideals P.
(3) M_P is free for all maximal ideals P.

7.5 Flatness and Standard Bases

In this section we show that standard bases can be characterized in terms of flatness. More precisely, for a suitable weighted ordering $>_w$ "approximating" the given ordering $>$, the ring $K[x]_{>_w}/I$ and the ring $K[x]_{>_w}/L(I)$ are fibres in a flat family $K[t] \to K[x,t]_{>_w}/J$, where $K[x]_{>_w}/L(I)$ is the special fibre and $K[x]_{>_w}/I$ is the general fibre. Even more, all fibres different from the special fibre are isomorphic to $K[x]_{>_w}/I$. In this family, each component of maximal dimension maps surjectively to the target, in particular, the family is faithfully flat. These properties are the background and the reason for many applications.

Let K be a field, and let $x = (x_1, \dots, x_n)$ and t be variables.

Theorem 7.5.1. *Let $>$ be any monomial ordering on $\mathrm{Mon}(x)$, $F \subset \mathrm{Mon}(x)$ a finite subset, and $I' \subset K[x]_>$ an ideal. Then there exist a weighted degree ordering $>_w$ on $\mathrm{Mon}(x,t)$, which is global in t and coincides with $>$ on F, such that the following holds: let $J \subset K[x,t]_{>_w}$ be the ideal generated by the weighted homogenization I^h of $I := I' \cap K[x]$ (w.r.t. the weights w and with homogenizing variable t), then the following holds:*

(1) $B := K[x,t]_{>_w}/J$ *is a flat $K[t]$-algebra.*

If $IK[x]_> \subsetneq K[x]_>$ then, for any maximal ideal $\mathfrak{m} \subset K[t]$, there exists a maximal ideal $M \subset B$ such that $M \cap K[t] = \mathfrak{m}$ and $\dim(B) = \dim(B_M)$. In particular, in this case B is faithfully flat over $K[t]$.

(2) $L(J) = L_>(I)K[x,t]$.

(3) $L_>(J|_{t=\lambda}) = L_>(I)$ *for all $\lambda \in K$.*

(4) $J|_{t=0} = L_>(I)K[x]_{>_w}$ *and* $J|_{t=1} = IK[x]_{>_w}$.

Moreover, the fibre $B \otimes_{K[t]} K[t]/\langle t - \lambda \rangle \cong K[x]_{>_w}/(J|_{t=\lambda})$ is isomorphic to $K[x]_{>_w}/IK[x]_{>_w}$, for all $\lambda \neq 0$.

Note that, for $>$ a mixed ordering, $K[x]_{>_w} \neq K[x]_>$, in general. For example, let $>$ be the product ordering (lp,ls), then $K[x_1, x_2]_> = (K[x_2]_{\langle x_2 \rangle})[x_1]$ is a ring in which the units are precisely the polynomials in $K[x_2]$ with a non-zero constant term. On the other hand, for any weight vector $w = (w_1, w_2)$, $w_1 > 0$, $w_2 < 0$, elements of the form $1 + x_1^{\alpha_1} x_2^{\alpha_2}$ with $w_1 \alpha_1 + w_2 \alpha_2 < 0$ are units in $K[x_1, x_2]_{>_w}$. Hence, $K[x_1, x_2]_> \subsetneq K[x_1, x_2]_{>_w}$. However, this does not occur for non-mixed orderings for which we obtain the following, stronger result:

Corollary 7.5.2. *If $>$ is a global or local monomial ordering on $\mathrm{Mon}(x)$, then Theorem 7.5.1 holds with the stronger assertion*

(4') $J|_{t=0} = L(I)K[x]_>$ *and* $J|_{t=1} = IK[x]_>$.

Moreover, the fibre $B \otimes_{K[t]} K[t]/\langle t - \lambda \rangle \cong K[x]_>/(J|_{t=\lambda})$ is isomorphic to $K[x]_>/IK[x]_>$, for all $\lambda \neq 0$.

Proof. We may assume that $\{1, x_1, \ldots, x_n\} \subset F$. Since $>_w$ coincides with $>$ on F this means, in particular, that the ordering $>_w$ is global (respectively local) if $>$ is global (respectively local). Hence, $K[x,t]_{>_w} = K[x,t]_>$. $\square$

To prove the theorem, we consider first the case that the ordering $>$ is a weighted degree ordering with weight vector $w = (w_1, \ldots, w_n) \in \mathbb{Z}^n$, $w_i \neq 0$ for $i = 1, \ldots, n$. We consider on $K[x,t]$ the weighted degree ordering with weight vector $(w_1, \ldots, w_n, 1)$ refined by the given ordering on $K[x]$ and denote it also by $>$. We re–formulate the theorem in terms of standard bases.

For $f \in K[x]$ let $f^h = t^{\mathrm{w-deg}(f)} f(x_1/t^{w_1}, \ldots, x_n/t^{w_n})$ be the (weighted) homogenization of f with respect to t. Moreover, let $f_1, \ldots, f_r \in K[x]$, having the properties

$$\mathrm{w-deg}\big(\mathrm{LM}(f_i)\big) > \mathrm{w-deg}\big(\mathrm{LM}(\mathrm{tail}(f_i))\big), \quad i = 1, \ldots, r.$$

This property implies that

$$f_i^h|_{t=0} = \mathrm{LT}(f_i), \qquad f_i^h|_{t=1} = f_i.$$

Let $I = \langle f_1, \ldots, f_r \rangle K[x]_>$ and $\widetilde{I} = \langle f_1^h, \ldots, f_r^h \rangle K[x,t]_>$.

Proposition 7.5.3. *With the above notations and assumptions, the following conditions are equivalent:*

(1) $\{f_1, \ldots, f_r\}$ *is a standard basis of I.*
(2) $\{f_1^h|_{t=\lambda}, \ldots, f_r^h|_{t=\lambda}\}$ *is a standard basis of $\widetilde{I}|_{t=\lambda}$ for all $\lambda \in K$.*
(3) $\{f_1^h, \ldots, f_r^h\}$ *is a standard basis of $\widetilde{I}$.*
(4) $K[x,t]_>/\widetilde{I}$ *is a flat $K[t]$–algebra.*

Moreover, if $I \neq K[x]_>$, then $K[x,t]_>/\widetilde{I}$ is faithfully flat over $K[t]$.

Proof. (1) $\Rightarrow$ (2). Let $\{f_1, \ldots, f_r\}$ be a standard basis of I, then $L(I) = \langle \mathrm{LM}(f_1), \ldots, \mathrm{LM}(f_r) \rangle$. The case $\lambda = 0$ is clear because $\widetilde{I}|_{t=0}$ is the ideal generated by the terms $f_1^h|_{t=0}, \ldots, f_r^h|_{t=0}$.

If $\lambda \neq 0$ then the K^*–action $\varphi_\lambda : K[x] \to K[x]$ defined by $\varphi_\lambda(x_i) = x_i/\lambda^{w_i}$ is a K–algebra isomorphism extending to an isomorphism of $K[x]_>$ also denoted by φ_λ.

Obviously, $\mathrm{LM}\big(\varphi_\lambda(g)\big) = \mathrm{LM}(g)$ for any non–zero polynomial $g \in K[x]$. This implies that $L(I) = L\big(\varphi_\lambda(I)\big)$ and, therefore, $\{\varphi_\lambda(f_1), \ldots, \varphi_\lambda(f_r)\}$ is a standard basis of $\varphi_\lambda(I)$. On the other hand, $\varphi_\lambda(f_i) = \frac{1}{\lambda^{\mathrm{w-deg}(f_i)}} f_i^h|_{t=\lambda}$. This implies $\widetilde{I}|_{t=\lambda} = \varphi_\lambda(I)$.

(2) $\Rightarrow$ (3). Let $h \in \widetilde{I}$, then we have to prove that $\mathrm{LM}(f_i^h) \mid \mathrm{LM}(h)$ for some i. Because $\widetilde{I}$ is homogeneous, we may assume that $h = t^\rho g^h$ for some $g \in I$. (2) implies that $\mathrm{LM}(f_i) \mid \mathrm{LM}(g)$ for some i, and, therefore, $\mathrm{LM}(f_i^h) = \mathrm{LM}(f_i)$ divides $\mathrm{LM}(h) = t^\rho \mathrm{LM}(g)$.

(3) $\Rightarrow$ (1) is similar. (3) $\Rightarrow$ (4) is a consequence of Exercise 7.3.19 because $\widetilde{I}K[x,t]_> = I^h K[x,t]_>$ is the homogenization of I (Exercise 1.7.5).

To prove (4) $\Rightarrow$ (3) assume that $\{f_1^h, \ldots, f_r^h\}$ is not a standard basis. We may assume that $h = \mathrm{NF}(\mathrm{spoly}(f_1^h, f_2^h) \mid \{f_1^h, \ldots, f_r^h\}) \neq 0$. Let $h = t^\rho G$ with (homogeneous) $G \in K[x,t]_>$, such that $t \nmid \mathrm{LM}(G)$. Since $t^\rho G \in \widetilde{I}$, we have, by flatness, $G \in \widetilde{I}$. Let $G = \sum_{i=1}^r h_i f_i^h$ then $\mathrm{LM}(G) = \mathrm{LM}(G|_{t=0})$. But

$$G\big|_{t=0} = \sum_{i=1}^r h_i\big|_{t=0} \cdot f_i^h\big|_{t=0} = \sum_{i=1}^r h_i\big|_{t=0} \cdot \mathrm{LT}(f_i) \in L(I),$$

which implies that $\mathrm{LM}(G|_{t=0})$ and, therefore, $\mathrm{LM}(h)$ is divisible by some $\mathrm{LM}(f_i^h) = \mathrm{LM}(f_i)$, in contradiction to the definition of h.

Now assume that $I \neq K[x]_>$. To prove faithful flatness, we have to show that, for each maximal ideal $\mathfrak{m}$ of $K[t]$, $(K[x,t]_>/\widetilde{I}) \otimes_{K[t]} K[t]/\mathfrak{m} \neq 0$ (cf. Exercise 7.3.13).

By passing to the algebraic closure of K, we may assume that K is algebraically closed (cf. Exercises 7.3.14 and 7.3.15). But then $\mathfrak{m} = \langle t - \lambda \rangle$ for some $\lambda \in K$. For $\lambda = 0$, we obtain

$$\left(K[x,t]_>/\widetilde{I}\right) \otimes_{K[t]} K[t]/\langle t \rangle = K[x]/L(I) \neq 0.$$

For $\lambda \neq 0$, we use, again, the K^*–action on $K[x]_>$, to obtain $\widetilde{I}|_{t=\lambda} = \varphi_\lambda(I)$. Therefore,

$$\left(K[x,t]_>/\widetilde{I}\right) \otimes_{K[t]} K[t]/\langle t - \lambda \rangle = K[x]_>/\varphi_\lambda(I) \cong K[x]_>/I \neq 0$$

for $\lambda \neq 0$. $\qquad\square$

Proof of Theorem 7.5.1. Let $I = \overline{Q}_1 \cap \cdots \cap \overline{Q}_r$ be an irredundant primary decomposition, and let $S_0, S_1, \ldots, S_r$ be standard bases of $I, \overline{Q}_1, \ldots, \overline{Q}_r$. We apply Corollary 1.7.9 and enlarge F if necessary, such that, for each monomial ordering $>_1$ on $\mathrm{Mon}(x)$ which coincides with $>$ on F, the following holds:

- $L_{>_1}(I) = L_>(I)$,
- $L_{>_1}(\overline{Q}_i) = L_>(\overline{Q}_i)$, $i = 1, \ldots, r$,
- $\mathrm{LM}_{>_1}(f) = \mathrm{LM}_>(f)$ for $f \in S_0 \cup \cdots \cup S_r$.

Using Lemma 1.2.11 and Exercise 1.7.15, we can choose an integer weight vector $w = (w_1, \ldots, w_n) \in \mathbb{Z}^n$, with $w_i > 0$ if $x_i > 1$, respectively $w_i < 0$ if $x_i < 1$, for all i, and a weighted degree ordering $>_w$ coinciding on F with $>$ such that

- w–$\deg(\mathrm{LM}_{>_w}(f)) >$ w–$\deg(\mathrm{LM}_{>_w}(\mathrm{tail}(f)))$ for $f \in S_0 \cup \cdots \cup S_r$.

Now consider on $K[x_1, \ldots, x_n, t]$ the weighted degree ordering defined by the weight vector $(w_1, \ldots, w_n, 1)$, refined with the ordering $>_w$ (and denote it also by $>_w$).

Let $S_i^h := \{f^h \mid f \in S_i\}$, $i = 0, \ldots, r$, and let $J := \langle S_0^h \rangle$ and $Q_i := \langle S_i^h \rangle$, $i = 1, \ldots, r$. Then $J = Q_1 \cap \cdots \cap Q_r$ is an irredundant primary decomposition (Exercise 4.1.9). Due to Proposition 7.5.3, $B := K[x,t]_{>w}/J$ and $K[x,t]_{>w}/Q_i$ are faithfully flat $K[t]$–algebras, $i = 1, \ldots, r$. We may assume that Q_1 defines a maximal dimensional component of B. Since the family $K[t] \to K[x,t]_{>w}/Q_1$ is faithfully flat, we can find, for every maximal ideal $\mathfrak{m} \subset K[t]$, a maximal ideal $M \subset B$ containing Q_1/J such that $M \cap K[t] = \mathfrak{m}$. This implies that $\dim(B) = \dim(B_M)$, which completes the proof of (1).

By construction, S_0^h is a standard basis of J, and $L(S_0^h) = L_>(I)K[x,t]$ implies (2). Moreover, (3) is a consequence of Proposition 7.5.3 (2).

(4) By construction, it is also clear that $J|_{t=0} = L_>(I)K[x]_{>w}$ and $J|_{t=1} = IK[x]_{>w}$. For the same reason, we have

$$K[x,t]_{>w}/J \otimes_{K[t]} K[t]/\langle t - \lambda \rangle \cong K[x]_{>w}/J|_{t=\lambda},$$

that is, $K[x]_{>w}/J|_{t=\lambda}$ is isomorphic to $K[x]_{>w}/IK[x]_{>w}$. □

Lemma 7.5.4. *Let $>$ be a monomial ordering on $\mathrm{Mon}(x)$ and $I \subset K[x]$ an ideal. Then there is a finite subset $F \subset \mathrm{Mon}(x)$ with the following property: for any monomial ordering $>_1$ on $\mathrm{Mon}(x)$, coinciding with $>$ on F, we have*

(1) $\dim_K(K[x]_>/IK[x]_>) = \dim_K(K[x]_{>_1}/IK[x]_{>_1})$,
(2) $\dim(K[x]_>/IK[x]_>) = \dim(K[x]_{>_1}/IK[x]_{>_1})$.

Proof. Let $I = Q_1 \cap \cdots \cap Q_r$ be a primary decomposition. By extending the set F (adding the monomials appearing in a system of generators for Q_i), we obtain that $Q_iK[x]_{>_1} = K[x]_{>_1}$ whenever $Q_iK[x]_> = K[x]_>$. Hence, we may assume that $Q_iK[x]_> \neq K[x]_>$ for $i = 1, \ldots, r$. Furthermore, we may choose a maximal ideal $M \subset K[x]$, $M \supset Q_1$ and $MK[x]_> \neq K[x]_>$, such that $\dim(K[x]_M/IK[x]_M) = \dim(K[x]_>/IK[x]_>) \geq \dim(K[x]_{M'}/IK[x]_{M'})$ for all maximal ideals M'.

Let $S_0, S_1, \ldots, S_r$ be standard bases of $M, Q_1, \ldots, Q_r$, respectively. We apply Corollary 1.7.9 to $S_0, \ldots, S_r$ and obtain a finite set $F \subset \mathrm{Mon}(x)$ such that for any ordering $>_1$ on $\mathrm{Mon}(x)$, coinciding with $>$ on F, the leading ideals of $M, Q_1, \ldots, Q_r$ with respect to $>$ and $>_1$ are the same.

To prove (1) we use (2) and only have to consider the case $\dim(Q_i) = 0$ for $i = 1, \ldots, r$. In this case, we have

$$K[x]_>/I = K[x]/I$$

because $Q_iK[x]_> \neq K[x]_>$ and all non–zerodivisors of $K[x]/I$ are units. But, by construction, we also have $K[x]_{>_1}/I = K[x]/I$ and, therefore, (1) holds.

To prove (2) note that, by construction, we have $Q_iK[x]_{>_1} \neq K[x]_{>_1}$ and $MK[x]_{>_1} \neq K[x]_{>_1}$. Therefore, $(K[x]_{>_1}/IK[x]_{>_1})_M = K[x]_M/IK[x]_M$. This implies that

$$\dim(K[x]_{>_1}/IK[x]_{>_1})_M = \dim(K[x]_>/IK[x]_>)_M.$$

Since $\dim(K[x]_M/IK[x]_M) \geq \dim(K[x]_{M'}/IK[x]_{M'})$ for all maximal ideals M', we obtain (2). □

The following two corollaries generalize Corollaries 5.3.9 and 5.3.14, respectively Corollary 5.3.17.

Corollary 7.5.5. *Let $>$ be any monomial ordering on $K[x]$, and let $I \subset K[x]$ be an ideal. Then*

$$\dim\big(K[x]_>/IK[x]_>\big) = \dim\big(K[x]/L(I)\big).$$

Proof. We may assume that $IK[x]_> \neq K[x]_>$ and $I = IK[x]_> \cap K[x]$. Let $f_1, \ldots, f_r$ be a standard basis of I with respect to $>$. We choose a finite set $F \subset \text{Mon}(x)$ such that each monomial ordering coinciding with $>$ on F satisfies the properties (1) and (2) of Lemma 7.5.4. Now we use Theorem 7.5.1 to obtain a weighted degree ordering $>_w$ on $\text{Mon}(x,t)$ and an ideal $J \subset K[x,t]$ such that $B := K[x,t]_{>_w}/J$ is a faithfully flat $K[t]$–algebra. Moreover, we know that for every maximal ideal $\mathfrak{m} \subset K[t]$ there exists a maximal ideal $M \subset B$ with $M \cap K[t] = \mathfrak{m}$ and $\dim(B) = \dim(B_M)$. By applying Exercise 7.4.6 to $\mathfrak{m} = \langle t \rangle$ and $\mathfrak{m} = \langle t - 1 \rangle$, we obtain

$$\dim\big(K[x]/L_>(I)\big) = \dim\big(K[x]/L_{>_w}(I)\big) = \dim(K[x]_{>_w}/IK[x]_{>_w})$$
$$= \dim(K[x]_>/IK[x]_>).$$ □

Corollary 7.5.6. *With the assumptions of Corollary 7.5.5,*

$$\dim_K\big(K[x]_>/IK[x]_>\big) = \dim_K\big(K[x]/L(I)\big).$$

Moreover, if $\dim_K(K[x]_>/I) < \infty$, then the monomials in $K[x] \setminus L(I)$ represent a K–basis of vector space $K[x]_>/I$.

Proof. It can easily be seen that the monomials of $K[x] \setminus L(I)$ are linearly independent modulo I (use a standard basis of I and Lemma 1.6.7). Hence, $\dim_K\big(K[x]/L(I)\big) = \infty$ implies $\dim_K(K[x]_>/I) = \infty$.

Suppose that $\dim_K\big(K[x]/L(I)\big) < \infty$. Then there exists a K–basis of $K[x]/L(I)$ consisting of finitely many monomials $x^\alpha \in K[x] \setminus L(I)$. We may assume that $I = IK[x]_> \cap K[x]$ and choose a finite set $F \subset \text{Mon}(x)$ such that each monomial ordering coinciding with $>$ on F satisfies the properties (1) and (2) of Lemma 7.5.4. Now we use Theorem 7.5.1, again, to obtain a weighted degree ordering $>_w$ on $\text{Mon}(x,t)$ and an ideal $J \subset K[x,t]$ such that $B := K[x,t]_{>_w}/J$ is a faithfully flat $K[t]$–algebra and $L_{>_w}(J) = L(I)K[x,t]$. Hence, the monomials of $K[x] \setminus L(I)$ generate $K[x,t]_{>_w}/J$ as $K[t]$–module (Exercise 7.5.1) and, therefore, this module is $K[t]$–free, as it is finitely generated. This implies that

$$\dim_K(K[x]_>/IK[x]_>) = \dim_K(K[x]_{>_w}/IK[x]_{>_w})$$
$$= \dim_K(K[x]_{>_w}/L_{>_w}(I)K[x]_{>_w}) = \dim_K(K[x]/L(I)).$$ □

SINGULAR Example 7.5.7 (flatness test).

In this example we study the ideal $I_0 = \langle x^3y + yz^2, xy^3 + x^2z \rangle \subset \mathbb{Q}[x, y, z]$. Its generators do not form a standard basis with respect to the degree reverse lexicographical ordering. Let $I = \langle x^3y + yz^2t, xy^3 + x^2zt \rangle \subset \mathbb{Q}[x, y, z, t]$ be the ideal generated by the homogenizations of the generators of I_0. We illustrate the relation between flatness and standard bases as shown above.

```
ring A=0,(x,y,z,t),dp;
ideal I0=x3y+yz2,xy3+x2z;
ideal I=homog(I0,t);
```

We compute $J = I : \langle t \rangle$ and see that $J \supsetneq I$. Therefore, $B := \mathbb{Q}[x, y, z, t]/I$ is not $\mathbb{Q}[t]$–flat, because it is not torsion free (Corollary 7.3.5), in particular, $I \neq I_0^h$:

```
ideal J=quotient(I,t);

prune(modulo(J,I));
//-> _[1]=t*gen(1)
//-> _[2]=-xy*gen(1)
```

Now we compute a standard basis of I_0 and use it to define $I_h := I_0^h$, the homogenization of I_0 with respect to t. We shall see that $C := \mathbb{Q}[x, y, z, t]/I_h$ is $\mathbb{Q}[t]$–flat, computing $I_h : \langle t \rangle = I_h$.

```
ideal I1=std(I0);
ideal Ih=homog(I1,t);
ideal L=quotient(Ih,t);

prune(modulo(L,J));
//-> _[1]=0
```

Finally, we apply the formula $\mathrm{Tor}_1^{\mathbb{Q}[t]}(\mathbb{Q}, B) \cong \mathrm{Tor}_1^{\mathbb{Q}[x,y,z,t]}(\mathbb{Q}[x, y, z], B)$ (cf. Exercise 7.1.8) to compute $\mathrm{Tor}_1^{\mathbb{Q}[t]}(\mathbb{Q}, B) \cong J/I$ (cf. Exercise 7.3.18) and $\mathrm{Tor}_1^{\mathbb{Q}[t]}(\mathbb{Q}, C) = 0$.

```
matrix Ph=t;
matrix Ps[1][2]=I;

Tor(1,Ps,Ph);              // Tor(1,Ps,Ph)=Q[x,y,z,t]/<xy,t>
//-> _[1]=t*gen(1)
//-> _[2]=xy*gen(1)

matrix Pl[1][4]=Ih;
Tor(1,Pl,Ph);              // Tor(1,Pl,Ph)=0
//-> _[1]=0
```

Hence, the Tor–criterion for flatness (Theorem 7.3.3) does also show that C is flat over $\mathbb{Q}[t]$.

Exercises

7.5.1. Let $>$ be a weighted degree ordering on $\mathrm{Mon}(x_1, \ldots, x_n, t)$, and let $I \subset K[x]$ be an ideal such that $IK[x]_> \cap K[x] = I$ and $\dim_K K[x]/L(I) < \infty$. Moreover, let $\{m_1, \ldots, m_s\} \subset K[x]$ be a monomial K–basis of $K[x]/L(I)$. Prove that $m_1, \ldots, m_s$ generate $K[x, t]_>/I^h K[x, t]_>$ as $K[t]$–module, where $I^h \subset K[x, t]$ denotes the weighted homogenization of I.

(Hint: Use Exercise 1.7.22 to choose a standard basis $G' \supset G$ of I such that REDNFBUCHBERGER$(- \mid G)$ terminates. Now use that the corresponding standard representation with respect to G is compatible with homogenization. Note that this implies, in particular, that I^h is generated by G^h. Use Exercise 1.7.23 to identify $K[x, t]_>/I^h K[x, t]_>$ with $K[x, t]/I^h$.)

7.5.2. With the notations of Theorem 7.5.1, let $>$ be a global degree ordering and I a homogeneous ideal. Prove that for all $\lambda \in K$ the Hilbert function of $K[x]_>/J|_{t=\lambda}$ is the same.

7.5.3. Formulate and prove Exercise 7.5.2 in the local case for the Hilbert–Samuel function.

In the following exercises, we extend the results of this section to modules.

7.5.4. Let $>$ be a module ordering on $K[x]^r = \bigoplus_{i=1}^r K[x]e_i$, $x = (x_1, \ldots, x_r)$. For a weight vector $w = (w_1, \ldots, w_{n+r}) \in \mathbb{Z}^{n+r}$, set

$$\mathrm{w\text{–}deg}(x^\alpha e_i) := w_1 \alpha_1 + \cdots + w_n \alpha_n + w_{n+i},$$

and let $w' = (w_1, \ldots, w_n)$.

Let $f_1, \ldots, f_k \in K[x]^r$ and I be the submodule of $(K[x]_>)^r$ generated by $f_1, \ldots, f_k$. Show that there exists a module ordering $>_w$ on $K[x]^r$, which induces the weighted degree ordering $>_{w'}$ on $K[x]$, and has the following properties:

(1) $L_{>_w}(I) = L_>(I)$,
(2) $\mathrm{LM}_{>_w}(f_i) = \mathrm{LM}_>(f_i)$,
(3) $\mathrm{w\text{–}deg}(\mathrm{LM}_{>_w}(f_i)) > \mathrm{w\text{–}deg}(\mathrm{LM}_{>_w}(\mathrm{tail}(f)))$,

for $i = 1, \ldots, k$.

7.5.5. Using the notations of Exercise 7.5.4, extend the ordering $>_w$ (respectively $>_{w'}$) to $K[x, t]^r$ (respectively $K[x, t]$), as in the proof of Theorem 7.5.1, and denote it also by $>_w$ (respectively $>_{w'}$). Let f^h be the homogenization of $f \in K[x]^r$ with respect to a new variable t, and let $\widetilde{I} = \langle f_1^h, \ldots, f_k^h \rangle K[x, t]_{>_{w'}}$. Prove Proposition 7.5.3 in this situation.

7.5.6. Formulate and prove an analog of Theorem 7.5.1 for modules.

7.5.7. Generalize Corollaries 7.5.5 and 7.5.6 to submodules of $K[x]^r$.

7.5.8. Consider the polynomial $f = 1 + x^n y^m$, $(n, m) \in \mathbb{N}^2 \setminus \{(0,0)\}$ in the ring `R=0,(x,y),(lp,ls)`; and in the ring `S=0,(x,y),M(A)`; with `intmat A[2][2]=u,v,1,0`; and $u > 0$, $v < 0$. Show that

(1) f is a unit in R if and only if $n = 0$;
(2) f is a unit in S if and only if $nu + mv < 0$.

Check this using `lead(f)` in SINGULAR for $(u, v) = (2, -1)$ and $n + m \leq 100$.

7.5.9. Write a SINGULAR procedure which gets, as input, a finite set F of monomials (in a SINGULAR basering $K[x_1, \dots, x_n]_>$) and which returns an integer vector $w = (w_1, \dots, w_n)$ satisfying $w_i < 0$ if $x_i < 1$, $w_i > 0$ if $x_i > 1$ and $> |_F =>_w |_F$.

7.5.10. Let $R = \big(\mathbb{Q}[x]_{\langle x \rangle}\big)[y, z]$ and $I \subset R^2$ be the submodule generated by $[x^3 y + y z^2, y]$, $[x y^3 + x^2 z, y + z]$.

(1) Find a monomial ordering $>$ on $\mathrm{Mon}(x, y, z)$ such that $\mathbb{Q}[x, y, z]_> = R$.
(2) Find a weight vector $w \in \mathbb{Z}^5$ satisfying the properties of Exercise 7.5.1.

7.5.11. Denote by I_0 the module $I \subset R^2$ of Exercise 7.5.10. Construct the submodules J and I^h and check flatness for the corresponding family of modules over $K[t]$ as in Example 7.5.7.

7.6 Koszul Complex and Depth

In this section we study regular sequences. We are especially interested in the length of a maximal regular sequence which can be characterized by the vanishing of certain homology groups of the Koszul complex.

Definition 7.6.1. Let A be a ring and M an A–module. $a_1, \dots, a_n \in A$ is called a *regular sequence* with respect to M, or, in short, an M–*sequence* if

(1) a_i is not a zerodivisor for $M/\langle a_1, \dots, a_{i-1} \rangle M$ for $i = 1, \dots, n$;
(2) $M \neq \langle a_1, \dots, a_n \rangle M$.

Remark 7.6.2. The permutation of a regular sequence need not be a regular sequence: let K be a field and $a_1 = x(y-1)$, $a_2 = y$, $a_3 = z(y-1) \in K[x, y, z]$. Then a_1, a_2, a_3 is a $K[x, y, z]$–sequence but a_1, a_3, a_2 is not (cf. Exercise 7.6.11).

We shall see that, for local rings, the permutation of a regular sequence is again a regular sequence. For checking this remark, we write a procedure to test a regular sequence.

SINGULAR Example 7.6.3 (regular sequences).
Let $>$ be any monomial ordering on $\mathrm{Mon}(x_1, \dots, x_m)$, let $R := K[x]_>$, let $N \subset R^n$ be a submodule, and let `f` $\subset R$ be an ideal, represented by the ordered system of polynomial generators $\{f_1, \dots, f_k\}$. Then the following procedure returns 1, if $f_1, \dots, f_k$ is an R^n/N–sequence and 0 otherwise.

```
proc isReg(ideal f, module N)
{
  int n=nrows(N);
  int i;
  while(i<ncols(f))
  {
    i++;
    N=std(N);
    if(size(reduce(quotient(N,f[i]),N))!=0){return(0);}
    N=N+f[i]*freemodule(n);
  }
  if (size(reduce(freemodule(n),std(N)))==0){return(0);}
  return(1);
}


ring R=0,(x,y,z),dp;
ideal f=x*(y-1),y,z*(y-1);
module N=0;
isReg(f,N);
//-> 1

f=x*(y-1),z*(y-1),y;
isReg(f,N);
//-> 0
```

From the output we read that $x(y-1), y, z(y-1)$ is a $Q[x, y, z]$–sequence, while $x(y-1), z(y-1), y$ is not a $Q[x, y, z]$–sequence.

Definition 7.6.4. Let A be a ring, $I \subset A$ an ideal and M an A–module. If $M \neq IM$, then the maximal length n of an M–sequence $a_1, \ldots, a_n \in I$ is called the I–depth of M and denoted by $\mathrm{depth}(I, M)$. If $M = IM$ then the I–depth of M is by convention ∞. If $(A, \mathfrak{m})$ is a local ring, then the $\mathfrak{m}$–depth of M is simply called the *depth* of M, that is, $\mathrm{depth}(M) := \mathrm{depth}(\mathfrak{m}, M)$.

Example 7.6.5.

(1) Let K be a field and $K[x_1, \ldots, x_n]$ the polynomial ring. Then

$$\mathrm{depth}(\langle x_1, \ldots, x_n \rangle, K[x_1, \ldots, x_n]) \geq n,$$

since $x_1, \ldots, x_n$ is an $\langle x_1, \ldots, x_n \rangle$–sequence (and we shall see later that it is $= n$).

(2) Let A be a ring, $I \subset A$ an ideal and M an A–module. Then the I–depth of M is 0 if and only if every element of I is a zerodivisor for M. Hence, $\mathrm{depth}(I, M) = 0$ if and only if I is contained in some associated prime ideal of M.

In particular, for a local ring $(A, \mathfrak{m})$, we have $\mathrm{depth}(\mathfrak{m}, A/\mathfrak{m}) = 0$.

The aim of this section is to give, for A a Noetherian ring and M a finitely generated A–module, a formula for the I–depth of M in terms of the homology $H_i(f_1, \ldots, f_n, M)$ of the Koszul complex: if $I = \langle f_1, \ldots, f_n \rangle$ then

$$\text{depth}(I, M) = n - \sup\{i \mid H_i(f_1, \ldots, f_n, M) \neq 0\}.$$

We shall also see that

$$\text{depth}(I, M) = \inf\{n \mid \text{Ext}_A^n(A/I, M) \neq 0\}.$$

In particular, we shall see that the I–depth is a geometric invariant, that is,

$$\text{depth}(I, M) = \text{depth}(\sqrt{I}, M).$$

Definition 7.6.6. Let A be a ring and $x_1, \ldots, x_n \in A$, $x := (x_1, \ldots, x_n)$. The *Koszul complex* $K(x)_\bullet$ is defined as follows:

(1) $K(x)_0 := A$, $K(x)_p := \langle 0 \rangle$ for $p \notin \{0, \ldots, n\}$.
(2) Let $1 \leq p \leq n$ then $K(x)_p := \bigoplus_{1 \leq i_1 < \cdots < i_p \leq n} Ae_{i_1 \ldots i_p}$, the free A–module with $\binom{n}{p}$ free generators $\{e_{i_1 \ldots i_p}\}_{1 \leq i_1 < \cdots < i_p \leq n}$.
(3) The differential $d_p : K(x)_p \to K(x)_{p-1}$ is defined by

$$d_p(e_{i_1 \ldots i_p}) = \sum_{\mu=1}^{p} (-1)^{\mu-1} x_{i_\mu} e_{i_1 \ldots i_{\mu-1} i_{\mu+1} \ldots i_p} \quad \text{if } p \in \{2, \ldots, n\},$$

$$d_1(e_i) = x_i,$$
$$d_p = 0 \quad \text{if } p \notin \{1, \ldots, n\}.$$

(4) For an A–module M we define

$$K(x, M)_\bullet := K(x)_\bullet \otimes_A M,$$

the *Koszul complex of M and x*.

Lemma 7.6.7.

(1) $K(x)_\bullet$ is a complex, that is, $d_p \circ d_{p+1} = 0$ for all p.
(2) Let $H_i(x, M)$ be the homology of $K(x, M)_\bullet$, that is,

$$H_i(x, M) := \text{Ker}(d_p \otimes 1_M) / (\text{Im}\, d_{p+1} \otimes 1_M),$$

then $H_0(x, M) = M/xM$ and $H_n(x, M) = \{m \in M \mid x_i m = 0 \ \forall i\}$.
(3) $K(x, M)_\bullet = K(x_1, M)_\bullet \otimes \cdots \otimes K(x_n, M)_\bullet$.

Proof. (1) and (2) are not difficult to see and left as an exercise.

To prove (3) we recall the definition of the tensor product of two complexes: let $(L_\bullet, d_\bullet)$, $(K_\bullet, d_\bullet)$ be two complexes of A-modules. Then

$$(L_\bullet \otimes K_\bullet)_n = \bigoplus_{p+q=n} L_p \otimes K_q$$

and for $x \in L_p$ and $y \in K_q$ we have

$$d_n(x \otimes y) = d_p(x) \otimes y + (-1)^p x \otimes d_q(y).$$

Using induction, it is enough to prove that

$$K(x_1, \ldots, x_n)_\bullet = K(x_1, \ldots, x_{n-1})_\bullet \otimes K(x_n)_\bullet.$$

Here $K(x_n)_\bullet$ is the complex $\cdots \to 0 \to A \to A \to 0$ with the differential being multiplication by x_n. This implies that

$$\bigl(K(x_1, \ldots, x_{n-1})_\bullet \otimes K(x_n)_\bullet\bigr)_p = K(x_1, \ldots, x_{n-1})_p \oplus K(x_1, \ldots, x_{n-1})_{p-1}.$$

Let $\{a_{i_1 \ldots i_p}\}_{1 \leq i_1 < \cdots < i_p \leq n-1}$ be bases of the free A-modules $K(x_1, \ldots, x_{n-1})_p$, $p \in \{0, \ldots, n-1\}$, and let $\{e_{i_1 \ldots i_p}\}_{1 \leq i_1 < \cdots < i_p \leq n}$ be a basis of $K(x_1, \ldots, x_n)_p$. We define isomorphisms

$$\varphi_p : K(x_1, \ldots, x_n)_p \xrightarrow{\cong} K(x_1, \ldots, x_{n-1})_p \oplus K(x_1, \ldots, x_{n-1})_{p-1}$$

by

$$\varphi_p(e_{i_1 \ldots i_p}) = \begin{cases} (a_{i_1 \ldots i_p}, 0) & \text{if } i_p < n \\ (0, a_{i_1 \ldots i_{p-1}}) & \text{if } i_p = n. \end{cases}$$

It is now straightforward to see that $\varphi = \{\varphi_p\}$ is, indeed, an isomorphism of complexes, and that $\varphi \otimes \mathrm{id}_M$ defines an isomorphism for the Koszul complex of M. $\qquad\square$

Corollary 7.6.8. *Let A be a ring, M an A–module and $x_1, \ldots, x_n \in A$. Let $x = (x_1, \ldots, x_n)$ and $y = (x_{\pi(1)}, \ldots, x_{\pi(n)})$ for a permutation π. Then the complexes $K(x, M)_\bullet$ and $K(y, M)_\bullet$ are isomorphic.*

Lemma 7.6.9. *Let A be a ring, $C_\bullet$ a complex and $x \in A$.*

(1) The sequence of complexes

$$0 \longrightarrow C_\bullet \longrightarrow C_\bullet \otimes K(x)_\bullet \longrightarrow C_\bullet(-1) \longrightarrow 0$$

is exact (with $C_\bullet(-1)_p = C_{p-1}$).

(2) The induced homology sequence

$$\cdots \longrightarrow H_p(C_\bullet) \longrightarrow H_p\bigl(C_\bullet \otimes K(x)_\bullet\bigr) \longrightarrow H_{p-1}(C_\bullet)$$
$$\xrightarrow{(-1)^{p-1}x} H_{p-1}(C_\bullet) \longrightarrow H_{p-1}\bigl(C_\bullet \otimes K(x)_\bullet\bigr) \longrightarrow \cdots$$

is exact.

(3) $x \cdot H_p\bigl(C_\bullet \otimes K(x)_\bullet\bigr) = 0$ for all $p \in \mathbb{Z}$.

Proof. (1) and (2) are simple consequences of the usual diagram chase, where we have to consider the following diagram

with $d'_p(a, b) = \big(d_p(a) + (-1)^{p-1}xb,\, d_{p-1}(b)\big)$.

To prove (3) let (a, b) be a representative of an element in $H_p(C_\bullet \otimes K(x)_\bullet)$, that is, $d_p(a) + (-1)^{p-1}xb = 0$ and $d_{p-1}(b) = 0$. But then we have

$$x \cdot (a, b) = (xa, xb) = d'_{p+1}\big(0, (-1)^p a\big),$$

that is, $x \cdot (a, b) \equiv 0 \bmod \operatorname{Im}(d'_{p+1})$. □

Corollary 7.6.10. *Let A be a ring, M an A–module and $x_1, \ldots, x_n \in A$.*

(1) Let $x := (x_1, \ldots, x_n)$, $x' := (x_1, \ldots, x_{i-1}, x_{i+1}, \ldots, x_n)$, then there exists an exact sequence

$$\ldots \to H_{p+1}(x, M) \to H_p(x', M) \xrightarrow{\cdot(-1)^p x_i} H_p(x', M) \to H_p(x, M) \to \ldots$$
$$\ldots \to H_1(x, M) \to M/\langle x'\rangle M \xrightarrow{\cdot x_i} M/\langle x'\rangle M \to M/\langle x\rangle M \to 0.$$

(2) $\langle x_1, \ldots, x_n\rangle \cdot H_p(x_1, \ldots, x_n, M) = 0$ for all $p \in \mathbb{Z}$.

Corollary 7.6.11. *Let A be a Noetherian ring, $I = \langle f_1, \ldots, f_n\rangle$ an ideal of A and M a finitely generated A–module. Assume that $M \neq IM$, and set $q := \sup\{i \mid H_i(f_1, \ldots, f_n, M) \neq 0\}$. Then every maximal M–sequence in I has length $n - q$. In particular,*

$$\operatorname{depth}(I, M) = n - \sup\{i \mid H_i(f_1, \ldots, f_n, M) \neq 0\}.$$

Proof. Let $x_1, \ldots, x_s \in I$ be a maximal M–sequence. We shall prove $s = n - q$ by induction on s. If $s = 0$ then every element of I is a zerodivisor for M. Then I is contained in an associated prime ideal P of M. Being associated prime of M, $P = (0 : m)$ for some non–zero $m \in M$. This implies $I \cdot m = 0$, that is, $m \in H_n(f_1, \ldots, f_n, M)$ (cf. Lemma 7.6.7 (2)). Therefore, $q = n$ and we have $s = n - q$. Assume now that $s > 0$, and consider the exact sequence $0 \to M \xrightarrow{x_1} M \to M/x_1M \to 0$. Corollary 7.6.10 implies that $I \cdot H_i(f_1, \ldots, f_n, M) = 0$ for all i. By Exercise 7.6.13, the sequence

$$0 \longrightarrow H_i(f_1, \ldots, f_n, M) \longrightarrow H_i(f_1, \ldots, f_n, M/x_1M)$$
$$\longrightarrow H_{i-1}(f_1, \ldots, f_n, M) \longrightarrow 0$$

is exact for all i. This implies especially that

$$0 = H_{q+1}(f_1, \ldots, f_n, M) \longrightarrow H_{q+1}(f_1, \ldots, f_n, M/x_1M)$$
$$\longrightarrow H_q(f_1, \ldots, f_n, M) \longrightarrow 0$$

is exact, hence $H_{q+1}(f_1, \ldots, f_n, M/x_1M) \neq 0$, and $H_i(f_1, \ldots, f_n, M/x_1M)$ vanishes for $i > q + 1$. But $x_2, \ldots, x_s$ is a maximal M/x_1M–sequence in I. Using induction we obtain $q + 1 = n - (s - 1)$ and, therefore, $q = n - s$. $\quad\square$

Corollary 7.6.12. *With the notations of Corollary 7.6.11 the following holds:*

(1) $f_1, \ldots, f_n$ is an M–sequence if and only if $\mathrm{depth}(I, M) = n$;
(2) let $J \subset A$ be an ideal and $f_1, \ldots, f_n$ an M–sequence in J, then

$$\mathrm{depth}(J, M/\langle f_1, \ldots, f_n \rangle M) = \mathrm{depth}(J, M) - n \,.$$

Example 7.6.13. Let A be one of the rings $K[x_1, \ldots, x_n]$, $K[x_1, \ldots, x_n]_\mathfrak{m}$, $K[[x_1, \ldots, x_n]]$, with $\mathfrak{m} := \langle x_1, \ldots, x_n \rangle$. Then $\mathrm{depth}(\mathfrak{m}, A) = n = \dim(A)$.

The example is a consequence of Example 7.6.5 and Corollary 7.6.12.

Theorem 7.6.14. *Let A be a ring, M an A–module and $x_1, \ldots, x_n \in A$.*

(1) If $x = (x_1, \ldots, x_n)$ is an M–sequence, then $H_p(x, M) = 0$ for all $p > 0$.
(2) If $(A, \mathfrak{m})$ is local, $x_1, \ldots, x_n \in \mathfrak{m}$ and $M \neq 0$ is finitely generated, then $H_1(x, M) = 0$ implies that $x = (x_1, \ldots, x_n)$ is an M–sequence.
(3) If A and M are $\mathbb{N}$–graded and $x_1, \ldots, x_n$ are homogeneous of positive degree, then $H_1(x, M) = 0$ and $M \neq 0$ implies that $x = (x_1, \ldots, x_n)$ is an M–sequence.

Proof. (1) is proved by induction on n. If $n = 1$ then, due to Lemma 7.6.7 (2), $H_1(x_1, M) = \{m \in M \mid x_1m = 0\}$. Thus, $H_1(x_1, M) = 0$ if and only if x_1 is M–regular. Assume now that $n > 1$ and consider the exact sequence (of Corollary 7.6.10 (1))

$$H_p(x_1, \ldots, x_{n-1}, M) \to H_p(x_1, \ldots, x_n, M) \to H_{p-1}(x_1, \ldots, x_{n-1}, M) \,.$$

If $x_1, \ldots, x_n$ is an M–sequence then $x_1, \ldots, x_{n-1}$ is an M–sequence and, for all $p \geq 1$, $H_p(x_1, \ldots, x_{n-1}, M) = 0$ by induction hypothesis. It follows that $H_p(x_1, \ldots, x_n, M) = 0$ for all $p > 1$. For $p = 1$, we can apply Corollary 7.6.10 (1), again, and the vanishing of $H_1(x_1, \ldots, x_n, M)$ follows from the fact that x_n is $M/\langle x_1, \ldots, x_{n-1} \rangle M$–regular.

To prove (2) assume that $H_1(x_1, \ldots, x_n, M) = 0$ and consider the exact sequence

$$H_1(x_1, \ldots, x_{n-1}, M) \xrightarrow{\pm x_n} H_1(x_1, \ldots, x_{n-1}, M) \to H_1(x_1, \ldots, x_n, M) = 0 \,,$$

again using Corollary 7.6.10 (1) (2). This implies that

$$x_n H_1(x_1, \ldots, x_{n-1}, M) = H_1(x_1, \ldots, x_{n-1}, M) \,.$$

We can use Nakayama's Lemma and obtain $H_1(x_1, \ldots, x_{n-1}, M) = 0$, because $x_n \in \mathfrak{m}$ and $H_1(x_1, \ldots, x_{n-1}, M)$ is finitely generated. Using induction we obtain that $x_1, \ldots, x_{n-1}$ is an M–sequence. Now

$$0 = H_1(x_1, \ldots, x_n, M) \longrightarrow H_0(x_1, \ldots, x_{n-1}, M) \xrightarrow{\pm x_n} H_0(x_1, \ldots, x_{n-1}, M)$$

is exact and $H_0(x_1, \ldots, x_{n-1}, M) = M/\langle x_1, \ldots, x_{n-1}\rangle M$. This implies that x_n is $M/\langle x_1, \ldots, x_{n-1}\rangle M$–regular and proves (2).

The proof of (3) is similar and left as an exercise. $\qquad\square$

Corollary 7.6.15. *Let $(A, \mathfrak{m})$ be a local ring, M a finitely generated A–module and $x_1, \ldots, x_n$ an M–sequence then $x_{\pi(1)}, \ldots, x_{\pi(n)}$ is an M–sequence for all permutations π.*

Proof. The corollary is an immediate consequence of Corollary 7.6.8 and Theorem 7.6.14. $\qquad\square$

Corollary 7.6.16. *Let A be a ring, M an A–module, $I \subset A$ an ideal.*

(1) If $x_1, \ldots, x_n \in I$ is an M–sequence, then $x_1^m, \ldots, x_n^m$ is an M–sequence for all $m \geq 1$.
(2) $\mathrm{depth}(I, M) = \mathrm{depth}(\sqrt{I}, M)$.

Proof. (2) is a consequence of (1) and the definition of I–depth.

(1) is proved by induction on n. The case $n = 1$ is trivial. Assume that $n > 1$ and that $x_1^m, \ldots, x_{n-1}^m$ is an M–sequence. We first prove that x_n is not a zerodivisor of $M/\langle x_1^m, \ldots, x_{n-1}^m\rangle M$. It is enough to check this in the localization at any prime ideal P. If $P \not\supset \langle x_1, \ldots, x_n\rangle$, then either x_n is a unit in A_P, or we have $M_P = \langle x_1^m, \ldots, x_{n-1}^m\rangle M_P$, that is, in this case x_n is a non–zerodivisor of $M_P/\langle x_1^m, \ldots, x_{n-1}^m\rangle M_P$.

For the remaining cases, we may assume that $(A, \mathfrak{m})$ is already local and $x_1, \ldots, x_n \in \mathfrak{m}$. Now $x_1, \ldots, x_n$ being an M–sequence implies that $x_1, \ldots, x_{n-1}, x_n^m$ is an M–sequence. Using Corollary 7.6.15, we obtain that $x_n^m, x_1, \ldots, x_{n-1}$ is an M–sequence. Repeating this n times we obtain that $x_n^m, \ldots, x_1^m$ is an M–sequence and Corollary 7.6.15 completes the proof. $\qquad\square$

SINGULAR Example 7.6.17 (Koszul Complex).

In this example, we should like to show how to compute the Koszul complex and its homology. The method is similar to the computation of Tor: let A be a ring and M an A–module given by the presentation $A^\ell \xrightarrow{\varphi} A^s \xrightarrow{\pi} M \to 0$. Let $x = (x_1, \ldots, x_n)$ and consider the following diagram:

$$\begin{array}{ccccccc}
& 0 & & 0 & & 0 & \\
& \uparrow & & \uparrow & & \uparrow & \\
\cdots \longrightarrow & K_{p+1}(x) \otimes M & \xrightarrow{d_{p+1}^M} & K_p(x) \otimes M & \xrightarrow{d_p^M} & K_{p-1}(x) \otimes M & \longrightarrow \cdots \\
& \uparrow & & \uparrow & & \uparrow & \\
\cdots \longrightarrow & K_{p+1}(x) \otimes A^s & \xrightarrow{d_{p+1}^{A^s}} & K_p(x) \otimes A^s & \xrightarrow{d_p^{A^s}} & K_{p-1}(x) \otimes A^s & \longrightarrow \cdots \\
& \uparrow & & \uparrow{\scriptstyle \varphi^{K_p}} & & \uparrow{\scriptstyle \varphi^{K_{p-1}}} & \\
\cdots \longrightarrow & K_{p+1}(x) \otimes A^\ell & \longrightarrow & K_p(x) \otimes A^\ell & \longrightarrow & K_{p-1}(x) \otimes A^\ell & \longrightarrow \cdots
\end{array}$$

As in the computation of Tor (cf. Proposition 7.1.3) it is an easy diagram chase to prove that

$$
H_p(x, M) := \operatorname{Ker}(d_p^M)/\operatorname{Im}(d_{p+1}^M)
$$
$$
= (d_p^{A^s})^{-1}\bigl(\operatorname{Im}(\varphi^{K_{p-1}})\bigr)/\bigl(\operatorname{Im}(d_{p+1}^{A^s}) + \operatorname{Im}(\varphi^{K_p})\bigr).
$$

First we need to compute the maps $d_i^{A^s} = d_i \otimes id_{A^s}$ and $\varphi^{K_p} = \varphi \otimes id_{K_p}$. To do this we have to choose suitable bases in $K_p(x)$ and order them linearly: we shall inductively order the sets $\{e_{i_1 \ldots i_p}\}_{1 \le i_1 < \cdots < i_p \le n}$, $p = 1, \ldots, n$, in the following manner. If $p = 1$ then the element e_{i_1} obtains the number i_1. Assume we already have a numbering for $p - 1$, then the number of $e_{i_1 \ldots i_p}$ is

$$
\sum_{\nu=1}^{i_1-1} \binom{n-\nu}{p-1} + \text{ number of } e_{i_2-i_1,\ldots,i_p-i_1} \text{ with respect to } n - i_1.
$$

We can use the following procedure:

Algorithm 7.6.18 (BASISNUMBER$\bigl(n, p, (i_1, \ldots, i_p)\bigr)$).

Input: n a positive integer, $p \in \{1, \ldots, n\}$ and a vector $(i_1, \ldots, i_p)$ of strictly increasing positive integers $\le n$.

Output: b, the number of $e_{i_1 \ldots i_p}$ with respect to the above numbering.

- if $p = 1$ then $b := i_1$;
  else $b := \sum_{\nu=1}^{i_1-1} \binom{n-\nu}{p-1} + \text{BASISNUMBER}\bigl(n - i_1, p - 1, (i_2 - i_1, \ldots, i_p - i_1)\bigr)$;
- return(b).

In the language of SINGULAR we obtain (since p is the size of the integer vector $\mathtt{v} = (i_1, \ldots, i_p)$, we only need two parameters):

```
proc basisNumber(int n,intvec v)
{
  int p=size(v);
  if(p==1){return(v[1]);}
```

```
    int j=n-1;
    int b;
    while(j>=n-v[1]+1)
    {
       b=b+binom(j,p-1);
       j--;
    }
    intvec w=v-v[1];
    w=w[2..size(w)];
    b=b+basisNumber(n-v[1],w);
    return(b);
}
```

The inverse function is given by the following procedure:

Algorithm 7.6.19 (BASISELEMENT(n, p, N)).

Input: n a positive integer, $p \in \{1, \ldots, n\}$ and $N \in \left\{1, \ldots, \sum_{\nu=1}^{n} \binom{n-\nu}{p-1}\right\}$.
Output: $(i_1, \ldots, i_p)$, the vector of strictly increasing integers in $\{1, \ldots, n\}$
         having number N with respect to the above numbering.

- if $p = 1$ then return(N);
  else choose s such that $\sum_{\nu=1}^{s-1} \binom{n-\nu}{p-1} < N \leq \sum_{\nu=1}^{s} \binom{n-\nu}{p-1}$;
    $i_1 := s$;
    $(j_1, \ldots, j_{p-1}) := $ BASISELEMENT$\left(n - i_1, p - 1, N - \sum_{\nu=1}^{s-1} \binom{n-\nu}{p-1}\right)$;
    $i_\nu := j_{\nu-1} + i_1$ for $\nu = 2, \ldots, p$;
    return$(i_1, \ldots, i_p)$.

The corresponding SINGULAR procedure is:

```
  proc basisElement(int n,int p,int N)
  {
     if(p==1){return(N);}
     int s,R;
     while(R<N)
     {
        s++;
        R=R+binom(n-s,p-1);
     }
     R=N-R+binom(n-s,p-1);
     intvec v=basisElement(n-s,p-1,R);
     intvec w=s,v+s;
     return(w);
  }
```

Now we compute the matrix describing $d_p : K(x)_p \to K(x)_{p-1}$, with respect
to the above bases.

```
proc KoszulMap(ideal x,int p)
{
  int n=size(x);
  int a=binom(n,p-1);
  int b=binom(n,p);
  matrix M[a][b];
  if(p==1){M=x;return(M);}
  int j,k;
  intvec v,w;
  for(j=1;j<=b;j++)
  {
    v=basisElement(n,p,j);
    w=v[2..p];
    M[basisNumber(n,w),j]=x[v[1]];
    for(k=2;k<p;k++)
    {
      w=v[1..k-1],v[k+1..p];
      M[basisNumber(n,w),j]=(-1)^(k-1)*x[v[k]];
    }
    w=v[1..p-1];
    M[basisNumber(n,w),j]=(-1)^(p-1)*x[v[p]];
  }
  return(M);
}
```

Finally, we compute the p–th Koszul homology $H_p(x, R^n/M)$ of the quotient R^n/M, M a submodule of R^n. It is presented via modulo (cf. SINGULAR Example 2.1.26). That is, the return value is a module hom such that matrix(hom) is a presentation matrix of $H_p(x, R^n/M)$ with the number of rows being minimized via prune. In particular, if $H_p(x, R^n/M) = 0$, then the return value is hom $= 0$, with nrows(hom) $= 0$.

```
proc KoszulHomology(ideal x, module M, int p)
{
  int n     =size(x);
  int a     =binom(n,p-1);
  int b     =binom(n,p);
  matrix N  =matrix(M);
  module ker=freemodule(nrows(N));
  if(p!=0)
  {
    module im=tensorMaps(unitmat(a),N);
    module f =tensorMaps(KoszulMap(x,p),unitmat(nrows(N)));
    ker       =modulo(f,im);
  }
  module im1=tensorMaps(unitmat(b),N);
```

```
    module im2=tensorMaps(KoszulMap(x,p+1),unitmat(nrows(N)));
    module hom=modulo(ker,im1+im2);
    hom         =prune(hom);
    return(hom);
}
```

As an example, we compute in the following the homology of the Koszul complex $K(x_1, x_2, x_3)_\bullet$ as complex of $\mathbb{Q}[x_1, x_2, x_3]$–modules, respectively as complex of $\mathbb{Q}[x_1, x_2, x_3]/\langle x_1 x_2 \rangle$–modules.

```
    LIB"matrix.lib";
    ring R=0,x(1..3),dp;
    ideal y=maxideal(1);
    module M=0;

    KoszulHomology(y,M,0);
    //-> _[1]=x(3)*gen(1)   _[2]=x(2)*gen(1)   _[3]=x(1)*gen(1)

    KoszulHomology(y,M,1);
    //-> _[1]=0
```

Hence, $H_0(x, \mathbb{Q}[x]) = \mathbb{Q}[x]/\langle x \rangle$, $x = (x_1, x_2, x_3)$, and $H_1(x, \mathbb{Q}[x]) = 0$. The latter is already clear by Theorem 7.6.14 (1).

```
    qring S=std(x(1)*x(2));
    module M=0;
    ideal x=maxideal(1);

    KoszulHomology(x,M,1);
    //-> _[1]=-x(3)*gen(1)   _[2]=-x(2)*gen(1)   _[3]=-x(1)*gen(1)

    KoszulHomology(x,M,2);
    //-> _[1]=0
```

From the output, we read $H_1(x, \mathbb{Q}[x]/\langle x_1 x_2 \rangle) = \mathbb{Q}[x]/\langle x \rangle$, $x = (x_1, x_2, x_3)$ and $H_2(x, \mathbb{Q}[x]/\langle x_1 x_2 \rangle) = 0$.

At the end of this section we describe the I–depth of a module M in terms of $\mathrm{Ext}_A^n(A/I, M)$. To prove this characterization we need the following lemma.

Lemma 7.6.20. *Let A be a Noetherian ring, M a finitely generated A–module, $I \subset A$ an ideal with $IM \neq M$. The following are equivalent:*

(1) $\mathrm{Ext}_A^i(N, M) = 0$ for all $i < n$ and all finitely generated A–modules N with $\mathrm{supp}(N) \subset V(I)$.

(2) $\mathrm{Ext}_A^i(A/I, M) = 0$ for all $i < n$.

(3) $\mathrm{Ext}_A^i(N, M) = 0$ for all $i < n$ and some finitely generated A–module N with $\mathrm{supp}(N) = V(I)$.

(4) I contains an M–sequence of length n.

Proof. That (1) implies (2) and that (2) implies (3) are obvious. To prove that (3) implies (4), let $n > 0$ and assume first that I contains only zerodivisors of M, that is, I is contained in an associated prime ideal $P = (0 : m)$, $m \neq 0$ an element of M. Then the map $A/P \to M$, defined by $1 \mapsto m$, is injective. Localizing at P, we obtain that $\mathrm{Hom}_{A_P}(k, M_P) \neq 0$ for $k = A_P/PA_P$. Now $P \in V(I) = \mathrm{supp}(N)$, that is, $N_P \neq 0$ and, hence, $N_P/PN_P = N \otimes_A k \neq 0$ (Lemma of Nakayama). This implies that $\mathrm{Hom}_k(N \otimes_A k, k) \neq 0$ and, therefore, we have a non–trivial A_P-linear map

$$N_P \longrightarrow N \otimes_A k \longrightarrow k \longrightarrow M_P \,,$$

that is, $\mathrm{Hom}_{A_P}(N_P, M_P) \neq 0$. This implies that $\mathrm{Hom}_A(N, M) \neq 0$, which contradicts (3) for $i = 0$. So we proved that I contains an M–regular element f. By assumption $M/IM \neq 0$, hence if $n = 1$ we are done. If $n > 1$, then we obtain from the exact sequence

$$0 \longrightarrow M \xrightarrow{f} M \longrightarrow M/fM \longrightarrow 0$$

that $\mathrm{Ext}_A^i(N, M/fM) = 0$ for $i < n - 1$. Using induction this implies that I contains an M/fM–regular sequence $f_2, \ldots, f_n$.

To prove (4) implies (1), let $f_1, \ldots, f_n \in I$ be an M–sequence and consider again the exact sequence

$$0 \longrightarrow M \xrightarrow{f_1} M \longrightarrow M/f_1M \longrightarrow 0 \,.$$

Using Exercise 7.1.3 we obtain the exact sequence

$$\cdots \to \mathrm{Ext}_A^i(N, M) \xrightarrow{f_1} \mathrm{Ext}_A^i(N, M) \to \mathrm{Ext}_A^i(N, M/f_1M) \to \cdots \,.$$

If $n = 1$ then we consider the first part of this sequence

$$0 \longrightarrow \mathrm{Hom}_A(N, M) \xrightarrow{f_1} \mathrm{Hom}_A(N, M) \,.$$

If $n > 1$ then we use induction to obtain $\mathrm{Ext}_A^i(N, M/f_1M) = 0$ for $i < n - 1$. This implies

$$0 \longrightarrow \mathrm{Ext}_A^i(N, M) \xrightarrow{f_1} \mathrm{Ext}_A^i(N, M)$$

is exact for $i < n$. Now $\mathrm{Ext}_A^i(N, M)$ is annihilated by elements of $\mathrm{Ann}(N)$ (Exercise 7.1.3).

On the other hand, by assumption, we have

$$\mathrm{supp}(N) = V\big(\mathrm{Ann}(N)\big) \subset V(I) \,.$$

This implies that $I \subset \sqrt{\mathrm{Ann}(N)}$. Therefore, a sufficiently large power of f_1 annihilates $\mathrm{Ext}_A^i(N, M)$. But we already saw that f_1 is a non–zerodivisor for $\mathrm{Ext}_A^i(N, M)$ and, consequently, $\mathrm{Ext}^i(N, M) = 0$ for $i < n$. $\qquad \square$

Corollary 7.6.21. *Let A be a Noetherian ring, M a finitely generated A–module and $I \subset A$ an ideal with $IM \neq M$, then*

$$\operatorname{depth}(I, M) = \inf\{n \mid \operatorname{Ext}_A^n(A/I, M) \neq 0\}.$$

Exercises

7.6.1. Prove (1) and (2) of Lemma 7.6.7.

7.6.2. Prove that the map $\varphi = \{\varphi_p\}$, given in the proof of Lemma 7.6.7 (3), is an isomorphism of complexes.

7.6.3. Prove (3) of Theorem 7.6.14.

7.6.4. Prove that in SINGULAR Example 7.6.17

$$H_p(x, M) = (d_p^{A^s})^{-1}\big(\operatorname{Im}(\varphi^{K_{p-1}})\big)/\big(\operatorname{Im}(d_{p+1}^{A^s}) + \operatorname{Im}(\varphi^{K_p})\big).$$

7.6.5. Use Corollary 7.6.21 to give another proof for the equality

$$\operatorname{depth}(I, M) = \operatorname{depth}(\sqrt{I}, M).$$

7.6.6. Let $(A, \mathfrak{m})$ be a local ring and $M \neq 0$ a finitely generated A–module. Let $x_1, \ldots, x_n \in \mathfrak{m}$ be an M–sequence. Prove that

$$\dim(M/\langle x_1, \ldots, x_n \rangle M) = \dim(M) - n.$$

7.6.7. Let A be a Noetherian ring, M a finitely generated A–module and $P \subset A$ a prime ideal. Prove that $\operatorname{depth}(P, M) \leq \operatorname{depth}(M_P)$.

7.6.8. Let $(A, \mathfrak{m})$ and $(B, \mathfrak{n})$ be local Noetherian rings such that $A \subset B$, $\mathfrak{n} \cap A = \mathfrak{m}$ and $\mathfrak{m}B$ is $\mathfrak{n}$–primary. Let M be a B–module which is finitely generated over A. Prove that $\operatorname{depth}_B(M) = \operatorname{depth}_A(M)$.

7.6.9. Let K be a field, $x = (x_1, \ldots, x_n)$ variables, and A a $K[x]$–algebra. Let M be an A–module. Prove that $H_i(x, M) = \operatorname{Tor}_i^{K[x]}(K, M)$.

7.6.10. Let A be a Noetherian ring, $x_1, \ldots, x_n$ an A–sequence. Prove that $\operatorname{ht}(\langle x_1 \rangle) < \operatorname{ht}(\langle x_1, x_2 \rangle) < \ldots < \operatorname{ht}(\langle x_1, \ldots, x_n \rangle)$.

7.6.11. Prove the statement in Remark 7.6.2.

7.6.12. Show that a faithfully flat (see Exercise 7.3.13) morphism between two rings maps a regular sequence to a regular sequence. In particular, if $(A, \mathfrak{m})$, $(B, \mathfrak{n})$ are local rings such that B is a flat A–algebra and $\mathfrak{n} \supset \mathfrak{m}B$, then any regular sequence of elements of A remains regular in B.

7.6.13. Let A be a ring, $x = (x_1, \ldots, x_n)$ a sequence of elements in A and $0 \to M \to N \to P \to 0$ an exact sequence of R–modules. Then, with the canonically induced maps, the sequence

$$\ldots \to H_i(x, M) \to H_i(x, N) \to H_i(x, P) \to H_{i-1}(x, M) \to \ldots$$

of homology modules is exact.

7.7 Cohen–Macaulay Rings

At the beginning of this section we shall see that, for a Noetherian local ring A and a finitely generated A–module, we always have the inequality

$$\operatorname{depth}(M) \leq \dim(M).$$

In the previous section we saw several examples with $\operatorname{depth}(M) = \dim(M)$. Modules with this property are called Cohen–Macaulay modules.

Definition 7.7.1. Let $(A, \mathfrak{m})$ be a local ring, M a finitely generated A–module. M is called a *Cohen–Macaulay module* if $M = 0$ or $M \neq 0$ and

$$\operatorname{depth}(M) = \dim(M).$$

If $\operatorname{depth}(M) = \dim(A)$ then M is called *maximal Cohen–Macaulay*. A is called a *Cohen–Macaulay ring* if it is a Cohen–Macaulay A–module.

Example 7.7.2. Regular local rings are Cohen–Macaulay (Corollary 5.6.18 and Proposition 5.6.17).

Proposition 7.7.3. *Let $(A, \mathfrak{m})$ be a Noetherian local ring and M a finitely generated A–module, then $\dim(A/P) \geq \operatorname{depth}(M)$ for all $P \in \operatorname{Ass}(M)$.*

To prove the proposition we need the following lemma:

Lemma 7.7.4. *Let $(A, \mathfrak{m})$ be a Noetherian local ring and M, N finitely generated A–modules different from zero, then*

$$\operatorname{Ext}_A^i(N, M) = 0 \ \text{ for } \ i < \operatorname{depth}(M) - \dim(N).$$

Proof of Proposition 7.7.3. Let $P \in \operatorname{Ass}(M)$, that is, $P = (0 : m)$ for a suitable $m \neq 0$ in M. This implies that $\operatorname{Hom}(A/P, M) \neq 0$ because $1 \mapsto m$ defines a non–trivial homomorphism. Hence, by Lemma 7.7.4, we obtain $0 \geq \operatorname{depth}(M) - \dim(A/P)$. $\square$

Proof of Lemma 7.7.4. Let $k = \operatorname{depth}(M)$ and $r = \dim(N)$. We prove the lemma by induction on r. If $r = 0$ then $\operatorname{Ann}(N)$ is $\mathfrak{m}$–primary. We use Lemma 7.6.20 and obtain $\operatorname{Ext}_A^i(N, M) = 0$ for $i < \operatorname{depth}(M)$. Now assume that $r > 0$. We choose a sequence $N = N_0 \supset N_1 \supset \cdots \supset N_n = \langle 0 \rangle$ such that $N_j/N_{j+1} \cong A/P_j$ for suitable prime ideals P_j (Exercise 4.1.15).

Now it is sufficient to prove that $\operatorname{Ext}_A^i(N_j/N_{j+1}, M) = 0$ for all j and $i < k - r$ because this implies $\operatorname{Ext}_A^i(N, M) = 0$ (cf. Exercise 7.1.3). Since $\dim(N_j/N_{j+1}) \leq \dim(N)$, we may assume that $N = A/P$, P a prime ideal. Since $\dim(A/P) > 0$, we can choose $x \in \mathfrak{m} \smallsetminus P$. We consider the exact sequence

$$0 \longrightarrow N \xrightarrow{x} N \longrightarrow N' \longrightarrow 0, \quad N' = A/\langle P, x \rangle.$$

Now $\dim(N') < r$ implies $\mathrm{Ext}_A^i(N', M) = 0$ for $i < k - r + 1$ (induction). On the other hand, for $i < k - r$ we have the exact sequence

$$0 \longrightarrow \mathrm{Ext}_A^i(N, M) \xrightarrow{x} \mathrm{Ext}_A^i(N, M) \longrightarrow \mathrm{Ext}_A^{i+1}(N', M) = 0.$$

This implies $\mathrm{Ext}_A^i(N, M) = x\,\mathrm{Ext}_A^i(N, M)$. Using Nakayama's Lemma we obtain $\mathrm{Ext}_A^i(N, M) = 0$. $\qquad\Box$

Theorem 7.7.5. *Let $(A, \mathfrak{m})$ be a Noetherian local ring and $M \neq 0$ a finitely generated A–module.*

(1) Let M be Cohen–Macaulay. Then $\dim(A/P) = \dim(M)$ for all $P \in \mathrm{Ass}(M)$. Moreover, if for $x \in A, \dim(M/xM) = \dim(M) - 1$, then x is M-regular.

(2) Let $x_1, \ldots, x_r \in \mathfrak{m}$ be an M–sequence, then M is Cohen–Macaulay if and only if $M/\langle x_1, \ldots, x_r\rangle M$ is Cohen–Macaulay.

(3) If M is Cohen–Macaulay, then M_P is Cohen–Macaulay for all prime ideals P and $\mathrm{depth}(P, M) = \mathrm{depth}_{A_P}(M_P)$ if $M_P \neq 0$.

Proof. (1) By definition, $\dim(M) = \sup\{\dim(A/P) \mid P \in \mathrm{Ass}(M)\}$. On the other hand, we have $\mathrm{depth}(M) \leq \inf\{\dim(A/P) \mid P \in \mathrm{Ass}(M)\}$ (Proposition 7.7.3), whence the first statement. Since $\dim(M/xM) < \dim(M) = \dim(A/P), x \notin P$ for all $P \in \mathrm{Ass}(M)$. Hence x is M-regular.

(2) We obtain $\mathrm{depth}(M/\langle x_1, \ldots, x_r\rangle M) = \mathrm{depth}(M) - r$, using Corollary 7.6.12. On the other hand, $\dim(M/\langle x_1, \ldots, x_r\rangle M) = \dim(M) - r$ (Exercise 7.6.6).

(3) Let P be a prime ideal. If $P \not\supset \mathrm{Ann}(M)$ then $M_P = 0$. Then M_P is Cohen–Macaulay by definition. If $P \supset \mathrm{Ann}(M)$ then $M_P \neq 0$. Now, because of Proposition 7.7.3, we have $\dim(M_P) \geq \mathrm{depth}(M_P)$ and, obviously, $\mathrm{depth}(M_P) \geq \mathrm{depth}(P, M)$. Therefore, it is enough to prove that

$$\dim(M_P) = \mathrm{depth}(P, M).$$

We use induction on $\mathrm{depth}(P, M)$. If $\mathrm{depth}(P, M) = 0$ then P is contained in an associated prime of M. But $P \supset \mathrm{Ann}(M)$ and, since all associated primes are minimal by (1), it follows that P is an associated prime of M. This implies $\dim(M_P) = 0$. Now assume that $\mathrm{depth}(P, M) > 0$ and choose $a \in P$ M-regular. Then

$$\mathrm{depth}(P, M/aM) = \mathrm{depth}(P, M) - 1$$

(Corollary 7.6.12). By (2), the quotient M/aM is Cohen–Macaulay and, by Nakayama's Lemma, $M/aM \neq 0$. If $(M/aM)_P = 0$, then $sM = aM$ for some $s \notin P$. Let $M = \langle m_1, \ldots, m_n\rangle$, then, for all $i = 1, \ldots, n$, there exist $h_{ij} \in P$ such that $sm_i = \sum_{j=1}^n h_{ij}m_j$, hence, $\det(h_{ij} - s\delta_{ij}) \in \mathrm{Ann}(M) \subset P$ and, therefore, $s \in P$, which is a contradiction. We conclude that $(M/aM)_P \neq 0$.

Now, by induction, we have $\dim(M/aM)_P = \operatorname{depth}(P, M/aM)$. On the other hand, a is M_P–regular and $(M/aM)_P = M_P/aM_P$. Using Exercise 7.6.6 once more, we obtain $\dim(M/aM)_P = \dim(M_P) - 1$. Finally, this implies $\dim(M_P) = \operatorname{depth}(P, M)$. $\qquad\square$

Corollary 7.7.6 (complete intersection rings[4] are Cohen-Macaulay). *Let K be a field and $f_1, \ldots, f_k$ be a regular sequence in $A = K[x]_{\langle x \rangle}$ (respectively $A = K[[x]]$), $x = (x_1, \ldots, x_n)$, then the quotient $A/\langle f_1, \ldots, f_k \rangle$ is a Cohen–Macaulay ring.*

Proposition 7.7.7. *Let $(A, \mathfrak{m})$ be a Noetherian local ring. Then $\operatorname{depth}(A) = \operatorname{depth}(\widehat{A})$. In particular, A is Cohen–Macaulay if and only if $\widehat{A}$ is Cohen–Macaulay.*

Proof. Let $\mathfrak{m} = \langle x_1, \ldots, x_n \rangle$, and let $K(x_1, \ldots, x_n, \widehat{A})_\bullet$ be the Koszul complex of the A–module $\widehat{A}$ and $x = (x_1, \ldots, x_n)$, considered in $\widehat{A}$. We already know that $\widehat{A}$ is flat A–algebra. This implies that $H_i(x, \widehat{A}) = H_i(x, A) \otimes_A \widehat{A}$. Now, the proposition is a consequence of Corollaries 7.6.11 and 6.3.6. $\qquad\square$

SINGULAR Example 7.7.8 (first test for Cohen–Macaulayness).
Let $(A, \mathfrak{m})$ be a local ring, $\mathfrak{m} = \langle x_1, \ldots, x_n \rangle$. Let M be an A–module given by a presentation $A^\ell \to A^s \to M \to 0$. To check whether M is Cohen–Macaulay we use that the equality

$$\dim\big(A/\operatorname{Ann}(M)\big) = \dim(M) = \operatorname{depth}(M)$$
$$= n - \sup\{i \mid H_i(x_1, \ldots, x_n, M) \neq 0\}.$$

is necessary and sufficient for M to be Cohen–Macaulay. The following procedure computes $\operatorname{depth}(\mathfrak{m}, M)$, where $\mathfrak{m} = \langle x_1, \ldots, x_n \rangle \subset A = K[x_1, \ldots, x_n]_>$ and M is a finitely generated A–module with $\mathfrak{m}M \neq M$, by using Corollary 7.6.11.

```
proc depth(module M)
{
   ideal m=maxideal(1);
   int n=size(m);
   int i;
   while(i<n)
   {
     i++;
     if(size(KoszulHomology(m,M,i))==0){return(n-i+1);}
   }
   return(0);
}
```

---

[4] A local ring A is called a *complete intersection ring* if there exists a regular local Noetherian ring R and a regular sequence $f_1, \ldots, f_k$ in R such that $A \cong R/\langle f_1, \ldots, f_k \rangle$.

Now the test for Cohen–Macaulayness is easy.

```
proc CohenMacaulayTest(module M)
{
  return(depth(M)==dim(std(Ann(M))));
}
```

The procedure returns 1 if M is Cohen–Macaulay and 0 if not.

As an application, we check that a complete intersection is Cohen–Macaulay (Corollary 7.7.6) and that $K[x,y,z]_{\langle x,y,z\rangle}/\langle xz,yz,z^2\rangle$ is not Cohen–Macaulay.

```
ring R=0,(x,y,z),ds;
ideal I=xz,yz,z2;
module M=I*freemodule(1);
CohenMacaulayTest(M);
//-> 0

I=x2+y2,z7;
M=I*freemodule(1);
CohenMacaulayTest(M);
//-> 1
```

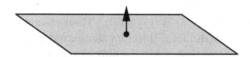

Fig. 7.1. The zero–set of $\langle xz,yz,z^2\rangle$, consisting of a plane with an embedded point (indicated by the arrow).

Cohen–Macaulay rings have several nice properties:

Proposition 7.7.9. Let $(A,\mathfrak{m})$ be a local Cohen–Macaulay ring, and let $x_1,\ldots,x_n \in \mathfrak{m}$, then the following conditions are equivalent:

(1) $x_1,\ldots,x_n$ is an A–sequence.
(2) $\mathrm{ht}(\langle x_1,\ldots,x_i\rangle) = i$ for $1 \le i \le n$.
(3) $\mathrm{ht}(\langle x_1,\ldots,x_n\rangle) = n$.
(4) $x_1,\ldots,x_n$ is part of a system of parameters of A.

Proof. The implication $(1) \Rightarrow (2)$ is a consequence of Remark 5.6.11, which states that $\mathrm{ht}(\langle x_1,\ldots,x_i\rangle) \le i$, and the fact that

$$0 < \mathrm{ht}(\langle x_1\rangle) < \mathrm{ht}(\langle x_1,x_2\rangle) < \cdots < \mathrm{ht}(\langle x_1,\ldots,x_i\rangle),$$

because $x_1,\ldots,x_i$ is an A–sequence (Exercise 7.6.10).

(2) $\Rightarrow$ (3) is trivial. (3) $\Rightarrow$ (4) is clear, at least, for the case $\dim(A) = n$. If $\dim(A) > n$ then we can choose $x_{n+1} \in \mathfrak{m}$ not being contained in any minimal prime ideal of $\langle x_1, \ldots, x_n \rangle$. Then $\mathrm{ht}(\langle x_1, \ldots, x_{n+1} \rangle) = n + 1$. We can continue this way and obtain a system of parameters.

(4) $\Rightarrow$ (1). We shall show that any system of parameters $x_1, \ldots, x_n$ is an A–sequence. Let $x_1, \ldots, x_n$ be a system of parameters, that is, $n = \dim(A)$ and $\sqrt{\langle x_1, \ldots, x_n \rangle} = \mathfrak{m}$. By Theorem 7.7.5 (1), we obtain $\dim(A/P) = n$ for all minimal prime ideals P of A. This implies that $x_1 \notin P$ for all minimal prime ideals. But then x_1 is A–regular and $A/\langle x_1 \rangle$ is an $(n-1)$–dimensional Cohen–Macaulay ring (Theorem 7.7.5 (2)). The images of $x_2, \ldots, x_n$ in $A/\langle x_1 \rangle$ form a system of parameters of $A/\langle x_1 \rangle$. Using induction on n we deduce that $x_1, \ldots, x_n$ is an A–sequence. $\qquad\square$

Corollary 7.7.10. *Let $(A, \mathfrak{m})$ be a Noetherian local Cohen–Macaulay ring, and let $I \subset A$ be an ideal. Then*

(1) $\mathrm{ht}(I) = \mathrm{depth}(I, A)$;
(2) $\mathrm{ht}(I) + \dim(A/I) = \dim(A)$;
(3) A *is equidimensional.*

Proof. If $\mathrm{ht}(I) = n$, then we can choose elements $x_1, \ldots, x_n \in I$ such that $\mathrm{ht}(\langle x_1, \ldots, x_i \rangle) = i$ for $1 \leq i \leq n$. Using the previous proposition, we see that $x_1, \ldots, x_n$ is an A–sequence. This implies $\mathrm{ht}(I) \leq \mathrm{depth}(I, A)$. Now let $x_1, \ldots, x_s \in I$ be an A–sequence then $s = \mathrm{ht}(\langle x_1, \ldots, x_s \rangle) \leq \mathrm{ht}(I)$. This proves (1).

By definition, $\mathrm{ht}(I) = \inf\{\mathrm{ht}(P) \mid P$ an associated prime ideal of $I\}$, and $\dim(A/I) = \sup\{\dim(A/P) \mid P$ an associated prime ideal of $I\}$. Hence, it is sufficient to prove that $\mathrm{ht}(P) = \dim(A) - \dim(A/P)$ for a prime ideal P.

Using (1) we obtain $\mathrm{ht}(P) = \mathrm{depth}(P, A)$. Let $x_1, \ldots, x_r \in P$ be an A–sequence, $r = \mathrm{ht}(P)$. Then $A/\langle x_1, \ldots, x_r \rangle$ is a $(\dim(A) - r)$–dimensional Cohen–Macaulay ring (Theorem 7.7.5 (2)).

We have $\mathrm{ht}(\langle x_1, \ldots, x_r \rangle) = r = \mathrm{ht}(P)$ (Proposition 7.7.9) and, therefore, P is a minimal associated prime for $\langle x_1, \ldots, x_r \rangle$. This implies

$$\dim(A/P) = \dim(A/\langle x_1, \ldots, x_r \rangle) = \dim(A) - \mathrm{ht}(P)$$

(Theorem 7.7.5 (1)) and proves (2).

(3) is an immediate consequence of Theorem 7.7.5 (1). $\qquad\square$

Corollary 7.7.11. *Let $(A, \mathfrak{m})$ be a Noetherian local Cohen–Macaulay ring, and let $I = \langle x_1, \ldots, x_r \rangle \subset A$ be an ideal of height r. Then all associated prime ideals of I have the same height.*

Proof. Assume $P \supset I$ is an associated prime. Localizing at P we may assume that $P = \mathfrak{m}$ (Theorem 7.7.5(3)). Using Proposition 7.7.9 we obtain that $x_1, \ldots, x_r$ is an A–sequence. This implies that A/I is a Cohen–Macaulay ring

(Theorem 7.7.5 (2)), hence $\dim(A/P) = \dim(A/I)$. Using Corollary 7.7.10 (2) we obtain $\mathrm{ht}(P) = \mathrm{ht}(I)$.   □

We now *globalize* the notion of a Cohen–Macaulay ring. Theorem 7.7.5 (3) gives rise to the following definition:

Definition 7.7.12. Let A be a Noetherian ring. A is called a *Cohen–Macaulay ring* if $A_{\mathfrak{m}}$ is a Cohen–Macaulay ring for all maximal ideals $\mathfrak{m}$ of A.

Theorem 7.7.13. *Let A be a Cohen–Macaulay ring, then $A[x_1, \ldots, x_n]$ is a Cohen–Macaulay ring.*

Proof. Using induction it is enough to prove the theorem for $n = 1$. Let P be a maximal ideal in $A[x]$ and $P \cap A = \mathfrak{m}$. Then $A[x]_P$ is a localization of $A_{\mathfrak{m}}[x]$. Therefore, we may assume that $(A, \mathfrak{m})$ is a local ring.

The quotient $A[x]/\mathfrak{m}A[x] \cong (A/\mathfrak{m})[x]$ is a principal ideal domain, and there exists a monic polynomial $f \in A[x]$ such that $P \equiv \langle f \rangle \bmod \mathfrak{m}A[x]$, that is, $P = \mathfrak{m}A[x] + \langle f \rangle$. Let $y_1, \ldots, y_n$ be a system of parameters for A which is an A–sequence by Proposition 7.7.9. Since $A[x]$ is a free A–module, $y_1, \ldots, y_n$ is an $A[x]$–sequence, too. On the other hand, f is (as monic polynomial) a non–zerodivisor mod $\langle y_1, \ldots, y_n \rangle A[x]$. This implies that $y_1, \ldots, y_n, f$ is an $A[x]$–sequence in P, especially $\mathrm{depth}(P, A[x]) \geq 1 + \mathrm{depth}(A)$. We conclude

$$\mathrm{depth}(A[x]_P) \geq \mathrm{depth}(P, A[x]) \geq 1 + \dim(A) = \dim(A[x]_P)$$

(cf. Exercise 3.3.8), that is, $A[x]_P$ is a Cohen–Macaulay ring.   □

Exercises

7.7.1. Let A be a local Cohen–Macaulay ring. Prove that $A[[x_1, \ldots, x_n]]$ is a Cohen–Macaulay ring.

7.7.2. Let A be a zero–dimensional ring. Prove that A is Cohen–Macaulay.

7.7.3. Let A be a reduced one–dimensional ring. Prove that A is Cohen–Macaulay.

7.7.4. Let A be a two–dimensional normal ring. Prove that A is Cohen–Macaulay.

7.7.5. Consider *Serre's conditions* (Ri), (Si), $i \geq 0$, for a Noetherian ring A:

(Ri) A_P is regular for every prime ideal P of height at most i.
(Si) $\mathrm{depth}(A_P) \geq \min\{\mathrm{ht}(P), i\}$ for every prime ideal P of A.

Show that A is Cohen–Macaulay if and only if (Si) holds for $i \geq 0$, and show that A is reduced if and only if (R0) and (S1) hold (see also Theorem 5.7.13).

7.7.6. Let $A = K[x_1, \ldots, x_n]/I$ be Cohen–Macaulay, with K a perfect field. Show that A is reduced if and only if $\dim(A/J) < \dim(A)$, where J is the ideal in A, generated by the $(n - \dim(A))$–minors of the Jacobian matrix $\left(\frac{\partial f_i}{\partial x_j}\right)$. (Geometrically, this means that a Cohen–Macaulay variety is reduced if and only if its singular locus has codimension at least 1.)

7.7.7. Which of the following rings are Cohen–Macaulay and which are not?

(1) $K[x, y, z]/\langle xy, yz \rangle$,
(2) $K[x, y, z]/\langle xy, yz, xz \rangle$,
(3) $K[x, y]/\langle xy, x^2 \rangle$,
(4) $K[x, y]/\langle x^2, y^2 \rangle$,
(5) $K[x, y]/\langle y^4 - y^2, xy^3 - xy, x^3y - xy, x^4 - x^2 \rangle$.

First use SINGULAR and then find a theoretical argument.

7.8 Further Characterization of Cohen–Macaulayness

In this section we shall first use finite ring extensions as, for instance, Noether normalizations to characterize Cohen–Macaulay rings. We prove that a local ring, which is finite over a regular local ring, is Cohen–Macaulay if and only if it is free. Further, we shall introduce the projective dimension $\mathrm{pd}_A(M)$ as the length of a minimal resolution of a finitely generated module M over a local ring A and prove the Auslander–Buchsbaum formula,

$$\mathrm{pd}_A(M) = \mathrm{depth}(A) - \mathrm{depth}(M),$$

which allows to check Cohen–Macaulayness over regular local rings by computing a free resolution.

Proposition 7.8.1. *Let* $(A, \mathfrak{m}) \to (B, \mathfrak{n})$ *be a local map of local rings (that is, $\mathfrak{m}B \subset \mathfrak{n}$), and assume that B is flat over A. Then*

(1) $\mathrm{depth}(B) = \mathrm{depth}(A) + \mathrm{depth}(B/\mathfrak{m}B)$.
(2) B *is Cohen–Macaulay if and only if A and $B/\mathfrak{m}B$ are Cohen–Macaulay.*

Proof. To prove (1), let $x_1, \ldots, x_r \in \mathfrak{m}$ be a maximal regular sequence and $y_1, \ldots, y_s \in \mathfrak{n}$ induce a maximal regular sequence in $B/\mathfrak{m}B$. Let $\tilde{x}_1, \ldots, \tilde{x}_r$ be the images of $x_1, \ldots, x_r$ in B. We shall prove that $\tilde{x}_1, \ldots, \tilde{x}_r, y_1, \ldots, y_s$ is a maximal regular sequence in $\mathfrak{n}$.

Using Exercise 7.6.12, we obtain that $\tilde{x}_1, \ldots, \tilde{x}_r$ is a regular sequence in $\mathfrak{n}$. Using Corollary 7.4.8, we obtain that y_1 is regular in B and $B/\langle y_1 \rangle$ is A–flat. Consider the exact sequence

$$0 \longrightarrow B \overset{\cdot y_1}{\longrightarrow} B \longrightarrow B/\langle y_1 \rangle \longrightarrow 0$$

then, for $\overline{A} := A/\langle x_1, \ldots, x_r \rangle$ we obtain

$$\mathrm{Tor}_1^A(\overline{A}, B/\langle y_1\rangle) \longrightarrow \overline{A}\otimes_A B \xrightarrow{\;\cdot y_1\;} \overline{A}\otimes_A B$$

is exact. But $\mathrm{Tor}_1^A(\overline{A}, B/\langle y_1\rangle) = 0$, since $B/\langle y_1\rangle$ is A–flat. This implies that y_1 is regular in $\overline{A}\otimes_A B = B/\langle \tilde{x}_1,\dots,\tilde{x}_r\rangle$. Continuing like this, we obtain that $y_1,\dots,y_s$ is a regular sequence in $B/\langle \tilde{x}_1,\dots,\tilde{x}_r\rangle$. It remains to prove that $\mathrm{depth}(B/\langle \tilde{x}_1,\dots,\tilde{x}_r,\ y_1,\dots,y_s\rangle) = 0$. But this is clear, since $\mathfrak{m}^a \subset \langle x_1,\dots,x_r\rangle$ and $\mathfrak{n}^b \subset \mathfrak{m}B + \langle y_1,\dots,y_s\rangle$ for suitable a,b, which implies that $\langle \tilde{x}_1,\dots,\tilde{x}_r,\ y_1,\dots,y_s\rangle$ is $\mathfrak{n}$–primary.

To prove (2), we use (1) and Exercise 7.4.6 and obtain

$$\dim(B) - \mathrm{depth}(B) = \dim(A) - \mathrm{depth}(A) + \dim(B/\mathfrak{m}B) - \mathrm{depth}(B/\mathfrak{m}B).$$

Since dimension is always greater or equal to depth, we obtain (2).   □

Theorem 7.8.2. *Let $(A,\mathfrak{m})$ be a regular local ring and $(B,\mathfrak{n})$ a Noetherian local A–algebra ($\mathfrak{m}B \subset \mathfrak{n}$).*

(1) B is flat over A if and only if $\mathrm{depth}(\mathfrak{m}B, B) = \dim(A)$.
(2) If B is Cohen–Macaulay, then B is flat over A if and only if $\dim(B) = \dim(A) + \dim(B/\mathfrak{m}B)$.

Proof. To prove (1), let $n = \dim(A)$ and $\mathfrak{m} = \langle x_1,\dots,x_n\rangle$. Then, as A is regular, $x_1,\dots,x_n$ is a regular sequence in A, which remains regular in B, since B is flat over A (see Exercise 7.6.12). This implies that $\mathrm{depth}(\mathfrak{m}B, B) = \dim(A)$ (Corollary 7.6.11).

If $\mathrm{depth}(\mathfrak{m}B, B) = \dim(A) = n$, then Corollary 7.6.11 implies that the homology $H_i(x_1,\dots,x_n,B)$ vanishes for $i > 0$. Now $K(x_1,\dots,x_n)_\bullet$ is a free resolution of $A/\mathfrak{m}$. This implies $\mathrm{Tor}_1^A(A/\mathfrak{m},B) = H_1(x_1,\dots,x_n,B) = 0$, and we can apply Theorem 7.4.2 to deduce that B is A–flat.

To prove (2) assume first that B is flat over A. Then Exercise 7.4.6 gives that $\dim(B) = \dim(A) + \dim(B/\mathfrak{m}B)$.

Conversely, assume that $\dim(B) = \dim(A) + \dim(B/\mathfrak{m}B)$. Let $y_1,\dots,y_s$ be a maximal regular sequence in $\mathfrak{m}B$, that is, $s = \mathrm{depth}(\mathfrak{m}B, B)$. We may extend this sequence to a maximal sequence $y_1,\dots,y_d$ in B. Because of Corollary 7.6.11 and since B is Cohen–Macaulay, we have $d = \dim(B)$. On the other hand, $\dim(B/\mathfrak{m}B) = d - s$ (Exercise 7.6.6). Finally, this implies $s = \mathrm{depth}(\mathfrak{m}B, B) = \dim(A)$ and, using (1), we obtain that B is flat.   □

Corollary 7.8.3. *Let $(A,\mathfrak{m})$ be a regular local ring and B a finite A–algebra. Assume that all localizations of B in maximal ideals have the same dimension. Then B is Cohen–Macaulay if and only if B is a free A–module.*

Proof. Let $\dim(A) = n$ and $\mathfrak{m} = \langle x_1,\dots,x_n\rangle$. Now assume that B is Cohen–Macaulay, that is $B_\mathfrak{n}$ is Cohen–Macaulay for all maximal ideals $\mathfrak{n}$ of B. By assumption, we know that $\dim(A) = \dim(B) = \dim(B_\mathfrak{n})$ for all maximal ideals $\mathfrak{n}$. By Theorem 7.8.2 (2), B is flat over A and, therefore, free (Corollary 7.3.8).

To prove the other direction, assume that B is a free A–module. Thus, $x_1, \ldots, x_n$, which is a regular sequence in A (since A is a regular local ring), remains a regular sequence in B by flatness. Therefore,

$$\text{depth}(B_\mathfrak{n}) \geq \dim(A) = \dim(B) = \dim(B_\mathfrak{n})$$

for all maximal ideals $\mathfrak{n} \subset B$. $\square$

Corollary 7.8.4. *Let* $B = K[x]_{(x)}/I$ *(resp.* $B = K[[x]]/I$*),* $x = (x_1, \ldots, x_n)$*, and let* $A \subset B$ *be a Noether normalization, then* B *is Cohen–Macaulay if and only if* B *is a free* A–module.

SINGULAR Example 7.8.5 (second test for Cohen–Macaulayness).
Let $A = K[x_1, \ldots, x_n]_{\langle x_1, \ldots, x_n \rangle}/I$. Using Noether normalization, we may assume that $A \supset K[x_{s+1}, \ldots, x_n]_{\langle x_{s+1}, \ldots, x_n \rangle} =: B$ is finite. We choose a monomial basis $m_1, \ldots, m_r \in K[x_1, \ldots, x_s]$ of $A\big|_{x_{s+1} = \cdots = x_n = 0}$.

Then $m_1, \ldots, m_r$ is a minimal system of generators of A as B–module. A is Cohen–Macaulay if and only if A is a free B–module, that is, there are no B–relations between $m_1, \ldots, m_r$, in other words, $\text{syz}_A(m_1, \ldots, m_r) \cap B^r = \langle 0 \rangle$. This test can be implemented in SINGULAR as follows:

```
proc isCohenMacaulay(ideal I)
{
    def A     = basering;
    list L    = noetherNormal(I);
    map phi   = A,L[1];
    I         = phi(I);
    int s     = nvars(basering)-size(L[2]);
    execute("ring B=("+charstr(A)+"),x(1..s),ds;");
    ideal m   = maxideal(1);
    map psi   = A,m;
    ideal J   = std(psi(I));
    ideal K   = kbase(J);
    setring A;
    execute("
      ring C=("+charstr(A)+"),("+varstr(A)+"),(dp(s),ds);");
    ideal I   = imap(A,I);
    qring D   = std(I);
    ideal K   = fetch(B,K);
    module N = std(syz(K));
    intvec v = leadexp(N[size(N)]);
    int i=1;
    while((i<s)&&(v[i]==0)){i++;}
    setring A;
    if(!v[i]){return(0);}
    return(1);
}
```

As the above procedure uses `noetherNormal` from `algebra.lib`, we first have to load this library.

```
LIB"algebra.lib";
ring r=0,(x,y,z),ds;
ideal I=xz,yz;
isCohenMacaulay(I);
//-> 0

I=x2-y3;
isCohenMacaulay(I);
//-> 1
```

We want to prove the Auslander–Buchsbaum formula, which is of fundamental importance for modules which allow a finite projective resolution. First we need a definition.

Definition 7.8.6. Let $(A, \mathfrak{m})$ be a Noetherian local ring and M be a finitely generated A–module. M has finite *projective dimension* if there exists an exact sequence (a free resolution)

$$0 \longrightarrow F_n \overset{\alpha_n}{\longrightarrow} F_{n-1} \longrightarrow \ldots \overset{\alpha_1}{\longrightarrow} F_0 \overset{\alpha_0}{\longrightarrow} M \longrightarrow 0 \,,$$

with finitely generated free A–modules F_i. The integer n is called the *length* of the resolution. The minimal length of a free resolution is called the *projective dimension* of M and is denoted by $\mathrm{pd}_A(M)$.

This notion is generalized to non–local rings: $\mathrm{pd}_A(M) = n$ if there exists an exact sequence (projective resolution)

$$0 \rightarrow P_n \overset{\alpha_n}{\longrightarrow} P_{n-1} \rightarrow \ldots \overset{\alpha_1}{\longrightarrow} P_0 \overset{\alpha_0}{\longrightarrow} M \rightarrow 0 \,,$$

with finitely generated projective A–modules P_i, and n is minimal with this property.

Note that in Chapter 2, Theorem 2.4.11, it is proved that all minimal free resolutions have the same length. Obviously, free modules have projective dimension zero.

Theorem 7.8.7 (Auslander–Buchsbaum formula). *Let $(A, \mathfrak{m})$ be a Noetherian local ring and M a finitely generated A–module of finite projective dimension. Then*

$$\mathrm{depth}(M) + \mathrm{pd}_A(M) = \mathrm{depth}(A) \,.$$

We shall postpone the proof and treat an example first.

SINGULAR Example 7.8.8 (3rd test for Cohen–Macaulayness).
We use the Auslander–Buchsbaum formula to compute the depth of M and
then use the procedure of SINGULAR Example 7.7.8 to test Cohen–Macaulayness (check if $\operatorname{depth}(M) = \dim(M) = \dim(A/\operatorname{Ann}(M)))$.

We assume that $A = K[x_1, \ldots, x_n]_{\langle x_1, \ldots, x_n \rangle}/I$ and compute a minimal free
resolution. Then $\operatorname{depth}(A) = n - \operatorname{pd}_{K[x_1, \ldots, x_n]_{\langle x_1, \ldots, x_n \rangle}}(A)$. If M is a finitely
generated A–module of finite projective dimension, then we compute a min-
imal free resolution of M and obtain $\operatorname{depth}(M) = \operatorname{depth}(A) - \operatorname{pd}_A(M)$.

```
proc projdim(module M)
{
    list l=mres(M,0);
    int i;
    while(i<size(l))
    {
      i++;
      if(size(l[i])==0){return(i-1);}
    }
}
```

Note that `projdim` assumes that $\operatorname{pd}_A(M) < \infty$. By Theorem 7.9.1 and Corollary 7.9.5 below, this is true if $A = K[x_1, \ldots, x_n]_>$ and if M is finitely generated. For quotient rings of such a ring, however, the resolution may not be
finite. In SINGULAR the command `mres(M,0)` (respectively `res(M,0), ...`)
has the effect that the resolution stops after n steps, where n is the number
of variables of the basering. [5]

Now it is easy to give another test for Cohen–Macaulayness.

```
proc isCohenMacaulay1(ideal I)
{
  int de=nvars(basering)-projdim(I*freemodule(1));
  int di=dim(std(I));
  return(de==di);
}


ring R=0,(x,y,z),ds;
ideal I=xz,yz;
isCohenMacaulay1(I);
//-> 0


I=x2-y3;
isCohenMacaulay1(I);
//-> 1
```

---

[5] `mres(M,k)` stops computation after k steps, where $k > 0$ is any positive integer.

```
I=xz,yz,xy;
isCohenMacaulay1(I);
//-> 1
kill R;
```

The following procedure checks whether the depth of M is equal to d. It uses the procedure Ann from primdec.lib.

```
proc CohenMacaulayTest1(module M, int d)
{
   return((d-projdim(M))==dim(std(Ann(M))));
}
```

```
LIB"primdec.lib";
ring R=0,(x,y,z),ds;
ideal I=xz,yz;
module M=I*freemodule(1);
CohenMacaulayTest1(M,3);
//-> 0
```

```
I=x2+y2,z7;
M=I*freemodule(1);
CohenMacaulayTest1(M,3);
//-> 1
```

To prove the Auslander–Buchsbaum formula, we need the following lemma:

Lemma 7.8.9. *Let* $(A, \mathfrak{m})$ *be a Noetherian local ring and*

$$0 \longrightarrow M_1 \longrightarrow M_2 \longrightarrow M_3 \longrightarrow 0$$

be an exact sequence of A–modules then

$$\operatorname{depth}(M_2) \geq \min\big(\operatorname{depth}(M_1), \operatorname{depth}(M_3)\big).$$

If the inequality is strict then $\operatorname{depth}(M_1) = \operatorname{depth}(M_3) + 1$.

Proof. If all three modules have positive depth, then we can find a non–zerodivisor $f \in \mathfrak{m} \subset A$ of M_1, M_2, M_3 (prime avoidance, Lemma 1.3.12), and the sequence

$$0 \longrightarrow M_1/fM_1 \longrightarrow M_2/fM_2 \longrightarrow M_3/fM_3 \longrightarrow 0$$

is exact (Snake Lemma). The depth drops by one if we divide by f (Corollary 7.6.12). Therefore, the proof can be reduced to the case that the depth of one of the M_i is zero. Suppose that $\operatorname{depth}(M_1) = 0$. But then $\operatorname{depth}(M_2) = 0$ because any non–zerodivisor of M_2 is a non–zerodivisor of

M_1. The lemma is proved in this case. Assume now that $\operatorname{depth}(M_2) = 0$. Suppose $\operatorname{depth}(M_1) > 0$ and $\operatorname{depth}(M_3) > 0$. Let $f \in \mathfrak{m} \subset A$ be a non–zerodivisor of M_1 and M_3 (prime avoidance, Lemma 1.3.12). From the Snake Lemma we obtain that f is a non–zerodivisor for M_2, too. This is a contradiction.

Finally, assume that $\operatorname{depth}(M_3) = 0$. If $\operatorname{depth}(M_2) > 0$, let $f \in \mathfrak{m}$ be a non–zerodivisor of M_2. This is also a non–zerodivisor for M_1 and, therefore, $\operatorname{depth}(M_1) \geq 1$. Using the Snake Lemma, we obtain an inclusion

$$\operatorname{Ker}(M_3 \xrightarrow{f} M_3) \subset M_1/fM_1 \,.$$

As $\operatorname{depth}(M_3) = 0$, we have $\operatorname{Ker}(f : M_3 \to M_3) \neq 0$. Any non–zerodivisor of M_1/fM_1 would give a non–zerodivisor of $\operatorname{Ker}(f : M_3 \to M_3) \subset M_3$. But this is not possible, and, therefore, $\operatorname{depth}(M_1) = 1$.   □

Proof of Theorem 7.8.7. If $\operatorname{depth}(A) = 0$, then there exists $x \in A$ different from zero satisfying $x\mathfrak{m} = 0$. Let

$$0 \longrightarrow F_n \xrightarrow{\pi_n} F_{n-1} \longrightarrow \dots \xrightarrow{\pi_1} F_0 \longrightarrow M \longrightarrow 0$$

be a minimal free resolution of M. If $n > 0$ we have $\pi_n(F_n) \subset \mathfrak{m}F_{n-1}$ (minimality of the resolution). Then $xF_n \hookrightarrow x\mathfrak{m}F_{n-1} = 0$, hence $F_n = 0$. This contradicts the minimality of the resolution. Therefore, $\operatorname{pd}_A(M) = 0$ and $\operatorname{depth}(M) = \operatorname{depth}(A)$.

Now we use induction on $\operatorname{depth}(A)$. Assume that $\operatorname{depth}(M) > 0$ and $\operatorname{depth}(A) > 0$. Let $f \in \mathfrak{m}$ be a non–zerodivisor of M and of A. The projective dimension is constant if we divide by f, that is, $\operatorname{pd}_{A/\langle f \rangle}(M/\langle f \rangle M) = \operatorname{pd}_A(M)$, but the depth drops by one. Moreover, by induction hypothesis, we have

$$\operatorname{pd}_{A/\langle f \rangle}(M/\langle f \rangle M) + \operatorname{depth}_{A/\langle f \rangle}(M/\langle f \rangle M) = \operatorname{depth}(A/\langle f \rangle) \,,$$

hence the statement.

Finally, assume $\operatorname{depth}(M) = 0$ and $\operatorname{depth}(A) > 0$. Then $\operatorname{pd}_A(M) > 0$, because, otherwise, M would be free and $\operatorname{depth}(M) > 0$. Let

$$0 \longrightarrow K \longrightarrow F \longrightarrow M \longrightarrow 0$$

be an exact sequence, F a finitely generated free A–module and $0 \neq K \subset \mathfrak{m}F$. We apply Lemma 7.8.9 and obtain $\operatorname{depth}(K) = 1$. Using the previous case we obtain $\operatorname{depth}(K) + \operatorname{pd}_A(K) = \operatorname{depth}(A)$. But $\operatorname{pd}_A(K) + 1 = \operatorname{pd}_A(M)$ and $\operatorname{depth}(K) = \operatorname{depth}(M) + 1$ implies $\operatorname{depth}(M) + \operatorname{pd}_A(M) = \operatorname{depth}(A)$.   □

Exercises

7.8.1. Let $(A, \mathfrak{m})$ be a zero–dimensional local ring. Let M be an A–module with $\operatorname{pd}_A(M) < \infty$. Prove that M is free.

7.8.2. Prove the Auslander–Buchsbaum formula for the case of graded rings and modules: let A be a finitely generated graded K–algebra with K a field, and let M be a finitely generated A–module with $\mathrm{pd}_A(M) < \infty$. Then

$$\mathrm{depth}(\mathfrak{m}, M) + \mathrm{pd}_A(M) = \mathrm{depth}(\mathfrak{m}, A),$$

where $\mathfrak{m}$ is the irrelevant ideal of A.

7.8.3. Let $I \subset K[x_1, \ldots, x_n]$ be a homogeneous ideal. Prove that the ring $K[x_1, \ldots, x_n]/I$ is Cohen–Macaulay if and only if $K[x_1, \ldots, x_n]_{\langle x_1, \ldots, x_n \rangle}/I$ is Cohen–Macaulay.

7.8.4. Show that the subring $A = K[s^3, s^2t, st^2, t^3]$ of $K[s,t]$ is Cohen–Macaulay.

(Hint: compute, by elimination, a homogeneous ideal $I \subset K[x_0, \ldots, x_3]$ such that $A \cong K[x_0, \ldots, x_3]/I$, and apply Exercise 7.8.3. Geometrically, this means to compute the homogeneous coordinate ring of the twisted cubic curve, which is the image of the morphism $\mathbb{P}^1 \to \mathbb{P}^3$, $(s : t) \mapsto (s^3 : s^2t : st^2 : t^3)$, cf. Appendix A.7.)

7.8.5. Show that the subring $A = K[s^4, s^3t, st^3, t^4]$ of $K[s,t]$ is not Cohen–Macaulay. A is the homogeneous coordinate ring of a smooth rational quartic curve in three–space.

(Hint: see Exercise 7.8.4.)

7.8.6. Check whether $K[x, y, z, w]_{\langle x,y,z,w \rangle}/\langle xz, xw, yz, yw \rangle$ is Cohen-Macaulay.

7.8.7. Let $A = K[x, y]_{\langle x,y \rangle}/\langle x \cdot y \rangle$ and M be the normalization of A. Prove that $\mathrm{pd}_A(M) = \infty$.

7.8.8. Let $(A, \mathfrak{m})$ be a Noetherian local ring, $k = A/\mathfrak{m}$ the residue field, and M a finitely generated A–module. Prove that

$$\mathrm{pd}_A(M) = \sup\{i \mid \mathrm{Tor}_i^A(M, k) \neq 0\}.$$

(Hint: use Exercise 7.1.9.)

7.8.9. Let $(A, \mathfrak{m})$ be a Noetherian local ring, $k = A/\mathfrak{m}$ the residue field and M a finitely generated A–module. Prove that $\mathrm{pd}_A(M) \leq \mathrm{pd}_A(k)$.

(Hint: use Theorem 7.8.7 and Exercise 7.8.8.)

7.8.10. Let $(A, \mathfrak{m})$ be a regular local ring, and let $(B, \mathfrak{n})$ be a local A–algebra, which is finitely generated as A–module. Then B is Cohen-Macaulay if and only if $\mathrm{pd}_A(B) = \dim(A) - \dim(A/\mathrm{Ann}_A(B))$.

7.8.11. Let $(A, \mathfrak{m})$ be a Noetherian local ring. An ideal $I \subset A$ is called *perfect*, if $\mathrm{depth}(I, A) = \mathrm{pd}_A(A/I)$. Prove the following statements:

(1) Let $I = \langle x_1, \ldots, x_r \rangle$, $x_1, \ldots, x_r$ be a regular sequence, then I is perfect.

(2) If A is Cohen–Macaulay and I a perfect ideal then A/I is Cohen–Macaulay.

7.8.12. Let $(A, \mathfrak{m})$ be a Noetherian local ring and M a finitely generated A–module. Let $P \in \mathrm{Ass}(M)$. Prove that $\mathrm{pd}_A(M) \geq \mathrm{depth}(P)$.

7.8.13. (*matrix factorization*) Let $(S, \mathfrak{m})$ be a regular local ring and $f \in \mathfrak{m}$. Let $R = S/\langle f \rangle$ and M be a maximal Cohen–Macaulay R–module, that is, $\mathrm{depth}(M) = \dim(R)$. Consider M via the canonical map $S \to R$ as an S–module. Prove the following statements:

(1) $\mathrm{pd}_S(M) = 1$.
(2) There exists a free resolution

$$0 \to S^n \xrightarrow{\varphi} S^n \to M \to 0 \,.$$

(3) There exists a linear map $\psi : S^n \to S^n$ such that $\varphi \circ \psi = f \cdot 1_{S^n}$ and $\psi \circ \varphi = f \cdot 1_{S^n}$. (φ, ψ) is called a *matrix factorization* of f.
(4) Let $\overline{\varphi} = \varphi \bmod \langle f \rangle$ and $\overline{\psi} = \psi \bmod f$, then the complex

$$\cdots \to R^n \xrightarrow{\overline{\varphi}} R^n \xrightarrow{\overline{\psi}} R^n \xrightarrow{\overline{\varphi}} R^n \xrightarrow{\overline{\psi}} R^n \to \cdots$$

is exact. In particular,

$$\cdots \to R^n \xrightarrow{\overline{\varphi}} R^n \xrightarrow{\overline{\psi}} R^n \xrightarrow{\overline{\varphi}} R^n \to M \to 0$$

is a *periodic free resolution* with periodicity 2.

(Hint: to prove (2) use (1) and the fact that M has rank 0 as an S–module. To prove (3), note that $fM = 0$ implies $fS^n \subset \varphi(S^n)$.)

7.8.14. Let $S = \mathbb{Q}[x, y]_{\langle x, y \rangle}$ and $f = x^2 + y^3$. Use SINGULAR to compute the first six modules of the periodic free resolution of the $S/\langle f \rangle$–module $\mathbb{Q} = S/\langle x, y \rangle$.

7.8.15. Let $S = \mathbb{Q}[x, y]_{\langle x, y \rangle}$ and $f = x^3 + y^5$, and let $\varphi = \begin{pmatrix} y & -x & 0 \\ 0 & y & -x \\ x & 0 & y^3 \end{pmatrix}$. Use SINGULAR to compute ψ such that (φ, ψ) is a matrix factorization of f.

7.9 Homological Characterization of Regular Rings

In this section we characterize regular rings by the property that all modules have finite projective dimension. With this characterization we can apply Hilbert's Syzygy Theorem to obtain that the localization of a regular ring at a prime ideal is regular, and that the rings $K[x_1, \ldots, x_n]_>$ are regular for every monomial ordering $>$.

Theorem 7.9.1 (Serre). *Let $(A, \mathfrak{m})$ be a Noetherian local ring. The following conditions are equivalent.*

(1) A is regular.
(2) $\mathrm{pd}_A(A/\mathfrak{m}) = \dim(A)$.
(3) $\mathrm{pd}_A(A/\mathfrak{m}) < \infty$.
(4) Every finitely generated A–module has finite projective dimension.

Proof. $(1) \Rightarrow (2)$: let A be regular of dimension n. Then there exists an A–sequence $x_1, \ldots, x_n$ such that $\mathfrak{m} = \langle x_1, \ldots, x_n \rangle$. Using Theorem 7.6.14 and Corollary 7.6.10, we obtain that the Koszul complex $K(x_1, \ldots, x_n)_\bullet$ is a minimal free resolution of $k = A/\mathfrak{m}$ of length n. This implies $\mathrm{pd}_A(A/\mathfrak{m}) = \dim(A)$.

The implication $(2) \Rightarrow (3)$ is trivial, and $(3) \Rightarrow (4)$ is a consequence of Exercise 7.8.9.

$(4) \Rightarrow (1)$: let $s := \mathrm{edim}(A)$. We have to prove that $s = \dim(A)$. Because $s \geq \dim(A)$, we may assume $s > 0$. We want to proceed by induction on s. Therefore, we need a non–zerodivisor. Assume that $\mathrm{depth}(A) = 0$, then Theorem 7.8.7 implies $\mathrm{pd}_A(A/\mathfrak{m}) = 0$, that is, $k = A/\mathfrak{m}$ is free, which is a contradiction to $s > 0$. We may choose a non–zerodivisor $x \in \mathfrak{m} \setminus \mathfrak{m}^2$. Let $B := A/\langle x \rangle$, then $\mathrm{edim}(B) = s - 1$.

We shall prove that $\mathrm{pd}_B(k) \leq \mathrm{pd}_A(k) + 1$. Once we have proved this, we obtain, by induction, that B is regular. This implies, using Proposition 5.6.17, that A is regular. It remains to prove that $\mathrm{pd}_B(k) \leq \mathrm{pd}_A(k) + 1$. Using the following exact sequence of B–modules

$$0 \longrightarrow \mathfrak{m}/\langle x \rangle \longrightarrow B \longrightarrow k \longrightarrow 0\,,$$

it is enough to prove that $\mathrm{pd}_B(\mathfrak{m}/\langle x \rangle) \leq \mathrm{pd}_A(k)$.

To do so, we show $\mathrm{pd}_B(\mathfrak{m}/x\mathfrak{m}) = \mathrm{pd}_A(k)$ and $\mathrm{pd}_B(\mathfrak{m}/\langle x \rangle) \leq \mathrm{pd}_B(\mathfrak{m}/x\mathfrak{m})$. The latter inequality is true, because the canonical surjection of B–modules $\mathfrak{m}/x\mathfrak{m} \to \mathfrak{m}/\langle x \rangle$ splits (Exercise 7.9.1). To prove $\mathrm{pd}_B(\mathfrak{m}/x\mathfrak{m}) = \mathrm{pd}_A(k)$, we use Exercise 7.8.8 and the fact that $\mathrm{Tor}_i^A(\mathfrak{m}, k) \cong \mathrm{Tor}_i^B(\mathfrak{m}/x\mathfrak{m}, k)$ for all i (Exercise 7.1.6). $\qquad \square$

Definition 7.9.2. Let A be a Noetherian ring, then the *global dimension* of A, $\mathrm{gldim}(A)$, is defined by[6]

$$\mathrm{gldim}(A) = \sup\{\mathrm{pd}_A(M) \mid M \text{ a finitely generated } A\text{–module}\}\,.$$

Lemma 7.9.3.

(1) Let $(A, \mathfrak{m})$ be a Noetherian local ring, then $\mathrm{gldim}(A) = \mathrm{pd}_A(A/\mathfrak{m})$.
(2) Let K be a field, then $\mathrm{gldim}(K[x_1, \ldots, x_n]) = n$.

---

[6] The restriction to finitely generated A–modules is not really necessary. The following theorem of Auslander is proved in [66]: $\mathrm{pd}_A(M) \leq n$ for all A–modules $M \iff \mathrm{pd}_A(M) \leq n$ for all finitely generated A–modules M.

Proof. (1) is a consequence of Exercise 7.8.9; (2) a consequence of Hilbert's Syzygy theorem (Theorem 2.5.15).   □

Lemma 7.9.4. *Let A be a Noetherian ring, and let $P \subset A$ be a prime ideal. If $\mathrm{pd}_A(A/P) < \infty$ then A_P is regular.*

Proof. By assumption, A/P has a finite free resolution

$$0 \longrightarrow F_r \longrightarrow \ldots \longrightarrow F_0 \longrightarrow A/P \longrightarrow 0\,.$$

Then $F_\bullet \otimes_A A_P$ is a free resolution of $(A/P) \otimes_A A_P = A_P/PA_P$. This implies that $\mathrm{pd}_{A_P}(A_P/PA_P) < \infty$ and, therefore, A_P is regular.   □

Corollary 7.9.5.

(1) Let A be a regular local ring, then A_P is regular for all prime ideals $P \subset A$.

(2) Let A be a Noetherian ring and $\mathrm{gldim}(A) < \infty$, then A is regular.[7]

(3) $K[x_1,\ldots,x_n]_>$ is a regular ring for every monomial ordering $>$.

Note that (3) follows from (2) and the constructive proof of Hilbert's Syzygy Theorem 2.5.15.

SINGULAR Example 7.9.6 (regularity test).
Lemma 7.9.4 gives another possibility to test whether $K[x]_P/I_P$ is regular for a prime ideal $P \subset K[x]$, $x = (x_1,\ldots,x_n)$: if $\mathrm{pd}_{K[x]/I}(K[x]/P) < \infty$ then $K[x]_P/I_P$ is regular. The test is applicable even if K is not perfect.

```
ring A=(2,a),(x,y),(c,dp);
qring B=std(x2+y2+a);
ideal I=x2+a,y;

print(mres(I,2));
//-> [1]:
//->    _[1]=y
//-> [2]:
//->    _[1]=0
```

From the output, we read that $\mathbb{F}_2(a)[x,y]_{\langle x^2+a,y\rangle}/\langle x^2 + y^2 + a\rangle$ is regular. Note that in this case the Jacobian matrix is the zero–matrix and the ground field is not perfect.

---

[7] A is called *regular* if A_P is regular for each prime ideal $P \subset A$. It is also true that a regular ring of finite Krull dimension has finite global dimension. The proof needs some knowledge about projective resolutions, which we did not develop here. This can be found in [159].

```
setring A;
qring C=std(x2+y2+1);
ideal I=x+1,y;

print(mres(I,2));
//-> [1]:
//->    _[1]=y
//->    _[2]=x+1
//-> [2]:
//->    _[1]=[y,x+1]
//->    _[2]=[x+1,y]
//-> [3]:
//->    _[1]=[x,y+1]
//->    _[2]=[x+1,y]
```

Hence, $\mathbb{F}_2(a)[x,y]_{\langle x+1,y\rangle}/\langle x^2+y^2+1\rangle$ is not regular.

Exercises

7.9.1. Let $(A, \mathfrak{m})$ be a Noetherian local ring, $x \in \mathfrak{m} \setminus \mathfrak{m}^2$ a non–zerodivisor. Prove that the canonical map $\mathfrak{m}/x\mathfrak{m} \to \mathfrak{m}/\langle x\rangle$ splits.
(Hint: choose a minimal system of generators $x = x_1,\ldots,x_s$ of $\mathfrak{m}$, and let $Q := \langle x_2,\ldots,x_s\rangle$. Prove first that $Q \cap \langle x\rangle \subset x\mathfrak{m}$ to obtain a map $Q/Q\cap\langle x\rangle \to \mathfrak{m}/x\mathfrak{m}$.)

7.9.2. Let A be a Noetherian ring. Prove that A is regular if and only if $A[x]$ is regular.

7.9.3. Let A be a regular local ring. Let $B \subset A$ be a Noetherian local ring such that A is a free B–module. Prove that B is a regular local ring.

7.9.4. Let $A = \mathbb{F}_3(a)[x,y]/\langle x^3+y^3+a\rangle$. Use SINGULAR to check whether $A_{\langle x^3+a,y\rangle}$ is a regular local ring.

7.9.5. Let $A = \mathbb{F}_2(a)[x,y]_{\langle x^2+a,y\rangle}/\langle x^2+y^2+a\rangle$. Use SINGULAR to compute $\mathrm{gldim}(A)$.

A. Geometric Background

Es ist die Freude an der Gestalt
in einem höheren Sinne,
die den Geometer ausmacht.

Alfred Clebsch

In this appendix we introduce a few concepts of algebraic geometry in order to illustrate some of the algebraic constructions used in this book. These illustrations are meant to stimulate the reader to develop his own geometric intuition on what is going on algebraically. Indeed, the connection between algebra and geometry has turned out to be very fruitful and both merged to become one of the leading areas in mathematics: algebraic geometry.

In order to provide some geometric understanding, we present, in this appendix, the very basic concepts of classical affine and projective varieties over an algebraically closed field K (say $\mathbb{C}$). However, we also introduce the modern counterparts, Spec and Proj, which bridge, quite naturally and with less assumptions, the canyon between algebra and geometry.

One word about the role of the ground field K. Algebraic geometers usually draw real pictures, think about them as complex varieties and perform computations over some finite field. We recommend following this attitude, which is justified by successful practice. Moreover, the modern language of schemes even allows one to formulate and prove geometric statements for arbitrary fields which coincide with the classical picture if the field is algebraically closed (cf. A.5).

For a deeper study of algebraic geometry, we recommend the book of Hartshorne [120], which is not only the standard reference book but also represents modern algebraic geometry in an excellent way, complemented perhaps by the red book of Mumford [182], and the books of Harris [119] and Brieskorn and Knörrer [29].

A.1 Introduction by Pictures

The basic problem of algebraic geometry is to understand the set of solutions $x = (x_1, \ldots, x_n) \in K^n$ of a system of polynomial equations

$$f_1(x_1, \ldots, x_n) = 0, \ldots, f_k(x_1, \ldots, x_n) = 0,$$

$f_i \in K[x] = K[x_1, \ldots, x_n]$ and K a field. The solution set is called an algebraic set or algebraic variety. The pictures in Figures A.1 – A.8 show examples of algebraic varieties.

However, algebraic sets really live in different worlds, depending on whether K is algebraically closed or not. For instance, the question whether the simple polynomial equation $x^n + y^n + z^n = 0$, $n \geq 3$, has any non–trivial solution in K, is of fundamental difference if we ask this for K to be $\mathbb{C}$, $\mathbb{R}$ or $\mathbb{Q}$. (For $\mathbb{C}$ we obtain a surface, for $\mathbb{R}$ we obtain a surface if n is odd but only $\{0\}$ if n is even, and for $\mathbb{Q}$ this is Fermat's problem, solved by A. Wiles in 1994.) Classical algebraic geometry assumes K to be algebraically closed. Real algebraic geometry is a field in its own and the study of varieties over $\mathbb{Q}$ belongs to arithmetic geometry, a merge of algebraic geometry and number theory. In this appendix we assume K to be algebraically closed.

Many of the problems in algebra, in particular, computer algebra, have a geometric origin. Therefore, we choose an introduction by means of some pictures of algebraic varieties, some of them being used to illustrate subsequent problems.

The pictures in this introduction, Figures A.1 – A.8, were not only chosen to illustrate the beauty of algebraic geometric objects but also because these varieties have had some prominent influence on the development of algebraic geometry and singularity theory.

The Clebsch cubic itself has been the object of numerous investigations in global algebraic geometry, the Cayley and the D_4–cubic also, but, moreover, since the D_4–cubic deforms, via the Cayley cubic, to the Clebsch cubic, these first three pictures illustrate deformation theory, an important branch of (computational) algebraic geometry.

The ordinary node, also called A_1–singularity (shown as a surface singularity), is the most simple singularity in any dimension. The Barth sextic illustrates a basic but very difficult and still (in general) unsolved problem: to determine the maximum possible number of singularities on a projective variety of given degree.

Whitney's umbrella was, at the beginning of stratification theory, an important example for the two Whitney conditions. We use the umbrella in Chapter 3 to illustrate that the algebraic concept of normalization may even lead to a parametrization of a singular variety, an ultimate goal in many contexts, especially for graphical representations. In general, however, such a parametrization is not possible, not even locally, if the variety has dimension larger than one. For curves, on the other hand, the normalization is always smooth and, hence, provides, at least locally, a parametrization. Indeed, computing the normalization of the ideal given by the implicit equations for the space curve in Figure A.8, we obtain the given parametrization. Conversely, the equations are derived from the parametrization by eliminating the variable t. Elimination of variables is perhaps the most important basic application of Gröbner bases.

Fig. A.1. The Clebsch Cubic.

Fig. A.2. The Cayley Cubic.

This is the unique cubic surface which has $\mathbb{S}_5$, the symmetric group of five letters, as symmetry group. It is named after its discoverer Alfred Clebsch and has the affine equation

$$81(x^3 + y^3 + z^3) - 189(x^2 y + x^2 z$$
$$+ xy^2 + xz^2 + y^2 z + yz^2) + 54xyz$$
$$+ 126(xy + xz + yz) - 9(x^2 + y^2 + z^2)$$
$$- 9(x + y + z) + 1 = 0.$$

There is a unique cubic surface which has four ordinary nodes (see Fig. A.5), usually called the Cayley cubic after its discoverer, Arthur Cayley. It is a degeneration of the Clebsch cubic, has $\mathbb{S}_4$ as symmetry group, and the projective equation is

$$z_0 z_1 z_2 + z_0 z_1 z_3 + z_0 z_2 z_3$$
$$+ z_1 z_2 z_3 = 0.$$

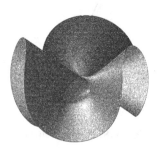

Fig. A.3. A cubic with a D_4–singularity.

Fig. A.4. The Barth Sextic.

Degenerating the Cayley cubic we get a D_4–singularity. The affine equation is

$$x(x^2 - y^2) + z^2(1 + z)$$
$$+ \tfrac{2}{5}xy + \tfrac{2}{5}yz = 0.$$

The equation for this sextic was found by Wolf Barth. It has 65 ordinary nodes, the maximal possible number for a sextic. Its affine equation is (with $c = \frac{1+\sqrt{5}}{2}$)

$$(8c+4)x^2 y^2 z^2 - c^4(x^4 y^2 + y^4 z^2 + x^2 z^4)$$
$$+ c^2(x^2 y^4 + y^2 z^4 + x^4 z^2)$$
$$- \frac{2c+1}{4}(x^2 + y^2 + z^2 - 1)^2 = 0.$$

Fig. A.5. An ordinary node.

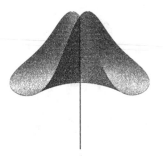

Fig. A.6. Whitney's Umbrella.

An ordinary node is the most simple singularity. It has the equation

$$x^2 + y^2 - z^2 = 0.$$

The Whitney umbrella is named after Hassler Whitney who studied it in connection with the stratification of analytic spaces. It has the affine equation

$$x^2 y - z^2 = 0.$$

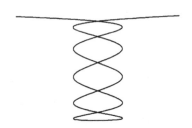

Fig. A.7. A 5–nodal plane curve of degree 11.

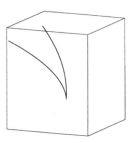

Fig. A.8. A space curve.

Deforming an A_{10}–singularity (normal form: $y^{11} - x^2 = 0$) we obtain a 5–nodal plane curve of degree 11:

$$32x^2 - 2097152y^{11} + 1441792y^9$$
$$- 360448y^7 + 39424y^5$$
$$- 1760y^3 + 22y - 1.$$

This space curve is given parametrically by $x = t^4$, $y = t^3$, $z = t^2$, or implicitly by

$$x - z^2 = y^2 - z^3 = 0.$$

Finally, the 5–nodal plane curve illustrates the singularities of plane curves, in particular, the deformation of a curve singularity into a nodal curve. Moreover, this kind of deformations, with the maximal number of nodes, also play a prominent role in the local theory of singularities. For instance, from this real picture we can read off the intersection form and, hence, the monodromy of the singularity A_{10} by the beautiful theory of A'Campo and Gusein-Zade. By means of standard bases in local rings, there exists a completely different, algebraic algorithm to compute the monodromy [206].

SINGULAR can be used to draw real pictures of plane curves or of surfaces in 3–space, that is, of hypersurfaces defined by polynomials in two or three variables. For this, SINGULAR calls the programme surf written by Stephan Endraß, which is distributed together with SINGULAR, but which can also be used as a stand–alone programme (unfortunately, up to now, surf runs only under Linux and Sun–Solaris). Drawing "nice" real pictures depends very much on the chosen equation, for example, the scaling of the variables, and the kind of projection chosen by the graphics system. It is, therefore, recommended to experiment with the concrete equations and with the parameters within surf. Here is the SINGULAR input for drawing the Whitney umbrella and a surprise surface:

SINGULAR Example A.1.1 (surface plot).

```
LIB "surf.lib";
ring r = 0,(x,y,z),dp;

poly f = -z2+yx2;                    //the Whitney umbrella
map phi = r,x,y,z-1/4x-2;
plot(phi(f));

f =(2*x^2+y^2+z^2-1)^3-(1/10)*x^2*z^3-y^2*z^3;
    // A surprise surface (equation due to Tore Norstrand)
phi = r,1/6z,1/6x,1/6y;             //rescaling
plot(phi(f));
```

Let us now discuss some geometric problems for which the book describes algebraic algorithms, which are implemented in SINGULAR.

Recall first the most basic, but also most important, applications of Gröbner bases to algebraic constructions (Sturmfels called these "Gröbner basics"). These can be found in Chapters 1, 2 and 5 of this book.

- Ideal (respectively module) membership problem,
- Intersection with subrings (elimination of variables),
- Intersection of ideals (respectively submodules),
- Zariski closure of the image of a map,
- Solvability of polynomial equations,
- Solving polynomial equations,
- Radical membership,
- Quotient of ideals,
- Saturation of ideals,
- Kernel of a module homomorphism,

- Kernel of a ring homomorphism,
- Algebraic relations between polynomials,
- Hilbert polynomial of graded ideals and modules.

The next questions and problems lead to algorithms which are slightly more (some of them much more) involved. They are, nevertheless, still very basic and quite natural. We should like to illustrate them by means of four simple examples, shown in the Figures A.9 and A.10, referred to as Examples (A) – (D). We recommend redoing the computations using SINGULAR, the appropriate commands can be found in the chapters we refer to.

(A) The Hypersurface $V(x^2 + y^3 - t^2 y^2)$.   (B) The Variety $V(xz, yz)$.

Fig. A.9. Examples (A) and (B).

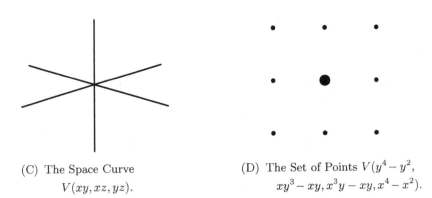

(C) The Space Curve   (D) The Set of Points $V(y^4 - y^2,$
$V(xy, xz, yz)$. $xy^3 - xy, x^3 y - xy, x^4 - x^2)$.

Fig. A.10. Examples (C) and (D).

Assume we are given an ideal $I \subset K[x] = K[x_1, \ldots, x_n]$, by a finite set of generators $f_1, \ldots, f_k \in K[x]$. Consider the following questions and problems:

(1) *Is $V(I)$ irreducible or may it be decomposed into several algebraic varieties? If so, find its irreducible components. Algebraically this means*

to compute a primary decomposition of I or of $\sqrt{I}$, the latter means to compute the associated prime ideals of I.

Example (A) is irreducible; Example (B) has two components (one of dimension 2 and one of dimension 1); Example (C) has three (one–dimensional), and Example (D) has nine (zero–dimensional) components. Primary decomposition is treated in Chapter 4.

(2) *Is I a radical ideal (that is, $I = \sqrt{I}$)? If not, compute its radical $\sqrt{I}$.*

In Examples (A)–(C) the ideal I is radical, while in Example (D) we have $\sqrt{I} = \langle y^3 - y, x^3 - x \rangle$, which is much simpler than I. In this example the central point corresponds to $V(\langle x, y \rangle^2)$ which is a *fat point*, that is, it is a solution of I of multiplicity $(= \dim_K K[x, y]/\langle x, y \rangle^2)$ larger than 1 (equal to 3). All other points have multiplicity 1, hence, the total number of solutions (counted with multiplicity) is 11. This is a typical example of the kind B. Buchberger (respectively W. Gröbner) had in mind at the time of writing his thesis, [32]. We show in Chapter 4, Section 4.5, how to compute the radical. In Corollary 5.3.17 we show how to compute the dimension as a K–vector space, respectively a K–basis, of $K[x_1, \ldots, x_n]/I$ if the quotient is finite dimensional.

(3) *A natural question to ask is, how independent are the generators $f_1, \ldots, f_k$ of I? That is, we ask for all relations $(r_1, \ldots, r_k) \in K[x]^k$ such that*

$$\sum_{i=1}^{k} r_i f_i = 0 \,.$$

These relations form a submodule of $K[x]^k$, which is called the *syzygy module* of $f_1, \ldots, f_k$ and is denoted by $\mathrm{syz}(I)$.[1] It is the kernel of the $K[x]$–linear module homomorphism

$$K[x]^k \longrightarrow K[x], \quad (r_1, \ldots, r_k) \longmapsto \sum_{i=1}^{k} r_i f_i \,.$$

Syzygies are introduced and computed in Chapter 2, Section 2.5.

(4) *More generally, we may ask for generators of the kernel of a $K[x]$–linear map $K[x]^r \to K[x]^s$ or, in other words, for solutions of a system of linear equations over $K[x]$.*

A direct geometric interpretation of syzygies is not so clear, but there are instances where properties of syzygies have important geometric consequences, cf. [205]. To compute the kernel of a module homomorphism, see Chapter 2, Section 2.8.7.

---

[1] In general, the notion $\mathrm{syz}(I)$ is a little misleading, because the syzygy module depends on the chosen system of generators for I, see Chapter 2, Remark 2.5.2.

In Example (A) $\mathrm{syz}(I) = 0$, in Example (B) $\mathrm{syz}(I) = \langle(-y, x)\rangle \subset K[x, y, z]^2$, in Example (C) $\mathrm{syz}(I) = \langle(-z, y, 0), (-z, 0, x)\rangle \subset K[x, y, z]^3$, and in Example (D) $\mathrm{syz}(I) \subset K[x, y, z]^4$ is generated by the vectors $(x, -y, 0, 0)$, $(0, 0, x, -y)$ and $(0, x^2 - 1, -y^2 + 1, 0)$.

(5) *A more geometric question is the following. Let $V(I') \subset V(I)$ be a subvariety. How can we describe $V(I) \setminus V(I')$? Algebraically, this amounts to finding generators for the ideal quotient[2]*

$$I : I' = \{f \in K[x] \mid fI' \subset I\}.$$

Geometrically, for I a prime ideal, $V(I : I')$ is the smallest variety containing $V(I) \setminus V(I')$ which is the (Zariski) closure of $V(I) \setminus V(I')$, a proof of this statement is given in Chapter 1, Section 1.8.9.

In Examples (B), (C) we compute the ideal quotients $\langle xz, yz\rangle : \langle x, y\rangle = \langle z\rangle$ and $\langle xy, xz, yz\rangle : \langle x, y\rangle = \langle z, xy\rangle$, which give, in both cases, equations for the complement of the z-axis $\{x = y = 0\}$. In Example (D) we obtain $I : \langle x, y\rangle^2 = \langle y(y^2 - 1), x(x^2 - 1), (x^2 - 1)(y^2 - 1)\rangle$, the corresponding zero-set being eight points, namely $V(I)$ without the central point.

See Chapter 1, Sections 1.8.8 and 1.8.9 for further properties of ideal quotients and for methods on how to compute them.

(6) *Geometrically important is the projection of a variety $V(I) \subset K^n$ onto a linear subspace K^{n-r}. Given generators $f_1, \ldots, f_k$ of $I \subset K[x_1, \ldots, x_n]$, we want to find generators for the (closure of the) image of $V(I) \subset K^n$ in $K^{n-r} = \{x \mid x_1 = \cdots = x_r = 0\}$. The closure of the image is defined by the ideal $I \cap K[x_{r+1}, \ldots, x_n]$, and finding generators for this intersection is known as eliminating $x_1, \ldots, x_r$ from $f_1, \ldots, f_k$.*

Projecting the varieties of Examples (A) – (C) to the (x, y)–plane is, in the first two cases, surjective and in the third case it gives the two coordinate axes in the (x, y)–plane. This corresponds to the fact that the intersection with $K[x, y]$ of the first two ideals is $\langle 0\rangle$, while the third one is $\langle xy\rangle$.

Projecting the nine points of Example (D) to the x–axis we obtain, by eliminating y, the polynomial $x^2(x - 1)(x + 1)$, describing the three image points. This example is discussed further in Example A.3.13. The geometric background of elimination is discussed in detail in A.2 and A.3. The algorithmic and computational aspects are presented in Chapter 1, Section 1.8.2.

(7) Another problem is related to the Riemann singularity removable theorem, which states that a function of a complex manifold, which is holomorphic and bounded outside a subvariety of codimension 1, is actually holomorphic everywhere. This statement is well–known for open subsets of $\mathbb{C}$. In higher dimensions there exists a second singularity removable theorem, which states that a function, which is holomorphic outside a

---

[2] The same definition applies if I, I' are submodules of $K[x]^k$.

subvariety of codimension 2 (no assumption on boundedness), is holomorphic everywhere. For singular complex algebraic varieties this is not true in general, but those for which the two removable theorems hold are called *normal*. Moreover, each reduced variety has a normalization, and there is a morphism with finite fibres from the normalization to the variety, which is an isomorphism outside the singular locus.

Given a variety $V(I) \subset K^n$, the problem is to find a normal variety $V(J) \subset K^m$ and a polynomial map $K^m \to K^n$ inducing the normalization map $V(J) \to V(I)$. It can be reduced to irreducible varieties (but need not be, as shown in Chapter 3, Section 3.6), and then the equivalent algebraic problem is to find the normalization of $K[x_1, \ldots, x_n]/I$, that is, the integral closure of $K[x]/I$ in the quotient field of $K[x]/I$, and to present this ring as an affine ring $K[y_1, \ldots, y_m]/J$ for some m and J.

For Examples (A) – (C) it can be shown that the normalization is smooth. In (B) and (C), it is actually the disjoint union of the smooth components. The corresponding rings are $K[x_1, x_2]$, $K[x_1, x_2] \oplus K[x_3]$, $K[x_1] \oplus K[x_2] \oplus K[x_3]$. (Use SINGULAR as in Chapter 3, Section 3.6.) The fourth example (D) has no normalization, as it is not reduced.

A related problem is to find, for a non–normal variety V, an ideal H such that $V(H)$ is the non–normal locus of V.

The normalization algorithm is described in Chapter 3, Section 3.6. There, we also present an algorithm to compute the non–normal locus.

In the examples above, the non–normal locus is equal to the singular locus.

(8) The significance of *singularities* appears not only in the normalization problem. The study of singularities is also called *local algebraic geometry* and belongs to the basic tasks of algebraic geometry. Nowadays, singularity theory is a whole subject on its own (cf. A.9 for a short introduction). A singularity of a variety is a point which has no neighbourhood in which the Jacobian matrix of the generators has constant rank.

In Example (A) the whole t–axis is singular, in the other three examples only the origin.

One task is to compute generators for the ideal of the singular locus, which is itself a variety. This is just done by computing subdeterminants of the Jacobian matrix, if there are no components of different dimensions. In general, however, we need, additionally, to compute either an equidimensional decomposition or annihilators of Ext groups.

For how to compute the singular locus see Chapter 5, Section 5.7.

In Examples (A) – (D), the singular locus is given by the ideals $\langle x, y \rangle$, $\langle x, y, z \rangle$, $\langle x, y, z \rangle$, $\langle x, y \rangle^2$, respectively.

(9) *Studying a variety $V(I)$, $I = \langle f_1, \ldots, f_k \rangle$, locally at a singular point, say the origin of K^n, means studying the ideal $IK[x]_{\langle x \rangle}$, generated by I in the local ring*

$$K[x]_{\langle x \rangle} = \left\{ \frac{f}{g} \;\middle|\; f, g \in K[x], \; g \notin \langle x_1, \ldots, x_n \rangle \right\}.$$

In this local ring the polynomials g with $g(0) \neq 0$ are units, and $K[x]$ is a subring of $K[x]_{\langle x \rangle}$.

Now all the problems we considered above can be formulated for ideals in $K[x]_{\langle x \rangle}$ and modules over $K[x]_{\langle x \rangle}$ instead of $K[x]$.

The geometric problems should be interpreted as properties of the variety in a neighbourhood of the origin, or more generally, the given point.

At first glance, it seems that computation in the localization $K[x]_{\langle x \rangle}$ requires computation with rational functions. It is an important fact that this is not necessary, but that basically the same algorithms which were developed for $K[x]$ can be used for $K[x]_{\langle x \rangle}$. This is achieved by the choice of a special ordering on the monomials of $K[x]$ where, loosely speaking, the monomials of lower degree are considered to be larger. A systematic study is given in Chapter 1.

In A.8 and A.9 we give a short account of local properties of varieties and of singularities.

All the above problems have algorithmic and computational solutions, which use, at some place, Gröbner basis methods. Moreover, algorithms for most of these have been implemented quite efficiently in several computer algebra systems. SINGULAR is also able to handle, in addition, local questions systematically.

A.2 Affine Algebraic Varieties

From now on, we always assume K to be an algebraically closed field, except when we specify K otherwise.

We start with the simplest algebraic varieties, affine varieties.

Definition A.2.1. $\mathbb{A}^n = \mathbb{A}^n_K$ denotes the *n–dimensional affine space* over K, the set of all *n–tuples* $x = (x_1, \ldots, x_n)$ with $x_i \in K$, together with its structure as affine space.

A set $X \subset \mathbb{A}^n_K$ is called an *affine algebraic set* or a *(classical) affine algebraic variety* or just an *affine variety* (over K) if there exist polynomials $f_\lambda \in K[x_1, \ldots, x_n]$, λ in some index set Λ, such that

$$X = V\big((f_\lambda)_{\lambda \in \Lambda}\big) = \{x \in \mathbb{A}^n_K \mid f_\lambda(x) = 0, \; \forall\, \lambda \in \Lambda\}.$$

X is then called the *zero–set* of $(f_\lambda)_{\lambda \in \Lambda}$.

If L is a non–algebraically closed field and f_λ are elements of $L[x]$, we may consider an algebraically closed field K containing L (for example, $K = \overline{L}$, the algebraic closure of L) and all statements apply to the f_λ considered as elements of $K[x]$.

Of course, X depends only on the ideal I generated by the f_λ, that is, $X = V(I)$ with $I = \langle f_\lambda \mid \lambda \in \Lambda \rangle_{K[x]}$. By the Hilbert basis theorem, 1.3.5, there are finitely many polynomials such that $I = \langle f_1, \ldots, f_k \rangle_{K[x]}$ and, hence, $X = V(f_1, \ldots, f_k)$.

If $X = V(f_1, \ldots, f_k)$ with $\deg(f_i) = 1$, then $X \subset \mathbb{A}^n$ is an affine linear subspace of $\mathbb{A}^n$. If $X = V(f)$ for a single polynomial $f \in K[x] \setminus \{0\}$ then X is called a *hypersurface* in $\mathbb{A}^n$. A hypersurface in $\mathbb{A}^2$ is called an *(affine) plane curve* and a hypersurface in $\mathbb{A}^3$ an *affine surface in 3-space*. Hypersurfaces in $\mathbb{A}^n$ of degree 2, 3, 4, 5, ... are called *quadrics, cubics, quartics, quintics,* ....

Figure A.11 shows pictures of examples with respective equations. These can be drawn using SINGULAR as indicated in the following example:

SINGULAR Example A.2.2 (surface plot).

```
ring r=0,(x,y,z), dp;
poly f= ...;
LIB"surf.lib";
plot(f);
```

The pictures shown in Figure A.11 give the correct impression that the varieties become more complicated if we increase the degree of the defining polynomial. However, the pictures are real, and it is quite instructive to see how they change if we change the coefficients of terms (in particular the signs). The quintic in Figure A.11 is the Togliatti quintic, which embellishes the cover of this book.

Lemma A.2.3. *For ideals* $I, I_i, I_\lambda \subset K[x_1, \ldots, x_n]$, Λ *any index set, we have*

(1) $\emptyset = V(\langle 1 \rangle)$, $\mathbb{A}^n_K = V(\langle 0 \rangle)$;
(2) $\bigcup_{i=1}^k V(I_i) = V\left(\bigcap_{i=1}^k I_i\right) = V\left(\prod_{i=1}^k I_i\right)$;
(3) $\bigcap_{\lambda \in \Lambda} V(I_\lambda) = V\left(\bigcup_{\lambda \in \Lambda} I_\lambda\right) = V\left(\sum_{\lambda \in \Lambda} I_\lambda\right)$;
(4) $V(I_1) \supset V(I_2)$ *if* $I_1 \subset I_2$;
(5) $V(I) = V(\sqrt{I})$;
(6) $V(I_1) = V(I_2)$ *if and only if* $\sqrt{I_1} = \sqrt{I_2}$.

The easy proof is left to the reader, (6) being a consequence of Hilbert's Nullstellensatz (Theorem 3.5.2). A direct consequence is

Lemma A.2.4.

(1) $\emptyset$, $\mathbb{A}^n$ *are affine varieties.*
(2) *The union of finitely many affine varieties is affine.*

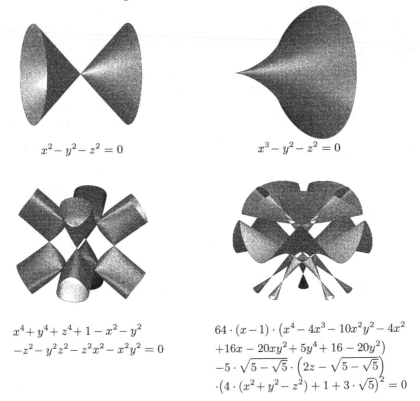

$$x^2 - y^2 - z^2 = 0 \qquad\qquad x^3 - y^2 - z^2 = 0$$

$$x^4 + y^4 + z^4 + 1 - x^2 - y^2$$
$$-z^2 - y^2 z^2 - z^2 x^2 - x^2 y^2 = 0$$

$$64 \cdot (x - 1) \cdot \left(x^4 - 4x^3 - 10x^2 y^2 - 4x^2\right.$$
$$\left. +16x - 20xy^2 + 5y^4 + 16 - 20y^2\right)$$
$$-5 \cdot \sqrt{5 - \sqrt{5}} \cdot \left(2z - \sqrt{5 - \sqrt{5}}\right)$$
$$\cdot \left(4 \cdot (x^2 + y^2 - z^2) + 1 + 3 \cdot \sqrt{5}\right)^2 = 0$$

Fig. A.11. Quadric, cubic, quartic, quintic hypersurfaces.

(3) The intersection of arbitrary many affine varieties is affine.

The lemma says that the affine algebraic sets in $\mathbb{A}^n$ are the closed sets of a topology. This topology is called the *Zariski topology* on $\mathbb{A}^n$. For an algebraic set $X \subset \mathbb{A}^n$, the induced topology is called the Zariski topology on X. Hence, an affine algebraic set is the same as a *(Zariski–)closed* (that is, closed in the Zariski topology) subset of an affine space.

If $Y \subset X$ is closed, we call Y also a *(closed) subvariety* of X. $X \subset \mathbb{A}^n$ is called *quasi–affine* if it is *locally closed*, that is, the intersection of an open and a closed subset.

Note that we can identify, as a set, $\mathbb{A}^n$ with $\mathbb{A}^1 \times \cdots \times \mathbb{A}^1$, but the Zariski topology on $\mathbb{A}^n$ is not the product of the Zariski topologies of $\mathbb{A}^1$ (for example, the only non–trivial closed sets in the product topology of $\mathbb{A}^1 \times \mathbb{A}^1$ are finite unions of points and lines).

Having defined the zero–set of an ideal, we define now the ideal of a set.

Definition A.2.5. For any set $X \subset \mathbb{A}^n$ define

$$I(X) := \left\{ f \in K[x_1, \ldots, x_n] \mid f|_X = 0 \right\},$$

the *(full vanishing) ideal of* X, where $f|_X : X \to K$ denotes the polynomial function of f restricted to X.

Lemma A.2.6. *Let $X \subset \mathbb{A}^n$ be a subset, $X_1, X_2 \subset \mathbb{A}^n$ affine varieties.*

(1) $I(X)$ is a radical ideal.
(2) $V\bigl(I(X)\bigr) = \overline{X}$ the Zariski closure of X in $\mathbb{A}^n$.
(3) If X is an affine variety, then $V\bigl(I(X)\bigr) = X$.
(4) $I(\overline{X}) = I(X)$.
(5) $X_1 \subset X_2$ if and only if $I(X_2) \subset I(X_1)$,
* $X_1 = X_2$ if and only if $I(X_1) = I(X_2)$.*
(6) $I(X_1 \cup X_2) = I(X_1) \cap I(X_2)$.
(7) $I(X_1 \cap X_2) = \sqrt{I(X_1) + I(X_2)}$.

The proof is left as an exercise.

It follows that, for a closed set X, the ideal $I(X)$ determines the affine algebraic set X (and vice versa), showing already a tight connection between ideals of $K[x]$ and affine algebraic sets in $\mathbb{A}^n_K$.

However, $I(X)$ is abstractly defined, we only know (by the Hilbert basis theorem) that it is finitely generated, but, given X, we do not know a set of generators of $I(X)$. Therefore, given any ideal I such that $X = V(I)$, we may ask how far does I differ from $I(X)$ or, how far can we recover I from its zero–set $V(I)$? Of course, $V(I) = V(\sqrt{I})$, that is, we can recover I at most up to radical.

Hilbert's Nullstellensatz (Theorem 3.5.2) says that, for algebraically closed fields, this is the only ambiguity. That is, if $I \subset K[x_1, \ldots, x_n]$ is an ideal, K is algebraically closed, and $X = V(I)$, then

$$I(X) = \sqrt{I}.$$

Note that the inclusion $\sqrt{I} \subset I(X)$ holds for any field. The other inclusion does not hold for K not algebraically closed. Consider, for example, $K = \mathbb{R}$, $I := \langle x^2 + y^2 + 1 \rangle \subset \mathbb{R}[x, y]$. Then $V(I) = \emptyset$, but $\sqrt{I} = I \subsetneq \mathbb{R}[x, y] = I(\emptyset)$.

As a consequence, we obtain, for K algebraically closed, an inclusion reversing bijection (HN refers to Hilbert's Nullstellensatz)

$$\{\text{affine algebraic sets in } \mathbb{A}^n_K\} \xleftrightarrow{\text{HN}} \{\text{radical ideals } I \subset K[x_1, \ldots, x_n]\}$$

$$X \longmapsto I(X)$$

$$V(I) \longleftarrow\!\shortmid I.$$

Corollary A.2.7. *Let $K[x] = K[x_1, \ldots, x_n]$. Then*

(1) $V(I) \neq \emptyset$ for any proper ideal $I \subsetneq K[x]$.
(2) If $\mathfrak{m} \subset K[x]$ is a maximal ideal then

$$\mathfrak{m} = \langle x_1 - p_1, \ldots, x_n - p_n \rangle K[x]$$

for some point $p = (p_1, \ldots, p_n) \in \mathbb{A}^n$. In particular, $V(\mathfrak{m}) = \{p\}$.

Proof. (1) $X = V(I) = \emptyset$ implies $I(X) = K[x]$ and, hence, $\sqrt{I} = K[x]$ by the Nullstellensatz, which implies $I = K[x]$.

(2) By (1) there exists some point $p = (p_1, \ldots, p_n) \in V(\mathfrak{m})$. The corresponding ideal $I_p := \langle x_1 - p_1, \ldots, x_n - p_n \rangle \subset K[x]$ is maximal and satisfies $V(I_p) = \{p\} \subset V(\mathfrak{m})$. Hence, $\mathfrak{m} \subset I_p$ and, therefore, $\mathfrak{m} = I_p$, since $\mathfrak{m}$ is maximal. $\qquad\square$

We see, in particular, that the bijection HN induces a bijection between points in $\mathbb{A}_K^n$ and maximal ideals in $K[x]$. (This will be the link between classical affine varieties and affine schemes defined in A.3.)

Moreover, we call an affine algebraic set $X \subset \mathbb{A}_K^n$ *irreducible* if $X \neq \emptyset$, and if it is not the union of two proper affine algebraic subsets. With this definition, the irreducible algebraic sets in $\mathbb{A}_K^n$ correspond to prime ideals in $K[x]$ (cf. Proposition 3.3.5).

Hence, for K an algebraically closed field, we have the following inclusion reversing bijections (with $K[x] = K[x_1, \ldots, x_n]$):

$$\{\text{affine algebraic sets in } \mathbb{A}_K^n\} \overset{\text{HN}}{\longleftrightarrow} \{\text{radical ideals in } K[x]\}$$
$$\cup \qquad\qquad\qquad\qquad \cup$$
$$\{\text{irreducible affine algebraic sets in } \mathbb{A}_K^n\} \longleftrightarrow \{\text{prime ideals in } K[x]\}$$
$$\cup \qquad\qquad\qquad\qquad \cup$$
$$\{\text{points of } \mathbb{A}_K^n\} \longleftrightarrow \{\text{maximal ideals in } K[x]\}.$$

In particular, the irreducible components of an affine algebraic set $V(I)$ are precisely $V(P_1), \ldots, V(P_r)$, where $P_1, \ldots, P_r$ are the minimal associated primes of I, that is, $V(I) = V(P_1) \cup \cdots \cup V(P_r)$ is the unique decomposition of $V(I)$ into irreducible affine algebraic sets, no one containing another.

Definition A.2.8. Let X be an affine algebraic set, then the affine ring

$$K[X] := K[x_1, \ldots, x_n]/I(X)$$

is called the *coordinate ring* of X, and the elements of $K[X]$ are called *regular functions* on X.

We define the *dimension* of an affine algebraic set X to be the dimension of its coordinate ring $K[X]$. In particular, if $X_1, \ldots, X_r$ are the irreducible components of X, then $\dim(X) = \max\{\dim(X_1), \ldots, \dim(X_r)\}$, by Lemma 3.3.9.

Definition A.2.9. Let $X \subset \mathbb{A}^n_K$, $Y \subset \mathbb{A}^m_K$ be affine algebraic sets. A map $f : X \to Y$ is called a *morphism* of algebraic sets if there exists a polynomial map $\widetilde{f} = (\widetilde{f}_1, \ldots, \widetilde{f}_m) : \mathbb{A}^n_K \to \mathbb{A}^m_K$, $\widetilde{f}_i \in K[x_1, \ldots, x_n]$, such that $f = \widetilde{f}|_X$. $\widetilde{f}$ is called a (polynomial) *representative of* f.

$\mathrm{Mor}(X, Y)$ denotes the set of all morphisms from X to Y. An *isomorphism* is a bijective morphism with f^{-1} also a morphism.

It is easy to see that the composition of morphisms is again a morphism. Moreover, morphisms are continuous in the Zariski topology: if $Z \subset Y$ is closed with $I(Z) = \langle g_1, \ldots, g_k \rangle_{K[Y]}$, then $f^{-1}(Z) = \{x \in X \mid g_i \circ \widetilde{f}(x) = 0\}$ is closed, too.

If $\widetilde{f} = (\widetilde{f}_1, \ldots, \widetilde{f}_m)$, $\widetilde{g} = (\widetilde{g}_1, \ldots, \widetilde{g}_m) : \mathbb{A}^n_K \to \mathbb{A}^m_K$ are two polynomial representatives of f then $(\widetilde{f}_i - \widetilde{g}_i)(x) = 0$ for all $x \in X$, hence, $\widetilde{f}_i - \widetilde{g}_i \in I(X)$. In particular, we obtain a bijection

$$\mathrm{Mor}(X, \mathbb{A}^1_K) = K[X].$$

More generally, $\mathrm{Mor}(X, Y) = \{\widetilde{f} \in K[x_1, \ldots, x_n]^m \mid \widetilde{f}(X) \subset Y\} \bmod I(X)$.

Since any algebraically closed field is infinite, $f|_{\mathbb{A}^n_K} = 0$ implies $f = 0$, hence the coordinate ring of $\mathbb{A}^n_K$ is $K[x] := K[x_1, \ldots, x_n]$. By the above remarks, we have

$$\mathrm{Mor}(X, \mathbb{A}^m_K) = K[X]^m$$

and, more generally, for any closed subvariety $Y \subset \mathbb{A}^m_K$,

$$\mathrm{Mor}(X, Y) = \{f \in K[X]^m \mid f(X) \subset Y\}.$$

In the following we point out that there is a tight relation between morphisms $X \to Y$ and K-algebra homomorphisms $K[Y] \to K[X]$: let $f : X \to Y$ be a morphism with polynomial representative $\widetilde{f} = (\widetilde{f}_1, \ldots, \widetilde{f}_m) : \mathbb{A}^n_K \to \mathbb{A}^m_K$. Then, for $g \in K[y] := K[y_1, \ldots, y_m]$, we set

$$f^*(g) := [g \circ \widetilde{f}],$$

the class of $g(\widetilde{f}_1, \ldots, \widetilde{f}_m)$ in $K[x_1, \ldots, x_n]/I(X)$, which is independent of the chosen representative $\widetilde{f}$. The latter defines a map $f^* : K[y] \to K[x]/I(X)$, which is easily checked to be a K-algebra homomorphism. Moreover, if $g \in I(Y)$, then $g \circ \widetilde{f}|_X = 0$ and, hence, $f^*(I(Y)) \subset I(X)$.

Altogether, we see that a morphism $f : X \to Y$ between algebraic sets $X \subset \mathbb{A}^n_K$ and $Y \subset \mathbb{A}^m_K$ induces a K-algebra homomorphism

$$f^* : K[Y] = K[y]/I(Y) \to K[x]/I(X) = K[X]$$

of the corresponding coordinate rings (in the opposite direction).

Conversely, let $\varphi : K[Y] = K[y]/I(Y) \to K[x]/I(X) = K[X]$ be a K-algebra homomorphism. Then we may choose any representatives $\widetilde{f}_i \in K[x]$ of $\varphi([y_i]) \in K[x]/I(X)$, $i = 1, \ldots, m$, and define the polynomial map

$$\widetilde{f} := (\widetilde{f}_1, \ldots, \widetilde{f}_m) : \mathbb{A}_K^n \to \mathbb{A}_K^m.$$

Since the possible choices for $\widetilde{f}_i$ differ only by elements of $I(X)$, the polynomial function $\widetilde{f}|_X : X \to \mathbb{A}_K^m$ is, indeed independent of the chosen representatives. Moreover, its image is contained in Y: consider the K–algebra homomorphism $\widetilde{\varphi} : K[y] \to K[x]$, defined by $\widetilde{\varphi}(y_i) := \widetilde{f}_i$, which is a lift of φ and satisfies $\widetilde{\varphi}(I(Y)) \subset I(X)$. It follows that, for each $x \in X$ and each $g \in I(Y)$, we have $g(\widetilde{f}_1(x), \ldots, \widetilde{f}_m(x)) = (\widetilde{\varphi}(g))(x) = 0$, showing that $\widetilde{f}(X) \subset Y$.

Altogether, we see that a K–algebra homomorphism $\varphi : K[Y] \to K[X]$ induces a morphism of algebraic sets

$$\varphi^\# := \widetilde{f}|_X : X \to Y.$$

As an example, consider $f : \mathbb{A}^1 \to X = V(y^2 - x^3) \subset \mathbb{A}^2$, $t \mapsto (t^2, t^3)$. The induced ring map is $\varphi = f^* : K[x,y]/\langle y^2 - x^3 \rangle \to K[t]$, induced by the K–algebra homomorphism $\widetilde{\varphi} : K[x,y] \to K[t]$, $x \mapsto t^2$, $y \mapsto t^3$, that is, $f^*(g) = g(t^2, t^3)$. φ induces $\varphi^\# : \mathbb{A}^1 \to X$, $t \mapsto (\widetilde{\varphi}(x), \widetilde{\varphi}(y)) = (t^2, t^3)$. Altogether, we see that $(f^*)^\# = f$ and $(\varphi^\#)^* = \varphi$. The following proposition shows that this is a general fact.

In SINGULAR, morphisms between affine varieties have to be represented by the corresponding ring maps, see Chapter 1, in particular Sections 1.1, 1.3 and 1.5 for definitions and examples.

Proposition A.2.10. *Let* $X \subset \mathbb{A}_K^n$, $Y \subset \mathbb{A}_K^m$, $Z \subset \mathbb{A}_K^p$ *be affine algebraic sets and* $K[X]$, $K[Y]$, $K[Z]$ *the corresponding coordinate rings.*

(1) $(\mathrm{id}_X)^* = \mathrm{id}_{K[X]}$, $(\mathrm{id}_{K[X]})^\# = \mathrm{id}_X$.

(2) For morphisms $X \xrightarrow{f} Y \xrightarrow{g} Z$ *of algebraic sets we have*

$$(g \circ f)^* = f^* \circ g^* : K[Z] \to K[X],$$

and for K–*algebra homomorphisms* $K[Z] \xrightarrow{\varphi} K[Y] \xrightarrow{\psi} K[X]$ *we have*

$$(\psi \circ \varphi)^\# = \varphi^\# \circ \psi^\# : X \to Z.$$

(3) $(f^*)^\# = f$ *for* $f : X \to Y$, *a morphism of algebraic sets;* $(\varphi^\#)^* = \varphi$ *for* $\varphi : K[Y] \to K[X]$, *a* K–*algebra homomorphism.*

All the statements are easy to check and left as an exercise.

Corollary A.2.11. *A morphism* $f : X \to Y$ *of affine varieties is an isomorphism if and only if* $f^* : K[Y] \to K[X]$ *is an isomorphism of* K–*algebras.*

In the language of categories and functors, Proposition A.2.10 says that associating

$$X \longmapsto K[X], \quad (f : X \to Y) \longmapsto (f^* : K[Y] \to K[X])$$

defines a (contravariant) functor from the category of affine algebraic varieties to the category of affine K-algebras. Indeed, this functor is an equivalence onto the full subcategory of all reduced affine K-algebras (by the Hilbert Nullstellensatz).

The following proposition gives a geometric interpretation of injective, respectively surjective, K-algebra homomorphisms.

Proposition A.2.12. *Let $f : X \to Y$ be a morphism of affine varieties and $f^* : K[Y] \to K[X]$ the corresponding map of coordinate rings. Then*

(1) f^* is surjective if and only if the image $f(X) \subset Y$ is a closed subvariety, and $f : X \to f(X)$ is an isomorphism.

(2) f^* is injective if and only if $f(X)$ is dense in Y, that is, $\overline{f(X)} = Y$.

We call $f : X \to Y$ a *closed embedding* or *closed immersion* if $f(X) \subset Y$ is closed and $f : X \to f(X)$ is an isomorphism; f is called *dominant* if $f(X)$ is dense in Y.

For the proof of Proposition A.2.12 we need an additional lemma. Moreover, we shall use that under the identification $\mathrm{Mor}(X, \mathbb{A}_K^1) = K[X]$ a morphism $g : X \to \mathbb{A}_K^1$ satisfies $g(x) = 0$ for all $x \in X$ if and only if $g = 0$ in $K[X]$.

If $Y \subset X \subset \mathbb{A}^n$ is any subset then we write

$$I(Y) := I_{\mathbb{A}^n}(Y) := \left\{ f \in K[x_1, \ldots, x_n] \mid f|_Y = 0 \right\},$$
$$I_X(Y) := \left\{ f \in K[X] \mid f|_Y = 0 \right\}.$$

Lemma A.2.13. *Let $f : X \to Y$ be a morphism of affine algebraic sets, then*

$$I_Y\big(f(X)\big) = I_Y\big(\overline{f(X)}\big) = \mathrm{Ker}(f^* : K[Y] \to K[X]).$$

Proof. For $g \in K[Y]$ we have $g\big(\overline{f(X)}\big) = \{0\}$ if and only if $g\big(f(X)\big) = \{0\}$, since g is continuous. Since $g\big(f(X)\big) = f^*(g)(X) = \{0\}$ if and only if $f^*(g)$ is the zero morphism, we obtain $I_Y\big(\overline{f(X)}\big) = I_Y\big(f(X)\big) = \mathrm{Ker}(f^*)$. $\square$

Proof of Proposition A.2.12. (1) Let $p \in \overline{f(X)}$ and $\mathfrak{m}_p = I_Y(\{p\}) \subset K[Y]$ the maximal ideal of p, then $\mathfrak{m}_p \supset I_Y\big(\overline{f(X)}\big) = \mathrm{Ker}(f^*)$, the latter equality being given by Lemma A.2.13. If f^* is surjective, then the induced map $f^* : K[Y]/\mathrm{Ker}(f^*) \to K[X]$ is an isomorphism, and the same holds for

$$f^* : K[Y]/\mathfrak{m}_p \xrightarrow{\cong} K[X]/f^*(\mathfrak{m}_p).$$

Hence, $f^*(\mathfrak{m}_p)$ is a maximal ideal and corresponds, by the Hilbert Nullstellensatz, to a unique point, $V\big(f^*(\mathfrak{m}_p)\big) = \{q\}$.

If $X \subset \mathbb{A}^n$, $Y \subset \mathbb{A}^m$, $p = (p_1, \ldots, p_m)$, and if f has the polynomial representative $\widetilde{f} = (\widetilde{f}_1, \ldots, \widetilde{f}_m) \in K[x]^m$ then $\mathfrak{m}_p = \langle y_1 - p_1, \ldots, y_m - p_m \rangle_{K[Y]}$,

$f^*(\mathfrak{m}_p) = \langle \tilde{f}_1 - p_1, \ldots, \tilde{f}_m - p_m \rangle_{K[X]}$. Hence, we obtain $\{q\} = V\big(f^*(\mathfrak{m}_p)\big) = \tilde{f}^{-1}(p) \cap X = f^{-1}(p)$, and it follows that $f(q) = p \in f(X)$ and that f is injective. We conclude that $f(X)$ is closed and that $f : X \to f(X)$ is bijective. Finally, since $f^* : K[f(X)] = K[Y]/I_Y\big(f(X)\big) \to K[X]$ is an isomorphism, f is an isomorphism, too, by Corollary A.2.11.

Conversely, if $f(X)$ is closed and if $f : X \to f(X)$ is an isomorphism then $f^* : K[Y]/I_Y\big(f(X)\big) \to K[X]$ is an isomorphism by Corollary A.2.11. Hence, $f^* : K[Y] \to K[X]$ is surjective.

(2) Using Lemma A.2.13, we have $\mathrm{Ker}(f^*) = 0$ if and only if $I_Y\big(\overline{f(X)}\big) = 0$ which is equivalent to $\overline{f(X)} = Y$.   □

For example, the projection $\mathbb{A}^2 \supset X = V(xy - 1) \xrightarrow{f} \mathbb{A}^1$, $(x,y) \mapsto x$, has the image $f(X) = \mathbb{A}^1 \setminus \{0\}$. For $g \in K[\mathbb{A}^1] = K[x]$ we have $f^*(g)(x) = g(x)$, and we see that f^* is not surjective ($[y] \in K[x,y]/\langle xy - 1\rangle \setminus \mathrm{Im}(f^*)$), hence, f is not a closed embedding. But f^* is injective, as it should be, since $f(X)$ is dense in $\mathbb{A}^1$.

The library `algebra.lib` contains procedures to test injectivity, asurjectivity and isomorphy of ring maps:

SINGULAR Example A.2.14 (injective, surjective).

```
LIB "algebra.lib";
ring R = 0,(x,y,z),dp;
qring Q = std(z-x2+y3);        // quotient ring R/<z-x2+y3>


ring S = 0,(a,b,c,d),dp;
map psi = R,a,a+b,c-a2+d3;     // a map from R to S,
                               // x->a, y->a+b, z->c-a2+d3
is_injective(psi,R);
//-> 1                         // psi is injective
is_surjective(psi,R);
//-> 0                         // psi is not surjective

qring T = std(ideal(d,c-a2+b3));// quotient ring
                               // S/<d,c-a2+b3>
map chi = Q,a,b,a2-b3;         // map Q --> T between two
                               // quotient rings,
                               // x->a, y->b, z->a2-b3
is_bijective(chi,Q);
//-> 1                         // chi is an isomorphism
```

Remark A.2.15. The reader might wonder whether there is an algebraic characterization of $f : X \to Y$ being surjective. Of course, f^* has to be injective but the problem is to decide whether $f(X)$ is closed in Y. In general, this is difficult and a simple algebraic answer does not exist.

By a theorem of Chevalley (cf. [120], [119]), $f(X)$ is *constructible*, that is, a finite union of locally closed subsets of Y. However, if f is a projective morphism of projective varieties, then $f(X)$ is closed. We prove this fact in Section A.7, where we also explain the origin of the points in $\overline{f(X)} \setminus f(X)$ (cf. Remark A.7.14).

For the further study of morphisms, we need to consider products of varieties. This will be especially useful for an algorithmic treatment and, hence, for computational aspects of images of affine varieties.

First we identify $\mathbb{A}^n \times \mathbb{A}^m$ with $\mathbb{A}^{n+m}$. In particular, we have the Zariski topology of $\mathbb{A}^{n+m}$ on $\mathbb{A}^n \times \mathbb{A}^m$ (which is *not* the product of the two Zariski topologies).

If $X = V(I) \subset \mathbb{A}^n$ and $Y = V(J) \subset \mathbb{A}^m$ are affine algebraic sets defined by ideals $I \subset K[x_1, \ldots, x_n]$ and $J \subset K[y_1, \ldots, y_m]$, then $X \times Y \subset \mathbb{A}^{n+m}$ is an affine algebraic set, since, as can easily be seen,

$$X \times Y = V(\langle I, J \rangle_{K[x,y]}).$$

Definition A.2.16. For a morphism $f : X \to Y$, we define the *graph* of f,

$$\Gamma_f := \left\{ (x,y) \in X \times Y \mid y = f(x) \right\}.$$

If $\widetilde{f} = (\widetilde{f}_1, \ldots, \widetilde{f}_m) \in K[x]^m$ is a polynomial representative of f, then

$$\Gamma_f = \left\{ (x,y) \in X \times Y \mid y_i - \widetilde{f}_i(x) = 0, \ i = 1, \ldots, m \right\},$$
$$= \left\{ (x,y) \in X \times \mathbb{A}^m \mid y_i - \widetilde{f}_i(x) = 0, \ i = 1, \ldots, m \right\}.$$

Hence, $\Gamma_f \subset \mathbb{A}^{n+m}$ is an affine algebraic set,

$$\Gamma_f = V\left(\langle I, \, y_1 - \widetilde{f}_1, \ldots, y_m - \widetilde{f}_m \rangle_{K[x,y]} \right).$$

Remark A.2.17. The projections $\mathrm{pr}_1 : X \times Y \to X$ and $\mathrm{pr}_2 : X \times Y \to Y$ are induced by the inclusions $j_1 : K[x] \hookrightarrow K[x,y]$ and $j_2 : K[y] \hookrightarrow K[x,y]$. For an ideal $I \subset K[x]$ with $V(I) = X$, and a morphism $f : X \to Y$ with polynomial representative $(\widetilde{f}_1, \ldots, \widetilde{f}_m)$, j_1 induces an isomorphism

$$K[x]/I \xrightarrow{\cong} K[x,y]/\langle I, \, y_1 - \widetilde{f}_1, \ldots, y_m - \widetilde{f}_m \rangle,$$

the inverse morphism being induced by $K[x,y] \to K[x]$, $x_i \mapsto x_i$, $y_j \mapsto \widetilde{f}_j$.

Back to geometry, we see that the projection $\mathrm{pr}_1 : X \times Y \to X$ induces an isomorphism $\pi_1 : \Gamma_f \to X$ of affine varieties: π_1 is the restriction of the projection $\mathbb{A}^n \times \mathbb{A}^m \to \mathbb{A}^n$, hence π_1 is a morphism, and the polynomial map $\mathbb{A}^n \to \mathbb{A}^n \times \mathbb{A}^m$, $x \mapsto (x, \widetilde{f}(x))$, induces an inverse to π_1. We obtain the following commutative diagram

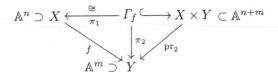

where π_2 is the restriction of the second projection $\mathrm{pr}_2 : X \times Y \to Y$. It follows that any morphism f of affine varieties can be represented as a composition of an inclusion and a projection.

The following lemma contains the *geometric meaning of elimination*. In particular, part (2) says that we can compute an ideal defining the closure of $f(X)$ by eliminating variables from an appropriate ideal J.

Lemma A.2.18.

(1) Let $f : X \to Y$ be a morphism of affine algebraic sets, then

$$I_Y\left(\overline{f(X)}\right) = I_{X \times Y}(\Gamma_f) \cap K[Y].$$

(2) Let $X \subset \mathbb{A}^n$, $Y \subset \mathbb{A}^m$ be affine algebraic sets, and let $I_X \subset K[x_1,\ldots,x_n]$ be any ideal with $X = V(I_X)$. Moreover, let $f : X \to Y$ be a morphism, induced by $\widetilde{f} = (\widetilde{f}_1,\ldots,\widetilde{f}_m) : \mathbb{A}^n \to \mathbb{A}^m$ with $\widetilde{f}_i \in K[x_1,\ldots,x_n]$. Define

$$J := \langle I_X, y_1 - \widetilde{f}_1, \ldots, y_m - \widetilde{f}_m \rangle_{K[x,y]}.$$

Then the closure of the image of f is given by

$$\overline{f(X)} = V(J \cap K[y_1,\ldots,y_m]).$$

Moreover, for $I_X = I(X)$ the full vanishing ideal, we obtain

$$I_{\mathbb{A}^{n+m}}(\Gamma_f) = \langle I(X), y_1 - \widetilde{f}_1, \ldots, y_m - \widetilde{f}_m \rangle_{K[x,y]},$$

and the full vanishing ideal of $\overline{f(X)} \subset \mathbb{A}^m$ is

$$I_{\mathbb{A}^m}\left(\overline{f(X)}\right) = I_{\mathbb{A}^{n+m}}(\Gamma_f) \cap K[y_1,\ldots,y_m].$$

Proof. (1) Note that Lemma A.2.13 gives $I_Y\left(\overline{f(X)}\right) = \mathrm{Ker}(f^*) = \mathrm{Ker}(\pi_2^*)$, since $f^* = (\pi_1^*)^{-1} \circ \pi_2^*$ and since π_1^* is an isomorphism.

Now the result follows, since we may consider $K[Y] = K[y]/I(Y)$ as a subalgebra of $K[X \times Y] = K[x,y]/\langle I(X), I(Y) \rangle_{K[x,y]}$ and since π_2^* is the canonical map $K[Y] \to K[X \times Y]/I_{X \times Y}(\Gamma_f)$.

For (2) note that $K[x]/I_X \to K[x,y]/J$ is an isomorphism with inverse induced by $x_i \mapsto x_i$, $y_i \mapsto \widetilde{f}_j$. Therefore, J is radical if and only if I_X is radical, and then $J \cap K[y]$ is a radical ideal, too.

Since $V(J) = \Gamma_f$, we have $\sqrt{J} = I_{\mathbb{A}^{n+m}}(\Gamma_f) \subset K[x,y]$ by Hilbert's Nullstellensatz. Therefore, $\sqrt{J \cap K[y]} = \sqrt{J} \cap K[y] = I_{\mathbb{A}^{n+m}}(\Gamma_f) \cap K[y]$, and (2) follows from (1). $\qquad\square$

A.3 Spectrum and Affine Schemes

Abstract algebraic geometry, as introduced by Grothendieck, is a far reaching generalization of classical algebraic geometry. One of the main points is that it allows the application of geometric methods to arbitrary commutative rings, for example, to the ring $\mathbb{Z}$. Thus, geometric methods can be applied to number theory, creating a new discipline called arithmetic geometry.

However, even for problems in classical algebraic geometry, the abstract approach has turned out to be very important.

For example, for polynomial rings over an algebraically closed field, affine schemes provide more structure than classical algebraic sets. In a systematic manner, the abstract approach allows nilpotent elements in the coordinate ring. This has the advantage of understanding and describing much better "dynamic aspects" of a variety, since nilpotent elements occur naturally in the fibre of a morphism, that is, when a variety varies in an algebraic family.

The abstract approach to algebraic geometry has, however, the disadvantage that it is often far away from intuition, although a geometric language is used. A scheme has many more points than a classical variety, even a lot of non–closed points. This fact, although against any "classical" geometric feeling, has, on the other hand, the effect that the underlying topological space of a scheme carries more information. For example, the abstract Nullstellensatz, which is formally the same as Hilbert's Nullstellensatz, holds without any assumption. However, since the geometric assumptions are much stronger than in the classical situation (we make assumptions on all prime ideals containing an ideal, not only on the maximal ideals), the abstract Nullstellensatz is more a remark than a theorem and Hilbert's Nullstellensatz is *not* a consequence of the abstract one. Nevertheless, the formal coincidence makes the formulation of geometric results in the language of schemes much smoother, and the relation between algebra and geometry is, even for arbitrary rings, as close as it is for classical algebraic sets defined by polynomials over an algebraically closed field.

At the end of Section A.5, we shall show how results about algebraic sets can, indeed, be deduced from results about schemes (in a functorial manner).

In the following we assume, as usual, all rings to be commutative with 1.

Definition A.3.1. Let A be a ring. Then

$$\operatorname{Spec}(A) := \{P \subset A \mid P \text{ is a prime ideal}\}$$

is called the (*prime*) *spectrum* of A, and

$$\operatorname{Max}(A) := \{\mathfrak{m} \subset A \mid \mathfrak{m} \text{ is a maximal ideal}\}$$

is called the *maximal spectrum* of A. For $X = \operatorname{Spec}(A)$ and $I \subset A$ an ideal

$$V(I) := \{P \in X \mid P \supset I\}$$

is called the *zero–set* of I in X. Note that $V(I) = \operatorname{supp}(A/I)$.

As for classical affine varieties, we have the following relations (recall that prime ideals are proper ideals):

(1) $V(\langle 1 \rangle) = \emptyset$, $V(\langle 0 \rangle) = X$;
(2) $\bigcup_{i=1}^{k} V(I_i) = V(\bigcap_{i=1}^{k} I_i)$;
(3) $\bigcap_{\lambda \in \Lambda} V(I_\lambda) = V(\bigcup_{\lambda \in \Lambda} I_\lambda) = V(\sum_{\lambda \in \Lambda} I_\lambda)$;
(4) if $I_1 \subset I_2$ then $V(I_1) \supset V(I_2)$;
(5) $V(I) = V(\sqrt{I})$;
(6) $V(I_1) = V(I_2)$ if and only if $\sqrt{I_1} = \sqrt{I_2}$.

Only the statements (2) and (6) are non–trivial. (2) follows, since, by definition, $P \in V(\bigcap_{i=1}^{k} I_i)$ if and only if $P \supset \bigcap_{i=1}^{k} I_i$, which means due to Lemma 1.3.12 that $P \supset I_i$ for some i, which is equivalent to $P \in \bigcup_{i=1}^{k} V(I_i)$. (6) will follow from Lemma A.3.3 and Theorem A.3.4 below.

Using the above properties (1)–(3) we can define on X the *Zariski topology* by defining the closed sets of X to be the sets $V(I)$ for $I \subset A$ an ideal. Note that this is formally the same definition as for classical varieties. $\mathrm{Max}(A)$ has the induced topology from $\mathrm{Spec}(A)$.

Definition A.3.2. Let $X = \mathrm{Spec}(A)$ and $Y \subset X$ any subset. The ideal

$$I(Y) := \bigcap_{P \in Y} P$$

is called the *(vanishing) ideal of Y in X*.

As in the classical case, we have the following

Lemma A.3.3. *Let $Y \subset X = \mathrm{Spec}(A)$ be a subset.*

(1) $I(Y)$ is a radical ideal.
(2) $V(I(Y))$ is the Zariski closure of Y in X.
(3) If Y is closed, then $V(I(Y)) = Y$.
(4) Let $\overline{Y}$ denote the Zariski closure of Y in X, then $I(\overline{Y}) = I(Y)$.

Proof. (1) If, for some $a \in A$, the n–th power a^n is in the intersection of prime ideals, then also a is in the intersection.

(2) $V(I(Y))$ is closed and contains Y. If $W = V(J)$ is closed in X and contains Y then, for each $a \in J$, we have $a \in P$ for all $P \supset J$, that is, for all $P \in W$. Hence, $a \in \bigcap_{P \in Y} P = I(Y)$. This implies $J \subset I(Y)$ and, therefore, $V(I(Y)) \subset V(J) = W$, showing that $V(I(Y))$ is the smallest closed subset of X containing Y.

Finally, (3) and (4) are consequences of (2). $\square$

The following analogue of Hilbert's Nullstellensatz is sometimes called the *abstract Nullstellensatz*, which holds for $\mathrm{Spec}(A)$, A any ring.

Theorem A.3.4. *Let* $X = \mathrm{Spec}(A)$, *and* $I \subset A$ *an ideal. Then*

$$I(V(I)) = \sqrt{I}.$$

Proof. The statement follows, since

$$I(V(I)) = \bigcap_{P \in V(I)} P = \bigcap_{\substack{P \in \mathrm{Spec}(A) \\ P \supset I}} P = \sqrt{I},$$

where the last equality follows from Exercise 3.3.1.   □

Note that a point $P \in \mathrm{Spec}(A)$ (or, more precisely, the set $\{P\} \subset \mathrm{Spec}(A)$) need not be closed. Indeed, by Lemma A.3.3 and Theorem A.3.4 we have

$$\overline{\{P\}} = V(P) = \{Q \in \mathrm{Spec}(A) \mid Q \supset P\}, \quad I(\{P\}) = I(\overline{\{P\}}) = P,$$

where $\overline{\{P\}}$ denotes the Zariski closure of $\{P\}$ in $\mathrm{Spec}(A)$.

Since P is always contained in some maximal ideal, it follows that $\{P\}$ is closed in $\mathrm{Spec}(A)$ if and only if P is a maximal ideal. Hence,

$$\mathrm{Max}(A) = \{P \in \mathrm{Spec}(A) \mid P \text{ is a closed point}\}.$$

Recall that a topological space X is called *irreducible* if $X \neq \emptyset$ and whenever $X = A_1 \cup A_2$, with $A_1, A_2 \subset X$ closed, then $X = A_1$ or $X = A_2$. X is called *reducible* if it is not irreducible.

The following lemma is left as an exercise.

Lemma A.3.5. *Let* A *be a ring and* $X = \mathrm{Spec}(A)$. *A closed subset* $Y \subset X$ *is irreducible if and only if* $I(Y) \subset A$ *is a prime ideal.*

We shall now define morphisms of spectra.

Let $\varphi : A \to B$ be a ring map. Since for a prime ideal $P \subset B$, the preimage $\varphi^{-1}(P)$ is a prime ideal in A, φ induces a map

$$\varphi^{\#} = \mathrm{Spec}(\varphi) : \ \mathrm{Spec}(B) \longrightarrow \mathrm{Spec}(A),$$
$$P \longmapsto \varphi^{-1}(P).$$

Note that the preimage $\varphi^{-1}(M)$ of a maximal ideal $M \subset B$ need not be a maximal ideal in A. Hence, in general, the map $\varphi^{\#}$ does *not* induce a map $\mathrm{Max}(B) \to \mathrm{Max}(A)$. However, if φ is integral, then $\varphi^{\#} : \mathrm{Max}(B) \to \mathrm{Max}(A)$ is defined by Lemma 3.1.9 (4).

Lemma A.3.6. *Let* $\varphi : A \to B$ *be a ring map. Then the induced map*

$$\varphi^{\#} : \mathrm{Spec}(B) \to \mathrm{Spec}(A)$$

is continuous. More precisely, we have $\varphi^{\#-1}(V(I)) = V(\varphi(I) \cdot B)$.

Proof. Let $I \subset A$ be an ideal, then

$$\varphi^{\#-1}(V(I)) = \varphi^{\#-1}(\{P \in \mathrm{Spec}(A) \mid P \supset I\})$$
$$= \{Q \in \mathrm{Spec}\, B \mid \varphi^{-1}(Q) \supset I\}$$
$$= \{Q \in \mathrm{Spec}\, B \mid Q \supset \varphi(I)\} = V(\varphi(I) \cdot B).$$

In particular, the preimages of closed sets in $\mathrm{Spec}(A)$ are closed.   □

Similarly to the classical case (cf. Lemma A.2.13), we have the following

Lemma A.3.7. *Let $\varphi : A \to B$ be a ring map. Then*

$$I(\varphi^{\#}(\mathrm{Spec}\, B)) = I(\overline{\varphi^{\#}(\mathrm{Spec}\, B)}) = \sqrt{\mathrm{Ker}(\varphi)}.$$

Proof. The first equality follows from Lemma A.3.3, the second from

$$I(\varphi^{\#}(\mathrm{Spec}\, B)) = \bigcap_{P \in \mathrm{Spec}\, B} \varphi^{-1}(P) = \varphi^{-1}\left(\bigcap_{P \in \mathrm{Spec}\, B} P \right) = \varphi^{-1}(\sqrt{\langle 0 \rangle}),$$

since, obviously, $\varphi^{-1}(\sqrt{\langle 0 \rangle}) = \sqrt{\mathrm{Ker}(\varphi)}$.   □

Proposition A.3.8. *Let $\varphi : A \to B$ be a ring map and consider the induced map $\varphi^{\#} : X = \mathrm{Spec}\, B \to \mathrm{Spec}(A) = Y$.*

(1) If φ is surjective then $\varphi^{\#}(X) = V(\mathrm{Ker}\,\varphi)$ and $\varphi^{\#} : X \to \varphi^{\#}(X)$ is a homeomorphism.

(2) If φ is injective then $\varphi^{\#}(X)$ is dense in Y.

(3) Let $\varphi : A \to A_{\mathrm{red}} = A/\sqrt{\langle 0 \rangle}$ be the canonical projection, then the induced map $\varphi^{\#} : \mathrm{Spec}(A_{\mathrm{red}}) \to \mathrm{Spec}(A)$ is a homeomorphism.

Proof. (1) φ induces an isomorphism $A/\mathrm{Ker}(\varphi) \to B$, hence, we have a bijection $\mathrm{Spec}\, B \to \mathrm{Spec}(A/\mathrm{Ker}(\varphi))$. However, we also have a bijection between prime ideals of $A/\mathrm{Ker}(\varphi)$ and prime ideals of A which contain $\mathrm{Ker}(\varphi)$. This shows that $\varphi^{\#}$ is a bijection $X \to V(\mathrm{Ker}(\varphi))$. It is also easy to see that $\varphi^{\#}$ and $\varphi^{\#-1}$ are continuous.

The remaining statements $(2), (3)$ follow from Lemma A.3.7 and (1).   □

Note that the converse of (1) and (2) need not be true if the rings are not reduced: for instance, $\varphi : K[x] \hookrightarrow K[x,y]/\langle y^2 \rangle$ is not surjective, but $\varphi^{\#}$ is a homeomorphism; $\varphi : K[x,y]/\langle y^2 \rangle \to K[x] = K[x,y]/\langle y \rangle$ is not injective, but again $\varphi^{\#}$ is a homeomorphism. However, if $\varphi^{\#}(X)$ is dense in Y, then, at least, $\mathrm{Ker}(\varphi)$ consists of nilpotent elements by Lemma A.3.7.

We have seen that $\mathrm{Spec}(A_{\mathrm{red}})$ and $\mathrm{Spec}(A)$ are homeomorphic. Hence, the topological space $\mathrm{Spec}(A)$ contains less information than A — the nilpotent elements of A are invisible in $\mathrm{Spec}(A)$. However, in many situations nilpotent elements occur naturally, and they are needed to understand the situation. The notion of a scheme takes care of this fact.

Definition A.3.9. The pair $\big(\mathrm{Spec}(A), A\big)$ with A a ring, is called an *affine scheme*. A *morphism* $f : \big(\mathrm{Spec}(A), A\big) \to \big(\mathrm{Spec}(B), B\big)$ of affine schemes is a pair $f = (\varphi^{\#}, \varphi)$ with $\varphi : B \to A$ a ring map and $\varphi^{\#} : \mathrm{Spec}(A) \to \mathrm{Spec}(B)$ the induced map. The map φ is sometimes denoted as f^{*}. An *isomorphism* is a morphism which has a two–sided inverse.

A *subscheme* of $\big(\mathrm{Spec}(A), A\big)$ is a pair $\big(\mathrm{Spec}(A/I), A/I\big)$, where $I \subset A$ is any ideal.

Although f is determined by φ we also mention $\varphi^{\#}$ in order to keep the geometric language. Usually we write $X = \mathrm{Spec}(A)$ to denote the scheme $(\mathrm{Spec}(A), A)$ and sometimes we write $|X|$ to denote the topological space $\mathrm{Spec}(A)$. As topological spaces, we have a canonical identification,

$$\mathrm{Spec}(A/I) = V(I) \subset \mathrm{Spec}(A),$$

and, usually, we shall not distinguish between $V(I)$ and $\mathrm{Spec}(A/I)$.

If $A = K[x_1, \ldots, x_n]/I$ then we call A also the *coordinate ring* of the affine scheme $\big(\mathrm{Spec}(A), A\big)$.

If $X_i = \big(\mathrm{Spec}(A/I_i), A/I_i\big)$, $i = 1, 2$, are two subschemes of the affine scheme $X = \big(\mathrm{Spec}(A), A\big)$ then we define the intersection and the union as

$$X_1 \cap X_2 := \big(\, \mathrm{Spec}(A/(I_1 + I_2)),\ A/(I_1 + I_2)\,\big),$$
$$X_1 \cup X_2 := \big(\, \mathrm{Spec}(A/(I_1 \cap I_2)),\ A/(I_1 \cap I_2)\,\big).$$

As topological spaces, we have, indeed,

$$\mathrm{Spec}(A/I_1 + I_2) = V(I_1) \cap V(I_2), \quad \mathrm{Spec}(A/I_1 \cap I_2) = V(I_1) \cup V(I_2).$$

As an example, consider $I := \langle y^2, xy \rangle \subset K[x, y]$. Then $V(I) \subset \mathbb{A}^2$ consists of the x–axis, but the affine scheme $X = \big(\mathrm{Spec}(K[x, y]/I), K[x, y]/I\big)$ has an *embedded fat point* at 0. Since $y \neq 0$ in $K[x, y]/I$, but $y^2 = 0$, this additional structure may be visualized as an infinitesimal direction, pointing in the y–direction (Fig. A.12).

Fig. A.12. A line with an infinitesimal direction pointing out of the line.

As in the classical case, we want to define products and the graph of a morphism: let $\varphi : C \to A$ and $\psi : C \to B$ be two ring maps, that is, A and B are C–algebras. Let $X = \mathrm{Spec}(A)$, $Y = \mathrm{Spec}(B)$ and $S = \mathrm{Spec}(C)$. Then the affine scheme

$$X \times_S Y := \big(\, \mathrm{Spec}(A \otimes_C B),\ A \otimes_C B \,\big),$$

together with the projection maps pr_1 and pr_2 defined below, is called the *fibre product* of X and Y over S. If $C = K$ is a field, then $S = \mathrm{Spec}(K)$ is a point, and we simply set

$$X \times Y := X \times_K Y := X \times_{\mathrm{Spec}(K)} Y.$$

The ring maps $A \to A \otimes_C B$, $a \mapsto a \otimes 1$, and $B \to A \otimes_C B$, $b \mapsto 1 \otimes b$ induce *projection maps*

$$\mathrm{pr}_1 : X \times_S Y \to X, \quad \mathrm{pr}_2 : X \times_S Y \to Y$$

such that the following diagram commutes

$$
\begin{array}{ccc}
X \times_S Y & \xrightarrow{\ \mathrm{pr}_2\ } & Y \\
{\scriptstyle \mathrm{pr}_1}\downarrow & & \downarrow{\scriptstyle \psi^\#} \\
X & \xrightarrow[\varphi^\#]{} & S.
\end{array}
$$

Moreover, given an affine scheme $Z = \mathrm{Spec}(D)$ and the following commutative diagram with solid arrows, there exists a unique dotted arrow such that everything commutes (all arrows being morphisms of affine schemes),

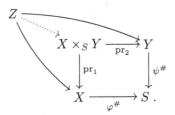

This is called the *universal property of the fibre product*, and it is a consequence of the universal property of tensor products (Proposition 2.7.11).

Note that for $A = K[x_1, \ldots, x_n]/I$, $B = K[y_1, \ldots, y_m]/J$, $X = \mathrm{Spec}(A)$, $Y = \mathrm{Spec}(B)$, we obtain

$$X \times Y = \mathrm{Spec}(K[x_1, \ldots, x_n, y_1, \ldots, y_n]/\langle I, J\rangle).$$

Now let $\varphi : B \to A$ be a C–algebra morphism. Then we have morphisms $f = (\varphi^\#, \varphi) : X \to Y$ and $\mathrm{id}_X : X \to X$ and, by the universal property of the fibre product, a unique morphism $X \to X \times_S Y$ such that the diagram

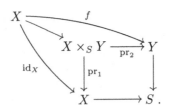

commutes. Note that the morphism $X \to X \times_S Y$ is given on the ring level by a ring map $\gamma_f : A \otimes_C B \to A$, $a \otimes b \mapsto a\varphi(b)$. This map is surjective and, hence, there exists a unique affine subscheme

$$\Gamma_f = \mathrm{Spec}\big((A \otimes_C B)/\mathrm{Ker}(\gamma_f)\big) \subset X \times_S Y,$$

called the *graph* of f.

Of course, since $(A \otimes_C B)/\mathrm{Ker}\,\gamma_f \cong A$, the morphism $X \to \Gamma_f$ is an isomorphism with $\pi_1 = \mathrm{pr}_1|_{\Gamma_f}$ as inverse. There is a commutative diagram

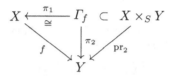

with π_1 and π_2 the restrictions of pr_1 and pr_2. Dually, we have a commutative diagram of ring maps

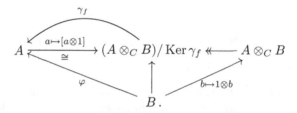

If $\varphi : B = K[y_1, \dots, y_m]/I_Y \to A = K[x_1, \dots, x_n]/I_X$ is a morphism of affine K–algebras, induced by $\widetilde{\varphi} : K[y] \to K[x]$, $y_i \mapsto \widetilde{f}_i$, with $\widetilde{\varphi}(I_Y) \subset I_X$, then

$$\gamma_f : A \otimes_K B = K[x,y]/\langle I_X + I_Y \rangle \longrightarrow K[x]/I_X = A$$

is given by $[x_i] \mapsto [x_i]$, $y_i \mapsto [\widetilde{f}_i]$, and γ_f has as kernel $J/\langle I_X, I_Y \rangle$ with

$$J := \langle I_X, y_1 - \widetilde{f}_1, \dots, y_m - \widetilde{f}_m \rangle_{K[x,y]}.$$

(Note that $I_Y \subset J$, since modulo the ideal $\langle y_1 - \widetilde{f}_1, \dots, y_m - \widetilde{f}_m \rangle$, we have $g \equiv g(\widetilde{f}_1, \dots, \widetilde{f}_m) = \widetilde{\varphi}(g) \in I_X$, for all $g \in I_Y$.)

Hence, in this case we have, for K an arbitrary field,

$$\Gamma_f = \mathrm{Spec}(K[x,y]/J).$$

Lemma A.3.10. *Let $f : X = \mathrm{Spec}(A) \to \mathrm{Spec}(B) = Y$ be as above, then*

$$\overline{f(X)} = \overline{\pi_2(\Gamma_f)} = V(J \cap K[y_1, \dots, y_m]).$$

Proof. The statement follows from Lemma A.3.7.   □

We shall explain, at the end of Section A.3, how this lemma generalizes Lemma A.2.18. It shows that, as topological space,

$$\overline{f(X)} = \operatorname{Spec}\big(K[y]/(J \cap K[y])\big)$$

and, hence, we can define a *scheme structure* on the closure of the image by defining the coordinate ring of $\overline{f(X)}$ as $K[y]/(J \cap K[y])$. The above lemma will be used to actually compute equations of the closure of the image of a morphism.

The ideal of $\overline{f(X)}$ in $\operatorname{Spec} B$ can be defined intrinsically, even for maps between arbitrary rings:

Lemma A.3.11. *Let $f : X = \operatorname{Spec}(A) \to Y = \operatorname{Spec}(B)$ be a morphism induced by the ring map $\varphi : B \to A$. Then*

$$\overline{f(X)} = V\big(\operatorname{Ann}_B(A)\big).$$

Proof. The proof follows easily from Lemma A.3.7, since

$$\operatorname{Ker}(\varphi) = \{b \in B \mid \varphi(b)A = \langle 0\rangle\} = 0 :_B A = \operatorname{Ann}_B(A). \qquad \square$$

Remark A.3.12. If $f : X \to Y$ is as in Lemma A.3.10, then

$$\operatorname{Ann}_B(A) = J \cap K[y],$$

that is, the structure defined on $\overline{f(X)}$ by eliminating x from J is the *annihilator structure*. Since annihilators are, in general, not compatible with base change, one has to be careful, for example, when computing multiplicities using ideals obtained by elimination. We illustrate this fact by an example.

Example A.3.13. Consider the set of nine points displayed in Figure A.13.

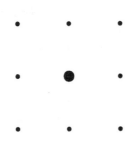

Fig. A.13. $V(y^4 - y^2, xy^3 - xy, x^3y - xy, x^4 - x^2)$.

Projecting these points to the x–axis, we obtain, by eliminating y, the polynomial $x^2(x-1)(x+1)$, describing the three image points. Set–theoretically, this is correct, however, it is not satisfactory if we wish to count multiplicities. For example, the two border points are the image of three points each, hence should appear with multiplicity three, but they appear only with multiplicity one.

In case the ideal is given by two polynomials we can use, instead of elimination, resultants (not discussed in this book, for a definition see, for example, the textbooks [85, 162]), which do count multiplicities correctly. For example, consider the polynomials $f = x(x-1)(x+1)$, $g = y(y-1)(y+1)$, defining together the same set as above, but all nine points being reduced. A SINGULAR analysis gives:

SINGULAR Example A.3.14 (elimination and resultant).

```
ring R = 0,(x,y),dp;
poly f = x*(x-1)*(x+1);
poly g = y*(y-1)*(y+1);

poly e = eliminate(ideal(f,g),y)[1];
factorize(e);
//-> [1]:                        // 3 linear factors,
//->    _[1]=1                    // each of multiplicity 1
//->    _[2]=x+1
//->    _[3]=x-1
//->    _[4]=x
//-> [2]:
//->    1,1,1,1

poly r = resultant(f,g,y);
factorize(r);
//-> [1]:                        // 3 linear factors,
//->    _[1]=1                    // each of multiplicity 3
//->    _[2]=x-1
//->    _[3]=x
//->    _[4]=x+1
//-> [2]:
//->    1,3,3,3
```

The resultant counts each image point with multiplicity 3, as each of this point has three preimage points, while elimination counts the image points only with multiplicity 1.

A.4 Projective Varieties

Affine varieties are the most important varieties as they are the building blocks for arbitrary varieties. Arbitrary varieties can be covered by open subsets which are affine varieties together with certain glueing conditions.

In modern treatments this glueing condition is usually coded in the notion of a sheaf, the structure sheaf of the variety. We are not going to introduce arbitrary varieties, since this would take us too deep into technical geometric constructions and too far away from commutative algebra.

However, there is one class of varieties which is the most important class of varieties after affine varieties and almost as closely related to algebra as affine ones. This is the class of the projective varieties.

What is the difference between affine and projective varieties? Affine varieties, for example $\mathbb{C}^n$, are in a sense open; travelling as far as we want, we can imagine the horizon — but we shall never reach infinity. On the other hand, projective varieties are closed (in the sense of compact, without boundary); indeed, we close up $\mathbb{C}^n$ by adding a "hyperplane at infinity" and, in this way, we domesticate infinity. The hyperplane at infinity can then be covered by finitely many affine varieties. In this way, finally, we obtain a variety covered by finitely many affine varieties, and we feel pretty well at home, at least locally.

However, the importance of projective varieties does not result from the fact that they can be covered by affine varieties, this holds for any variety. The important property of projective varieties is that they are closed, hence there is no escape to infinity. The simplest example demonstrating this are two parallel lines which do not meet in $\mathbb{C}^2$ but do meet in the projective plane $\mathbb{P}^2(\mathbb{C})$. This is what a perspective picture suggests, two parallel lines meeting at infinity.

This fact has many important consequences, the most important being probably *Bézout's theorem* (Theorem A.8.17), which says that two projective varieties $X, Y \subset \mathbb{P}_K^n$ (where K is an algebraically closed field) of complementary dimension, that is, $\dim(X) + \dim(Y) = n$, and without common component, meet in exactly $\deg(X) \cdot \deg(Y)$ points, if we count the intersection points with appropriate multiplicities. The degree $\deg(X)$ is a global invariant and can be defined using the Hilbert polynomial (see Definitions 5.3.3 and A.8.8), while the correct definition of the local multiplicities (for arbitrary singularities), for which Bézout's theorem holds, is a non–trivial task (cf. [120] for special cases and [82], [184] for arbitrary varieties). The Bézout's theorem, and many other "projective theorems", do not hold for affine varieties and, with respect to this geometric point of view, affine and projective varieties are completely different.

From an algebraic point of view, the difference is not that large, at least at first glance. Affine varieties in $\mathbb{A}_K^n$ are the zero–set of arbitrary polynomials in n variables, while projective varieties in $\mathbb{P}_K^n$ are the zero–set of homogeneous polynomials in $n+1$ variables. For any affine variety in $\mathbb{A}_K^n$ we can consider the projective closure in $\mathbb{P}_K^n$, obtained by homogenizing the ideal with the help of an extra variable. However, homogenizing an ideal is not completely trivial, except for hypersurfaces, since the homogenized ideal is not generated by an arbitrary homogenized set of generators. But we can compute generators for the homogenized ideal by homogenizing a Gröbner basis (with respect to a degree ordering).

Thus, we are able to compute equations for the projective closure of an affine variety. For homogeneous ideals we can compute the degree, the Hilbert polynomial, the graded Betti numbers and many more, all being important invariants of projective varieties. However, one has to be careful in choosing the correct ordering. Special care has to be taken with respect to elimination. This will be discussed in Section A.7.

Again, let K denote an algebraically closed field.

Definition A.4.1. Let V be a finite dimensional vector space over K. We define the *projective space* of V, denoted by $\mathbb{P}_K(V) = \mathbb{P}(V)$, as the space of lines in V going through 0.

More formally, two elements $v, w \in V \smallsetminus \{0\}$ are called equivalent, $v \sim w$, if and only if there exists a $\lambda \in K^*$ such that $v = \lambda w$. The equivalence classes are just the lines in V through 0 and, hence,

$$\mathbb{P}(V) := (V \smallsetminus \{0\})/ \sim \ .$$

We denote by $\pi : (V \smallsetminus \{0\}) \to \mathbb{P}(V)$ the canonical projection, mapping a point $v \in V \smallsetminus \{0\}$ to the line through 0 and v.

$$\mathbb{P}^n := \mathbb{P}^n_K := \mathbb{P}(K^{n+1})$$

is called the *projective n–space* over K.

An equivalence class of $(x_0, \ldots, x_n) \in K^{n+1} \smallsetminus \{0\}$, that is, a line through 0 and $(x_0, \ldots, x_n)$ is called a *point of* $\mathbb{P}^n$, which we denote as

$$p = \pi(x_0, \ldots, x_n) =: (x_0 : \ldots : x_n) \in \mathbb{P}^n \,,$$

and $(x_0 : \ldots : x_n)$ are called *homogeneous coordinates* of p.

Note that for $\lambda \in K^*$, $(\lambda x_0 : \ldots : \lambda x_n)$ are also homogeneous coordinates of p, that is, $(x_0 : \ldots : x_n) = (\lambda x_0 : \ldots : \lambda x_n)$.

Any linear isomorphism of K–vector spaces $V \cong W$, induces a bijection $\mathbb{P}(V) \cong \mathbb{P}(W)$, and the latter is called a *projective isomorphism*. In particular, if $\dim_K(V) = n + 1$, then $\mathbb{P}(V) \cong \mathbb{P}^n$ and, therefore, it is sufficient to consider $\mathbb{P}^n$.

Note that $f \in K[x_0, \ldots, x_n]$ defines a polynomial function $f : \mathbb{A}^{n+1} \to K$, but not a function $\mathbb{P}^n \to K$ since $f(\lambda x_0, \ldots, \lambda x_n) \neq f(x_0, \ldots, x_n)$, in general. However, if f is a *homogeneous polynomial* of degree d, then for $\lambda \in K$,

$$f(\lambda x_0, \ldots, \lambda x_n) = \lambda^d f(x_0, \ldots, x_n)$$

since $(\lambda x)^\alpha = (\lambda x_0)^{\alpha_0} \cdot \ldots \cdot (\lambda x_n)^{\alpha_n} = \lambda^d x^\alpha$ for any $\alpha = (\alpha_0, \ldots, \alpha_n)$ with $|\alpha| = d$. In particular, $f(x_0, \ldots, x_n) = 0$ if and only if $f(\lambda x_0, \ldots, \lambda x_n) = 0$

for all λ. Hence, the zero–set in $\mathbb{P}^n$ of a homogeneous polynomial is well-defined.

In the following, x denotes $(x_0, \ldots, x_n)$, and $K[x]_d$ denotes the vector space of homogeneous polynomials of degree d. Note that any $f \in K[x] \smallsetminus \{0\}$ has a unique *homogeneous decomposition*

$$f = f_{d_1} + \cdots + f_{d_k}, \quad f_i \in K[x]_i \smallsetminus \{0\}.$$

Definition A.4.2. A set $X \subset \mathbb{P}_K^n$ is called a *projective algebraic set* or a *(classical) projective variety* if there exists a family of homogeneous polynomials $f_\lambda \in K[x_0, \ldots, x_n]$, $\lambda \in \Lambda$, such that

$$X = V\big((f_\lambda)_{\lambda \in \Lambda}\big) := \{p \in \mathbb{P}_K^n \mid f_\lambda(p) = 0 \text{ for all } \lambda \in \Lambda\}.$$

X is called the *zero–set* of $(f_\lambda)_{\lambda \in \Lambda}$ in $\mathbb{P}^n$.

As in the affine case, X depends only on the ideal $I = \langle f_\lambda \mid \lambda \in \Lambda \rangle_{K[x]}$, which is finitely generated, since $K[x_0, \ldots, x_n]$ is Noetherian. Note that the polynomials f_λ may have different degree for different λ. If $X = V(f_1, \ldots, f_k)$ and all f_i are homogeneous of degree 1, then $X \subset \mathbb{P}^n$ is a projective linear subspace of $\mathbb{P}^n$, that is $X \cong \mathbb{P}^m$ with $m = n - \dim_K \langle f_1, \ldots, f_k \rangle_K$. If $X = V(f)$ for a single homogeneous polynomial $f \in K[x_0, \ldots, x_n]$ of degree $d > 0$, then X is called a *projective hypersurface in $\mathbb{P}^n$ of degree d*.

The polynomial ring $K[x]$, $x = (x_0, \ldots, x_n)$, has a canonical grading, where the *homogeneous component of degree d*, $K[x]_d$, consists of the homogeneous polynomials of degree d (see Section 2.2).

Recall that an ideal $I \subset K[x]$ is called homogeneous if it can be generated by homogeneous elements. Hence, projective varieties in $\mathbb{P}^n$ are the zero–sets $V(I)$ of homogeneous ideals $I \subset K[x_0, \ldots, x_n]$.

As in the affine case (cf. Lemma A.2.4) we have, using Exercise 2.2.4,

Lemma A.4.3.

(1) $\emptyset$, $\mathbb{P}^n$ are projective varieties.
(2) The union of finitely many projective varieties is projective.
(3) The intersection of arbitrary many projective varieties is projective.

The *Zariski topology* on $\mathbb{P}^n$ is defined by taking as closed sets the projective varieties in $\mathbb{P}^n$. The Zariski topology on a projective variety $X \subset \mathbb{P}^n$ is the induced topology. An open subset of a projective variety is called a *quasi-projective variety*.

Again, we call a projective variety $X \subset \mathbb{P}^n$ *irreducible* if it is irreducible as a topological space, that is, if $X \neq \emptyset$, and if it is not the union of two proper projective algebraic subsets.

As in the affine case, there is a tight connection between (irreducible) projective varieties and homogeneous (prime) ideals, the reason being the projective Nullstellensatz.

Definition A.4.4. For any non–empty set $X \subset \mathbb{P}^n_K$ define

$$I(X) := \langle f \in K[x_0, \dots, x_n] \mid f \text{ homogeneous, } f|_X = 0 \rangle \subset K[x_0, \dots, x_n],$$

the *(full) homogeneous (vanishing) ideal* of $X \subset \mathbb{P}^n_K$. The quotient ring

$$K[X] := K[x_0, \dots, x_n]/I(X)$$

is called the *homogeneous coordinate ring* of X. We set

$$K[X]_d := K[x_0, \dots, x_n]_d / \big(I(X) \cap K[x_0, \dots, x_n]_d\big).$$

If $X = V(I) \subset \mathbb{P}^n$ is projective, $I \subset K[x_0, \dots, x_n]$ a homogeneous ideal, then the projective Nullstellensatz (see below) compares I and the full homogeneous vanishing ideal $I(X)$ of X. Of course, we can recover I from X only up to radical, but there is another ambiguity in the homogeneous case, namely $V(\langle 1 \rangle) = \emptyset = V(\langle x_0, \dots, x_n \rangle)$. We define

$$I(\emptyset) := \langle x_0, \dots, x_n \rangle = \bigoplus_{d > 0} K[x]_d =: K[x]_+$$

which is called the *irrelevant ideal*.

Remark A.4.5. The irrelevant ideal can be used to adjust the degrees of the defining equations of a projective variety. Since $V(f) = V(x_0 f, \dots, x_n f) \subset \mathbb{P}^n$, we see that a projective variety defined by a polynomial of degree d is also the zero–set of polynomials of degree $d + 1$. Hence, if $X \subset \mathbb{P}^n$ is the zero–set of homogeneous polynomials of degree $d_i \leq d$, then X is also the zero–set of homogeneous polynomials having all the same degree d (but with more equations).

If $X \subset \mathbb{P}^n$ is a projective algebraic set and $\pi : K^{n+1} \smallsetminus \{0\} \to \mathbb{P}^n$ the projection, we may also consider the affine variety

$$CX := \pi^{-1}(X) \cup \{0\} = \{(x_0, \dots, x_n) \in \mathbb{A}^{n+1} \mid (x_0 : \dots : x_n) \in X\} \cup \{0\},$$

which is the union of lines through 0 in $\mathbb{A}^{n+1}$, corresponding to the points in X if $X \neq \emptyset$, and $C\emptyset = \{0\}$. CX is called the *affine cone* of X (see the symbolic picture in Figure A.14, respectively Figure A.15). It is the affine variety in $\mathbb{A}^{n+1}$ defined by the homogeneous ideal $I \subset K[x_0, \dots, x_n]$.

Now consider $I(CX)$, the vanishing ideal of the affine variety $CX \subset \mathbb{A}^{n+1}$. If $f = f_{d_1} + \dots + f_{d_k} \in I(CX)$ is the homogeneous decomposition of f, then, for any $(x_0, \dots, x_n) \in CX \smallsetminus \{0\}$, we have

$$f(\lambda x_0, \dots, \lambda x_n) = \lambda^{d_1} f_{d_1}(x_0, \dots, x_n) + \dots + \lambda^{d_k} f_{d_k}(x_0, \dots, x_n) = 0,$$

for all $\lambda \in K$. Hence, since K is algebraically closed and has infinitely many elements, all homogeneous components $f_{d_1}, \dots, f_{d_k}$ of f are in $I(CX)$ (this

$$CV \subset \mathbb{A}^{n+1}$$

Fig. A.14. The affine cone of a projective variety V.

need not be true if K is finite). By Lemma 2.2.7, $I(CX)$ is a homogeneous ideal.

Slightly more general, let $\mathbb{P}^n \subset \mathbb{P}^{n+1}$ be a projective hyperplane, let $X \subset \mathbb{P}^n$ be a projective variety, and let $p \in \mathbb{P}^{n+1} \setminus \mathbb{P}^n$. Then the *projective cone over* X with vertex p is the union of all projective lines $\overline{pq}$, q running through all points of X. We denote it by $\overline{C_pX}$, it is a projective variety in $\mathbb{P}^{n+1}$. If X is given by homogeneous polynomials $f_i(x_1, \ldots, x_{n+1})$ and if $p = (1 : 0 : \ldots : 0)$, then $\overline{C_pX}$ is given by the same polynomials, considered as elements of $K[x_0, \ldots, x_{n+1}]$. An affine or a projective cone over some variety X with vertex some point $p \notin X$ is simply called a *cone*.

Theorem A.4.6 (Projective Nullstellensatz). *Let K be an algebraically closed field and $I \subset K[x_0, \ldots, x_n]$ a homogeneous ideal. Then*

(1) $V(I) = \emptyset$ if and only if $\langle x_0, \ldots, x_n \rangle \subset \sqrt{I}$;
(2) if $V(I) \neq \emptyset$ then $I(V(I)) = \sqrt{I}$.

Moreover, there is an inclusion reversing bijection

$$
\left\{ \begin{array}{c} \textit{projective algebraic} \\ \textit{sets in } \mathbb{P}_K^n \end{array} \right\} \longleftrightarrow \left\{ \begin{array}{c} \textit{homogeneous radical ideals} \\ I \subset \langle x_0, \ldots, x_n \rangle \end{array} \right\}
$$
$$
\cup \qquad\qquad\qquad\qquad \cup
$$
$$
\left\{ \begin{array}{c} \textit{irreducible projective} \\ \textit{algebraic sets in } \mathbb{P}_K^n \end{array} \right\} \longleftrightarrow \left\{ \begin{array}{c} \textit{homogeneous prime ideals} \\ I \subsetneq \langle x_0, \ldots, x_n \rangle \end{array} \right\}
$$
$$
\cup \qquad\qquad\qquad\qquad \cup
$$
$$
\left\{ \textit{points in } \mathbb{P}_K^n \right\} \longleftrightarrow \left\{ \begin{array}{c} \textit{homogeneous maximal ideals} \\ I \subsetneq \langle x_0, \ldots, x_n \rangle \end{array} \right\}.
$$

Proof. Let $V_a(I) \subset \mathbb{A}^{n+1}$, respectively $V(I) \subset \mathbb{P}^n$, denote the affine, respectively projective, variety defined by I.

(1) We have $V(I) = \emptyset$ if and only if $V_a(I) \subset \{0\}$, which is equivalent to $\langle x_0, \ldots, x_n \rangle \subset I(V_a(I))$. But the latter ideal equals $\sqrt{I}$, due to the usual Nullstellensatz.

(2) The considerations before Theorem A.4.6 imply that $I(V_a(I))$ is generated by all *homogeneous* polynomials vanishing on $V_a(I)$, hence, we have

$I\big(V_a(I)\big) = I\big(V(I)\big)$ and, therefore, everything follows from the affine Null-stellensatz. For example, the point $(p_0 : \ldots : p_n)$ with $p_0 \neq 0$ corresponds to the ideal $\langle p_0 x_1 - p_1 x_0, \ldots, p_0 x_n - p_n x_0 \rangle$.   □

As an example, look at the affine cone of the cuspidal cubic, given by $f = y^3 + x^2 z = 0$. By Remark A.4.5, this is also given as the zero–set of the ideal $I = \langle xf, yf, zf \rangle$. However, the latter ideal is not radical. We check this and draw the surface using SINGULAR:

SINGULAR Example A.4.7 (projective Nullstellensatz).

```
ring R = 0,(x,y,z),dp;
poly f = x2z+y3;
ideal I = maxideal(1)*f;     //<xf,yf,zf>
LIB"primdec.lib";
radical(I);
//-> _[1]=y3+x2z

LIB"surf.lib";
plot(f);                     // cf. Figure A.15
```

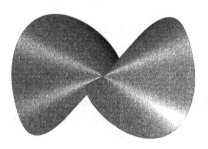

Fig. A.15. Affine cone of cuspidal cubic.

Moreover, any projective algebraic set $X = V(I) \subset \mathbb{P}^n$ has a unique minimal decomposition into irreducible components: let $P_1, \ldots, P_r$ denote the minimal associated primes of I (which are homogeneous ideals by Exercise 4.1.8), then $X = V(P_1) \cup \cdots \cup V(P_r)$ is the unique decomposition of X into irreducible projective varieties, no one containing another.

Let us now see how to cover $\mathbb{P}^n$ by affine charts.

Definition A.4.8. In $\mathbb{P}^n_K$, with homogeneous coordinates $(x_0 : \ldots : x_n)$, let

$$H_i := V(x_i) = \{(x_0 : \ldots : x_n) \in \mathbb{P}^n \mid x_i = 0\}$$

denote the *i–th hyperplane at infinity*, $i = 0, \ldots, n$, and let

$$U_i := D(x_i) := \mathbb{P}^n \smallsetminus H_i = \{(x_0 : \ldots : x_n) \in \mathbb{P}^n \mid x_i \neq 0\}$$

denote the *i–th (canonical) affine chart* of $\mathbb{P}^n$. Usually, $H_0 = V(x_0)$ is denoted by H_∞ and called the *hyperplane at infinity*, and U_0 is called the *affine chart* of $\mathbb{P}^n$.

In an obvious way, H_i can be identified with $\mathbb{P}^{n-1}$, and we shall see below that U_i can be identified with $\mathbb{A}^n$. Of course, U_i and H_i depend on the chosen coordinates.

Definition A.4.9. For a homogeneous polynomial $f \in K[x_0, \ldots, x_n]$, let

$$f^a(x_1, \ldots, x_n) := f(1, x_1, \ldots, x_n) \in K[x_1, \ldots, x_n]$$

denote the *affinization* or *dehomogenization* of f. For an arbitrary polynomial $g \in K[x_1, \ldots, x_n]$ of degree d, let

$$g^h(x_0, x_1, \ldots, x_n) := x_0^d \cdot g\left(\frac{x_1}{x_0}, \ldots, \frac{x_n}{x_0}\right) \in K[x_0, \ldots, x_n]$$

denote the *homogenization* of g (with respect to x_0). x_0 is called the *homogenizing variable*.

Remark A.4.10.

(1) Instead of substituting $x_0 = 1$, we can substitute $x_i = 1$ and obtain the affinization with respect to x_i. Since any monomial $x_1^{\alpha_1} \cdot \ldots \cdot x_n^{\alpha_n}$ of g satisfies $|\alpha| = \alpha_1 + \cdots + \alpha_n \leq d$,

$$x_0^d \left(\frac{x_1}{x_0}\right)^{\alpha_1} \cdot \ldots \cdot \left(\frac{x_n}{x_0}\right)^{\alpha_n} = x_0^{d-|\alpha|} x_1^{\alpha_1} \cdot \ldots \cdot x_n^{\alpha_n}$$

is a monomial of degree d. That is, g^h is a homogeneous polynomial of degree d.

(2) $(g^h)^a = g$ for any $g \in K[x_1, \ldots, x_n]$.

(3) Let $f \in K[x_0, \ldots, x_n]$ be homogeneous and $f = x_0^p g(x_0, \ldots, x_n)$ such that x_0 does not divide g, then $(f^a)^h = g$. Hence, we have $x_0^e \cdot (f^a)^h = f$ for some e, and $(f^a)^h = f$ if and only if $x_0 \nmid f$. For example, if $f = x^2 z^2 + y^3 z$, and if z is the homogenizing variable, then $z \cdot (f^a)^h = f$.

Lemma A.4.11. *Let $U_i := D(x_i)$, then $\{U_i\}_{i=0,\ldots,n}$ is an open covering of $\mathbb{P}^n$ and*

$$\varphi_i : U_i \longrightarrow \mathbb{A}^n, \quad (x_0 : \ldots : x_n) \longmapsto \left(\frac{x_0}{x_i}, \ldots, \frac{\widehat{x_i}}{x_i}, \ldots, \frac{x_n}{x_i} \right)$$

is a homeomorphism ($\widehat{}$ means that the respective component is deleted). The map

$$\varphi_j \circ \varphi_i^{-1} : \mathbb{A}^n \setminus V(x_j) \longrightarrow \mathbb{A}^n \setminus V(x_i),$$

$$(x_1, \ldots, x_n) \longmapsto \frac{1}{x_j}(x_1, \ldots, x_i, 1, x_{i+1}, \ldots, \widehat{x_j}, \ldots, x_n)$$

describes the coordinate transformation on the intersection $U_i \cap U_j$, $i < j$.

Proof. The first statement is clear, since for $p = (x_0 : \ldots : x_n) \in \mathbb{P}^n$ there exists an i with $x_i \neq 0$, hence $p \in U_i$. It is also clear that φ_i is well–defined and bijective, with $\varphi_i^{-1}(x_1, \ldots, x_n) = (x_1 : \ldots : x_{i-1} : 1 : x_i : \ldots : x_n)$.

It remains to show that φ_i and φ_i^{-1} are continuous, that is, the preimages of closed sets are closed. We show this for $i = 0$ and set $\varphi := \varphi_0$, $U = U_0$.

If $Y \subset U$ is closed, then there exists a closed set $X = V(f_1, \ldots, f_k) \subset \mathbb{P}^n$, $f_i \in K[x_0, \ldots, x_n]$ homogeneous, such that $Y = U \cap X$. Then the image $\varphi(Y) = V(f_1^a, \ldots, f_k^a) \subset \mathbb{A}^n$ is closed.

Conversely, let $W = V(g_1, \ldots, g_\ell) \subset \mathbb{A}^n$ be closed, $g_i \in K[x_1, \ldots, x_n]$. Since $(g_j^h)^a = g_j$, $\varphi^{-1}(W) = V(g_1^h, \ldots, g_\ell^h) \cap U$ is closed in U. □

We usually identify U_i with $\mathbb{A}^n$ and $V(x_i)$ with $\mathbb{P}^n$. More generally, if $X \subset \mathbb{P}^n$ is a projective variety, we identify $X \cap U_i$ with $\varphi_i(X \cap U_i) \subset \mathbb{A}^n$ and, in this sense, $X \cap U_i$ is an affine variety. We specify this when we define morphisms of quasi–projective varieties in Section A.6.

Example A.4.12. Consider the projective cubic curve C in $\mathbb{P}^2$, given by the homogeneous equation $x^2 z - y^3 = 0$.

In $U_0 := \{x \neq 0\}$, we have the affine equation (setting $x = 1$), $z - y^3 = 0$, in $U_1 := \{y \neq 0\}$, we have $x^2 z - 1 = 0$ and in $U_2 := \{z \neq 0\}$, $x^2 - y^3 = 0$.

The line $H_0 := \{x = 0\}$ meets C in $(0 : 0 : 1) \in U_2$, $H_1 := \{y = 0\}$, meets C in $(1 : 0 : 0)$ and $(0 : 0 : 1)$, while $H_2 := \{z = 0\}$, meets C in $(1 : 0 : 0)$. The local pictures in the neighbourhoods of 0 in U_0, U_1, U_2 are displayed in Figure A.16.

We can also represent the global curve C by a symbolic picture in all three coordinate neighbourhoods (cf. Figure A.17).

We define the *dimension* of the projective variety X as

$$\dim(X) = \dim(CX) - 1,$$

where CX is the affine cone of X.

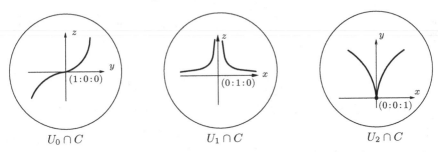

Fig. A.16. Local pictures of the projective cubic $x^2 z - y^3 = 0$.

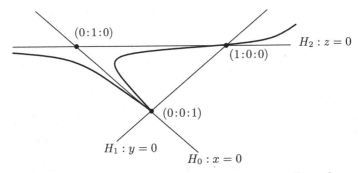

Fig. A.17. A symbolic picture of the projective cubic $x^2 z - y^3 = 0$.

Lemma A.4.13. *If $X \subset \mathbb{P}^n$ is a projective variety, then*

$$\dim X = \max\{\dim \varphi_i(X \cap U_i) \mid i = 0, \dots, n\}.$$

If $X = V(f_1, \dots, f_k)$, $f_i \in K[x_0, \dots, x_n]$ homogeneous, then $\varphi(X \cap U_i)$ is the affine variety in $\mathbb{A}^n$ defined by $f_1|_{x_i=1}, \dots, f_k|_{x_i=1} \in K[x_0, \dots, \widehat{x}_i, \dots, x_n]$, where $f_j|_{x_i=1} := f_j(x_0, \dots, x_{i-1}, 1, x_{i+1}, \dots, x_n)$.

The proof is left as an exercise.

So far, we related a projective variety $X \subset \mathbb{P}^n$ to affine varieties. We did this in two different ways, by considering the affine cone $CX \subset \mathbb{A}^{n+1}$ and the affine pieces $X \cap U_i \subset \mathbb{A}^n$, where we fixed coordinates $x_1, \dots, x_n$ of $\mathbb{A}^n$ and $x_0, \dots, x_n$ of $\mathbb{A}^{n+1}$, respectively $\mathbb{P}^n$. In particular, we have, for $i = 1, \dots, n$,

$$X = (X \cap U_i) \cup (X \cap V(x_i))$$

where $X \cap U_i \subset \mathbb{A}^n$ is affine and $X \cap V(x_i) \subset \mathbb{P}^{n-1}$ is projective. Sometimes we call $X \cap U_0$ the *affine part* of X and $X_\infty := X \cap V(x_0)$ the *part at infinity*.

Now we go the opposite way and associate to any affine variety X in $\mathbb{A}^n$ a projective variety $\overline{X}$ in $\mathbb{P}^n$, such that its affine part is equal to X.

Definition A.4.14. Let $x_0, \ldots, x_n$ be homogeneous coordinates on $\mathbb{P}^n$, let $U_0 = D(x_0) \subset \mathbb{P}^n$ be the affine chart, and let $\varphi_0 : U_0 \to \mathbb{A}^n$ be as before.

(1) Let $V \subset \mathbb{A}^n$ be an affine variety. The closure of $\varphi_0^{-1}(V)$ in $\mathbb{P}^n$ in the Zariski topology is called the *projective closure* of V in $\mathbb{P}^n$ and denoted by $\overline{V}$. Let $V_\infty := \overline{V} \cap H_\infty$ denote the part at infinity of V (respectively $\overline{V}$), where $H_\infty := V(x_0)$ denotes the *hyperplane at infinity*.

(2) If $I \subset K[x_1, \ldots, x_n]$ is an ideal, then

$$I^h := \langle f^h \mid f \in I \rangle \subset K[x_0, \ldots, x_n]$$

is called the *homogenization of I*. For $g \in K[x_0, \ldots, x_n]$ let

$$g|_{x_0=0} := g(0, x_1, \ldots, x_n) \in K[x_1, \ldots, x_n],$$

and set

$$I_\infty := \langle g|_{x_0=0} \mid g \in I^h \rangle \subset K[x_1, \ldots, x_n].$$

(3) If $X \subset \mathbb{P}^n$ is projective then $X^a = \varphi_0(X \cap U_0) \subset \mathbb{A}^n$ is called the *affine part* of X. If $J \subset K[x_0, \ldots, x_n]$ is a homogeneous ideal, the ideal

$$J^a := J|_{x_0=1} := \langle f^a \mid f \in J \rangle$$

is called the *affinization* of J.

The following lemma shows that I^h, respectively I_∞, describes the projective closure, respectively the part at infinity of $V(I)$, and that the affinization J^a describes the affine part of $V(J)$. Of course, an analogous result holds for $J|_{x_i=1}$, the dehomogenization of J with respect to x_i.

Recall that $I(V)$ denotes the full vanishing ideal of the affine, respectively projective, variety V.

Lemma A.4.15.

(1) Let $I \subset K[x_1, \ldots, x_n]$ be an ideal and $V = V(I) \subset \mathbb{A}^n$. Then

$$I(\overline{V}) = I(V)^h, \ V(I^h) = \overline{V} \ and \ I(V_\infty) = \sqrt{I(V)_\infty}, \ V(I_\infty) = V_\infty.$$

(2) If $J \subset K[x_0, \ldots, x_n]$ is a homogeneous ideal and $X = V(J) \subset \mathbb{P}^n$. Then

$$I(X^a) = I(X)^a, \ V(J^a) = X^a.$$

Proof. (1) By definition, $\overline{V}$ is the intersection of all projective varieties containing $\varphi_0^{-1}(V)$ and, hence, the intersection of all projective hypersurfaces containing $\varphi_0^{-1}(V)$. Therefore, $I(\overline{V})$ is generated by all homogeneous polynomials f satisfying $f|_V = 0$, that is, $f^a \in I(V)$. Hence, if $x_0 g \in I(\overline{V})$, then $g \in I(\overline{V})$.

This implies that $I(\overline{V})$ is generated by homogeneous elements f which are not divisible by x_0. By Remark A.4.10, for such elements we have $f = (f^a)^h \in I(V)^h$, that is, $I(\overline{V}) \subset I(V)^h$.

Conversely, $I(V)^h \subset I(\overline{V})$, since, by continuity, $f^h|_{\overline{V}} = 0$ for $f^h \in I(V)^h$. If I is any ideal with $V(I) = V$, then $V(I^h)$ is projective and contains $\varphi_0^{-1}(V)$, hence $\overline{V} \subset V(I^h)$. For the other inclusion, let $p \in V(I^h)$. Since $V(I^h) \cap U_0 = V = \overline{V} \cap U_0$, we only have to consider points p of the form $(0 : x_1 : \ldots : x_n) \in V(I^h)_\infty$. Then $g^h(p) = 0$ for all $g \in I$. Since $I(V)^m \subset I$ for some m by Hilbert's Nullstellensatz, $0 = (f^m)^h(p) = (f^h(p))^m$ and, hence, $f^h(p) = 0$ for all $f \in I(V)$, that is, $p \in V\big(I(V)^h\big) \subset \overline{V}$ by the first part. This implies also $V(I_\infty) = V_\infty = V\big(I(V)_\infty\big)$ and, hence, by Hilbert's Nullstellensatz $\sqrt{I_\infty} = I(V_\infty) = \sqrt{I(V)_\infty}$.

(2) is obvious. $\square$

It can easily be seen that it is not sufficient to homogenize an arbitrary set of generators of I in order to obtain a set of generators of I^h (cf. Example A.4.17). However, if $f_1, \ldots, f_k$ is a Gröbner basis of I with respect to a degree ordering, then $I^h = \langle f_1^h, \ldots, f_k^h \rangle$, cf. Exercise 1.7.4, and we have

$$\overline{V(I)} = V(f_1^h, \ldots, f_k^h), \qquad V(I)_\infty = V(f_1^h|_{x_0=0}, \ldots, f_k^h|_{x_0=0}).$$

In contrast, if $J = \langle f_1, \ldots, f_k \rangle$, with f_i arbitrary homogeneous generators, then $J^a = \langle f_1^a, \ldots, f_k^a \rangle$.

Corollary A.4.16. *Let $V \subset \mathbb{A}^n$ be an affine variety, $\overline{V} \subset \mathbb{P}^n$ its projective closure and $H_\infty := \mathbb{P}^n \setminus \mathbb{A}^n$ the hyperplane at infinity. Then*

(1) V is irreducible if and only if $\overline{V}$ is irreducible.
(2) If $V = V_1 \cup \cdots \cup V_s$ is the decomposition of V into irreducible components, then $\overline{V} = \overline{V}_1 \cup \cdots \cup \overline{V}_s$ is the decomposition of $\overline{V}$ into irreducible components.
(3) The map $V \mapsto \overline{V}$ induces a bijection

$$\{\text{affine varieties in } \mathbb{A}^n\} \longleftrightarrow \left\{ \begin{array}{l} \text{projective varieties in } \mathbb{P}^n \text{ with} \\ \text{no component contained in } H_\infty \end{array} \right\}.$$

Proof. (1) follows, since I is a prime ideal if and only if I^h is a prime ideal (Exercise 4.1.9 (1)). (2) follows easily from (1) and, using (2), we may assume in (3) that V is irreducible. For any irreducible projective variety $W \subset \mathbb{P}^n$, we have $W \subset H_\infty$ if and only if $x_0 \in I(W)$. This implies (3), since $x_0 \notin I^h$ for any ideal $I \subset K[x_1, \ldots, x_n]$. $\square$

Example A.4.17. Consider the ideal $I := \langle x^3 + z^2, x^3 + y^2 \rangle \subset K[x, y, z]$. Then $I^h \ni z^2 - y^2 \notin J = \langle x^3 + z^2 u, x^3 + y^2 u \rangle$, where u is the homogenizing variable. Indeed, $V(I)$ is a curve in $\mathbb{A}^3$, its projective closure $V(I^h)$ should meet $H_\infty = \{u = 0\}$ only in finitely many points, while $V(J)$ contains the line $\{u = x = 0\}$ as component in H_∞.

We verify this, using SINGULAR:

SINGULAR Example A.4.18 (projective closure).

```
ring R = 0,(x,y,z,u),dp;
ideal I = x3 + z2, x3 + y2;
ideal J = homog(I,u);
J;
//-> J[1]=x3+z2u
//-> J[2]=x3+y2u

ideal Ih = homog(std(I),u);
Ih;                          //the homogenization of I
//-> Ih[1]=y2-z2
//-> Ih[2]=x3+y2u

LIB "primdec.lib";
minAssGTZ(J,1);              //the minimal associated primes
//-> [1]:
//->    _[1]=u
//->    _[2]=x
//-> [2]:
//->    _[1]=y+z
//->    _[2]=x3+z2u
//-> [3]:
//->    _[1]=y-z
//->    _[2]=x3+z2u
```

Note that we obtain in $V(J)$ the extra component $\{u = x = 0\}$ at infinity, as J is not equal to I^h. $\overline{V} = V(I^h)$ meets H_∞ in $(0 : 1 : \pm1 : 0)$.

A.5 Projective Schemes and Varieties

Let us give a short account of *projective schemes*, the projective counterpart of affine schemes.

Definition A.5.1. Let $A = \bigoplus_{d \geq 0} A_d$ be a graded ring and A_+ the homogeneous ideal $\bigoplus_{d > 0} A_d$, which is called the *irrelevant ideal* of A. We set

$$\text{Proj}(A) := \{P \subset A \mid P \text{ a homogeneous prime ideal}, A_+ \not\subset P\}.$$

The elements of $\text{Proj}(A)$ are called *relevant prime ideals* of A, and $\text{Proj}(A)$ is called the *projective*, or *homogeneous, spectrum* of A. For $X = \text{Proj}(A)$ and $I \subset A$ a homogeneous ideal, the set

$$V(I) := \{P \in X \mid P \supset I\}$$

is called the *zero–set of I* in X.

We have $V(A_+) = \emptyset$, $V(\langle 0 \rangle) = \text{Proj}(A)$, and, as for $\text{Spec}(A)$, the sets $V(I)$ are the closed sets of a topology, the *Zariski topology* on $\text{Proj}(A)$. This is also the induced topology from $\text{Proj}(A) \subset \text{Spec}(A)$.

From now on, let R be a Noetherian ring and $A = R[x_0, \ldots, x_n]$ the polynomial ring over R with A_d the homogeneous polynomials of degree d. Then $\text{Proj}(A)$ is called the n–dimensional *projective space over R* and is denoted by $\mathbb{P}^n_R$.

If $I \subset A = R[x_0, \ldots, x_n]$ is a homogeneous ideal, then $V(I)$ coincides with $\text{Proj}(A/I)$ under the map sending $P \in V(I)$ to its residue class modulo I. This bijection is even a homeomorphism, and we shall consider $\text{Proj}(A/I)$ as a closed subspace of $\mathbb{P}^n_R$.

So far, we have defined $\mathbb{P}^n_R$ only as a topological space. To define a scheme structure on $\mathbb{P}^n_R$ or, more generally, on the closed subspace $\text{Proj}(A/I)$ of $\mathbb{P}^n_R$ we *cannot*, as we did in the affine case, define it as a pair of the topological space $\text{Proj}(A/I)$ together with the homogeneous coordinate ring A/I. This would distinguish between schemes which we want to be the same. Namely, if $\mathfrak{m} = \langle x_0, \ldots, x_n \rangle \subset A$ is the irrelevant ideal, then $\mathfrak{m}^k I$ and I define the same zero–sets in $\mathbb{P}^n_R$. But we also want $\text{Proj}(A/I)$ and $\text{Proj}(A/\mathfrak{m}^k I)$ to define the same projective scheme. For example, we do not want to distinguish between $\text{Proj}(A/\mathfrak{m})$ and $\text{Proj}(A/\mathfrak{m}^2)$, since both are the empty scheme. Of course, $\text{Spec}(A/I)$ and $\text{Spec}(A/\mathfrak{m}^k I)$ are different, but the difference is located in the vertex of the affine cone, which is irrelevant for the projective scheme.

What we do is, we define the scheme structure of a projective variety locally: for $f \in A$ homogeneous of positive degree we set $V(f) := V(\langle f \rangle)$ and call its complement

$$D(f) := \mathbb{P}^n_R \smallsetminus V(f)$$

a *principal open set* in $\mathbb{P}^n_R$. As a set, $D(f)$ coincides with the homogeneous prime ideals not containing f. Intersecting them with

$$(A_f)_0 = \left\{ \frac{h}{f^n} \;\middle|\; h \in R[x_1, \ldots, x_n] \text{ homogeneous of degree } n \cdot \deg(f) \right\},$$

the subring of A_f of homogeneous elements of degree 0, they correspond to all prime ideals of $(A_f)_0$. That is, we have

$$D(f) = \text{Spec}\big((A_f)_0\big),$$

which is an affine scheme. Since each $V(I)$ can be written as a finite intersection $V(I) = V(f_1) \cap \ldots \cap V(f_r)$ with $f_1, \ldots, f_r$ a system of homogeneous generators for I, it follows that the open sets $D(f)$ are a basis of the Zariski topology of $\mathbb{P}^n_R$.

We define the *scheme structure* on $\mathbb{P}^n_R$ by the collection of the affine schemes $\text{Spec}\big((A_f)_0\big)$, where f varies over the homogeneous elements of A of positive degree. To specify the scheme structure, we can actually take any open covering by affine schemes. In particular, it is sufficient to consider the covering of $\mathbb{P}^n_R$ by the *canonical charts*

$$U_i := D(x_i) = \mathrm{Spec}\big((R\,[x_0,\ldots,x_n]_{x_i})_0\big),$$

$i = 0,\ldots,n$, where

$$(R\,[x_0,\ldots,x_n]_{x_i})_0 = R\left[\frac{x_0}{x_i},\ldots,\frac{x_{i-1}}{x_i},\frac{x_{i+1}}{x_i},\ldots,\frac{x_n}{x_i}\right]$$

is isomorphic to the polynomial ring $R\,[y_1,\ldots,y_n]$ in n variables over R.

Now let $X = \mathrm{Proj}(A/I) \subset \mathbb{P}^n_R$ be a closed subset. Then X is covered by the open subsets

$$D(x_i) \cap X \cong \mathrm{Spec}\big(((A/I)_{x_i})_0\big),$$

where

$$((A/I)_{x_i})_0 \cong R\,[x_0,\ldots,x_{i-1},x_{i+1},\ldots,x_n]/(I|_{x_i=1}).$$

Then the *scheme* $X = \mathrm{Proj}(A/I)$ is given by the topological space $|X|$ and the *scheme structure*, which is specified by the collection of affine schemes $\mathrm{Spec}\big(((A/I)_{x_i})_0\big)$, $i = 0,\ldots,n$. Any scheme $X \subset \mathbb{P}^n_R$, given in this way, is called a *projective subscheme of* $\mathbb{P}^n_R$. A *projective scheme over* R is a closed subscheme of some $\mathbb{P}^n_R$ and a *quasi-projective scheme* is an open subset of a projective scheme with the induced scheme structure. A projective scheme X is called *reduced* if the affine schemes $\mathrm{Spec}\big(((A/I)_{x_i})_0\big)$, that is, the rings $((A/I)_{x_i})_0$ are reduced for $i = 0,\ldots,n$.

With the above definition, $\mathrm{Proj}(A/I)$ and $\mathrm{Proj}(A/\mathfrak{m}^k I)$ coincide. In particular, in contrast to the case of affine schemes (where, by definition, the ideals in A correspond bijectively to the subschemes of $\mathrm{Spec}(A)$), there is no bijection between homogeneous ideals in $\mathfrak{m}$ and projective subschemes of $\mathbb{P}^n_R$, even if R is an algebraically closed field.

The following lemma clarifies the situation and gives a geometric interpretation of saturation (considered in Section 1.8.9).

Lemma A.5.2. *Let R be a Noetherian ring, and let $I, J \subset A = R[x_0,\ldots,x_n]$ be two homogeneous ideals. Then $\mathrm{Proj}(A/I)$ and $\mathrm{Proj}(A/J)$ define the same projective subscheme of $\mathbb{P}^n_R$ if and only if the saturations of I and J coincide. Hence, there is a bijection*

$$\left\{\begin{array}{c} \textit{projective subschemes} \\ X \textit{ of } \mathbb{P}^n_R \end{array}\right\} \longleftrightarrow \left\{\begin{array}{c} \textit{saturated homogeneous ideals} \\ I \subset \mathfrak{m} := \langle x_0,\ldots,x_n\rangle \end{array}\right\}$$

Recall that the *saturation* of the homogeneous ideal $I \subset R\,[x_0,\ldots,x_n]$ is

$$\mathrm{sat}(I) := I : \mathfrak{m}^\infty = \{f \in A \mid \mathfrak{m}^r f \in I \text{ for some } r \geq 0\},$$

$\mathfrak{m} = \langle x_0,\ldots,x_n\rangle$ (see Section 1.8.9). I is called *saturated* if $\mathrm{sat}(I) = I$. We leave the proof of the lemma as an exercise.

SINGULAR Example A.5.3 (projective subscheme, saturation).

```
ring R=0,(x,y,z),dp;
ideal M=maxideal(1);
ideal I=(x2z+y3)*M^2;
I;
//-> I[1]=x2y3+x4z
//-> I[2]=xy4+x3yz
//-> I[3]=xy3z+x3z2
//-> I[4]=y5+x2y2z
//-> I[5]=y4z+x2yz2
//-> I[6]=y3z2+x2z3

LIB"elim.lib";
sat(I,M);
//-> [1]:
//->    _[1]=y3+x2z
//-> [2]:
//->    2
```

We obtain that I and $\langle y^3 + x^2 z \rangle$ define the same subscheme of $\mathbb{P}^2_{\mathbb{C}}$.

If $X = \mathrm{Proj}(A/I)$ is a projective scheme, then $\mathrm{Spec}(A/I)$ is called the *affine cone of* X. As in the classical case, we define the dehomogenization of a homogeneous polynomial, the homogenization of an arbitrary polynomial and the projective closure of an affine scheme. We leave the details to the reader.

To summarize the discussion so far, we see that a projective scheme $X \subset \mathbb{P}^n_K$ (K an algebraically closed field) can be given algebraically in three different ways, each way having a clearly distinguished geometric meaning.

(1) We give X by a homogeneous radical ideal $I \subset \langle x_0, \ldots, x_n \rangle$.
    This means, by the projective Nullstellensatz, to give X *set–theoretically*. That is, we specify X as a classical projective variety or, in the language of schemes, we specify X with its reduced structure.
(2) We give X by a homogeneous saturated ideal $I \subset \langle x_0, \ldots, x_n \rangle$.
    This is, by the above lemma, equivalent to specifying X as a projective scheme *scheme–theoretically*.
(3) We give X by an arbitrary homogeneous ideal $I \subset \langle x_0, \ldots, x_n \rangle$.
    This means that we do not only specify X with its scheme–structure but we do also specify the affine cone over X scheme–theoretically. This is sometimes called the *arithmetic structure* of X.

Note that in Section 4.5, respectively Section 1.8.9, we have shown how the radical, respectively the saturation, of a polynomial ideal can be computed effectively.

So far we have not yet defined morphisms between projective varieties, respectively projective schemes, since this is more involved than in the affine case. It is not true that such morphisms are induced by morphisms of the homogeneous coordinate rings, rather we have to refer to a covering by open affine pieces. This will be explained in detail in Section A.7.

Nevertheless, already, we give here the relation between varieties and schemes in a functorial way, referring to Section A.7 for the definition of morphisms between quasi–projective schemes, used in (2) below.

Theorem A.5.4.

(1) There is a functor F from the category of classical affine varieties to the category of affine schemes, defined as follows: for X an affine variety, set $F(X) := \mathrm{Spec}(K[X])$ where $K[X]$ is the coordinate ring of X and for $f : X \to Y$ a morphism of affine varieties set $\varphi := f^* : K[Y] \to K[X]$ and $F(f) := (\varphi^\#, \varphi) : \mathrm{Spec}(K[X]) \to \mathrm{Spec}(K[Y])$ the corresponding morphism of affine schemes. F has the following properties:

a) F is fully faithful, that is, the map

$$\mathrm{Mor}(X, Y) \to \mathrm{Mor}\big(\mathrm{Spec}(K[X]), \mathrm{Spec}(K[Y])\big), \quad f \mapsto F(f),$$

is bijective. In particular, X and Y are isomorphic if and only if the schemes $\mathrm{Spec}(K[X])$ and $\mathrm{Spec}(K[Y])$ are isomorphic.

b) For any affine variety X, the topological space $|X|$ is homeomorphic to the set of closed points of $\mathrm{Spec}(K[X])$ which is dense in $\mathrm{Spec}(K[X])$.

c) The image of F is exactly the set of reduced affine schemes of finite type over K, that is, schemes of the form $\mathrm{Spec}(A)$ with A a reduced affine ring over K.

(2) The functor F extends to a fully faithful functor from quasi–projective varieties to quasi–projective schemes over K. For any quasi–projective variety X, the topological space $|X|$ is homeomorphic to the (dense) set of closed points of $F(X)$. Irreducible varieties correspond to irreducible schemes. The image under F of the set of projective varieties is the set of reduced projective schemes over K.

The proof of (1) is an easy consequence of the results of Section A.1 and A.2. For a detailed proof we refer to [120, Propositions II.2.6 and II.4.10]. Note that varieties in [120] are irreducible, but the generalization to reducible varieties is straightforward.

Although classical varieties are not schemes in a strict sense, however, we can use the above theorem to transfer results about schemes to results about varieties. For example, the statement of Lemma A.3.10, saying that the closure of the image $\overline{f(X)}$ equals $V(J \cap K[y_1, \ldots, y_m])$, holds for the whole scheme, in particular, it holds when restricting ourselves to the subset of closed points. We conclude that the statement of Lemma A.3.10 holds for classical varieties, too, provided we work over an algebraically closed field K. In other words, Lemma A.2.18 is an immediate consequence of Lemma A.3.10.

A.6 Morphisms Between Varieties

The definition of morphisms between projective varieties is more complicated than for affine varieties. To see this, let us make a naive try by simply using homogeneous polynomials. Let $X \subset \mathbb{P}^n$ be a projective variety, and let $f_0, \ldots, f_m$ be homogeneous polynomials of the same degree d in $K[x_0, \ldots, x_n]$.

If $p = (x_0 : \ldots : x_n) \in X$ is a point with $f_i(p) \neq 0$ for at least one i then $\big(f_0(p) : \ldots : f_m(p)\big)$ is a well-defined point in $\mathbb{P}^m$. Thus, $f = (f_0 : \ldots : f_m)$ defines a map $X \to \mathbb{P}^m$, provided that $X \cap V(f_0, \ldots, f_m) = \emptyset$. However, by Bézout's theorem, the assumption $X \cap V(f_0, \ldots, f_m) = \emptyset$ is very restrictive as this is only possible if $\dim(X) + \dim V(f_0, \ldots, f_m) < n$. In particular, since $\dim V(f_0, \ldots, f_m) \geq n - m - 1$, the naive approach excludes morphisms $X \to \mathbb{P}^m$ with $\dim X > m$.

The naive approach is, in a sense, too global. The good definition of morphisms between projective and, more generally, quasi–projective varieties uses the concept of a regular function, where regularity is a local condition (not to be confused with regularity of a local ring). Since any affine variety is open in its projective closure, it is quasi–projective. Thus, we shall obtain a new definition of morphisms between affine varieties which turns out to be equivalent to the previous one, given in Section A.2.

Definition A.6.1. Let $X \subset \mathbb{P}_K^n$ and $Y \subset \mathbb{P}_K^m$ be quasi–projective varieties.

(1) A function $f : X \to K$ is called *regular at a point* $p \in X$ if there exists an open neighbourhood $U \subset X$ of p and homogeneous polynomials $g, h \in K[x_0, \ldots, x_n]$ of the same degree such that, for each $q \in U$, we have

$$f(q) = \frac{g(q)}{h(q)}, \quad h(q) \neq 0.$$

f is called *regular* on X if it is regular at each point of X. $\mathcal{O}(X)$ denotes the K–algebra of regular functions on X.

(2) A *morphism* $f : X \to Y$ is a continuous map such that for each open set $V \subset Y$ and for each regular function $g : V \to K$ the composition

$$g \circ f : f^{-1}(V) \to K$$

is a regular function on $f^{-1}(V)$.

Note that in (1), if g and h both have degree d, then

$$\frac{g(\lambda q)}{h(\lambda q)} = \frac{\lambda^d g(q)}{\lambda^d h(q)} = \frac{g(q)}{h(q)},$$

for all $\lambda \in K^*$, that is, the quotient g/h is well–defined on $\mathbb{P}^n$. If $X \subset \mathbb{A}^n$ is quasi–affine, we can, equivalently, define $f : X \to K$ to be regular at p, if there exists an open neighbourhood U of p in $\mathbb{A}^n$ and polynomials $g, h \in K[x_1, \ldots, x_n]$, not necessarily homogeneous, such that $h(q) \neq 0$ and $f(q) = g(q)/h(q)$ for all $q \in U$.

If we identify K with $\mathbb{A}_K^1$, that is, if we endow K with the Zariski topology, then a regular function is a continuous map $X \to \mathbb{A}^1$. Therefore, the regular functions $X \to K$ are just the morphisms $X \to \mathbb{A}^1$.

It is also quite easy to see that a map $f : X \to Y$ is a morphism if and only if there exist open coverings $\{U_i\}$ of X and $\{V_i\}$ of Y with $f(U_i) \subset V_i$ such that $f|_{U_i} : U_i \to V_i$ is a morphism.

It is now easy to check that the quasi–projective varieties and, hence, also the projective varieties, together with the morphisms defined above, are a category. This means, essentially, that the composition of two morphisms is again a morphism of quasi–projective varieties.

A morphism $f : X \to Y$ between quasi–projective varieties is an *isomorphism* if it has an inverse, that is, there exists a morphism $g : Y \to X$ such that $f \circ g = \mathrm{id}_Y$ and $g \circ f = \mathrm{id}_X$. $f : X \to Y$ is called a *closed immersion* or an *embedding* if $f(X)$ is a closed subvariety of Y and if $f : X \to f(X)$ is an isomorphism.

Note that an isomorphism is a homeomorphism but the converse is not true, as the map $\mathbb{A}^1 \to V(x^2 - y^3) \subset \mathbb{A}^2$, $t \mapsto (t^3, t^2)$ shows.

So far, affine varieties were always given as subvarieties of some $\mathbb{A}^n$. Having the notion of morphisms of quasi–projective varieties, we extend this by saying that a quasi–projective variety is *affine* if it is isomorphic to an affine subvariety of some $\mathbb{A}^n$.

An important example is the complement of a hypersurface in an affine variety: if $X \subset \mathbb{A}^n$ is affine, $f \in K[x_1, \ldots, x_n]$, then $X \smallsetminus V(f)$ is affine. Indeed, the projection $\mathbb{A}^{n+1} \to \mathbb{A}^n$, $(x_1, \ldots, x_n, t) \mapsto (x_1, \ldots, x_n)$, induces an isomorphism from the affine subvariety $V(I, tf(x) - 1) \subset \mathbb{A}^{n+1}$ onto $X \smallsetminus V(f)$.

We should like to emphasize that isomorphic projective varieties need not have isomorphic homogeneous coordinate rings as this was the case for affine varieties. The simplest example is the *Veronese embedding* of $\mathbb{P}^1$ in $\mathbb{P}^2$,

$$\nu_2 : \mathbb{P}^1 \longrightarrow \mathbb{P}^2, \ (x_0 : x_1) \longmapsto (y_0 : y_1 : y_2) = (x_0^2 : x_0 x_1 : x_1^2),$$

which maps $\mathbb{P}^1$ isomorphic to the non–singular quadric $\nu_2(\mathbb{P}^1)$ defined by the homogeneous polynomial $y_1^2 - y_0 y_1$. But the coordinate rings $K[x_0, x_1]$ and $K[y_0, y_1, y_2]/\langle y_1^2 - y_0 y_1 \rangle$ are not isomorphic (show this).

Example A.6.2. More generally, the *d–tuple Veronese embedding* is given by

$$\nu_d : \mathbb{P}^n \longrightarrow \mathbb{P}^N, \ p \longmapsto \big(M_0(p) : \ldots : M_N(p)\big),$$

where $M_0, \ldots, M_N$, $N = \binom{n+d}{d}$, are all monomials $x_0^{\alpha_0} \cdots x_n^{\alpha_n}$ in $x_0, \ldots, x_n$ of degree $|\alpha| = d$, in, say, lexicographical order. Note that for $V = V(f) \subset \mathbb{P}^n$ a hypersurface of degree d, the image $\nu_d(V) \subset \mathbb{P}^N$ is the intersection of $\nu_d(\mathbb{P}^n)$ with a linear hyperplane in $\mathbb{P}^N$.

Two projective varieties $X, Y \subset \mathbb{P}^n$ are called *projectively equivalent* if there exists a linear automorphism of $K[x_0, \ldots, x_n]$, inducing an isomorphism of the homogeneous coordinate rings $K[X]$ and $K[Y]$. Projective equivalence is an important equivalence relation which implies isomorphy, but it is stronger.

As we have seen, morphisms between projective varieties are not as easy to describe as between affine varieties. For morphisms to $\mathbb{A}^n$ this is, however, still fairly simple, as the following lemma shows.

Lemma A.6.3. *Let X be a quasi–projective variety, and let $Y \subset \mathbb{A}^n$ be quasi–affine. The component functions*

$$x_i : \mathbb{A}^n \longrightarrow K, \quad (x_1, \ldots, x_n) \longmapsto x_i,$$

define regular functions $x_i|_Y : Y \to K$ and a map $f : X \to Y$ is a morphism if and only if the component function $f_i = x_i \circ f$ is regular for $i = 1, \ldots, n$.

We leave the proof of the lemma to the reader.

It is now also easy to see that the canonical morphism

$$\varphi_i : U_i = \mathbb{P}^n \smallsetminus V(x_i) \longrightarrow \mathbb{A}^n,$$

given by $(x_0 : \ldots : x_n) \mapsto (x_0/x_i, \ldots, \widehat{1}, \ldots, x_n/x_i)$, is indeed an isomorphism of affine varieties.

Lemma A.6.3 already shows that morphisms between affine varieties in the sense of Definition A.2.9 (that is, represented by a polynomial map) are morphisms in the sense of Definition A.6.1. To see that both definitions do coincide, we still need a better understanding of regular functions on an affine variety.

Let $X \subset \mathbb{A}^n$ be an affine, respectively $X \subset \mathbb{P}^n$ a projective, variety, then the principal open sets

$$D(f) = X \smallsetminus V(f) = \{x \in X \mid f(x) \neq 0\},$$

$f \in K[X]$, form a basis of the Zariski topology on X (since $K[X]$ is Noetherian, see also Section A.5). We have $D(fg) = D(f) \cap D(g)$, and $D(f)$ is empty if and only if $f = 0$, due to the affine, respectively projective, Nullstellensatz.

If $U \subset X$ is open and $f : U \to K$ is a regular function U, then it follows that U can be covered by finitely many principal open sets, $U = \bigcup_{i=1}^k D(h_i)$, such that $f = g_i/h_i$ on $D(h_i)$ for some g_i, $h_i \in K[X]$, where g_i and h_i are homogeneous of the same degree if X is projective.

In general, it is not true that f has, on all of U, a representation as g/h with $g(p) \neq 0$ for all $p \in U$. However, for U a principal open set, this is true.

Proposition A.6.4. *Let* X *be an affine (respectively projective) variety, and let* $h \in K[X]$ *be non–zero (respectively homogeneous of positive degree). Then every regular function* $f : D(h) \to K$ *is of the form* $f = g/h^d$ *on* $D(h)$, *for some* $d > 0$ *and* $g \in K[X]$ *(respectively with* g *homogeneous of degree* $d \deg(h)$, *if* X *is projective).*

Proof. Let $D(h) = \bigcup_{i=1}^{s} D(h_i)$ such that $f|_{D(h_i)} = g_i/h_i$. Then $g_i h_j = g_j h_i$ on $D(h_i) \cap D(h_j) = D(h_i h_j)$, hence, we have $h_i h_j (g_i h_j - g_j h_i) = 0$ on X and $g_i h_j = g_j h_i$ on $D(h_i h_j)$. Replacing, for all i, h_i by h_i^2 and g_i by $g_i h_i$, we may assume that $g_i h_j = g_j h_i$ on X, hence this equality holds in $K[X]$.

Since $D(h) = \bigcup_i D(h_i)$, we obtain $V(h) = X \smallsetminus D(h) = V(h_1, \ldots, h_s)$ and the affine, respectively projective, Nullstellensatz implies $h \in \sqrt{\langle h_1, \ldots, h_s \rangle}$, that is, $h^d = \sum_{i=1}^{s} k_i h_i$ for some $d > 0$ and $k_i \in K[X]$; if X is projective, the k_i are homogeneous of appropriate degree.

Set $g := \sum_{i=1}^{s} g_i k_i$. Then $h^d \cdot g_j = \sum_{i=1}^{s} k_i h_i g_j = \sum_{i=1}^{s} (k_i g_i) h_j = g \cdot h_j$, hence, $f = g_j/h_j = g/h^d$ on $D(h_j)$ for each j, that is, $f = g/h^d$ on $D(h)$. $\square$

For a further understanding of morphisms between projective varieties, we need to consider rational functions.

Definition A.6.5. Let X be a quasi–projective variety. A *rational function* on X is given by a pair (U, f) where $U \subset X$ is open and dense, and $f : U \to K$ is a regular function on U. Two pairs, (U, f) and (V, g), are equivalent if $f|_W = g|_W$ for an open dense subset $W \subset U \cap V$. An equivalence class is called a *rational function* on X. The set of rational functions on X is denoted by $R(X)$.

Given a representative (U, f) of a rational function on X, there exists a maximal open set $\widetilde{U}$ such that $U \subset \widetilde{U} \subset X$ and a regular function $\widetilde{f} : \widetilde{U} \to K$ such that $\widetilde{f}|_U = f$. Hence, a rational function is uniquely determined by an open dense set $U \subset X$ and a regular function $f : U \to K$ such that f has no extension to a regular function on a larger open set in X. U is then called the *definition set* and $X \smallsetminus U$ the *pole set* of f. Hence, a rational function is a function on its definition set, while it is not defined on its pole set.

For example, x_i/x_0 is a rational function on $\mathbb{P}^n$, $i = 1, \ldots, n$, with pole set $V(x_0)$ and definition set $D(x_0)$.

We define the addition and multiplication of rational functions on X by taking representatives. Thus, $R(X)$ is a K–algebra.

Theorem A.6.6.

(1) Let X *be a quasi–projective variety and* $X = X_1 \cup \cdots \cup X_r$ *the decomposition of* X *into irreducible components. Then the map*

$$R(X) \longrightarrow R(X_1) \oplus \cdots \oplus R(X_r), \quad f \longmapsto (f|_{X_1}, \ldots, f|_{X_r})$$

is an isomorphism of K–*algebras.*

(2) *Let X be an affine or a projective variety, and let $f \in R(X)$. Then there exists a non–zerodivisor $g \in K[X]$, homogeneous of positive degree if X is projective, such that $D(g)$ is contained and dense in the definition set of f and*

$$f = \frac{h}{g^d} \text{ on } D(g)$$

for some $d > 0$, $h \in K[X]$, and h homogeneous of degree $d \cdot \deg(g)$ if X is projective.

(3) *If X is an irreducible affine or projective variety, then $R(X)$ is a field.*

    a) *If X is affine, then $R(X) \cong K(X)$ where $K(X) = \operatorname{Quot}(K[X])$ is the quotient field of $K[X]$.*

    b) *If X is projective, then $R(X) \cong K(X)_0$, where $K(X)_0$ is the subfield of $K(X)$ of homogeneous elements of degree 0, that is,*

$$K(X)_0 = \left\{ \frac{f}{g} \;\middle|\; f, g \in K[X] \text{ homogeneous of same degree, } g \neq 0 \right\}.$$

Proof. (1) is straightforward.

(2) We use the fact that for $U \subset X$, open and dense, there exists a principal open set $D(g) \subset U$, which is dense in X. Indeed, if $U = X \smallsetminus A$, and if $X_1, \ldots, X_r$ are the irreducible components of X, then $I(A) \not\subset I(X_i)$ for all i, and, hence, by prime avoidance (Lemma 1.3.12), $I(A) \not\subset I(X_1) \cup \cdots \cup I(X_r)$, that is, there is some $g \in I(A)$ such that $g \notin I(X_i)$ for all i. Such a g is the desired non–zerodivisor (homogeneous of positive degree if X is projective) and $f = h/g^d$ by Proposition A.6.4.

(3) Let $f/g \in K(X)$, respectively $K(X)_0$, then $D(g)$ is dense in X, since X is irreducible and, hence, we have a map to $R(X)$ which is clearly injective. Surjectivity follows from (2). $\qquad\Box$

Note that the dimension of an irreducible affine algebraic set X is equal to the transcendence degree of the field extension $K \hookrightarrow K(X)$ (Theorem 3.5.1).

If X is a reducible quasi–projective variety with irreducible components $X_1, \ldots, X_r$, then the above theorem implies that $R(X)$ is a direct sum of fields. More precisely,

$$R(X) \cong \operatorname{Quot}(K[X_1]) \oplus \cdots \oplus \operatorname{Quot}(K[X_r])$$

for X an affine variety, respectively

$$R(X) \cong \operatorname{Quot}(K[X_1])_0 \oplus \cdots \oplus \operatorname{Quot}(K[X_r])_0$$

for X a projective variety, and the right–hand side is isomorphic to the total quotient ring $\operatorname{Quot}(K[X])$, respectively $\operatorname{Quot}(K[X])_0$ (see Example 1.4.6 and Exercise 1.4.8).

With these preparations we can identify the ring $\mathcal{O}(X)$ of regular functions on affine and projective varieties.

Theorem A.6.7.

(1) Let $X \subset \mathbb{A}^n$ be an affine variety, then $\mathcal{O}(X)$ is isomorphic to the affine coordinate ring $K[X] = K[x_1, \ldots, x_n]/I(X)$.

(2) Let $X \subset \mathbb{P}^n$ be a connected (for example, irreducible) projective variety, then $\mathcal{O}(X) = K$.

The theorem shows a principal difference between affine and projective varieties: affine varieties have plenty of regular functions, while on a projective variety we have only regular functions which are constant on each connected component.

Proof. To see (1), note that each polynomial defines a regular function on X. Hence, we have a canonical map $K[x_1, \ldots, x_n] \to \mathcal{O}(X)$ with kernel $I(X)$.

To see that the injective map $K[X] \to \mathcal{O}(X)$ is surjective, let $f \in \mathcal{O}(X)$ and apply Proposition A.6.4 to $h = 1$.

(2) Let $X \subset \mathbb{P}^n$ be projective and $f \in \mathcal{O}(X)$. By Proposition A.6.4, we can write f on the dense principal open set $U_i := D(x_i)$ as

$$f|_{U_i} = \frac{g_i}{x_i^{d_i}}, \quad g_i \in K[X]_{d_i}.$$

By Theorem A.6.6, we may consider $R(X)$ and, in particular, $\mathcal{O}(X)$, as subring of the total ring of fractions $\mathrm{Quot}(K[X])$. Then the above equality means that $f = g_i/x_i^{d_i} \in \mathrm{Quot}(K[X])$, that is, $x_i^{d_i} f \in K[X]_{d_i}$, for $i = 0, \ldots, n$.

Each monomial $x_0^{\alpha_1} \cdot \ldots \cdot x_n^{\alpha_n}$ of degree $N = |\alpha| \geq d_0 + \cdots + d_k$ satisfies $\alpha_i \geq d_i$ for at least one i. Therefore, $K[X]_N \cdot f \subset K[X]_N$. Multiplying, successively, both sides with f, we see that

$$K[X]_N \cdot f^q \subset K[X]_N \text{ for all } q > 0.$$

Let $h \in K[X]$ be a non–zerodivisor of degree 1 (for example, a generic linear combination $h := \lambda_0 x_0 + \cdots + \lambda_n x_n$), then $h^N f^q \in K[X]_N$ for all $q \geq 0$ and, hence,

$$(K[X])[f] \subset \frac{1}{h^N} K[X].$$

Since $1/h^N K[X]$ is a finitely generated $K[X]$–module, and since $K[X]$ is Noetherian, the subring $(K[X])[f]$ is finite over $K[X]$ and, therfore, f satisfies an integral relation (by Proposition 3.1.2)

$$f^m + a_1 f^{m-1} + \cdots + a_m = 0, \quad a_i \in K[X].$$

Since $f \in \mathrm{Quot}(K[X])_0$, we can replace in this equation the a_i by their homogeneous components of degree 0 and see that f is integral, hence algebraic, over $K[X]_0 = K$.

If X is irreducible, then $\mathrm{Quot}(K[X])$ is a field and, since K is algebraically closed, it follows that f belongs to K. Otherwise, the same reasoning gives that f is constant on each irreducible component, and the result follows from continuity of f. $\qquad\square$

Morphisms in concrete terms

We can now describe morphisms $f : X \to Y$ between varieties in concrete terms. If $Y \subset \mathbb{A}^m$, respectively $Y \subset \mathbb{P}^m$, then f may be considered as a morphism to $\mathbb{A}^m$, respectively $\mathbb{P}^m$, such that $f(X) \subset Y$. We, therefore, have to distinguish two main cases, $Y = \mathbb{A}^m$ and $Y = \mathbb{P}^m$.

(1) *Morphisms to $\mathbb{A}^m$*: By Lemma A.6.3, a morphism $f : X \to \mathbb{A}^m$ is given by regular component functions $f_i : X \to K$, $i = 1, \ldots, m$.

If $X \subset \mathbb{A}^n$ is affine, then, by Theorem A.6.7 (1), $f_i = \widetilde{f}_i|_X$, where $\widetilde{f}_i$ is a polynomial in $K[x_1, \ldots, x_n]$ representing $f_i \in K[X]$.

If $X \subset \mathbb{P}^n$ is projective, then f is constant on each connected component of X by Theorem A.6.7 (2), and $f(X)$ consists of finitely many points.

(2) *Morphisms to $\mathbb{P}^m$*: Consider a map $f : X \to \mathbb{P}^m$, and let $V_i := D(y_i)$ be the canonical charts of $\mathbb{P}^m$. Then f is a morphism if and only if the restriction $f : f^{-1}(V_i) \to V_i \cong \mathbb{A}^m$ is a morphism for all $i = 0, \ldots, m$. Hence, we can give a morphism f by giving morphisms $g_i : U_i \to V_i \cong \mathbb{A}^m$, where $\{U_i\}$ is an open covering of X, such that g_i and g_j coincide on $U_i \cap U_j$, taking into account the coordinate transformation $V_i \cong V_j$. Since this is not very practical, we prefer to describe morphisms to $\mathbb{P}^m$ by specifying homogeneous polynomials.[3]

To see that this is possible, assume that, for each irreducible component X_i of X, $f(X_i) \not\subset H_\infty = V(y_0) \subset \mathbb{P}^m$ (this can always be achieved by a linear change of the homogeneous coordinates $y_0, \ldots, y_m$ of $\mathbb{P}^m$). Then $f^{-1}(V_0) \cap X_i \neq \emptyset$ and, hence, $X' := f^{-1}(V_0)$ is dense in X and the restriction of f to X' is a morphism $f' = (f_1', \ldots, f_m') : X' \to \mathbb{A}^m$ with f_i' regular functions on X'. Each f_i' defines a rational function f_i on X and, hence, on the projective closure $\overline{X} \subset \mathbb{P}^n$. By Theorem A.6.6, f_i is of the form $h_i/g_i^{d_i}$ with $g_i, h_i \in K[x_0, \ldots, x_n]$ homogeneous, $\deg(h_i) = d_i \deg(g_i)$ and $D(g_i) \cap X$ dense in X. Set $g := g_1^{d_1} \cdot \ldots \cdot g_m^{d_m}$ and $\widehat{g}_i := g/g_i^{d_i}$. Consider the homogeneous polynomials $\widetilde{f}_0 := g$, $\widetilde{f}_1 := h_1 \widehat{g}_1, \ldots, \widetilde{f}_m = h_m \widehat{g}_m$, are all homogeneous of the same degree. Hence the polynomial map $\widetilde{f}$,

$$x = (x_0 : \ldots : x_n) \longmapsto \left(\widetilde{f}_0(x) : \ldots : \widetilde{f}_m(x) \right),$$

is a morphism to $\mathbb{P}^m$, defined on $\mathbb{P}^n \setminus V(\widetilde{f}_0, \ldots, \widetilde{f}_m)$, which coincides on $D(\widetilde{f}_0)$ with f. Since $D(\widetilde{f}_0) = D(g_1) \cap \cdots \cap D(g_m)$ is dense in X, f coincides with $\widetilde{f}$ on $X \setminus V(\widetilde{f}_0, \ldots, \widetilde{f}_m)$, the definition set of $\widetilde{f}|_X$, which is, again, dense in X. Since f is continuous, f is uniquely determined by $\widetilde{f}$. Hence, any morphism $X \to \mathbb{P}^m$ can be given by homogeneous polynomials $\widetilde{f}_0, \ldots, \widetilde{f}_m$ of the same degree, satisfying that $(\widetilde{f}_0 : \ldots : \widetilde{f}_m)$ extends from $X \setminus V(\widetilde{f}_0, \ldots, \widetilde{f}_m)$ to a morphism defined on X. It may, indeed, happen that $X \cap V(\widetilde{f}_0, \ldots, \widetilde{f}_m) \neq \emptyset$ (cf. Example A.6.9).

---

[3] More conceptual but less concrete is the description of f in terms of sections of some invertible sheaf [120].

Conversely, if $\widetilde{f}_0, \ldots, \widetilde{f}_m$ are homogeneous of same degree, then, certainly, $\widetilde{f} = (\widetilde{f}_0 : \ldots : \widetilde{f}_m)$ defines a morphism $X \to \mathbb{P}^m$ if $X \cap V(\widetilde{f}_0, \ldots, \widetilde{f}_m) = \emptyset$.

SINGULAR Example A.6.8 (morphisms of projective varieties).
We consider $X = V(z^3 - xy^2 + y^3) \subset \mathbb{P}^2$, and want to define a morphism $\phi : X \to \mathbb{P}^2$, given by the homogeneous polynomials $\widetilde{f}_0 = xz$, $\widetilde{f}_1 = xy$ and $\widetilde{f}_2 = x^2 + yz$. We have to check that $X \cap V(\widetilde{f}_0, \widetilde{f}_1, \widetilde{f}_2) = \emptyset$.

```
ring R=0,(x,y,z),dp;
ideal I=z3-xy2+y3;
ideal J=xz,xy,x2+yz;
dim(std(I+J));
//-> 0
```

From the result we read that, indeed, $X \cap V(\widetilde{f}_0, \widetilde{f}_1, \widetilde{f}_2) = \emptyset$.

```
map phi=R,J;
phi(I);
//-> _[1]=x6+x3y3+3x4yz-x3y2z+3x2y2z2+y3z3
```

The latter means that the image of X is a curve of degree 6, given by the polynomial $x^6 + x^3y^3 + 3x^4yz - x^3y^2z + 3x^2y^2z^2 + y^3z^3$.

However, if $X \cap V(\widetilde{f}_0, \ldots, \widetilde{f}_m) \neq \emptyset$, then $\widetilde{f}$ defines, in general, only a rational map (cf. Definition A.7.11). It defines a morphism $X \to \mathbb{P}^m$ if and only if X can be covered by open sets U_j such that $U_j \supset \widetilde{f}^{-1}(V_j)$, and such that for each i, the function $\widetilde{f}_i/\widetilde{f}_j$, which is regular on $\widetilde{f}^{-1}(V_j)$, extends to a regular function on U_j (since $\widetilde{f}$ is not defined everywhere, the preimages $\widetilde{f}^{-1}(V_j)$ need not cover X).

To see how this can happen, consider $(x_0 : \ldots : x_n) \mapsto (x_1 : \ldots : x_n)$, which defines a morphism $\pi : \mathbb{P}^n \smallsetminus \{p_0\} \to \mathbb{P}^{n-1}$, $p_0 := (1 : 0 : \ldots : 0)$. Geometrically, this is the *projection from a point*, here p_0, to $\mathbb{P}^{n-1}$. A point $p \in \mathbb{P}^n \smallsetminus \{p_0\}$ is mapped to the intersection point of the projective line $\overline{p_0 p}$ through p_0 and p with the hyperplane at infinity, $H_\infty = \mathbb{P}^{n-1}$. Certainly, if $n \geq 2$ then π cannot be extended to a continuous map $\mathbb{P}^n \to \mathbb{P}^{n-1}$, since otherwise $\pi(p_0) = \pi(\overline{p_0 p}) = \pi(p)$ for each $p \in \mathbb{P}^n \smallsetminus \{p_0\}$.

If $X \subset \mathbb{P}^n$ is a subvariety and $p_0 \notin X$, then the restriction of π is a morphism from X to $\mathbb{P}^{n-1}$. But even for $p_0 \in X$, π may define a morphism on all of X. We show this by a concrete example.

Example A.6.9. Let $X \subset \mathbb{P}^2$ be the curve defined by $x^2 - y^2 - yz = 0$, and consider the map π defined by $X \ni (x : y : z) \mapsto (x : y) \in \mathbb{P}^1$.

π is the projection from $q = (0 : 0 : 1)$ to the projective line $\mathbb{P}^1 = \{z = 0\}$. Since $q \in X$, π is not defined at q, and we have to analyse the situation in the charts $V_0 = \{x \neq 0\}$ and $V_1 = \{y \neq 0\}$ of the image $\mathbb{P}^1$. We have

$$\pi^{-1}(V_0) = X \smallsetminus \{(0 : 0 : 1), (0 : 1 : -1)\},$$
$$\pi^{-1}(V_1) = X \smallsetminus \{(0 : 0 : 1)\}.$$

On $\pi^{-1}(V_0)$ the regular function y/x coincides with

$$\frac{y}{x} = \frac{xy}{x^2} = \frac{yx}{y(y+z)} = \frac{x}{y+z}$$

which is a regular function on $X \smallsetminus \{(0:1:-1)\}$, in particular in $(0:0:1)$. Setting $\pi(0:0:1) := (1:0)$, we see that π defines a morphism from X to $\mathbb{P}^1$. On $X \smallsetminus \{(0:0:1)\}$, the projection π is given by $(x:y:z) \mapsto (x:y)$ and on $X \smallsetminus \{(0:1:-1)\}$ by $(y+z:x)$. This fact is illustrated by the picture in Figure A.18.

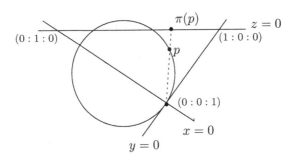

Fig. A.18. Projecting $X = V(x^2 - y^2 - yz)$ from the point $(0:0:1) \in X$ to the projective line $\{z = 0\}$.

A.7 Projective Morphisms and Elimination

We introduce projective morphisms and prove the "main theorem of elimination theory", which says that the image of a closed subvariety under a projective morphism is again closed. Then we discuss in some detail the geometric meaning of elimination in the context of projective morphisms and, more generally, in the context of rational maps.

In order to be able to compute images under projective morphisms, we need, as in the affine case, products of projective varieties and graphs. For simplicity, we work in the framework of varieties, although there is no essential difficulty in establishing the results for algebraic schemes over an algebraically closed field.

To show that the product of projective varieties is again projective is less straightforward than for affine varieties. The easiest way is to use the Segre embedding.

Definition A.7.1. The *Segre embedding* of $\mathbb{P}^n \times \mathbb{P}^m$ is the map

$$\sigma : \mathbb{P}^n \times \mathbb{P}^m \longrightarrow \mathbb{P}^N, \quad N = (n+1)(m+1) - 1,$$

$$\big((x_0 : \ldots : x_n), (y_0 : \ldots : y_m)\big) \longmapsto (x_0 y_0 : \ldots : x_i y_j : \ldots : x_n y_m),$$

with the image of a point being the pairwise product of the x_i and y_j sorted, say, lexicographically.

Let z_{ij}, $0 \leq i \leq n$, $0 \leq j \leq m$, denote the homogeneous coordinates of $\mathbb{P}^N$. One can use the local description of σ below to check that σ is injective and the image is $\Sigma_{n,m} := \sigma(\mathbb{P}^n \times \mathbb{P}^m)$, the zero–set of the quadratic equations

$$z_{ij}z_{kl} - z_{il}z_{kj} = 0\,,$$

hence, a projective subvariety of $\mathbb{P}^N$ with

$$I\left(\Sigma_{n,m}\right) = \langle z_{ij}z_{kl} - z_{il}z_{kj} \mid 0 \leq i < k \leq n,\ 0 \leq j < l \leq m \rangle\,.$$

Definition A.7.2. We identify $\mathbb{P}^n \times \mathbb{P}^m$ with the image $\Sigma_{n,m}$ and, thus, endow $\mathbb{P}^n \times \mathbb{P}^m$ with the structure of a projective variety, which is called the *product of $\mathbb{P}^n$ and $\mathbb{P}^m$*. A *subvariety of $\mathbb{P}^n \times \mathbb{P}^m$* is a subvariety of $\mathbb{P}^N$ contained in $\Sigma_{n,m}$.

If f is homogeneous of degree d in z_{ij}, then we can replace z_{ij} by $x_i y_j$ and obtain a bihomogeneous polynomial in x_i and y_j of bidegree (d,d). Here $f \in K[x_0, \ldots, x_n, y_0, \ldots, y_m]$ is called *bihomogeneous of bidegree (d,e)* if every monomial $x^\alpha y^\beta$ appearing in f satisfies $|\alpha| = d$ and $|\beta| = e$.

A bihomogeneous polynomial has a well–defined zero–set in $\mathbb{P}^n \times \mathbb{P}^m$, and it follows that the closed sets in the Zariski topology or, equivalently, the projective subvarieties of $\mathbb{P}^n \times \mathbb{P}^m$ are the zero–sets of bihomogeneous polynomials of arbitrary bidegrees. Indeed, by Remark A.4.5, the zero–set of a bihomogeneous polynomial of bidegree (d,e) with $e < d$, is also the zero–set of bihomogeneous polynomials of bidegree (d,d).

For example, denote by $U_i = D(x_i)$, $i = 0, \ldots, n$, respectively $V_j = D(y_j)$, $j = 0, \ldots, m$, the canonical affine charts in $\mathbb{P}^n$, respectively $\mathbb{P}^m$. Then the x_i, respectively the y_j, are bihomogeneous polynomials of degree $(1,0)$, respectively $(0,1)$, the products $x_i y_j$ are bihomogeneous of degree $(1,1)$, and the sets $U_i \times V_j = \mathbb{P}^n \times \mathbb{P}^m \smallsetminus V(x_i y_j) \cong \mathbb{A}^{n+m}$ are open in $\mathbb{P}^n \times \mathbb{P}^m$. It follows that $\{U_i \times V_j\}_{i,j}$ is an open, affine covering of $\mathbb{P}^n \times \mathbb{P}^m$.

We use the Segre embedding $\sigma : \mathbb{P}^n \times \mathbb{P}^m \to \mathbb{P}^N$ also to define the product of two arbitrary quasi–projective varieties $X \subset \mathbb{P}^n$, $Y \subset \mathbb{P}^m$. Namely, the image $\sigma(X \times Y)$ is a quasi–projective subvariety of $\mathbb{P}^N$ and, by identifying $X \times Y$ with $\sigma(X \times Y)$, we define on the set $X \times Y$ the structure of a quasi–projective variety. Moreover, if X, respectively Y, are closed in $\mathbb{P}^n$, respectively $\mathbb{P}^m$, hence projective, then the image $\sigma(X \times Y)$ is closed in $\mathbb{P}^N$. Therefore, $X \times Y$ is a projective variety.

In affine coordinates $(x_1, \ldots, x_n)$ on $U_0 \subset \mathbb{P}^n$ and $(y_1, \ldots, y_m)$ on $V_0 \subset \mathbb{P}^n$ the Segre embedding is (up to permutation of coordinates) just the map $\mathbb{A}^{n+m} \to \mathbb{A}^N$,

$$(x_1, \ldots, x_n, y_1, \ldots, y_m) \longmapsto (x_1, \ldots, x_n, y_1, \ldots, y_m, x_1 y_1, \ldots, x_n y_m)\,,$$

hence, the image is isomorphic to the graph of the map $\mathbb{A}^{n+m} \to \mathbb{A}^{nm}$,

$$(x_1, \ldots, x_n, y_1, \ldots, y_m) \longmapsto (x_1 y_1, \ldots, x_i y_j, \ldots, x_n y_m)$$

between affine spaces. Using the universal properties of the product and of the graph of affine varieties from Section A.2, respectively A.3, one can show that the product of two quasi–projective varieties X, Y has the following *universal property*: the projections $\pi_X : X \times Y \to X$ and $\pi_Y : X \times Y \to Y$ are morphisms of quasi–projective varieties, and for any quasi–projective variety Z and morphisms $f : Z \to X$ and $g : Z \to Y$ there exists a unique morphism $f \times g : Z \to X \times Y$ such that the following diagram commutes:

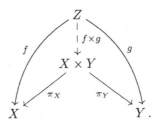

Example A.7.3. The Segre map $\sigma : \mathbb{P}^2 \times \mathbb{P}^1 \to \mathbb{P}^5$,

$$(x : y : z,\ s : t) \longmapsto (xs : xt : ys : yt : zs : zt) = (z_0 : \cdots : z_5)$$

has as image, the *Segre threefold* $\Sigma_{2,1}$, given in $\mathbb{P}^5$ by the equations

$$z_0 z_3 - z_1 z_2 = z_0 z_5 - z_1 z_4 = z_2 z_5 - z_3 z_4 = 0,$$

which may be defined as

$$\Sigma_{2,1} = \left\{ z \in \mathbb{P}^5 \ \middle|\ \mathrm{rank} \begin{pmatrix} z_0 & z_2 & z_4 \\ z_1 & z_3 & z_5 \end{pmatrix} < 2 \right\}.$$

If X is the quadric $x^2 - y^2 - yz = 0$ in $\mathbb{P}^2$, then the product $X \times \mathbb{P}^1$ can be defined in $\mathbb{P}^2 \times \mathbb{P}^1$ by the two bihomogeneous polynomials $s^2(x^2 - y^2 - yz)$ and $t^2(x^2 - y^2 - yz)$. Hence, $\sigma(X \times \mathbb{P}^1) \subset \Sigma_{2,1} \subset \mathbb{P}^5$ is given by the two additional quadratic equations $z_0^2 - z_2^2 - z_2 z_4 = z_1^2 - z_3^2 - z_3 z_5 = 0$.

As in the affine case, the notion of a product can be generalized to the fibre product of two morphisms.

Let $f : X \to S$ and $g : Y \to S$ be two morphisms of quasi–projective varieties, then there exists a quasi–projective variety $X \times_S Y$ and morphisms $\mathrm{pr}_1 : X \times_S Y \to X$, $\mathrm{pr}_2 : X \times_S Y \to Y$ such that the following diagram commutes:

$$
\begin{array}{ccc}
X \times_S Y & \xrightarrow{\ \mathrm{pr}_2\ } & Y \\
{\scriptstyle \mathrm{pr}_1}\downarrow & & \downarrow{\scriptstyle g} \\
X & \xrightarrow[\ f\ ]{} & S.
\end{array}
$$

$(X \times_S Y, \mathrm{pr}_1, \mathrm{pr}_2)$ is called the *fibre product* of X and Y over S. It has the same universal property as explained in Section A.3, but with Z a quasi–projective scheme.

$X \times_S Y$ can be implemented as a closed subvariety of $X \times Y$ with pr_1 and pr_2 induced by the projections. In particular, $X \times_S Y$ is projective if X and Y are projective.

The existence of the fibre product can be shown as follows: cover S by affine varieties and cover their preimages in X and Y by affine varieties, too, then glue the corresponding affine fibre products (cf. [120]). To discuss this in concrete terms, let $X \subset \mathbb{P}^n$ and $Y \subset \mathbb{P}^m$ be projective and $S = \mathbb{P}^k$. Assume that f and g are given by homogeneous polynomials $\widetilde{f}_0, \ldots, \widetilde{f}_k$ and $\widetilde{g}_0, \ldots, \widetilde{g}_k$, where the $\widetilde{f}_i$, respectively the $\widetilde{g}_j$, have the same degree, with $U = X \smallsetminus V(\widetilde{f}_0, \ldots, \widetilde{f}_k)$, respectively $V = Y \smallsetminus V(\widetilde{g}_0, \ldots, \widetilde{g}_k)$, dense in X, respectively Y. Assume also that no component of X, respectively Y, is mapped to $V(z_\ell) \subset \mathbb{P}^k$, $\ell = 0, \ldots, k$ (see the discussion in A.6). Then the set

$$W := \left\{ (p, q) \in U \times V \mid \left(\widetilde{f}_0(p) : \cdots : \widetilde{f}_k(p) \right) = \left(\widetilde{g}_0(q) : \cdots : \widetilde{g}_k(q) \right) \right\}$$

is well–defined, and over the open chart $\{z_\ell \neq 0\}$ of $\mathbb{P}^k$, this set is given by the equations $\widetilde{f}_i / \widetilde{f}_\ell = \widetilde{g}_i / \widetilde{g}_\ell$, $i = 0, \ldots, k$. Hence, W is closed in $U \times V$ and given by the bihomogeneous equations $\widetilde{f}_\ell \widetilde{g}_i - \widetilde{g}_\ell \widetilde{f}_i = 0$, $i, \ell = 0, \ldots, k$. The closure of W in $X \times Y$ is $X \times_S Y$, and we have

$$X \times_S Y \subset Z := \left(X \times Y \right) \cap V\left(\widetilde{f}_i \widetilde{g}_j - \widetilde{g}_i \widetilde{f}_j \mid 0 \leq i < j \leq k \right).$$

The inclusion may be strict, since $V(\widetilde{f}_0, \ldots, \widetilde{f}_k) \times Y$ and $X \times V(\widetilde{g}_0, \ldots, \widetilde{g}_k)$ are contained in Z. To compute the ideal of $X \times_S Y$, we have to saturate the ideal of Z by $\langle \widetilde{f}_0, \ldots, \widetilde{f}_k \rangle$ and by $\langle \widetilde{g}_0, \ldots, \widetilde{g}_k \rangle$.

The above construction makes sense for f and g rational maps. An arbitrary closed subvariety of $X \times Y$ is called a *correspondence* between X and Y (cf. [181]).

As in Section A.3, we obtain, as a special case of the fibre product, the existence of the graph of a morphism. If $f : X \to Y$ is a morphism, then the *graph of f*, $\Gamma_f := \{(x, y) \in X \times Y \mid f(x) = y\}$, is a closed subvariety of $X \times Y$. The projection $\mathrm{pr}_1 : \Gamma_f \to X$ is an isomorphism. If $X = \mathbb{P}^n$, $Y = \mathbb{P}^m$ and $f = (f_0 : \ldots : f_m)$ defines a rational map, then the saturation $\langle \{f_i y_j - f_j y_i\} \rangle : \langle \{f_i y_j\} \rangle^\infty$ describes Γ_f.

Example A.7.4. Let $X \subset \mathbb{P}^2$ be the quadric $x^2 - y^2 - yz = 0$, and $\pi : X \to \mathbb{P}^1$ the projection to the first two coordinates, as in Example A.6.9. The product $X \times \mathbb{P}^1$ is described in $\mathbb{P}^2 \times \mathbb{P}^1$ by $x^2 - y^2 - yz = 0$ (bidegree $(2, 0)$) and Γ_π by the additional equation $xt - ys = 0$, where $(x : y : z, s : t)$ denote the coordinates of $\mathbb{P}^2 \times \mathbb{P}^1$.

Definition A.7.5. Let $f : X \to Y$ be a morphism of quasi–projective varieties. Then f is called a *projective morphism* if there exists a commutative diagram

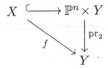

with $X \hookrightarrow \mathbb{P}^n \times Y$ as closed embedding and pr_2 the projection.

If $X \subset \mathbb{P}^n$ is a projective variety, then each morphism $f : X \to Y$, Y a quasi–projective variety, is projective: it factors through the closed embedding $X \cong \Gamma_f \hookrightarrow \mathbb{P}^n \times Y$ and the projection to the second factor, where Γ_f denotes the graph of f.

Theorem A.7.6 (Main Theorem of Elimination Theory).

Let Y be any quasi–projective variety and $\pi : \mathbb{P}^n \times Y \to Y$ the projection on the second factor. Then π is a closed map, that is, the image of any closed set in $\mathbb{P}^n \times Y$ is closed in Y.

Proof. Let $\{U_i\}_{i \in I}$ be an open affine covering of Y. Then $A \subset Y$ is closed if and only if, for all $i \in I$, the intersection $A \cap U_i$ is closed in U_i. Hence, we may assume that Y is affine, that is, Y is a closed subset of some $\mathbb{A}^m$. But then it is obviously sufficient to consider the case $Y = \mathbb{A}^m$.

Let $X \subset \mathbb{P}^n \times \mathbb{A}^m$ be closed. Then $X = V(f_1, \ldots, f_k)$, where the f_i are homogeneous polynomials of degree d_i in the variables $x = (x_0, \ldots, x_n)$, but not necessarily homogeneous in $y = (y_1, \ldots, y_m)$.

We show that $\pi(X)$ is the zero–set of certain determinants, hence closed: if $K[x]_d \subset K[x]$ denotes the vector space of homogeneous polynomials of degree d, then $K[x]_d \otimes_K K[y]$ is a free $K[y]$–module of finite rank. For $d \geq 0$, define a $K[y]$–linear map between free modules ($K[x]_i := 0$ for $i < 0$),

$$A_d : \left(\bigoplus_{j=1}^{k} K[x]_{d-d_j} \right) \otimes_K K[y] \longrightarrow K[x]_d \otimes_K K[y] \subset K[x,y] \,,$$

$$(g_1(x), \ldots, g_k(x)) \otimes 1 \longmapsto \sum_{j=1}^{k} g_j(x) f_j(x,y) \,,$$

and consider the set

$$S_d := \left\{ y \in Y \mid \mathrm{rank}\, A_d(y) \leq \dim_K K[x]_d - 1 \right\} ,$$

which is exactly the set of all $y \in Y$ such that $A_d(y)$ is not surjective.

Since $A_d(y)$ is, with respect to fixed bases, a matrix with polynomial entries, S_d is a zero–set of appropriate subdeterminants of A_d and, hence, a closed subset of Y.

Finally, we show that $\pi(X) = \bigcap_{d=0}^{\infty} S_d$. We actually prove the equivalence of the following four statements (for $y \in Y$ fixed):

(a) $y \in \pi(X)$, that is, $\pi^{-1}(y) \cap X \neq \emptyset$,

(b) $V\big(f_1(x,y),\dots,f_k(x,y)\big) \neq \emptyset$,

(c) $A_d(y)$ is not surjective for all d,

(d) $y \in \bigcap_{d=0}^{\infty} S_d$.

The equivalence of (a) and (b), respectively (c) and (d), being obvious, it remains to show the equivalence of (b) and (c). Clearly, for any fixed y and d, the map $A_d(y)$ is surjective if and only if $\langle x_0,\dots,x_n\rangle^d \subset \langle f_1(x,y),\dots,f_k(x,y)\rangle$. On the other hand, by the projective Nullstellensatz, the latter is satisfied for some d if and only if the variety $V\big(f_1(x,y),\dots,f_k(x,y)\big)$ is empty. $\qquad\square$

As a corollary, we obtain

Theorem A.7.7 (image of projective morphisms is closed).
Let $f : X \to Y$ be a projective morphism of quasi–projective varieties. Then the image $f(X)$ is a closed subvariety of Y.

In particular, if $X \subset \mathbb{P}^n$ is a projective variety and $f : X \to Y$ a morphism to a quasi–projective variety Y, then $f(X)$ is a closed subvariety of Y.

Note that Theorem A.7.7 provides a new proof of Theorem A.6.7 (2) about regular functions on a projective variety. Since any regular function f on X may be considered as a morphism to $\mathbb{A}^1$, and since the closed subsets of $\mathbb{A}^1$ are just finitely many points, it follows that f is constant on each connected component of X.

Geometric interpretation of elimination

The above theorem allows a geometric interpretation of what we really compute when we do elimination.

Case 1: To start with, let $X \subset \mathbb{P}^n \times \mathbb{A}^m$ be a closed subvariety. As shown in the proof of Theorem A.7.6, X is given by an ideal $I \subset K[x,y]$, generated by polynomials $f_1,\dots,f_k \in K[x,y]$, homogeneous in $x = (x_0,\dots,x_n)$, and arbitrary in $y = (y_1,\dots,y_m)$. Consider the projection $\pi : \mathbb{P}^n \times \mathbb{A}^m \to \mathbb{A}^m$. By Theorem A.7.7, the image $\pi(X)$ is closed in $\mathbb{A}^m$.

What do we compute when we eliminate $x_0,\dots,x_n$ from I ?

In general, perhaps nothing. To see this, let $V_a(I) \subset \mathbb{A}^{n+1} \times \mathbb{A}^m$ be the affine variety defined by I (the "affine cone with respect to x") and let $V_p(I)$ denote the "partially projective" variety defined by I in $\mathbb{P}^n \times \mathbb{A}^m$. π denotes in both cases the projection to $\mathbb{A}^m$. Eliminating x from I computes $I \cap K[y]$. We have $V(I \cap K[y]) = \overline{\pi(V_a(I))} \subset \mathbb{A}^m$, by Lemma A.2.13. This may be strictly larger than the image $\pi\big(V_p(I)\big) = \pi(X)$, we may even have $\pi\big(V_a(I)\big) = \mathbb{A}^m$.

Namely, $V_p(I) = V_p(\langle x\rangle^r I)$ for any $r \geq 0$ (see Remark A.4.5), but since $\{0\} \times \mathbb{A}^m \subset V_a(\langle x\rangle^r I)$ we have $\pi\big(V_a(\langle x\rangle^r I)\big) = \mathbb{A}^m$ for $r > 0$. Algebraically, this means that $I \cap K[y] = \langle 0\rangle$ if I is of the form $\langle x\rangle^r I'$ for some ideal I' with $r > 0$ (Lemma A.2.18). Hence, if $I = \langle x\rangle^r I'$, then eliminating x from I gives $\langle 0\rangle$.

Nevertheless, we can compute the image $\pi(X)$ by elimination.

Lemma A.7.8. *With the notations from above,*

$$\pi(X) = V\left(\bigcap_{i=0}^{n}(I|_{x_i=1}) \cap K[y]\right).$$

Proof. Let $U_i = \{x_i \neq 0\} \cong \mathbb{A}^n \times \mathbb{A}^m \subset \mathbb{P}^n \times \mathbb{A}^m$, then

$$\pi(X) = \bigcup_{i=0}^{n} \pi(X \cap U_i) = \bigcup_{i=0}^{n} \overline{\pi(X \cap U_i)} \subset \mathbb{A}^m,$$

since $\pi(X)$ is closed, by Theorem A.7.7. Hence, by Lemmas A.2.18 and A.2.3,

$$\pi(X) = \bigcup_{i=0}^{n} V\left(I|_{x_i=1} \cap K[y]\right) = V\left(\bigcap_{i=0}^{n}(I|_{x_i=1}) \cap K[y]\right). \qquad \square$$

The following lemma implies that eliminating x from I, indeed, defines the image $\pi(X)$ if I is saturated with respect to $x_0, \ldots, x_n$.

Lemma A.7.9. *Let $I \subset K[x, y]$, $x = (x_0, \ldots, x_n)$, $y = (y_1, \ldots, y_m)$, be an ideal, which is homogeneous with respect to x, then*

$$\left(\bigcap_{i=0}^{n} I|_{x_i=1}\right) \cap K[y] = (I : \langle x_0, \ldots, x_n\rangle^\infty) \cap K[y].$$

We notice that $I : \langle x_0, \ldots, x_n\rangle^\infty = I : \langle x_0^\infty, \ldots, x_n^\infty\rangle$, the latter being by definition $\bigcup_{d \geq 0} I : \langle x_0^d, \ldots, x_n^d\rangle$, and that we may use $I : \langle x_0^\infty, \ldots, x_n^\infty\rangle$ for computations.

Proof. If $f \in I : \langle x\rangle^\infty \cap K[y]$, then, for each i, $x_i^d f \in I$ for some d and, hence, $f = (x_i^d f|_{x_i=1}) \in I|_{x_i=1}$ for all i.

Conversely, if $f \in (\bigcap_{i=0}^{n} I|_{x_i=1})$, then, for each i, there exists some $F_i \in I$ such that $f = F_i|_{x_i=1}$. By Lemma 2.2.7, each x–homogeneous part of F_i is in I. Hence, multiplying the parts with appropriate powers of x_i, we can assume that F_i is, indeed, x–homogeneous. But then $F_i = x_i^{d_i} f^{h(i)}$, where the superscript $h(i)$ denotes the x–homogenization with x_i as homogenizing variable.

If, additionally, $f \in K[y]$, then it is already x–homogeneous, and we obtain $x_i^{d_i} f = x_i^{d_i} f^{h(i)} \in I$ for all i. This implies that, for sufficiently large d, we have $\langle x\rangle^d f \in I$, hence $f \in I : \langle x\rangle^\infty \cap K[y]$. $\qquad \square$

Algorithms to compute intersections of ideals and ideal quotients were described in Section 1.8. Since these computations may be expensive, it is worth looking for situations where they can be avoided. This happens, for example, if no irreducible component of X is contained in the hyperplane $\{x_i = 0\}$, then $V(I|_{x_i=1} \cap K[y]) = \pi(X)$. In concrete examples, it may even be worthwhile to change coordinates instead of computing the ideal quotient.

Remark A.7.10. The statement (and proof) of Lemma A.7.9 is even simpler if the ideal is of the form I^h of some ideal $I \subset K[x_1, \ldots, x_n, y_1, \ldots, y_m]$, where I^h denotes the homogenization with respect to the x–variables with x_0 the homogenizing variable. Then

$$I \cap K[y] = I^h \cap K[y].$$

Namely, since $I \cap K[y]$ is x–homogeneous, $I \cap K[y] \subset I^h$. The other inclusion $I^h \cap K[y] \subset I \cap K[y]$ follows, since $I = I^h|_{x_0=1}$. Moreover, since $x_0^k f \in I^h$ for some k if and only if $f \in I^h$ (by definition of I^h), we have

$$I^h = I^h : \langle x_0, \ldots, x_n \rangle^\infty.$$

Combining this with Lemmas A.7.9 and A.7.8, we obtain the following: let $X = V(I) \subset \mathbb{A}^n \times \mathbb{A}^m$, and let $\overline{X}$ be the closure of X in $\mathbb{P}^n \times \mathbb{A}^m$. Then

$$\pi(\overline{X}) = \pi(V(I^h)) = V(I \cap K[y]) = V(I^h \cap K[y]).$$

Recall that it is, in general, not sufficient to homogenize the generators of I with respect to x in order to obtain generators for I^h (see also Exercise 1.7.4).

Case 2: Let $X \subset \mathbb{P}^n \times \mathbb{P}^m$ be a projective subvariety. Then X is given by an ideal $I \subset K[x, y]$ generated by polynomials f_i, $i = 1, \ldots, k$, which are bihomogeneous in $x = (x_0, \ldots, x_n)$ and $y = (y_0, \ldots, y_m)$, and the image $\pi(X) \subset \mathbb{P}^m$ is closed by Theorem A.7.6.

Let $V_a(I) \subset \mathbb{P}^n \times \mathbb{A}^{m+1}$ be the affine cone with respect to y. It is easy to see that $\pi(V_a(I)) \subset \mathbb{A}^{m+1}$ is a cone and that the corresponding projective variety in $\mathbb{P}^m$ coincides with $\pi(X)$. By Lemmas A.7.8 and A.7.9, we obtain $\pi(V_a(I)) = V_a((I : \langle x \rangle^\infty) \cap K[y]) \subset \mathbb{A}^{m+1}$. Now $(I : \langle x \rangle^\infty) \cap K[y]$ is a homogeneous ideal in $K[y]$, and we obtain

$$\pi(X) = V((I : \langle x \rangle^\infty) \cap K[y]).$$

The same remark as above shows that $I \cap K[y]$ may be 0, hence, in general, it does not describe $\pi(X)$. Remark A.7.10 applies, too.

Computing the image of projective morphisms and rational maps.

Consider a morphism $f : X \to \mathbb{P}^m$ with $X \subset \mathbb{P}^n$ projective. Then the graph $\Gamma_f \subset X \times \mathbb{P}^n \subset \mathbb{P}^n \times \mathbb{P}^m$ is closed and $f(X) = \pi(\Gamma_f) \subset \mathbb{P}^m$. Thus to compute a homogeneous ideal defining the image $f(X) \subset \mathbb{P}^m$, we can apply the method described above to an ideal defining Γ_f in $\mathbb{P}^n \times \mathbb{P}^m$. However, there is a simpler method which we are going to describe now (cf. Corollary A.7.13).

Any morphism $f : X \to \mathbb{P}^m$ can be described by homogeneous polynomials $f_0, \ldots, f_m \in K[x]$, all of the same degree, subject to some conditions which guarantee that, at the points $x \in X \cap V(f_0, \ldots, f_m)$, $(f_0 : \ldots : f_m)$ can be extended to a morphism in some open neighbourhood of x (see the discussion at the end of Section A.6).

Slightly more general, let us consider rational maps.

Definition A.7.11. Let X, Y be two quasi–projective varieties. A *rational map* $f : X \dashrightarrow Y$ is an equivalence class of pairs (U, g), where $U \subset X$ is open and dense and where $g : U \to Y$ is a morphism. Two pairs (U, g) and (V, h) are equivalent if there exists an open dense subset $W \subset U \cap V$ such that $g|_W = h|_W$.

A rational map $f : X \dashrightarrow Y$ is called *birational* if there exist open dense sets $U \subset X$, $V \subset Y$ such that $f : U \to V$ is an isomorphism.

A rational map is denoted by a dotted arrow, since it is not defined everywhere. As for rational functions, there exists a maximal open dense set $U \subset X$ such that $f : U \to Y$ is a morphism. f is not defined on $X \setminus U$, which is called the *indeterminacy set* or *pole set* of the rational map; U is called the *definition set*. Note that, in general, f cannot be extended in any reasonable way to all of X, even if $Y = \mathbb{P}^m$. There is, in general, no way to map points from $X \setminus U$ to "points at infinity".

A rational map $X \dashrightarrow \mathbb{A}^m$ is, up to a choice of coordinates of $\mathbb{A}^m$, the same as an m–tuple of rational functions on X.

A rational map $X \dashrightarrow \mathbb{P}^m$ can be given either by $m + 1$ rational functions g_i/h_i where g_i and h_i are homogeneous of the same degree or, multiplying g_i/h_i with the least common multiple of $h_0, \ldots, h_m$, by an $(m + 1)$–tuple $(f_0 : \ldots : f_m)$, where the f_i are homogeneous polynomials of the same degree, such that $X \setminus V(f_0, \ldots, f_m)$ is dense in X.

Since the ring of rational functions on X is the direct sum of the fields of rational functions of the irreducible components of X (Theorem A.6.6), we can give a rational map, separately, on each irreducible component (with no condition on the intersection of components).

Case 3: Let $f_0, \ldots, f_m \in K[x_0, \ldots, x_n]$ define a rational map $f : \mathbb{P}^n \dashrightarrow \mathbb{P}^m$, the f_i being homogeneous of the same degree, let $X \subset \mathbb{P}^n$ be closed, and set $X_0 := X \setminus V(f_0, \ldots, f_m)$. Then $f(X_0) \subset \mathbb{P}^m$ is well–defined, though not necessarily closed. We describe how to compute the closure $\overline{f(X_0)}$.

Let $I \subset K[x_0, \ldots, x_n]$ be a homogeneous ideal with $V(I) = X$, and let $X_a = V_a(I)$ denote the affine cone of X in $\mathbb{A}^{n+1}$. Moreover, let f_a be the affine morphism $(f_0, \ldots, f_m) : \mathbb{A}^{n+1} \to \mathbb{A}^{m+1}$. Since the polynomials f_i are homogeneous of the same degree, it follows easily that $f_a(X_a) \subset \mathbb{A}^{m+1}$ is a cone and that $\pi\big(f_a(X_a) \setminus \{0\}\big) = f(X_0)$, where $\pi : \mathbb{A}^{m+1} \setminus \{0\} \to \mathbb{P}^m$ denotes the canonical projection. Hence, any homogeneous ideal in $K[y_0, \ldots, y_m]$ describing the closure of $f_a(X_a)$ in $\mathbb{A}^{m+1}$ describes $\overline{f(X_0)}$. By Lemma A.2.18,

$$\overline{f_a(X_a)} = V_a\big(\langle I,\, f_0 - y_0, \ldots, f_m - y_m \rangle \cap K[y]\big).$$

This does also hold if $X \subset V(f_0, \ldots, f_m)$, since then f is not defined anywhere on X, that is, $X_0 = \emptyset$, and $f(X) = \emptyset$. On the other hand, $X \subset V(f_0, \ldots, f_m)$ implies $\langle f_0, \ldots, f_m \rangle^r \subset I$ for some r by the Nullstellensatz. Using the isomorphism $K[x]/I \cong K[x, y]/\langle I, y_1 - f_1, \ldots, y_m - f_m \rangle$, we obtain an inclusion $\langle y_0, \ldots, y_m \rangle^r \subset \langle I, y_1 - f_1, \ldots, y_m - f_m \rangle \cap K[y]$, hence the latter ideal defines the empty set in $\mathbb{P}^m$. Thus, we have shown:

Proposition A.7.12. *Let $f = (f_0 : \ldots : f_m) : \mathbb{P}^n \dashrightarrow \mathbb{P}^m$ be a rational map, with $f_i \in K[x_0, \ldots, x_n]$ being homogeneous polynomials of the same degree. Moreover, let $I \subset K[x_0, \ldots, x_n]$ be a homogeneous ideal, defining the projective variety $X := V(I) \subset \mathbb{P}^n$, and let $X_0 := X \smallsetminus V(f_0, \ldots, f_m)$. Then*

$$\overline{f(X_0)} = V(\langle I, f_0 - y_0, \ldots, f_m - y_m \rangle \cap K[y_0, \ldots, y_m]) \subset \mathbb{P}^m.$$

Corollary A.7.13. *Let $X = V(I) \subset \mathbb{P}^n$ be a projective variety, and assume that $f : X \to \mathbb{P}^m$ is a morphism, then*

$$f(X) = V(\langle I, f_0 - y_0, \ldots, f_m - y_m \rangle \cap K[y_0, \ldots, y_m]).$$

Proof. f can be described by homogeneous polynomials $f_0, \ldots, f_m \in K[x]$ of the same degree, such that $X_0 := X \smallsetminus V(f_1, \ldots, f_m)$ is dense in X (see Page 494). Since f is continuous, it follows that

$$\overline{f(X_0)} \supset f(\overline{X_0}) = f(X),$$

which is closed in $\mathbb{P}^m$ and contains $f(X_0)$. Hence, $\overline{f(X_0)} = f(X)$, and the statement follows from Proposition A.7.12. $\qquad\square$

Let us return to morphisms of affine varieties $f : X \to \mathbb{A}^m$, where $X \subset \mathbb{A}^n$ is affine. As we saw in Section A.2, the closure of the image $f(X)$ is the zero–set of the ideal $J = \langle I, y_1 - f_1, \ldots, y_m - f_m \rangle \cap K[y]$, where $y = (y_1, \ldots, y_m)$ are coordinates of $\mathbb{A}^m$, $f_1, \ldots, f_m \in K[x]$, $x = (x_1, \ldots, x_n)$, are the component functions of f, and where $I \subset K[x]$ is an ideal describing X.

Remark A.7.14. In view of the previous discussion, we are able to explain where the points in $\overline{f(X)} \smallsetminus f(X)$ come from: since f factors as

$$f : X \xrightarrow{\cong} \Gamma_f \subset \mathbb{A}^n \times \mathbb{A}^m \xrightarrow{\pi} \mathbb{A}^m,$$

the image $f(X)$ coincides with the image of the graph Γ_f under the projection π. Now let $\overline{\Gamma_f}$ be the closure of Γ_f in $\mathbb{P}^n \times \mathbb{A}^m$. It is easy to see that $\overline{\Gamma_f} = V(J^h)$, where J^h is the homogenization of J with respect to $x_1, \ldots, x_n$ and with the new homogenizing variable x_0 (see Exercise 1.7.4 on how to compute J^h). By Remark A.7.10, we obtain

$$\overline{f(X)} = V(J \cap K[y]) = V(J^h \cap K[y]) = \pi(\overline{\Gamma_f}).$$

That is, the closure of $f(X)$ is the image of $\overline{\Gamma_f}$ under the projective map π. Hence, in general, the image $f(X)$ is not closed, because the presages of points in $\overline{f(X)} \smallsetminus f(X)$ escaped to infinity, that is, to $\overline{\Gamma_f} \smallsetminus \Gamma_f = V(J^h|_{x_0=0})$ in $\mathbb{P}^n \times \mathbb{A}^m$. Therefore,

$$\overline{f(X)} \smallsetminus f(X) = \pi(\overline{\Gamma_f}) \smallsetminus \pi(\Gamma_f) \subset \pi(V(J^h|_{x_0=0})).$$

Note that we had to take the closure of Γ_f in $\mathbb{P}^n \times \mathbb{A}^m$, but not the closure of X in $\mathbb{P}^n$. Namely, if $f(X)$ is not a finite set, then $f : X \to \mathbb{A}^m$ does not extend to a morphism $\overline{X} \to \mathbb{A}^m$, not even to $\overline{X} \to \mathbb{P}^m$ (consider, for instance, the projection $\mathbb{A}^2 \to \mathbb{A}^1$).

Example A.7.15. Look at the parametrization of the cuspidal cubic

$$\mathbb{A}^1 \longrightarrow \mathbb{A}^2, \quad t \longmapsto (t^2, t^3).$$

We have $\overline{f(X)} = V(x^3 - y^2) \subset \mathbb{A}^2$, and $\Gamma_f = V(x - t^2, y - t^3) \subset \mathbb{A}^1 \times \mathbb{A}^2$. In order to compute the closure $\overline{\Gamma_f} \subset \mathbb{P}^1 \times \mathbb{A}^2$, we compute a Gröbner basis of the ideal $\langle x - t^2, y - t^3 \rangle \subset K[t, x, y]$ with respect to a global monomial ordering satisfying $t^\alpha x^{\beta_0} y^{\beta_1} > t^{\alpha'} x^{\beta_0} y^{\beta_1}$ if $\alpha > \alpha'$. We obtain, by homogenizing this Gröbner basis with respect to t and with the homogenizing variable s,

$$\overline{\Gamma_f} = V(x^3 - y^2, \, ty - sx^2, \, tx - sy, \, t^2 - s^2 x) \subset \mathbb{P}^1 \times \mathbb{A}^2.$$

Let us do this using SINGULAR.

SINGULAR Example A.7.16 (projective elimination).

```
ring R = 0,(t,s,x,y),(dp(1),dp);
ideal I = x-t2,y-t3; //ideal of the graph of f

eliminate(I,t);
//-> _[1]=x3-y2      //ideal of the closure of f(X)

ideal J = std(I);    //Groebner basis w.r.t. a good ordering
J;
//-> J[1]=x3-y2
//-> J[2]=ty-x2
//-> J[3]=tx-y
//-> J[4]=t2-x
```

In order to homogenize only with respect to t, with s as homogenizing variable, we map to a ring R1, where x, y are considered as parameters, and homogenize in this new ring:

```
ring R1 = (0,x,y),(t,s),dp;
ideal Jh = homog(imap(R,J),s);

setring R;           //go back to R
ideal Jh= imap(R1,Jh);
Jh;                  //ideal of the closure of the graph of f
//-> Jh[1]=x3-y2
//-> Jh[2]=ty-sx2
//-> Jh[3]=tx-sy
//-> Jh[4]=t2-s2x

std(subst(Jh,t,1,s,0)); //points at infinity of the closure
//-> _[1]=1              //of the graph of f
```

We see that the closure of the graph of f in $\mathbb{P}^1 \times \mathbb{A}^2$ has no points at infinity, hence, in this case the image $f(X)$ is closed.

In the following simple example $f(X)$ is not closed: let f be the projection of $X = V(xt - 1, y) \subset \mathbb{A}^3$ to the (x, y)–plane:

```
ring S = 0,(t,s,x,y),(dp(1),dp);
ideal I = xt-1,y;              //ideal of affine hyperbola

eliminate(I,t);
//-> _[1]=y                    //ideal of the projection
```

By the above, the closure of the image $f(X)$ equals the image of the closure of $X \subset \mathbb{A}^3 = \mathbb{A}^1 \times \mathbb{A}^2$ in $\mathbb{P}^1 \times \mathbb{A}^2$ under the projection $(t, x, y) \mapsto (x, y)$. We compute

```
ideal J = std(I);
ring S1 = (0,x,y),(t,s),dp;   //homogenize J as ideal of
ideal Jh = homog(imap(S,J),s); //polynomials in t only

setring S;                     //go back to original ring
ideal Jh = imap(S1,Jh);
Jh;
//-> Jh[1]=y
//-> Jh[2]=tx-s

std(subst(Jh,t,1,s,0));        //intersection with infinity
//-> _[1]=y
//-> _[2]=x
```

Hence, the closure of X in $\mathbb{P}^1 \times \mathbb{A}^2$, with coordinates $(t:s; x, y)$, meets infinity at $(1:0; 0, 0)$ which is not a point of X.

Remark A.7.17. So far, we explained how to compute the image of a map by elimination. We now pose the opposite question. Let $I \subset K[x_0, \ldots, x_n]$ be a homogeneous ideal defining the projective variety $X = V(I) \subset \mathbb{P}^n$. What do we compute, when we eliminate, say, $x_0, \ldots, x_{r-1}$ from I? That is, what is the zero–set of $J = I \cap K[x_r, \ldots, x_n]$?

First, notice that J is homogeneous, that is, it defines a projective variety $Y = V(J) \subset \mathbb{P}^{n-r}$. We claim that Y is the closure of the image of X under the projection from the linear subspace $H = V(x_r, \ldots, x_n) \cong \mathbb{P}^{r-1}$ to $\mathbb{P}^{n-r}$. Here, we define the *projection from H to $\mathbb{P}^{n-r}$* as

$$\pi_H : \mathbb{P}^n \smallsetminus H \longrightarrow \mathbb{P}^{n-r}, \quad (x_0 : \ldots : x_n) \longmapsto (x_r : \ldots : x_n).$$

Geometrically, this is the map which sends $p \in \mathbb{P}^n \smallsetminus H$ to the intersection of $V(x_0, \ldots, x_r) \cong \mathbb{P}^{n-r}$ with the subspace $\overline{pH} \cong \mathbb{P}^r$, the union of all lines in $\mathbb{P}^n$ through p and any point of H. Similarly, we can define the projection from any other $(r - 1)$–dimensional linear subspace to a complementary $\mathbb{P}^{n-r}$.

Since π_H defines a rational map from $\mathbb{P}^n$ to $\mathbb{P}^{n-r}$, the claim is a direct consequence of Proposition A.7.12, and we obtain

$$V(I \cap K[x_r, \ldots, x_n]) = \overline{\pi_H(X_0)}.$$

If $X \cap H = \emptyset$, then the projection π_H is defined everywhere on X, hence, defines a morphism $\pi_H : X \to \mathbb{P}^{n-r}$, and then the image $\pi_H(X)$ is closed.

If $r = 1$, then H is the point $p_0 = (1 : 0 : \ldots : 0)$ and π_{p_0} is the projection from the point p_0 to $\mathbb{P}^{n-1}$. That is, a point $q \in \mathbb{P}^n \smallsetminus \{p_0\}$ is sent to the intersection of the projective line $\overline{p_0 q}$ with $V(x_0) = H_\infty = \mathbb{P}^{n-1}$. We saw in Example A.6.9 that even if $p_0 \in X$, π_{p_0} may define a morphism from X to $\mathbb{P}^{n-1}$, and then $\pi_{p_0}(X)$ is closed in $\mathbb{P}^{n-1}$.

Note that the result may be quite different if we project from different points, as is shown in the following example.

Example A.7.18. Consider the *twisted cubic* in threespace, which is the closure in $\mathbb{P}^3$ of the image of $\mathbb{A}^2 \to \mathbb{A}^3$, $t \mapsto (t, t^2, t^3)$. The image in $\mathbb{A}^3$ is given by eliminating t from $\langle x_1 - t, x_2 - t^2, x_3 - t^3 \rangle$:

SINGULAR Example A.7.19 (degree of projection).

```
ring R = 0,(x(0..3),t),dp;
ideal I = x(1)-t,x(2)-t2,x(3)-t3;
ideal J = eliminate(I,t);
J;
//-> J[1]=x(2)^2-x(1)*x(3)
//-> J[2]=x(1)*x(2)-x(3)
//-> J[3]=x(1)^2-x(2)
```

Hence, the image is given by $x_2^2 - x_1 x_3 = x_1 x_2 - x_3 = x_1^2 - x_2 = 0$.

Homogenizing gives the ideal $I = \langle x_2^2 - x_1 x_3, x_1 x_2 - x_3 x_0, x_1^2 - x_2 x_0 \rangle$ of X in $\mathbb{P}^3$. Projecting X from $(1:0:0:0)$ to $\mathbb{P}^2$, that is, eliminating x_0 from I, gives the quadric polynomial $x_2^2 - x_1 x_3$ for the image, which has one degree less than X (see Definition A.8.8). Projecting from $(0:1:0:0)$ gives, by eliminating x_1, the cubic polynomial $x_2^3 - x_0 x_3^2$, hence, the degree of the image equals the degree of X. This is due to the fact that $(1:0:0:0) \in X$, while $(0:1:0:0) \notin X$.

```
ideal Jh = homog(std(J),x(0));
eliminate(Jh,x(0));
//-> _[1]=x(2)^2-x(1)*x(3)

eliminate(Jh,x(1));
//-> _[1]=x(2)^3-x(0)*x(3)^2
```

To summarize the above discussion, we assume that we have an ideal (arbitrary, homogeneous, partially homogeneous, or bihomogeneous) and eliminate some of the variables.

What do we compute geometrically when we eliminate?

(1) Let $I = \langle g_1, \ldots, g_k \rangle \subset K[x_1, \ldots, x_n]$ and $X = V(I) \subset \mathbb{A}^n$.

   a) By eliminating $x_1, \ldots, x_r$ from I, we compute an ideal I' describing the closure of the image of X under the projection $\pi : \mathbb{A}^n \to \mathbb{A}^{n-r}$ on the last $n - r$ coordinates (that is, $I' = I \cap K[x_{r+1}, \ldots, x_n]$ and $V(I') = \overline{\pi(X)}$).

   b) Let $f_1, \ldots, f_m \in K[x_1, \ldots, x_n]$ and $f = (f_1, \ldots, f_m) : \mathbb{A}^n \to \mathbb{A}^m$, then eliminating $x_1, \ldots, x_n$ from the ideal

   $$J = \langle I, y_1 - f_1, \ldots, y_m - f_m \rangle \subset K[x_1, \ldots, x_n, y_1, \ldots, y_m]$$

   computes the closure of the image, $\overline{f(X)} \subset \mathbb{A}^m$.

   Moreover, if $J^h \subset K[x_0, \ldots, x_n, y_1, \ldots, y_m]$ is the homogenization of J with respect to $x_1, \ldots, x_n$ and with homogenizing variable x_0,[4] then $V(J^h)$ is the closure in $\mathbb{P}^n \times \mathbb{A}^m$ of $V(J) \subset \mathbb{A}^n \times \mathbb{A}^m$. Moreover, $V(J) \cong X$ and $\overline{f(X)} = \pi(V(J^h))$, where π denotes the projection $\pi : \mathbb{P}^n \times \mathbb{A}^m \to \mathbb{A}^m$.

(2) Let $I = \langle f_1, \ldots, f_k \rangle$, $f_i \in K[x_0, \ldots, x_n, y_1, \ldots, y_m]$ being homogeneous in $x_0, \ldots, x_n$ and being arbitrary in $y_1, \ldots, y_m$. Let $X = V(I) \subset \mathbb{P}^n \times \mathbb{A}^m$ and $\pi : \mathbb{P}^n \times \mathbb{A}^m \to \mathbb{A}^m$ the projection.

   a) By eliminating $x_0, \ldots, x_n$ from I, we compute an ideal describing $\overline{\pi(V_a(I))} \subset \mathbb{A}^m$, where $V_a(I) \subset \mathbb{A}^{n+1} \times \mathbb{A}^m$ is the affine variety defined by I. Note that $\overline{\pi(V_a(I))}$ contains $\pi(X)$, but it may be strictly larger.

   b) Eliminating $x_0, \ldots, x_n$ from $I : \langle x_0, \ldots, x_n \rangle^\infty$ computes an ideal describing $\pi(X)$ (which is closed).

   c) If $J \subset K[x_1, \ldots, x_n, y_1, \ldots, y_n]$ is arbitrary and if $I = J^h$ denotes the ideal in $K[x_0, \ldots, x_n, y_1, \ldots, y_n]$, obtained by homogenizing J with respect to $x_1, \ldots, x_n$, then $\pi(X)$ is described by eliminating $x_0, \ldots, x_n$ from I, that is, $\pi(X) = V(I \cap K[y_1, \ldots, y_n])$.

(3) If I is as in (2), but also homogeneous in y_i (that is, the f_i are bihomogeneous in x and y), $X = V(I) \subset \mathbb{P}^n \times \mathbb{P}^{m-1}$, and π the projection $\mathbb{P}^n \times \mathbb{P}^{m-1} \to \mathbb{P}^{m-1}$. Then a), b) and c) of (2) hold.

(4) Let $f_0, \ldots, f_m \in K[x_0, \ldots, x_n]$ be homogeneous polynomials of the same degree, let $X \subset \mathbb{P}^n$ be a projective variety given by a homogeneous ideal $I \subset K[x_0, \ldots, x_n]$ and let $f = (f_0 : \ldots : f_m) : \mathbb{P}^n \dashrightarrow \mathbb{P}^m$ be the rational map defined by the f_i. Eliminating $x_0, \ldots, x_n$ from

$$\langle I, f_0 - y_0, \ldots, f_m - y_m \rangle \subset K[x_0, \ldots, x_n, y_0, \ldots, y_m]$$

computes $\overline{f(X_0)}$, where $X_0 := X \smallsetminus V(f_0, \ldots, f_m)$. If f is a morphism, then $\overline{f(X_0)} = f(X)$.

---

[4] Note that J^h can be computed from J by computing a Gröbner basis for J with respect to an ordering satisfying $x^\alpha y^\beta > x^{\alpha'} y^{\beta'}$ if $|\alpha| > |\alpha'|$.

(5) Let $I \subset K[x_0, \ldots, x_n]$ be a homogeneous ideal and $X = V(I) \subset \mathbb{P}^n$. Eliminating $x_0, \ldots, x_{r-1}$ from I computes $\overline{\pi_H(X)}$, where $\pi_H : \mathbb{P}^n \dashrightarrow \mathbb{P}^{n-r}$ is the projection from $H = V(x_r, \ldots, x_n) \cong \mathbb{P}^{r-1}$. If $X \cap H = \emptyset$, then π_H is a morphism and the image $\pi_H(X)$ is closed in $\mathbb{P}^{n-r}$.

A.8 Local Versus Global Properties

So far, we have discussed global properties of affine or projective varieties. Algebraically, these properties are coded in the affine or homogeneous coordinate ring. When talking about local properties, we mean those properties which hold in an arbitrary small neighbourhood of a given point. Since any projective variety is covered by open affine varieties, we can, without loss of generality, restrict our attention to affine varieties. However, of special interest are results which combine local and global properties.

Given an affine variety X and a point $p \in X$, then the neighbourhoods of p in X are complements of closed subvarieties not containing the point p. Recall from Section A.6 that the coordinate ring of X is isomorphic to $\mathcal{O}(X)$, the ring of regular functions on X. Restricting the regular functions to smaller and smaller neighbourhoods of p, we obtain, in the limit, a ring of germs at p of regular functions, the local ring of X at p. It turns out that this ring is just the localization of the coordinate ring at the maximal ideal defining the closed point p. As we have seen in Chapter 1, we can effectively compute in such rings.

If $X = \mathrm{Spec}(A)$ is an affine scheme, and $\mathfrak{m} \in X$ a maximal ideal of A, then the localization $A_{\mathfrak{m}}$ is the straightforward generalization of the local ring of a classical variety. Hence, $A_{\mathfrak{m}}$ should be considered as the ring carrying information about X, which is valid in an arbitrary small neighbourhood of the closed point $\mathfrak{m}$ in X, in the Zariski topology.

So far we considered "local" with respect to the Zariski topology, which is a rather coarse topology. When working with complex varieties or schemes we also have the (usual) Euclidean topology. The local rings arising from this topology are discussed in the next section on singularities.

In this section, a variety means a classical quasi–projective variety. Eventually, we mention the corresponding definitions and statements for arbitrary quasi–projective schemes.

The following definition applies to any variety.

Definition A.8.1. Let X be a variety and p a point of X.

A *germ of a regular function at p* is an equivalence class of pairs (U, f) where U is an open neighbourhood of p and f a regular function on U. Two such pairs (U, f) and (V, g) are equivalent if there exists an open neighbourhood $W \subset U \cap V$ of p such that $f|_W = g|_W$.

The *local ring of* X *at* p is the ring of germs of regular functions of X at p and is denoted by $\mathcal{O}_{X,p}$. We set $\dim_p X := \dim \mathcal{O}_{X,p}$. X is called *pure dimensional* or *equidimensional* if $\dim_p X$ is independent of $p \in X$.

Proposition A.8.2. *Let* $X \subset \mathbb{A}^n$ *be an affine variety with coordinate ring* $K[X] = K[x_1, \ldots, x_n]/I(X)$, *and let* $p \in X$, *then the following holds:*

(1) $\mathcal{O}_{X,p}$ *is a local ring with maximal ideal*

$$\mathfrak{m}_p := \mathfrak{m}_{X,p} := \{ f \in \mathcal{O}_{X,p} \mid f(p) = 0 \},$$

the vanishing ideal of p.

(2) $\mathcal{O}_{X,p} \cong K[X]_{\mathfrak{m}_p}$.

(3) $\dim \mathcal{O}_{X,p} = \max\{ \dim X_i \mid p \in X_i \}$ *where the* X_i *are the irreducible components of* X.

Proof. (1) Let f be a regular function with $f(p) \neq 0$. Then, since f is continuous, $f(q) \neq 0$ for all q in a neighbourhood U of p. Hence, $(U, 1/f)$ is an inverse of (U, f) and, therefore, every element $f \in \mathcal{O}_{X,p} \smallsetminus \mathfrak{m}_p$ is a unit and $\mathfrak{m}_p$ is the only maximal ideal of $\mathcal{O}_{X,p}$.

To see (2), note that we have $\mathcal{O}_{X,p} = \{(U, f) \mid f$ regular on $U\}/\sim$, and

$$K[X]_{\mathfrak{m}_p} = \left\{ \frac{f}{g} \,\middle|\, f, g \in K[X], \, g(p) \neq 0 \right\}.$$

Now, let $f/g \in K[X]_{\mathfrak{m}_p}$. Since g is continuous, $g \neq 0$ in a neighbourhood U of p, hence, f/g defines a regular function on U. Therefore, we obtain a map $K[X]_{\mathfrak{m}_p} \to \mathcal{O}_{X,p}$, which is easily seen to be bijective.

(3) The canonical map $j : K[X] \to K[X]_{\mathfrak{m}_p}$, $f \mapsto f/1$, induces a bijection between the set of prime ideals in $K[X]_{\mathfrak{m}_p}$ and the set of all prime ideals in $K[X]$ which are contained in $\mathfrak{m}_p$, via $P \mapsto j^{-1}(P)$ and $Q \mapsto QK[X]_{\mathfrak{m}_p}$.

Hence, $\dim K[X]_{\mathfrak{m}_p}$ equals the maximal length of chains of prime ideals in $K[X]$ contained in $\mathfrak{m}_p$. But each maximal chain of prime ideals starts with a minimal associated prime I_0 of $I(X)$, which corresponds to an irreducible component of X. The latter contains p, since $I_0 \subset \mathfrak{m}_p$. $\qquad\square$

Example A.8.3. Consider the reducible variety X defined by $zx = zy = 0$, consisting of the line $\{x = y = 0\}$ and the plane $\{z = 0\}$ (see Figure A.19).

Let $0 := (0, 0, 0)$, $q := (1, 0, 0)$ and $p := (0, 0, 1)$, then the geometric picture (Figure A.19) suggests $\dim_0 X = \dim_q X = 2$ and $\dim_p X = 1$. We check this for the respective local rings: $\mathcal{O}_{X,0} = K[x, y, z]_{\langle x,y,z \rangle}/\langle xz, yz \rangle$, which is a local ring of dimension 2. As x is a unit in $K[x, y, z]_{\langle x-1,y,z \rangle}$, the local ring $\mathcal{O}_{X,q}$ is isomorphic to $K[x, y, z]_{\langle x-1,y,z \rangle}/\langle z \rangle$, which has dimension 2. In $K[x, y, z]_{\langle x,y,z-1 \rangle}$ we have z as a unit, hence, $\mathcal{O}_{X,p} \cong K[x, y, z]_{\langle x,y,z-1 \rangle}/\langle x, y \rangle$ has dimension 1.

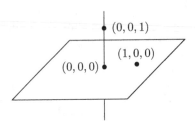

Fig. A.19. The variety $V(\langle xz, yz \rangle)$.

SINGULAR Example A.8.4 (local and global dimension).

We confirm the latter by using SINGULAR:

```
ring R = 0,(x,y,z),dp;   //global affine ring
ideal I = xz,yz;
dim(std(I));
//-> 2                   //dimension of affine variety V(I)


ring r = 0,(x,y,z),ds;   //localization of R at (0,0,0)
ideal I = imap(R,I);
dim(std(I));
//-> 2                   //dimension of V(I) at (0,0,0)


map phi1 = r,x-1,y,z;    //maps the point (1,0,0) to (0,0,0)
ideal I1 = phi1(I);
dim(std(I1));
//-> 2                   //dimension of V(I) at (1,0,0)


map phi2 = r,x,y,z-1;    //maps the point (0,0,1) to (0,0,0)
ideal I2 = phi2(I);
dim(std(I2));
//-> 1                   //dimension of V(I) at (0,0,1)
```

For projective varieties, we can describe the local rings also directly with the help of the homogeneous coordinate ring.

Lemma A.8.5. *Let $X \subset \mathbb{P}^n$ be a projective variety, let $p \in X$ be a point, and let $K[X] = K[x_0, \ldots, x_n]/I(X)$ be the homogeneous coordinate ring of X. Then*

$$\mathcal{O}_{X,p} = K[X]_{(\mathfrak{m}_p)} := \left\{ \frac{f}{g} \; \middle| \; \begin{array}{l} f, g \in K[X] \; homogeneous \; of \\ the \; same \; degree, \; g(p) \neq 0 \end{array} \right\}.$$

We leave the proof, which is similar to the affine case, as an exercise.

The above statements are used to define local rings for schemes.

Definition A.8.6. Let X be a quasi–projective scheme, $p \in X$, and let $U = \operatorname{Spec}(A)$ be an affine neighbourhood of p. If P denotes the prime ideal of A corresponding to p, then the *local ring of X at p* is $\mathcal{O}_{X,p} := A_P$, the localization of A at P.

Of course, $p = P$, but we have chosen different letters to distinguish between geometry and algebra.

Let $A = K[x_1, \ldots, x_n]/I$ be an affine K–algebra, K algebraically closed and $X = \operatorname{Spec}(A)$. Then any closed point p corresponds to a maximal ideal $\mathfrak{m}_p$, which is of the form $\mathfrak{m}_p = \langle x_1 - p_1, \ldots, x_n - p_n \rangle / I$ (Corollary A.2.7) and, hence, $A/\mathfrak{m}_p = K$. Therefore, any $f \in A$ determines a function

$$\{\text{closed points of } X\} = \operatorname{Max}(A) \longrightarrow K,$$

$$p \longmapsto f(p) := \text{class of } f \text{ in } A/\mathfrak{m}_p.$$

However, due to nilpotent elements in A, $p \mapsto f(p)$ may be the zero–function, even if $f \neq 0$ in A. The simplest example is $A = K[x]/\langle x^2 \rangle$ with 0 as the only closed point, $\mathfrak{m}_0 = \langle x \rangle$, $x(0) = 0$, but $x \neq 0$ in A.

Hence, if X is a non–reduced scheme, we cannot define the local rings as germs of continuous functions. Indeed, since $\operatorname{Max}(A)$ is dense in $\operatorname{Spec}(A)$, every continuous function which is 0 on $\operatorname{Max}(A)$ must be 0 on $\operatorname{Spec}(A)$, identically.

Local properties of varieties are of particular interest in the neighbourhood of singular points.

Definition A.8.7. Let X be a variety or scheme and $p \in X$. Then p is called a *singular point of X*, or X is called *singular at p*, if the local ring $\mathcal{O}_{X,p}$ is not a regular local ring. Otherwise, p is called a *regular point*, or *non–singular point* of X. X is called *regular* if it is regular at each point p of X.

The definition applies, if X is a scheme, to closed and non–closed points. Recall that a local ring A with maximal ideal $\mathfrak{m}$ is regular, if $\dim(A) = \operatorname{edim}(A)$, where $\operatorname{edim}(A) = \dim_{A/\mathfrak{m}}(\mathfrak{m}/\mathfrak{m}^2)$ is the embedding dimension. This is the definition which works in general, unfortunately, it is not very geometric.

A good geometric interpretation of regular points is given by the *Jacobian criterion* (cf. Corollary 5.6.14), which is valid for varieties over an algebraically closed field. If $X \subset \mathbb{A}^n$ is affine with ideal $I(X) = \langle f_1, \ldots, f_k \rangle$ and $p \in X$, then X is regular at p if and only if

$$\dim_p X = n - \operatorname{rank}\left(\frac{\partial f_i}{\partial x_j}(p)\right).$$

That is, the linear parts of the power series expansions of $f_1, \ldots, f_k$ at p define a linear subspace of the same dimension as X at p. By the implicit function theorem for formal power series (Theorem 6.2.17), there exists an analytic coordinate change at p, that is, an automorphism φ of the power series ring $K[[x_1, \ldots, x_n]]$ (but, in general, not of the polynomial ring) such that $\varphi(I(X)K[[x_1, \ldots, x_n]]) = \langle x_1, \ldots, x_r \rangle K[[x_1, \ldots, x_n]]$, where r is the rank of the Jacobian matrix at p. This means that X is regular at p if and only if formally at the point p, X is analytically isomorphic to a linear subspace.

We shall discuss regular and singular points in the framework of local analytic geometry in the next section on singularities.

The Jacobian criterion is also the basis for computing the *singular locus*

$$\mathrm{Sing}(X) = \big\{ p \in X \mid X \text{ is singular at } p \big\}$$

of a variety X. If X is equidimensional, this is particularly simple. We just have to compute the $(n - \dim X)$–minors of the Jacobian matrix $\left(\frac{\partial f_i}{\partial x_j} \right)$, see Corollary 5.7.5.

If X is not equidimensional, then $\mathrm{Sing}(X)$ is the union of the singular loci of the irreducible components of X and the locus of the pairwise intersections of the components. Hence, $\mathrm{Sing}(X)$ can be computed from a primary decomposition of X and then applying the Jacobian criterion to each irreducible component. In practice, it is, however, cheaper to compute an equidimensional than a prime decomposition. Methods to compute an ideal describing the singular locus without any decomposition by using either fitting ideals or Ext (cf. Exercises 7.3.24 and 7.3.25) also exist. Algorithms, describing the singular locus, are given in Chapter 5, Section 5.7.

We are particularly interested in distinguishing singular points by invariants. Invariants can be numbers, groups, vector spaces or other varieties which reflect certain properties of the variety at p and which do not change under isomorphisms of the local ring $\mathcal{O}_{X,p}$. Ideally, one could hope to classify all singular points on a given variety by a discrete set of invariants. However, this is only possible for the most simple singularities, the ADE–singularities (see Section A.9).

One warning is perhaps in order. One should not expect the classification of singularities to be simpler than the classification of, say, projective varieties. For example, if $X \subset \mathbb{P}^n$ is projective, then the affine cone $CX \subset \mathbb{A}^{n+1}$ has a singularity at 0 and a classification of this singularity, up to local isomorphism (which must be linear as CX is a cone), implies a classification of X up to projective equivalence. Hence, the classification of singularities includes the projective classification of projective varieties, but the singularities arising this way are only the homogeneous ones, which is a small subclass of all singularities.

In this section we discuss only one local invariant, the multiplicity, and compare it with a global invariant, the degree.

The multiplicity was introduced in Section 5.5, using the Hilbert–Samuel function. This is, perhaps, algebraically the most elegant way but not very geometric. Algebraically, as well as geometrically, the multiplicity is the local counterpart of the degree, see Section 5.3.

Definition A.8.8.

(1) Let X be any variety and $p \in X$. The *multiplicity of X at p* is the multiplicity of the local ring $\mathcal{O}_{X,p}$ in the sense of Definition 5.5.2, and denoted by $\mathrm{mult}(X, p)$.

(2) Let $X \subset \mathbb{P}^n$ be a projective variety, then the *degree* of X is the degree of the homogeneous coordinate ring $K[X]$ in the sense of Definition 5.3.3, and denoted by $\deg(X)$.

Note that mult is invariant under local isomorphisms of varieties (that is, isomorphisms of their local rings), while deg is not invariant under isomorphisms of projective varieties. The degree is only invariant under projective equivalence or, geometrically speaking, it is an invariant of the embedding.

Example A.8.9. Consider the d–tuple *Veronese embedding* of Example A.6.2. One shows that ν_d is an isomorphism from $\mathbb{P}^n$ onto the image $V_d := \nu_d(\mathbb{P}^n)$, which is called the *Veronese variety*. The hypersurface $M_0(x) := \{x_0^d = 0\}$ in $\mathbb{P}^n$ is mapped isomorphically onto the hyperplane $\{z_0 = 0\}$ in $\mathbb{P}^N$. $M_0(x)$ has degree d, $\{z_0 = 0\}$ degree 1. Moreover, one can show that V_d has degree d^n in $\mathbb{P}^N$, while $\mathbb{P}^n$ certainly has degree 1 in $\mathbb{P}^n$.

It is useful to compare the notions of degree and multiplicity for the most simple case, the case of a hypersurface. If $f \in K[x_1, \ldots, x_n]$ is a squarefree polynomial with $f(0) = 0$, and if $X \subset \mathbb{A}^n$ is the hypersurface defined by $f = 0$, then $\mathrm{mult}(X, 0) = \mathrm{ord}(f)$, the smallest degree of a monomial appearing in f (Corollary 5.5.9), while $\deg(X) = \deg(\overline{X}) = \deg(f)$, the largest degree of a monomial appearing in f (Lemma 5.3.5). Here, $\overline{X} \subset \mathbb{P}^n$ is the projective closure of X, defined by $f^h = 0$, where $f^h \in K[x_0, \ldots, x_n]$ denotes the homogenization of f.

We start with a geometric interpretation of the degree as the number of intersection points with a sufficiently general hyperplane. What is meant here by "sufficiently general" will be explained in the proof.

Proposition A.8.10. *Let $X \subset \mathbb{P}^n$ be a projective variety of dimension d, and let $X_1, \ldots, X_r, X_{r+1}, \ldots, X_s$ be the irreducible components of X, ordered such that $\dim X_i = d$ for $i = 1, \ldots, r$ and $\dim X_i < d$ for $i = r + 1, \ldots, s$. Then*

$$\deg(X) = \#(X \cap H) = \sum_{i=1}^{r} \#(X_i \cap H),$$

where $H \subset \mathbb{P}^n$ is a sufficiently general projective hyperplane of dimension $n - d$ and $\#$ denotes the cardinality.

Proof. By Lemma 5.3.11, we have $\deg(X) = \sum_{i=1}^{r} \deg(X_i)$. Hence, we can assume X to be irreducible with homogeneous coordinate ring $K[x_0, \ldots, x_n]/P$, where P is a homogeneous prime ideal.

Choose a homogeneous Noether normalization $X \to \mathbb{P}^d$ (Theorem 3.4.1): after a linear change of coordinates, we have an inclusion

$$K[x_0, \ldots, x_d] \subset K[x_0, \ldots, x_n]/P$$

together with irreducible homogeneous polynomials $g_{d+1}, \ldots, g_n \in P$,

$$g_i = x_i^{p_i} + \sum_{j=0}^{p_i-1} c_{ij}(x_0, \ldots, x_{i-1}) x_i^j \,,$$

such that $\mathrm{Quot}(K[x_0, \ldots, x_n]/P) = \mathrm{Quot}(K[x_0, \ldots, x_d])[x_{d+1}]/\langle g_{d+1}\rangle$, especially, $p_{d+1} = \deg(X)$, by Proposition 5.3.10.

Set $f := g_{d+1}$, which defines a hypersurface $V(f) \subset \mathbb{P}^{d+1}$ of degree $\deg(X)$. Then, for a sufficiently general line L, the restriction $f|_L$ has exactly $\deg(X)$ simple roots, that is, the intersection $V(f) \cap L$ consists of precisely $\deg(X)$ points.

We show now that the projection $\pi : X \to V(f)$ is birational, and that the preimage of a general line, $H = \pi^{-1}(L)$ intersects X in $\deg(X)$ points.

Let $\Delta \in K[x_0, \ldots, x_d]$ be the discriminant of f (with respect to x_{d+1}). Δ is homogeneous and vanishes at a point $a = (a_0 : \ldots : a_d) \in \mathbb{P}^d$ if and only if $f(a_0, \ldots, a_d, x_{d+1})$ has a multiple root.[5]

By Lemma 3.5.12 (applied as in the proof of Theorem 3.5.10), we have an inclusion

$$K[x_0, \ldots, x_n]/P \subset \frac{1}{\Delta} K[x_0, \ldots, x_{d+1}]/\langle f\rangle \,.$$

In particular, there exist polynomials $q_{d+2}, \ldots, q_n \in K[x_0, \ldots, x_{d+1}]$ such that $x_i \Delta - q_i \in P$, for $i = d+2, \ldots, n$, and

$$\langle f, x_{d+2}\Delta - q_{d+2}, \ldots, x_n\Delta - q_n\rangle K[x_0, \ldots, x_n]_\Delta = P K[x_0, \ldots, x_n]_\Delta \,.$$

Hence, for any point $(x_0 : \ldots : x_d) \in \mathbb{P}^d \smallsetminus V(\Delta)$ we have

$$(x_0 : \ldots : x_n) \in V(P) \iff f(x_0, \ldots, x_{d+1}) = 0 \quad \text{and}$$

$$x_i = \frac{q_i(x_0, \ldots, x_{d+1})}{\Delta(x_0, \ldots, x_d)}, \quad i = d+2, \ldots, n.$$

This shows that the projection $\pi : \mathbb{P}^n \to \mathbb{P}^{d+1}$ induces an isomorphism

---

[5] Another way to describe the zero–set of the discriminant is given by considering the projection $V(f) \to \mathbb{P}^d$. Then $V(\Delta)$ is the image of the set of critical points $C = V(f, \partial f/\partial x_{d+1})$ in $\mathbb{P}^d$ (which we know already to be closed by Theorem A.7.7). Note that the polynomial $\Delta' \in K[x_0, \ldots, x_d]$, obtained by eliminating x_{d+1} from $\langle f, \partial f/\partial x_{d+1}\rangle$, has the same zero–set as Δ and satisfies $\langle \Delta\rangle \subset \langle \Delta'\rangle$, however, the inclusion may be strict.

$$\pi : X \setminus V(\Delta) \xrightarrow{\;\cong\;} V(f) \setminus V(\Delta) \,.$$

Therefore, if $a = (a_0 : \ldots : a_d) \in \mathbb{P}^d \setminus V(\Delta)$, then the line

$$L := \{(a : x_{d+1}) \mid x_{d+1} \in K\}$$

intersects $V(f)$ in $\deg(X)$ distinct points, and the hyperplane

$$H := \pi^{-1}(L) = \{(a : x_{d+1} : \ldots : x_n) \mid x_{d+1}, \ldots, x_n \in K\}$$

intersects X in exactly $\deg(X)$ points, too. $\qquad\qquad\square$

In the above proof we showed, moreover,

Proposition A.8.11. *Let $X \subset \mathbb{P}^n$ be an irreducible d–dimensional projective variety, then, for general homogeneous coordinates $x_0 : \ldots : x_n$, there exists a hypersurface $V(f) \subset \mathbb{P}^{d+1} = V(x_{d+2}, \ldots, x_n)$ such that the projection $\pi : X \to V(f)$ is birational. Moreover, $\deg(X) = \deg(f)$.*

SINGULAR Example A.8.12 (degree of projective variety).
Consider the *rational normal curve* C of degree r in $\mathbb{P}^r$, which is the projective closure of the image of the morphism

$$\mathbb{P}^1 \ni (s : t) \longmapsto (s^r : s^{r-1}t : \ldots : t^r) \in \mathbb{P}^r \,.$$

The homogeneous ideal of C is the kernel of the ring map

$$K[x_0, \ldots, x_r] \longrightarrow K[s, t], \quad x_i \longmapsto s^{r-i}t^i \,.$$

We compute the degree of C and count the number of intersection points of C with a general hyperplane:

```
LIB"random.lib";
int r = 5;
ring R = 0,x(0..r),dp;

ring S = 0,(s,t),dp;
ideal I = maxideal(r);  //s^r, s^(r-1)*t,..., s*t^(r-1), t^r
ideal zero;
map phi = R,I;          //R --> S, x(i) --> s^(r-i)*t^i

setring R;
ideal I = preimage(S,phi,zero); //kernel of map phi R --> S
I = std(I);             //ideal of rational normal curve C
dim(I);                 //dimension of affine cone over C
//-> 2
degree(I);              //degree of C is 5
//-> 5
```

```
ideal L = sparsepoly(1,1,0,10); //a general linear form
                                //?sparsepoly; explains the syntax
ideal CL = std(I+L+(x(0)-1)); //ideal of intersection of C
                                //with L=0
                                //in affine chart x(0)=1
vdim(CL);                       //number of intersection points is 5
//-> 5                          //in affine chart x(0)=1
```

Note that `vdim` counts the number of intersection points with multiplicity. By Proposition A.8.10, this number should coincide with the degree of C. In order to count the points set theoretically, we compute the radical of CL. Since C is smooth (check this), a general hyperplane meets C in simple points, hence CL should coincide with the radical of CL and `vdim` should give the same number. We check this:

```
LIB"primdec.lib";
vdim(std(radical(CL)));
//-> 5
```

Let us now consider the multiplicity. It has a similar geometric interpretation as the degree. It counts the number of intersection points of a variety with a general hyperplane of the right dimension in a sufficiently small neighbourhood of the given point. However, this interpretation is only valid in the Euclidean topology, not in the Zariski topology. The algebraic reason for this fact is that the Noether normalization of an algebraic local ring may fail. The Noether normalization holds for affine algebras (Theorem 3.4.1) and for analytic algebras (Theorem 6.2.16) but, in general, not for the localization of an affine algebra (Exercise 3.4.7).

Therefore, we assume in the following discussion that $K = \mathbb{C}$, and we use the *Euclidean topology*.

Proposition A.8.13. *Let $X \subset \mathbb{C}^n$ be a variety of dimension d and $p \in X$. We denote by $X_1, \ldots, X_r, X_{r+1}, \ldots, X_s$ the irreducible components of X passing through p, such that $\dim X_i = d$ for $i = 1, \ldots, r$ and $\dim X_i < d$ for $i = r + 1, \ldots, s$. Then, for any sufficiently small (Euclidean) neighbourhood U of p and for any sufficiently general hyperplane $H \subset \mathbb{C}^n$ of dimension $n - d$ and sufficiently close to p (but not passing through p), we have*

$$\mathrm{mult}(X, p) = \#(X \cap H \cap U) = \sum_{i=1}^{r} \#(X_i \cap H \cap U).$$

The proof is completely analogous to the previous one, by using an analytic Noether normalization (cf. Theorem 6.2.16, Exercise 6.2.1 for the formal case). Moreover, we need a Weierstraß polynomial for generic coordinates,

$$f = x_{d+1}^p + \sum_{j=0}^{p-1} c_j(x_1, \ldots, x_d) x_{d+1}^j \,,$$

where $c_j(x_1, \ldots, x_d)$ are convergent power series and, thus, define holomorphic functions in a neighbourhood of $(p_1, \ldots, p_d) \in \mathbb{C}^d$. This follows from the Weierstraß preparation theorem for convergent power series (cf. [108], [128], for example). Now everything else works as in the global case.

Note that $\operatorname{mult}(X, p) = \operatorname{mult}(C_p(X), p)$, where $C_p(X)$ is the *tangent cone* of X at p, which is defined as follows: if $X \subset \mathbb{C}^n$ is defined by an ideal $I \subset \mathbb{C}[x_1, \ldots, x_n]$, then $C_p(X)$ is defined by $\operatorname{In}_p(I)$. Here $\operatorname{In}_p(I)$ denotes the ideal generated by the initial forms of all $f \in I$, written as polynomials in $y := x - p$ (cf. Proposition 5.5.12). The tangent cone can be computed according to Lemma 5.5.11.

Geometrically, the tangent cone is the union of all limits of secants $\overline{p p_i}$ with $p_i \in X \setminus \{p\}$ a sequence of points converging to p. The direction of the hyperplane H in Proposition A.8.13 (equivalently, the choice of the generic coordinates) is predicted by the tangent cone. Choose a hyperplane H_0 through p, which is transversal to $C_p(X)$, that is, $H_0 \cap C_p(X) = \{p\}$. Then a small, sufficiently general displacement of H_0 intersects $U \cap X$ in exactly $\operatorname{mult}(X, p)$ points.

Example A.8.14. Let X be the cuspidal cubic given by $f := x^3 - y^2 = 0$, and let $L_\varepsilon := \{x + by = \varepsilon\}$ be a general line. Then $f|_{L_\varepsilon} = (\varepsilon - by)^3 - y^2$ has two zeros close to 0 (check this), and we have $\operatorname{ord}(f) = \operatorname{mult}(X, 0) = 2$ (cf. Figure A.20). The tangent cone is the x–axis $\{y = 0\}$, and the line L_ε is a small displacement of the line $L_0 = \{x + by = 0\}$, which is transversal to the tangent cone. A small displacement of the line $\{y = 0\}$ meets X in three complex points (check this), although we see only one point in the real picture.

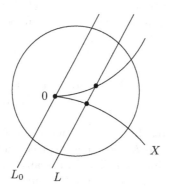

Fig. A.20. $\operatorname{mult}(X, 0) = \#(L_\varepsilon \cap X) = 2$.

SINGULAR Example A.8.15 (multiplicity and tangent cone).
Although the above example is very simple, let us demonstrate the commands
to compute the multiplicity and the tangent cone. Moreover, we compute the
intersection multiplicity of the curve $\{f = 0\}$ with a general line L_o through 0
(which is transversal to the tangent cone) and with the special line L_1 (which
coincides with the tangent cone). In order to compute the correct multiplicity,
we have to work with a local degree ordering (the procedure `tangentcone`
works for any ordering).

```
ring r = 0,(x,y),ds;
ideal f = x3-y2;
mult(std(f));               //the multiplicity of f at 0
//-> 2


LIB"sing.lib";
tangentcone(f);             //the tangent cone of f at 0
//-> _[1]=y2

ideal Lo = random(1,100)*x + random(1,100)*y;
                            //a general line through 0
vdim(std(f+Lo));            //intersection multiplicity of
//-> 2                      //f and Lo at 0

ideal L1 = y;               //the special line y=0
vdim(std(f+L1));            //intersection multiplicity of
//-> 3                      //f and L1 at 0
```

When we want to compute the number of intersection points of $\{f = 0\}$ with
a small displacement L_e of L_o (where e is a small number), we have to be
careful: in the local ring r the polynomial defining the displaced line is a
unit, hence, the intersection multiplicity is 0. Thus, we have to use a global
ordering. However, counting the intersection number in a global ring gives
the total intersection number in affine space of $\{f = 0\}$ with the given line
(which is 3 for L_o as well as for L_e), and not only in a small Euclidean
neighbourhood of 0 (which is 2).

The only thing we can do is to solve the system given by $f = L_e = 0$
numerically and then "see", which points are close to 0 (which is, in general,
a guess).

Let X be the parabola $\{x - y^2 = 0\}$ with $\mathrm{mult}(X, 0) = 1$. The tangent
cone is the y–axis, equal to L_0. A small displacement of L_0 intersects X in
more than $\mathrm{mult}(X, 0)$ points. The line L_0' is transversal to the tangent cone,
L_ε' intersects X in one point close to 0 (cf. Figure A.21).

What we discussed in the above examples was the intersection multiplicity
of a curve with a line. We generalize this to the intersection multiplicity of

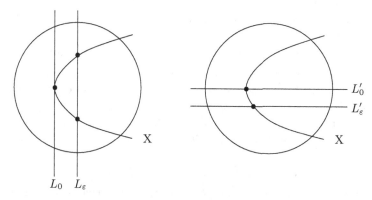

Fig. A.21. $\mathrm{mult}(X,0) = 1$, $\#X \cap L_\varepsilon = 2$, $\#X \cap L'_\varepsilon = 1$.

two plane curves. In the following, K is again an algebraically closed field (of arbitrary characteristic).

Definition A.8.16.

(1) Let $f, g \in K[x,y]$ be of positive degree, $p = (p_1, p_2) \in V(f) \cap V(g) \subset \mathbb{A}^2_K$, and let $\mathfrak{m}_p = \langle x - p_1, y - p_2 \rangle$ be the maximal ideal of p. We define

$$i(f,g;p) := \dim_K \left(K[x,y]_{\mathfrak{m}_p} / \langle f, g \rangle \right),$$

and call it the *intersection multiplicity* of f and g at p.

(2) If $C = \mathrm{Spec}(K[x,y]/\langle f \rangle)$ and $D = \mathrm{Spec}(K[x,y]/\langle g \rangle)$, then

$$i(C, D; p) := i(f, g; p)$$

is called the *intersection multiplicity* of C and D at p.

(3) If $C, D \subset \mathbb{A}^2_K$ are classical affine plane curves and $I(C) = \langle f \rangle$, $I(D) = \langle g \rangle$, then $i(C, D; p) := i(f, g; p)$ is called the *intersection multiplicity* of C and D at p.[6]

Note that in (2) and (3) the generator of the (principal) ideal $\langle f \rangle$, respectively $\langle g \rangle$, is unique up to multiplication by a non–zero constant. Hence, the intersection multiplicity $i(C, D; p)$ is well–defined. Note also that $i(f, g; p)$ is finite if and only if f and g have no common factor which vanishes at p. This follows from Krull's principal ideal theorem (cf. Corollary 5.6.10) and Corollary 5.3.17.

Now let $F, G \in K[z, x, y]$ be homogeneous polynomials of positive degree, let $p = (p_0 : p_1 : p_2) \in V(F) \cap V(G) \subset \mathbb{P}^2_K$, and let $\mathfrak{m}_p = \langle p_0 x - p_1 z, p_0 y - p_2 z \rangle$

---

[6] The difference here is just that f and g are not allowed to have multiple factors which are allowed in (1) and (2).

be the homogeneous ideal of p. Assume that $p_0 \neq 0$, then it is easy to see that

$$i(F, G; p) := \dim_K \big(K[z, x, y]_{\mathfrak{m}_p} / \langle F, G \rangle \big) = i(f, g; p) \,,$$

where $f = F^a = F|_{z=1}$ and $g = G^a = G|_{z=1}$ are the affinizations of F and G. $i(F, G; p)$ is called the *intersection multiplicity* of F and G at p. Note that, by definition, it is a purely local invariant.

If $C = \mathrm{Proj}(K[z, x, y]/\langle f \rangle)$ and $D = \mathrm{Proj}(K[z, x, y]/\langle g \rangle)$, respectively, if $C, D \subset \mathbb{P}^2_K$ are classical projective plane curves with $I(C) = \langle F \rangle$, $I(D) = \langle G \rangle$, then we set

$$i(C, D; p) := i(F, G; p) \,,$$

and call it the *intersection multiplicity* of the projective plane curves C and D at p.

It is clear that $i(F, G; p) = 0$ if and only if f or g is a unit in the local ring $K[x, y]_{\langle x-p_1, y-p_2 \rangle}$, and the latter is equivalent to $p \notin V(F) \cap V(G)$. Moreover, $i(F, G; p) < \infty$ for all $p \in \mathbb{P}^2_K$ if and only if F and G have no common non-constant factor. It is also clear that $i(f^r, g^s; p) \geq i(f, g; p)$ for $r, s \geq 1$, hence, $i(F_{\mathrm{red}}, G_{\mathrm{red}}; p) \leq i(F, G; p)$, where $F_{\mathrm{red}}, G_{\mathrm{red}}$ denote the squarefree parts of F, G, respectively. Hence, the scheme–theoretic intersection multiplicity is at least the intersection multiplicity of the reduced curves.

Now we can formulate one of the most important theorems in projective geometry, combining local and global invariants:

Theorem A.8.17 (Bézout's Theorem). *Let $F, G \in K[z, x, y]$ be polynomials of positive degree, without common non–constant factor. Then the intersection $V(F) \cap V(G)$ is a finite set and*

$$\deg(F) \cdot \deg(G) = \sum_{p \in V(F) \cap V(G)} i(F, G; p) \,.$$

Since the right–hand side is zero if and only if $V(F) \cap V(G)$ is empty, we obtain:

Corollary A.8.18. *Let $C, D \in \mathbb{P}^2_K$ be two projective plane curves without common components, $C = V(F)$, $D = V(G)$, then*

$$\deg(F) \cdot \deg(G) \geq \#(C \cap D) \geq 1 \,.$$

Proof of Theorem A.8.17. Since F and G have no common factor, F is a non-zerodivisor in $K[z, x, y]/\langle G \rangle$. By Lemma 5.3.5, $\deg(G) = \deg(K[z, x, y]/\langle G \rangle)$, and, by Proposition 5.3.6, we obtain

$$\deg(K[z, x, y]/\langle F, G \rangle) = \deg(F) \cdot \deg(G) \,.$$

By Krull's principal ideal theorem 5.6.8, $\dim(K[z, x, y]/\langle F, G \rangle) = 1$. Now Theorem 5.3.7 implies that the Hilbert polynomial of $K[z, x, y]/\langle F, G \rangle$ is constant and, by definition, this constant is $\deg(K[z, x, y]/\langle F, G \rangle)$. In particular, the projective variety $V(F, G) \subset \mathbb{P}^2_K$ has dimension 0, and we may assume (after a linear change of coordinates) that $V(F, G) \cap V(z) = \emptyset$ and F and G are not divisible by z.

Let $f = F^a = F|_{z=1}$ and $g = G^a = G|_{z=1} \in K[x, y]$ be the affinizations of F and G. Then, by Remark 5.3.16, the Hilbert function of $K[z, x, y]/\langle F, G \rangle$ coincides with the affine Hilbert function of $K[x, y]/\langle f, g \rangle$ and, hence,

$$\deg(K[z, x, y]/\langle F, G \rangle) = \dim_K(K[x, y]/\langle f, g \rangle) \,.$$

Let $V(f) \cap V(g) = \{p_1, \ldots, p_r\}$, $p_i = (p_{i1}, p_{i2})$, and let $\mathfrak{m}_i = \langle x - p_{i1}, y - p_{i2} \rangle$ be the corresponding maximal ideals, $i = 1, \ldots, r$. Then, by the Chinese remainder theorem (cf. Exercise 1.3.13),

$$K[x, y]/\langle f, g \rangle \cong \bigoplus_{i=1}^{r} K[x, y]/Q_i \,,$$

where the Q_i are $\mathfrak{m}_i$–primary ideals. Now $K[x, y]/Q_i \cong K[x, y]_{\mathfrak{m}_i}/\langle f, g \rangle$, and the result follows.   □

A.9 Singularities

In Section A.8 we have already defined singular points as points p of a variety X where the local ring $\mathcal{O}_{X,p}$ is not regular. In particular, "singular" is a local notion, where "local" so far was mainly considered with respect to the Zariski topology. However, since the Zariski topology is so coarse, small neighbourhoods in the Zariski topology might not be local enough. If our field K is $\mathbb{C}$, then we may use the Euclidean topology and we can study singular points p in arbitrary small ε–neighbourhoods (as we did already at the end of Section A.8). But then we must also allow more functions, since the regular functions at p (in the sense of Definition A.6.1) are always defined in a Zariski neighbourhood of p. Thus, instead of considering germs of regular functions at p, we consider germs of complex analytic functions at $p = (p_1, \ldots, p_n) \in \mathbb{C}$. The ring of these functions is isomorphic to the ring of convergent power series $\mathbb{C}\{x_1 - p_1, \ldots, x_n - p_n\}$, which is a local ring and contains the ring $\mathbb{C}[x_1, \ldots, x_n]_{\langle x_1 - p_1, \ldots, x_n - p_n \rangle}$ of regular functions at p.

For arbitrary (algebraically closed) fields K, we cannot talk about convergence and then a substitute for $\mathbb{C}\{x_1 - p_1, \ldots, x_n - p_n\}$ is the formal power series ring $K[[x_1 - p_1, \ldots, x_n - p_n]]$. Unfortunately, with formal power series, we cannot go into a neighbourhood of p; formal power series are just not defined there. Therefore, when talking about geometry of singularities, we consider $K = \mathbb{C}$ and convergent power series. Usually, the algebraic statements which hold for convergent power series do also hold for formal power series (but are easier to prove since we need no convergence considerations). We just mention in passing that there is, for varieties over general fields, another notion of "local" with étale neighbourhoods and Henselian rings (cf. [143]) which is a geometric substitute of convergent power series over $\mathbb{C}$. For $I \subset \mathbb{C}[x]$, $x = (x_1, \ldots, x_n)$, an ideal, we have inclusions of rings

$$\mathbb{C}[x]/I \subset \mathbb{C}[x]_{\langle x \rangle}/I\mathbb{C}[x]_{\langle x \rangle} \subset \mathbb{C}\{x\}/I\mathbb{C}\{x\} \subset \mathbb{C}[[x]]/I\mathbb{C}[[x]] \, .$$

To distinguish the different points of view, we make the following definition:

Definition A.9.1. Let $X \subset \mathbb{C}^n$ be an algebraic variety and $p \in X$. The *analytic germ* (X, p) of X at p is an equivalence class of open neighbourhoods of p in X, in the Euclidean topology, where any two open neighbourhoods of p in X are equivalent.

The *analytic local ring* of X at p is the ring of germs of complex analytic functions at $p = (p_1, \ldots, p_n)$,

$$\mathcal{O}_{X,p}^{an} := \mathbb{C}\{x_1 - p_1, \ldots, x_n - p_n\}/I(X) \cdot \mathbb{C}\{x_1 - p_1, \ldots, x_n - p_n\} \, .$$

If $X \subset \mathbb{A}_K^n$, K an arbitrary algebraically closed field, let

$$\mathcal{O}_{X,p} = K[x_1, \ldots, x_n]_{\langle x_1 - p_1, \ldots, x_n - p_n \rangle}$$

be the *algebraic local ring* of X at the (closed) point $p = (p_1, \ldots, p_n)$ and

$$\mathcal{O}_{X,p}^{an} := \widehat{\mathcal{O}}_{X,p} = K[[x_1 - p_1, \ldots, x_n - p_n]]/I(X) \cdot K[[x_1 - p_1, \ldots, x_n - p_n]]$$

be the *analytic local ring* of X at p, $\widehat{}$ denoting the $\langle x_1 - p_1, \ldots, x_n - p_n \rangle$-adic completion.

We call the analytic germ (X, p), and the analytic local ring $\mathcal{O}_{X,p}^{an}$, also a *singularity*.

In the following, we write $K\langle x_1, \ldots, x_n \rangle$ to denote either $K[[x_1, \ldots, x_n]]$ or $\mathbb{C}\{x_1, \ldots, x_n\}$.

For analytic singularities, regular (in the sense of non–singular) points have a very nice interpretation. By the Jacobian criterion and the implicit function theorem, p is a regular point of the complex variety X if, in a small Euclidean neighbourhood of p, X is a complex manifold. Algebraically, this is equivalent to $\mathcal{O}_{X,p}^{an}$ being isomorphic to $K\langle x_1, \ldots, x_d \rangle$, as analytic algebra, which holds for convergent and formal power series (but not for regular functions in the sense of Definition A.6.1).

Many invariants of singularities, which are defined in the convergent, re-spectively formal, power series ring, can be computed in the localization $\mathbb{C}[x]_{\langle x \rangle}$ but this is not always the case, as the following shows.

Consider the singularity at $0 = (0,0)$ of the plane curve $y^2 - x^2(1 + x) = 0$ (cf. Figure A.22).

The picture shows that in a small neighbourhood of 0 (with respect to the Euclidean topology) the curve has two irreducible components, meeting transversally, but in the affine plane the curve is irreducible.

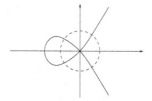

Fig. A.22. The variety $V\left(y^2 - x^2(1+x)\right)$.

To see this algebraically, let us consider $f = y^2 - x^2(1 + x)$ as an element of $\mathbb{C}\{x, y\}$. We have a decomposition

$$f = \left(y - x\sqrt{1 + x}\right)\left(y + x\sqrt{1 + x}\right)$$

with $y \pm x\sqrt{1 + x} \in \mathbb{C}\{x, y\}$, that is, f can be factorized in $\mathbb{C}\{x, y\}$ into two non–trivial factors. The zero–set of these factors corresponds to the two components of $V(f)$ in a small neighbourhood of 0. This is also the factorization in $\mathbb{C}[[x, y]]$, since the factorization is unique. However, f is irreducible in $\mathbb{C}[x, y]$, and even in $\mathbb{C}[x, y]_{\langle x,y\rangle}$. Otherwise, there would exist $g, h \in \mathbb{C}[x, y]_{\langle x,y\rangle}$ satisfying $f = (y + xg)(y + xh)$, hence $g = -h$ and $g^2 = 1 + x$. But, since $1 + x$ is defined everywhere, g^2 and, hence, g must be a polynomial, which is impossible since g^2 has degree 1. We can imagine the irreducibility of f in $\mathbb{C}[x, y]_{\langle x,y\rangle}$ also geometrically, since this corresponds to the irreducibility of the curve $\{f = 0\}$ in an arbitrary small neighbourhood of 0 with regard to the Zariski topology. Such a neighbourhood consists of the curve minus finitely many points different from 0. Since the curve is a complex curve and since a connected open subset of $\mathbb{C}$ minus finitely many points is irreducible (here, the above real picture is misleading), we can "see" the irreducibility of f in $\mathbb{C}[x, y]_{\langle x,y\rangle}$ (this argument can actually be turned into a proof).

The above example shows that the Zariski neighbourhoods are too big for certain purposes, or, algebraically, the *algebraic local rings* $K[x]_{\langle x\rangle}/I \cdot K[x]_{\langle x\rangle}$ are too small and we have to work with the *analytic local rings*, $K\langle x\rangle/IK\langle x\rangle$. The basic theorem for the study of $\mathbb{C}\{x\}$ and $K[[x]]$ is the Weierstraß preparation theorem (which does not hold for $K[x]_{\langle x\rangle}$) and which is treated in Section 6.2 for formal power series.

Computationally, however, we can, basically, treat only $K[x]$ and $K[x]_{\langle x\rangle}$ or factor rings of those in SINGULAR. In particular, we cannot put a polynomial into Weierstraß normal form, nor factorize it in $K[[x]]$, effectively in SINGULAR. We are only able to do so approximately, up to a given order (not to mention the problem of coding an infinite, but not algebraic, power series).

Nevertheless, it has turned out that many invariants of singularities can be computed in $K[x]_{\langle x\rangle}$, an algebraic reason for this being the fact that $K[x]_{\langle x\rangle} \subset K[[x]]$ is faithfully flat (Corollary 7.4.6).

We illustrate this with a few examples, the simplest is stated in the following lemma:

Lemma A.9.2. *Let $\mathcal{O}_{X,p}$ be the algebraic local ring of a variety at $p \in X$ and let $I \subset \mathcal{O}_{X,p}$ be an ideal such that $\dim_K(\mathcal{O}_{X,p}/I) < \infty$. Then, as local k-algebras, $\mathcal{O}_{X,p}/I \cong \mathcal{O}_{X,p}^{an}/I\mathcal{O}_{X,p}^{an}$. In particular, both vector spaces have the same dimension and a common basis represented by monomials.*

To see this, we may assume $p = 0$. Then $\langle x_1, \ldots, x_n \rangle^s \subset I\mathcal{O}_{X,0}$ for some s, by assumption, and $\mathcal{O}_{X,0}/\langle x_1, \ldots, x_n \rangle^s = \mathcal{O}_{X,0}^{an}/\langle x_1, \ldots, x_n \rangle^s$. The same result holds for submodules $I \subset \mathcal{O}_{X,0}^r$ of finite K-codimension. However, if $I \subset \mathcal{O}_{X,0}$ is generated by polynomials $f_1, \ldots, f_k$, then there may be no s such that $\langle x_1, \ldots, x_n \rangle^s \subset \langle f_1, \ldots, f_k \rangle \subset K[x_1, \ldots, x_n]$ (this inclusion holds if and only if $V(f_1, \ldots, f_k) = \{0\}$) and, therefore,

$$\dim_K \mathcal{O}_{X,0}/I \neq \dim_K K[x_1, \ldots, x_n]/\langle f_1, \ldots, f_k \rangle,$$

in general.

Important examples are given by the Milnor number and by the Tjurina number of an isolated hypersurface singularity.

Definition A.9.3.

(1) We say that $f \in K[x]$, $x = (x_1, \ldots, x_n)$, has an *isolated critical point* at p if p is an isolated point of $V(\partial f/\partial x_1, \ldots, \partial f/\partial x_n)$. Similarly, we say that p is an *isolated singularity* of f, or of the hypersurface $V(f) \subset \mathbb{A}_K^n$, if p is an isolated point of $V(f, \partial f/\partial x_1, \ldots, \partial f/\partial x_n)$.[7]

(2) We call the number

$$\mu(f, p) := \dim_K \left(K\langle x_1 - p_1, \ldots, x_n - p_n \rangle \Big/ \left\langle \frac{\partial f}{\partial x_1}, \ldots, \frac{\partial f}{\partial x_n} \right\rangle \right)$$

the *Milnor number*, and

$$\tau(f, p) := \dim_K \left(K\langle x_1 - p_1, \ldots, x_n - p_n \rangle \Big/ \left\langle f, \frac{\partial f}{\partial x_1}, \ldots, \frac{\partial f}{\partial x_n} \right\rangle \right)$$

the *Tjurina number of f at p*. We write $\mu(f)$ and $\tau(f)$ if $p = 0$.

If p is an isolated critical point of f, then $\mathfrak{m}_p := \langle x_1 - p_1, \ldots, x_n - p_n \rangle$ is a minimal associated prime of $\langle \partial f/\partial x_1, \ldots, \partial f/\partial x_n \rangle$ by the Hilbert Nullstellensatz and, therefore, $\mathfrak{m}_p^s \subset \langle \partial f/\partial x_1, \ldots, \partial f/\partial x_n \rangle \cdot K[x]_{\mathfrak{m}_p}$ for some s. It follows that the Milnor number $\mu(f, p)$ is finite and, similarly, if p is an isolated singularity of $V(f)$, then the Tjurina number $\tau(f, p)$ is finite, too.

By Lemma A.9.2 we can compute the Milnor number $\mu(f)$, resp. the Tjurina number $\tau(f)$, by computing a standard basis of $\langle \partial f/\partial x_1, \ldots, \partial f/\partial x_n \rangle$, respectively $\langle f, \partial f/\partial x_1, \ldots, \partial f/\partial x_n \rangle$ with respect to a local monomial ordering and then apply the SINGULAR command vdim.

---

[7] Examples of isolated and non-isolated hypersurface singularities are shown in Figure A.23.

Isolated Singularities

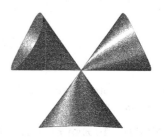

$$A_1 : x^2 - y^2 + z^2 = 0 \qquad\qquad D_4 : z^3 - zx^2 + y^2 = 0$$

Non–isolated singularities

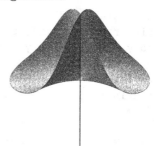

$$A_\infty : x^2 - y^2 = 0 \qquad\qquad D_\infty : y^2 - zx^2 = 0$$

Fig. A.23. Isolated and non–isolated singularities.

We can use the interplay between local and global orderings to check the existence of critical points and of singularities outside 0. For this we use the (easy) facts:

- $\mu(f, p) = 0$ if and only if p is a non–critical point of f, that is,

$$p \notin V\left(\frac{\partial f}{\partial x_1}, \dots, \frac{\partial f}{\partial x_n}\right) =: \mathrm{Crit}(f),$$

- $\tau(f, p) = 0$ if and only if p is a non–singular point point of $V(f)$, that is,

$$p \notin V\left(f, \frac{\partial f}{\partial x_1}, \dots, \frac{\partial f}{\partial x_n}\right) =: \mathrm{Sing}(f).$$

Note that we have the following equalities for the *total Milnor number*, respectively the *total Tjurina number*, of f:

$$\dim_K \left(K[x_1, \ldots, x_n] \Big/ \left\langle \frac{\partial f}{\partial x_1}, \ldots, \frac{\partial f}{\partial x_n} \right\rangle \right) = \sum_{p \in \mathrm{Crit}(f)} \mu(f, p),$$

$$\dim_K \left(K[x_1, \ldots, x_n] \Big/ \left\langle f, \frac{\partial f}{\partial x_1}, \ldots, \frac{\partial f}{\partial x_n} \right\rangle \right) = \sum_{p \in \mathrm{Sing}(f)} \tau(f, p),$$

Moreover, note that a singular point of f is a critical point of f which lies on the hypersurface $\{f = 0\}$.

SINGULAR Example A.9.4 (Milnor and Tjurina number).
We compute the local and the total Milnor, respectively Tjurina, number and check in this way, whether there are further critical, respectively singular, points outside 0.

```
LIB "sing.lib";
ring r = 0,(x,y,z),ds;          //local ring
poly f = x7+y7+(x-y)^2*x2y2+z2;
milnor(f);
//-> 28                         //Milnor number at 0
tjurina(f);
//-> 24                         //Tjurina number at 0


ring R = 0,(x,y,z),dp;          //affine ring
poly f = x7+y7+(x-y)^2*x2y2+z2;
milnor(f);
//-> 36                         //total Milnor number
tjurina(f);
//-> 24                         //total Tjurina number
```

We see that the difference between the total and the local Milnor number is 8; hence, f has eight critical points (counted with their respective Milnor numbers) outside 0.

On the other hand, since the total Tjurina number coincides with the local Tjurina number, $V(f) \subset \mathbb{A}^3$ has no other singular points except 0.

The most simple singularities of a hypersurface are (ordinary) nodes: a critical point $p = (p_1, \ldots, p_n)$ of $f \in K[x_1, \ldots, x_n]$ is called a *node* or an A_1-*singularity* if there exist analytic coordinates $y_1, \ldots, y_n$ centred at p such that $f(y_1, \ldots, y_n) = y_1^2 + \cdots + y_n^2$ (that is, there exist $\varphi_i \in K[[y_1, \ldots, y_n]]$, $\varphi_i(0) = 0$, $i = 1, \ldots, n$, such that substituting x_i by $\varphi_i + p_i$, $i = 1, \ldots, n$, induces an isomorphism $\varphi : K[[x_1 - p_1, \ldots, x_n - p_n]] \to K[[y_1, \ldots, y_n]]$ with $\varphi(f) = y_1^2 + \cdots + y_n^2$). In this definition, we assume that $\mathrm{char}(K) \neq 2$.

By the *Morse lemma* (cf. [164], [108], [128]), a critical point p of f is a node if and only if the Hessian at p is nondegenerate, that is, if and only if

$$\det \left(\frac{\partial^2 f}{\partial x_i \partial x_j}(p) \right) \neq 0.$$

Moreover, as is not difficult to see, this is also equivalent to $\mu(f,p) = 1$ and also to $\tau(f - f(p),p) = 1$ (in characteristic 0). Hence, we can count the number of nodes of a function as in the following example:

SINGULAR Example A.9.5 (counting nodes).
Consider f from Example A.9.4 and use the rings defined there. We compute the ideal nn of critical points which are not nodes.

```
setring R;
ideal j = jacob(f);              //ideal of critical locus
poly h = det(jacob(j));          //det of Hessian of f
ideal nn = j,h;                  //ideal of non-nodes
vdim(std(nn));
//-> 27
setring r;                       //go back to local ring
ideal nn = jacob(f),det(jacob(jacob(f)));
vdim(std(nn));
//-> 27
```

The computation in the affine ring R shows that there are, perhaps several, non–nodes. The ideal of non–nodes of f is generated by $\partial f/\partial x_1, \ldots, \partial f/\partial x_n$ and by the determinant of the Hessian $\det\left(\partial^2 f/(\partial x_i \partial x_j)\right)$. The SINGULAR command vdim(std(nn)) counts the number of non–nodes, each non–node p being counted with the multiplicity

$$\dim_K \left(K\langle x_1 - p_1, \ldots, x_n - p_n \rangle \middle/ \left\langle \frac{\partial f}{\partial x_1}, \ldots, \frac{\partial f}{\partial x_n}, \det\left(\frac{\partial^2 f}{\partial x_i \partial x_j}\right) \right\rangle \right)$$

(which is equal to or less than the local Milnor number, since the determinant of the Hessian reduces the multiplicity). The computation in the local ring r shows, however, that 0 is the only non–node, since the multiplicity of the non–nodes is 27 in both cases. Hence, all the critical points in $\mathbb{A}^3 \setminus \{0\}$ of f are nodes and there are $36 - 28 = 8$ of them (the singularity at 0 is, of course, not a node, since it has Milnor number 29).

As the analytic local ring of a singularity is a quotient ring of a power series ring, singularity theory deals with power series rather than with polynomials. It is, however, a fundamental fact that isolated hypersurface singularities are *finitely determined* by a sufficiently high *jet* (power series expansion up to a sufficiently high order). That is, if $f \in K\langle x_1, \ldots, x_n \rangle$ has an isolated singularity at 0, then there exists a $k > 0$ such that any $g \in K\langle x_1, \ldots, x_n \rangle$, having the same k–jet as f, is *right equivalent* to g (that is, there exists an automorphism φ of $K\langle x_1, \ldots, x_n \rangle$ such that $\varphi(f) = g$). We say that f is k–*determined* in this situation, and the minimal such k is called the *determinacy* of f. Hence, if f has an isolated singularity, then we can replace it by its k–jet (which is a polynomial), without changing the singularity.

The finite determinacy has important consequences, theoretical as well as computational ones. A typical application is the construction of analytic coordinate transformations $\varphi = \sum_{\nu \geq 1} \varphi_\nu$ of $K\langle x_1, \ldots, x_n \rangle$, where φ_ν is homogeneous of degree ν and which are constructed degree by degree. Usually, this process will not stop, but if f is k–determined, then we know that $\left(\sum_{\nu=1,\ldots,k} \varphi_\nu \right)(f)$ is right equivalent to f.

As a general estimate, we have that $f \in K\langle x_1, \ldots, x_n \rangle$ is $\big(\mu(f) + 1\big)$–determined, provided $\mathrm{char}(K) = 0$ (cf. [128], [108]).

Using this estimate, the Morse lemma cited above is an easy consequence: it follows that for $f \in \langle x_1, \ldots, x_n \rangle^2$ and $\mu(f) = 1$, f is right equivalent to its 2–jet and, hence, a node.

The estimate $\mu(f) + 1$ for the determinacy is, in general, not very good. Instead, we have a much better estimate given by the following theorem (cf. [128], [108]):

Theorem A.9.6. *Let $f \in \mathfrak{m} = \langle x_1, \ldots, x_n \rangle \subset K\langle x_1, \ldots, x_n \rangle$, where K is a field of characteristic 0. Then*

$$\mathfrak{m}^{k+1} \subset \mathfrak{m}^2 \cdot \left\langle \frac{\partial f}{\partial x_1}, \ldots, \frac{\partial f}{\partial x_n} \right\rangle$$

implies that f is k–determined.

In particular, if $\mathfrak{m}^k \subset \mathfrak{m} \cdot \langle \partial f/\partial x_1, \ldots, \partial f/\partial x_n \rangle$, then f is k–determined. Of course, we can compute, by the method of Section 1.8.1, the minimal k satisfying $\mathfrak{m}^{k+1} \subset I := \mathfrak{m}^2 \cdot \langle \partial f/\partial x_1, \ldots, \partial f/\partial x_n \rangle$, by computing a standard basis G of I and then a normal form for $\mathfrak{m}^i$ with respect to G, for increasing i.

However, this can be avoided due to a very useful feature of SINGULAR, the so–called *highest corner* (cf. Definition 1.7.11). The *highest corner* of an ideal I is the minimal monomial m (with respect to the monomial ordering) such that $m \notin I$. In case of a local degree ordering the highest corner exists if and only if $\dim_K K\langle x_1, \ldots, x_n \rangle/I$ is finite (Lemma 1.7.14). If we compute a standard basis of I, then SINGULAR computes automatically the highest corner, if it exists. The command `highcorner(I);` returns it. Hence, if we compute a standard basis of I with respect to a local degree ordering and if the monomial m is the highest corner of I, then $\mathfrak{m}^{\deg(m)+1} \subset I$. We obtain

Corollary A.9.7. *Let $\mathrm{char}(K) = 0$, and let $f \in \mathfrak{m} \subset K\langle x_1, \ldots, x_n \rangle$ have an isolated singularity. Moreover, let $m_i \in \mathrm{Mon}(x_1, \ldots, x_n)$ be the highest corner of $\mathfrak{m}^i \cdot \langle \partial f/\partial x_1, \ldots, \partial f/\partial x_n \rangle$, $i = 0, 1, 2$ with respect to a local degree ordering. Then f is $\deg(m_i) + 2 - i$ determined.*

SINGULAR Example A.9.8 (estimating the determinacy).
We compute the highest corner of $\langle x, y \rangle^i \cdot \langle \partial f/\partial x, \partial f/\partial y \rangle$, $i = 0, 1, 2$, for the E_7–singularity $f = x^3 + xy^3 \in \mathbb{C}\{x, y\}$ and estimate the determinacy.

```
ring r = 0,(x,y),ds;
poly f = x3+xy3;
ideal j = jacob(f);
ideal j1 = maxideal(1)*j;
ideal j2 = maxideal(2)*j;
j  = std(j);
deg(highcorner(j));
//-> 4
j1 = std(j1);
deg(highcorner(j1));
//-> 4
j2 = std(j2);
deg(highcorner(j2));
//-> 5
```

Corollary A.9.7 implies, using the ideal j, that f is 6–determined. If we use the ideals j1, respectively j2, it follows that f is 5–determined. One can show that E_7 is even 4–determined, hence Corollary A.9.7 provides only an estimate for the determinacy (which, nevertheless, is quite good in general).

We should like to finish this section with two further applications of standard bases in local rings: classification of singularities and deformation theory. There are many more, some of them are in the SINGULAR libraries and we refer to the examples given there.

In a tremendous work, V.I. Arnold started, in the late sixties, the classification of hypersurface singularities up to right equivalence. His work culminated in impressive lists of normal forms of singularities and, moreover, in a determinator for singularities which allows the determination of the normal form for a given power series ([5]). This work of Arnold has found numerous applications in various areas of mathematics, including singularity theory, algebraic geometry, differential geometry, differential equations, Lie group theory and theoretical physics. The work of Arnold was continued by many others, we just mention C.T.C. Wall [224].

Most prominent is the list of *ADE* or *simple* or *Kleinian singularities*, which have appeared in surprisingly different areas of mathematics, and still today, new connections of these singularities to other areas are being discovered (see, for example, [62], [101]). Here is the list of ADE–singularities for algebraically closed fields of characteristic 0 (for the classification in positive characteristic see [52]). The names come from their relation to the simple Lie groups of type A, D and E.

$$A_k : x_1^{k+1} + x_2^2 + x_3^2 + \cdots + x_n^2 , \quad k \geq 1 ,$$
$$D_k : x_1(x_1^{k-2} + x_2^2) + x_3^2 + \cdots + x_n^2 , \ k \geq 4 ,$$
$$E_6 : x_1^4 + x_2^3 + x_3^2 + \cdots + x_n^2 ,$$
$$E_7 : x_2(x_1^3 + x_2^2) + x_3^2 + \cdots + x_n^2 ,$$
$$E_8 : x_1^5 + x_2^3 + x_3^2 + \cdots + x_n^2 .$$

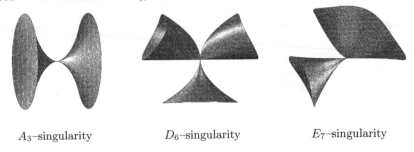

A_3–singularity D_6–singularity E_7–singularity

Fig. A.24. Some two–dimensional simple singularities.

Arnold introduced the concept of "modality", related to Riemann's idea of moduli, into singularity theory and classified all singularities of modality ≤ 2 (and also of Milnor number ≤ 16), and many more. The ADE–singularities are just the singularities of modality 0. Singularities of modality 1 are the three *parabolic singularities*:

$$\widetilde{E}_6 = P_8 = T_{333} \ : \ x^3 + y^3 + z^3 + axyz, \ a^3 + 27 \neq 0 \,,$$
$$\widetilde{E}_7 = X_9 = T_{244} \ : \ x^4 + y^4 + ax^2y^2, \qquad a^2 \neq 4 \,,$$
$$\widetilde{E}_8 = J_{10} = T_{236} \ : \ x^3 + y^5 + ax^2y^2, \qquad 4a^3 + 27 \neq 0 \,,$$

the 3–indexed series of *hyperbolic singularities*

$$T_{pqr} : x^p + y^q + z^r + axyz, \quad a \neq 0, \quad \frac{1}{p} + \frac{1}{q} + \frac{1}{r} < 1 \,,$$

and 14 exceptional families, cf. [5].

The proof of Arnold for his determinator is, to a great part, constructive and has been partly implemented in SINGULAR, cf. [140]. Although the whole theory and the proofs deal with power series, everything can be reduced to polynomial computations, since we deal with isolated singularities, which are finitely determined, as explained above.

An important initial step in Arnold's classification is the *generalized Morse lemma*, or *splitting lemma*, which says that

$$f \circ \varphi(x_1, \ldots, x_n) = x_1^2 + \cdots + x_r^2 + g(x_{r+1}, \ldots, x_n) \,,$$

for some analytic coordinate change φ and some power series $g \in \mathfrak{m}^3$, if the rank of the Hessian matrix of f at 0 is r.

The determinacy allows the computation of φ up to sufficiently high order and the polynomial g. This has been implemented in SINGULAR and is a cornerstone in classifying hypersurface singularities.

In the following example we use SINGULAR to obtain the singularity $T_{5,7,11}$ from a database A_L ("Arnold's list"), make some coordinate change and determine then the normal form of the complicated polynomial after coordinate change.

SINGULAR Example A.9.9 (classification of singularities).

```
LIB "classify.lib";
ring r = 0,(x,y,z),ds;
poly f = A_L("T[5,7,11]"); f;
//-> xyz+x5+y7+z11

map phi = r, x+z,y-y2,z-x;
poly g = phi(f); g;
//-> -x2y+yz2+x2y2-y2z2+x5+5x4z+10x3z2+10x2z3+5xz4+z5+y7
//-> -7y8+21y9-35y10-x11+35y11+11x10z-55x9z2+165x8z3
//-> -330x7z4+462x6z5-462x5z6+330x4z7-165x3z8+55x2z9
//-> -11xz10+z11-21y12+7y13-y14

quickclass(g);
//-> Singularity R-equivalent to :   T[k,r,s]=T[5,7,11]
//-> normal form : xyz+x5+y7+z11
//-> xyz+x5+y7+z11
```

Beyond classification by normal forms, the construction of moduli spaces for singularities, for varieties or for vector bundles is a pretentious goal, theoretically as well as computational. First steps towards this goal for singularities were undertaken in [19] and [76].

Let us finish with a few remarks about *deformation theory*. Consider a singularity $(X, 0)$ given by power series $f_1, \ldots, f_k \in K\langle x_1, \ldots, x_n \rangle$. The idea of deformation theory is to perturb the defining functions in a controlled way, that is, we consider power series $F_1(t, x), \ldots, F_k(t, x)$ with $F_i(0, x) = f_i(x)$, where $t \in S$ may be considered as a small parameter of a parameter space S (containing 0).

For $t \in S$, the power series $f_{i,t}(x) = F_i(t, x)$ define a singularity X_t, which is a perturbation of $X = X_0$ for $t \neq 0$ close to 0. It may be hoped that X_t is simpler than X_0, but still contains enough information about X_0. For this hope to be fulfilled, it is, however, necessary to restrict the possible perturbations of the equations to *flat* perturbations, which are called *deformations* (cf. Chapter 7 for the notion of flatness).

By a theorem of Grauert [98], for any isolated singularity $(X, 0)$ there exists a *semi–universal* (or *miniversal*) *deformation*, which contains essentially all information about all deformations of $(X, 0)$.

For an isolated hypersurface singularity $f(x_1, \ldots, x_n)$, the semi–universal deformation is given by

$$F(t, x) = f(x) + \sum_{j=1}^{\tau} t_j g_j(x) ,$$

where $1 =: g_1, g_2, \ldots, g_\tau$ represent a K–basis of the Tjurina algebra

$$K\langle x_1, \ldots, x_n \rangle \Big/ \Big\langle f, \frac{\partial f}{\partial x_1}, \ldots, \frac{\partial f}{\partial x_n} \Big\rangle \,,$$

τ being the Tjurina number.

To compute $g_1, \ldots, g_\tau$ we only need to compute a standard basis of the ideal $\langle f, \partial f/\partial x_1, \ldots, \partial f/\partial x_n \rangle$ with respect to a local ordering and then compute a basis of $K[x]$ modulo the leading monomials of the standard basis. For complete intersections, that is, singularities whose analytic local rings are complete intersection rings, we have similar formulas.

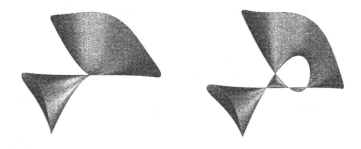

Fig. A.25. Deformation of an E_7–singularity in four A_1–singularities.

For non–hypersurface singularities, the semi–universal deformation is much more complicated and, up to now, no finite algorithm is known in general. However, there exists an algorithm to compute this deformation up to arbitrary high order (cf. [145], [158]), which is implemented in SINGULAR.

As an example, we calculate the semi–universal deformation of the normal surface singularity, being the cone over the rational normal curve C of degree 4, parametrized by $t \mapsto (t, t^2, t^3, t^4)$. Homogeneous equations for the cone over C are given by the 2×2–minors of the matrix:

$$m := \begin{pmatrix} x & y & z & u \\ y & z & u & v \end{pmatrix} \in \mathrm{Mat}(2 \times 4, K[x, y, z, u, v])\,.$$

SINGULAR Example A.9.10 (deformation of singularities).

```
LIB "deform.lib";
ring r = 0,(x,y,z,u,v),ds;
matrix m[2][4] = x,y,z,u,y,z,u,v;
ideal f = minor(m,2);  //ideal of 2x2 minors of m
versal(f);             //computes semi-universal deformation
setring Px;            //data are contained in the ring Px
```

```
Fs;
//-> Fs[1,1]=-u2+zv+Bu+Dv
//-> Fs[1,2]=-zu+yv-Au+Du
//-> Fs[1,3]=-yu+xv+Cu+Dz
//-> Fs[1,4]=z2-yu+Az+By
//-> Fs[1,5]=yz-xu+Bx-Cz
//-> Fs[1,6]=-y2+xz+Ax+Cy
Js;
//-> Js[1,1]=BD
//-> Js[1,2]=-AD+D2
//-> Js[1,3]=-CD
```

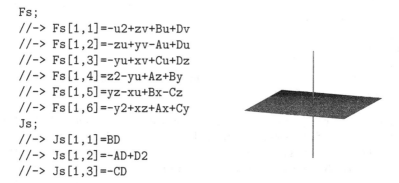

The ideal $J_s = \langle BD, AD - D^2, CD \rangle \subset K[A, B, C, D]$ defines the required base space, which consists of a 3–dimensional component $(D = 0)$ and a transversal 1–dimensional component $(B = C = A - D = 0)$. This was the first example, found by Pinkham, of a base space of a normal surface singularity having several components of different dimensions.

The full versal deformation is given by the canonical map (Fs and Js as above) $K[[A, B, C, D]]/J_s \to K[[A, B, C, D, x, y, z, u, v]]/\langle J_s, F_s \rangle$.

Although, in general, the equations for the versal deformation are formal power series, in many cases of interest (as in the example above) the algorithm terminates and the resulting ideals are polynomial.

Finally, let us compute the discriminant of the semi-universal deformation of the hypersurface singularity A_3.

Fig. A.26. *Swallow tail*, the discriminant of the semi-universal deformation of the hypersurface singularity A_3.

```
ring s = 0,x,ds;
poly f = x4;
versal(f);
setring Px;
Fs;                                //the semi-universal deformation
//-> Fs[1,1]=x4+Ax2+Bx+C
```

The *discriminant* of F_s is the ideal describing the locus of points (A, B, C) in the base space of the semi–universal deformation such that the polynomial $x^4 + Ax^2 + Bx + C$ has multiple roots. We compute it by projecting the singular locus of the semi–universal deformation to the parameter space.

```
ideal sing = Fs,diff(Fs,x);        //the singular locus
eliminate(sing,x);                 //the discriminant
//-> _[1]=256C3-27B4+144AB2C-128A2C2-4A3B2+16A4C
```

Let us plot the discriminant.

```
LIB"surf.lib";
plot(256C3-27B4+144AB2C-128A2C2-4A3B2+16A4C);
```

B. Polynomial Factorization

In this section we shall give a short introduction to univariate and multivariate polynomial factorization. The aim is to present the basic algorithms for polynomial factorization over finite fields $\mathbb{F}_q$, the rational numbers $\mathbb{Q}$, and algebraic extensions of these fields. Finally we explain absolute factorization, i.e. factorization over the algebraic closure of a field.

Factorization of polynomials from $K[x_1, \ldots, x_n]$ is, besides Gröbner basis computation, the most important tool for computational commutative algebra and algebraic geometry. It is used, for example, in any primary decomposition algorithm. In contrast to Gröbner basis methods, algorithms for polynomial factorization depend in an essential way on the ground field K. On the other hand, we face again the problem of intermediate coefficient swell when working over non finite fields. Therefore, factorization over $\mathbb{Q}$ is reduced to factorization over finite fields, together with Hensel lifting. This works quite well since checking of the result is easy (in contrast to Gröbner basis computation).

Multivariate factorization over K is reduced to univariate factorization by substituting all variables except one by a randomly chosen constant and going back to several variables again by Hensel lifting.

In the last section we explain absolute factorization of multivariate polynomials f from $K[x_1, \ldots, x_n]$ where K is a field of characteristic zero. That is, we compute a suitable algebraic field extension L of K and the factorization of f into irreducible factors in $L[x_1, \ldots, x_n]$ which is the factorization of f in $\overline{K}[x_1, \ldots, x_n]$, where $\overline{K}$ is an algebraic closure of K.

We start with the univariate case and with a partial factorization (square-free factorization and distinct degree factorization) which uses only gcd–computations and is therefore fast and hence indispensible as preprocessing.

We assume the reader to be familiar with the elementary notions and results of factorization in unique factorization domains. In particular, every non–constant polynomial $f \in K[x_1, \ldots, x_n]$ factorizes as $f = f_1 \cdot \ldots \cdot f_r$ with $f_i \in K[x_1, \ldots, x_n] \setminus K$ irreducible. This factorization is unique up to multiplication with non–zero constants and permutation of the irreducible factors f_i.

B.1 Squarefree Factorization

Let K be a field of characteristic p. In this chapter we will explain how to decompose a univariate polynomial $g \in K[x]$ as a product $g = \prod_{i=1}^{k} g_{(i)}^{i}$ of powers of pairwise coprime[1] squarefree factors $g_{(1)}, \ldots, g_{(k)}$.

Definition B.1.1. (1) $g \in K[x]$ is called *squarefree* if g is not constant and if it has no non–constant multiple factor, that is, each irreducible factor of g appears with multiplicity 1.

(2) Let $g \in K[x]$, $g = \prod_{i=1}^{k} g_{(i)}^{i}$ is called the *squarefree factorization* of g if $g_{(1)}, \ldots, g_{(k)}$ are squarefree, and those $g_{(i)}$, which are non–constant are pairwise coprime.

It follows from the existence and uniqueness of the factorization of f into irreducible factors that the squarefree factorization exists and the squarefree factors are unique up to multiplication by a non–zero constant.

Example B.1.2. Let $g = x^2(x+1)^2(x+3)^4(x^2+1)^5 \in \mathbb{Q}[x]$ then $g_{(1)} = g_{(3)} = 1$ and $g_{(2)} = x(x+1)$, $g_{(4)} = x+3$, $g_{(5)} = x^2+1$.

As the case of char $K = 0$ is an easy exercise using some of the same ideas as in the following proposition we concentrate from this point on on the case of fields of positive characteristic.

Proposition B.1.3. *Let* $f \in \mathbb{F}_q[x]$ *be non–constant with* $q = p^r$ *and* p *prime. Then* f *is squarefree if and only if* $f' \neq 0$ *and* $\gcd(f, f') = 1$.

Proof. If $f' = 0$ then $f = \sum_{j=0}^{s} a_j x^{pj}$. Since p–th roots exist in $\mathbb{F}_q$, i.e. $a_j = b_j^p$ for suitable $b_j \in \mathbb{F}_q$, this implies $f = \sum b_j^p x^{pj} = (\sum b_j x^j)^p$.

Let $f' \neq 0$ and h an irreducible polynomial with $h | \gcd(f, f')$. Then $f = h \cdot g$ and $h | (f' = h'g + hg')$. If $h' = 0$ then h is a p–th power and hence f is not squarefree. Otherwise $h | (h'g)$ and, since h is irreducible, $h|g$. This implies $h^2 | f$ and again f is not squarefree. Conversely, if $f = h^2 \cdot g$ then $h|f'$ and $\gcd(f, f') \neq 1$.   □

The following lemma is the basis for the algorithm to compute the square-free decomposition.

Lemma B.1.4. *Let* $f \in \mathbb{F}_q[x]$ *and* $f' \neq 0$. *If* $f = \prod_{i=1}^{k} f_{(i)}^{i}$ *is the squarefree decomposition then*

$$g := \gcd(f, f') = \prod_{p|j} f_{(j)}^{j} \prod_{i=2, p \nmid i}^{k} f_{(i)}^{i-1}.$$

---

[1] f and g are called *coprime* if $\gcd(f, g) = 1$.

Proof. The fact that $g|f$ is clear. The product rule implies

$$f' = g \cdot \sum_{i=2,p\nmid i}^{k} \left(i \cdot f'_{(i)} \cdot \prod_{j \neq i, p\nmid j} f_{(j)} \right).$$

It remains to prove that $\gcd(f, \sum_{i=2,p\nmid i}^{k} i f'_{(i)} \prod_{j \neq i, p\nmid j} f_{(j)}) = 1$.

Let h be irreducible such that $h | \frac{f}{g}$ and $h | \frac{f'}{g}$. There exist l such that $p \nmid l$ and $h | f_{(l)}$ and $h \nmid f_{(j)}$ for $j \neq l$ since the $f_{(j)}$ are pairwise coprime. This implies that $h | l \cdot f'_{(l)} \prod_{j \neq l, p\nmid j} f_{(j)}$ and therefore $h | f'_{(l)}$. But $f_{(l)}$ is squarefree and this is impossible. $\qquad\square$

Lemma B.1.5. *Let $f \in \mathbb{F}_q[x]$ and $f = \prod_{i=1}^{k} f_{(i)}^i$ be the squarefree decomposition. Let $g_0 = \gcd(f, f')$ and $w_0 = \frac{f}{g_0}$, and define inductively for $i \leq k$*

$$w_i = \gcd(g_{i-1}, w_{i-1}) \quad and \quad g_i = \frac{g_{i-1}}{w_i}.$$

Then

$$w_\nu = \prod_{p\nmid i, i \geq \nu+1} f_{(i)} \quad and \quad g_\nu = \prod_{p|i} f_{(i)}^i \cdot \prod_{p\nmid i, i \geq \nu+2} f_{(i)}^{i-\nu-1},$$

especially

$$w_k = 1, \qquad g_k = \prod_{p|i} f_{(i)}^i, \quad and \quad \frac{w_{i-1}}{w_i} = f_{(i)} \; if \; p \nmid i .$$

Proof. The proof is left as exercise. $\qquad\square$

By Lemma B.1.5 we get the squarefree factors $f_{(i)}$ of f by gcd–computations if $p \nmid i$. If $p | i$ then the factors $f_{(i)}$ are contained in g_k which is of the form $(\widetilde{g})^p$ for some $\widetilde{g}$ and we can apply the lemma to $\widetilde{g}$. This leads to the following algorithm.

Algorithm B.1.6 (SQUAREFREEDECO (f)).

Input: $f \in \mathbb{F}_q[x]$ monic
Output: a list of monic polynomials $f_1, \ldots, f_k \in \mathbb{F}_q[x]$

such that $f = \prod_{i=1}^{k} f_{(i)}^i$ is the squarefree decomposition

- $i = 0; L = \emptyset$;
- if $(f' = 0)$
  compute $g \in \mathbb{F}_q[x]$ such that $f = g^p$;
  $\{g_1, \ldots, g_s\} = $ SQUAREFREEDECO(g);
  return $\{(\underbrace{1, \ldots, 1}_{p-1}, g_1, \underbrace{1, \ldots, 1}_{p-1}, g_2, 1, \ldots, g_s)\}$

- $g_0 = \gcd(f, f'); \ \omega_0 = \frac{f}{g_0};$
- while $(w_i \neq 1)$
  $i = i + 1;$
  $w_i = \gcd(f_{i-1}, w_{i-1}); \ g_i = \frac{g_{i-1}}{w_i};$
  $L[i] = \frac{w_{i-1}}{w_i};$
- if $(g_i = 1)$
  return $L;$
- $\{h_1, \ldots, h_s\} = \text{SQUAREFREEDECO}(g_i)$
  For $j = 1$ to s
  $L[j] = L[j] \cdot h_j;$
  return $L;$

We give now an example to show how the algorithm works.

Example B.1.7. $f = x(x+1)^2(x^2+x+1)^3 \in \mathbb{F}_2[x]$

- $f' = (x+1)^4(x^2+x+1)^2$
- $g_0 = \gcd(f, f') = (x+1)^2(x^2+x+1)^2 \qquad w_0 = \frac{f}{g_0} = x(x^2+x+1)$
- $w_1 = \gcd(g_0, w_0) = x^2+x+1 \qquad g_1 = \frac{g_0}{w_1} = (x+1)^2(x^2+x+1)$
  $L[1] = \frac{w_0}{w_1} = x$
- $w_2 = \gcd(g_1, w_1) = x^2+x+1 \qquad g_2 = \frac{g_1}{w_2} = (x+1)^2$
  $L[2] = \frac{w_1}{w_2} = 1$
- $w_3 = \gcd(g_2, w_2) = 1 \qquad g_3 = \frac{g_2}{w_3} = (x+1)^2$
  $L[3] = \frac{w_2}{w_3} = x^2+x+1$
- $i = 3 \quad$ and $\quad g_3 = (x+1)^2$
  $\text{SQUAREFREEDECO}(g_3) = \{1, x+1\}$
  $L[1] = x \cdot 1 = x$
  $L[2] = 1 \cdot (x+1) = x+1$
  $L[3] = x^2+x+1$

B.2 Distinct Degree Factorization

In this section we describe an (optional) preprocessing step for the factorization. We show how to factor a given squarefree polynomial $f \in \mathbb{F}_q[x]$ into a product $f = f_{[1]} \cdot \ldots \cdot f_{[m]}$, where $f_{[i]}$ is the product of all irreducible factors of f of degree i.

Recall that two finite fields $\mathbb{F}_q$ and $\mathbb{F}_{q'}$ satisfy: $\mathbb{F}_q$ can be embedded as a subfield of $\mathbb{F}_{q'}$ iff, for some prime number p, $q = p^d, q' = p^{d'}$ and $d|d'$. Recall also that $\mathbb{F}_{q^d}$ is the splitting field of $x^{q^d} - x \in \mathbb{F}_q[x]$.

Lemma B.2.1. *The polynomial* $x^{q^d} - x$ *is the product of all monic, irreducible polynomials* $g \in \mathbb{F}_q[x]$ *whose degree is a divisor of* d.

Proof. It is easy to see that $x^{q^d} - x$ is squarefree; as $\frac{d}{dx}(x^{q^d} - x) = -1$ this is a consequence of B.1.3. Let $g \in \mathbb{F}_q[x]$ be monic and irreducible of degree e. We have to prove that $e|d$ if and only if $g|(x^{q^d} - x)$. Assuming $e|d$, then $\mathbb{F}_{q^e} = \mathbb{F}_q[x]/\langle g \rangle$ can be embedded as a subfield of $\mathbb{F}_{q^d}$. If $\alpha \in \mathbb{F}_{q^e}$ is a root of g then $\alpha^{q^d} - \alpha = 0$. This implies that $g|(x^{q^d} - x)$. Assume now that $g|(x^{q^d} - x)$. Then there is a subset $L \subset \mathbb{F}_{q^d}$ such that $g = \prod_{a \in L}(x - a)$. Let $\alpha \in \mathbb{F}_{q^e} = \mathbb{F}_q[x]/\langle g \rangle$ be a zero of g then $\mathbb{F}_q(\alpha) = \mathbb{F}_q[x]/\langle g \rangle$. $\mathbb{F}_q(\alpha) \subset \mathbb{F}_{q^d}$ implies $e|d$. $\square$

Corollary B.2.2. *Let $f \in \mathbb{F}_q[x]$ be squarefree. Then $\gcd(f, x^{q^d} - x) = \prod_{i|d} f_{[i]}$, where $f_{[i]}$ denotes the product of all irreducible factors of f of degree $i \geq 1$.*

Example B.2.3. In $\mathbb{F}_2[x]$ we have

$$x^{16} - x =$$
$$\underbrace{x(x + 1)}_{f_{[1]}} \underbrace{(x^2 + x + 1)(x^4 + x + 1)}_{f_{[2]}} \underbrace{(x^4 + x^3 + 1)(x^4 + x^3 + x^2 + x + 1)}_{f_{[4]}}.$$

We obtain the following algorithm to compute a squarefree polynomial into factors of distinct degrees.

Algorithm B.2.4 (DISTINCTDEGFAC (f)).

Input: $f \in \mathbb{F}_q[x]$ monic, squarefree

Output: $(f_{[d_1]}, d_1), \ldots, (f_{[d_2]}, d_2)$ such that $f = \prod_{i=1}^{s} f_{[d_i]}$ and
 $f_{[d_i]}$ is the product of all irreducible factors of f of degree d_i

- $d = 0;\ R = \emptyset$;
- $g_0 = f,\ h_0 = x$;
- while $(d \leq \frac{\deg(g_d)}{2} - 1)$
  $d = d + 1$;
  $h_d = h_{d-1}^q \mod g_{d-1}$;
  $f_{[d]} = \gcd(g_{d-1}, h_d - x)$;
  if $(f_{[d]} \neq 1)$
  $R = R \cup \{(f_{[d]}, d)\}$;
  $g_d = \frac{g_{d-1}}{f_{[d]}}$;
- If $(g_d \neq 1)$
  $R = R \cup \{(g_d, \deg(g_d))\}$;
- Return R;

Example B.2.5. Let $f = x^{15} - 1 \in \mathbb{F}_{11}[x]$

- $g_0 = f = x^{15} - 1, h_0 = x$
- $d = 1,\ h_1 = x^{11},\ f_{[1]} = \gcd(x^{15} - 1, x^{11} - x) = x^5 - 1$
  $g_1 = \frac{g_0}{f_{[1]}} = x^{10} + x^5 + 1,\ R = \{(x^5 - 1, 1)\}$

- $d = 2$ $h_2 = x^{121}$ mod $(x^{10} + x^5 + 1) = x$
  $f_{[2]} = \gcd(x^{10} + x^5 + 1, 0) = x^{10} + x^5 + 1$
  $g_2 = \frac{g_1}{f_{[2]}} = 1$, $R = \{(x^5 - 1, 1), (x^{10} + x^5 - 1, 2)\}$.

We obtain that $f = (x^5 - 1)(x^{10} + x^5 + 1)$ has 5 factors of degree 1 and 5 factors of degree 2.

B.3 The Algorithm of Berlekamp

In the previous section we described partial factorization methods, based on gcd–computation. In this section we describe the algorithm of Berlekamp and its improvement by Cantor and Zassenhaus which computes a complete factorization of polynomials in $\mathbb{F}_q[x]$.

Let $f \in \mathbb{F}_q[x]$ be a squarefree, non–constant, monic polynomial. Let $f = f_1 \cdot \ldots \cdot f_s$ be the irreducible decomposition of f, which we want to compute. The key tool in Berlekamp's algorithm is the Frobenius map $F : \mathbb{F}_q[x] \to \mathbb{F}_q[x]$, $F(h) = h^q$, which reduces the factorization problem to linear algebra. F is an $\mathbb{F}_q$–linear endomorphism of $\mathbb{F}_q[x]$. It induces an endomorphism

$$\phi : \mathbb{F}_q[x]/\langle f \rangle \to \mathbb{F}_q[x]/\langle f \rangle,$$

$\phi(\bar{h}) = \bar{h}^q - \bar{h}$, of the finite dimensional $\mathbb{F}_q$–vector space $\mathbb{F}_q[x]/\langle f \rangle$.

Lemma B.3.1. *With the notations above we have*

(1) $\dim_{\mathbb{F}_q}(\mathrm{Ker}(\phi)) = s$.
(2) Let $h \in \mathbb{F}_q[x]$ represent $\bar{h} \in \mathrm{Ker}(\phi)$ then

$$f = \prod_{a \in \mathbb{F}_q} \gcd(f, h - a) .$$

(3) Let $1 = h_1, h_2, \ldots, h_s \in \mathbb{F}_q[x]$ represent a basis of $\mathrm{Ker}(\phi)$. Then, for all $1 \le i < j \le s$, there exist $k \in \{2, \ldots, s\}$ and $a, b \in \mathbb{F}_q, a \ne b$, such that

$$f_i | (h_k - a) \text{ and } f_j | (h_k - b) .$$

In particular $\gcd(f, h_k - a)$ is a proper factor of f.

Proof. We use the Chinese remainder theorem (Exercise 1.3.13) and obtain an isomorphism

$$\chi : \mathbb{F}_q[x]/\langle f \rangle \xrightarrow{\sim} \mathbb{F}_q[x]/\langle f_1 \rangle \oplus \ldots \oplus \mathbb{F}_q[x]/\langle f_s \rangle,$$
$$\chi(h \mod \langle f \rangle) = (h \mod \langle f_1 \rangle, \ldots, h \mod \langle f_s \rangle).$$

If $d_i = \deg(f_i)$ then $\mathbb{F}_q[x]/\langle f_i \rangle \cong \mathbb{F}_{q^{d_i}}$.

Now $\bar{h} = h \mod \langle f \rangle \in \text{Ker}(\phi)$ if and only if $\bar{h} = \bar{h}^q$. This is the case if and only if $(h \mod \langle f_i \rangle)^q = h \mod \langle f_i \rangle$, i.e. $h^q - h \in \langle f_i \rangle$, for $1 \le i \le s$. But in $\mathbb{F}_{q^{d_i}}$ we have $\alpha^q = \alpha$ if and only if $\alpha \in \mathbb{F}_q$. This implies that

$$\text{Ker}(\phi) = \chi^{-1}(\mathbb{F}_q \oplus \ldots \oplus \mathbb{F}_q)$$

which is an s–dimensional $\mathbb{F}_q$–vector space. This proves (1).

To prove (2) let $h \in \mathbb{F}_q[x]$ represent $\bar{h} \in \text{Ker}(\phi)$ and $\chi(\bar{h}) = (a_1, \ldots, a_s)$. Then, as discussed in the proof of (1), we have $a_i \in \mathbb{F}_q$. This implies $f_i | (h - a_i)$ hence $f_i | \prod_{a \in \mathbb{F}_q} (h - a)$. Since the $f_1, \ldots, f_s$ are pairwise coprime we obtain that $f | \prod_{a \in \mathbb{F}_q} (h - a)$. On the other hand the $h - a, a \in \mathbb{F}_q$, are pairwise coprime. Hence, each irreducible factor f_i occurs in precisely one of the $h - a$. This implies

$$f = \prod_{a \in \mathbb{F}_q} \gcd(f, h - a).$$

To prove (3), consider f_i, f_j for $i < j$. Since $\{\bar{h}_1, \ldots, \bar{h}_s\}$ is a basis of $\text{Ker}(\phi) = \chi^{-1}(\mathbb{F}_q \oplus \ldots \oplus \mathbb{F}_q)$ there exist $k \ge 2$ such that $\chi(h_k) = (a_1, \ldots, a_s)$ and $a_i \ne a_j$. But $f_i | (h_k - a_i)$ and $f_j | (h_k - a_j)$. This proves (3) with $a = a_i$ and $b = a_j$. □

The lemma is the basis of Berlekamp's algorithm.

Algorithm B.3.2 (BERLEKAMP(f)).

Input: $f \in \mathbb{F}_q[x]$ squarefree, non-constant and monic
Output: $f_1, \ldots, f_s$ the irreducible factors of f

- Compute representatives $1, h_2, \ldots, h_s \in \mathbb{F}_q[x]$ of a basis of $\text{Ker}(\phi)$
- $R = \{f\}$; $I = \emptyset$;
- while ($\#R + \#I < s$)
  choose $g \in R$;
  $h = 1; i = 1$;
  while (($i < s$) and (($h = 1$ or ($h = g$)))
  $i = i + 1$;
  for $a \in \mathbb{F}_q$ do
  $h = \gcd(g, h_i - a)$;
  if ($h \ne g$)
  $I = I \cup \{h\}$;
  $R = (R \setminus \{g\}) \cup \{\frac{g}{h}\}$;
  else
  $R = R \setminus \{g\}$;
  $I = I \cup \{g\}$;
- return $R \cup I$;

Let us consider the following example.

Example B.3.3. Let $f = x^8 + x^6 + x^4 + x^3 + 1 \in \mathbb{F}_2[x]$ then $\{1, x, \ldots, x^7\}$ is a basis of $\mathbb{F}_2[x]/\langle f \rangle$. With respect to this basis ϕ has the following matrix.

$$\begin{pmatrix} 0 & 0 & 0 & 0 & 1 & 1 & 0 & 1 \\ 0 & 1 & 0 & 0 & 0 & 0 & 0 & 1 \\ 0 & 1 & 1 & 0 & 0 & 1 & 1 & 0 \\ 0 & 0 & 0 & 1 & 1 & 1 & 0 & 1 \\ 0 & 0 & 1 & 0 & 0 & 1 & 1 & 1 \\ 0 & 0 & 0 & 0 & 0 & 0 & 1 & 1 \\ 0 & 0 & 0 & 1 & 1 & 0 & 0 & 0 \\ 0 & 0 & 0 & 0 & 0 & 0 & 1 & 1 \end{pmatrix}$$

We obtain that $h_1 = 1$ and $h_2 = x + x^2 + x^5 + x^6 + x^7$ induce a basis of $\text{Ker}(\phi)$. Now $\gcd(f, h_2) = x^6 + x^5 + x^4 + x + 1$ and $\gcd(f, h_2 - 1) = x^2 + x + 1$. This implies

$$f = (x^6 + x^5 + x^4 + x + 1)(x^2 + x + 1)$$

which is the decomposition of f into irreducible factors.

Remark B.3.4. (1) If q is a power of 2, say $q = 2^r$ for some r, then $x^q - x = \text{Tr}(x)(\text{Tr}(x) + 1)$ with $\text{Tr}(x) = \sum_{i=0}^{r-1} x^{2^i}$.
(2) If q is odd, then $x^q - x = x(x^{\frac{q-1}{2}} + 1)(x^{\frac{q-1}{2}} - 1)$.

Applying these observations to k, we obtain the following corollary which is the basis for the probabilistic algorithm of Cantor and Zassenhaus.

Corollary B.3.5. *Let f be squarefree and let $h \in \mathbb{F}_q[x]$ satisfy $h^q - h \in \langle f \rangle$, that is, h induces an element $\bar{h} \in \text{Ker}(\phi)$.*

(1) If $q = 2^r$, then

$$f = \gcd(f, \text{Tr}(h)) \cdot \gcd(f, \text{Tr}(h) + 1).$$

(2) If q is odd, then

$$f = \gcd(f, h) \cdot \gcd(f, h^{\frac{q-1}{2}} + 1) \cdot \gcd(f, h^{\frac{q-1}{2}} - 1).$$

It is not difficult to see that for a monic squarefree polynomial f with $s \geq 2$ irreducible factors in case (1) the probabiliby of $\gcd(f, \text{Tr}(h))$ being non–trivial is at least $\frac{1}{2}$. Similarly the decomposition of (2) is non–trivial with probability at least $\frac{1}{2}$.

Using this, we obtain the improvement of Berlekamp's algorithm by Cantor and Zassenhaus. This algorithm is usually more efficient due to fewer gcd–computations.

Algorithm B.3.6 (BERLEKAMPCANTORZASSENHAUS(f)).

Input: $f \in \mathbb{F}_q[x]$ squarefree, non–constant and monic
Output: $f_1, \ldots, f_s$ the irreducible factors of f

- Compute representatives $1, h_2, \ldots, h_s \in \mathbb{F}_q[x]$ of a basis of $\mathrm{Ker}(\phi)$
- define the function $T(h) = \begin{cases} h^{\frac{q-1}{2}} - 1 & \text{if } q \text{ odd} \\ \mathrm{Tr}(h) & \text{if } q \text{ even} \end{cases}$
- $R = \{f\}$;
- while $(\#R < s)$
    choose $g \in R$ at random
    choose $c = (c_1, \ldots, c_s) \in \mathbb{F}_q^s \setminus \{0\}$ at random
    $h = c_1 + c_2 h_2 + \cdots + c_s h_s$;
    $w = \gcd(g, T(h))$;
    if $(w \neq 1$ and $w \neq g)$
        $R = (R \setminus \{g\}) \cup \{w, \frac{g}{w}\}$;
- return R

Example B.3.7. Let $f = x^6 - 3x^5 + x^4 - 3x^3 - x^2 - 3x + 1 \in \mathbb{F}_{11}[x]$

- $1, h_2 = x^4 + x^3 + x^2 + x, h_3 = x^5 - 2x^3 - 4x^2$ induces a basis of $\mathrm{Ker}(\phi)$
- $R = \{f\}$
- random choice $h = 3 - 2h_2 + 5h_3$
- $\gcd(f, h^5 - 1) = x^5 - 4x^4 + 5x^3 + 3x^2 - 4x + 1 = w$
- $R = \{x + 1, w\}$
- random choice $h = 2 + 3h_2 + 4h_3$
- $\gcd(w, h^5 - 1) = 1$
- random choice $h = 1 + 3h_2 - 4h_3$
- $\gcd(w, h^5 - 1) = x^2 + 5x + 3$
- $R = \{x + 1, x^2 + 5x + 3, x^3 + 2x^2 + 3x + 4\}$

We obtain as factorization

$$f = (x + 1)(x^2 + 5x + 3)(x^3 + 2x^2 + 3x + 4).$$

B.4 Factorization in $\mathbb{Q}[x]$

In this chapter we shall show how to relate the problem of factorization in $\mathbb{Q}[x]$ with factorization in $\mathbb{F}_p[x]$ for a suitable p. First of all let us consider an example, showing that this relation is by no means obvious.

Example B.4.1. The polynomial $x^4 + 1$ is irreducible in $\mathbb{Z}[x]$ and $\mathbb{Q}[x]$ but reducible in $\mathbb{F}_p[x]$ for all prime numbers p.

Proof. Obviously $x^4 + 1$ has no linear factor in $\mathbb{Q}[x]$ because its zero's are not real. Moreover it is easy to see that $x^4 + 1 = (x^2 + ax + b)(x^2 + cx + d)$ has no solution for a, b, c, d in $\mathbb{Q}$. This implies that $x^4 + 1$ is irreducible in $\mathbb{Z}[x]$, and hence in $\mathbb{Q}[x]$ by the lemma of Gauss (Proposition B.4.3).

In $\mathbb{F}_2[x]$ we have $x^4 + 1 = (x+1)^4$. To prove reducibility for the odd primes we fix a prime number $p > 2$ and apply the theory of the previous sections. The polynomial $x^4 + 1$ is squarefree by Proposition B.1.3 since $\gcd(x^4 + 1, 4x^3) = 1$. Let us consider the endomorphism introduced in B.3

$$\phi : \mathbb{F}_p[x]/\langle x^4 + 1 \rangle \rightarrow \mathbb{F}_p[x]/\langle x^4 + 1 \rangle,$$

$\phi(\bar{h}) = \bar{h}^p - \bar{h}$. To prove that $x^4 + 1$ is reducible, we have to prove that $\mathrm{Ker}(\phi)$ is non–trivial. We choose the basis $\{1, x, x^2, x^3\}$ in $\mathbb{F}_p[x]/\langle x^4 + 1 \rangle$. With respect to this basis we compute the matrix of ϕ.

If $p \equiv 1(8)$ then $x^p \equiv x \mod x^4 + 1$ and therefore ϕ is the zero–map. A basis of $\mathrm{Ker}(\phi)$ is $\{1, x, x^2, x^3\}$. If $p \equiv 1(4)$ and $p \not\equiv 1(8)$ then $x^p \equiv -x \mod x^4 + 1$ and we obtain

$$\begin{pmatrix} 0 & 0 & 0 & 0 \\ 0 & -2 & 0 & 0 \\ 0 & 0 & 0 & 0 \\ 0 & 0 & 0 & -2 \end{pmatrix}$$

as the matrix of ϕ. A basis of $\mathrm{Ker}(\phi)$ is $\{1, x^2\}$. If $p = 4r+3$ then $x^p \equiv (-1)^r x^3 \mod x^4 + 1$ and we obtain

$$\begin{pmatrix} 0 & 0 & 0 & 0 \\ 0 & -1 & 0 & (-1)^r \\ 0 & 0 & -2 & 0 \\ 0 & (-1)^r & 0 & -1 \end{pmatrix}$$

as the matrix of ϕ. A basis of $\mathrm{Ker}(\phi)$ is $\{1, x^3 + (-1)^r x\}$.
We deduce the following splitting of $x^4 + 1$:

If $p \equiv 1(8)$ then $x^4 + 1 = \prod_{i=1}^{4} (x - a_i)$, where $a_i^4 = -1$ for $1 \leq i \leq 4$.

If $p \equiv 1(4)$ and $p \not\equiv 1(8)$ then $x^4 + 1 = (x^2 + a)(x^2 - a)$, where $a^2 = -1$.

If $p = 4r + 3$ then $x^4 + 1 = (x^2 - (-1)^r ax - (-1)^r)(x^2 + (-1)^r ax - (-1)^r)$, where $a^2 = -(-1)^r \cdot 2$.   □

Next we will show that irreducibility in $\mathbb{Z}[x]$ and $\mathbb{Q}[x]$ is (modulo integer factors) the same.

Definition B.4.2. $f = \sum_{\nu=0}^{n} a_\nu x^\nu \in \mathbb{Z}[x]$ is called a *primitive polynomial* if $\gcd(a_0, \ldots, a_n) = 1$. The integer $\gcd(a_0, \ldots, a_n)$ is called the *content* of f. It is unique up to sign. Any $g \in \mathbb{Z}[x] \smallsetminus \{0\}$ can be uniquely written as $g = cf$ where $c \in \mathbb{Z}$ is the content of g and $f \in \mathbb{Z}[x]$ is primitive with $a_n > 0$. f is called the *primitive part* of g.

Of course, every $g \in \mathbb{Z}[x] \setminus \{0\}$ can be made primitive by dividing it by its content.

Proposition B.4.3 (Gauss). *Let $f \in \mathbb{Z}[x]$ be a primitive polynomial. Then f is irreducible in $\mathbb{Z}[x]$ if and only if f is irreducible in $\mathbb{Q}[x]$.*

Proof. Obviously f irreducible in $\mathbb{Q}[x]$ and f primitive implies $f \in \mathbb{Z}[x]$ irreducible. Conversely, assume $f \in \mathbb{Z}[x]$ being irreducible but $f = f_1 \cdot f_2$ in $\mathbb{Q}[x]$, where $\deg(f_i) > 0$. We can choose $n_1, n_2 \in \mathbb{Q}$ such that $n_i f_i \in \mathbb{Z}[x]$ are primitive polynomials. Now consider in $\mathbb{Q}[x]$ the equation $(n_1 n_2) f = (n_1 f_1) \cdot (n_2 f_2)$. Since $n_i f_i \in \mathbb{Z}[x]$ we have $n_1 n_2 f \in \mathbb{Z}[x]$.

Since $\mathbb{Z}[x]$ is a unique factorization domain and $n_1 f_1, n_2 f_2$ are primitive it follows $n_1 n_2 = 1$. This is a contradiction. $\qquad\square$

The following example illustrates the way we want to factor polynomials in $\mathbb{Z}[x]$. We use Hensel lifting (cf. Proposition B.4.10), that is, for a fixed prime p, we try to lift a given factorization over $\mathbb{Z}/p$, to $\mathbb{Z}/p^k$ for increasing k.

Example B.4.4. Let $F = x^5 - 5x^4 + 10x^3 - 10x^2 + 1 \in \mathbb{Z}[x]$. We consider $f = F(\mathrm{mod}\ 2) = x^5 + x^4 + 1$ in $\mathbb{F}_2[x]$. Here $f' = x^4$ and hence f is squarefree. With respect to the basis $1, x, x^2, x^3, x^4$ of $\mathbb{F}_2[x]/\langle f\rangle$ we obtain the following matrix for ϕ

$$\begin{pmatrix} 0 & 0 & 0 & 1 & 1 \\ 0 & 1 & 0 & 1 & 1 \\ 0 & 1 & 1 & 0 & 1 \\ 0 & 0 & 0 & 1 & 1 \\ 0 & 0 & 1 & 1 & 0 \end{pmatrix}.$$

This implies that $\{1, x^4 + x^3 + x^2\}$ is a basis of $\mathrm{Ker}(\phi)$. We compute that $\gcd(x^5 + x^4 + 1, x^4 + x^3 + x^2) = x^2 + x + 1$, which implies $x^5 + x^4 + 1 = (x^2 + x + 1)(x^3 + x + 1)$.

Let $g_1 = x^2 + x + 1$ and $h_1 = x^3 + x + 1$. Note that in $\mathbb{F}_2[x]$ we have $1 = (x + 1)h_1 + x^2 g_1$, that is, h_1 and g_1 are coprime.

We consider g_1 and h_1 as polynomials in $\mathbb{Z}[x]$ and obtain $F - g_1 h_1 = -6x^4 + 8x^3 - 12x^2 - 2x$. In order to lift the decomposition $f = g_1 h_1$ in $\mathbb{Z}/2[x]$ to a decomposition $F(\mathrm{mod}\ 4) = g_2 h_2$ in $\mathbb{Z}/4[x]$ we make the Ansatz (using that $n \in \mathbb{Z}$ can be written as finite sum $n = \pm \sum \varepsilon_i 2^i, \varepsilon_i \in \{0, 1\}$)

$$g_2 = g_1 + 2s$$
$$h_2 = h_1 + 2t$$

with $s, t \in \mathbb{Z}[x]$. We can lift the factorization if and only if $f = g_2 h_2(\mathrm{mod}\ 4)$, that is

$$\frac{F - g_1 h_1}{2} \equiv g_1 t + h_1 s \quad \mathrm{mod}\ 2.$$

This implies $s = 0$ and $t = x^2 + x$ and we obtain

$$g_2 = x^2 + x + 1, \quad h_2 = x^3 + 2x^2 + 3x + 1.$$

We make the Ansatz
$$g_3 = g_2 + 4s$$
$$h_3 = h_2 + 4t$$

to lift the factors to $\mathbb{Z}/8[x]$. We obtain as condition for lifting

$$\frac{F - g_2 h_2}{4} = -2x^4 + x^3 - 4x^2 - x \equiv g_2 t + h_2 s \quad \text{mod } 2.$$

This implies $s = x$ and $t = x^2$ and hence $g_3 = x^2 + 5x + 1, h_3 = x^3 + 6x^2 + 3x + 1$ in $\mathbb{Z}/8[x]$. In order to have unique representatives in $\mathbb{Z}[x]$ with coefficients in $[-4, 4)$ we choose $g_3 = x^2 - 3x + 1, h_3 = x^3 - 2x^2 + 3x + 1$. These are the true factors of F, which we can easily check by multiplying g_3 with h_3.

In this example everything worked well. The following two examples show that not every prime p is good (Example B.4.5) and that F may have more factors over $\mathbb{Z}/p^k$ than over $\mathbb{Z}$. In this case we have to check whether products of factors over $\mathbb{Z}/p^k$ are factors over $\mathbb{Z}$. Moreover, if $f = a_n x^n + \cdots + a_1 x + a_0 \in \mathbb{Z}[x]$ is not normalized, a_n may become a unit in $\mathbb{Z}/p$ and then we have to distribute the integer factors of a_n to the lifted factors in an appropriate manner (Example B.4.6).

Example B.4.5. $F = x^4 + 3x^3 - 13x^2 + 6x - 30$ has the following properties:

(1) F is squarefree and primitive in $\mathbb{Z}[x]$
(2) F is not squarefree in $\mathbb{F}_3[x]$ (that is, 3 is not a good prime)
(3) F is squarefree in $\mathbb{F}_5[x]$, namely $F = x(x - 2)(x^2 + 2)$
(4) The factors lift to $F = (x + 45)(x - 42)(x^2 + 2)$ in $\mathbb{Z}/5^4[x]$ and $45, 42, 2 \in [-\frac{1}{2}5^4, \frac{1}{2}5^4]$.
(5) $x^2 + 2$ is a factor of F in $\mathbb{Z}[x]$ but $x + 45$ and $x - 42$ are not. We have that $F = (x^2 + 2)(x^2 + 3x - 15)$, with $x^2 + 3x - 15$ irreducible in $\mathbb{Z}[x]$ and representing $(x + 45)(x - 42)$ in $\mathbb{Z}/5^4[x]$.

Example B.4.6. $F = 12x^3 + 10x^2 - 36x + 35$ has the following properties:

(1) It is squarefree and primitive in $\mathbb{Z}[x]$, but not normalized
(2) $F = 2x(x^2 + 2)$ in $\mathbb{F}_5[x]$
(3) $F = 12(x + 315)(x^2 - 210x + 103)$ in $\mathbb{Z}/5^4[x]$.

(4) By checking all possibilities we find that the factors 2 resp. 6 of 12 (which is a unit in $\mathbb{Z}/5^4$) have to be distributed as factors of the first resp. second irreducible factor of F over $\mathbb{Z}/5^4$.
Then we get $F = (2x + 5)(6x^2 - 10x + 7)$ in $\mathbb{Z}[x]$.

The idea of Berlekamp and Zassenhaus is as in the above examples: Reduce the factorization problem for f in $\mathbb{Z}[x]$ to factorization in $\mathbb{Z}/p[x]$ and lift the result to $\mathbb{Z}/p^N[x]$ with the choice of the coefficients in $[-\frac{1}{2}p^N, \frac{1}{2}p^N)$, a bound for N is given in Proposition B.4.9. First of all we have to make f primitive (by dividing by the content) and then apply the squarefree decomposition of section B.1. Then we apply the following algorithm of Berlekamp and Zassenhaus.

Algorithm B.4.7 (UNIVARIATEFACTORIZE).

Input: $f \in \mathbb{Z}[x]$, $\deg(f) > 1$, f primitive and squarefree
Output: $f_1, \ldots, f_s \in \mathbb{Z}[x]$ irreducible, $\deg(f_i) \geq 1$ and $f = f_1 \cdot \ldots \cdot f_s$.

(1) choose a prime p such that $\bar{f} := f \mod p\mathbb{Z}[x]$ is squarefree of the same degree as f (a random prime number will do)
(2) compute the irreducible factors $\bar{g}_1, \ldots, \bar{g}_s$ of $\bar{f}$ in $\mathbb{F}_p[x]$ (using Berlekamp's algorithm from section B.3)
(3) choose an integer N such that the coefficients of each factor of f in $\mathbb{Z}[x]$ are integers in $[-\frac{1}{2}p^N, \frac{1}{2}p^N)$ (choose N to be the bound from Proposition B.4.9)
(4) lift the irreducible factorization $\bar{g}_1 \cdot \ldots \cdot \bar{g}_s$ to a factorization $\tilde{g}_1 \cdot \ldots \cdot \tilde{g}_s$ of f in $\mathbb{Z}/p^N[x]$ (using Hensel–lifting B.4.10)
(5) Take products of a subset of the (unique) representatives $g_1, \ldots, g_s \in \mathbb{Z}[x]$ of $\tilde{g}_1, \ldots, \tilde{g}_s$ with coefficients in $[-\frac{1}{2}p^N, \frac{1}{2}p^N)$ in an appropriate way to obtain the true factors (cf. Algorithm B.4.11).

We shall now discuss the details of steps (3) - (5) of the above algorithm. First we give bounds for the coefficients of the factors of f.

Definition B.4.8. Let $f = \sum_{i=0}^{n} a_i x^i \in \mathbb{C}[x]$ and denote by the *quadratic norm* of f $\|f\|_2 := \sqrt{\sum_{i=0}^{n} |a_i|^2}$ and by $M(f) := |a_m| \prod_{i=1}^{n} \max\{1, |z_i|\}$ the *measure* of f, where $z_1, \ldots, z_n \in \mathbb{C}$ are the roots of f.

Proposition B.4.9. *Let $g = \sum_{i=0}^{k} b_i x^i \in \mathbb{Z}[x]$ be a factor of $f = \sum_{i=0}^{n} a_i x^i$, then*
$$|b_i| \leq \binom{k}{[\frac{k}{2}]} \cdot \|f\|_2 \text{ for all } i.$$

Proof. The coefficients of g are sums of products of the roots of f multiplied by a_n. Let L_i be the number of products of roots contributing to b_i then $|b_i| \leq M(f) \cdot L_i$. Now $L_i \leq \binom{k}{i}$ and $\binom{k}{i} \leq \binom{k}{[\frac{k}{2}]}$. It remains to prove that $M(f) \leq \|f\|_2$. Let $z_1, \ldots, z_n$ be the roots of f and assume $|z_i| > 1$ for $i = 1, \ldots, l$, then $M(f) = |a_n| \prod_{i=1}^{l} |z_i|$. Let
$$h := a_n \prod_{i=1}^{l} (\bar{z}_i x - 1) \prod_{i=l+1}^{n} (x - z_i) = \sum_{i=0}^{n} h_i x^i$$

then $|h_n| = |a_n| \prod_{j=1}^{l} |\bar{z}_j| = M(f)$. Especially $\|h\|_2 \geq |h_n| = M(f)$. To finish the proof we will show that $\|h\|_2 = \|f\|_2$ holds. Let $p = \sum_{i=0}^{n} p_i x^i \in \mathbb{C}[x]$ and $z \in \mathbb{C}$ then

$$
\begin{aligned}
\|(\bar{z}x - 1)p\|_2^2 &= \sum_{i=0}^{n+1} (\bar{z}p_{i-1} - p_i)(z\bar{p}_{i-1} - \bar{p}_i) \\
&= \sum_{i=0}^{n} \bar{z}z\bar{p}_i p_i + \sum_{i=0}^{n} \bar{p}_i p_i - \sum_{i=0}^{n+1} (\bar{z}p_{i-1}\bar{p}_i + z\bar{p}_{i-1}p_i) \\
&= \sum_{i=0}^{n+1} (zp_i - p_{i-1})(\bar{z}\bar{p}_i - \bar{p}_{i-1}) \\
&= \|(z - x)p\|_2^2
\end{aligned}
$$

Applying this equality l times to h we obtain $\|h\|_2 = \|f\|_2$. $\qquad \square$

Next we prove that we can always lift a factorization from $\mathbb{Z}/p[x]$ to $\mathbb{Z}/p^N[x]$.

Proposition B.4.10 (Hensel Lifting).
Let p be a prime number and k a non–negative integer. Let $f, g_k, h_k, u, v \in \mathbb{Z}[x]$, $f = \sum_{i=0}^{n} a_i x^i$, with the following properties:

(1) $p \nmid a_n$,
(2) $f \equiv a_n g_k h_k \mod p^k$,
(3) $u g_k + v h_k \equiv 1 \mod p$,
(4) $\deg(u) < \deg(h_k)$ and $\deg(v) < \deg(g_k)$,
(5) g_k, h_k are monic.

Then there exist monic polynomials $g_{k+1}, h_{k+1} \in \mathbb{Z}[x]$ such that

$$f \equiv a_n g_{k+1} h_{k+1} \mod p^{k+1}$$

and

$$g_{k+1} \equiv g_k \mod p^k, \quad h_{k+1} \equiv h_k \mod p^k.$$

Moreover, g_{k+1}, h_{k+1} are uniquely determined $\mod p^{k+1}$.

Proof. We consider the following Ansatz (with $s, t \in \mathbb{Z}[x]$):

$$g_{k+1} = g_k + p^k \cdot s, \quad h_{k+1} = h_k + p^k \cdot t$$

Let $d := \frac{f - a_n g_k h_k}{p^k}$ then we have to solve

$$d \equiv g_k t + h_k s \mod p,$$

where condition (3) implies that $u d g_k + v d h_k \equiv d \mod p$.

Using pseudo–division with remainder by the monic polynomial $h_k \in \mathbb{Z}[x]$, we find $z, t \in \mathbb{Z}[x]$ such that $ud = h_k z + t$, $\deg(t) < \deg(h_k)$ and define $s := vd + zg_k$. Then $\deg(s) < \deg(g_k)$ since $tg_k + sh_k \equiv d(p)$ and $\deg(d) < \deg(f) = \deg(g_k) + \deg(h_k)$. The fact that t, s are uniquely determined mod p implies that g_{k+1}, h_{k+1} are uniquely determined mod p^{k+1}. $\qquad \square$

Finally we give an algorithm to find the "true" factors.

Algorithm B.4.11 (TRUEFACTORS).

Input: $f = \sum_{\nu=0}^{n} a_\nu x^\nu \in \mathbb{Z}[x]$ squarefree and primitive,

   p prime, $p \nmid a_n$,

   $N \in \mathbb{N}, p^N \geq 2\binom{n}{[\frac{n}{2}]}\|f\|_2$

   $g_1, \ldots, g_l \in \mathbb{Z}[x], \sum_{i=0}^{l} \deg(g_i) = \deg(f), f \equiv a_n g_1 \cdot \ldots \cdot g_l \mod p^N$,

Output: Irreducible polynomials $f_1, \ldots, f_r \in \mathbb{Z}[x]$ such that $f = f_1 \cdot \ldots \cdot f_r$.

(1) $d = 1; c = a_n; M = \{1, \ldots, l\}; g = f$; Result=$\emptyset$;

(2) while $(d \leq \frac{\#M}{2})$
   - $S = \{T \subset M \mid \#T = d\}$
   - while $((S \neq \emptyset)$ and $(d \leq \frac{\#M}{2}))$
     - choose $T \in S$
     - $S = S \setminus \{T\}$
     - compute $\bar{h} = c \cdot \prod_{i \in T} g_i$ in $\mathbb{Z}/p^N[x]$ and represent it by its unique representative $h \in \mathbb{Z}[x]$ with coefficients in $[-\frac{1}{2}p^N, \frac{1}{2}p^N)$
     - if $(h|c \cdot g)$
       $h :=$ primitive part of h
       Result = Result $\cup\{h\}$
       $g = \frac{g}{h}$
       $M = M \setminus T$
       $S = S \setminus \{D \subset M | D \cap T \neq \emptyset\}$
   - $d = d + 1$
(3) if $(\deg(g) > 0)$ Result = Result $\cup \{h\}$
(4) return (Result)

B.5 Factorization in Algebraic Extensions

Let K be a field and $K(\alpha) \supset K$ an algebraic extension. In this chapter we show how to reduce the factorization over $K(\alpha)$ to the factorization over K.

   Let us recall first the definition and the basic properties of the resultant of two polynomials.

   Let $f = \sum_{\nu=0}^{n} a_\nu x^\nu$ and $g = \sum_{\nu=0}^{m} b_\nu x^\nu$ two polynomials of degree n respectively m, i.e. $a_n \neq 0$ and $b_m \neq 0$. Then the resultant of f and g is

$$\text{Res}(f,g) := \det \begin{pmatrix} a_0 & \cdots & a_m & & & \\ & \ddots & & \ddots & & \\ & & a_0 & \cdots & a_m & \\ b_0 & \cdots & b_n & & & \\ & \ddots & & \ddots & & \\ & & b_0 & \cdots & b_n & \end{pmatrix} \begin{matrix} \left.\right\} n \text{ rows} \\ \\ \left.\right\} m \text{ rows} \end{matrix}$$

By definition $\text{Res}(f,g) \in \mathbb{F}[a_0,\ldots,a_n,b_0,\ldots,b_m]$ where $\mathbb{F}$ is a prime field for K.

As an example we consider

(1) $\text{Res}(x^2 - 2x + 1, 3x + 7) = \det \begin{pmatrix} 1 & -2 & 1 \\ 3 & 7 & 0 \\ 0 & 3 & 7 \end{pmatrix} = 100$

(2) $\text{Res}(x^2 - 2x + 1, 5) = \det \begin{pmatrix} 5 & 0 \\ 0 & 5 \end{pmatrix} = 25.$

SINGULAR Example B.5.1 (resultant).

```
ring R=0,x,dp;
resultant(x2-2x+1,3x+7,x);
//->100
```

The resultant is characterized by the following properties, a proof of which can e.g. be found in [192].

Proposition B.5.2.

(1) $\text{Res}(f,g) = (-1)^{nm} \text{Res}(g,f)$

(2) $\text{Res}(f,g) = g^n,\ if\ m = 0.$

(3) $\text{Res}(f,g) = a_n^{m-\deg(r)} \text{Res}(f,r)\ \ if\ \ g = qf + r,\ \deg(r) < \deg(f).$

Corollary B.5.3.

(1) There exist polynomials $a, b \in K[x]$ such that $af + bg = \text{Res}(f,g)$.

(2) Let $\alpha_1,\ldots,\alpha_n$ be the zeros of f and $\beta_1,\ldots,\beta_m$ the zeros of g (in some extension field of K) then

$$\text{Res}(f,g) = a_n^m b_m^n \prod_{i,j}(\alpha_i - \beta_j) = a_n^m \prod_i g(\alpha_i) = b_m^n \prod_i f(\beta_i)$$

(3) $\text{Res}(f_1 f_2, g) = \text{Res}(f_1, g) \cdot \text{Res}(f_2, g)$

Definition B.5.4. Let $f \in K[x]$ be the minimal polynomial of the extension $K(\alpha)$ of K. Let $\beta \in K(\alpha)$ and $b \in K[x]$ with $\beta = b(\alpha)$. We define the *norm* of β to be $\text{Norm}(\beta) = \text{Res}(b, f) \in K$. Moreover, for $g \in K(\alpha)[z]$ and $G \in K[x,z]$ with $G(\alpha) = g$ we define[2] $\text{Norm}\,(g) = \text{Res}_x(G, f) \in K[z]$.

---

[2] We consider G as element in $K(z)[x]$ and therefore $\text{Res}(G, f)$ is well–defined and an element in $K[z]$.

Remark B.5.5. It is not difficult to see that the definition is independent of the choice of b (resp. G). E.g. the uniqueness of Norm (β) follows since (by Corollary B.5.3 (3))

$$\text{Norm}(\beta) = \prod_{i=1}^{n} b(\alpha_i),$$

where $\alpha_1, \ldots, \alpha_n$ are the roots of f in an algebraic closure of K.

The norm is (by Corollary B.5.3 (3)) a multiplicative map $K(\alpha)[z] \rightarrow K[z]$ since $\text{Norm}(g) = \prod_{i=1}^{n} G(\alpha_i, z)$, where $\alpha_1, \ldots, \alpha_n$ are the roots of f.

Proposition B.5.6. *Let $g \in K(\alpha)[z]$ be irreducible then $\text{Norm}(g)$ is a power of an irreducible element in $K[z]$.*

Proof. Let f be the minimal polynomial of the extension $K(\alpha) \supset K$ and denote its roots by $\alpha_1 = \alpha, \alpha_2, \ldots, \alpha_n$. Assume $\text{Norm}(g) = h_1(z) \cdot h_2(z), h_1, h_2 \in K[z]$ relatively prime. Since $g | \text{Norm}(g)$ and g is irreducible we obtain $g | h_i$ for a suitable i. We may assume that $g | h_1$. Let $h_1 = g_1 g$. Let $\sigma_i : K(\alpha) \rightarrow K(\alpha_i)$ be the map defined by $\sigma(\alpha) = \alpha_i$. Then $h_1 = \sigma_i(h_1) = \sigma_i(g_1) \cdot \sigma_i(g)$ for all i. But $\text{Norm}(g) = \prod_{i=1}^{n} \sigma_i(g)$ implies that $h_1 h_2 = \text{Norm}(g)$ divides h_1^n which is a contradiction. $\qquad\square$

Theorem B.5.7. *Let $g \in K(\alpha)[z]$ and assume that $\text{Norm}(g)$ is squarefree. Let $\text{Norm}(g) = \prod_{i=1}^{k} h_i$ be a factorization in $K[z]$, then*

$$g(z) = \prod_{i=1}^{k} gcd(g(z), h_i(z))$$

is a factorization of $g(z)$ in $K(\alpha)[z]$.

Proof. Let $g = \prod_{i=1}^{k} g_i$ be the factorization of g in $K(\alpha)[z]$, then $\text{Norm}(g) = \prod_{i=1}^{k} \text{Norm}(g_i)$. Since $\text{Norm}(g)$ is squarefree we may assume by Proposition B.5.6 that $\text{Norm}(g_i) = h_i$, $i = 1, \ldots, k$. Let $\alpha = \alpha_1, \alpha_2, \ldots, \alpha_n$ be the roots of f then $h_i = \text{Norm}(g_i) = \prod_{j=1}^{n} G_i(\alpha_j, z)$ with $G_i(\alpha, z) = g_i(z)$ as in Definition B.5.4. This implies $g_i | h_i$.

Assume $g_j | h_i$ for some $j \neq i$. Then $\text{Norm}(g_j) | \text{Norm}(h_i)$. But, since $h_i \in K[z]$ does not depend on x, $\text{Norm}(h_i) = \text{Res}(h_i, f) = h_i^n$ by Proposition B.5.2 (2), and $\text{Norm}(g_j) = h_j$. This implies $h_j | h_i^n$ which is a contradiction. $\qquad\square$

Proposition B.5.8. *Let $g \in K(\alpha)[z]$ be squarefree. Then $\text{Norm}(g(z - s \cdot \alpha))$ is squarefree except for finitely many of $s \in K$.*

Proof. Let $\text{Norm}(g) = \prod_{i=1}^{r} g_i^i$ be the squarefree factorization in $K[z]$. Since g is assumed to be squarefree and $g | \text{Norm}(g)$ we obtain $g | \prod_{i=1}^{r} g_i$. Let $p := \prod_{i=1}^{r} g_i$ and $\beta_1, \ldots, \beta_t$ be the roots of p in a suitable extension field of K. Since p is squarefree the roots are pairwise distinct. Let $s \in K$ and denote by $a_s(z) := \text{Norm}\,(p(z - s\alpha)) = \prod_{i=1}^r \text{Norm}(g_i(z - s\alpha))$.

Since $\text{Norm}(g_i(z - s\alpha)) = \text{Res}_x(G_i(z - sx), f) = \prod_j G_i(z - s\alpha_j)$, where α_i are the roots of the minimal polynomial f of the field extension $K(\alpha) \supset K$, we get from Corollary B.5.3 that $a_s(z) = \prod_{i,j}(z - (s\alpha_j + \beta_i))$. Therefore a_s has multiple roots if and only if $s\alpha_j + \beta_i = s\alpha_k + \beta_l$ for suitable i,j,k,l, i.e. for the finitely many $s = \frac{\beta_l - \beta_i}{\alpha_j - \alpha_k}$.

But $\text{Norm}\,(g(z - s \cdot \alpha)) \mid p(z - s\alpha)$ and $p(z - s\alpha) \mid a_s(z)$. This implies that for all but finitely many $s \in K$ the polynomial $g(z - s\alpha)$ has a squarefree norm. $\square$

Remark B.5.9. (1) If $\deg(f) = n$ and $\deg(g) = k$, the proof of Proposition B.5.8 shows that for $\#K > k^n \cdot n^2$ there exists an $s \in K$ such that $\text{Norm}(g(z - s\alpha))$ is squarefree.

(2) For small fields, however, it is possible that $\text{Norm}(g(z - s\alpha))$ is not squarefree for every $s \in K$. In this case one has to pass to a suitable extension field $L \supset K$ to obtain this property. We will illustrate this in the following example where we compute the factorization in an algebraic extension.

SINGULAR Example B.5.10. Let $K = \mathbb{F}_2$ and let $K(\alpha) = \mathbb{F}_2[x]/f$ where $f = x^2 + x + 1$.

The aim is to factorize the polynomial $g(z) = z^2 + z + 1$ in $K(\alpha)[z]$. We have $g(z - s \cdot \alpha) = g(z) + s$ for $s \in K = \mathbb{F}_2$. This implies $\text{Norm}(g(z - s\alpha)) = \text{Norm}(g(z) + s) = (g(z) + s)^2$ is not squarefree.

We choose $L = \mathbb{F}_2(\beta)$, $\beta^5 + \beta^3 + \beta^2 + \beta + 1 = 0$ as field extension. Let $h = y^5 + y^3 + y^2 + y + 1$ then $K(\beta) = \mathbb{F}_2[y]/h$.

We consider now $g(z + \beta\alpha) = z^2 + z + \alpha^2\beta^2 + \alpha\beta + 1 \in \mathbb{F}_2(\alpha, \beta)$ and $G = z^2 + z + x^2y^2 + xy + 1$ its representative [3] in $\mathbb{F}_2[x, y, z]$.

As one can see later in the SINGULAR–computation $r := \text{Norm}(g(z + \beta\alpha)) = z^4 + (\beta^2 + \beta + 1)z^2 + (\beta^2 + \beta)z + \beta^4 + \beta^3 + \beta + 1$. Note that $g(z + \beta\alpha) \in \mathbb{F}_2(\alpha, \beta)[z]$ and its Norm r is an element of $\mathbb{F}_2(\beta)[z]$. If we compute the norm of r we obtain $\text{Norm}(r) = z^{20} + z^{18} + z^{16} + z^8 + z^6 + z + 1 \in \mathbb{F}_2[z]$. This norm is squarefree and factorizes as

$$\text{Norm}(r) = (z^{10} + z^9 + z^7 + z^6 + z^3 + z^2 + 1)(z^{10} + z^9 + z^7 + z^5 + z^2 + z + 1).$$

Taking the gcd of r and the factors of $\text{Norm}(r)$ we obtain a factorization of r by Theorem B.5.7,

$$r = (z^2 + (\beta + 1)z + \beta^2 + \beta + 1)(z^2 + (\beta + 1)z + \beta^2 + 1).$$

We can again apply Theorem B.5.7 to get a factorization of $g(z + \beta\alpha)$. However, we want to factorize $g(z)$. Therefore we use the inverse transformation

---

[3] $\mathbb{F}_2[x, y]/\langle f, h \rangle = \mathbb{F}_2(\alpha, \beta)$. We can compute in this extension field of $\mathbb{F}_2$ without choosing a primitive element just computing in the qring defined by $\langle f, g \rangle$. The computation of the gcd is then replaced by the corresponding Gröbner basis computation.

to $z \mapsto z + \beta\alpha$ (which is the same here) and we obtain a factorization of $r(z + \beta\alpha)$. Taking the gcd of g and these factors we finally obtain the factorization of g as $g = (z + \alpha + 1)(z + \alpha)$.

```
ring R=2,(z,x,y),lp;
poly f=x2+x+1;          //minimal polynomial of the field
                        //extension
poly g=z2+z+1;          //polynomial to factorize
resultant(f,g,x);
//->z4+z2+1             //the norm of the polynomial g
resultant(f,subst(g,z,z+x),x);
//->z4+z2               //the norm of g(z+x)

poly h=y5+y3+y2+y+1;    //the minimal polynomial for the new
                        //field extension
poly G=subst(g,z,z+xy); //the transformed polynomial
poly r=resultant(f,G,x);
r;                      //the norm r of the transformed
                        //polynomial
//->z4+z2y2+z2y+z2+zy2+zy+y4+y3+y+1

poly s=resultant(h,r,y);
s;                      //the norm of r
//->z20+z18+z16+z8+z6+z+1

factorize(s);

//->[1]:                //the factors of the norm of r
//->    _[1]=1
//->    _[2]=z10+z9+z7+z6+z3+z2+1
//->    _[3]=z10+z9+z7+z5+z2+z+1
//->[2]:
//->    1,1,1

ring S=(2,y),(z,x),lp;
minpoly=y5+y3+y2+y+1;
poly r=imap(R,r);
poly f1=gcd(r,z10+z9+z7+z6+z3+z2+1); //the gcd with the first
                                     //factor
f1;
//->(y2+1)*z2+(y3+y2+y+1)*z+(y4+y3+y+1);
f1=simplify(f1,1);                   //we normalize f1
f1;
//->z2+(y+1)*z+(y2+y+1)
poly f2=gcd(r,z10+z9+z7+z5+z2+z+1);  //the gcd with the second
```

```
                                        //factor
f2=simplify(f2,1);
f2;
//->z2+(y+1)*z+(y2+1)
r-f1*f2;                    //test the result
//->0
f1=subst(f1,z,z+x*y);      //the inverse transformation
f1=simplify(f1,1);
f1;
//->z2+(y+1)*z+(y2)*x2+(y2+y)*x+(y2+y+1)
f2=subst(f2,z,z+x*y);      //the inverse transformation
f2=simplify(f2,1);
f2;
//->z2+(y+1)*z+(y2)*x2+(y2+y)*x+(y2+1)

setring R;
ideal I=std(ideal(f,h));
qring T=I;
option(redSB);
poly f1=imap(S,f1);
poly f2=imap(S,f2);
poly g=imap(R,g);
std(ideal(f1,g));          //gcd of g with the first factor

//->_[1]=z+x+1

std(ideal(f2,g));          //gcd of g with the second factor
//->_[1]=z+x

std((x+z+1)*(x+z));        //test the result
//->_[1]=z2+z+1            //our original f
```

For an infinite field K we obtain the following algorithm for the factorization in the algebraic extension $K(\alpha)$.

Algorithm B.5.11 (FACTORIZATION).

Input: $f \in K(\alpha)[z]$ squarefree
Output: $f_1, \ldots, f_k \in K(\alpha)[z]$ irreducible such that $f = f_1 \cdot \ldots \cdot f_k$.

(1) Choose a random element $s \in K$ and compute $g = \text{Norm}(f(z - s\alpha))$
(2) If g is not squarefree go to (1)
(3) factorize $g = g_1 \cdot \ldots \cdot g_k$ in $K[z]$
(4) For $i = 1$ to k compute $f_i = \gcd(f, g_i)$
(5) return $\{f_1, \ldots, f_k\}$.

B.6 Multivariate Factorization

In this chapter we show how to reduce the factorization of multivariate polynomials in $K[x_1, \ldots, x_n]$ to the case of one variable. The idea is similar to the reduction of the factorization in $\mathbb{Z}[x]$ to the factorization in $\mathbb{Z}/p[x]$. We choose a so-called main variable, say x_n and a suitable point $a = (a_1, \ldots, a_{n-1}) \in K^{n-1}$. Let $\mathfrak{m}_a \in K[x_1, \ldots, x_{n-1}]$ be the maximal ideal corresponding to a. We factorize $f(a, x_n)$ in $K[x_n] = (K[x_1, \ldots, x_{n-1}]/\mathfrak{m}_a)[x_n]$ and use Hensel lifting to lift the factors to $(K[x_1, \ldots, x_{n-1}]/\mathfrak{m}_a^N)[x_n]$ for sufficiently large N. We choose (unique) representatives of these liftings and combine them to obtain the true factors. Let us start with an example.

Example B.6.1.

$$f = x^4 + (-z + 3)x^3 + (z^3 + (y - 3)z - y^2 - 13)x^2$$
$$+ (-z^4 + (y^2 + 3y + 15)z + 6)x$$
$$+ yz^4 + 2z^3 + z(-y^3 - 15y) - 2y^2 - 30.$$

We chose x as main variable, $a = (0, 0)$, $\mathfrak{m}_a = \langle y, z \rangle$ and factorize $f(x, 0, 0)$. We obtain

$$f(x, 0, 0) = x^4 + 3x^3 - 13x^2 + 6x - 30 = g_1 \cdot h_1$$

with $g_1 = x^2 + 2$ and $h_1 = x^2 + 3x - 15$.
We want to lift the factorization $f = g_1 h_1 \pmod{\mathfrak{m}_a}$ to $f = g_i h_i \pmod{\mathfrak{m}_a^i}$ for increasing i (Hensel lifting).
We have

$$f - g_1 h_1 \pmod{\mathfrak{m}_a^2} = -zx^3 - 3zx^2 + 15zx$$
$$= -zx \cdot h_1 .$$

For the Hensel lifting we obtain

$$h_2 = h_1 \text{ and } g_2 = g_1 - zx = x^2 - zx + 2.$$

In the next step we have

$$f - g_2 h_2 \bmod \mathfrak{m}_a^3 = (yz - y^2)x^2 + 3yzx - 15yz - 2y^2$$
$$= yzg_2 - y^2 \cdot h_2 .$$

Therefore we obtain

$$h_3 = h_2 - y^2 = x^2 + 3x - 15 - y^2$$
$$g_3 = g_2 + yz = x^2 - zx + 2 + yz.$$

In the following step we have

$$f - g_3 h_3 \bmod \mathfrak{m}_a^4 = z^3 x^2 + 2z^3 = z^3 g_3 \bmod \mathfrak{m}_a^4.$$

This implies $g_4 = g_3$ and $h_4 = x^2 + 3x - 15 - y^2 + z^3$. One checks that $f = g_4 h_4$ by multiplication.

SINGULAR Example B.6.2 (multivariate factorization).

```
ring R=0,(x,y,z),dp;
poly f=x4+(-z+3)*x3+(z3+(y-3)*z-y2-13)*x2
        +(-z4+(y2+3y+15)*z+6)*x
        +yz4+2z3+z*(-y3-15y)-2y2-30;
factorize(f);
//->[1]:
//->    _[1]=1
//->    _[2]=x2-xz+yz+2
//->    _[3]=z3+x2-y2+3x-15
//->[2]:
//->    1,1,1
```

Example B.6.3. This example shows that, similar to Example B.4.5, f may factor modulo $\mathfrak{m}_a$ but not f itself, which is detected after finitely many Hensel liftings.

Factorisation in $K[x, y]/\langle y \rangle$

$f = x^2 - (y + 1)^3$

$f(x, 0) = x^2 - 1 = (x + 1)(x - 1) = g_1 h_1$

Step 1: Lifting

$f = g_1 h_1 \mod \langle y \rangle^2 = -3y = -\frac{3}{2}yg_1 + \frac{3}{2}yh_1$

$g_2 = x + 1 + \frac{3}{2}y, \qquad h_2 = x - 1 - \frac{3}{2}y$

Step 2: Lifting

$f - g_2 h_2 \mod \langle y \rangle^3 = -\frac{3}{4}y^2 = -\frac{3}{8}y^2(x + 1) + \frac{3}{8}y^2(x - 1)$

$g_3 = x + 1 + \frac{3}{2}y + \frac{3}{8}y^2, \qquad h_3 = x - 1 - \frac{3}{2}y - \frac{3}{8}y^2$

Step 3: Lifting

$f - g_3 h_3 \mod \langle y \rangle^4 = \frac{1}{8}y^3 = \frac{1}{16}y^3(x + 1) - \frac{1}{16}y^3(x - 1)$

$g_4 = x + 1 + \frac{3}{2}y + \frac{3}{8}y^2 - \frac{1}{16}y^3, \qquad h_4 = x - 1 - \frac{3}{2}y - \frac{3}{8}y^2 + \frac{1}{16}y^3$

Testing the product of $g_4 h_4$:

$f \neq g_4 h_4$ implies f is irreducible. We can stop now since $\deg(f) = 3$.

SINGULAR Example B.6.4.

We compute this example with SINGULAR.

```
ring R=0,(x,y),dp;
poly f=x2-(y+1)^3;
factorize(f);
//->[1]:
//->    _[1]=-1
//->    _[2]=y3-x2+3y2+3y+1
//->[2]:
//->    1,1
```

Example B.6.5. $f = (xy + 1)(x + y) \in \mathbb{F}_2[x, y]$ has the following (bad) properties:

(1) $f(x,0) = x$, i.e. the degree drops,
(2) $f(x,1) = (x+1)^2$ is not squarefree.

That is, substituting y by any element of $\mathbb{F}_2$ has bad properties. Therefore we have to substitute y by an element of a field extension of $\mathbb{F}_2$.

Consider f over $\mathbb{F}_4 = \mathbb{F}_2(\alpha)$, $\alpha^2 + \alpha + 1 = 0$. Substituting y by α we get $f(x,\alpha) = (\alpha x + 1)(x + \alpha) = \alpha(x + \alpha + 1)(x + \alpha)$. It is squarefree of the same degree as f. Now we choose $g_1 = \alpha(x + \alpha + 1) = \alpha x + 1$ and $h_1 = x + \alpha$ as factors of the factorization of f in $\mathbb{F}_2(\alpha)[x,y]/\langle y - \alpha \rangle$.

We obtain $f - g_1 h_1 = (y - \alpha)x^2 + (y - \alpha)^2 x + y - \alpha$, i.e. $f - g_1 h_1$ mod $\langle y - \alpha \rangle^2 = (y - \alpha)(\alpha x + 1) + (xy - \alpha x)(x + \alpha) = (y - \alpha)g_1 + (xy - \alpha x)h_1$.

We apply Hensel lifting (cf. the proof of Proposition B.6.9) and get $g_2 = \alpha x + 1 + xy - \alpha x = xy + 1$, $h_2 = x + \alpha + y - \alpha = x + y$.

The examples gave already an idea of the following algorithm. Let K be a field and assume we can factorize univariate polynomials over K.

Algorithm B.6.6 (Extended Zassenhaus).

Input: $f \in K[x_1, \ldots, x_n]$ primitive and squarefree and $\deg_{x_i}(f) \geq 1$ for all $i = 1, \ldots, n$.
Output: $f_1 \ldots, f_s \in K[x_1, \ldots, x_n]$ irreducible such that $f = f_1 \cdot \ldots \cdot f_s$

- Choose a main variable[4], say $x = x_n$ and let $f = \sum_{\nu=0}^{k} f_\nu x^\nu, f_k \neq 0, f_\nu \in K[x_1, \ldots, x_{n-1}]$.
- Choose a point $a = (a_1, \ldots, a_{n-1}) \in K^{n-1}$ such that
    $f_k(a) \neq 0$,
    $f(a, x)$ is squarefree[5].
- Choose N to be the total degree of $F := f(y_1 + a_1, \ldots, y_{n-1} + a_{n-1}, x)$ as polynomial in $y = (y_1, \ldots, y_{n-1})$ with coefficients in $K[x]$.
- Compute the irreducible factorization of $f(a, x)$ in $K[x]$.
- Apply Hensel lifting: lift the irreducible factorization in $K[x] = (K[y]/\langle y \rangle)[x]$ to $(K[y]/\langle y \rangle^{N+1})[x]$ such that $f \equiv \bar{g}_1 \cdot \ldots \cdot \bar{g}_e$ mod $\langle y \rangle^{N+1}$(cf. Proposition B.6.9).

---

[4] A good choice of the main variable is very important for the performance of the algorithm. There are several heuristics: $f(a, x)$ should have low degree, f should be monic with respect to x if possible, etc.

[5] In case of an infinite field K a random choice of a will have these properties. In case of a finite field it may happen that there is no $a \in K$ satisfying these conditions. In this situation one has to pass to a field extension with sufficiently many elements, see Example B.6.5. To avoid dense polynomials, the point a should satisfy an algebraic relation with only few terms.

- Combine the (unique) representatives $g_1, \ldots, g_e \in K[x, y]$ of the $\bar{g}_1, \ldots, \bar{g}_e$ of total degree $\leq N$ in an appropriate way to obtain the true factors[6]. Then obtain the irreducible factors of f by substituting y_i by $x_i - a_i$.

Remark B.6.7. (1) The algorithm is correct since the degree of the factors is bounded by the degree of f and since the Hensel lifting is unique (cf. Proposition B.6.9).

(2) In the last step one has to take products of the unique representatives of the factors $\mathrm{mod}\langle y\rangle^{N+1}$, obtained by Hensel lifting, to obtain the true factors of f

(3) Similar to the univariate case one can reduce the factorization of multivariate polynomials to the factorization of primitive, squarefree multivariate polynomials: factorize the content$(f) = \gcd(f_0, \ldots, f_k)$ and the primitive part of f ($= f/\mathrm{content}(f)$) separately. The squarefree decomposition is similar to the univariate case.

We prove now that the choice of a point $a \in K^{n-1}$ with the required properties is always possible. The set of those points is Zariski open in K^n.

Lemma B.6.8. *Let K be an infinite field and $f = \sum_{\nu=0}^{k} f_\nu x_n^\nu$ squarefree, $f_k \neq 0$ and $f_\nu \in K[x_1, \ldots, x_{n-1}]$. Let D be the discriminant of f with respect to x_n. Then $f(a, x_n)$ is squarefree for $a \in K^{n-1}$ if and only if $\frac{\partial f}{\partial x_n}(a) \neq 0$ and $D(a) \neq 0$.*

Proof. This is an immediate consequence of Proposition B.1.3.   □

Proposition B.6.9. *(Hensel Lifting). Let K be a field and let $R = K[y]$, $y = (y_1, \ldots, y_m)$. Let $f = \sum_{\nu=0}^{s} f_\nu x^\nu \in R[x]$, $f_s \notin \langle y\rangle$ and $g_k, h_k \in R[x]$ such that*

(a) $f \equiv g_k h_k \mod \langle y\rangle^k$
(b) $\deg_x(f) = \deg_x(g_k) + \deg_x(h_k)$
(c) For all i there exist $s_i, t_i \in K[x]$ such that [7]
  - $s_i g_k + t_i h_k \equiv x^i \mod \langle y\rangle$
  - $\deg(s_i) < \deg(h_k)$, $\deg(t_i) \leq \deg(g_k)$.

Then there are $g_{k+1}, h_{k+1} \in R[x]$ uniquely determined up to units in $R/\langle y\rangle^{k+1}$ with the following properties:

(1) $f \equiv g_{k+1} h_{k+1} \mod \langle y\rangle^{k+1}$
(2) $g_{k+1} \equiv g_k \mod \langle y\rangle^k$, $h_{k+1} \equiv h_k \mod \langle y\rangle^k$.

Moreover the conditions (b) and (c) are satisfied for g_{k+1} and h_{k+1}.

---

[6] The number of factors of $f(a, x)$ may by far exceed the one of f. To guess the correct combination of the factors one can consider the factorization of $f(a, x)$ with different points $a \in K^{n-1}$ and compare them. If f is monic then each irreducible factor of f corresponds to some product of the (monic) g_i. In general it is more complicated.

[7] This condition is satisfied if and only if $\gcd(g_k \mod \langle y\rangle, h_k \mod \langle y\rangle) = 1$.

Proof. Let $g_k h_k - f = \sum_{j=0}^{s} b_j x_j$ and define $g_{k+1} := g_k - \sum_{j=0}^{s} b_j t_j$ and $h_{k+1} = h_k - \sum_{j=0}^{s} b_j s_j$. The condition (a) implies $b_i \in \langle y \rangle^k$ for all i. This implies (2).

Condition (c) implies $\deg_x(h_k) = \deg_x(h_{k+1})$ and $\deg_x(g_{k+1}) \leq \deg_x(g_k)$. Now (1) holds because

$$g_{k+1} h_{k+1} \equiv g_k h_k - \sum_{j=0}^{s} b_j(s_j g_k + t_j h_k) \equiv f \quad \mathrm{mod} \ \langle y \rangle^{k+1}.$$

This implies (by (c)) that $\deg_x(f) = \deg_x(g_{k+1}) + \deg_x(h_{k+1})$. Finally

$$s_i g_{k+1} + t_i h_{k+1} = s_i g_k + t_i h_k - \sum_{j=0}^{s} b_j(s_i t_j + t_i s_j) \equiv x^i \quad \mathrm{mod} \ \langle y \rangle.$$

To prove uniqueness assume that $f \equiv g_{k+1} h_{k+1} \equiv \bar{g}_{k+1} \bar{h}_{k+1} \ \mathrm{mod} \ \langle y \rangle^{k+1}$ and $\bar{g}_{k+1} \equiv g_k \ \mathrm{mod} \ \langle y \rangle^k$, $\bar{h}_{k+1} \equiv h_k \ \mathrm{mod} \ \langle y \rangle^k$. This implies $g_{k+1} + \bar{g} = \bar{g}_{k+1}$ and $h_{k+1} + \bar{h} = \bar{h}_{k+1}$ for suitable, $\bar{h}, \bar{g} \in \langle y \rangle^k$, and $\bar{g} h_{k+1} + \bar{h} g_{k+1} \equiv 0 \ \mathrm{mod} \ \langle y \rangle^{k+1}$

From condition (c) for g_{k+1} and h_{k+1} we deduce that $\bar{g} \equiv \xi \cdot g_{k+1} + \eta h_{k+1}$ $\mathrm{mod} \ \langle y \rangle^{k+1}$ and $\bar{h} = \lambda g_{k+1} + \mu h_{k+1} \ \mathrm{mod} \ \langle y \rangle^{k+1}$ for suitable $\xi, \eta, \lambda, \mu \in \langle y \rangle^k$. This implies that $(\xi + \mu) g_{k+1} h_{k+1} + \eta h_{k+1}^2 + \lambda g_{k+1}^2 \equiv 0 \ \mathrm{mod} \ \langle y \rangle^{k+1}$.

Let $g_{k+1} = \sum_{|r| \leq k} g_{k+1}^{(r)} y^r$ and $h_{k+1} = \sum_{|r| \leq k} h_{k+1}^{(r)} y^r$ for suitable $g_{k+1}^{(r)}, h_{k+1}^{(r)} \in K[x]$. Assume that $\xi + \mu = \sum_{|\alpha|=k} c_\alpha y^\alpha \ \mathrm{mod} \ \langle y \rangle^{k+1}, \eta = \sum_{|\alpha|=k} d_\alpha y^\alpha \ \mathrm{mod} \ \langle y \rangle^{k+1}$ and $\lambda = \sum_{|\alpha|=k} e_\alpha y^d \ \mathrm{mod} \ \langle y \rangle^{k+1}$. This implies that $c_\alpha g_{k+1}^{(0)} h_{k+1}^{(0)} + d_\alpha h_{k+1}^{(0)^2} + e_\alpha g_{k+1}^{(0)^2} = 0$ for all $\alpha, |\alpha| = k$. But (c) implies $\gcd (g_{k+1}^{(0)}, h_{k+1}^{(0)}) = 1$. This implies $c_\alpha = d_\alpha = e_\alpha = 0$, for all α and therefore $\lambda \in \langle y \rangle^{k+1}$ and $\xi + \mu \in \langle y \rangle^{k+1}$. Then we have $\bar{g}_{k+1} \equiv (1 + \xi) g_{k+1}$ $\mathrm{mod} \ \langle y \rangle^{k+1}$ and $\bar{h}_{k+1} = (1 - \xi) h_{k+1} \ \mathrm{mod} \ \langle y \rangle^{k+1}$. $\qquad \square$

Remark B.6.10. Our aim in this appendix was to explain only the main ideas for computing the factorization. These ideas can also be used to compute the multivariate gcd. If K is a field, then the univariate gcd in $K[x]$ can be computed by using the Euclidean algorithm. The reduction of the multivariate case to the univariate case is the same as in the factorization. If $K = \mathbb{Q}$ the Euclidean algorithm can produce coefficients with large numerators and denominators. Therefore it is better to consider the problem over $\mathbb{Z}$ and use modular techniques. If we want to factorize a polynomial $f \in \mathbb{Z}[x_1, \ldots, x_n]$ or compute the gcd of two such polynomials it is more efficient to reduce the problem to the computation in $\mathbb{Z}/p[x_i] = \mathbb{Z}[x_1, \ldots, x_n]/\langle p, I \rangle$ for a suitable prime p, a main variable x_i, and an ideal $I = \langle x_1 - a_1, \ldots, \widehat{x_i}, \ldots, x_n - a_n \rangle$. Then use a combination of the Hensel liftings B.6.8 and B.4.10 to lift the result to $\mathbb{Z}[x_1, \ldots, x_n]/\langle p^l, I^k \rangle$ for suitable l, k. Finally, products of the factors give the true factors resp. we obtain the true gcd. This method was proposed by Zassenhaus. The corresponding gcd–computation based on this idea is called EZGCD (extended Zassenhaus gcd).

A significantly more efficient algorithm was proposed by Wang ([229], [230]). The corresponding algorithm to compute the gcd was called EEZ-GCD (Enhanced EZGCD).

Here the Hensel lifting from $\mathbb{Z}/p[x_1]$ to $\mathbb{Z}[x_1, \ldots, x_n]/\langle p^l, I^k \rangle$ (x_1 the main variable) is replaced by single–variable Hensel liftings step by step. From $\mathbb{Z}/p[x_1]$ to $\mathbb{Z}/p^l[x_1]$, from $\mathbb{Z}/p^l[x_1]$ to $\mathbb{Z}/p^l[x_1, x_2], \ldots,$ to $\mathbb{Z}/p^l[x_1, \ldots, x_n]$. For more details see [85].

To illustrate this remark let us describe the EZGCD algorithm.

Algorithm B.6.11 (EZGCD).

Input: $f, g \in \mathbb{Z}[x_1, \ldots, x_n]$
Output: $gcd(f, g)$

(1) Choose a main variable[8] x_i and let $x = x_i$ and $y = (y_1, \ldots, y_{n-1}) = (x_1, \ldots, \widehat{x_i}, \ldots, x_n)$.

(2) Compute f_c (resp. g_c) the content of f (resp. g) with respect to x and set $f = f/f_c$, $g = g/g_c$. If $(\deg_x(g) > \deg_x(f))$ exchange f and g.

(3) Compute $a = \text{EZGCD}(f_c, g_c)$ in $\mathbb{Z}[y_1, \ldots, y_{n-1}]$.

(4) Choose a random prime p such that the leading coefficients of f and g with respect to x do not vanish mod p.

(5) Choose randomly an evaluation[9] point $b = (b_1, \ldots, b_{n-1}), 0 \le b_i < p$, such that the leading coefficients of f and g with respect to x do not vanish at b mod p. If this is not possible go to (4).

(6) Use the Euclidean algorithm to compute $c = \gcd(f(b, x), g(b, x))$ in $\mathbb{Z}/p[x]$, $d = \deg(c)$.

(7) If $d = 0$ return(a).

(8) Choose a random prime[10] $q \ne p$ such that the leading coefficients of f and g with respect to x do not vanish mod q.

(9) Choose at random an evaluation point $\bar{b} = (\bar{b}_1, \ldots, \bar{b}_{n-1}), 0 \le \bar{b}_i < q$, such that the leading coefficients of f and g with respect to x do not vanish at $\bar{b}$ mod q. If this is not possible go to (8).

(10) Use the Euclidean algorithm to compute $e = \gcd(f(\bar{b}, x), g(\bar{b}, x))$ in $\mathbb{Z}/q[x]$.

(11) If $\deg(e) = 0$ return(a)

(12) If $(\deg(e) < d)$ then $p = q, c = e, b = \bar{b}, d = \deg(e)$ and go to (8).

(13) If $(\deg(e) > d)$ go to (8)

(14) If $d = \deg_x(g)$

---

[8] A suitable choice is to minimize the degree of f and g with respect to this main variable.

[9] To avoid density b should contain as many zeros as possible. In practice we will try special values first and use random values after several failures (cf. the following example B.6.12).

[10] To double check the answer mod p. This double check has the effect that, in practice, we almost never have to go back to step (8) again after expensive calculations.

- if $g|f$ return(ag)
- $d = d - 1$ go to (8)

(15) If $\gcd(\frac{f(b,x)}{c}, c) = 1$ use Hensel lifting to lift the product $f(b,x) = \frac{f(b,x)}{c}\cdot c$ of coprime factors over $\mathbb{Z}/p[x]$ to $\mathbb{Z}[x,y]$. Let C be the primitive part of the lifting corresponding to c. If $C|f$ and $C|g$ return (aC) else go to (8).

(16) If $\gcd(\frac{g(b,x)}{c}, c) = 1$ use Hensel lifting to lift the product $g(b,x) = \frac{g(b,x)}{c} \cdot c$ of coprime factors over $\mathbb{Z}/p[x]$ to $\mathbb{Z}[x,y]$. Let C be the primitive part of the lifting corresponding to c.
If $C|f$ and $C|g$ return(aC) else go to (8).

(17) Choose[11]$\alpha, \beta \in \mathbb{Z}$, $0 \le \alpha, \beta < p$ such that $\gcd(\frac{\alpha f(b,x)+\beta g(b,x)}{c}, c) = 1$ and use Hensel lifting to lift the product $\frac{\alpha f(b,x)+\beta g(b,x)}{c} \cdot c$ of coprime factors over $\mathbb{Z}/p[x]$ to $\mathbb{Z}[x,y]$. Let C be the primitive part of the lifting corresponding to c. If $C|f$ and $C|g$ return (a, C) else go to (8).

To illustrate the algorithm consider the following example.

SINGULAR Example B.6.12 (EZGCD).

```
ring R=0,(x,y),dp;
poly f=3x2y+4xy2+y3+3x2+6xy+y2+3x;
poly g=2x2y+xy2-y3-2x2-y2-2x;

//we decide to use x as main variable

ring S=(0,y),x,dp;
poly f=imap(R,f); poly g=imap(R,g);
f;
//->(3y+3)*x2+(4y2+6y+3)*x+(y3+y2)
g;
//->(2y-2)*x2+(y2-2)*x+(-y3-y2)

//we can choose as prime p=5 and as evaluation point y=0

ring T1=5,x,dp;
map phi=R,x,0;
poly f=phi(f); poly g=phi(g);
f;
//->-2x2-2x
g;
//->-2x2-2x

ring T2=7,x,dp;
```

---

[11] Since $\deg_x(\gcd(\frac{f(b,x)}{c}, c)) \ge 1$ and $\deg_x(\gcd(\frac{g(b,x)}{c}, c)) \ge 1$ there are infinitely many α, β with this property. We can choose them by trial and error.

```
map psi=R,x,0;
poly f=psi(f); poly g=psi(g);
f;
//->3x2+3x
g;
//->-2x2-2x
//in both cases we obtain x2+x as gcd which has the same
//degree as g, but g does not divide f and step (14) of the
//algorithm implies that we should try again with (8)
//note: here y=0 was a bad choice

setring T1;
phi=R,x,2;
f=phi(f); g=phi(g);    //we substitute y by 2
gcd(f,g);
//->x-2

setring T2;
psi=R,x,2;
f=psi(f); g=psi(g);
gcd(f,g);
//->x+3

//now we come to step (15) in the algorithm and
//Hensel lifting gives x+y+1

setring R;
gcd(f,g);              //the SINGULAR built-in function
//->x+y+1
```

B.7 Absolute Factorization

Let K be a field of characteristic 0, $\overline{K}$ its algebraic closure and assume we are able to compute the multivariate factorization over algebraic extensions of K (our main example is $K = \mathbb{Q}$). In this chapter we explain how to compute the *absolute factorization* of a polynomial $f \in K[x_1, \ldots, x_n]$, that is, to compute the irreducible factors (and their multiplicities) of f in $\overline{K}[x_1, \ldots, x_n]$. To solve this problem we may assume that f is irreducible in $K[x_1, \ldots, x_n]$.

There exist several approaches to solve this problem (cf. [59], the part written by Chèze and Galligo, or [42]). We concentrate on the algorithm implemented by G. Lecerf in SINGULAR.

The idea of this algorithm is to find an algebraic field extension $K(\alpha)$ of K and a smooth point of the affine variety $V(f)$ in $K(\alpha)^n$. Then an (absolutely) irreducible factor of f will be defined over $K(\alpha)$ which can be computed by using the factorization over $K(\alpha)$ described in Section B.5. The idea is based on the following theorem:

Theorem B.7.1. *Let $f \in K[x_1, \ldots, x_n]$ be irreducible and $a \in K^n$ a smooth point of $V(f) \subseteq \overline{K}^n$. Then f is absolutely irreducible, i.e. irreducible in $\overline{K}[x_1, \ldots, x_n]$.*

Proof. Let $f = f_1 \cdot \ldots \cdot f_t$ be the factorization of f in $\overline{K}[x_1, \ldots, x_n]$. We may assume that $f_1(a) = 0$. Assume that $t > 1$. This implies that $f_i \notin K[x_1, \ldots, x_n]$ for all i. Now a being a smooth point of $V(f)$ implies that $f_i(a) \neq 0$ for $i > 1$. We may choose $\alpha \in \overline{K}$ such that $f_i \in K(\alpha)[x_1, \ldots, x_n]$ for all i. Since $f_1 \notin K[x_1, \ldots, x_n]$ there exist $\sigma \in \mathrm{Gal}_K(K(\alpha))$ such that $\sigma(f_1) \neq f_1$.

But $\sigma(f) = f = \sigma(f_1) \cdot \ldots \cdot \sigma(f_t)$ implies that there is $i \neq 1$ such that $\sigma(f_1) = c \cdot f_i$ for some non–zero constant $c \in K(\alpha)$.

This implies $0 = \sigma(f_1(a)) = c \cdot f_i(a)$ which is a contradiction. We obtain $t = 1$ and f is absolutely irreducible.   □

If we apply the theorem to $K(\alpha)$ we obtain:

Corollary B.7.2. *Let $f \in K[x_1, \ldots, x_n]$ be irreducible, $\alpha \in \overline{K}$ and $a \in K(\alpha)^n$ a smooth point of $V(f) \subseteq \overline{K}^n$, then at least one absolutely irreducible factor of f is defined over $K(\alpha)$.*

For f irreducible it is not difficult to find α and a smooth point of $V(f)$ in $K(\alpha)^n$ (use Lemma B.6.8). We deduce that for irreducible $f \in K[x_1, \ldots, x_n]$, $\deg_{x_n}(f) > 0$, $f(a, x_n)$ is squarefree for almost all $a \in K^{n-1}$.

Lemma B.7.3. *Let $f \in K[x_1, \ldots, x_n]$ be irreducible, $\deg_{x_n}(f) > 0$ and $f(a, x_n)$ squarefree for some $a \in K^{n-1}$. Let $g(x_n)$ be an irreducible factor of $f(a, x_n)$ and $\alpha \in \overline{K}$ with $g(\alpha) = 0$. Then $(a, \alpha) \in K(\alpha)^n$ is a smooth point of $V(f) \subset \overline{K}^n$.*

Proof. Obviously $(a, \alpha) \in V(f)$. Let $f(a, x_n) = g(x_n) \cdot h(x_n)$. Since $f(a, x_n)$ is squarefree $h(\alpha) \neq 0$. Now $\frac{\partial f}{\partial x_n}(a, \alpha) = \frac{\partial g}{\partial x_n}(\alpha) \cdot h(\alpha) \neq 0$ and therefore (a, α) is a smooth point in $V(f)$.   □

We describe now an algorithm to compute the factorization of a multivariate polynomial over the algebraic closure of the ground field K with $\mathrm{char}(K) = 0$.

Algorithm B.7.4 (ABSFACTORIZATION).

Input: $f \in K[x_1, \ldots, x_n]$ irreducible, $\deg(f) = d$.
Output: A list of two polynomials g, q with $g \in K[x_1, \ldots, x_n, z]$, $q \in K[z]$
   irreducible such that for $\alpha \in \overline{K}$ with $q(\alpha) = 0$ the following holds:
   $g(x, \alpha)$ irreducible in $\overline{K}[x_1, \ldots, x_n]$ and $f = \prod_{\sigma \in \mathrm{Gal}_K(K(\alpha))} g(x, \sigma(\alpha))$.

(1) Choose i such that $\deg_{x_i}(f) > 0$ and minimal with this property. We may assume $i = n$.
(2) Choose a random point $a = (a_1, \ldots, a_{n-1}) \in K^{n-1}$ such that $f(a, x_n)$ has a simple irreducible factor, say $q(x_n)$ (using univariate factorization in $K[x_n]$). Let $r = \deg(q)$.
(3) If $(r = 1)$ return (f, z).
(4) Let $K(\beta) = K[z]/q(z)$ and factorize f over $K(\beta)[x_1, \ldots, x_n]$. Let $h \in K[x, z]$ be a representative of an irreducible factor of minimal degree.
(5) If $(d = r \cdot \deg(h))$ return $(h(x, z), q(z))$.
    Note: if this holds r is the degree of a minimal field extension $K(\beta)$ of K s.t. f has an irreducible factor h over $K(\beta)$ satisfying $d = r \cdot \deg(h)$. But in general h may already be defined over a smaller field extension.
(6) Choose, by Algorithm B.7.8, an extension $K(\alpha) \subset K(\beta)$ of K given by a minimal polynomial $p(z) \in K[z]$ of minimal degree $(= d/\deg(h))$ such that h is defined over $K(\alpha)[x_1, \ldots, x_n]$ given by a representative $h_1 \in K[x_1, \ldots, x_n, z]$.
(7) return (h_1, p).

SINGULAR Example B.7.5 (absolute factorization).

```
LIB"absfact.lib";


ring R1=0,(x,y),dp;
poly f=(x2+y2)^3*(x3+y2)^5*(x4+4)^2;
def S1 =absFactorize(f);
setring(S1);
absolute_factors;


//->[1]:
//->   _[1]=1            //coefficient
//->   _[2]=x+(a-2)      //factor of x4+4
//->   _[3]=x+(a+2)      //factor of x4+4
//->   _[4]=x+(a)*y      //factor of x2+y2
//->   _[5]=x3+y2
//->[2]:
//->   1,2,2,3,5         //the multiplicities of the factors
//->[3]:
//->   _[1]=(a)          //the minimal polynomials corresponding
//->   _[2]=(a2-2a+2)    //to the factors above
//->   _[3]=(a2+2a+2)
//->   _[4]=(a2+1)
//->   _[5]=(a)
//->[4]:
//->   19                //total number of irreducible factors of f


ring R2=0,(x,y,z,w),dp;
```

```
poly f=(x2+y2+z2)^2+w4;
def S2 =absFactorize(f);
setring(S2);
absolute_factors;

//->[1]:
//->    _[1]=1
//->    _[2]=-x2-y2-z2+(a)*w2
//->[2]:
//->    1,1
//->[3]:
//->    _[1]=(a)
//->    _[2]=(a2+1)
//->[4]:
//->    2
```

Finally let us make some remarks on step (6) in the algorithm. Here one has to construct a primitive element α of the field extension L of K generated by the coefficients of $h(x, \beta)$, and the corresponding minimal polynomial. Theoretically, this could be done using a generic linear combination of the coefficients of h. Then $h(x, \beta)$ has to be rewritten in terms of α. In practise, this method is not recommended.

In the implementation in SINGULAR this problem is solved in a different way, using the Rothstein–Trager algorithm (cf. [42], [150]) which computes the absolute partial fraction decomposition. We explain now the idea of G. Lecerf (cf. [150]) to find the corresponding minimal field extension.

In step (6) of Algorithm B.7.4 we are in the following situation with $x' = (x_1, \ldots, x_{n-1})$ and $x = ((x', x_n)$:

We have $f \in K[x]$ irreducible, $a \in K^{n-1}$, $q \in K[z]$ a simple factor of $f(a, z)$ of degree r, $h \in K[x, z]$ with $\deg_z h < r$ and $\deg_{x_n} h \geq 1$, representing a proper factor[12] of f over $K(\beta) := K[z]/q$. There exist roots $\gamma_1, \ldots, \gamma_s \in \overline{K}$ of $q, s \leq r$ such that $f_i := h(z = \gamma_i)$[13] gives the irreducible decomposition $f_1 \cdot \ldots \cdot f_s$ of f over $\overline{K}[x]$.

We want to find an irreducible polynomial $p \in K[z]$ of degree s and a polynomial $k \in K[x, z]$ such that $f = \prod_{\alpha, p(\alpha)=0} k(z = \alpha)$. Let $w \in K[x, z]$ such that $\deg_z w < r$ and $hw = f \frac{\partial h}{\partial x_n}$ mod q and write $w = g_1 + g_2 z + \ldots + g_r z^{r-1}$. Then we have $w(z = \gamma) = \frac{f}{h(z=\gamma)} \frac{\partial h}{\partial x_n}(z = \gamma)$ for all roots γ of q.

Proposition B.7.6 (Lecerf). *With these notations let $c = (c_1, \ldots, c_r) \in K^r$ be a generic point and $g = c_1 g_1 + \ldots + c_r g_r$. Let p be the monic squarefree part of* $\operatorname{Res}_{x_n}(f(a, x_n), z \frac{\partial f}{\partial x_n}(a) - g(a, x_n))$ *and $k \in K[x, z], \deg_z k < s$, be a*

---

[12] Note that this implies that no proper factor of f in $\overline{K}[x]$ is an element of $\overline{K}[x']$.

[13] Here, and in the following, this notation means that z is substituted by γ_i in h.

representative of $\gcd(f, z\frac{\partial f}{\partial x_n} - g)$ *in* $(K[z]/p)[x]$. *Then* $p \in K[z]$ *is irreducible of degree* s *and* $f = \prod\limits_{\alpha, p(\alpha)=0} k(z = d)$.

Proof. Consider the $\overline{K}$–subvectorspace of $\overline{K}[x]$

$$\langle g_1, \dots, g_r \rangle_{\overline{K}} = \left\langle \left\{ \frac{f}{h(z=\gamma)} \frac{\partial h}{\partial x_n}(z = \gamma) \mid q(\gamma) = 0 \right\} \right\rangle_{\overline{K}}$$

$$= \left\langle \frac{f}{f_1}\frac{\partial f_1}{\partial x_n}, \dots, \frac{f}{f_s}\frac{\partial f_s}{\partial x_n} \right\rangle_{\overline{K}}.$$

There are $\rho_{ij} \in \overline{K}$ such that $g_j = \sum\limits_{i=1}^{s} \rho_{ij} \frac{f}{f_i}\frac{\partial f_i}{\partial x_n}$. The matrix (ρ_{ij}) has maximal rank s because $\left\{ \frac{f}{f_i}\frac{\partial f_i}{\partial x_n} \mid 1 \le i \le s \right\}$ are linearly independent over $\overline{K}$ as we shall prove in Theorem B.7.7. Let $\rho_i = (\rho_{i1}, \dots, \rho_{ir})$ then it is not difficult to see that the set of all $c \in K^r$ such that the scalar products $(c, \rho_1), \dots, (c, \rho_r)$ are pairwise distinct contains a non–empty Zariski open subset $U \subseteq K^r$. Let $c = (c_1, \dots, c_r) \in U$ and $g := \sum\limits_{j=1}^{r} c_j g_j = \sum\limits_{i=1}^{s} (c, \rho_i)\frac{f}{f_i}\frac{\partial f_i}{\partial x_n}$ then

$$\frac{g}{f} = \sum_{i=1}^{s} (c, \rho_i)\frac{\frac{\partial f_i}{\partial x_n}}{f_i}.$$

The classical theorem of Rothstein–Trager (Theorem B.7.7) applied to $K(x')[x_n]$ implies

$$f_i = \gcd(f, (c, \rho_i)\frac{\partial f}{\partial x_n} - g)$$

and[14]

$$\mathrm{Res}_{x_n}(f, z\frac{\partial f}{\partial x_n} - g) = \mathrm{Res}_{x_n}(f, \frac{\partial f}{\partial x_n}) \prod_{i=1}^{s} (z - (c, \rho_i))^{d_i} .$$

for suitable $d_i \in \mathbb{N}$.

This implies, with $b(x') = \mathrm{Res}_{x_n}(f, \frac{\partial f}{\partial x_n})$,

$$b(a) \cdot \prod_{i=1}^{s} (z - (c, \rho_i))^{d_i} = \mathrm{Res}_{x_n}(f(a, x_n), z\frac{\partial f}{\partial x_n}(a, x_n) - g(a, x_n)) \in K[z]$$

and especially $p := \prod_{i=1}^{s}(z - (c, \rho_i))$ (which is a priori an element of $\overline{K}[z]$) is in $K[z]$.   □

---

[14] The classical theorem of Rothstein–Trager requires monic polynomials but we can extend it to non–monic polynomials as well. Let $\hat{f}, \tilde{f}_i \in K(x')[x_n]$ be monic and $f = e\hat{f}, f_i = e_i\tilde{f}_i, e, e_i \in K[x']$, then Theorem B.7.7 implies $\tilde{f}_i = \gcd_{x_n}(\tilde{f}, (c, \rho_i)\frac{\partial \tilde{f}}{\partial x_n} - \frac{g}{e})$. But f is irreducible over $K[x]$ and has no factors in $\overline{K}[x']$ implies $f_i = \gcd(f, (c, \rho_i)\frac{\partial f}{\partial x_n} - g)$. Similarly the next equation follows.

It remains to prove the classical Rothstein–Trager theorem:

Theorem B.7.7 (Rothstein–Trager). *Let $f \in K[y]$ be a univariate monic polynomial of degree $d \geq 1$ such that $\delta := \mathrm{Res}(f, f') \neq 0$.*

Let $g \in K[y]$ and $\deg(g) \leq d - 1$. Let $\rho_1, \ldots, \rho_r \in \overline{K}$ be the roots of $Q := \mathrm{Res}_y(f(y), zf'(y) - g(y)) \in K[z]$ and $f_i := \gcd(f, \rho_i f' - g)$ be monic of degree $d_i > 0$. Then the following hold

(1) $f = f_1 \cdot \ldots \cdot f_r$,

(2) $\rho_i = \frac{g(\beta)}{f'(\beta)}$ for all roots β of f_i,

(3) $\frac{g}{f} = \sum\limits_{i=1}^{r} \rho_i \frac{f_i'}{f_i}$,

(4) $Q = \delta \prod\limits_{i=1}^{r} (z - \rho_i)^{d_i}$,

(5) the set $\{\frac{f_i'}{f_i} | i = 1, \ldots, r\}$ is linearly independent over $\overline{K}$.

Proof. $\delta \neq 0$ implies $\gcd(f_i, f_j) = 1$ for $i \neq j$ and therefore $\prod_{i=1}^{r} f_i \mid f$. Let $\beta \in \overline{K}$ be a root of f then $f'(\beta) \neq 0$. Let $\rho := \frac{g(\beta)}{f'(\beta)}$, then $\mathrm{Res}_y(f(y), \rho f'(y) - g(y)) = 0$. This implies $\rho \in \{\rho_1, \ldots, \rho_r\}$. Let $\rho = \rho_i$ then β is a root of f_i. This implies (1) (otherwise β would be a root of another factor of f) and (2).

Now let $\widehat{f_i} := \frac{f}{f_i}$ and consider $H := g - \sum\limits_{i=1}^{r} \rho_i \widehat{f_i} f_i'$. Let β be a root of f and $f_i(\beta) = 0$ then $H(\beta) = g(\beta) - \rho_i \widehat{f_i}(\beta) f_i'(\beta)$ and as before we obtain $H(\beta) = 0$. This implies $H \equiv 0$ because $\deg(H) < d$ and $\delta \neq 0$. This implies (3).

To prove (4) we use (1) and the properties of the resultant,

$$\mathrm{Res}_y(f(y), zf'(y) - g(y)) = \prod_{i=1}^{r} \mathrm{Res}_y(f_i(y), zf'(y) - g(y)) \ .$$

By (3) we have $g = \sum_{i=1}^{r} \rho_i f_i' \widehat{f_i}$. This implies that the remainder of g of divison by f_i is the same as the remainder of $\rho_i f_i' \widehat{f_i}$ of division by f_i which is the same as the remainder of $\rho_i f'$ of division by f_i. We obtain $\mathrm{Res}_y(f_i(y), zf'(y) - g(y)) = \mathrm{Res}_y(f_i(y), (z - \rho_i)f'(y))$. This proves (4) since $\mathrm{Res}_y(f(y), zf'(y) - g(y)) = \mathrm{Res}_y(f, f') \prod_{i=1}^{r}(z - \rho_i)^{d_i}$.

To prove (5) let $\sum_{i=1}^{r} c_i \frac{f_i'}{f_i} = 0$ for suitable $c_i \in \overline{K}$. This implies $\sum_{i=1}^{r} c_i f_i' \widehat{f_i} = 0$. Let β be a root of f_i then $c_i f_i'(\beta) \widehat{f_i}(\beta) = 0$ but $f_i'(\beta) \widehat{f_i}(\beta) \neq 0$, hence (5). $\square$

Proposition B.7.6 leads to the following algorithm.

Algorithm B.7.8 (ROTHSTEIN–TRAGER–LECERF).

Input: $f \in K[x]$ irreducible, $a \in K^{n-1}$,
          $q \in K[z]$ a simple irreducible factor of $f(a, z)$ of degree r,
          $h \in K[x, z]$ representing a proper factor of f over $K(\beta) := K[z]/q$,
          i.e. $\deg_z(h) < r, \deg_{x_n}(h) \geq 1$ and $\deg_x(h) < \deg(f)$.

Output: k, p such that $p \in K[z]$ irreducible of degree $s = \frac{\deg(f)}{\deg_x(h)}$
          $k \in K[x, z]$ with $\deg_z(k) < s$, $\deg_x(k) = \deg_x(h)$ such that k represents a factor of f over $K(\alpha) := K[z]/p$.

(1) $s = \frac{\deg(f)}{\deg_x(h)}$

(2) If $\deg(q) = s \quad \text{return}(h, q)$

(3) Choose $w = \sum\limits_{i=1}^{r} g_i z^{i-1} \in K[x, z]$, $\deg_z w < r$ such that
    $$hw = f \frac{\partial h}{\partial x_n} \quad \text{mod } q$$

(4) Choose $c_1, \dots, c_r \in K$ at random and let $g := \sum\limits_{i=1}^{r} c_i g_i$

(5) Let p be the monic squarefree part of
    $\text{Res}_{x_n}(f(a, x_n), z \frac{\partial f}{\partial x_n}(a, x_n) - g(a, x_n))$

(6) If $\deg(p) \neq s$ go to (4)

(7) In $(K[z]/p)[x]$ compute $k = \gcd(f, z \frac{\partial f}{\partial x_n} - g)$

(8) return (k, p).

C. SINGULAR — A Short Introduction

De computer is niet de steen, maar de slijpsteen der wijzen.
(The computer is not the philosophers' stone but the philosophers' whetstone.)
Hugo Battus, Rekenen op taal (1983).

In this section we shall give a short introduction to the computer algebra system SINGULAR. In all the chapters of this book there are already many examples for the interactive use of SINGULAR and the possibility to write own programmes in the SINGULAR programming language. For more details we refer to the SINGULAR Manual, which is offered as an online help for SINGULAR, and can be found as a postscript file in the distribution. See also the SINGULAR examples in the distribution. We start with instructions on how to obtain SINGULAR, explain the first steps on how to work with SINGULAR and, finally, give an overview about the data types and functions of SINGULAR.

C.1 Downloading Instructions

SINGULAR is available, free of charge, as a binary programme for most common hardware and software platforms. Release versions of SINGULAR can be downloaded through ftp from our FTP site

```
ftp://www.mathematik.uni-kl.de/pub/Math/Singular/,
```

or, using your favourite WWW browser, from

```
http://www.singular.uni-kl.de/download.html
```

There are detailed installation instruction available on this webpage or on the included CD–ROM. On Windows they will start with the autostart feature. Otherwise you can just open `index.html` in the root directory of the CD–ROM.

C.2 Getting Started

SINGULAR can either be run in a text terminal or within Emacs. To start SINGULAR in its text terminal user interface, enter `Singular` (or `Singular-<version>`) at the system prompt. The SINGULAR banner appears which, among others, reports the version (and the compilation date, search path, location of programme and libraries, etc.):

```
                    SINGULAR                     /
A Computer Algebra System for Polynomial Computations   /   version <version>
                                              0<
     by: G.-M. Greuel, G. Pfister, H. Schoenemann     \   May 2007
FB Mathematik der Universitaet, D-67653 Kaiserslautern \
```

To start SINGULAR in its Emacs user interface enter `ESingular` at the system prompt.

While using Windows you can just double–click the icons, as you are used from other programmes.

Generally, we recommend using SINGULAR in its Emacs interface, since this offers many more features and is more convenient to use than the ASCII–terminal interface.

To exit SINGULAR type `quit;`, `exit;` or `$` (or, when running within Emacs preferably type `CTRL-C $`). To exit during computation type `CTRL-C`.

There are a few important notes which one should keep in mind:

- Every command has to be terminated by a ; (semicolon) followed by a $\boxed{\text{RETURN}}$.
- The online help is accessible by means of `help;` or `?;`.

SINGULAR is a special purpose system for polynomial computations. Hence, most of the powerful computations in SINGULAR require the prior definition of a ring. The most important rings are polynomial rings over a field, local-izations thereof, or quotient rings of such rings modulo an ideal. However, some simple computations with integers (machine integers of limited size) and manipulations of strings are available without a ring.

Once SINGULAR is started, it awaits an input after the prompt >. Every statement has to be terminated by ; . If you type `37+5;` followed by a Return, then 42 will appear on the screen. We always start the SINGULAR output with `//->` to have it clearly separated from the input.

```
37+5;
//-> 42
```

All objects have a type, for example, integer variables are defined by the word `int`. An assignment is done by the symbol = .

```
int k = 2;
```

Testing for equality, respectively inequality, is done using ==, respectively !=, (or <>), where 0 represents the boolean value FALSE, any other integer value represents TRUE.

```
k == 2;
//-> 1
k != 2;
//-> 0
```

The value of an object is displayed by simply typing its name.

```
k;
//-> 2
```

On the other hand, the output is suppressed if an assignment is made.

```
int j;
j = k+1;
```

The last displayed (!) result is always available with the special symbol _.

```
2*_;      // two times the value of k displayed above
//-> 4
```

Starting with // denotes a comment and the rest of the line is ignored in calculations, as seen in the previous example. Furthermore, SINGULAR maintains a history of the previous lines of input, which may be accessed by CTRL-P (previous) and CTRL-N (next) or the arrows on the keyboard.

To edit SINGULAR input we have the following possibilities:

TAB: automatic name completion.
CTRL-B: moves the cursor to the left.
CTRL-F: moves the cursor to the right.
CTRL-D: deletes the symbol in the cursor.
CTRL-P: gives the preceding line in the history.
RETURN: sends the current line to the SINGULAR–Parser.

For more commands see the SINGULAR Manual.

The SINGULAR Manual is available online by typing the command help;. Explanations on single topics, for example, on intmat, which defines a matrix of integers, are obtained by

```
help intmat;
```

Next, we define a 3×3 matrix of integers and initialize it with some values, row by row from left to right:

```
intmat m[3][3] = 1,2,3,4,5,6,7,8,9;
```

A single matrix entry may be selected and changed using square brackets [and].

```
m[1,2]=0;
m;
//-> 1,0,3,
//-> 4,5,6,
//-> 7,8,9
```

To calculate the trace of this matrix, we use a for loop. The curly brackets {, respectively }, denote the beginning, respectively end, of a block. If you define a variable without giving an initial value, as the variable tr in the example below, SINGULAR assigns a default value for the specific type. In this case, the default value for integers is 0. Note that the integer variable j has already been defined above.

```
int tr;
for ( j=1; j <= 3; j++ ) { tr=tr + m[j,j];}
tr;
//-> 15
```

j++ is equivalent to j=j+1. To count the even and the odd entries of the second row of this matrix, we use the if and else and a while loop.

```
j=0;
int even,odd;
while(j<3)
{
   j++;
   if((m[2,j] mod 2)==0){even++;}
   else                 {odd++;}
}
even;
//-> 2
odd;
//-> 1
```

Variables of type string can also be defined and used without a ring being active. Strings are delimited by " (quotation–marks). They may be used to comment the output of a computation or to give it a nice format. If a string contains valid SINGULAR commands, it can be executed using the function execute. The result is the same as if the commands were written on the command line. This feature is especially useful to define new rings inside procedures.

```
"example for strings:";
//-> example for strings:
string s="The element of m ";
s = s + "at position [2,3] is:";    //+ concatenates strings
s , m[2,3] , ".";
//-> The element of m at position [2,3] is: 6.
```

```
s="m[2,1]=0; m;";
execute(s);
//-> 1,0,3,
//-> 0,5,6,
//-> 7,8,9
```

This example shows that expressions can be separated by , (comma) giving a list of expressions. SINGULAR evaluates each expression in this list and prints all results separated by spaces.

To read data from a file, or to write it to a file, we can use the commands read and write. For example, the following SINGULAR session creates a file hallo.txt and writes 2 and example to this file.

```
int a = 2;
write("hallo.txt",a,"example");
```

This data can be read from the file again (as strings):

```
read("hallo.txt");
//-> 2
//-> example
```

If we want to execute commands from a file, we have to write

```
execute(read("hallo.txt"));
```

or, in short,

```
<"hallo.txt";
```

If, for instance, in hallo.txt we have the command

```
int j=1; j;
```

we obtain

```
//-> 1
```

To calculate with objects as polynomials, ideals, matrices, modules, and polynomial vectors, a ring has to be defined first:

```
ring r = 0,(x,y,z),lp;
```

The definition of a ring consists of three parts: the first part determines the ground field, the second part determines the names of the ring variables, and the third part determines the monomial ordering to be used. So the example above declares a polynomial ring called r with a ground field of characteristic 0 (that is, the rational numbers) and ring variables called x, y, and z. The lp at the end means that the lexicographical ordering is used.

The default ring in SINGULAR is $\mathbb{Z}/32003[x, y, z]$ with degree reverse lexicographical ordering:

```
ring s;
s;
//-> //    characteristic : 32003
//-> //    number of vars : 3
//-> //         block   1 : ordering dp
//-> //                   : names    x y z
//-> //         block   2 : ordering C
```

Defining a ring makes this ring the current active basering, so each ring definition above switches to a new basering.

The basering is now s. If we want to calculate in the ring r, which we defined first, we have to switch back to it. This can be done by using

```
setring r;
```

Once a ring is active, we can define polynomials. A monomial, say x^3, may be entered in two ways: either using the power operator ^, saying x^3, or in shorthand notation without operator, saying x3. Note that the shorthand notation is forbidden if the name of the ring variable consists of more than one character. Note also that SINGULAR always expands brackets and automatically sorts the terms with respect to the monomial ordering of the basering.

```
poly f = x3+y3+(x-y)*x2y2+z2;
f;
//-> x3y2-x2y3+x3+y3+z2
```

C.3 Procedures and Libraries

SINGULAR offers a comfortable programming language, with a syntax close to C. So it is possible to define procedures which combine several commands to form a new one. Procedures are defined with the keyword **proc** followed by a name and an optional parameter list with specified types. Finally, a procedure may return values using the command **return**.

Assume we want to compute for $f = x^3y^2 - x^2y^3 + x^3 + y^3 + z^2 \in \mathbb{Q}[x, y, z]$ the vector space dimension $\dim_{\mathbb{Q}}(\mathbb{Q}[x, y, z]/\langle \partial f/\partial x, \partial f/\partial y, \partial f/\partial z \rangle)$, the so-called *global Milnor number* of f (see also Page 527). Then we type

```
ring  R=0,(x,y,z),dp;
poly  f=x3y2-x2y3+x3+y3+z2;
ideal J=diff(f,x),diff(f,y),diff(f,z);
```

or

```
ideal J=jacob(f);
```

Then we have to compute a standard basis of J:

```
ideal K=std(J);
```

We obtain the Milnor number as

```
vdim(K);
//-> 12.
```

As a procedure, we can write

```
proc Milnor (poly h)
{
  ideal J=jacob(h);
  ideal K=std(J);
  int    d=vdim(K);
  return(d);
}
```

or, in a more compact form:

```
proc Milnor (poly h)
{
 return(vdim(std(jacob(h))));
}
```

Note: if you have entered the first line of the procedure and pressed RETURN, SINGULAR prints the prompt . (dot) instead of the usual prompt > . This shows that the input is incomplete and SINGULAR expects more lines. After typing the closing curly bracket, SINGULAR prints the usual prompt, indicating that the input is now complete.

Then call the procedure:

```
Milnor(f);
//-> 12
```

If we want to compute the local Milnor number of f at 0 (cf. Definition A.9.3), then we can use the same procedure, but we have to define a local monomial ordering on R, for example, `ring R=0,(x,y,z),ds;`, and then define f, J, K, as above.

The distribution of SINGULAR contains several libraries, which extend the functionality of SINGULAR. Each of these libraries is a collection of useful procedures based on the kernel commands. The command `help all.lib;` lists all libraries together with a one–line explanation.

One of these libraries is `sing.lib` which already contains a procedure called `milnor` to calculate the Milnor number not only for hypersurfaces but, more generally, for complete intersection singularities.

Libraries are loaded with the command LIB. Some additional information during the process of loading is displayed on the screen, which we omit here.

```
LIB "sing.lib";
```

As all input in SINGULAR is case sensitive, there is no conflict with the previously defined procedure `Milnor`, but the result is the same.

```
milnor(f);
//-> 12
```

Let us compute the local Milnor number of f. We use the command `imap` to map f from R to the new basering with a local ordering:

```
ring S=0,(x,y,z),ds;
poly f=imap(R,f);
milnor(f);
//-> 4
```

(Since the global Milnor number is the sum of all local Milnor numbers at all critical points of f, this shows that f must have further critical points outside 0.)

The procedures in a library have a help part, which is displayed by typing

```
help milnor;
```

as well as some examples, which are executed by

```
example milnor;
```

Likewise, the library itself has a help part, to show a list of all the functions available for the user, which are contained in the library.

```
help sing.lib;
```

The output of the help commands is omitted here.

We want to give some more examples to explain how to write procedures. First of all we give an example of a procedure to compute the prime numbers up to a given bound to explain the use of loops. We use the sieve of Eratosthenes:

We test the integers from 2 to n whether they are divisible by the primes which are already in the list L. If this is not the case we include this number in L and continue.

```
proc primList(int n)
{
  int i,j;
  list L;
  for(i=2;i<=n;i++)
  {
    j=1;
    while(j<=size(L))
    {
      if((i mod L[j])==0){break;}
      j++;
```

```
    }
    if(j==size(L)+1){L[size(L)+1]=i;}
  }
  return(L);
}

primList(10);
//->[1]:   [2]:   [3]:   [4]:
      2      3      5      7
```

Finally we will explain a recursive procedure call. Let us program the generalized Euclidean algorithm. The generalized Euclidean algorithm is applied to two vectors x, y in $\mathbb{Z}^n$ as follows. Let $x = (x_1, \ldots, x_n), y = (y_1, \ldots, y_n)$. We perform the Euclidean algorithm for x_1, y_1 but do the operations for the whole vectors.

Let $x_1 - qy_1 = r, 0 \leq r < |y_1|$ then we compute in the first step the vector $x - qy$. We continue with y and $x - qy$. If the condition c to stop is true we return y.

Algorithm C.3.1 (GENERALEUCLID(x, y, c)).

Input: $x, y \in \mathbb{Z}^n, c$ a condition to stop
Output: $z \in \mathbb{Z}^n$

(1) $z = y$
(2) if (c) return(z)
(3) $q = x_1$ div y_1
(4) $z = x - qy$
(5) $x = y$
(6) $y = z$
(7) go to (1)

If we use the procedure recursively it looks like that:

Algorithm C.3.2 (GENERALEUCLID(x, y, c)).

- if (c) return (y)
- return (generalEuclid $(y, x - (x_1$ div $y_1)y, c)$)

Before we come to the corresponding SINGULAR example we will make a remark on the use of this procedure.

(1) If $x, y \in \mathbb{Z}$ and the condition c is $(x \equiv 0 \mod y)$ then we have the classical Euclidean algorithm.
(2) If $x = (a, 1, 0), y = (b, 0, 1)$ and the condition c is $(x_1 \equiv 0 \mod y_1)$ then for the result $y = (y_1, y_2, y_3)$ we have $\gcd(a, b) = y_1 = ay_2 + by_3$.
(3) If $x = (p, 0), y = (b, a), \gcd(b, p) = 1$, and the condition c is as before $(x_1 \equiv 0 \mod y_1)$ then the result $(y_1, y_2) = (1, ab^{-1} \mod p)$.

(4) If $x = (p,0), y = (q,1), p$ a prime $p \nmid q$, and the condition c is $(2y_1^2 <$ p or $y_1 = 0)$ then in every step of the algorithm $y = (y_1, y_2)$ has the property $\frac{y_1}{y_2}$ mod $p = q$. The result (y_1, y_2) has also the property $2y_i^2 < p, i = 1, 2$ if $y_1 \neq 0$. This is the so–called Farey[1] fraction corresponding to q and is uniqueley determined.

In SINGULAR we obtain the following procedures

```
proc generalEuclid(intvec x,intvec y,int c)
{
   if(c==0)
   {
      if((x[1] mod y[1])==0){return(y);}
   }
   else  //the case of Farey fractions
   {
      if((2*y[1]^2<c)||(y[1]==0)){return(y);}
   }
   return(generalEuclid(y,x-(x[1] div y[1])*y,c));
}
```

```
intvec x=28;
intvec y=36;
generalEuclid(x,y,0);
//->4                     4=gcd(28,36)
```

```
x=28,1,0;
y=36,0,1;
generalEuclid(x,y,0);
//->4,4,-3                4=4*28-3*36
```

```
x=37,0;
y=3,4;
generalEuclid(x,y,0);
//->1,-48                 -48 mod 37=26=4/3 mod 37
```

---

[1] The set $F_n := \{\frac{a}{b} \in \mathbb{Q} \mid \gcd(a,b) = 1), 0 \leq |a| \leq n, 0 < b \leq n\}$ is called Farey frations of order n for a prime number p. Let $\mathbb{Q}_p = \{\frac{a}{b} \in \mathbb{Q} \mid \gcd(a,b) = \gcd(b,p) = 1\}$ and denote by $f_p : \mathbb{Q}_p \to \mathbb{F}_p = \mathbb{Z}/p$ the canonical map defined by $f_p(\frac{a}{b}) = (a \mod p) \cdot (b \mod p)^{-1}$. If $2n^2 < p$ then $F_n \subset \mathbb{Q}_p$ and $f_p|F_n$ is injective. In this case the Euclidean algorithm computes for $q \in f_p(F_n)$ the uniquely determined $\frac{a}{b} \in F_n$ such that $f_p(\frac{a}{b}) = q$. For more details and proofs cf. [118]. The Farey fractions and the algorithm to compute them are used to compute a Gröbnerbasis of an ideal in $\mathbb{Q}[x_1, \ldots, x_n]$ computing it over $\mathbb{Z}/p$ for several primes and lifting the coefficients using Chinese remainder theorem and Farey fractions (cf. [4]). This is implemented in SINGULAR in modstd.lib.

```
x=37,0;
y=26,1;
generalEuclid(x,y,37);
//->4,3                    26 = 4/3 mod 37
```

More Examples of procedures can be found in Section 3.7 and Section 4.8.

C.4 Data Types

In this section, we give an overview of all data types in SINGULAR. For more details see the Manual.

def: Objects may be defined without a specific type: they get their type from the first assignment to them. For example, `ideal i=x,y,z; def j=i^2;` defines the ideal `i^2` with the name `j`.

ideal: Ideals are represented as lists of polynomials, which generate the ideal: `ideal I=x2-1,xy;`. Like polynomials, they can only be defined or accessed with respect to a basering.

ideal operations are:
+   addition (concatenation of the generators and simplification),
*   multiplication (with ideal, poly, vector, module; simplification in case of multiplication with ideal),
^   exponentiation (by a nonnegative integer).

ideal related functions are:
`char_series, coeffs, contract, diff, degree, dim, eliminate, facstd, factorize, fglm, finduni, groebner, highcorner, homog, hilb, indepSet, interred, intersect, jacob, jet, kbase, koszul, lead, lift, liftstd, lres, maxideal, minbase, minor, modulo, mres, mstd, mult, ncols, preimage, qhweight, quotient, reduce, res, simplify, size, sortvec, sres, std, stdfglm, stdhilb, subst, syz, vdim, weight`

int: Variables of type int represent the machine integers and are, therefore, limited in their range (for example, the range is between -2147483647 and 2147483647 on 32–bit machines): `int i=1;` .

int operations are:
++   changes its operand to its successor, is itself not an int expression
--   changes its operand to its predecessor, is itself not an int expression
+   addition
-   negation or subtraction
*   multiplication
/   integer division (omitting the remainder), rounding toward 0

div	integer division (omitting the remainder >=0)
%	integer modulo (the remainder of the division /)
mod	integer modulo (the remainder of the division div), always nonnegative
^, **	exponentiation (exponent must be nonnegative)
<, >, <=, >=, ==,<>	comparison

An assignment j=i++; or j=i--; is not allowed, in particular it does not change the value of j.

int related functions are:

char, deg, det, dim, extgcd, find, gcd, koszul, memory, mult, ncols, npars, nrows, nvars, ord, par, pardeg, prime, random, regularity, rvar, size, trace, var, vdim.

boolean expressions are:

A boolean expression is really an int expression used in a logical context: an int expression (<> 0 evaluates to *TRUE* (represented by 1), 0 represents *FALSE*).

boolean operations are:

and (logical), may also be written as &&,

or (logical), may also be written as ||,

not (not logical), may also be written as !

intmat: Integer matrices are matrices with integer entries. Integer matrices do not belong to a ring, they may be defined without a basering being defined: intmat im[2][3]=1,2,3,4,5,6;.

intmat operations are:

+	addition with intmat or int; the int is converted into a diagonal intmat
−	negation or subtraction with intmat or int; the int is converted into a diagonal intmat
*	multiplication with intmat, intvec, or int; the int is converted into a diagonal intmat
div,/	division of entries in the integers (omitting the remainder)
%, mod	entries modulo int (remainder of the division)
<>, ==	comparison

intmat related functions are:

betti, det, ncols, nrows, random, size, transpose, trace.

intvec: Variables of type intvec are lists of integers: intvec iv=1,2,3,4;.

intvec operations are:

+	addition with intvec or int (component–wise)
−	change of sign or subtraction with intvec or int (component–wise)

    *   multiplication with int (component–wise)

`/, div`   division by int (component–wise)

`%, mod`   modulo (component–wise)

`<>, ==, <=, >=, >, <`

        comparison (done lexicographically)

intvec related functions are:

`hilb,` `indepSet, leadexp, nrows, qhweight, size, sortvec, transpose,` `weight`.

link: Links are the communication channels of SINGULAR, that is, something SINGULAR can write to and/or read from. Currently, SINGULAR supports four different link types:

- ASCII links,
- MPfile links,
- MPtcp links,
- DBM links.

link related functions are:

`close, dump, getdump, open, read, status, write, kill.`

Via ASCII links, data, which can be converted to a string, can be written into files for storage or communication with other programmes. The data is written in plain text format. The output format of polynomials is done w.r.t. the value of the global variable **short**. Reading from an ASCII link returns a string — conversion into other data is up to the user. This can be done, for example, using the command **execute**.

MP (Multi Protocol) links give the possibility to store and communicate data in the binary MP format: read and write access is very fast compared to ASCII links.

MPtcp links give the possibility to exchange data in the binary MP format between two processes, which may run on the same or on different computers.

DBM links provide access to data stored in a data base.

list: Lists are arrays whose elements can be of any type (including ring and qring). If one element belongs to a ring the whole list belongs to that ring:

`list L=5,"Hallo";`.

Note that a list stores the objects themselves and not the names. Hence, if L is a list, `L[1]` for example has no name. A name, say R, can be created for `L[1]` by `def R=L[1];`.

list operations are:

    +   concatenation

`delete`   deletes one element from list, returns new list

`insert`   inserts or appends a new element to list, returns a new list

list related functions are:

`bareiss, betti, delete, factstd, factorize, insert, lres, minres, mres,` `names, res, size, sres.`

map: Maps are ring maps from a ring into the basering: `map g=R,x,y,z;`.

map operations are:

composition of maps. If, for example, `f` and `g` are maps, then `f(g)` is a map expression giving the composition of `f` and `g`.

matrix: Objects of type matrix are matrices with polynomial entries. Like polynomials they can only be defined or accessed with respect to a basering: `matrix m[2][2]=x,1,y,xy;`.

matrix related functions are:

`bareiss, coef, coeffs, det, diff, jacob, koszul, lift, liftstd, minor, ncols, nrows, print, size, subst, trace, transpose, wedge`.

module: Modules are submodules of a free module over the basering with basis `gen(1)`, `gen(2)`, ... They are represented by lists of vectors, which generate the submodule. Like vectors they can only be defined or accessed with respect to a basering. If M is a submodule of R^n, R the basering, generated by vectors $v_1, \ldots, v_k$, then $v_1, \ldots, v_k$ may be considered as the generators of relations of R^n/M between the canonical generators `gen(1)`, ..., `gen(n)`. Hence, any finitely generated R–module can be represented in SINGULAR by its module of relations:
`module mo=[x,yz,y],[xy,1,3];`.

module operations are:

+   addition (concatenation of the generators and simplification)

*   multiplication with ideal or poly.

module related functions are:

`coeffs, degree, diff, dim, eliminate, freemodule, groebner, hilb, homog, interred, intersect, jet, kbase, lead, lift, liftstd, lres, minbase, modulo, mres, mult, ncols, nrows, print, prune, qhweight, quotient, reduce, res, simplify, size, sortvec, sres, std, subst, syz, vdim, weight`.

number: Numbers are elements from the coefficient field (or ground field). They can only be defined or accessed with respect to a basering, which determines the coefficient field: `number n=4/6;`.

number operations are:

+	addition
–	change of sign or subtraction
*	multiplication
/	division
^, **	power, exponentiation (by an integer)
<=, >=, ==, <>	comparison.

number related functions are:

`cleardenom, impart, numerator, denominator, leadcoef, par, pardeg, parstr, repart`.

poly: Polynomials are the basic data for all main algorithms in SINGULAR. They consist of finitely many terms, which are combined by the usual polynomial operations. Polynomials can only be defined or accessed with respect to a basering, which determines the coefficient type, the names of the indeterminants and the monomial ordering.

```
ring r=32003,(x,y,z),dp;
poly f=x3+y5+z2;
```

poly operations are:

+	addition
−	change of sign or subtraction
*	multiplication
/	division by a monomial, non divisible terms yield 0
^, **	power by an integer
<, <=, >, >=, ==, <>	comparison (w.r.t. the monomial ordering)

poly related functions are:

cleardenom, coef, coeffs, deg, det, diff, extgcd, factorize, finduni, gcd, homog, jacob, lead, leadcoef, leadexp, leadmonom, jet, ord, qhweight, reduce, rvar, simplify, size, subst, trace, var, varstr.

proc: Procedures are sequences of SINGULAR commands in a special format. They are used to extend the set of SINGULAR commands with user defined commands. Once a procedure is defined it can be used like any other SINGULAR command. Procedures may be defined by either typing them on the command line or by loading them from a file. A file containing procedure definitions, which comply with certain syntax rules is called a library. Such a file is loaded using the command LIB.

qring: SINGULAR offers the opportunity to calculate in quotient rings, that is, rings modulo an ideal. The ideal has to be given as a standard basis: qring Q=I;.

resolution: The resolution type is intended as an intermediate representation, which internally retains additional information obtained during computation of resolutions. It furthermore enables the use of partial results to compute, for example, Betti numbers or minimal resolutions. Like ideals and modules, a resolution can only be defined w.r.t. a basering.

resolution related functions are:

betti, lres, minres, mres, res, sres.

ring: Rings are used to describe properties of polynomials, ideals, etc. Almost all computations in SINGULAR require a basering.

ring related functions are:

charstr, keepring, npars, nvars, ordstr, parstr, qring, setring, varstr.

ring operations are:

+   construct a new ring $k[X, Y]$ from $k[X]$ and $k[Y]$.

string: Variables of type `string` are used for output (almost every type can be "converted" to `string`) and for creating new commands at run-time. They are also return values of certain interpreter related functions. String constants consist of a sequence of any characters (including newline!) between a starting " and a closing " . There is also a string constant `newline`, which is the newline character. The + sign "adds" strings, "" is the empty string. Strings may be used to comment the output of a computation or to give it a nice format. Strings may also be used for intermediate conversion of one type into another.

A comma between two strings makes an expression list out of them (such a list is printed with a separating blank in between), while a + concatenates strings.

string operations are:

+   concatenation

<=, >=, ==, <>   comparison (lexicographical with respect to the ASCII encoding)

string related functions are:

`charstr, execute, find, names, nameof, option, ordstr, parstr, read, size, sprintf, typeof, varstr.`

vector: Vectors are elements of a free module over the basering with basis `gen(1), gen(2), ...` . Each vector belongs to a free module of rank equal to the greatest index of a generator with non–zero coefficient. Since generators with zero coefficients need not be written, any vector may be considered also as an element of a free module of higher rank. (For example, if `f` and `g` are polynomials then `f*gen(1)+g*gen(3)+gen(4)` may also be written as `[f,0,g,1]` or as `[f,0,g,1,0]`.) Like polynomials they can only be defined or accessed with respect to the basering. Note that the elements of a vector have to be surrounded by square brackets (`[ , ]`).

vector operations are:

+   addition

−   change of sign or subtraction

/   division by a monomial, not divisible terms yield 0

<, <=, >, >=, ==, <>   comparison of leading terms w.r.t. monomial ordering

vector related functions are:

`cleardenom, coeffs, deg, diff, gen, homog, jet, lead, leadcoef, leadexp, leadmonom, nrows, ord, reduce, simplify, size, subst.`

C.5 Functions

This section gives a short reference of all functions, commands and special variables of the SINGULAR kernel (that is, all built–in commands).
The general syntax of a function is

[target =] function_name (<arguments>);

If no target is specified, the result is printed. In some cases (for example, export, keepring, kill, setring, type) the brackets enclosing the arguments are optional. For the commands help, continue, break, quit and exit brackets are not allowed.

attrib: attrib (name) displays the attribute list of the object called name. attrib (name , string) returns the value of the attribute string of the variable name. If the attribute is not defined for this variable, attrib returns the empty string. attrib (name , string , expression) sets the attribute string of the variable name to the value expression.
An attribute may be described by any string. Some of these are used by the kernel of SINGULAR and referred to as reserved attributes.
Reserved attributes are: isSB, isHomog, isCI, isCM, rank, withSB, withHilb, withRes, withDim, withMult.

bareiss: bareiss(M) applies the sparse Gauß-Bareiss algorithm to a module (or with type conversion to a matrix) M with an "optimal" pivot strategy. The vectors of the module are the columns of the matrix, hence elimination takes place with respect to rows. With only one parameter a complete elimination is done. Result is a list: the first entry is a module with a minimal independent set of vectors (as a matrix lower triangular), the second entry an intvec with the permutation of the rows w.r.t. the original matrix, that is, a k at position l indicates that row k was carried over to the row l.
The further parameters control the algorithm. bareiss(M, i, j) does not attempt to diagonalize the last i rows in the elimination procedure and stops computing when the remaining number of vectors (columns) to reduce is at most j.

betti: betti(L) computes the graded Betti numbers of a minimal resolution of R^n/M, if R denotes the basering and M a homogeneous submodule of R^n and the argument represents a resolution of R^n/M. The entry d of the intmat at place (i,j) is the minimal number of generators in degree $i + j$ of the j-th syzygy module (= module of relations) of R^n/M (the 0–th (resp. 1–st) syzygy module of R^n/M is R^n (resp. M)).The argument is considered to be the result of a res/sres/mres/nres/lres command. This implies that a zero is only allowed (and counted) as a generator in the first module. For the computation betti uses only the initial monomials. This could lead to confusing results for a non-homogeneous input.

bracket: bracket(p,q) computes the Lie bracket $[p, q] = pq - qp$ of p with q. It uses special routines, based on the Leibniz rule.

char: char(R) returns the characteristic of the coefficient field of the ring R.

char_series: char_series(I) computes a matrix. The rows of the matrix represent the irreducible characteristic series (cf. [167]) of the ideal I with respect to the current ordering of variables.

charstr: charstr(R) returns the description of the coefficient field of the ring R as a string.

cleardenom: cleardenom(f) multiplies the polynomial f, respectively vector f, by a suitable constant to cancel all denominators of the coefficients and then divides it by its content.

close: close(L) closes the link L.

coef: coef(f,m) determines the monomials in f divisible by one of the variables appearing in m (which is a product of ring variables) and the coefficients of these monomials as polynomials in the remaining variables.

coeffs: coeffs(J,z) develops each polynomial of J as a univariate polynomial in the given ring variable z, and returns the coefficients as a $k \times d$ matrix M, where $d-1$ is the maximal z–degree of all occurring polynomials; k is the number of generators if J is an ideal ($k = 1$, if J is a polynomial). If J is a vector or a module this procedure is repeated for each component and the resulting matrices are appended. An (optional) third argument T is used to return the matrix T of coefficients such that matrix(J) $= TM$.

coeffs(M,K,p) returns a matrix A of coefficients with $KA = M$ such that the entries of A do not contain any variable from p. Here K is a set of monomials, respectively vectors, with monomial entries, in the variables appearing in p, p is a product of variables (if this argument is not given, then the product of all ring variables is taken as default argument).

M is supposed to consist of elements of (respectively have entries in) a finitely generated module over a ring in the variables not appearing in p. K should contain the generators of M over this smaller ring.

If K does not contain all M, then $KA = M'$ where M' is the part of M corresponding to the monomials of K.

contract: contract(I,J) contracts each of the n elements of the second ideal J by each of the m elements of the first ideal I, producing an $m \times n$–matrix. Contraction is defined on monomials by:

$$\text{contract}(x^A, x^B) := \begin{cases} x^{B-A}, & \text{if } B \geq A \text{ component–wise} \\ 0, & \text{otherwise} \end{cases}$$

where A and B are the multi–exponents of the ring variables represented by x. contract is extended bilinearly to all polynomials.

defined: defined(name) returns a value $\neq 0$ *(TRUE)* if there is a user-defined object with this name, and 0 *(FALSE)* otherwise.

deg: deg(f) returns the maximal (weighted) degree of the terms of a polynomial or a vector f; the weights are the weights used for the first block of the ring ordering. deg(0) is -1.

degree: `degree(I)` computes the (weighted) degree of the projective variety, respectively sheaf over the projective variety, defined by the ideal, respectively module, generated by the leading monomials of the input. This is equal to the (weighted) degree of the projective variety, respectively sheaf over the projective variety, defined by the ideal, respectively module, if the input is a standard basis with respect to a (weighted) degree ordering.

delete: `delete(L,n)` deletes the n-th element from the list L.

det: `det(M)` computes the determinant of the matrix M.

diff: `diff(f,x)` computes the partial derivative of the polynomial f by the ring variable x.

dim: `dim(I)` computes the dimension of the ideal, respectively module, generated by the leading monomials of the given generators of the ideal I, respectively module. This is also the dimension of the ideal if it is represented by a standard basis.

division: computes a division with remainder. `division(I,J)` returns a list `T,R,U` where `T` is a matrix, `R` is an ideal, respectively a module, and `U` is a diagonal matrix of units such that `matrix(I)*U=matrix(J)*T+matrix(R)` is a standard representation for the (weak) normal form `R` of `I` with respect to a standard basis of `J`.

dump: `dump(L)` dumps (that is, writes in one "message" or "block") the state of the SINGULAR session (that is, all defined variables and their values) to the link L (which must be either an ASCII or MP link) such that a `getdump` can retrieve it later on.

eliminate: `eliminate(I,m)` eliminates variables occurring as factors of the second argument m from an ideal I, respectively module, by intersecting it with the subring not containing these variables. `eliminate` does not need a special ordering nor a standard basis as input.

Since elimination is expensive, for homogeneous input it might be useful first to compute the Hilbert function of the ideal (first argument) with a fast ordering (for example, `dp`). Then make use of it to speed up the computation: a Hilbert–driven elimination uses the intvec provided as the third argument.

Remark: in a noncommutative algebra, not every subset of a set of variables generates a proper subalgebra. But if it is so, there may be cases, when no elimination is possible. In these situations error messages will be reported.

envelope: `envelope(r)` creates an enveloping algebra of a given algebra r, that is $A^{env} = A \otimes_K A^{opp}$, where A^{opp} is the opposite algebra of A.

eval: evaluates (quoted) expressions. Within a quoted expression, the quote can be "undone" by an `eval` (that is, each `eval` "undoes" the effect of exactly one quote). Used only when receiving a quoted expression from an MP link, with `quote` and `write` to prevent local evaluations when writing to an MP link.

ERROR: immediately interrupts the current computation, returns to the top level, and displays the argument `string_expression` as error message.

This should be used as an emergency, respectively failure, exit within procedures.

example: `example topic;` computes an example for `topic`. Examples are available for all SINGULAR kernel and library functions.

execute: executes a string containing a sequence of SINGULAR commands.

exit: exits (quits) SINGULAR, works also from inside a procedure or from an interrupt.

extgcd: computes extended gcd: the first element is the greatest common divisor of the two arguments, the second and third are factors such that if `list L=extgcd(a,b);` then $L[1]=a*L[2]+b*L[3]$. Polynomials must be univariate to apply `extgcd`.

facstd: `facstd(I)` returns a list of ideals computed by the factorizing Gröbner basis algorithm. The intersection of these ideals has the same zero–set as the ideal I, that is, the radical of the intersection coincides with the radical of the input ideal. In many (but not all!) cases this is already a decomposition of the radical of the ideal. (Note, however, that, in general, no inclusion between the input and output ideals holds.) A second, optional, argument can be a list of polynomials, which define non–zero constraints. Hence, the intersection of the output ideal has a zero–set, which is the (closure of the) complement of the zero–set of the second argument in the zero–set of the first argument.

factorize: `factorize(f)` computes the irreducible factors (as an ideal) of the polynomial f together with or without the multiplicities (as an intvec) depending on the optional second argument:

  0:   returns factors and multiplicities, first factor is a constant.
       (`factorize(f)` is a short notation for `factorize(f,0)`).

  1:   returns non–constant factors (no multiplicities).

  2:   returns non–constant factors and multiplicities.

fetch: `fetch(R,I)` maps the object I defined over the ring R to the basering. `fetch` is the canonical map between rings and qrings: the i–th variable of the source ring R is mapped to the i–th variable of the basering. The coefficient fields must be compatible.

Compared to `imap`, `fetch` uses the position of the ring variables, not their names.

fglm: `fglm(R,I)` computes for the ideal I in the ring R a reduced Gröbner basis in the basering, by applying the so–called FGLM (Faugère, Gianni, Lazard, Mora) algorithm. The main application is to compute a lexicographical Gröbner basis from a reduced Gröbner basis with respect to a degree ordering. This can be much faster than computing a lexicographical Gröbner basis directly. The ideal must be zero–dimensional and given as a reduced Gröbner basis in the ring R.

fglmquot: `fglmquot(I,p)` computes a reduced Gröbner basis of the ideal quotient $I : \langle p \rangle$ of a zero–dimensional ideal I and an ideal generated by a polynomial p, by using FGLM–techniques. The ideal must be zero–

dimensional and given as a reduced Gröbner basis in the given ring. The polynomial must be reduced with respect to the ideal.

find: returns the first position of a substring in a string or 0 (if not found), starts the search at the position given in the (optional) third argument.

finduni: `finduni(I)` returns an ideal, which is contained in the given ideal I such that the i-th generator is a univariate polynomial in the i-th ring variable. The polynomials have minimal degree with respect to this property. The ideal must be zero–dimensional and given as a reduced Gröbner basis in the current ring.

fprintf: `fprintf(l,fmt,...)` performs output formatting. The second argument is a format control string. Additional arguments may be required, depending on the content of the control string. A series of output characters is generated as directed by the control string; these characters are written to the link l. The control string fmt is simply text to be copied, except that the string may contain conversion specifications.

freemodule: `freemodule(n)` creates the free module of rank n generated by `gen(1),...,gen(n)`.

frwalk: `frwalk(R,I)` computes for the ideal I in the ring R a Groebner basis in the current ring, by applying the fractal walk algorithm. The main application is to compute a lexicographical Gröbner basis from a reduced Gröbner basis with respect to a degree ordering. This can be much faster than computing a lexicographical Gröbner basis directly.

gcd: computes the greatest common divisor of two integers or two polynomials.

gen: `gen(i)` is the i-th free generator of a free module.

getdump: `getdump(L)` reads the content of the entire file, respectively link, L and restores all variables from it. For ASCII links, `getdump` is equivalent to an `execute(read(L))` command. For MP links, `getdump` should only be used on data, which were previously dumped.

groebner: `groebner(I)` computes the standard basis of the argument I (ideal or module), by a heuristically chosen method: Possibilities are `std`, `slimgb`, `fglm` or `stdhilb`. If a second argument `wait` is given, then the computation proceeds at most `wait` seconds. That is, if no result could be computed in `wait` seconds, then the computation is interrupted, 0 is returned, a warning message is displayed, and the global variable `groebner_error` is defined.

help: `help topic;` displays online help information for `topic` using the currently set help browser. If no `topic` is given, the title page of the manual is displayed.

- `?` may be used instead of `help`.
- `topic` can be an index entry of the SINGULAR manual or the name of a (loaded) procedure, which has a help section.
- `topic` may contain wildcard characters (that is, * characters).

- If a (possibly "wildcarded") topic cannot be found (or uniquely matched) a warning is displayed and no help information is provided.
- If topic is the name of a (loaded) procedure whose help section has changed w.r.t. the help available in the manual then, instead of displaying the respective help section of the manual in the help browser, the "newer" help section of the procedure is simply printed to the terminal.
- The browser in which the help information is displayed can be set either with the command–line option
  `--browser=<browser>`
  or, if SINGULAR is already running, with the command
  `system("--browser", "<browser>")`.
  Use the command
  `system("browsers");`
  for a list of all available browsers.

highcorner: `highcorner(I)` returns the smallest monomial not contained in the ideal, respectively module, generated by the initial terms of the given generators of I. If the generators are a standard basis, this is also the smallest monomial not contained in the ideal, respectively module I. If the ideal, respectively module, is not zero–dimensional, 0 is returned. Hence, `highcorner` is always 1 or 0 for global monomial orderings.

hilb: computes the (weighted) Hilbert series of the ideal, respectively module, defined by the leading terms of the generators of the given ideal, respectively module. If `hilb(I)` is called with one argument (the ideal or module I), then the first and second Hilbert series, together with some additional information, are displayed. If `hilb(I,n)` is called with two arguments, then the n–th Hilbert series is returned as an intvec, where $n = 1, 2$ is the second argument. More precisely, if `hilb(I,n)` $= v_0, \ldots, v_d, 0$, then the n–th Hilbert series ($Q(t)$, respectively $G(t)$) is $\sum_{i=0}^{d} v_i t^i$, cf. page 316.
If a weight vector w is a given as third argument, then the Hilbert series is computed with respect to these weights w (by default all weights are set to 1).
The last entry of the returned intvec is not part of the actual Hilbert series, but is used in the Hilbert driven standard basis computation.
If the input is homogeneous with respect to the weights and a standard basis, the result is the (weighted) Hilbert series of the original ideal, respectively module.

homog: `homog(I)` tests for homogeneity: returns 1 for homogeneous input, 0 otherwise. `homog(I,t)` homogenizes polynomials, vectors, generators of ideals or modules I by multiplying each monomial with a suitable power of the given ring variable t (which must have weight 1).

imap: is the map between rings and qrings with compatible ground fields, which is the identity on variables and parameters of the same name and 0 otherwise. `imap(R,I)` maps I defined over the ring R to the basering.

Compared with fetch, imap uses the names of variables and parameters. Unlike map and fetch, imap can map parameters to variables.

impart: returns the imaginary part of a number in a complex ground field, returns 0 otherwise.

indepSet: indepSet(I) computes a maximal set U of independent variables of the ideal I given by a standard basis. If v is the result then $v[i]$ is 1 if and only if the i-th variable of the ring, $x(i)$, is an independent variable. Hence, the set U consisting of all variables $x(i)$ with $v[i] = 1$ is a maximal independent set.

insert: insert(L,I) inserts a new element I into a list L at the first place or (if called with three arguments) after the given position.

interred: interred(I) interreduces a set of polynomials or vectors I .

intersect: intersect(I,J, ...) computes the intersection of the ideals, respectively modules, $I, J, \ldots$.

jacob: jacob(f) computes the Jacobian ideal, respectively Jacobian matrix, generated by all partial derivatives of the input f.

janet: janet(I) computes the Janet basis of the ideal I, resp. a standard basis if 1 is given as the second argument.

jet: jet(f,k) deletes from the first argument, f, all terms of degree larger than the second argument, k. If a third argument, w, of type intvec is given, the degree is replaced by the weighted degree defined by w. jet is independent of the given monomial ordering.

kbase: kbase(I) computes a vector space basis (consisting of monomials) of the quotient ring by the ideal, respectively of a free module by the module, I, in case it is finite dimensional and if the input is a standard basis with respect to the ring ordering.

With two arguments: computes the part of a vector space basis of the respective quotient with degree of the monomials equal to the second argument. Here, the quotient does not need to be finite dimensional.

Note: in the noncommutative case, a ring modulo an ideal has a ring structure if and only if the ideal is two-sided.

kill: deletes objects.

killattrib: deletes the attribute given as the second argument

koszul: koszul(d,n) computes a matrix of the Koszul relations of degree d of the first n ring variables. koszul(d,id) computes a matrix of the Koszul relations of degree d of the generators of the ideal id. koszul(d,n,id) computes a matrix of the Koszul relations of degree d of the first n generators of the ideal id.

laguerre: laguerre(p,n,m) computes all complex roots of the univariate polynomial p using Laguerre's algorithm. The second argument, n, defines the precision of the fractional part if the ground field is the field of rational numbers, otherwise it will be ignored. The third argument (can be 0, 1 or 2) gives the number of extra runs for Laguerre's algorithm (with corrupted roots), leading to better results.

lead: `lead(I)` returns the leading term(s) of a polynomial, a vector, respectively of the generators of an ideal or module I with respect to the monomial ordering.

leadcoef: `leadcoef(f)` returns the leading coefficient of a polynomial or a vector f with respect to the monomial ordering.

leadexp: `leadexp(f)` returns the exponent vector of the leading monomial of a polynomial or a vector f. In the case of a vector the last component is the index in the vector.

leadmonom: `leadmonom(f)` returns the leading monomial of a polynomial or a vector f as a polynomial or vector, whose coefficient is one.

LIB: reads a library of procedures from a file. If the given filename does not start with . or / and cannot be located in the current directory, each directory contained in the SearchPath (SINGULARPATH) for libraries is searched for a file of this name.

lift: `lift(m,sm)` computes the transformation matrix, which expresses the generators of a submodule in terms of the generators of a module. More precisely, if m denotes the module (or ideal), if sm denotes the submodule (or subideal), and if T denotes the transformation matrix returned by lift, then `matrix(sm)*U=matrix(m)*T`, where U is a diagonal matrix of units. U is always the unity matrix if the basering is a polynomial ring (not a power series ring). U is stored in the optional third argument.

liftstd: `liftstd(m,T)` returns a standard basis of an ideal or module and the transformation matrix from the given ideal, respectively module, to the standard basis. That is, if m is the ideal or module, sm the standard basis returned by `liftstd`, and T the transformation matrix, then `matrix(sm)=matrix(m)*T`.

listvar: lists all (user–)defined names in the current namespace:
- `listvar():`   all currently visible names except procedures,
- `listvar(type):`   all currently visible names of the given type,
- `listvar(ring_name):`   all names, which belong to the given ring,
- `listvar(name):`   the object with the given name,
- `listvar(all):`   all names except procedures,
- `listvar(proc):`   all names of currently available library procedures.

lres: computes a free resolution of a homogeneous ideal using La Scala's algorithm. It can be used in the same way as `res`.

maxideal: `maxideal(i)` returns the i–th power of the maximal ideal generated by all ring variables (`maxideal(i)=1` for $i \leq 0$).

memory: returns statistics concerning the memory management:
- `memory(0)` is the number of active (used) bytes,
- `memory(1)` is the number of bytes allocated from the operating system,
- `memory(2)` is the maximal number of bytes ever allocated from the operating system during the current SINGULAR session.

To monitor the memory usage during ongoing computations the option `mem` should be set.

minbase: `minbase(I)` returns a minimal set of generators of an ideal, respectively module I, if the input is either homogeneous or if the ordering is local.

minor: `minor(M,r)` returns the set of all minors (=subdeterminants) of the given size r of a matrix M. The optional third argument must be a standard basis. If a third argument is given, the computations will be performed modulo that ideal.

minres: `minres(L)` minimizes a free resolution L of an ideal or module.

modulo: `modulo(h1,h2)` returns generators of the kernel of the map $R^k \to R^\ell/h_2$ induced by h_1; k is the number of generators of h_1, that is, it represents $h_1/(h_1 \cap h_2) \cong (h_1 + h_2)/h_2$ where h_1 and h_2 are considered as submodules of the same free module R^ℓ ($\ell = 1$ for ideals).

monitor: `monitor("xxx.txt", "io")` controls the recording of all user input and/or programme output into the file `xxx.txt`. The second argument describes what to log: `"i"` means input, `"o"` means output, `"io"` for both. The default for the second argument is `"i"`. Each `monitor` command closes a previous monitor file and opens the file given by the first string expression. `monitor ("")` turns off recording.

mpresmat: `mpresmat(I,n)` computes the multipolynomial resultant matrix of the ideal I. It uses the sparse resultant matrix method of Gelfand, Kapranov and Zelevinsky (second parameter = 0) or the resultant matrix method of Macaulay (second parameter = 1).

When using the resultant matrix method of Macaulay the input system must be homogeneous. The number of elements in the input system must be the number of variables in the basering, plus one.

mres: computes a minimal free resolution of an ideal or module M by the standard basis method. More precisely, let $A = \mathtt{matrix}(M)$, then `mres(M,k)` computes a free resolution of $\mathrm{Coker}(A) = F_0/M$

$$\ldots \to F_2 \xrightarrow{A_2} F_1 \xrightarrow{A_1} F_0 \to F_0/M \to 0\,,$$

where the columns of the matrix A_1 are a minimal set of generators of M if the basering is local or if M is homogeneous. If k is not zero then the computation stops after k steps and returns a list of modules $M_i = \mathtt{module}(A_i), i = 1 \ldots k$.

`mres(M,0)` returns a resolution consisting of, at most, $n+2$ modules, where n is the number of variables of the basering. Let `list L=mres(M,0);` then `L[1]` consists of a minimal set of generators of the input, `L[2]` consists of a minimal set of generators for the first syzygy module of `L[1]`, etcetera.

mstd: `mstd(I)` returns a list whose first entry is a standard basis for the ideal, respectively module I. If the monomial ordering is global then the second entry is both a generating set for the ideal, respectively module, and a subset of the standard basis. If, additionally, the input is homogeneous then the second entry is a minimal generating set for I.

mult: `mult(I)` computes the multiplicity of the monomial ideal, respectively module I, generated by the leading monomials of the input. If the input is a standard basis of an ideal, respectively module, with respect to a local degree ordering then it returns the multiplicity of this ideal, respectively module, (in the sense of Samuel, with respect to the maximal ideal of the basering).

nameof: returns the name of an expression as string.

names: `names();` returns the names of all user–defined variables, which are ring independent (this includes the names of procedures), `names(R);` returns the names belonging to the ring R.

ncalgebra: executed in the commutative basering R, say, in k variables $x_1, \ldots, x_k$, `ncalgebra(C,D)` creates the noncommutative extension of R subject to relations $\{x_j x_i = c_{ij} \cdot x_i x_j + d_{ij}, 1 \leq i < j \leq k\}$, where c_{ij} and d_{ij} must be put into two strictly upper triangular matrices C with entries c_{ij} from the ground field of R and D with polynomial entries d_{ij} from R. If $\forall i < j$, $c_{ij} = n$, one can input a number n instead of the matrix C. If $\forall i < j$, $d_{ij} = p$, one can input a poly p instead of the matrix D.

ncols: `ncols(M)` returns the number of columns of a matrix M or an intmat or the number of given generators of the ideal, including zeros.

npars: `npars(R)` returns the number of parameters of a ring R.

nres: computes a free resolution of an ideal or module M, which is minimized from the second module onwards (by the standard basis method).

nrows: `nrows(M)` returns the number of rows of a matrix M, an intmat or an intvec, respectively the minimal rank, of a free module in which the given module or vector lives (the index of the last non–zero component).

nvars: `nvars(R)` returns the number of variables of a ring R.

open: `open(L)` opens the link L.

oppose: `oppose(r,p)` for a given object p in the given ring r, creates its opposite object in the opposite ring (r^{opp} is assumed to be the current ring).

opposite: `opposite(r)` creates an opposite algebra of a given algebra r.

option: lists all set options. `option(option_name)` sets an option. To disable an option, use the prefix `no`. The state of all options is dumped to an intvec by `option(get)`. The state of all options from an intvec (produced by `option(get)`) is restored by `option(set,intvec_expression)`. The following options are used to manipulate the behaviour of computations and act like boolean switches. Notice that some options are ring dependent and reset to their default values on a change of the current basering:

none: turns off all options.

returnSB: the functions `syz`, `intersect`, `quotient`, `modulo` return a standard base instead of a generating set if `returnSB` is set. This option should not be used for `lift`.

fastHC: tries to the find the highest corner of the staircase as fast as possible during a standard basis computation (only used for local orderings).

intStrategy: avoids division of coefficients during standard basis computations. This option is ring dependent. By default, it is set for rings with characteristic 0 or parameters.

notRegularity: disables the regularity bound for **nres** and **mres**.

notSugar: disables the sugar strategy during standard basis computation.

notBuckets: disables the bucket representation of polynomials during standard basis computations. This option usually decreases the memory usage but increases the computation time. It should only be set for memory critical standard basis computations.

prot: shows protocol information indicating the progress during the following computations: **facstd, fglm, groebner, lres, mres, minres, mstd, res, sres, std, stdfglm, stdhilb, syz**.

redSB: computes reduced standard bases S (up to normalization) in any standard basis computation in rings with global monomial orderings. Warning: since, for efficiency reasons, SINGULAR prefers to compute with integers, the leading coefficients are not necessarily 1. Use **simplify(S,1)** to obtain a reduced standard basis.

redTail: reduction of the tails of polynomials during standard basis computations. This option is ring dependent. By default, it is set for rings with global degree orderings and not set for all other rings.

redThrough: for inhomogeneous input, polynomial reductions during standard basis computations are never postponed, but always finished through. This option is ring dependent. By default, it is set for rings with global degree orderings and not set for all other rings.

sugarCrit: uses criteria similar to the homogeneous case to keep more useless pairs.

weightM: automatically computes suitable weights for the weighted ecart and the weighted sugar method.

ord: ord(f) returns the (weighted) degree of the initial term of the polynomial or a vector f; the weights are the weights used for the first block of the ring ordering. ord(0) is −1.

ordstr: ordstr(R) returns the description of the monomial ordering of the ring R as string.

par: par(n) returns the n–th parameter of the basering.

pardeg: pardeg(p) returns the degree of a number p considered as a polynomial in the ring parameters.

parstr: parstr(R) returns the list of parameters of the ring R as a string. parstr(R,n) returns the name of the n–th parameter. If the ring name is omitted, the basering is used, thus parstr(n) is equivalent to parstr(basering,n).

preimage: `preimage(R,phi,I)` returns the preimage of the ideal I under the map `phi`. The monomial ordering must be global. The second argument `phi` has to be a map from the basering to the given ring R (or an ideal defining such a map), and the ideal has to be an ideal in the given ring R. To compute the kernel of a map, the preimage of the zero–ideal has to be computed. Use preimageLoc from `ring.lib` for the non–global case.

Remark: In the noncommutative case, it is implemented only for maps $A \to B$, where A is a commutative ring.

prime: `prime(n)` returns the largest prime smaller or equal to n; returns 2 for all arguments smaller than 3 and 32003 for arguments ≥ 32004.

print: `print(E)` prints the expression E to the terminal and has no return value. `print(E,F)` prints the expression E in a special format. The format string F determines which format to use to generate the output:

"betti"   The Betti numbers are printed in a matrix–like format where the entry d in row i and column j is the minimal number of generators in degree $i + j$ of the j–th syzygy module of R^n/M (the 0–th, respectively 1–st, syzygy module of R^n/M are R^n, respectively M).

"%s"   returns `string(` expression `)`.

"%2s"   similar to "%s", except that newlines are inserted after every comma and at the end.

"%l"   similar to "%s", except that each object is embraced by its type such that it can be directly used for "cutting and pasting"

"%2l"   similar to "%l", except that newlines are inserted after every comma and at the end.

"%;"   returns the string equivalent to typing `expression;`.

"%t"   returns the string equivalent to typing `type expression;`.

"%p"   returns the string equivalent to typing `print(expression);`.

"%b"   returns the string equivalent to typing `print(expression, "betti");`.

prune: `prune(M)` returns the module M minimally embedded in a free module such that the corresponding factor modules are isomorphic.

qhweight: `qhweight(I)` computes the weight vector of the variables for a quasihomogeneous ideal I. If the input is not weighted homogeneous, an intvec of zeros is returned.

quote: prevents expressions from evaluation. Used only in connection with write to MPfile links, prevents evaluation of an expression before sending it to another SINGULAR process. Within a quoted expression, the quote can be "undone" by an `eval` (that is, each `eval` "undoes" the effect of exactly one quote).

quotient: computes the ideal quotient, respectively module quotient. Let R be the basering, I, J ideals in R, and M, N modules in R^n. Then $quotient(I,J) = \{a \in R \mid aJ \subset I\}$, $quotient(M,J) = \{b \in R^n \mid bJ \subset M\}$ and $quotient(M,N) = \{a \in R \mid aN \subset M\}$. In the noncommutative case, quotient can only be used for two-sided ideals (bimodules), otherwise the result may have no meaning.

random: random(a,b) returns a random integer between the integer a and the integer b. random(a,b,c) returns a $b \times c$ random intmat. The absolute value of the entries of the matrix is smaller than or equal to the integer a.

read: read(L) reads data from a link.

For ASCII links, the content of the entire file is returned as one string. If the ASCII link is the empty string, read reads from standard input. For MP links, one expression is read from the link and returned after evaluation. For MPtcp links the read command blocks as long as there is no data to be read from the link. The status command can be used to check whether or not there is data to be read. For DBM links, a read with one argument returns the value of the next entry in the data base, and a read with two arguments returns the value to the key given as the second argument from the data base.

reduce: reduce(I,J) reduces a polynomial, vector, ideal or module I to its normal form with respect to an ideal or module J represented by a standard basis.[2] Returns 0 if and only if the polynomial (respectively vector, ideal, module) is an element (respectively subideal, submodule) of the ideal (respectively module). reduce(I,J,1) does no tail reduction. reduce(I,J,U) reduces $U^{-1}I$ modulo J. This works only for zero dimensional ideals (respectively modules) J and gives a reduced normal form; U has to be a diagonal matrix with units on the diagonal. One may give a degree bound in the fourth argument with respect to a weight vector in the fifth argument in order have a finite computation. If some of the weights are zero, the procedure may not terminate!

regularity: regularity(L) computes the regularity of a homogeneous ideal, respectively module, from a minimal resolution L.

Let $0 \to \bigoplus_a K[x]e_{a,n} \to \ldots \to \bigoplus_a K[x]e_{a,0} \to I \to 0$ be a minimal resolution of I considered with homogeneous maps of degree 0. The regularity is the smallest number s with the property $\deg(e_{a,i}) \le s + i$ for all i.

If the input to the commands res and mres is homogeneous, then the regularity is computed and used as a degree bound during the computation unless option(notRegularity); is given.

repart: returns the real part of a number from a complex ground field, returns its argument otherwise.

res: res(M,k) computes a (possibly minimal) free resolution of an ideal or module M using a heuristically chosen method. The second argument k specifies the length of the resolution. If it is not positive then k is assumed to be the number of variables of the basering. If a third argument is given, the returned resolution is minimized.

---

[2] If J is not a standard basis, then a warning is displayed (the result has in general no invariant meaning).

reservedName: `reservedName()` prints a list of all reserved identifiers, `reservedName("xyz")` tests whether the string xyz is a reserved identifier.

resultant: `resultant(f,g,x)` computes the resultant of f and g with respect to the variable x.

ringlist: `ringlist(r)` decomposes a ring/qring into a list of 4 (or 6 in the noncommutative case) components:

1 the field description in the following format:
   - for Q, Z/p: the characteristic, type int (0 or prime number)
   - for real, complex: a list of: the characteristic, type int (always 0) the precision, type list (2 integers: external, internal precision) the name of the imaginary unit, type string
   - for transcendental or algebraic extensions: described as a ringlist (that is, as list L with 4 entries:
     L[1] the characteristic,
     L[2] the names of the parameters,
     L[3] the monomial ordering for the ring of parameters (default: lp),
     L[4] the minimal polynomial (as ideal))
2 the names of the variables (a list L of strings: L[i] is the name of the i-th variable)
3 the monomial ordering (a list L of lists): each block L[i] consists of
   - the name of the ordering (string)
   - parameters specifying the ordering and the size of the block (intvec : typically the weights for the variables [default: 1])
4 the quotient ideal.
   In the noncommutative case, two additional fields appear:
5 square matrix C with nonzero upper triangle, containing structural coefficients of a G-algebra (this corresponds to the matrix C from the definition of G-algebras)
6 square matrix D, containing structural polynomials of a G-algebra (this corresponds to the matrix D from the definition of G-algebras). ¿From a list L of such structure, a new ring S may be defined by the command def S=ring(L).

rvar: `rvar(x)` returns the number of the variable x if the name is a ring variable of the basering or if the string is the name of a ring variable of the basering; returns 0 if not.

setring: `setring(S)` changes the basering to the (already defined) ring S.

simplex: `simplex(M,m,n,m1,m2,m3)` perform the simplex algorithm for the tableau given by the input:
   - M matrix of numbers :first row describing the objective function (maximize problem), the remaining rows describing constraints;
   - n = number of variables;
   - m = total number of constraints;
   - m_1 = number of $\leq$ constraints ($rows\ 2\ ...\ m_1 + 1\ of\ M$);
   - m_2 = number of $\geq$ constraints ($rows\ m_1 + 2\ ...\ m_1 + m2 + 1\ of\ M$);

- m_3 = number of =constraints.

The following assumptions are made:

- ground field is of type (real,N), $N \geq 4$;
- the matrix M is of size m x n;
- $m = m_1 + m_2 + m_3$;
- the entries M[2,1] ,..., M[m+1,1] are non-negative;
- the variables x(i) are non-negative;
- a row b, a(1) ,..., a(n) corresponds to b+a(1)x(1)+...+a(n)x(n);
- for a $\leq$, $\geq$, or = constraint: add "in mind" ≥ 0, ≤ 0, or =0.

The output is a list L with

- L[1] = matrix
- L[2] = int: 0 = finite solution found; 1 = unbounded; -1 = no solution; -2 = error occurred;
- L[3] = intvec : L[3][k] = number of variable which corresponds to row k+1 of L[1];
- L[4] = intvec : L[4][j] = number of variable which is represented by column j+1 of L[1] ("non-basis variable");
- L[5] = int : number of constraints (= m);
- L[6] = int : number of variables (= n).

The solution can be read from the first column of L[1] as is done by the procedure simplexOut in solve.lib.

simplify: simplify(f,n) returns the "simplified" first argument f depending on the simplification rule given by n. The simplification rules are sums of the following basic rules:

1	normalize (make leading coefficients 1),
2	erase zero generators (respectively columns),
4	keep only the first one of identical generators (respectively columns),
8	keep only the first one of generators (respectively columns), which differ only by a factor in the ground field,
16	keep only those generators (respectively columns) whose leading monomials differ,
32	keep only those generators (respectively columns) whose leading monomials are not divisible by other ones.

size: depends on the type of argument: size(M) for an ideal or module M returns the number of (non–zero) generators. For a string, intvec, list or resolution it returns the length, that is, the number of characters, entries or elements. For a polynomial or vector it returns the number of monomials. For a matrix or intmat it returns the number of entries. For a ring it returns the number of elements in the ground field (for $\mathbb{Z}/p$ and algebraic extensions) or -1.

slimgb: slimgb(I) returns a Groebner basis of an ideal I with respect to the monomial ordering of the basering, which has to be global. Note: It is designed to keep polynomials slim (short with small coefficients).

sortvec: `sortvec(I)` computes the permutation v, which orders the ideal, respectively module, I by its leading monomials, starting with the smallest, that is, $I(v[i]) < I(v[i+1])$ for all i if the leading monomials of the generators of I are different.

sres: computes a free resolution of an ideal or module with Schreyer's method. The ideal, respectively module, has to be a standard basis. More precisely, let M be given by a standard basis and A_1 =`matrix`(M). Then `sres(M,k)` computes a free resolution of $\mathrm{Coker}(A_1) = F_0/M$

$$\ldots \to F_2 \xrightarrow{A_2} F_1 \xrightarrow{A_1} F_0 \to F_0/M \to 0 \,.$$

If the int expression k is not zero then the computation stops after k steps and returns a list of modules (given by standard bases), M_i =`module`(A_i), $i = 1 \ldots k$. `sres(M,0)` returns a list of n modules, where n is the number of variables of the basering. Even if `sres` does not compute a minimal resolution, the `betti` command gives the true betti numbers! In many cases of interest `sres` is much faster than any other known method. Let `list L=sres(M,0);` then `L[1]=M` is identical to the input, `L[2]` is a standard basis with respect to the Schreyer ordering of the first syzygy module of `L[1]`, etc. ($L[i] = M_i$ in the notations from above.)

status: returns the status of the link as asked for by the second argument. If a third argument is given, the result of the comparison to the status string is returned: (`status(1,s1)==s2`) is equivalent to `status(1,s1,s2)`. If a fourth integer argument (say, i) is given and if `status(1,s1,s2)` yields 0, then the execution of the current process is suspended (the process is put to "sleep") for approximately i microseconds, and afterwards the result of another call to `status(1,s1,s2)` is returned. The latter is useful for "polling" the `read` status of MPtcp links such that busy loops are avoided. Note that on some systems, the minimum time for a process to be put to sleep is one second. The following string expressions are allowed:

`"name"`   the name string given by the definition of the link (usually the filename)

`"type"`   returns `"ASCII"`, `"MPfile"`, `"MPtcp"` or `"DBM"`

`"open"`   returns `"yes"` or `"no"`

`"openread"`
           returns `"yes"` or `"no"`

`"openwrite"`
           returns `"yes"` or `"no"`

`"read"`   returns `"ready"` or `"not ready"`

`"write"`   returns `"ready"` or `"not ready"`

`"mode"`   (depending on the type of the link and its status)
           `""`,`"w"`,`"a"`,`"r"` or `"rw"`

std: `std(I)` returns a standard basis of the ideal or module I with respect to the monomial ordering of the basering. Use an optional second argument of type intvec as Hilbert series if the ideal, respectively module, is homogeneous (Hilbert driven standard basis computation). If the ideal is quasihomogeneous with respect to weights w and if the Hilbert series is computed w.r.t. to these weights, then use w as third argument. Use an optional second argument of type poly, respectively vector, to construct the standard basis from an already computed one (given as the first argument) and one additional generator (the second argument). For global orderings, use the groebner command instead, which heuristically chooses the "best" algorithm to compute a Gröbner basis. To view the progress of computations, use option(prot).

stdfglm: stdfglm(I) computes the standard basis of the 0–dimensional ideal I in the basering via fglm (from the ordering given as the second argument to the ordering of the basering). If no second argument is given, "dp" is used.

stdhilb: stdhilb(I) computes a standard basis of the ideal I in the basering, via a Hilbert driven standard basis computation. It contains the computation of the Hilbert function of the homogenized ideal I^h.

subst: subst(f,x,m) substitutes the ring variable x by the term (a polynomial of length at most 1) m.
Use map for substitutions by polynomials.

system: interface to internal data and the operating system.

syz: syz(I) computes the first syzygy (that is, the module of relations of the given generators) of the ideal, respectively module, I.

trace: trace(A) returns the trace of the matrix A.

transpose: transpose(A) transposes the matrix A.

twostd: twostd(I) returns a left Groebner basis of the two-sided ideal, generated by the input, treated as a set of two-sided generators.

type: prints the name, level, type and value of a variable. To display the value of an expression, it is sufficient to type the expression followed by ; .

typeof: returns the type of an expression as string. Possible types are: "ideal", "int", "intmat", "intvec", "list", "map", "matrix", "module", "number", "none", "poly", "proc", "qring", "resolution", "ring", "string", "vector".

uressolve: uressolve(I,a,b,c) computes all complex roots of the zero–dimensional ideal I. Makes either use of the multipolynomial resultant of Macaulay ($a = 1$), which works only for homogeneous ideals, or uses the sparse resultant of Gelfand, Kapranov and Zelevinsky ($a = 0$).
The third argument b defines the precision of the fractional part if the ground field is the field of rational numbers, otherwise it will be ignored. The fourth argument c (can be 0, 1 or 2) gives the number of extra runs of Laguerre's algorithm, leading to better results.

vandermonde: vandermonde(p,v,d) computes the (unique) polynomial of degree d with prescribed values $v_1, \ldots, v_N$ at the points

$$p_0 = (p_{0,1}, \ldots, p_{0,n}), \ldots, p_{N-1} = (p_{N-1,1}, \ldots, p_{N-1,n}),$$

where n denotes the number of ring variables and $N := (d+1)^n$.
The returned polynomial is $\sum c_{\alpha_1 \ldots \alpha_n} \cdot x_1^{\alpha_1} \cdot \ldots \cdot x_n^{\alpha_n}$, where the coefficients $c_{\alpha_1 \ldots \alpha_n}$ are the solution of the (transposed) Vandermonde system of linear equations

$$\sum_{\alpha_1 + \ldots + \alpha_n \leq d} c_{\alpha_1 \ldots \alpha_n} \cdot p_{k-1,1}^{\alpha_1} \cdot \ldots \cdot p_{k-1,n}^{\alpha_n} = v_k, \quad k = 1, \ldots, N.$$

The ground field has to be the field of rational numbers.

var: var(n) returns the n-th ring variable.

varstr: varstr(R) returns the list of the names of the ring variables as a string: varstr(R,n) returns the name of the n-th ring variable. If the ring name is omitted, the basering is used, thus varstr(n) is equivalent to varstr(basering,n).

vdim: vdim(I) computes the vector space dimension of the ring, respectively free module, modulo the ideal, respectively module, I generated by the initial terms of the given generators. If the generators form a standard basis, this is the same as the vector space dimension of the ring, respectively free module, modulo the ideal, respectively module I. If the ideal, respectively module, is not zero–dimensional, -1 is returned.
Note: In the noncommutative case, a ring modulo an ideal has a ring structure if and only if the ideal is two-sided.

weightKB: weightKB(M,i,L) computes the part of a vector space basis of the quotient defined by M with weighted degree of the monomials equal to i. L is a list containing the information about the weights:

L[1]   for all variables (positive),

L[2]   only for module for the generators.

wedge: wedge(M,n) computes the n-th exterior power of the matrix M.

weight: weight(I) computes an "optimal" weight vector for an ideal, respectively module I, which may be used as weight vector for the variables in order to speed up the standard basis algorithm. If the input is weighted homogeneous, a weight vector for which the input is weighted homogeneous is found.

write: writes data to a link.
If the link is of type ASCII, all expressions are converted to strings (and separated by a newline character) before they are written. As a consequence, only such values, which can be converted to a string can be written to an ASCII link. For MP links, ring–dependent expressions are written together with a ring description. To prevent an evaluation of the expression before it is written, the quote command (possibly together with eval) can be used. A write call blocks (that is, does not return to the prompt), as long as a MPtcp link is not ready for writing. For DBM links, write with three arguments inserts the first string as key and the second string as value into the DBM data base. Called with two arguments, it deletes the entry with the key specified by the string from the data base.

C.6 Control Structures

A sequence of commands surrounded by curly brackets is a so–called *block*. Blocks are used in SINGULAR to define procedures and to collect commands belonging to `if`, `else`, `for` and `while` statements and to the `example` part in libraries. Even if the sequence of statements consists of only a single command it has to be surrounded by curly brackets! Variables, which are defined inside a block, are local to that block. Note that there is no ending semicolon at the end of the block.

break: leaves the innermost `for` or `while` block.

breakpoint: sets a breakpoint at the beginning of the specified procedure or at the given line. **Note:** Line number 1 is the first line of a library (for procedures from libraries), respectively the line with the `{`. A line number of -1 removes all breakpoints from that procedure.

continue: skips the rest of the innermost `for` or `while` loop and jumps to the beginning of the block. This command is only valid inside a `for` or a `while` construction.

else: executes the false block if the boolean expression of the `if` statement is false. This command is only valid in combination with an `if` command.

export: converts a local variable of a procedure to a global one that is the identifier is moved from the current package to package Top. Objects defined in a ring are not automatically exported when exporting the ring.

export to: `exportto(p,n)` transfers an identifier n in the current package into the package p. p can be Current, Top or any other identifier of type package. (Objects defined in a ring are not automatically exported when exporting the ring.)

for: repetitive, conditional execution of a command block. In `for(i=1; i<7;i++)`, the init command `i=1` is executed first. Then the boolean expression `i<7` is evaluated. If its value is *TRUE* the block is executed, otherwise the `for` statement is complete. After each execution of the block, the iterate command `i++` is executed and the boolean expression is evaluated. This is repeated until the boolean expression evaluates to *FALSE*. The command `break;` leaves the innermost `for` construct.

if: executes true block if the boolean condition is true. If the `if` statement is followed by an `else` statement and the boolean condition is false, then false block is executed.

importfrom: `importfrom(p,n)` creates a new identifier n in the current package which is a copy of the one specified by n in the package p. p can be Top or any other identifier of type package.

load: `load(s)` reads a library of procedures from the file s. In contrast to the command LIB, load can also handle dynamic modules.

quit: quits SINGULAR; works also from inside a procedure. The commands `quit` and `exit` are synonymous.

return: returns the result(s) of a procedure and can only be used inside a procedure. Note that the brackets are required even if no return value is given, `return()`.

while: repetitive, conditional execution of a block. In `while(i<7)`, the boolean expression `i<7` is evaluated and if its value is *TRUE*, the block is executed. This is repeated until the boolean expression evaluates to *FALSE*. The command `break` leaves the innermost `while` construction. ·

˜(break point): sets a break point. Whenever SINGULAR reaches the command `˜`; in a sequence of commands it prompts for input. The user may now input lines of SINGULAR commands. The line length cannot exceed 80 characters. SINGULAR proceeds with the execution of the command following `˜`; as soon as it receives an empty line.

C.7 System Variables

System variables can be used to modify the default behaviour of SINGULAR.

degBound: The standard basis computation is stopped if the (weighted) total degree exceeds `degBound`. `degBound` should not be used for a global ordering with inhomogeneous input. Reset this bound by setting `degBound` to 0.

echo: Input is echoed if `echo >= voice`. `echo` is a local setting for a procedure and defaulted to 0. `echo` does not affect the output of commands.

minpoly: describes the coefficient field of the current basering as an algebraic extension with the minimal polynomial equal to `minpoly`. Setting the `minpoly` should be the first command after defining the ring.

The minimal polynomial has to be specified in the syntax of a polynomial. Its variable is not one of the ring variables, but the algebraic element, which is adjoined to the field. Algebraic extensions in SINGULAR are only possible over the rational numbers or over $\mathbb{Z}/p$, p a prime number.

SINGULAR does not check whether the given polynomial is irreducible! It can be checked in advance with the function `factorize`.

multBound: The standard basis computation is stopped if the ideal is zero–dimensional in a ring with local ordering and its multiplicity (`mult`) is lower than `multBound`. Reset this bound by setting `multBound` to 0.

noether: The standard basis computation in local rings cuts off all monomials above (in the sense of the monomial ordering) the monomial `noether` during the computation. Reset `noether` by setting `noether` to 0.

printlevel: sets the debug level for `dbprint`. If `printlevel >= voice` then `dbprint` is equivalent to `print`, otherwise nothing is printed.

short: the output of monomials is done in the short manner, if `short` is non–zero. A C–like notation is used, if short is zero. Both notations may be used as input. The default depends on the names of the ring variables (0 if there are names of variables longer than one character, 1 otherwise).

Every change of the basering sets short to the previous value for that ring. In other words, the value of the variable short is "ring-local".

timer:

1. the CPU time (i.e, user and system time) used for each command is printed if timer is set to a positive value , if this time is bigger than a (customizable) minimal time and if printlevel+1 $\geq$ voice (which is by default true on the SINGULAR top level, but not true while procedures are executed).

2. yields the used CPU time since the start-up of SINGULAR in a (customizable) resolution.

   The default setting of timer is 0, the default minimal time is 0.5 seconds, and the default timer resolution is 1 (i.e., the default unit of time is one second).

TRACE: sets level of debugging.

TRACE=0 no debugging messages are printed.

TRACE=1 messages about entering and leaving of procedures are displayed.

TRACE=3 messages about entering and leaving of procedures together with line numbers are displayed.

TRACE=4 each line is echoed and the interpretation of commands in this line is suspended until the user presses RETURN.

TRACE   is defaulted to 0. It does not affect the output of commands.

voice: shows the nesting level of procedures.

C.8 Libraries

C.8.1 Standard-lib

LIBRARY: standard.lib   Procedures, which are always loaded at Start-up

PROCEDURES:
```
stdfglm(ideal[,ord])     standard basis of ideal via fglm [and ordering ord]
stdhilb(ideal[,h])       standard basis of ideal using the Hilbert function
groebner(ideal/module)   standard basis using a heuristically chosen method
res(ideal/module,[i])    free resolution of ideal or module
sprintf(fmt,...)         returns formatted string
fprintf(link,fmt,..)     writes formatted string to link
printf(fmt,...)          displays formatted string
weightkb(s,d,v);         degree d part of kbase(s) with respect to weights v
```

C.8.2 General purpose

LIBRARY: general.lib   Elementary Computations of General Type

PROCEDURES:
```
A_Z("a",n);              string a,b,... of n comma separated letters
```

ASCII([n,m]);	string of printable ASCII characters (number n to m)
binomial(n,m[,../..]);	n choose m (type int), [type string/type number]
cr_roots(n,m);	complex and real (positive) m-th roots of number n
chebychev(n);	n-th Chebychev polynomial
deleteSublist(iv,l);	delete entries given by iv from list l
factorH(p);	factorizes with good choice of principal variable
factorial(n[,../..]);	n factorial (=n!) (type int), [type string/number]
fibonacci(n[,p]);	n-th Fibonacci number [char p]
kmemory([n[,v]]);	active [allocated] memory in kilobyte
killall();	kill all user-defined variables
number_e(n);	compute exp(1) up to n decimal digits
number_pi(n);	compute pi (area of unit circle) up to n digits
primecoeffs(J[,q]);	prime factors $\leq$ min(p,32003) of coeffs of J
primefactors(n[,p]);	prime factors $\leq$ min(p,32003) of n
primes(n,m);	intvec of primes p, n $\leq$ p $\leq$ m
product(../..[,v]);	multiply components of vector/ideal/...[indices v]
sort(ideal/module);	sort generators according to monomial ordering
sum(vector/id/..[,v]);	add components of vector/ideal/...[with indices v]
timeStd(i,d);	std(i) if computation finished after d seconds else i
timeFactorize(p,d);	works as timeStd with factorization
watchdog(i,cmd);	only wait for result of command cmd for i seconds
which(command);	search for command and return absolute path, if found

LIBRARY: inout.lib   Printing and Manipulating In- and Output

PROCEDURES:

allprint(list);	print list if ALLprint is defined, with pause if ¿0
lprint(poly/...[,n]);	display poly/... fitting to pagewidth [size n]
pmat(matrix[,n]);	print form-matrix [first n chars of each column]
rMacaulay(string);	read Macaulay'1 output and return its Singular format
show(any);	display any object in a compact format
showrecursive(id,p);	display id recursively with respect to variables in p
split(string,n);	split given string into lines of length n
tab(n);	string of n space tabs
writelist(...);	write a list into a file and keep the list structure
pause([prompt]);	stop the computation until user input

LIBRARY: poly.lib   Procedures for Manipulating Polys, Ideals, Modules

AUTHORS: Olaf Bachmann,   obachman@mathematik.uni-kl.de,
         Gert-Martin Greuel,   greuel@mathematik.uni-kl.de,
         Anne Frühbis-Krüger,   anne@mathematik.uni-kl.de

PROCEDURES:

hilbPoly(I);	Hilbert polynomial of basering/I
cyclic(int);	ideal of cyclic n-roots
katsura([i]);	katsura [i] ideal
freerank(poly/...)	rank of coker(input) if coker is free else -1
is_zero(poly/...);	int, =1 resp. =0 if coker(input) is 0 resp. not
lcm(ideal);	lcm of given generators of ideal
maxcoef(poly/...);	maximal length of coefficient occurring in poly/...
maxdeg(poly/...);	int/intmat = degree/s of terms of maximal order

maxdeg1(poly/...);	int = [weighted] maximal degree of input
mindeg(poly/...);	int/intmat = degree/s of terms of minimal order
mindeg1(poly/...);	int = [weighted] minimal degree of input
normalize(poly/...);	normalize poly/... such that leading coefficient is 1
rad_con(p,I);	check radical containment of poly p in ideal I
content(f);	content of polynomial/vector f
numerator(n);	numerator of number n
denominator(n)	denominator of number n
mod2id(M,iv);	conversion of a module M to an ideal
id2mod(i,iv);	conversion inverse to mod2id
subrInterred(i1,i2,iv);	interred w.r.t. a subset of variables
substitute(I):	substitute in I variables by monomials
newtonDiag(p);	Newton diagram of the polynomial p

LIBRARY: random.lib   Creating Random and Sparse Matrices, Ideals, Polys

PROCEDURES:

genericid(i[,p,b]);	generic sparse linear combinations of generators of i
randomid(id,[k,b]);	random linear combinations of generators of id
randommat(n,m[,id,b]);	nxm matrix of random linear combinations of id
sparseid(k,u[,o,p,b]);	ideal of k random sparse poly's of degree d
sparsematrix(n,m,o[,.]);	nxm sparse matrix of polynomials of degree ≤ o
sparsemat(n,m[,p,b]);	nxm sparse integer matrix with random coefficients
sparsepoly(u[,o,p,b]);	random sparse polynomial, terms of degree in [u,o]
sparsetriag(n,m[,.]);	nxm sparse lower-triang intmat with random coeffs
triagmatrix(n,m,o[,.]);	nxm sparse lower-triang matrix of poly's of deg ≤ o
randomLast(b);	random transformation of the last variable
randomBinomial(k,u,..);	binomial ideal, k random generators of deg ≥ u

LIBRARY: ring.lib   Manipulating Rings and Maps

PROCEDURES:

changechar("R",c[,r]);	make a copy R of basering [ring r] with new char c
changeord("R",o[,r]);	make a copy R of basering [ring r] with new ord o
changevar("R",v[,r]);	make a copy R of basering [ring r] with new vars v
defring("R",c,n,v,o);	define a ring R in specified char c, n vars v, ord o
defrings(n[,p]);	define ring Sn in n vars, char 32003 [p], ord ds
defringp(n[,p]);	define ring Pn in n vars, char 32003 [p], ord dp
extendring("R",n,v,o);	extend given ring by n vars v, ord o and name it R
fetchall(R[,str]);	fetch all objects of ring R to basering
imapall(R[,str]);	imap all objects of ring R to basering
mapall(R,i[,str]);	map all objects of ring R via ideal i to basering
ord_test(R);	test whether ordering of R is global, local or mixed
ringtensor("R",s,t,..);	create ring R, tensor product of rings s,t,...
ringweights(r);	intvec of weights of ring variables of ring r
preimageLoc(I);	computes the preimage of I for non-global orderings
rootofUnity(n);	the minimal polynomial of the n-th primitive root of unity

C.8.3 Linear algebra

LIBRARY: linalg.lib   Algorithmic Linear Algebra

AUTHORS: Ivor Saynisch,   ivs@math.tu-cottbus.de
         Mathias Schulze,   mschulze@mathematik.uni-kl.de
PROCEDURES:

inverse(A);	the inverse matrix of A
inverse_B(A);	list(matrix Inv,poly p), Inv·A = p·En (using busadj(A))
inverse_L(A);	list(matrix Inv,poly p), Inv·A = p·En (using lift)
sym_gauss(A);	symmetric Gaussian algorithm
orthogonalize(A);	Gram-Schmidt orthogonalization
diag_test(A);	test whether A can be diagonalized
busadj(A);	coeffs of Adj(En·t−A) and coeffs of det(En·t−A)
charpoly(A,v);	characteristic polynomial of A (using busadj(A))
adjoint(A);	adjoint of A (using busadj(A))
det_B(A);	determinant of A (using busadj(A))
gaussred(A);	Gaussian reduction: P·A = U·S, S row reduced form of A
gaussred_pivot(A);	Gaussian reduction: P·A = U·S, uses row pivoting
gauss_nf(A);	Gaussian normal form of A
mat_rk(A);	rank of constant matrix A
U_D_O(A);	P·A = U·D·O, P,D,U,O = permutation, diagonal, lower-, upper-triangular matrix
pos_def(A,i);	test symmetric matrix for positive definiteness
hessenberg(M);	Hessenberg form of M
eigenvals(M);	eigenvalues with multiplicities of M
minipoly(M);	minimal polynomial of M
spnf(sp);	normal form of spectrum sp
spprint(sp);	print spectrum sp
jordan(M[,opt]);	eigenvalues, Jordan block sizes, transformation matrix
jordanbasis	Jordan basis and weight filtration of M
jordanmatrix(l);	Jordan matrix with eigenvalues, Jordan block sizes
jordannf(M);	**Jordan normal form of constant square matrix M**

LIBRARY: matrix.lib   Elementary Matrix Operations

PROCEDURES:

compress(A);	matrix, zero columns from A deleted
concat(A1,A2,..);	matrix, concatenation of matrices A1,A2,...
diag(p,n);	matrix, nxn diagonal matrix with entries poly p
dsum(A1,A2,..);	matrix, direct sum of matrices A1,A2,...
flatten(A);	ideal, generated by entries of matrix A
genericmat(n,m[,id]);	generic nxm matrix [entries from id]
is_complex(c);	1 if list c is a complex, 0 if not
outer(A,B);	matrix, outer product of matrices A and B
power(A,n);	matrix/intmat, n-th power of matrix/intmat A
skewmat(n[,id]);	generic skew-symmetric nxn matrix [entries from id]
submat(A,r,c);	submatrix of A with rows/cols specified by intvec r/c
symmat(n[,id]);	generic symmetric nxn matrix [entries from id]
tensor(A,B);	matrix, tensor product of matrices A and B
unitmat(n);	unit square matrix of size n
gauss_col(A);	transform a matrix into col-reduced Gauß normal form
gauss_row(A);	transform a matrix into row-reduced Gauß normal form

```
addcol(A,c1,p,c2);    add p·(c1-th col) to c2-th column of matrix A, p poly
addrow(A,r1,p,r2);    add p·(r1-th row) to r2-th row of matrix A, p poly
multcol(A,c,p);       multiply c-th column of A with poly p
multrow(A,r,p);       multiply r-th row of A with poly p
permcol(A,i,j);       permute i-th and j-th columns
permrow(A,i,j);       permute i-th and j-th rows
rowred(A[,any]);      reduction of matrix A with elementary row-operations
colred(A[,any]);      reduction of matrix A with elementary col-operations
rm_unitrow(A);        remove unit rows and associated columns of A
rm_unitcol(A);        remove unit columns and associated rows of A
headstand(A,i,j);     A[n-i+1,m-j+1]=A[i,j]
```

C.8.4 Commutative algebra

LIBRARY: absfact.lib   Absolute factorization for characteristic 0

AUTHORS: Wolfram Decker,   decker@math.uni-sb.de
         Gregoire Lecerf,   lecerf@math.uvsq.fr
         Gerhard Pfister,   pfister@mathematik.uni-kl.de

OVERVIEW:
A library for computing the absolute factorization of multivariate polynomials f
with coefficients in a field K of characteristic zero. Using Trager's idea, the implemented algorithm computes an absolutely irreducible factor by factorizing over
some finite extension field L (which is chosen such that V(f) has a smooth point
with coordinates in L). Then a minimal extension field is determined making use
of the Rothstein-Trager partial fraction decomposition algorithm. See [Cheze and
Lecerf, Lifting and recombination techniques for absolute factorization].

PROCEDURES:
  absFactorize(f);   absolute factorization of poly

LIBRARY: algebra.lib   Compute with Algebras and Algebra Maps

AUTHORS: Gert-Martin Greuel,   greuel@mathematik.uni-kl.de,
         Agnes Eileen Heydtmann, agnes@math.uni-sb.de,
         Gerhard Pfister,   pfister@mathematik.uni-kl.de
PROCEDURES:
  algebra_containment();   query of algebra containment
  module_containment();   query of module containment over a subalgebra
  inSubring(p,I);   test whether poly p is in subring generated by I
  algDependent(I);   computes algebraic relations between generators of I
  alg_kernel(phi,R);   computes the kernel of the ring map phi
  is_injective(phi,R);   test for injectivity of ring map phi
  is_surjective(phi,R);   test for surjectivity of ring map phi
  is_bijective(phi,R);   test for bijectivity of ring map phi
  noetherNormal(id);   Noether normalization of ideal id
  mapIsFinite(R,phi,I);   query for finiteness of map phi: R → basering/I
  finitenessTest(I,z);   find variables which occur as pure power in lead(I)
```
```

LIBRARY: ehv.lib   Procedures for Primary Decomposition of Ideals

AUTHORS: Kai Dehmann, dehmann@mathematik.uni-kl.de;
OVERVIEW:
Algorithms for primary decomposition and radical-computation based on
the ideas of Eisenbud, Huneke, and Vasconcelos.
PROCEDURES:

equiMaxEHV(I);	equidimensional part of I
equiRadEHV(I [,Strategy]);	equidimensional radical of I
radEHV(I [,Strategy]);	radical of I
decompEHV(I);	decomposition of a zero-dimensional I
AssEHV(I [,Strategy]);	associated primes of I
minAssEHV(I [,Strategy]);	minimal associated primes of I
localize(I,P,1);	the contraction of the ideal generated by I
componentEHV(I,P,L [,Strategy]);	a P-primary component for I
primdecEHV(I [,Strategy]);	a minimal primary decomposition of I

LIBRARY: elim.lib   Elimination, Saturation and Blowing up

PROCEDURES:

blowup0(j[,s1,s2]);	create presentation of blownup ring of ideal j
elim(id,n,m);	variable n..m eliminated from id (ideal/module)
elim1(id,p);	p=product of vars to be eliminated from id
nselect(id,n[,m]);	select generators not containing nth [..m-th] variable
sat(id,j);	saturated quotient of ideal/module id by ideal j
select(id,n[,m]);	select generators containing all variables n...m
select1(id,n[,m]);	select generators containing one variable n...m

LIBRARY: grwalk.lib   Groebner Walk Conversion Algorithms

AUTHOR: I Made Sulandra
PROCEDURES:

fwalk(ideal[,intvec]);	standard basis of ideal via fractalwalk alg
twalk(ideal[,intvec]);	standard basis of ideal via Tran's alg
awalk1(ideal[,intvec]);	standard basis of ideal via the first alt. alg
awalk2(ideal[,intvec]);	standard basis of ideal via the second alt. alg
pwalk(ideal[,intvec]);	standard basis of ideal via perturbation walk alg
gwalk(ideal[,intvec]);	standard basis of ideal via groebnerwalk alg

LIBRARY: homolog.lib   Procedures for Homological Algebra

AUTHORS: Gert-Martin Greuel,   greuel@mathematik.uni-kl.de,
         Bernd Martin,   martin@math.tu-cottbus.de
         Christoph Lossen,   lossen@mathematik.uni-kl.de
PROCEDURES:

cup(M);	$\cup : \text{Ext}^1(M',M) \times \text{Ext}^1(M') \to \text{Ext}^2(M')$
cupproduct(M,N,P,p,q);	$\cup : \text{Ext}^p(M',N) \times \text{Ext}^q(N',P') \to \text{Ext}^{p+q}(M',P')$
depth(I,M);	depth(I,M'), I ideal, M module, M'=coker(M)

Ext_R(k,M);	$\text{Ext}^k(\text{M'},\text{R})$, M module, R basering, M'=coker(M)
Ext(k,M,N);	$\text{Ext}^k(\text{M'},\text{N'})$, M'=coker(M), N'=coker(N)
fitting(M,n);	n-th Fitting ideal of M'=coker(M), M module, n int
flatteningStrat(M);	Flattening stratification of M'=coker(M), M module
Hom(M,N);	Hom(M',N'), M'=coker(M), N'=coker(N)
homology(A,B,M,N);	ker(B)/im(A), homology of complex $R^k \xrightarrow{A} M' \xrightarrow{B} N'$
isCM(M);	test if coker(M) is Cohen-Macaulay, M module
isFlat(M);	test if coker(M) is flat, M module
isLocallyFree(M,r);	test if coker(M) is locally free of constant rank r
isReg(I,M);	test if I is coker(M)-sequence, I ideal, M module
kernel(A,M,N);	ker(A:M' → N') M,N modules, A matrix
kohom(A,k);	Hom(R^k,A), A matrix over basering R
kontrahom(A,k);	Hom(A,R^k), A matrix over basering R
KoszulHomology(I,M,n);	n-th Koszul homology H_n(I,coker(M)), I=ideal
tensorMod(M,N);	Tensor product of modules M'=coker(M), N'=coker(N)
Tor(k,M,N);	Tor`k(M',N'), M,N modules, M'=coker(M), N'=coker(N)

LIBRARY: intprog.lib   Integer Programming with Gröbner Basis Methods

AUTHOR: Christine Theis, ctheis@math.uni-sb.de
PROCEDURES:
solve_IP(..);   procedures for solving integer programming problems

LIBRARY: lll.lib   Integral LLL-Algorithm (see also [151])

AUTHOR: Alberto Vigneron-Tenorio, alberto.vigneron@uca.es
         Alfredo Sanchez-Navarro, alfredo.sanchez@uca.es
PROCEDURES:
LLL(..);   Integral LLL-Algorithm

LIBRARY: mprimdec.lib   Procedures for Primary Decomposition of Modules

AUTHORS: Alexander Dreyer, adreyer@web.de

REMARK:
These procedures are implemented to be used in characteristic 0. They also work
in positive characteristic. In small characteristic and for algebraic extensions, the
procedures via Gianni, Trager, Zacharias may not terminate.

PROCEDURES:

separator(l);	computes a list of separators of prime ideals
PrimdecA(N[,i]);	decomposition via Shimoyama/Yokoyama (S/Y)
PrimdecB(N,p);	primary decomposition for pseudo-primary ideals
modDec(N[,i]);	minimal primary decomposition via S/Y
zeroMod(N[,check]);	minimal zero-dimensional primary decomposition via Gianni, Trager and Zacharias (GTZ)
GTZmod(N[,check]);	minimal primary decomposition
dec1var(N[,check[,ann]]);	primary decomposition for one variable
annil(N);	the annihilator of R^n/N in the basering
splitting(N[,check[,ann]]);	splitting to simpler modules
primTest(i[,p]);	tests whether i is prime or homogeneous

```
preComp(N,check[,ann]);        enhanced version of splitting
indSet(i);                     lists with varstrings of(in)dependent variables
GTZopt(N[,check[,ann]]);       a faster version of GTZmod
zeroOpt(N[,check[,ann]]);      a faster version of zeroMod
clrSBmod(N);                   extracts an minimal SB from a SB
minSatMod(N,I);                minimal saturation of N w.r.t. I
specialModulesEqual(N1,N2);    checks if N1 is contained in N2 or vice versa
stdModulesEqual(N1,N2);        checks for equality of standard bases
modulesEqual(N1,N2);           checks for equality of modules
getData(N,l[,i]);              extracts oldData and computes the remaining data
```

LIBRARY: modstd.lib   Groebner basis of ideals

AUTHORS: A. Hashemi,   Amir.Hashemi@lip6.fr
         G. Pfister   pfister@mathematik.uni-kl.de
         H. Schoenemann   hannes@mathematik.uni-kl.de

NOTE:
A library for computing the Groebner basis of an ideal in the polynomial ring over
the rational numbers using modular methods. The procedures are inspired by the
following paper: Elizabeth A. Arnold: Modular Algorithms for Computing Groebner
Bases, Journal of Symbolic Computation, April 2003, Volume 35, (4), p. 403-419.

PROCEDURES:
modStd(I);   compute a standard basis of I using modular methods
modS(I,L);   liftings to Q of standard bases of I mod p for p in L

LIBRARY: compregb.lib   comprehensive Groebner systems

AUTHOR: Akira Suzuki   sakira@kobe-u.ac.jp

OVERVIEW:
A simple algorithm to compute *Comprehensive Groebner Bases* using Groebner
Bases by Akira Suzuki and Yosuke Sato.

PROCEDURES:
cgs(polys,vars,pars,R1,R2);   comprehensive Groebner systems
base2str(G);   pretty print of the result G

LIBRARY: mregular.lib   Castelnuovo-Mumford Regularity of CM-Schemes

AUTHORS: Isabel Bermejo,   ibermejo@ull.es
         Philippe Gimenez,   pgimenez@agt.uva.es
         Gert-Martin Greuel,   greuel@mathematik.uni-kl.de

OVERVIEW:
A library for computing the *Castelnuovo–Mumford regularity* of a subscheme of the
projective n-space that DOES NOT require the computation of a minimal graded
free resolution of the saturated ideal defining the subscheme. The procedures are
based on [24], and [25]. The algorithm assumes the variables to be in Noether
position.

PROCEDURES:
regIdeal(id,[,e]);	regularity of homogeneous ideal id
depthIdeal(id,[,e]);	depth of S/id with S=basering, id homogeneous ideal
satiety(id,[,e]);	saturation index of homogeneous ideal id
regMonCurve(li);	regularity of projective monomial curve defined by li
NoetherPosition(id);	Noether normalization of ideal id
is_NP(id);	checks whether variables are in Noether position
is_nested(id);	checks whether monomial ideal id is of nested

LIBRARY: noether.lib   Noether normalization of an ideal

AUTHORS: A. Hashemi,   Amir.Hashemi@lip6.fr

OVERVIEW:
A library for computing the Noether normalization of an ideal that DOES NOT require the computation of the dimension of the ideal. It checks whether an ideal is in Noether position. A modular version of these algorithms is also provided. The procedures are based on a paper of Amir Hashemi 'Efficient Algorithms for Computing Noether Normalization' Submitted to: Special Issue of Mathematics in Computer Science on Symbolic and Numeric Computation. This library computes also Castelnuovo-Mumford regularity and satiety of an ideal. A modular version of these algorithms is also provided. The procedures are based on a paper of Amir Hashemi 'Computation of Castelnuovo-Mumford regularity and satiety' Submitted to: IS-SAC 2007.

PROCEDURES:
NPos_test(id);	checks whether monomial ideal id is in Noether position
modNPos_test(id);	checks the same by modular methods
NPos(id);	Noether normalization of ideal id
modNPos(id);	Noether normalization of ideal id by modular methods
nsatiety(id);	Satiety of ideal id
modsatiety(id)	Satiety of ideal id by modular methods
regCM(id);	Castelnuovo-Mumford regularity of ideal id
modregCM(id);	the same by modular methods

LIBRARY: normal.lib   Normalization of Affine Rings

AUTHORS: Gert-Martin Greuel,   greuel@mathematik.uni-kl.de,
         Gerhard Pfister,   pfister@mathematik.uni-kl.de
PROCEDURES:
normal(I[,"wd"]);	computes the normalization of basering/I, respectively the normalization of basering/I and the delta-invariant
HomJJ(L);	presentation of End`R(J) as affine ring, L a list
genus(I);	computes the genus of the projective curve defined by I
primeClosure(L)	integral closure of R/p, p prime, L a list
closureFrac(L)	write poly in integral closure as element of Q(R/p)

LIBRARY: primdec.lib   Primary Decomposition and Radical of Ideals

AUTHORS: Gerhard Pfister,   pfister@mathematik.uni-kl.de (GTZ),
         Wolfram Decker,   decker@math.uni-sb.de   (SY),

Hans Schönemann,   hannes@mathematik.uni-kl.de   (SY)
Santiago Laplagne, laplagn@dm.uba.ar   (GTZ)

OVERVIEW:
Algorithms for primary decomposition based on ideas of Gianni,Trager and Zacharias, [90], (implementation by G. Pfister), respectively based on ideas of Shimoyama and Yokoyama [213] (implementation by W. Decker and H. Schönemann). The procedures are implemented to be used in characteristic 0. They also work in positive characteristic >> 0. In small characteristic and for algebraic extensions, primdecGTZ and minAssGTZ may not terminate, while primdecSY and minAssChar may not give a complete decomposition. Algorithms for the computation of the based on the ideas of Krick, Logar [139] and Kemper (implementation by G. Pfister).

PROCEDURES:

Ann(M);	annihilator of module R^n/M
primdecGTZ(I);	complete primary decomposition via Gianni,Trager,Zacharias
primdecSY(I...);	complete primary decomposition via Shimoyama-Yokoyama
minAssGTZ(I);	the minimal associated primes via Gianni,Trager,Zacharias
minAssChar(I...);	the minimal associated primes using characteristic sets
testPrimary(L,k);	tests the result of the primary decomposition
radical(I);	computes the radical of I via Krick/Logar and Kemper
radicalEHV(I);	computes the radical of I via Eisenbud,Huneke,Vasconcelos
equiRadical(I);	the radical of the equidimensional part of the ideal I
prepareAss(I);	list of radicals of the equidimensional components of I
equidim(I);	weak equidimensional decomposition of I
equidimMax(I);	equidimensional locus of I
equidimMaxEHV(I);	equidimensional locus of I via Eisenbud,Huneke,Vasconcelos
zerodec(I);	zero-dimensional decomposition via Monico
absPrimdec GTZ(I);	the absolute prime components of I

LIBRARY: primitiv.lib   Computing a Primitive Element

AUTHOR: Martin Lamm, lamm@mathematik.uni-kl.de
PROCEDURES:

primitive(ideal i);	find minimal polynomial for a primitive element
primitive_extra(i);	find primitive element for two generators
splitring(f,R[,L]);	define ring extension with name R and switch to it

LIBRARY: reesclos.lib   Procedures to Compute Integral Closure of an Ideal

AUTHOR: Tobias Hirsch, hirsch@math.tu-cottbus.de

OVERVIEW:
A library to compute the integral closure of an ideal I in a R=k[x(1),...,x(n)] using the Rees–Algebra R[It] of I. It computes the integral closure of R[It] (in the same manner as done in the library 'normal.lib'), which is a graded subalgebra of R[t]. The degree k component is the integral closure of the k-th power of I.

PROCEDURES:

ReesAlgebra(I);	computes Rees-Algebra of an ideal I
normalI(I[,p[,r]]);	computes integral closure of an ideal I using R[It]

LIBRARY: toric.lib   Standard Basis of Toric Ideals

AUTHOR: Christine Theis,   ctheis@math.uni-sb.de
PROCEDURES:
  toric_ideal(A,..);   computes the toric ideal of A
  toric_std(ideal I); standard basis of I by a specialized Buchberger algorithm

LIBRARY: resolve.lib   Resolution of singularities (Algorithm of Villamayor)

AUTHORS: A. Fruehbis-Krueger,   anne@mathematik.uni-kl.de,
           G. Pfister,   pfister@mathematik.uni-kl.de
MAIN PROCEDURES:
  blowUp(J,C[,W,E]) blowing up of the variety V(J) in V(C)
  blowUp2(J,C)   blowing up of the variety V(J) in V(C)
  Center(J[,W,E])   computes 'Villamayor'-center for blow up
  resolve(J)   computes the desingularization of the variety V(J)
PROCEDURES FOR PRETTY PRINTING OF OUTPUT:
  showBO(BO)   prints the content of a BO in more human readable
                     form
  presentTree(L)   prints the final charts in more human readable form
  showDataTypes()   prints help text for output data types
AUXILIARY PROCEDURES:
  createBO(J,W,E)   creates basic object from input data
  CenterBO(BO)   computes the center for the next blow-up
  Delta(BO)   apply the Delta-operator of [Bravo,Encinas,Villamayor]
  DeltaList(BO)   list of results of Delta^0 to Delta^bmax

LIBRARY: reszeta.lib   Zeta-function of Denef and Loeser

AUTHORS: A. Fruehbis-Krueger,   anne@mathematik.uni-kl.de,
           G. Pfister,   pfister@mathematik.uni-kl.de
MAIN PROCEDURES:
  intersectionDiv(L) intersection form, genera of exceptional divisors
  spectralNeg(L)   negative spectral numbers
  discrepancy(L)   computes discrepancy of given resolution
  zetaDL(L,d)   computes Denef-Loeser zeta function
AUXILIARY PROCEDURES:
  collectDiv(L[,iv]) identify exceptional divisors in different charts
  abstractR(L)   pass from embedded to non-embedded resolution

LIBRARY: sagbi.lib Subalgebras bases Analogous to Groebner bases for ideals

AUTHORS: Gerhard Pfister,   pfister@mathematik.uni-kl.de,
         Anen Lakhal,   alakhal@mathematik.uni-kl.de
PROCEDURES:
  proc reduction(p,I);   Perform one step subalgebra reduction
  proc sagbiSPoly(I);   S-polynomial of the Subalgebra
  proc sagbiNF(id,I);   iterated S-reductions
  proc sagbi(I);   SAGBI basis for the Subalgebra defined by I

proc sagbiPart(I);   partial SAGBI basis

LIBRARY: sheafcoh.lib   Procedures for Computing Sheaf Cohomology

AUTHORS: Wolfram Decker,   decker@math.uni-sb.de,
          Christoph Lossen, lossen@mathematik.uni-kl.de
          Gerhard Pfister,   pfister@mathematik.uni-kl.de
PROCEDURES:
  truncate(phi,d);   truncation of coker(phi) at d
  CM_regularity(M);   Castelnuovo-Mumford regularity of coker(M)
  sheafCohBGG(M,l,h);   cohomology of sheaf associated to coker(M)
  sheafCoh(M,l,h);   cohomology of sheaf associated to coker(M)
  dimH(i,M,d);   compute h^i(F(d)), F sheaf associated to coker(M)
AUXILIARY PROCEDURES:
  displayCohom(B,l,h,n);   display intmat as Betti diagram (with zero rows)

C.8.5 Singularities

LIBRARY: alexpoly.lib   Resolution Graph and Alexander Polynomial

AUTHOR: Fernando Hernando Carrillo, hernando@agt.uva.es
          Thomas Keilen,   keilen@mathematik.uni-kl.de

OVERVIEW:
A library for computing the resolution graph of a plane curve singularity f, the
total multiplicities of the total transforms of the branches of f alongthe exceptional
divisors of a minimal good resolution of f, the Alexander polynomial of f, and the
zeta function of its monodromy operator.

PROCEDURES:
  resolutiongraph(f);   resolution graph f
  totalmultiplicities(f);   resolution graph, multiplicities of f
  alexanderpolynomial(f);   Alexander polynomial of f
  semigroup(f);   calculates generators for the semigroup of f
  multseq2charexp(v);   multiplicity sequence to characteristic exponents
  charexp2multseq(v);   characteristic exponents to multiplicity sequence
  charexp2generators(v);   characteristic exponents to the semigroup
  charexp2inter(c,e);   charact. exp. to intersection matrix
  charexp2conductor(v);   characteristic exponents to conductor
  charexp2poly(v,a);   calculates f with characteristic exponents v
  tau_es2(f);   equisingular Tjurina number of f

LIBRARY: arcpoint.lib   Truncations of arcs at a singular point

AUTHOR: Nadine Cremer   cremer@mathematik.uni-kl.de

OVERVIEW:
An arc is given by a power series in one variable, say t, and truncating it at a
positive integer i means cutting the t-powers i. The set of arcs truncated at order
-bound- is denoted Tr(i). An algorithm for computing these sets (which happen to
be constructible) is given in [Lejeune-Jalabert, M.: Courbes tracées sur un germe
d'hypersurface, American Journal of Mathematics, 112 (1990)]. Our procedures for

computing the locally closed sets contributing to the set of truncations rely on this algorithm.

PROCEDURES:

nashmult(f,bound);	the sequence of Nash Multiplicities
removepower(I);	removes powers of variable
idealsimplify(I,maxiter);	further simplification of I
equalJinI(I,J);	tests if two ideals I and J are equal

LIBRARY: classify.lib   Arnold Classifier of Singularities

AUTHOR: Kai Krüger,   krueger@mathematik.uni-kl.de
           Corina Baciu,   baciu@mathematik.uni-kl.de

OVERVIEW:
A library for classifying isolated hypersurface singularities w.r.t. right equivalence, based on the determinator of singularities by V.I. Arnold.

PROCEDURES:

basicinvariants(f);	computes Milnor number, determinacy-bd. and crk of f
classify(f);	normal form of poly f determined with Arnold's method
corank(f);	computes the corank of f (i.e. of the Hessian of f)
Hcode(v);	coding of intvec v according to the number repetitions
init_debug([n]);	print trace and debugging information depending on int n
internalfunctions();	display names of internal procedures of this library
milnorcode(f[,e]);	Hilbert poly of [e-th] Milnor algebra coded with Hcode
morsesplit(f);	residual part of f after applying the splitting lemma
quickclass(f)	normal form of f determined by invariants (milnorcode)
singularity(s,[]);	normal form of singularity given by its name s and index
swap(a,b);	returns b,a
A_L(s/f)	shortcut for quickclass(f) or normalform(s)
normalform(s);	normal form of singularity given by its name s
debug_log(lev,[]);	print trace and debugging information w.r.t level

LIBRARY: deform.lib   Miniversal Deformation of Singularities and Modules

AUTHOR: Bernd Martin,   martin@math.tu-cottbus.de
PROCEDURES:

versal(Fo[,d,any]);	miniversal deformation of isolated singularity Fo
mod_versal(Mo,I,[,d,any]);	miniversal deformation of module Mo mod ideal I
lift_kbase(N,M);	lifting N into standard kbase of M
lift_rel_kb(N,M[,kbM,p]);	relative lifting N into a kbase of M

LIBRARY: equising.lib   Equisingularity Stratum of a Family of Plane Curves

AUTHOR: Andrea Mindnich,   mindnich@mathematik.uni-kl.de
PROCEDURES:

esStratum(F[,m]);	computes the equisingularity stratum of the family F
isEquising(F[,m]);	tests if a given deformation is equisingular

LIBRARY: gmspoly.lib   Gauss-Manin System of Tame Polynomials

AUTHOR: Mathias Schulze, email: mschulze@mathematik.uni-kl.de

OVERVIEW:
A library to compute invariants related to the Gauss-Manin system of a cohomologically tame polynomial

PROCEDURES:
  isTame(f);   test if the polynomial f is tame
  goodBasis(f);   a good basis of the Brieskorn lattice of a cohomologically tame f

LIBRARY: gmssing.lib   Gauss-Manin System of Isolated Singularities

AUTHOR: Mathias Schulze, email: mschulze@mathematik.uni-kl.de

OVERVIEW:
A library to compute invariants related to the the Gauss-Manin system of an isolated hypersurface singularity

PROCEDURES:
gmsring(t,s);	Gauss-Manin system of t with variable s
gmsnf(p,K);	Gauss-Manin normal form of p
gmscoeffs(p,K);	Gauss-Manin basis representation of p
bernstein(t);	roots of the Bernstein polynomial of t
monodromy(t);	Jordan data of complex monodromy of t
spectrum(t);	singularity spectrum of t
sppairs(t);	spectral pairs of t
vfilt(t);	V-filtration of t on Brieskorn lattice
vwfilt(t);	weighted V-filtration of t on Brieskorn lattice
tmatrix(t);	matrix of t w.r.t. good basis of Brieskorn lattice
endvfilt(V);	endomorphism V-filtration on Jacobian algebra
sppnf(a,w[,m]);	spectral pairs normal form of (a,w[,m])
sppprint(spp);	print spectral pairs spp
spadd(sp1,sp2);	sum of spectra sp1 and sp2
spsub(sp1,sp2);	difference of spectra sp1 and sp2
spmul(sp0,k);	linear combination of spectra sp
spissemicont(sp[,opt]);	semicontinuity test of spectrum sp
spmilnor(sp);	Milnor number of spectrum sp
spgeomgenus(sp);	geometrical genus of spectrum sp
spgamma(sp);	gamma invariant of spectrum sp

LIBRARY: hnoether.lib   Hamburger-Noether (Puiseux) Development

AUTHOR: Martin Lamm,   lamm@mathematik.uni-kl.de
           Christoph Lossen lossen@mathematik.uni-kl.de

OVERVIEW:
A library for computing the Hamburger–Noether, respectively Puiseux, development of a plane curve singularity following [39]. The library contains also procedures for computing the (topological) numerical invariants of plane curve singularities.

MAIN PROCEDURES:
```
hnexpansion(f);            Hamburger-Noether (H-N) expansion of f
develop(f[,n]);            H-N development of irreducible curves
extdevelop(hne,n);         extension of the H-N development hne of f
param(hne [,x]);           a parametrization of f (input=output(develop))
displayHNE(hne);           display H-N development as an ideal
invariants(hne);           invariants of f, e.g. the characteristic exponents
multsequence(hne);         sequence of multiplicities
displayInvariants(hne);    display invariants of f
intersection(hne1,hne2);   intersection multiplicity of two curves
displayMultsequence(hne);  display sequence of multiplicities
is_irred(f);               test for irreducibility
delta(f);                  delta-invariant of f
newtonpoly(f)              local Newton polygon of f
is_NND(f)                  test whether f is Newton non-degenerate
```
AUXILIARY PROCEDURES:
```
puiseux2generators(m,n);   convert Puiseux pairs to generators of semigroup
separateHNE(hne1,hne2);    number of quadratic transf. needed for separation
squarefree(f);             a squarefree divisor of the poly f
allsquarefree(f,l);        the maximal squarefree divisor of the poly f
further_hn_proc();         show further procedures useful for
interactive use
stripHNE(hne);             reduce amount of memory consumed by hne
```

LIBRARY: kskernel.lib   Kernel of the Kodaira--Spencer map

AUTHOR: Tetyana Povalyaeva
PROCEDURES:
```
KSker(p,q);           kernel of the Kodaira-Spencer map
KSlinear(M);          matrix of linear terms of the kernel
KScoef(i,j,P,Q,qq);   coefficient of the given term in the matrix
```

OVERVIEW:
computes the kernel of the Kodaira-Spencer map of a versal deformation of an irreducible plane curve singularity as a matrix.

LIBRARY: KVequiv.lib   Procedures related to K˙V-Equivalence

AUTHOR: Anne Fruehbis-Krueger, anne@mathematik.uni-kl.de
PROCEDURES:
```
derlogV(iV);                     derlog(V(iV))
KVtangent(I,rname,dername,k)      K˙V tangent space to given singularity
KVversal(KVtan,I,rname,idname)    K˙V versal family
KVvermap(KVtan,I)                 section inducing K˙V versal family
lft_vf(I,rname,idname)            liftable vector fields
```

LIBRARY: mondromy.lib   Monodromy of an Isolated Hypersurface Singularity

AUTHOR: Mathias Schulze, mschulze@mathematik.uni-kl.de

OVERVIEW:
Library to compute the monodromy of an isolated hypersurface singularity. It uses
an algorithm by Brieskorn [28] to compute a connection matrix of the meromorphic
Gauß-Manin connection up to arbitrarily high order, and an algorithm of Gerard
and Levelt [87] to transform it to a simple pole.

PROCEDURES:

detadj(U);	determinant and adjoint matrix of square matrix U
invunit(u,n);	series inverse of polynomial u up to order n
jacoblift(f);	lifts f^κ in jacob(f) with minimal κ
monodromyB(f[,opt]);	monodromy of isolated hypersurface singularity f
H2basis(f);	basis of Brieskorn lattice H''

LIBRARY: qhmoduli.lib   Moduli Spaces of Semi-Quasihomogeneous Singularities

AUTHOR: Thomas Bayer, bayert@in.tum.de

OVERVIEW:
Compute equations for the moduli space of an isolated semi-quasihomogeneous
hypersurface singularity with fixed principal part (based on [105]).

PROCEDURES:

ArnoldAction(f,[G,w]);	induced action of G˙f on T˙
ModEqn(f);	equations of the moduli space for principal part f
QuotientEquations(G,A,I);	equations of Variety(I)/G w.r.t. action 'A'
StabEqn(f);	equations of the stabilizer of f
StabEqnId(I,w);	equations of the stabilizer of the qhom. ideal I
StabOrder(f);	order of the stabilizer of f
UpperMonomials(f,[w]);	upper basis of the Milnor algebra of f
Max(data);	maximal integer contained in 'data'
Min(data);	minimal integer contained in 'data'

LIBRARY: sing.lib   Invariants of Singularities

AUTHORS: Gert-Martin Greuel, greuel@mathematik.uni-kl.de,
         Bernd Martin,   martin@math.tu-cottbus.de

PROCEDURES:

codim (id1, id2);	vector space dimension of of id2/id1 if finite
deform(i);	infinitesimal deformations of ideal i
dim_slocus(i);	dimension of singular locus of ideal i
is_active(f,id);	is poly f an active element mod id? (id ideal/module)
is_ci(i);	is ideal i a complete intersection?
is_is(i);	is ideal i an isolated singularity?
is_reg(f,id);	is poly f a regular element mod id? (id ideal/module)
is_regs(i[,id]);	are gen's of ideal i regular sequence modulo id?
milnor(i);	Milnor number of ideal i; (assume i is ICIS in nf)
nf_icis(i);	generic combinations of generators; get ICIS in nf
qhspectrum(f,w);	spectrum numbers of w-homogeneous polynomial f
slocus(i);	ideal of singular locus of ideal i
tangentcone(i);	tangent cone of ideal i
Tjurina(i);	SB of Tjurina module of ideal i (assume i is ICIS)
tjurina(i);	Tjurina number of ideal i (assume i is ICIS)
T_1(i);	T^1-module of ideal i

```
T_2((i));              T^2-module of ideal i
T_12(i);               T^1- and T^2-module of ideal i
locstd(I);             SB of I for local degree orderings without
                       cancelling units
```

LIBRARY: space_curve.lib

AUTHOR: Viazovska Maryna, viazovsk@mathematik.uni-kl.de
PROCEDURES:
```
BlowingUp(f,I,1);      BlowingUp of V(I) at the point 0;
CurveRes(I);           Resolution of V(I)
CurveParam(I);         Parametrization of algebraic branches of V(I)
WSemigroup(X,b);       Weierstrass semigroup of the curve
```

LIBRARY: spcurve.lib   Deformations and Invariants of CM-codim 2 Singularities

AUTHOR: Anne Frühbis-Krüger, anne@mathematik.uni-kl.de
PROCEDURES:
```
isCMcod2(i);           presentation matrix of the ideal i, if i is CM
CMtype(i);             Cohen-Macaulay type of the ideal i
matrixT1(M,n);         1-st order deformation T1 in matrix description
semiCMcod2(M,T1);      semiuniversal deformation of maximal minors of M
discr(sem,n);          discriminant of semiuniversal deformation
qhmatrix(M);           weights if M is quasihomogeneous
relweight(N,W,a);      relative matrix weight of N w.r.t. weights (W,a)
posweight(M,T1,i);     deformation of coker(M) of non-negative weight
KSpencerKernel(M);     kernel of the Kodaira-Spencer map
```

LIBRARY: spectrum.lib   Singularity Spectrum for Nondegenerate
Singularities

AUTHOR: Stefan Endraß
PROCEDURES:
```
spectrumnd(poly[,1]);  spectrum of a nondegenerate isolated singularity
```

C.8.6 Invariant theory

LIBRARY: ainvar.lib   Invariant Rings of the Additive Group (see also [113])

AUTHORS: Gerhard Pfister,   pfister@mathematik.uni-kl.de,
         Gert-Martin Greuel, greuel@mathematik.uni-kl.de
PROCEDURES:
```
invariantRing(m..);    compute ring of invariants of (K,+)-action given by m
derivate(m,f);         derivation of f with respect to the vectorfield m
actionIsProper(m);     tests whether action defined by m is proper
reduction(p,I);        SAGBI reduction of p in the subring generated by I
completeReduction();   complete SAGBI reduction
localInvar(m,p..);     invariant polynomial under m computed from p,...
furtherInvar(m..);     compute further invariants of m from the given ones
sortier(id);           sorts generators of id by increasing leading terms
```

LIBRARY: finvar.lib   Invariant Rings of Finite Groups

AUTHOR: Agnes E. Heydtmann,   agnes@math.uni-sb.de
          Simon A. King,   king@mfo.de

OVERVIEW:
A library for computing polynomial invariants of finite matrix groups and generators
of related varieties. The algorithms are based on B. Sturmfels, G. Kemper and W.
Decker et al.

MAIN PROCEDURES:

invariant_ring();	generators of the invariant ring (i.r.)
invariant_ring_random();	generators of the i.r., randomized alg.
primary_invariants();	primary invariants (p.i.)
primary_invariants_random();	primary invariants, randomized alg.

AUXILIARY PROCEDURES:

cyclotomic();	cyclotomic polynomial
group_reynolds();	finite group and Reynolds operator (R.o.)
molien();	Molien series (M.s.)
reynolds_molien();	Reynolds operator and Molien series
partial_molien();	partial expansion of Molien series
evaluate_reynolds();	image under the Reynolds operator
invariant_basis();	basis of homogeneous invars of a degree
invariant_basis_reynolds();	as invariant basis(), with R.o.
primary_char0();	primary invariants in char 0
primary_charp();	primary invariant in char p
primary_char0_no_molien();	p.i., char 0, without Molien series
primary_charp_no_molien();	p.i., char p, without Molien series
primary_charp_without();	p.i., char p, without R.o. or Molien series
primary_char0_random();	primary invariants in char 0, randomized
primary_charp_random();	primary invariants in char p, randomized
primary_char0_no_molien_random();	p.i., char 0, without M.s., randomized
primary_charp_no_molien_random();	p.i., char p, without M.s., randomized
primary_charp_without_random();	p.i., char p, without R.o. or M.s., random.
power_products();	exponents for power products
secondary_char0();	secondary (s.i.) invariants in char 0
secondary_charp();	secondary invariants in char p
secondary_no_molien();	secondary invariants, without M.s.
secondary_and_irreducibles_no_molien();	s.i. & irreducible s.i., without M.s.
secondary_not_cohen_macaulay();	s.i. when invariant ring not CM
orbit_variety();	ideal of the orbit variety
relative_orbit_variety();	ideal of a relative orbit variety
image_of_variety();	ideal of the image of a variety

LIBRARY: rinvar.lib   Invariant Rings of Reductive Groups

AUTHOR: Thomas Bayer,   tbayer@in.tum.de

OVERVIEW:
Implementation based on Derksen's algorithm. Written in the frame of the diploma

thesis (advisor: Prof. Gert-Martin Greuel) 'Computations of moduli spaces of semi-quasihomogeneous singularities and an implementation in Singular'

PROCEDURES:

HilbertSeries(I, w);	Hilbert series of the ideal I w.r.t. weight w
HilbertWeights(I, w);	weighted degrees of the generators of I
ImageVariety(I, F);	ideal of the image variety F(variety(I))
ImageGroup(G, F);	ideal of G w.r.t. the induced representation
InvariantRing(G, Gaction);	generators of the invariant ring of G
InvariantQ(f, G, Gaction);	decide if f is invariant w.r.t. G
LinearizeAction(G, Gaction);	linearization of the action 'Gaction' of G
LinearActionQ(action,s,t);	decide if action is linear in var(s..nvars)
LinearCombinationQ(base, f);	decide if f is in the linear hull of 'base'
MinimalDecomposition(f,s,t);	minimal decomposition of f (like coef)
NullCone(G,act);	ideal of the nullcone of the action 'act' of G
ReynoldsImage(RO,f);	image of f under the Reynolds operator 'RO'
ReynoldsOperator(G, Gaction);	Reynolds operator of the group G
SimplifyIdeal(I[,m,s]);	simplify the ideal I (try to reduce variables)

LIBRARY: stratify.lib   Algorithmic Stratification for Unipotent Group-Actions

AUTHOR: Anne Frühbis-Krüger,   anne@mathematik.uni-kl.de

OVERVIEW:
This library provides an implementation of the algorithm of Greuel and Pfister introduced in the article "Geometric quotients of unipotent group actions".

PROCEDURES:

prepMat(M,wr,ws,step);	list of sub matrices corresp. to given filtration
stratify(M,wr,ws,step);	algorithmic stratification (main procedure)

C.8.7 Symbolic-numerical solving

LIBRARY: digimult.lib   Satisfiability of prop. logical expressions

AUTHORS: Michael Brickenstein, bricken@mathematik.uni-kl.de

OVERVIEW:
Various algorithms for verifying digital circuits, including SAT-Solvers

PROCEDURES:

satisfiable(I);	returns 1, if system is satisfiable

LIBRARY: ntsolve.lib   Real Newton Solving of Polynomial Systems

AUTHORS: Wilfred Pohl,   pohl@mathematik.uni-kl.de
         Dietmar Hillebrand
PROCEDURES:

nt_solve(G,ini,[..]);	find one real root of 0-dimensional ideal G
triMNewton(G,a,[..]);	find one real root for 0-dim triangular system G

LIBRARY: presolve.lib   Pre-Solving of Polynomial Equations

AUTHORS: Gert-Martin Greuel,   greuel@mathematik.uni-kl.de,
PROCEDURES:

degreepart(id,d1,d2);	elements of id of total degree $\geq$ d1 and $\leq$ d2
elimlinearpart(id);	linear part eliminated from id
elimpart(id[,n]);	partial elimination of vars [among first n vars]
elimpartanyr(i,p);	factors of p partially eliminated from i in any ring
fastelim(i,p[..]);	fast elimination of factors of p from i [options]
findvars(id[..]);	ideal of variables occurring in id [more information]
hilbvec(id[,c,o]);	intvec of Hilbert series of id [in char c and ord o]
linearpart(id);	elements of id of total degree $\leq$ 1
tolessvars(id[,]);	maps id to new basering having only vars occurring in id
solvelinearpart(id);	reduced std-basis of linear part of id
sortandmap(id,s1,s2);	map to new basering with vars sorted w.r.t. complexity
sortvars(id[n1,p1..]);	sort vars w.r.t. complexity in id [different blocks]
valvars(id[..]);	valuation of vars w.r.t. to their complexity in id
idealSplit(id,tF,fS);	radical of the intersection of the ideals =radical(id)

LIBRARY: realrad.lib   Computation of real radicals

AUTHOR : Silke Spang

OVERVIEW:
Algorithms about the computation of the *real radical* of an arbitrary ideal over the
rational numbers and transcendental extensions thereof

PROCEDURES:

realpoly(f);	Computes the real part of the univariate polynomial f
realzero(j);	Computes the real radical of the zerodimensional ideal j
realrad(j);	Computes the real radical of an arbitrary ideal over transcendental extension of the rational numbers

LIBRARY: solve.lib   Complex Solving of Polynomial Systems

AUTHOR: Moritz Wenk,   wenk@mathematik.uni-kl.de
         Wilfred Pohl,   pohl@mathematik.uni-kl.de
PROCEDURES:

laguerre_solve(p,[..]);	find all roots of univariate polynomial p
solve(i,[..]);	all roots of 0-dim. ideal i using triangular sets
ures_solve(i,[..]);	find all roots of 0-dimensional ideal i with resultants
mp_res_mat(i,[..]);	multipolynomial resultant matrix of ideal i
interpolate(p,v,d);	interpolate poly from evaluation points p and results v
fglm_solve(i,[..]);	find roots of 0-dim. ideal using FGLM and lex'solve
lex_solve(i,p,[..]);	find roots of reduced lexicographic standard basis
triangLf_solve(l,[..]);	find roots using triangular sys. (factorizing Lazard)
triangM_solve(l,[..]);	find roots of given triangular system (Moeller)
triangL_solve(l,[..]);	find roots using triangular system (Lazard)
triang_solve(l,p,[..]);	find roots of given triangular system
simplexOut(L);	print solution L of simplex in nice format

LIBRARY: triang.lib   Decompose Zero-dimensional Ideals into Triangular Sets

AUTHOR: Dietmar Hillebrand
PROCEDURES:
triangL(G);   Decomposition of (G) into triangular systems (Lazard).
triangLfak(G);   Decomp. of (G) into tri. systems plus factorization.
triangM(G[,.]);   Decomposition of (G) into triangular systems (Möller).
triangMH(G[,.]);   Decomp. of (G) into tri. syst. with disjoint varieties.

LIBRARY: zeroset.lib   Procedures For Roots and Factorization

AUTHOR: Thomas Bayer, bayert@in.tum.de

OVERVIEW:
Algorithms for finding the zero–set of a zero–dimensional ideal in $Q(a)[x_1, ..., x_n]$.
Roots and Factorization of univariate polynomials over $Q(a)[t]$ where a is an algebraic number. Written in the frame of the diploma thesis (advisor: Prof. Gert-
Martin Greuel) 'Computations of moduli spaces of semiquasihomogeneous singularities and an implementation in Singular'. This library is meant as a preliminary
extension of the functionality of Singular for univariate factorization of polynomials over simple algebraic extensions in characteristic 0. Subprocedures with postfix
'Main' require that the ring contains a variable 'a' and no parameters, and the ideal
'mpoly', where 'minpoly' from the basering is stored.

PROCEDURES:
EGCD(f,g);   gcd over an algebraic extension field of Q
Factor(f);   factorization of f over an algebraic extension field
Quotient(f,g);   quotient q of f w.r.t. g (in f = q·g + remainder)
Remainder(f,g);   remainder of the division of f by g
Roots(f);   computes all roots of f in an extension field of Q
SQFRNorm(f);   norm of f (f must be squarefree)
ZeroSet(I);   zero-set of the 0-dim. ideal I

AUXILIARY PROCEDURES:
EGCDMain(f,g);   gcd over an algebraic extension field of Q
FactorMain(f);   factorization of f over an algebraic extension field
InvertNumberMain(c);   inverts an element of an algebraic extension field
QuotientMain(f,g);   quotient of f w.r.t. g
RemainderMain(f,g);   remainder of the division of f by g
RootsMain(f);   computes all roots of f, might extend the groundfield
SQFRNormMain(f);   norm of f (f must be squarefree)
ContainedQ(data,f);   f in data
SameQ(a,b);   a = b (list a,b)

LIBRARY: rootsmr.lib Counting the number of real roots of polynomial systems

AUTHOR: Enrique A. Tobis, etobis@dc.uba.ar

OVERVIEW:
Routines for counting the number of real roots of a multivariate polynomial system. Two methods are implemented: deterministic computation of the number of

roots, via the signature of a certain bilinear form (nrRootsDeterm); and a rational univariate projection, using a pseudorandom polynomial (nrRootsProbab). It also includes a command to verify the correctness of the pseudorandom answer. References: Basu, Pollack, Roy, Algorithms in Real Algebraic Geometry; Springer, 2003.

PROCEDURES:

nrRootsProbab(I)	Number of real roots of 0-dim ideal (probabilistic)
nrRootsDeterm(I)	Number of real roots of 0-dim ideal (deterministic)
symsignature(m)	Signature of the symmetric matrix m
sturmquery(h,B,I)	Sturm query of h on V(I)
matbil(h,B,I)	Matrix of the bilinear form on R/I associated to h
matmult(f,B,I)	Matrix of multiplication by f (m·f) on R/I in the basis B
tracemult(f,B,I)	Trace of m·f (B is an ordered basis of R/I)
coords(f,B,I)	Coordinates of f in the ordered basis B
randcharpoly(B,I,n)	Pseudorandom charpoly of univ. projection, n optional
verify(p,B,i)	Verifies the result of randcharpoly
randlinpoly(n)	Pseudorandom linear polynomial, n optional
powersums(f,B,I)	Powersums of the roots of a char polynomial
symmfunc(S)	Symmetric functions from the powersums S
univarpoly(l)	Polynomial with coefficients from l
qbase(i)	Like kbase, but the monomials are ordered

LIBRARY: rootsur.lib   Counting number of real roots of univariate polynomial

AUTHOR: Enrique A. Tobis, etobis@dc.uba.ar

OVERVIEW:
Routines for bounding and counting the number of real roots of a univariate polynomial, by means of several different methods, namely Descartes' rule of signs, the Budan-Fourier theorem, Sturm sequences and Sturm-Habicht sequences. The first two give bounds on the number of roots. The other two compute the actual number of roots of the polynomial. There are several wrapper functions, to simplify the application of the aforesaid theorems and some functions to determine whether a given polynomial is univariate. References: Basu, Pollack, Roy, Algorithms in Real Algebraic Geometry; Springer, 2003.

PROCEDURES:

isuni(p)	Checks whether a polynomial is univariate
whichvariable(p)	The only variable of a univariate monomial (or 0)
varsigns(p)	Number of sign changes in a list
boundBuFou(p,a,b)	Bound for number of real roots of poly p
boundposDes(p)	Bound for the number of positive real roots of p
boundDes(p)	Bound for the number of real roots of poly p
allrealst(p)	Checks whether all the roots of a poly are real
maxabs(p)	A bound for the maximum absolute value of a root
allreal(p)	Checks whether all the roots of a poly are real
sturm(p,a,b)	Number of real roots of a poly on an interval
sturmseq(p)	Sturm sequence of a polynomial
sturmha(p,a,b)	Number of real roots of a poly in (a,b)
sturmhaseq(p)	A Sturm-Habicht Sequence of a polynomial
reverse(l)	Reverses a list
nrroots(p)	The number of real roots of p

LIBRARY: signcond.lib Routines for computing realizable sign conditions

AUTHOR: Enrique A. Tobis, etobis@dc.uba.ar

OVERVIEW:
Routines to determine the number of solutions of a multivariate polynomial system which satisfy a given sign configuration. References: Basu, Pollack, Roy, Algorithms in Real Algebraic Geometry; Springer, 2003.

PROCEDURES:

signcnd(P,I)	The sign conditions realized by polynomials of P on a V(I)
psigncnd(P,l)	Pretty prints the output of signcnd (1)
firstoct(I)	The number of elements of V(I) with every coordinate ¿ 0

C.8.8 Visualization

LIBRARY: latex.lib   Typesetting of Singular-Objects in LaTeX2e

AUTHOR: Christian Gorzel,   gorzelc@math.uni-muenster.de
PROCEDURES:

closetex(fnm);	writes closing line for LaTeX-document
opentex(fnm);	writes header for LaTeX-file fnm
tex(fnm);	calls LaTeX2e for LaTeX-file fnm
texdemo([n]);	produces a file explaining the features of this lib
texfactorize(fnm,f);	creates string in LaTeX-format for factors of poly f
texmap(fnm,m,r1,r2);	creates string in LaTeX-format for map m: r1 → r2
texname(fnm,s);	creates string in LaTeX-format for identifier
texobj(l);	creates string in LaTeX-format for any (basic) type
texpoly(f,n[,l]);	creates string in LaTeX-format for poly
texproc(fnm,p);	creates string in LaTeX-format of text from proc p
texring(fnm,r[,l]);	creates string in LaTeX-format for ring/qring
rmx(s);	removes .aux and .log files of LaTeX-files
xdvi(s);	calls xdvi for dvi-files

LIBRARY: surf.lib   Procedures for Graphics with surf

AUTHOR: Hans Schönemann, hannes@mathematik.uni-kl.de,
        the programme surf is written by Stefan Endraß

NOTE:
To use this library requires the programme surf to be installed. surf is only available for Linux PCs and Sun workstations. You can download surf either from
                http://sourceforge.net/projects/surf
or from ftp://www.mathematik.uni-kl.de/pub/Math/Singular/utils/.

PROCEDURES:
plot(I,[...]);   plots plane curves and surfaces

LIBRARY: `surfex.lib`   Procedures for visualizing Surfaces.

AUTHOR: Oliver Labs
NOTE:
This library uses the program surf and surfex. surfex was written by Oliver
Labs and others, mainly Stephan Holzer. surf was written by Stefan Endrass
and others.
This software is used for producing raytraced images of the surfaces.
You can download surfex from   `http://www.surfex.AlgebraicSurface.net`.
surfex is a front-end for surf which aims to be easier to use than the original tool.

PROCEDURES:

`plotRotated();`	Plot the surface given by the polynomial p
`plotRot();`	Similar to plotRotated
`plotRotatedList();`	Plot the varieties given by a list of polynomials
`plotRotatedDirect();`	Plot the varieties given by a list directely

C.8.9 Coding theory

LIBRARY: `brnoeth.lib`   Brill-Noether Algorithm, Weierstraß-SG and AG-codes

AUTHORS: Jose Ignacio Farran Martin, ignfar@eis.uva.es
          Christoph Lossen,   lossen@mathematik.uni-kl.de

OVERVIEW:
Implementation of the Brill–Noether algorithm for solving the Riemann–Roch problem and applications in Algebraic Geometry codes. The computation of Weierstraß
semigroups is also implemented. The procedures are intended only for plane (singular) curves defined over a prime field of positive characteristic.

MAIN PROCEDURES:

`Adj_div(f);`	computes the conductor of a curve
`NSplaces(h,A);`	computes non-singular places up to given degree
`BrillNoether(D,C);`	computes a vector space basis of the linear system L(D)
`Weierstrass(P,m,C);`	computes the Weierstraß semigroup of C at P up to m
`extcurve(d,C);`	extends the curve C to an extension of degree d
`AGcode_L(G,D,E);`	computes the evaluation AG code with divisors G and D
`AGcode_Omega(G,D,E);`	computes the residual AG code with divisors G and D
`prepSV(G,D,F,E);`	preprocessing for the basic decoding algorithm
`decodeSV(y,K);`	decoding of a word with the basic decoding algorithm

AUXILIARY PROCEDURES:

`closed_points(I);`	computes the zero-set of a zero-dim. ideal in 2 vars
`dual_code(C);`	computes the dual code
`sys_code(C);`	computes an equivalent systematic code
`permute_L(L,P);`	applies a permutation to a list

C.8.10 System and Control theory

LIBRARY: `control.lib` Algebraic analysis tools for System and Control Theory

AUTHORS: Oleksandr Iena   yena@mathematik.uni-kl.de
          Markus Becker   mbecker@mathematik.uni-kl.de
          Viktor Levandovskyy   levandov@mathematik.uni-kl.de

MAIN PROCEDURES:
```
control(R);      analysis of controllability-related properties of R
controlDim(R);   analysis of controllability-related properties of R
autonom(R);      analysis of autonomy-related properties of R
autonomDim(R);   analysis of autonomy-related properties of R
```

COMPONENT PROCEDURES:
```
leftKernel(R);      a left kernel of R
rightKernel(R);     a right kernel of R
leftInverse(R);     a left inverse of R
rightInverse(R);    a right inverse of R
smith(M);           a Smith form of a module M
colrank(M);         a column rank of M as of matrix
genericity(M);      analysis of the genericity of parameters
canonize(L);        Groebnerification for modules
iostruct(R);        computes an IO-structure of behaviour by a module R
findTorsion(R, I);  submodule of R, annihilated by the ideal I
```

AUXILIARY PROCEDURES:
```
controlExample(s);  set up an example from the mini database inside
view();             well-formatted output of lists, modules and matrices
```

C.8.11 Teaching

LIBRARY: teachstd.lib   Procedures for Teaching Standard Bases

AUTHOR: Gert-Martin Greuel, greuel@mathematik.uni-kl.de

NOTE:
The library is intended to be used for teaching purposes only. The procedures are implemented exactly as described in the book 'A SINGULAR Introduction to Commutative Algebra' by G.-M. Greuel and G. Pfister. Sufficiently high printlevel allows to control each step.

PROCEDURES:
```
ecart(f);             ecart of f
tail(f);              tail of f
sameComponent(f,g);   test for same module component of lead(f) and lead(g)
leadmonomial(f);      leading monomial as poly (also for vectors)
monomialLcm(m,n);     lcm of monomials m and n as poly (also for vectors)
spoly(f[,1]);         s-polynomial of f [symmetric form]
minEcart(T,h);        element g ∈ T of minimal ecart s.t. LM(g) divides LM(h)
NFMora(i);            normal form of i w.r.t Mora algorithm
prodcrit(f,g);        test for product criterion
chaincrit(f,g,h);     test for chain criterion
pairset(G);           pairs form G neither satisfying prodcrit nor chaincrit
updatePairs(P,S,h);   pairset P enlarged by not useless pairs (h,f), f in S
standard(id);         standard basis of ideal/module
localstd(id);         local standard basis of id using Lazard's method
```

LIBRARY: weierstr.lib   Procedures for the Weierstraß Theorems

AUTHOR: Gert-Martin Greuel,   greuel@mathematik.uni-kl.de
PROCEDURES:
weierstr_div(g,f,d);   perform Weierstrass division of g by f up to degree d
weierstr_prep(f,d);   perform Weierstrass preparation of f up to degree d
lastvar_general(f);   make f general of finite order w.r.t. last variable
general_order(f);   compute integer b s.t. f is x`n-general of order b

LIBRARY: aksaka.lib   Primality testing after Agrawal, Saxena, Kayal

AUTHOR: Christoph Mang
OVERVIEW:
Algorithms for primality testing in polynomial time based on the ideas of Agrawal,
Saxena and Kayal.
PROCEDURES:
schnellexpt(a,m,n)   a^m for numbers a,m;
log2(n)   logarithm to basis 2 of n
PerfectPowerTest(n)   checks if there are a,b such that a^b=n
wurzel(r)   square root of number r
euler(r)   phi-function of euler
coeffmod(f,n)   poly f modulo number n (coefficients mod n)
powerpolyX(q,n,a,r)   (poly a)^q modulo (poly r,number n)
ask(n)   ASK-Algorithm; deterministic Primality test

LIBRARY: atkins.lib   Procedures for Teaching Cryptography

AUTHOR: Stefan Steidel, Stefan.Steidel@gmx.de

NOTE:
The library contains auxiliary procedures to compute the elliptic curve primality
test of Atkin and the Atkin's Test itself. The library is intended to be used for
teaching purposes but not for serious computations. Sufficiently high printLevel
allows to control each step, thus illustrating the algorithms at work.

PROCEDURES:
newTest(L,D)   checks if number D already exists in list L
bubblesort(L)   sorts elements of the list L
disc(N,k)   generates a list of negative discriminants
Cornacchia(d,p)   computes solution (x,y) for $x^2+d*y^2=p$
CornacchiaModified(D,p) computes solution (x,y) for $x^2+D*y^2=4p$
maximum(L)   computes the maximal number contained in L
expo(z,k)   computes exp(z)
jOft(t,k)   computes the j-invariant of t
round(r)   rounds r to the nearest number out of Z
HilbertClassPoly(D,k)   computes the Hilbert Class Polynomial
rootsModp(p,P)   computes roots of the polynomial P modulo p
wUnit(D)   computes the number of units in Q(sqr(D))
Atkin(N,K,B)   tries to prove that N is prime

LIBRARY: crypto.lib   Procedures for Teaching Cryptography

AUTHOR: Gerhard Pfister, pfister@mathematik.uni-kl.de

NOTE:
The library contains procedures to compute the discrete logarithm, primaly-tests, factorization included elliptic curve methodes. The library is intended to be used for teaching purposes but not for serious computations. Sufficiently high printlevel allows to control each step, thus illustrating the algorithms at work.

PROCEDURES:

decimal(s);	number corresponding to the hexadecimals
exgcdN(a,n)	compute s,t,d such that d=gcd(a,n)=s*a+t*n
eexgcdN(L)	T with sum L[i]*T[i]=T[n+1]=gcd(L[1],...,L[n])
gcdN(a,b)	compute gcd(a,b)
lcmN(a,b)	compute lcm(a,b)
powerN(m,d,n)	compute m^d mod n
chineseRem(T,L)	compute x such that x = T[i] mod L[i]
Jacobi(a,n)	the generalized Legendre symbol of a and n
primList(n)	the list of all primes below n
intPart(x)	the integral part of a rational number
intRoot(m)	the integral part of the square root of m
squareRoot(a,p)	the square root of a in Z/p, p prime
solutionsMod2(M)	basis solutions of Mx=0 over $Z/2$
powerX(q,i,I)	q-th power of the i-th variable modulo I
babyGiant(b,y,p)	discrete logarithm x: b^x=y mod p
rho(b,y,p)	discrete logarithm x: b^x=y mod p
MillerRabin(n,k)	probabilistic primaly-test of Miller-Rabin
SolowayStrassen(n,k)	probabilistic primaly-test of Soloway-Strassen
PocklingtonLehmer(N,[])	primaly-test of Pocklington-Lehmer
PollardRho(n,k,a,[])	Pollard's rho factorization
pFactor(n,B,P)	Pollard's p-factorization
quadraticSieve(n,c,B,k)	quadratic sieve factorization
isOnCurve(N,a,b,P)	P is on the curve $y^2z=x^3+a*xz^2+b*z^3$
ellipticAdd(N,a,b,P,Q)	P+Q, addition on elliptic curves
ellipticMult(N,a,b,P,k)	k*P on elliptic curves
ellipticRandomCurve(N)	generates $y^2z=x^3+a*xz^2+b*z^3$
ellipticRandomPoint(N,a,b)	random point on $y^2z=x^3+a*xz^2+b*z^3$
countPoints(N,a,b)	number of points of $y^2=x^3+a*x+b$ over Z/N
ellipticAllPoints(N,a,b)	points of $y^2=x^3+a*x+b$ over Z/N
ShanksMestre(q,a,b,[])	number of points of $y^2=x^3+a*x+b$ over Z/N
Schoof(N,a,b)	number of points of $y^2=x^3+a*x+b$ over Z/N
generateG(a,b,m)	m-th division polynomial of $y^2=x^3+a*x+b$
factorLenstraECM(N,S,B,[])	Lenstra's factorization
ECPP(N)	primaly-test of Goldwasser-Kilian

LIBRARY: hyperelliptic.lib   Procedures for Teaching Cryptography

AUTHOR: Markus Hochstetter, markushochstetter@gmx.de

NOTE:
This library provides procedures for computing with divisors in the jacobian of *hyperelliptic curves*. In addition procedures are available for computing the rational

representation of divisors and vice versa. The library is intended to be used for teaching and demonstrating purposes but not for efficient computations.

PROCEDURES:

ishyper(h,f)	test, if y^2+h(x)y=f(x) is hyperelliptic
isoncurve(P,h,f)	test, if point P is on C: y^2+h(x)y=f(x)
chinrestp(b,moduli)	compute polynom x, s.t. x=b[i] mod moduli[i]
norm(a,b,h,f)	norm of a(x)-b(x)y in IF[C]
multi(a,b,c,d,h,f)	(a(x)-b(x)y)*(c(x)-d(x)y) in IF[C]
ratrep (P,h,f)	returns polynomials a,b, s.t. div(a,b)=P
divisor(a,b,h,f,[])	computes divisor of a(x)-b(x)y
gcddivisor(p,q)	gcd of the divisors p and q
semidiv(D,h,f)	semireduced divisor of the pair of polys D[1], D[2]
cantoradd(D,Q,h,f)	adding divisors of the hyperell. curve y^2+h(x)y=f(x)
cantorred(D,h,f)	returns reduced divisor which is equivalent to D
double(D,h,f)	computes 2*D on y^2+h(x)y=f(x)
cantormult(m,D,h,f)	computes m*D on y^2+h(x)y=f(x)

C.8.12 Non–commutative

LIBRARY: central.lib   Computation of central elements of GR-algebras

AUTHOR: Oleksandr Motsak   motsak@mathematik.uni-kl.de

OVERVIEW:
A library for computing elements of the center and centralizers of sets of elements in GR-algebras.

KEYWORDS: center; centralizer; reduce; centralize; PBW

PROCEDURES:

centralizeSet(F, V)	v.s. basis of the centralizer of F within V
centralizerVS(F, D)	v.s. basis of the centralizer of F
centralizerRed(F, D[, N])	reduced basis of the centralizer of F
centerVS(D)	v.s. basis of the center
centerRed(D[, k])	reduced basis of the center
center(D[, k])	reduced basis of the center
centralizer(F, D[, k])	reduced bais of the centralizer of F
sa_reduce(V)	's.a. reduction' of elements
sa_poly_reduce(p, V)	's.a. reduction' of p
inCenter(T)	checks the centrality of T
inCentralizer(T, S)	checks whether T, S commutes
isCartan(p)	checks whether p is a Cartan element
applyAdF(Basis, f)	images under the k-linear map Ad`f
linearMapKernel(Images)	kernel of a linear map given by images
linearCombinations(Basis, C)	k-linear combinations of elements
variablesStandard()	set of algebra generators
variablesSorted()	heuristically sorted set of generators
PBW_eqDeg(Deg)	PBW monomials of given degree
PBW_maxDeg(MaxDeg)	PBW monomials up to given degree
PBW_maxMonom(MaxMonom)	PBW monomials up to given maximal

LIBRARY: dmod.lib   Algorithms for algebraic D-modules

AUTHORS: Viktor Levandovskyy,   levandov@risc.uni-linz.ac.at
         Jorge Martin Morales,   jorge@unizar.es

THEORY:
Given a polynomial ring $R = K[x_1, ..., x_n]$ and a polynomial F in R, one is interested in the ring $R[1/F^s]$ for a natural number s. In fact, the ring $R[1/F^s]$ has a structure of a D(R)-module, where $D(R)$ is a Weyl algebra. Constructively, one needs to find a left ideal $I = I(F^s)$ in $D(R)$, such that $K[x_1, ..., x_n, 1/F^s]$ is isomorphic to $D(R)/I$ as a $D(R)$-module.

We provide two implementations:
1) the classical Ann F^s algorithm from Oaku and Takayama (J. Pure
   Applied Math., 1999) and
2) the newer Ann F^s algorithm by Briancon and Maisonobe (Remarques sur
   l'ideal de Bernstein associe a des polynomes, preprint, 2002).
PROCEDURES:
annfsOT(F[,eng]);   compute Ann F^s for a poly F (algorithm of Oaku-Takayama)
annfsBM(F[,eng]);   compute Ann F^s for a poly F (Briancon-Maisonobe)
minIntRoot(P,fact); minimal integer root of a maximal ideal P
reiffen(p,q);   create a poly, describing a Reiffen curve
arrange(p);   create a poly, describing a generic hyperplane arrangement
isHolonomic(M);   check whether a module is holonomic
convloc(L);   replace global orderings with local in the ringlist L

LIBRARY: gkdim.lib   Procedures for calculating the Gelfand-Kirillov dimension

AUTHORS: Lobillo, F.J.,   jlobillo@ugr.es
         Rabelo, C.,   crabelo@ugr.es
SUPPORT: 'Metodos algebraicos y efectivos en grupos cuanticos',
         BFM2001-3141, MCYT, Jose Gomez-Torrecillas (Main researcher).
PROCEDURES:
  GKdim(M);   Gelfand-Kirillov dimension

LIBRARY: involut.lib   Procedures for Computations with Involutions

AUTHORS: Oleksandr Iena,   yena@mathematik.uni-kl.de,
         Markus Becker,   mbecker@mathematik.uni-kl.de,
         Viktor Levandovskyy,   levandov@mathematik.uni-kl.de

THEORY:
Involution is an antiisomorphism of a noncommutative algebra with the property that applied an involution twice, one gets an identity. Involution is linear with respect to the ground field. In this library we compute linear involutions, distinguishing the case of a diagonal matrix (such involutions are called homothetic) and a general one.

NOTE:
This library provides algebraic tools for computations and operations with algebraic involutions and linear automorphisms of noncommutative algebras.

PROCEDURES:
findInvo();	computes linear involutions on a basering;
findInvoDiag();	computes homothetic (diagonal) involutions;
findAuto();	computes linear automorphisms of a basering;
ncdetection();	computes an ideal, presenting an involution map.

LIBRARY: ncalg.lib   Definitions of important GR-algebras

AUTHORS: Viktor Levandovskyy,   levandov@mathematik.uni-kl.de,
            Oleksandr Motsak,   motsak@mathematik.uni-kl.de

CONVENTIONS:
This library provides pre-defined important noncommutative algebras. For universal enveloping algebras of finite dimensional Lie algebras sl_n, gl_n, g_2 etc. there are functions makeUsl, makeUgl, makeUg2 etc. There are quantized enveloping algebras $U_q(sl_2)$ and $U_q(sl_3)$ (via functions makeQsl2, makeQsl3) and non-standard quantum deformation of so_3, accessible via makeQso3 function. For bigger algebras we suppress the output of the (lengthy) list of non-commutative relations and provide only the number of these relations instead.

PROCEDURES:
makeUsl2([p])	create U(sl`2) in the variables (e,f,h) in char p
makeUsl(n[,p])	create U(sl`n) in char p
makeUgl(n,[p])	create U(gl`n) in the variables (e`i`j (1¡i,j¡n)) in char p
makeUso5([p])	create U(so`5) in the variables (x(i),y(i),H(i)) in char p
makeUso6([p])	create U(so`6) in the variables (x(i),y(i),H(i)) in char p
makeUso7([p])	create U(so`7) in the variables (x(i),y(i),H(i)) in char p
makeUso8([p])	create U(so`8) in the variables (x(i),y(i),H(i)) in char p
makeUso9([p])	create U(so`9) in the variables (x(i),y(i),H(i)) in char p
makeUso10([p])	create U(so`10) in the variables (x(i),y(i),H(i)) in char p
makeUso11([p])	create U(so`11) in the variables (x(i),y(i),H(i)) in char p
makeUso12([p])	create U(so`12) in the variables (x(i),y(i),H(i)) in char p
makeUsp1([p])	create U(sp`1) in the variables (x(i),y(i),H(i)) in char p
makeUsp2([p])	create U(sp`2) in the variables (x(i),y(i),H(i)) in char p
makeUsp3([p])	create U(sp`3) in the variables (x(i),y(i),H(i)) in char p
makeUsp4([p])	create U(sp`4) in the variables (x(i),y(i),H(i)) in char p
makeUsp5([p])	create U(sp`5) in the variables (x(i),y(i),H(i)) in char p
makeUg2([p])	create U(g`2) in the variables (x(i),y(i),Ha,Hb) in char p
makeUf4([p])	create U(f`4) in the variables (x(i),y(i),H(i)) in char p
makeUe6([p])	create U(e`6) in the variables (x(i),y(i),H(i)) in char p
makeUe7([p])	create U(e`7) in the variables (x(i),y(i),H(i)) in char p
makeUe8([p])	create U(e`8) in the variables (x(i),y(i),H(i)) in char p
makeQso3([n])	create U`q(so`3) in the presentation of Klimyk
makeQsl2([n])	preparation for U`q(sl`2) as factor-algebra;
makeQsl3([n])	preparation for U`q(sl`3) as factor-algebra;
Qso3Casimir(n [,m])	Casimir elements of U`q(so`3)
GKZsystem(A, sord, alg [,v])	Gelfand-Kapranov-Zelevinsky system

LIBRARY: ncdecomp.lib   Decomposition of a module into central characters

AUTHORS: Viktor Levandovskyy,   levandov@mathematik.uni-kl.de.

OVERVIEW:
This library presents algorithms for the central character decomposition of a module, i.e. a decomposition into generalized weight modules with respect to the center. Based on ideas of O. Khomenko and V. Levandovskyy (see the article [153] in the References for details).

PROCEDURES:
CentralQuot(M,G);	central quotient M:G,
CenCharDec(I,C);	decomposition of I into central characters w.r.t. C
IntersectWithSub(M,Z);	intersection of M with the subalgebra.

LIBRARY: nctools.lib   General tools for noncommutative algebras

AUTHORS: Levandovskyy V.,   levandov@mathematik.uni-kl.de,
            Lobillo, F.J.,   jlobillo@ugr.es,
            Rabelo, C.,   crabelo@ugr.es,
            Motsak, O.,   motsak@mathematik.uni-kl.de.
MAIN PROCEDURES:
Gweights(r);	compute weights for a compatible ordering
weightedRing(r);	change the ordering to a weighted one
ndcond();	the ideal of non-degeneracy conditions
Weyl([p]);	create Weyl algebra structure in a basering
makeWeyl(n, [p]);	return n-th Weyl algebra
makeHeisenberg(N, [p,d]);	return n-th Heisenberg algebra
Exterior();	return qring, the exterior algebra of a basering,
findimAlgebra(M,[r]);	create finite dimensional algebra
SuperCommutative([b,e,Q]);	the super-commutative algebra over a basering,
rightStd(I);	compute a right Groebner basis of an ideal,

LIBRARY: perron.lib   computation of algebraic dependences

AUTHORS: Oleksandr Motsak, motsak@mathematik.uni-kl.de.
PROCEDURES:
perron(L[, D]);	relations between pairwise commuting polynomials

LIBRARY: qmatrix.lib   Quantum matrices, quantum minors and symmetric groups

AUTHORS: Lobillo, F.J.,   jlobillo@ugr.es
         Rabelo, C.,   crabelo@ugr.es
MAIN PROCEDURES:
quantMat(n, [p]);	generates the quantum matrix ring of order n;
qminor(u, v, nr);	calculate a quantum minor of a quantum matrix

AUXILIARY PROCEDURES:

SymGroup(n);	generates an intmat containing S(n),
LengthSymElement(v);	calculates the length of the element v of S(n)
LengthSym(M);	calculates the length of each element of M

LIBRARY: ratgb.lib   Groebner bases in Ore localizations

AUTHOR: Viktor Levandovskyy,   levandov@risc.uni-linz.ac.at
PROCEDURES:
ratstd(ideal I, int n);   compute Groebner basis in Ore localization

C.9 SINGULAR and Maple

In this section we give two examples how to use SINGULAR as a support for a Maple session and one example showing how to use Maple in a SINGULAR session. The first example is the trivial way, the second is based on a simplified version of a script of G. Kemper.[3]

Assume we are in a Maple session and want to compute a Gröbner basis with SINGULAR of the ideal $I = \langle x^{10} + x^9 y^2, y^8 - x^2 y^7 \rangle$ in characteristic 0 with the degree reverse lexicographical ordering dp.

The first solution is to write the polynomials to the file singular_input (already in the SINGULAR language). This is done by the following:

```
f:=x^10+x^9*y^2;
g:=y^8-x^2*y^7;

interface(prettyprint=0);
interface(echo=0);
writeto( singular_input );
lprint('ideal I = ');
f, g ;
lprint(';');
writeto(terminal);
```

The resulting file looks like:

```
ideal I =
x^10+x^9*y^2, y^8-x^2*y^7
;
```

Now we can start SINGULAR, and perform the following

---

[3] Warning: The scripts run only on Unix like operating systems.

```
ring R=0,(x,y),dp;
< "singular_input";
short=0;            // output in Maple format
ideal J=std(I);
write(":w maple_input",J);
```

This Singular session writes the computed Gröbner basis (in Maple format)
to the file maple_input:

```
x^2*y^7-y^8,x^9*y^2+x^10,x^12*y+x*y^11,x^13-x*y^12,y^14+x*y^12,
x*y^13+y^12
```

A more advanced solution is given by the following procedure. Here I is a
list of polynomials, P is the characteristic of the ground field, and tord is a
string specifying the Singular ordering (for instance dp, lp, ...).[4]

```
SINGULARlink:=proc(I,P,tord)
    local i,j,path,vars,F,p,ele;

    F:=map(expand,I);
    vars:=indets(F);
    if P>0 then p:=P else p:=0 fi;

    path:="Trans";
    if assigned(pathname) then path:=cat(pathname,path) fi;
    if system("mkdir ".path)<>0 then
        ERROR("Couldn't make the Transfer-directory")
    fi;

    # produce input for Singular ...
    writeto(cat(path,"/In"));

    # Define the ring (with term order)
    lprint('ring R ='));
    lprint(''.p.','');
    lprint('(x(1..'.(nops(vars)).')),');
    lprint('('.tord.');'');

    # Define the ideal ...
    lprint('ideal I =');
    for i to nops(F) do
        if i>1 then lprint(',') fi;
        ele:=subs([seq(vars[j]=x(j), j=1..nops(vars))],F[i]);
        if type(ele,monomial) then lprint(ele)
```

---

[4] This procedure works with Maple V Release 5. In older versions of Maple,
string expression were enclosed in a pair of back quotes ' ' instead of " ";
moreover, the nullary operator was denoted by " instead of %. The directory EXAMPLES/ on the enclosed CD contains two versions of the procedure
– one for Maple V Release 5 and one for Maple V Release 3 (with the
old syntax). A modified worksheet for Maple V, Release 10 is available at
http://www.singular.uni-kl.de/interfaces.html.

```
            else # split into summands, otherwise line might get chopped!
                lprint(op(1,ele));
                for j from 2 to nops(ele) do
                    lprint('+',op(j,ele))
                od
            fi
    od;
    lprint(';');

    lprint('short=0;');
    lprint('ideal b = std(I);');
    lprint('write(\"'.path.'/Out\",\"[\");');
    lprint('write(\"'.path.'/Out\",b);');
    lprint('write(\"'.path.'/Out\",\"];\");');
    lprint('quit;');
    writeto(terminal);

    # Call Singular ...
    system("Singular < ".path."/In > ".path."/temp");
    if %<>0 then
        system("'rm' -r ".path);
        ERROR("Something went wrong while executing Singular")
    fi;

    # Retrieve the results ...
    read(cat(path,"/Out"));
    F:=%;
    system("'rm' -r ".path);
    F:=subs([seq(x(i)=vars[i],i=1..nops(vars))],F);
    F:=map(expand,F);
    F
end:
```

Let's apply this procedure to an example:

```
f:=x^10+x^9*y^2;
g:=y^8-x^2*y^7;
J:=SINGULARlink([f,g],0,"dp"):

interface(prettyprint=0);
J;
#-> [x^2*y^7-y^8, x^10+x^9*y^2, x^12*y+x*y^11, x^13-x*y^12,
#->  y^14+x*y^12, x*y^13 +y^12]
```

Assume now that we are in a SINGULAR session and want to use Maple to factorize a polynomial. This can be done by using the following SINGULAR procedure:

```
proc maple_factorize(poly p)
{
    int saveshort=short;
    short=0;
    string in="maple-in."+string(system("pid"));
    string out="maple-out."+string(system("pid"));
```

```
link l=in;
write(l,"res:=factors("+string(p)+");");
write(l,"interface(prettyprint=0);");
write(l,"interface(echo=0);");
write(l,"writeto('"+out+"');");
write(l,"lprint('ideal fac=');");
write(l,"lprint(op(1,res));");
write(l,"for i to nops(op(2,res)) do");
write(l,"  lprint(',');");
write(l,"  lprint(op(1,op(i,op(2,res))));");
write(l,"od;");
write(l,"lprint(';');");
write(l,"lprint('intvec multies=');");
write(l,"lprint('1');");
write(l,"for i to nops(op(2,res)) do");
write(l,"  lprint(',');");
write(l,"  lprint(op(2,op(i,op(2,res))));");
write(l,"od;");
write(l,"lprint(';');");
write(l,"lprint('list res=fac,multies;');");
write(l,"writeto(terminal);");
write(l,"quit;");
int dummy=system("sh","maple <"+in+" > dummy");
if (dummy <> 0) { ERROR("something went wrong"); }
string r=read(out);
execute(r);
return(res);
}
```

Here comes an example:

```
ring R = 0,(x,y),dp;
poly f = 5*(x-y)^2*(x+y);

maple_factorize(f);
//-> [1]:
//->    _[1]=5
//->    _[2]=x+y
//->    _[3]=-x+y
//-> [2]:
//->    1,1,2
```

C.10 SINGULAR and Mathematica

We show by an example how to use SINGULAR as support for a Mathematica session.[5] Assume we are in a Mathematica session and want to compute a

---

[5] Warning: The scripts run only on Unix like operating systems. M. Kauers and V. Levandovskyy developed a Mathematica package to call SINGULAR from Mathematica. It is available from RICAM Linz www.risc.uni-linz.ac.at/research/combinat/software/Singular and also at www.singular.uni-kl.de/interfaces.html.

Gröbner basis with SINGULAR of the ideal $\langle x^{10} + x^9 y^2, y^8 - x^2 y^7 \rangle$ in characteristic 0 with the degree reverse lexicographical ordering dp.

```
SINGULARlink[J_List,P_Integer,tord_String] := Module[
  {i,vars,F,p,subst,varnames,SINGULARin,SINGULARout},
  F=J;
  vars=Variables[F];
  (* Substitution of Variable names *)
  varnames = Table[ ToExpression[ "x[" <> ToString[i] <> "]"],
                    {i, 1, Length[vars]} ];
  subst = Dispatch[MapThread[Rule, {vars, varnames} ]];
  F = F /. subst;
  singF = ToString[F // InputForm];
  singF = StringReplace[ singF, { "{"->"", "}"->"", ", "->",\n",
                                  "]"->")", "["->"(" } ];
  If[P>0, p=P, (*Else*) p=0];
  (* Prepare Singular input string *)
  SINGULARin = "ring R =" <> ToString[p] <> ", (" <>
               "x(1.." <> ToString[Length[vars]] <> ")), (" <>
               tord <> ");\n";
  (* Define the ideal... *)
  SINGULARin = SINGULARin <> "ideal I =" <> singF <> " ;\n" <>
               "short=0;\n" <>
               "ideal b = std(I);\n" <>
               "write(\".tmp.sing.mathematica\",b );\n" <>
               "quit;\n";

  (* Call Singular ... *)
  SINGULARin // OutputForm >> "!Singular -q";
  SINGULARout=ReadList[".tmp.sing.mathematica",String];
  >>"!rm .tmp.sing.mathematica";
  SINGULARout= StringReplace[SINGULARout , { ")"->"]",
                                             "("->"[" } ];
  SINGULARout= "{" <> SINGULARout <> "}";
  SINGULARout= ToExpression[SINGULARout];
  subst = Dispatch[MapThread[Rule, {varnames, vars} ]];
  SINGULARout=SINGULARout/. subst;
  SINGULARout
]
```

Let's apply this

```
f=x^10+x^9*y^2;
g=y^8-x^2*y^7;
J=SINGULARlink[{f,g},0,"dp"];

J // InputForm
(*-> //InputForm=
    {x^2*y^7 - y^8, x^10 + x^9*y^2, x^12*y + x*y^11,
     x^13 - x*y^12, x*y^12 + y^14, y^12 + x*y^13}
*)
```

C.11 SINGULAR and MuPAD

Finally, we give an example how to use SINGULAR as support for a MuPAD session.[6] Assume we are in a MuPAD session and want to compute a Gröbner basis with SINGULAR of the ideal $\langle x^{10} + x^9 y^2, y^8 - x^2 y^7 \rangle$ in characteristic 0 with the degree reverse lexicographical ordering dp.

```
SINGULARlink := proc(L:DOM_LIST, P:DOM_INT, tord)
  local i, path, ipath, opath, tpath, fd, vars, F, p;

begin
  F := map(L, expand);
  vars := indets(F);

  if P > 0 then
    p := P;
  else
    p := 0;
  end_if;

  // Create the directory where "communication files" are stored
  path := "Trans";

  if singpath <> hold(singpath) then
    path := pathname(singpath, path) ;
  end_if;

  if system("mkdir ".path) <> 0 then
    error("Could not make the Transfer-directory")
  end_if;

  // produce input for Singular ...
  ipath := pathname(path)."In" ;
  if version() = [2, 0, 0] then
    fd := fopen(Text, ipath, Write) ;
  else // MuPAD 2.5
    fd := fopen(ipath, Write, Text) ;
  end_if ;

  // Define the ring (with term order)
  fprint(NoNL, fd, "ring R = ");
  fprint(NoNL, fd, p,", (");
  fprint(NoNL, fd, vars[i], ", " ) $i=1..nops(vars)-1;
  fprint(NoNL, fd, vars[nops(vars)], "), ");
  fprint(Unquoted, fd, "( ",tord," ); ");

  // Define the ideal ...
  fprint(Unquoted, fd, "ideal I =");
  fprint(Unquoted, fd, F[i], ", " ) $i=1..nops(F)-1;
  fprint(Unquoted, fd, F[nops(F)], ";");
```

---

[6] We should like to thank Torsten Metzner for providing the scripts. Warning: The scripts run only on Unix like operating systems.

```
fprint(Unquoted, fd, "short=0;");
fprint(Unquoted, fd, "ideal b = std(I);");
opath := pathname(path)."Out" ;
fprint(Unquoted, fd, "write(\"".opath."\",\"[\");");
fprint(Unquoted, fd, "write(\"".opath."\",b);");
fprint(Unquoted, fd, "write(\"".opath."\",\"];\");");
fprint(Unquoted, fd, "quit;");
fclose(fd);

// Call Singular ...
tpath := pathname(path)."temp" ;
if system("Singular < ".ipath."> ".tpath) <> 0 then
  system("rm -r ".path);
  error("Something went wrong while executing Singular");
end_if;

// Retrieve the results ...
F := read(opath, Quiet):
system("rm -r ".path);
F := map(F,expand);
end_proc:
```

Now apply this procedure:

```
f:=x^10+x^9*y^2;
g:=y^8-x^2*y^7;
J:=SINGULARlink([f,g],0,"dp"):
output::tableForm(J,",");
//-> x^2*y^7 - y^8,x^10 + x^9*y^2,x*y^11 + x^12*y,x^13 - x*y^12,
//-> y^14 + x*y^12,y^12 + x*y^13
```

Assume now that we are in a SINGULAR session and want to use MuPAD to factorize a polynomial. This can be done by using the following SINGULAR procedure:

```
proc mupad_factorize(poly p)
{
  int saveshort = short;
  short = 0;
  string in  = "mupad-in."+string(system("pid"));
  string out = "mupad-out."+string(system("pid"));
  link l = in;
  write(l, "res:=factor("+string(p)+");");
  write(l, "if version() = [2, 0, 0] then ");
  write(l, "fd := fopen(Text, \""+out+"\", Write) ;");
  write(l, "else");
  write(l, "fd := fopen(\""+out+"\", Write, Text) ;");
  write(l, "end_if;");

  write(l,"fprint(Unquoted, fd, \"ideal fac=\");");
  write(l,"fprint(Unquoted, fd, op(res,1));");
  write(l,"for i from 2 to nops(res) step 2 do");
  write(l," fprint(Unquoted, fd, \",\", op(res,i));");
```

```
write(1,"end_for;");
write(1,"fprint(Unquoted, fd, \";\");");

write(1,"fprint(Unquoted, fd, \"intvec multies=\");");
write(1,"fprint(Unquoted, fd, \"1\");");
write(1,"for i from 3 to nops(res) step 2 do");
write(1,"  fprint(Unquoted, fd, \",\", op(res,i));");
write(1,"end_for;");
write(1,"fprint(Unquoted, fd, \";\");");

write(1,"fprint(Unquoted, fd, \"list res=fac,multies;\");");
write(1,"fclose(fd);");
write(1,"quit;");
int dummy = system("sh","mupad <"+in+" > dummy");
if (dummy <> 0) { ERROR("something went wrong"); }
string r=read(out);
execute(r);
return(res);
}
```

We apply this procedure to an example:

```
ring R = 0,(x,y),dp;
poly f = 5*(x-y)^2*(x+y);

mupad_factorize(f);
//-> [1]:
//->    _[1]=5
//->    _[2]=x+y
//->    _[3]=-x+y
//->[2]:
//->    1,1,2
```

C.12 SINGULAR and GAP

Of course we can use GAP's functionality in a SINGULAR session in the same way as in the preceding sections.

For the opposite direction there is a GAP package called "singular" written by Marco Costantini and Willem A. de Graaf. This package allows the GAP user to access functions of SINGULAR from within GAP, and to apply these functions to the GAP objects. With this package, the user keeps working with GAP and, if he needs a function of SINGULAR that is not present in GAP, he can use this function via "SingularInterface".

The interface is expected to work with every version of GAP 4, every (not very old) version of SINGULAR, and on every platform, on which both GAP and SINGULAR run. It may be found at

http://www.gap-system.org/Packages/singular.html or
http://www.singular.uni-kl.de/interfaces.html.

Since there is a detailed documentation on the homepage, we will confine ourselves to give a short example:

We compute a Groebner basis (that will be returned as a list of polynomials) of an ideal I of a polynomial ring. As term ordering, SINGULAR will use the value of TermOrdering of the polynomial ring containing I. If this value is not set, then the degree reverse lexicographical ordering ("dp") will be used.

```
gap> LoadPackage("singular");
gap> R:= PolynomialRing( Rationals, ["x","y","z"] : old );;
gap> x:= R.1;; y := R.2;; z := R.3;;
gap> r:= [x*y*z-x^2*z,x^2*y*z-x*y^2*z-x*y*z^2,x*y-x*z-y*z];;
gap> I:= Ideal( R, r );
<two-sided ideal in PolynomialRing(..., [ x, y, z ]),
 (3 generators)>
gap> GroebnerBasis( I );
[x*y-x*z-y*z,x^2*z-x*z^2-y*z^2,x*z^3+y*z^3,-x*z^3+y^2*z^2
 -y*z^3]
```

C.13 SINGULAR and SAGE

SAGE [249] is a new computer algebra system and mathematics software distribution developed under the lead of William Stein by a worldwide team of developers.

The SAGE developers describe SAGE as a free distribution of open source math software with new functionality that fills in gaps in what is available elsewhere, and provides a unified interface to most math software: to Axiom, GAP, Macaulay2, Magma, Maple, Mathematica, MATLAB, Pari, SINGULAR, etc. .

Singular is shipped with SAGE and is used to provide functionality for global and local commutative algebra, for non–commutative algebra and for symbolic–numerical polynomial solving.

SAGE communicates with several computer algebra systems by interacting with their command line using the pexpect Python package. As a result Singular can be used from the SAGE command line in several ways.

First, there is an object oriented interface to 'native' Singular objects, which allows to perform almost every calculation Singular is capable of.

```
sage: r = singular.ring(0,'(x,y,z)','dp')
sage: f = singular('x + 2*y + 2*z - 1')
sage: g = singular('x^2 + 2*y^2 + 2*z^2 - x')
sage: h = singular('2*x*y + 2*y*z - y')
sage: I = singular.ideal(f,g,h)
sage: I.std()
```

```
//->x+2*y+2*z-1,
//->10*y*z+12*z^2-y-4*z,
//->4*y^2+2*y*z-y,
//->210*z^3-79*z^2+7*y+3*z
```

Alternatively, commands can be passed to the Singular interpreter directly, allowing SAGE to perform exactly every calculation Singular is capable of.

```
sage: ret = singular.eval('ring r = 0,(x,y,z),dp')
sage: ret = singular.eval('poly f= x + 2*y + 2*z - 1')
sage: ret = singular.eval('poly g = x^2 + 2*y^2 + 2*z^2 - x')
sage: ret = singular.eval('poly h = 2*x*y + 2*y*z - y')
sage: ret = singular.eval('ideal i = f,g,h')
sage: print singular.eval('std(i)')


//->_[1]=x+2y+2z-1
//->_[2]=10yz+12z2-y-4z
//->_[3]=4y2+2yz-y
//->_[4]=210z3-79z2+7y+3z
```

This functionality is used by several 'native' SAGE objects. For instance, if a Gröbner basis is to be calculated in SAGE Singular is used by default:

```
sage: P.<x,y,z> = PolynomialRing(QQ,3)
sage: I = sage.rings.ideal.Katsura(P,3)
sage: I.groebner_basis() # calls Singular in background


//->[x + 2*y + 2*z - 1, \
//->10*y*z + 12*z^2 - y - 4*z, \
//->5*y^2 - 3*z^2 - y + z, \
//->210*z^3 - 79*z^2 + 7*y + 3*z]
```

However, the communication channel via string parsing of SINGULAR in- and output may be pretty slow depending on the (size of the) task at hand.

To use SINGULAR 's multivariate polynomial arithmetic, SAGE links directly against a shared library called libSINGULAR which is derived from SINGULAR by the SAGE developers[7]. As of SAGE 2.7 multivariate polynomial arithmetic over Q, $GF(p)$, and $GF(p^n)$ is implemented this way providing SAGE with the very fast multivariate polynomial arithmetic of SINGULAR . However, in this shared library mode no support for the SINGULAR command-line interpreter is provided which means that only a basic subset of SINGULAR 's capabilities are available, i.e. those written in C/C++. All of SINGULAR 's capabilities are however available through the aforementioned pexpect interface and conversion methods are provided.

---

[7] We would like to thank Martin Albrecht for improving the Singular interface and making it possible to compile SINGULAR as a shared library.

```
# uses Singular in shared library mode
sage: P.<x,y,z> = PolynomialRing(QQ,3)
sage: I = sage.rings.ideal.Katsura(P)
# calls Singular via the pexpect interface in background
sage: I.groebner_basis()
//->[x + 2*y + 2*z - 1, \
//->10*y*z + 12*z^2 - y - 4*z, \
//->5*y^2 - 3*z^2 - y + z, \
//->210*z^3 - 79*z^2 + 7*y + 3*z]
sage: f = I.gens()[0]

# shared library implementation -> pexpect interface
sage: g = f._singular_()

# pexpect interface implementation -> shared library mode
sage: P(g)
//->x + 2*y + 2*z - 1
```

References

1. Adams, W.W.; Loustaunau, P.: An Introduction to Gröbner Bases. Vol. 3. Graduate Studies in Mathematics, Providence, RI, AMS (1996).
2. Altman, A.; Kleiman, S.: Introduction to Grothendieck duality theory. SLN 146, Springer (1970).
3. Apel, J.: Gröbnerbasen in nicht kommutativen Algebren und ihre Anwendung. Dissertation, Universität Leipzig (1988).
4. Arnold, E.A.: Modular algorithms for computing Gröbner bases. J. of Symbolic Computations 35, 403–419, (2003).
5. Arnold, V.I.; Gusein-Zade, S.M., Varchenko, A.N.. Singularities of Differential Maps. Vol. I, Birkhäuser (1985).
6. Atiyah, M.F.; Macdonald,I.G.: Introduction to Commutative Algebra. Addison–Wesley, London (1969).
7. Aubry, P.; Lazard, D.; Moreno Maza M.: On the Theories of Triangular Sets. J. Symb. Comp. 28, 105–124 (1999).
8. Bachmann, O; Gray, S.; Schönemann, H.: A Proposal for Syntactic Data Integration for Math Protocols. In: Erich Kaltofen, Markus Hitz (eds.) Proc. of Second International Symposium on Parallel Symbolic Computation, PASCO '97, Maui, HI, USA, July, 1997. ACM Press, New York, 165–175 (1997).
9. Bachmann, O; Gray, S.; Schönemann, H.: MPP: A Framework for Distributed Polynomial Computations. In: Lakshman, Y.N. (ed.), Proc, of 1996 International Symposium on Symbolic and Algebraic Computation, ISSAC '96, Zürich, Switzerland, July, 1996. ACM Press, New York, 103–109 (1996).
10. Bachmann, O.; Schönemann,H.: Monomial Operations for Computations of Gröbner Bases. In: Proceedings ISSAC'98, 309–316.
11. Baciu, C.: The Classification of the Hypersurface Singularities using the Milnor Code. Diplomarbeit, Kaiserslautern (2001).
12. Bandman, T.; Greuel, G.-M.; Grunewald, F.; Kunyavskii, B.; Pfister, G.; Plotkin, E.: Two-variable identities for finite solvable groups, COMPTES RENDUS DE L'ACADEMIE DES SCIENCES, Ser. I337(2003)581–586.
13. Bandman, T.; Greuel, G.-M.; Grunewald, F.; Kunyavskii, B.; Pfister, G.; Plotkin, E.: Engel-Like Identities Characterizing Finite Solvable Groups, Compositio Math. 142 (2006), No. 3, 734–764.
14. Bayer, D.: The Division Algorithm and the Hilbert Scheme. Ph.D. Thesis, Harvard University, Cambridge, MA (1982).
15. Bayer, D.; Galligo, A.; Stillman, M.: Gröbner Bases and Extension of Scalars. In: Proceedings Comput. Algebraic Geom. and Commut. Algebra, 198–215, Cortona, Italy, Cambridge University Press (1991).
16. Bayer, D.; Mumford, D.: What can be Computed in Algebraic Geometry? In: Proceedings Comput. Algebraic Geom. and Commut. Algebra, 1–48, Cortona, Italy, Cambridge University Press (1991).

17. Bayer, D.; Stillman, M.: On the Complexity of Computing Syzygies. J. Symb. Comp. 6, 135–147 (1988).
18. Bayer, D.; Stillman, M.: Computation of Hilbert Functions. J. Symb. Comp. 14, 31–50 (1992).
19. Bayer, T.: Computations of Moduli Spaces for Semiquasihomogeneous Singularities and an Implementation in SINGULAR. Diplomarbeit, Kaiserslautern (1999).
20. Becker, T.; Grobe, R.; Niermann, M.: Radicals of Binomial Ideals. Journal of Pure and Applied Algebra 117 & 118, 41–79 (1997).
21. Becker, E.; Marinari, M.G.; Mora, T.; Traverso, C.: The Shape of the Shape Lemma. In: von zur Gathen, J., Giesbrecht, M. (eds.): Proceedings of the International Symposium on Symbolic and Algebraic computation. Association for Computing Machinery. New York, 129–133 (1994).
22. Becker, T.; Weispfenning, V.: Gröbner Bases, A Computational Approach to Commutative Algebra. Graduate Texts in Mathematics 141, Springer (1993).
23. Becker, E.; Wörmann, T.: Radical Computations of Zero–dimensional Ideals and real Root Counting. Mathematics and Computers in Simulation 42, 561–569 (1996).
24. Bermejo, I.; Gimenez, P.: On Castelnuovo–Mumford Regularity of Projective Curves. Proc. Amer. Math. Soc. 128 (5) (2000).
25. Bermejo, I.; Gimenez, P.: Computing the Castelnuovo–Mumford regularity of some subschemes of $\mathbb{P}^n$ using quotients of monomial ideals, JPAA 164 (2001)
26. Bigatti, A.; Caboara, M.; Robbiano, L.: On the Computation of Hilbert-Poincaré Series. Appl. Algebra Eng. Comm. Comput. 2, 21–33 (1991).
27. Boege, R.; Gebauer, R.; Kredel, H.: Some Examples for Solving Systems of Algebraic Equations by Calculating Gröbner Bases. J. Symb. Comp. 2, 83–89 (1987).
28. Brieskorn, E.: Die Monodromie der isolierten Singularitäten von Hyperflächen. Manuscripta Math. 2, 103–161 (1970).
29. Brieskorn, E.; Knörrer, H.: Plane algebraic curves. Birkhäuser (1986).
30. Bruns, W.; Herzog, J.: Cohen–Macaulay Rings. Cambridge University Press, Cambridge (1993).
31. Bruns, W.; Vetter, U.: Determinantal rings. SLN 1327, Springer (1988).
32. Buchberger, B.: Ein Algorithmus zum Auffinden der Basiselemente des Restklassenringes nach einem nulldimensionalen Polynomideal. PhD Thesis, University of Innsbruck, Austria (1965).
33. Buchberger, B.: Ein algorithmisches Kriterium für die Lösbarkeit eines algebraischen Gleichungssystems. Äqu. Math. 4, 374–383 (1970).
34. Buchberger, B.: A Criterion for Detecting Unnecessary Reductions in the Construction of Gröbner Bases. In: EUROSAM'79, An International Symposium on Symbolic and Algebraic Manipulation (E.W.Ng,ed.). Lecture Notes in Comput. Sci., Vol. 72, 3–21, Springer (1979).
35. Buchberger, B.: Gröbner bases — an Algorithmic Method in Polynomial Ideal Theory. In: Recent trends in Multidimensional System Theory, N.B. Bose, ed., 184–232, Reidel (1985).
36. Buchberger, B.: Applications of Gröbner Bases in Non–linear Computational Geometry. In Trends in Computer Algebra, R. Janssen Ed., Springer LNCS Vol. 296, 52–80 (1987).
37. Buchberger, B.; Winkler, F.: Gröbner Bases and Applications. LNS 251, Cambridge University Press, 109–143 (1998).
38. Bueso, J.; Gomez-Torrecillas, J.; Verschoren, A.: Algorithmic methods in non–commutative algebra. Applications to quantum groups. Kluwer Academic Publishers (2003).

39. Campillo, A.: Algebroid Curves in Positive Characteristic. SLN 813, Springer (1980).
40. Canny, J.F.; Kaltofen, E.; Yagati, L.: Solving Systems of Non–linear Polynomial Equations Faster. Proc. Int. Symp. Symbolic Algebraic Comput. ISSAC'92, ACM Press 121–128 (1989).
41. Caviness, B.F:: Computer Algebra: Past and Future. J. Symb. Comp. 2, 217–236 (1986).
42. Chèze, G.; Lecerf, G.: Lifting and recombination techniques for absolute factorization. Manuscript, Université de Versailles Saint-Quentin-en-Yvelines, (2005).
43. Chrusciel, P.; Greuel, G.-M.; Meinel, R.; Szybka, S.: The Ernst equation and ergosurfaces. Preprint, http://arxiv.org/abs/gr-qc/0603041, 1–23 (2006).
44. Chyzak, F.; Salvy, B.: Non–commutative Elimination in Ore Algebras Proves Multivariate Identities. J. Symbolic Computation, 26(2):187–227 (1998).
45. Cohen, H.: A Course in Computational Algebraic Number Theory. Springer (1995).
46. Cohen, H.; Frey, G.: Handbook of Elliptic and Hyperelliptic Curve Cryptography. Chapman & Hall/CRC (2006).
47. Cohen, A.M.; Cuypers, H.; Sterk, H.: Some Tapas of Computer Algebra. Springer (1999).
48. Cojocaru, S.; Pfister, G.; Ufnarovski, V. (Eds.):Computational Commutative and Non-commutative Algebraic Geometry, NATO Science Series III, Computer and Systems Sciences 196 (2005)
49. Collins, G.E.; Mignotte, M.; Winkler, F.: Arithmetic in Basic Algebraic Domains. In: Buchberger, B.; Collins, G.E.; Loos, E. (eds.): Computer Algebra, Symbolic and Algebraic Computation, 2nd edn., 189–220. Springer (1983).
50. Corso, A.; Huneke, C.; Vasconcelos, W.: On the Integral Closure of Ideals. Manuscripta Math. 95, 331–347 (1998).
51. Coutinho, S.C.: A primer of algebraic $\mathcal{D}$–modules. Cambridge Univ. Press (1995).
52. Cox, D.; Little, J.; O'Shea, D.: Ideals, Varieties and Algorithms. Springer (1992).
53. Cox, D.; Little, J.; O'Shea, D.: Using Algebraic Geometry. Springer (1998).
54. Czapor, S.R.: Solving Algebraic Equations. Combining Buchberger's Algorithm with Multivariate Factorization. J. Symb. Comp. 7, 49–53 (1989).
55. Davenport, J.H.; Siret, Y.; Tournier, E.: Computer Algebra, Systems and Algorithms for Algebraic Computation. Academic Press, London (1988).
56. Decker, W.; Greuel, G.-M.; de Jong, T.; Pfister, G.: The Normalization: a new Algorithm, Implementation and Comparisons. In: Proceedings EURO-CONFERENCE Computational Methods for Representations of Groups and Algebras (1.4. – 5.4.1997), Birkhäuser, 177–185 (1999).
57. Decker, W.; Greuel, G.-M.; Pfister, P.: Primary Decomposition: Algorithms and Comparisons. In: Algorithmic Algebra and Number Theory, Springer, 187–220 (1998).
58. Decker, W.; Lossen, Chr.: Computing in Algebraic Geometry; A quick start using SINGULAR. Springer, (2006).
59. Dickenstein, A.; Emiris, I.Z.: Solving Polynomial Equations; Foundations, Algorithms, and Applications. Algorithms and Computations in Mathematics, Vol. 41, Springer, (2005).
60. Dixmier, J.: Enveloping Algebras. AMS (1996).
61. Dreyer, A: Primary Decomposition of Modules. Diplomarbeit, Kaiserslautern (2001).

652   References

62. Durfee, A.H.: Fifteen Characterizations of Rational Double Points and Simple Critical Points. Enseign. Math., II. Ser. 25, 132–163 (1979).
63. Duval, D.: Absolute Factorization of Polynomials: a Geometric Approach. SIAM J. Comput. 20, 1–21 (1991).
64. Ebert, G.L.: Some comments on the modular approach to Gröbner bases. ACM SIGSAM Bulletin 17, 28–32, (1983).
65. Eisenbud, D.: Homological Algebra on a Complete Intersection, with an Application to Group Representations. Transactions of the AMS 260 (1), 35–64 (1980).
66. Eisenbud, D.: Commutative Algebra with a View toward Algebraic Geometry. Springer (1995).
67. Eisenbud, D.; Huneke, C.; Vasconcelos, W.: Direct Methods for Primary Decomposition. Invent. Math. 110, 207–235 (1992).
68. Eisenbud, D.; Grayson, D.; Stillman, M., Sturmfels, B.: Computations in Algebraic Geometry with Macaulay2. Springer, (2001).
69. Eisenbud, D.; Riemenschneider, O.; Schreyer, F.-O.: Resolutions of Cohen–Macaulay Algebras. Math. Ann. 257 (1981).
70. Eisenbud, D.; Robbiano, L.: Open Problems in Computational Algebraic Geometry and Commutative Algebra. In: Computational Algebraic Geometry and Commutative Algebra, Cortona 1991, Cambridge University Press 49–71 (1993).
71. Eisenbud, D.; Sturmfels, B.: Finding Sparse Regular Sequences. J. Pure Appl. Algebra 94, No. 2, 143–157 (1994).
72. Faugère, C.: A new Efficient Algorithm for Computing Gröbner Bases (F_4). Journal of Pure and Applied Algebra 139, 61–88 (1999).
73. Faugère, C.; Gianni, P.; Lazard, D.; Mora, T.: Efficient Computation of Zero–dimensional Gröbner Bases by Change of Ordering. J. Symb. Comp. 16, 329–344 (1993).
74. Fröberg, R.: An Introduction to Gröbner Bases. John Wiley & Sons, Chichester (1997).
75. Frühbis-Krüger, A.: Moduli Spaces for Space Curve Singularities. Dissertation, Kaiserslautern (2000).
76. Frühbis-Krüger, A.: Computing Moduli Spaces for Space Curve Singularities. JPAA 164, 165–178 (2001).
77. Frühbis-Krüger, A.: Partial Standard Bases for Families. Proc. of the ICMS (2002).
78. Frühbis-Krüger, A.; Pfister, G.: Some Applications of Resolution of Singularities from a Practical Point of View, Proceedings of the Conference Computational Commutative and Non-commutative Algebraic Geometry, Chisinau 2004, NATO Science Series III, Computer and Systems Sciences 196 (2005),104–117
79. Frühbis-Krüger, A.; Pfister, G.: Auflösung von Singularitäten, DMV-Mitteilungen 13–2 (2005),89–105
80. Frühbis-Krüger, A.; Pfister, G.: Algorithmic Resolution of Singularities, In: C. Lossen, G. Pfister: Singularities and Computer Algebra. LMS Lecture Notes 324, Cambridge University Press 157–184 (2006)
81. Frühbis-Krüger, A.; Terai, N.: Bounds for the Regularity of Monomial Ideals. LE Mathematiche, Vol. LIII, 83–97 (1998).
82. Fulton, W.: Intersection Theory. Springer (1984).
83. Galligo, A: Théorème de Division et Stabilité en Géométrie Analytique Locale. Ann. Inst. Fourier 29, 107–184 (1979).
84. von zur Gathen, J.; Gerhard, J.: Modern Computer Algebra. Cambridge University Press (1999).

85. Geddes, K.O.; Czapor, S.R.; Labahn, G.: Algorithms for Computer Algebra. Kluwer Academic Publishers Group. XVIII, 585 p., (1992).

86. Gelfand, I.; Kapranov, M.; Zelevinski, A.: Discriminants, Resultants and Multidimensional Determinants. Birkhäuser (1994).

87. Gerard, R.; Levelt, A.H.M.: Invariants Mésurant l'Irrégularité en un Point Singulier des Systèmes d'Equations Différentielles Linéaires. Ann. Inst. Fourier, Grenoble 23 (1), 157–195 (1973).

88. Gianni, P.: Properties of Gröbner Bases under Specialization. In: EURO-CAL'87, J.H. Davenport, Ed., Springer LNCS, Vol. 378, 293–297 (1989).

89. Gianni, P.; Miller, V.; Trager, B.: Decomposition of Algebras. ISSAC 88, Springer LNC 358, 300–308.

90. Gianni, P.; Trager, B.; Zacharias, G.: Gröbner Bases and Primary Decomposition of Polynomial Ideals. J. Symb. Comp. 6, 149–167 (1988).

91. Giovini, A.; Mora, T.; Niesi, G.; Robbiano, L., Traverso, C.: "One Sugar Cube, Please" or Selection Strategies in the Buchberger Algorithm. In: Proceedings Int. Symposium on Symbolic and Algebraic Computation ISSAC'91 (S.M. Watt ed.), ACM Press, New York, 49–54 (1991).

92. Giusti, M.: Some Effectivity Problems in Polynomial Ideal Theory. EUROSAM 84, Symbolic and Algebraic Computation. In: Proc. Int. Symp., Cambridge/Engl. 1984, Lecture Notes Comput. Sci. 174, 159–171 (1984).

93. Gordan, P.: Les Invariants des Formes Binaires. J. de Math. Pure et Appl. 6, 141–156 (1900).

94. Gräbe, H.-G.: The Tangent Cone Algorithm and Homogenization. Journal of Pure and Applied Algebra 97, 303–312 (1994).

95. Gräbe, H.-G.: Algorithms in Local Algebra. J. Symb. Comp. 19, 545–557 (1995).

96. Gräbe, H.: On lucky primes. Journal of Symbolic Computation 15, 199–209, (1994).

97. Grassmann, H.; Greuel, G.-M.; Martin, B.; Neumann, W.; Pfister, G.; Pohl, W.; Schönemann, H.; Siebert, T.: Standard Bases, Syzygies and their Implementation in SINGULAR. In: Beiträge zur angewandten Analysis und Informatik, Shaker, Aachen, 69–96 (1994).

98. Grauert, H.: Über die Deformation isolierter Singularitäten analytischer Mengen. Invent. Math. 15, 171–198 (1972).

99. Grauert, H.; Remmert, R.: Analytische Stellenalgebren. Grundl. 176, Springer (1971).

100. Gregory, R.T.; Krishnamurthy, E.V.: Methods and Applications of Errorfree Computations. Springer (1984).

101. Greuel, G.-M.: Deformation und Klassifikation von Singularitäten und Moduln. Jahresber. Deutsch. Math.–Verein. Jubiläumstagung 1990, 177–238 (1992).

102. Greuel, G.-M.: Applications of Computer Algebra to Algebraic Geometry, Singularity Theory and Symbolic–Numerical Solving. ECM 2000, Progress in Math. 202, 169–188, Birkhäuser Verlag (2001).

103. Greuel, G.-M.: Computer Algebra and Algebraic Geometry — Achievements and Perspectives. J. Symb. Comp. 30, 253–289 (2000).

104. Greuel, G.-M.: A Computer Algebra Solution to a Problem in Finite groups. Revista Matematica Iberoamericana 19, No. 2 (2003), 413–424.

105. Greuel, G.-M.; Hertling, C.; Pfister, Gerhard: Moduli Spaces of Semiquasihomogeneous Singularities with fixed Principal Part. J. Algebraic Geometry 6, 169–199 (1997).

106. Greuel, G.-M.; Kröning, H.: Simple Singularies in Positive Characteristic. Math. Z. 203, 339–354 (1990).

107. Greuel, G.-M.; Lossen, C.; Schulze, M.: Three Algorithms in Algebraic Geometry, Coding Theory and Singularity Theory. In: C. Ciliberto et al (eds.), Applications of Algebraic Geometry to Coding Theory, Physics and Computation, Kluver (2001).

108. Greuel, G.-M.; Lossen, C.; Shustin, E.: Introduction to Singularities and Deformations. Springer (2006).

109. Greuel, G.-M.; Martin, B.; Pfister, G.: Numerische Charakterisierung quasihomogener Gorenstein–Kurvensingularitäten. Math. Nachr. 124, 123–131 (1985).

110. Greuel, G.-M.; Pfister, G.: Moduli for Singularities. In: Singularities London Math. Soc. LN 201, 119–146 (1991).

111. Greuel, G.-M.; Pfister, G.: Advances and Improvements in the Theory of Standard Bases and Syzygies. Arch. Math. 66, 163–196 (1996).

112. Greuel, G.-M.; Pfister, G.: Gröbner bases and Algebraic Geometry. In: Proceedings of the Conference 33 Years of Gröbner bases (Linz). Lecture Notes of London Math. Soc., 251, 109–143 (1998).

113. Greuel, G.-M.; Pfister, G.: Geometric Quotients of Unipotent Group Actions II. In: Singularities. The Brieskorn Anniversary Volume, 27–36, Progress in Mathematics 162, Birkhäuser (1998).

114. Greuel, G.-M.; Pfister, G.: Computer Algebra And Finite Groups, In Proceedings of the ICMS, Beijing 2002, 4–14.

115. Greuel, G.-M.; Pfister, G.: SINGULAR and Applications, Jahresbericht der DMV 108 (4), 167–196, (2006).

116. Greuel, G.-M.; Pfister, G.; Schönemann, H.: SINGULAR Reference Manual. Reports On Computer Algebra Number 12, May 1997, Centre for Computer Algebra, University of Kaiserslautern from www.singular.uni-kl.de.

117. Gröbner, W.: Über die algebraischen Eigenschaften der Integrale von linearen Differentialgleichungen mit konstanten Koeffizienten. Monatsh. der Math. 47, 247–284 (1939).

118. Hardy, G.H.; Wright, E.M.: An Introduction to the Theory of Numbers. Oxford University Press (1954).

119. Harris, J.: Algebraic Geometry, A First Course. Springer (1992).

120. Hartshorne, R.: Algebraic Geometry, Vol. 52, Graduate Texts in Mathematics, New York, Springer (1977).

121. Hermann, G.: Die Frage der endlich vielen Schritte in der Theorie der Polynomideale. Math. Annalen 95, 736–788 (1926).

122. Hillebrand, D.: Triangulierung nulldimensionaler Ideale — Implementierung und Vergleich zweier Algorithmen. Diplomarbeit, Dortmund (1999).

123. Hironaka, H.: Resolution of Singularities of an Algebraic Variety over a Field of Characteristic Zero. Ann. of Math. 79, 109–326 (1964).

124. Hirsch, T.: Die Berechnung des ganzen Abschlusses eines Ideals mit Hilfe seiner Rees–Algebra. Diplomarbeit, Kaiserslautern (2000).

125. Iglesias, A.; Takayama, N.: Mathematical Software - ICMS 2006. LNCS 4151.

126. Jansen, C.; Lux, K.; Parker, R.; Wilson, R.: An Atlas of Brauer Characters. Oxford University Press (1995).

127. De Jong, T.: An Algorithm for Computing the Integral Closure. J. Symb. Comp. 26, 273–277 (1999).

128. De Jong, T.; Pfister, G.: Local Analytic Geometry. Vieweg (2000).

129. Kalkbrener, M.: Algorithmic Properties of Polynomial Rings. J. Symb. Comp. 26, 525–581 (1998).

130. Kalkbrener, M.: Prime Decomposition of Radicals in Polynomial Rings. J. Symb. Comp. 18, 365–372 (1994).

131. Kandri–Rody, A.; Kapur, D.: Algorithms for Computing Gröbner Bases of Polynomial Ideals over Various Euclidean Rings. EUROSAM 84, Symbolic and Algebraic Computation, Proc. Int. Symp., Cambridge/Engl. 1984, Lecture Notes Comput. Sci. 174, 195–206 (1984).

132. Kandri–Rody, A.; Weispfenning, V.: Non–commutative Gröbner bases in algebras of solvable type. J. Symbolic Computation, 9(1):1–26 (1990)

133. Knuth, D.E.: The art of computer programming I (1997, 3rd ed.), II (1997, 3rd ed.), III (1998, 3rd ed.), Addison-Wesley.

134. Kornerup, P.; Gregory, R.: Mapping integers and Hensel codes onto Farey fractions. Bit 23, 9–20, (1983).

135. Kredel, H.: Solvable polynomial rings. Shaker (1993).

136. Kredel, H.; Weispfenning, V.: Computing Dimension and Independent Sets for Polynomial Ideals. J. Symb. Comp. 6, 231–248 (1988).

137. Kreuzer, M.; Robbiano, L.: Computational Commutative Algebra 1. Springer (2000).

138. Kreuzer, M.; Robbiano, L.: Computational Commutative Algebra 2. Springer (2005).

139. Krick, T.; Logar, A.: An Algorithm for the Computation of the Radical of an Ideal in the Ring of Polynomials. AAECC9, Springer LNCS 539, 195–205 (1991).

140. Krüger, K.: Klassifikation von Hyperflächensingularitäten. Diplomarbeit, Kaiserslautern (1997).

141. Kunz, E.: Einführung in die kommutative Algebra und algebraische Geometrie. Vieweg (1980). English Translation: Introduction to Commutative Algebra and Algebraic Geometry. Birkhäuser (1985).

142. Kurke, H.; Mostowski, T.; Pfister, G.; Popescu, D.; Roczen, M.: Die Approximationseigenschaft lokaler Ringe. SLN 634, Springer (1978).

143. Kurke, H.; Pfister, G.; Roczen, M.: Henselsche Ringe und algebraische Geometrie. Berlin (1975).

144. Lang, S.: Algebra. Addison–Wesley (1970).

145. Laudal, O.A.: Formal Moduli of Algebraic Structures. SLN 754, Springer (1979).

146. Laudal, O.A.; Pfister G.: Local moduli and singularities. SLN 1310, Springer (1988).

147. Lazard, D.: A new Method for Solving Algebraic Systems of Positive Dimension. Discr. Appl. Math. 33, 147–160 (1991).

148. Lazard, D.: Gröbner Bases, Gaussian Elimination and Resolution of Systems of Algebraic Equations. Proceedings of EUROCAL 83, Lecture Notes in Computer Science 162, 146–156 (1983).

149. Lazard, D.: Solving Zero–dimensional Algebraic Systems. J. Symb. Comp. 13, 117–131 (1992).

150. Lecerf, G.; A quick Implementation of the Absolute Factorization, preprint (2006).

151. Lenstra, A.K.; Lenstra, H.W. Jr.; Lovász, L.: Factoring Polynomials with Rational Coefficients. Math. Ann. 261, 515–534 (1982).

152. Levandovskyy, V.: Non–commutative computer algebra for polynomial algebras: Gröbner bases, applications and implementation. Doctoral Thesis, Universität Kaiserslautern (2005). Available from: http://kluedo.ub-uni-kl.de/volltexte/2005/1883/.

153. Levandovskyy, V.: PBW Bases, Non–Degenerate Conditions and Applications. In: Buchweitz, R.-O.; Lenzing, H. (eds.): Representation of algebras and related topics. Proceedings of the ICRA X conference, Vol. 45, 229–246. AMS. Fields Institute Communications (2005).

154. Li, H.: Noncommutative Gröbner bases and filtered–graded transfer. Springer (2002).
155. Lossen, C.; Pfister, G. (Eds.): Singularities and Computer Algebra. LMS Lecture Notes 324 , Cambridge University Press (2006)
156. Lossen, C.; Pfister, G.: Aspects of Gert-Martin Greuel's Mathematical Work. In: C. Lossen, G. Pfister: Singularities and Computer Algebra. LMS Lecture Notes 324, Cambridge University Press (2006), xvii-xxxvi
157. Macaulay, F.S.: Some Properties of Enumeration in the Theory of Modular Systems. In: Proc. London Math. Soc. 26, 531–555 (1939).
158. Martin, B.: Computing Versal Deformations with SINGULAR. In: Algorithmic Algebra and Number Theory, Springer, 283–294 (1998).
159. Matsumura, H.: Commutative Ring Theory. Cambridge Studies in Advanced Math. 8 (1986).
160. Matsumura, H.: Commutative Algebra (second edition). Mathematical Lecture Notes Series, Benjamin (1980).
161. Mayr, E.W.; Meyer, A.R.: The Complexity of the Word Problem for Commutative Semigroups and Polynomial Ideals. Adv. Math. 46, 305–329 (1982).
162. Mignotte, M.: Mathematics for Computer Algebra. Springer (1991).
163. Miller, Ezra; Sturmfels, Bernd: Cobinatorial Commutative Algebra. Graduate Texts in Mathematics 227, Springer (2005).
164. Milnor, T.: Singular Points of Complex Hypersurfaces. Ann. of Math. Studies 61, Princeton (1968).
165. Mines, R.; Richman, F.; Reitenburg, W.: A Course in Constructive Algebra. Springer, New York (1988).
166. Miola, A.; Mora, T.: Constructive Lifting in Graded Structures: A Unified View of Buchberger and Hensel Methods. J. Symb. Comp. 6, 305–322 (1988).
167. Mishra, B.: Algorithmic Algebra. Springer (1993).
168. Möller, H.M.: On the Construction of Gröbner Bases using Syzygies. J. Symb. Comp. 6, 345–360 (1988).
169. Möller, H.M.: On Decomposing Systems of Polynomial Equations with Finitely many Solutions. Appl. Algebra Eng. Commun. Comput. 4, 217–230 (1993).
170. Möller, H.M.: Solving of Algebraic Equations — an Interplay of Symbolical and Numerical Methods. Multivariate Approximation, Recent Trends and Results, Akademie Verlag, 161–176 (1997).
171. Möller, H.M.: Gröbner Bases and Numerical Analysis. In: B. Buchberger and F. Winkler (Eds.): Gröbner Bases and Applications. Cambridge University Press, LNS 251, 159–178 (1998).
172. Motsak, O.: Computation of the central elements and centralizers of sets of elements in non–commutative polynomial algebras. Diploma thesis, Kaiserslautern 2006.
173. Möller, H.M.; Mora, T.: Upper and Lower Bounds for the Degree of Gröbner Bases. In: EUROSAM 84, J. Fitch Ed., Springer LNCS, Vol. 174, 172–183 (1984).
174. Möller, H.M.; Mora, T.: New Constructive Methods in Classical Ideal Theory. J. Algebra 100, 138–178 (1986).
175. Möller, H.M.; Mora, T.: Computational Aspects of Reduction Strategies to Construct Resolutions of Monomial Ideals. In: Proceedings AAECC 2, Lecture Notes in Comput. Sci. 228 (1986).
176. Mora, T.: An Algorithm to Compute the Equations of Tangent Cones. In: Proceedings EUROCAM 82, Lecture Notes in Comput. Sci. (1982).
177. Mora, T.: La Queste del Saint Graal: A Computational Approach to Local Algebra. Discrete Applied Math. 33, 161–190 (1991).

178. Mora, T.: Gröbner Bases in non–commutative Algebras. In: Proc. of the Internatinal Symposium on Symbolic and Algebraic Computation (ISSAC '88), 150–161, LNCS 358 (1989).
179. Mora, T.; Pfister, G.; Traverso, C.: An Introduction to the Tangent Cone Algorithm. In: Issues in non–linear geometry and robotics, JAI Press (6), 199–270 (1992).
180. Mora, T.; Robbiano, L.: The Gröbner Fan of an Ideal. J. Symb. Comp. 6, 183–208 (1988).
181. Mumford, D.: Algebraic Geometry I, Complex Projective Varieties. Springer (1976).
182. Mumford, D.: The Red Book of Varieties and Schemes. LNM 1358, Springer (1989).
183. Nagata, M.: Local Rings. Wiley (1962).
184. O'Caroll, L.; Flenner, H.; Vogel, W.: Joins and Intersections. Springer (1999).
185. Pauer, F.: On lucky ideals for Gröbner bases computations. Journal of Symbolic Computation 14, 471–482, (1992).
186. Peskine, C.: An Algebraic Introduction to Complex Projective Geometry I. Commutative Algebra. Cambridge studies in advanced mathematics 47 (1996).
187. Pfister, G.: The Tangent Cone Algorithm and some Applications to Algebraic Geometry. Proc. of the MEGA Conference 1990. Birkhäuser Vol. 94, 401–409 (1991).
188. Pfister, G.: A Problem in Group Theory Solved by Computer Algebra, In Proceedings of the Conference Commutative Algebra, Singularities and Computer Algebra, Sinaia 2002, Kluwer Academic Publishers 2003, 217–223.
189. Pfister, G.: On Modular Computation of Standard Basis, to appear in Analele Stiintific e ale Univ. Ovidius, Mathematical Series, (2007).
190. Pfister, G.; Schönemann, H.: Singularities with exact Poincarè Complex but not Quasihomogeneous. Revista Mathematica Madrid (2) 2y3, 161–171 (1989).
191. Radu, N.: Inele Locale. Bucharest (1968).
192. Rédei, L.: Algebra, Akademische Verlagsgesellschaft Geest & Portig K.-G., Leipzig (1959)
193. Renschuch, B.: Elementare und praktische Idealtheorie. Berlin (1979).
194. Ritt, J.F.: Differential equations from the algebraic standpoint. Colloquium Publications XIV, AMS, (1932).
195. Ritt, J.F.: Differential algebra. Colloquium Publications XXXIII, AMS, (1950).
196. Robbiano, L.: Term Orderings on the Polynomial Ring. In: Proceedings EUROCAL 85, Lecture Notes in Computer Science 204, Springer (1985).
197. Robbiano, L.: On the Theory of Graded Structures. J. Symb. Comp. 2, 139–170 (1986).
198. Robbiano, L.; Sweedler, M.: Subalgebra Bases. In: Commutative Algebra, Proceedings Workshop, Salvador/Brazil 1988, Lecture Notes in Math., Vol. 1430, 61–87, Springer (1990).
199. Saito, M.; Sturmfels, B., Takayama, N.: Gröbner Deformations of Hypergeometric Differential Equations. Springer (2000).
200. Sasaki, T.; Takeshima, T.: A modular method for Gröbner–basis construction over $\mathbb{Q}$ and solving system of algebraic equations. Journal of Information Processing 12, 371–379, (1989).
201. Scheja, G.; Storch, U.: Lehrbuch der Algebra Teil 1 und 2. B.G. Teubner, Stuttgart (1988).
202. Schmidt, J.: Algorithmen zur Berechnung größter gemeinsamer Teiler im univariaten Polynomring über den ganzen Zahlen und deren effiziente Implementierung in Factory. Diplomarbeit, Kaiserslautern (1999).

203. Schreyer, F.-O.: Die Berechnung von Syzygien mit dem verallgemeinerten Weierstrasschen Divisionssatz. Diplomarbeit, Hamburg (1980).
204. Schreyer. F.-O.: A Standard Basis Approach to Syzygies of Canonical Curves. J. Reine Angew. Math., 421, 83–123 (1991).
205. Schreyer, F.-O.: Syzygies of Canonical Curves and Special Linear Series. Math. Ann., 275 (1986).
206. Schulze, M.: Computation of the Monodromy of an Isolated Hypersurface Singularity. Diploma Thesis, Kaiserslautern (1999).
207. Schulze, M.: Algorithms for the Gauss–Manin Connection. J. Symb. Comp. 32, 549–564 (2001).
208. Schulze, M.: Algorithmic Gauss–Manin Connection. Algorithms to Compute Hodge–theoretic Invariants of Isolated Hypersurface Singularities. Dissertation, Kaiserslautern (2002).
209. Seidenberg, A.: Constructions in Algebra. Trans. Amer. Math. Soc. 197, 273–313 (1974).
210. Seidenberg, A.: Construction of the Integral Closure of a Finite Integral Domain II. Proceedings Amer. Math. Soc. 52, 368–372 (1975).
211. Serre, J.-P.: Algèbre locale multiplicités. SLN 11, Springer (1975).
212. Shannon, D.; Sweedler, M.: Using Gröbner Bases to Determine Algebra Membership, Split Surjective Algebra Homomorphisms Determine Birational Equivalence. J. Symb. Comp. 6, 267–273 (1988).
213. Shimoyama, T.; Yokoyama, K.: Localization and Primary Decomposition of Polynomial ideals. J. Symb. Comp. 22, 247–277 (1996).
214. Stillman, M.; Tsai, H.: Using SAGBI Bases to Compute Invariants. Journal of Pure and Applied Algebra 139, 285–302 (1999).
215. Stobbe, R.: Darstellung algebraischer Kurven mittels Computergraphik. Diplomarbeit, Kaiserslautern (1992).
216. Stolzenberg, G.: Constructive Normalization of an Algebraic Variety. Bull. AMS., 74, 595–599 (1968).
217. Sturmfels, B.: Algorithms in Invariant Theory. In: Texts and Monographs in Symbolic Computation. Springer (1993).
218. Sturmfels, B.: Gröbner Bases and Convex Polytopes. AMS University Lecture Series 8 (1995).
219. Traverso, C.: Hilbert functions and the Buchberger algorithm. J. Symb. Comp. 22, No. 4, 355–376 (1996).
220. Trinks, W.: Über B. Buchbergers Verfahren, Systeme algebraischer Gleichungen zu lösen. J. Number Theory, 10, 475–488 (1978).
221. Valentine, F.A: Convex Sets. Krieger (1976).
222. Vasconcelos, W.V.: Arithmetic of Blowup Algebras. Lecture Note 195, Cambridge University Press (1994).
223. Vasconcelos, W.V.: Computational Methods in Commutative Algebra and Algebraic Geometry. Springer (1998).
224. Wall, C.T.C.: Classification of Unimodal Isolated Singularities of Complete Intersections. In: Singularities, Arcata 1981 (Ed. Orlik, P.), Proceedings Sympos. Pure Math. 40 (2), 625–640 (1983).
225. Wang, D.: Irreducible Decomposition of Algebraic Varieties via Characteristic Sets and Gröbner Bases. Computer aided Geometric Design 9, 471–484 (1992).
226. Wang, D.: Decomposing Polynomial Systems into Simple Systems. J. Symb. Comp. 25, 295–314 (1998).
227. Wang, D.: Computing Triangular Systems and Regular Systems. J. Symb. Comp. 30, 221–236 (2000).
228. Wang, D.: Elimination Methods. In: Texts and Monographs in Symbolic Computation. Springer (2001).

229. Wang, P.S.: An Improved Multivariate Polynomial Factorizing Algorithm, Math. Comp. 32, 1215–1231 (1978)
230. Wang, P.S.: The EEZ-GCD Algorithm, ACM SIGSAM Bull. 14, 50–60 (1980)
231. Weispfenning, V.: Constructing Universal Gröbner Bases. Applied Algebra, Algebraic Algorithms and Error-correcting Codes, Proc. 5th Int. Conference, AAECC-5, Menorca, Spain, 1987, Lect. Notes Comput. Sci. 356, 408–417 (1989).
232. Weispfenning, V.: Comprehensive Gröbner Bases. J. Symb. Comp. 14, 1–29 (1992).
233. Wichmann, T.: Der FGLM–Algorithmus: verallgemeinert und implementiert in SINGULAR. Diplomarbeit, Kaiserslautern (1997).
234. Winkler, F.: A p-adic approach to the computation of Gröbner bases. Journal of Symbolic Computation 6, 287–304, (1987).
235. Winkler, F.: Polynomial Algorithms in Computer Algebra. In: Texts and Monographs in Symbolic Computation. Springer (1996).
236. Wu, W.T.: Basic Principles of Mechanical Theorem Proving in Elementary Geometries. J. of System Science and Mathematical Science, 4(3), 207–235 (1984).
237. Yoshino, Y.: Cohen–Macaulay Modules over Cohen–Macaulay Rings. Cambridge University Press (1990).
238. Zariski, O.; Samuel, P.: Commutative Algebra. Vol. I, II, Springer (1960).

Computer Algebra Systems

239. ASIR (Noro, M.; Shimoyama, T.; Takeshima, T.): http://www.asir.org/.
240. CoCoA (Robbiano, L.): A System for Computation in Algebraic Geometry and Commutative Algebra. Available from cocoa.dima.unige.it/cocoa
241. Macaulay (Bayer, D.; Stillman, M.): A System for Computation in Algebraic Geometry and Commutative Algebra. Available via ftp from zariski.harvard.edu.
242. Macaulay 2 (Grayson, D.; Stillman, M.): A Computer Software System Designed to Support Research in Commutative Algebra and Algebraic Geometry. Available from http://math.uiuc.edu/Macaulay2.
243. Macsyma (Macsyma Inc.): http://www.scientek.com/macsyma/main.htm
244. Magma (Computational Algebra Group within School of Maths and Statistics of University of Sydney): http://magma.maths.usyd.edu.au/magma
245. Maple (Waterloo Maple Inc.): http://www.maplesoft.com/.
246. MATHEMATICA (Wolfram Research Inc.): http://www.wolfram.com/.
247. MuPAD (MuPAD Research Group and Sciface Software GmbH & Co. KG): http://www.mupad.de/.
248. Reduce: http://www.uni-koeln.de/REDUCE.
249. SAGE (Mathematics Software (Verson 2.7), (Stein, W.): http://www.sagemath.org/
250. SINGULAR (Greuel, G.-M.; Pfister, G.; Schönemann, H.): A Computer Algebra System for Polynomial Computations. Centre for Computer Algebra, University of Kaiserslautern, free software under the GNU General Public Licence (1990–2007). http://www.singular.uni-kl.de.

Glossary

Index

G–algebra, 90

A_1–singularity, 528
absfact.lib, 566, 611
AbsFactorization, 565
absolute factorization, 564
A'Campo, 447
addition, 109
additive function, 148
ADE–singularity, 531
admissible, 101
affine, 489
  algebraic set, 452
    dimension of, 456, 492
  chart, 478
  cone, 475, 484, 486
  Hilbert function, 330
  part, 480, 481
  ring, 23, 487
  scheme, 467
  space
    n–dimensional, 452
  surface, 453
  variety, 452
affinization, 478, 481
AG–codes, 630
ainvar.lib, 623
aksaka.lib, 632
alexpoly.lib, 618
algDependent, 86, 611
algebra, 3, 110
  enveloping, 91
  G–, 90
  GR–, 99
  map, 110
  of solvable type, 90
  opposite, 97
  PBW, 90

algebra.lib, 20, 86, 87, 216, 433, 460, 611
algebra_containment, 87, 611
algebraic
  dependence, 86
  local ring, 524, 525
  set, 444
    affine, 452, 455
    irreducible, 456, 474
    projective, 474
  variety, 444
algebraically dependent, 86
algebras of solvable type, 90
alg_kernel, 20, 611
A–linear, 110
analytic
  function, 523
  germ, 524
  K–algebra, 363
  local ring, 524, 525
Ann, 341, 435, 616
annihilator, 27, 119, 201, 203, 613, 616
  structure, 470
arcpoint.lib, 618
arithmetic
  geometry, 463
  structure, 486
Arnold, 531, 619
Artin–Rees, 332
Artinian
  K–algebra, 30
  ring, 30
ascending chain condition, 22
ascending set, 286
associated prime, 259, 449
  embedded, 259
  minimal, 226

Algorithms

SINGULAR–Examples